AF876235

Ecology and Conservation of Mountaintop Grasslands in Brazil

Geraldo Wilson Fernandes
Editor

Ecology and Conservation of Mountaintop Grasslands in Brazil

Springer

Editor
Geraldo Wilson Fernandes
Biologia Geral
Universidade Federal de Minas Gerais
Belo Horizonte, Minas Gerais
Brazil

and

Biology Department
Stanford University
Stanford
USA

ISBN 978-3-319-29807-8 ISBN 978-3-319-29808-5 (eBook)
DOI 10.1007/978-3-319-29808-5

Library of Congress Control Number: 2016931429

© Springer International Publishing Switzerland 2016
This work is subject to copyright. All rights are reserved by the Publisher, whether the whole or part of the material is concerned, specifically the rights of translation, reprinting, reuse of illustrations, recitation, broadcasting, reproduction on microfilms or in any other physical way, and transmission or information storage and retrieval, electronic adaptation, computer software, or by similar or dissimilar methodology now known or hereafter developed.
The use of general descriptive names, registered names, trademarks, service marks, etc. in this publication does not imply, even in the absence of a specific statement, that such names are exempt from the relevant protective laws and regulations and therefore free for general use.
The publisher, the authors and the editors are safe to assume that the advice and information in this book are believed to be true and accurate at the date of publication. Neither the publisher nor the authors or the editors give a warranty, express or implied, with respect to the material contained herein or for any errors or omissions that may have been made.

Printed on acid-free paper

This Springer imprint is published by SpringerNature
The registered company is Springer International Publishing AG Switzerland

This book is dedicated to my children, Stephannie and Arthur.
I hope that the stories about the exuberant life in the quartzitic and ferrous mountains that I have shared with them in the field has given them an appreciation for this paramount ecosystem. I wish that they emulate those experiences for the next generations to come.

Foreword

This book offers a feast of information on a little-known Brazilian ecosystem. Rupestrian grasslands occur on extremely infertile soils derived from quartzites and ironstones. They are tropical grassy systems embedded in cerrado, the Brazilian savanna, but with some striking affinities to temperate Cape fynbos and Australian heathlands. I first encountered rupestrian grasslands (RG) on a visit to the Sierra de Cipo to view a restoration project. As in South Africa, restoration of the Brazilian montane grasslands had proved inordinately difficult. I had assumed strongly convergent ecologies of the grasslands starting with intense competition from densely rooted grasses. I was completely wrong. The grasses had sparse roots and bare soil patches were ubiquitous. The vegetation, though grassy, had some elements strikingly similar to Cape fynbos. Many shrubs had an ericoid appearance with tiny evergreen leaves but were from tropical families, not temperate ones. Though C4 grasses dominated, many graminoids were from other families in the Poales (Cyperaceae, Eriocaulaceae, Xyridaceae, Bromeliaceae) which, like fynbos Restionaceae, seem to thrive where grasses struggle. As in South Africa, a fire had produced a stunning display of fire-stimulated flowering herbs. But unlike South African grasslands, decades of fire suppression at one site had produced no successional trend to closed forests. Whereas lowland cerrado on oxisols appears to be fire-maintained, the rupestrian grasslands seem to exist because of distinctive soil properties, not fire. It's hard to recall a day with so many preconceptions demolished so quickly.

This book explores these extraordinary grasslands in rich detail. Though embedded in cerrado, they have some striking affinities (and clear differences) with the heathland systems in winter rainfall climates of South Africa and Australia. The similarities are functional, rather than phylogenetic. There are no Proteaceae, Ericaceae, or Restionaceae, just plants that look like them. Like the southern heathlands, the RGs are remarkably rich in plant species accounting for nearly half the diversity of the cerrado, the world's most species—rich savannas. There are also high levels of endemism (30 %) and many species have tiny distribution ranges. RGs resemble heathlands in nutrient acquisition strategies with many species lacking mycorrhizal associates; yet cluster roots are absent. RGs also resemble heathlands in the preponderance of autochory in striking contrast to the dominance of zoochory in closed forest patches; yet myrmecochory is absent.

The genius of this book is first to recognise the distinct nature of rupestrian grasslands within Brazil, and then to make the links with ecosystems on similar geological substrates elsewhere in the world. The result is a book that deserves a wide readership. Brazil has a venerable ecological tradition starting with Warming and others in the 19th century. Though much research has been published in Portuguese, language is not the only barrier to wider knowledge of these systems. For far too long, grassy biomes have been neglected in favour of tropical forests. This neglect is coming to an end as researchers begin to delve into the lighter side of tropical ecology. Most of the papers cited in this volume are from the 2000s and many from the 2010s. The excitement of discovery is tangible in many of the chapters. New questions are being asked: on the origins, history, affinities, functioning of RGs and their relationships with lowland cerrado and Atlantic forests. What, one wonders, grew on these hostile habitats before the C4 grass revolution of the late Miocene? Several chapters explore predictions of Steve Hopper's OCBIL theory (old climatically buffered infertile landscapes) providing context for novel comparisons with tropical and extra-tropical ecosystems elsewhere in the world. Readers interested in the human ecology of RGs are also well served. There are chapters on the history of land use, traditional and novel uses of the plants from craftwork made with members of the Eriocaulaceae to the potential for novel antibiotics revealed by studies of the rich fungal symbiotic flora.

While RGs have escaped the massive transformation of cerrado to croplands on clay-rich oxisols, their future is far from secure. Besides the general uncertainties of climate change, the extraordinary vegetation of ironstone outcrops is clearly threatened by future mining. On the quartzites, poorly conceived afforestation plans, the spread of alien invasive trees (Pines and eucalypts) and African grasses, and the vexed question of appropriate fire regimes in a legislative environment that has suppressed fires until recently, are discussed as concerns for the future of RGs. As regards the threats posed by increasing tourism, this book will undoubtedly exacerbate the problem. The many photos in the book of the magnificent plants and the habitats in which they grow will surely lead to an influx of yet more visitors keen to explore the rupestrian grasslands for themselves.

William J. Bond
South African Environmental Observation Network
National Research Foundation and Emeritus Professor
Department of Biological Sciences, University of Cape Town
Cape Town, South Africa

Acknowledgments

The number of people to be acknowledged for their help with this book edition is very high, but among them, I must express most of my gratitude to my students, primarily those who worked on the rupestrian grasslands and brought to light some of the information shown in this book. Many of them are authors of chapters in this book. What I have learned was mostly with and from them. Among the students, I am truly grateful to Daniel Negreiros, Barbara Nunes, Hernani Almeida, Patricia Angrisano, Leticia Sena, Patricia Costa, and Daniela Marchesotti for helping with the editing of many chapters. I thank also all authors for accepting the challenges to write chapters. I am also indebted to many of them who also spent additional time patience to review several chapters. Reviews were kindly done by: Aldicir Scariot, Brazil; Ana Neves, Brazil; Armando Aguirre, Mexico; Benjamim Wilder, USA; Elise Buisson, France; Neil Cobb, USA; Abel Conceição, Brazil; Jose Fragoso, USA; Christina Lara, Canada; Gerhard Overback, Brazil; Silvana Marten, Mexico; Beth Morrison, USA; Daniel Negreiros, Brazil; Ramigo Aguilar, Argentina; Ramón Perea, Spain; John Schroeder, USA; and Robert Whelam, USA. I also thank the Conselho Nacional de Pesquisas—CNPq, Fundação de Desenvolvimento da Pesquisa do Estado de Minas Gerais—FAPEMIG, and Reserva Vellozia for sponsoring many research grants that have resulted in several ideas presented in this book. I also thank a special UFMG internal sabbatical leave at the IEAT-UFMG and at Stanford University, which were fundamental to further elaborate and polish the ideas that culminated in the design and publication of this book. Finally, I thank my graduate supervisor and friend Rogério Parentoni Martins for having introduced me to the studies of plant animal interactions and for taking me to my first field trip to the quartz mountains of Serra do Cipó. It did change it all.

Contents

About the Book

This book is a pioneer attempt to bring forward the first synthesis on the rupestrian grassland ecosystem. It brings to light the state of the art information on this ecosystem geological genesis, soil formation, environmental filters, biodiversity, species interactions, and their fine-tuned adaptations to survive and succeed in this harsh environment. The book also brings to light the several human dimensions of the rupestrian grasslands, including the anthropogenic threats that may irreversibly impact this ecosystem and ecosystem services. Whenever possible, the book also addresses several aspects related to conservation of this ecosystem and relates those to human pressure and climate change. While most of the knowledge is concentrated on the quartzitic grasslands of the Espinhaço Mountains, we have witnessed a growing interest and science being produced on the ironstone rupestrian grasslands, which are mostly confined to the Iron Quadrangle and to the Carajás region in northern Brazil. In spite of the exuberance and uniqueness of life found in this ecosystem, the rupestrian grasslands are being converted into novel ecosystems, fragmented and invaded by exotic species at a fast pace. These have resulted in losses of its biodiversity, ecosystem services, and unique identity with still unknown consequences to society. While ecological restoration is still in its infancy and threats augmenting, climate models suggest that threats to the rupestrian grasslands shall worsen in the future. In this way, urgent conservation efforts are needed to save the rupestrian grassland 'nanoforests' and its organisms.

Part I
Foreword 1 + Photo

Students interested in the physical and biological environments of the rupestrian grassland are now exposed to a large array of information not available a few years ago. We now have several sources and alternatives for finding the most recent information, thanks to modern technology which has made them much more accessible. The literature on the rupestrian grassland was very poor just a few decades ago when most of it was in unpublished monographs, dissertations, theses, and reports. The number of articles on the geological history, geography, hydrology, and climate of the rupestrian grasslands was few in comparison with what has been produced in the last two decades.

Part I presents an updated evaluation of the physical environment of the rupestrian grassland and includes novel information on water resources. A separation between the physical factors and the biological factors is not attempted, as in nature they are mostly associated.

The magnitude of the rupestrian grassland starts with the geological events that produced it. The geological history of these altimontane grasslands is not solely of interest to geologists and geographers, or even of mining engineers, but is also needed by biologists, land managers, and even the general public to develop a deeper appreciation for the forces of the Earth that gave rise to the rupestrian grassland millions of years ago. To that end, we must understand the processes that eroded these gigantic mountains and made them short in comparison with others throughout the world. The combination of slope, climate, and rock properties gave rise to many different soils. On top of these mountains, one of the most unique floras of the world developed, the rupestrian grassland. Life on these poor nutrient soils is very tough, where temperature and water are key elements in the survival of both plants and animals. One ecosystem service provided by the rupestrian grassland that is of major relevance to life is water. The rupestrian grassland contains the springs of major rivers that occur in this semiarid region of Brazil, including the San Francisco River and the Rio Doce watershed, where cities with millions of people depend on the crystal pure water. This part covers it all, bringing to light plenty of new information as well as opening space for more studies.

Chapter 1
The Megadiverse Rupestrian Grassland

G. Wilson Fernandes

Abstract The Rupestrian Grassland is an ancient ecosystem characterized by high herbaceous species richness, high endemism, and unique species compositions. The vegetation and habitats in the Rupestrian Grassland are maintained by edapho-climatic factors that limit tree growth and distribution, frequent fires, and, possibly, herbivory. The synergism of the environmental filters caused by these factors and interactions amongst species are the drivers of biodiversity and ecosystem services in this megadiverse environment. But this natural heritage is under threat due to human induced disturbances and by climate change. This book brings to light the most updated synthesis on this unique and paramount ecosystem.

Contents

1.1 The Harsh and Unique Rupestrian Grassland

Rupestrian grasslands (also known as *campos rupestres*, *campos altitudinais, cangas,* rupestrian fields, rupestrian savannas, rocky savannas, rocky fields, ferruginous rocky fields, ironstone fields, and altitudinal fields) are old-growth

G.W. Fernandes (✉)
Ecologia Evolutiva & Biodiversidade/DBG, ICB/Universidade Federal de Minas Gerais, CP 486, Belo Horizonte, MG 31270-901, Brazil
e-mail: gw.fernandes@gmail.com

G.W. Fernandes
Department of Biology, Stanford University, Stanford, CA 94305, USA

© Springer International Publishing Switzerland 2016

G.W. Fernandes (ed.), *Ecology and Conservation of Mountaintop Grasslands in Brazil*, DOI 10.1007/978-3-319-29808-5_1

grasslands; i.e., ancient ecosystems characterized by high herbaceous species richness, high endemism, and unique species compositions. They originate from quartzite, sandstone, or ironstone (i.e. banded iron formations such as itabirites in the case of *cangas*). The Espinhaço mountain chain, a relict of an ancient sea floor and deserts, is where most of the quartzite rupestrian grasslands occur (Barbosa and Sad 1973). This mountain chain extends almost continuously for over 1200 km from northeast to southeast Brazil (Chap. 2, 23). The ironstone layers of cangas are well fragmented and occur in the southern portion of the Espinhaço chain (Iron Quadrangle) as well as in northern Brazil in some hills of the Carajás region. Ironstone rupestrian grasslands represent relicts of an ancient sedimentary river and/or lacustrine basin (Tolbert et al. 1973).

Given the nature of the mother rock that gave rise to rupestrian grasslands, the soils experience heavy drainage and/or leaching, and are exceptionally nutrient impoverished. Those of quartzitic origin very often have a high aluminum content (e.g. Benites et al. 2007; Carvalho et al. 2014; Le Stradic et al. 2015), while those of ironstone origin, where surface soil may be a thin film isolated from true soils by a compact ironstone layer (or crust), present high levels of heavy metals (e.g., Jacobi et al. 2007).

The vegetation that evolved in the Rupestrian Grassland is maintained by edapho-climatic factors that limit tree growth and distribution, by frequent fires, and, possibly, by herbivory. To survive the strong filters imposed by the harsh environment many herbs, forbs, and shrubs of the Rupestrian Grassland have developed a huge array of morphological, anatomical and physiological traits, phenological behaviors, and specific trophic and symbiotic relationships (e.g., Rizzini 1979; Giulietti et al. 1997; Belo et al. 2013; Nishi et al. 2013). Unique specialized underground organs enable many species to resprout repeatedly after damage (Warming 1892; Ribeiro and Fernandes 2000; Barbosa et al. 2014; Negreiros et al. 2014), while poikilohydric mosses, ferns, Cyperaceae, and Velloziaceae of canga outcrops survive to almost complete desiccation (Jacobi et al. 2007). The diversity of habitat types is also enormous in the Rupestrian Grassland. In quartzitic grasslands, sandy, stony, and waterlogged grassland habitats, rocky outcrops, gallery forests, and relict hilltop forest patches, locally known as *capões de mata,* are the most common habitats. Immersed in this mosaic in the Espinhaço Range one can also find patches of Cerrado, at least up to 1350 meters in elevation (see Benites et al. 2007) (Fig. 1.1). Likewise, in the Ironstone rupestrian grasslands many different habitats are found side by side (e.g., Dorr 1969; Jacobi et al. 2007). While a rich variety of herbs occur over the compact ironstone layer, small woodland patches with trees, shrubs and vines are restricted to large depressions, places where the ironstone layer is deeply fragmented and often with caves produced by erosion, allowing the growth of roots that may reach and spread through the soil bellow (Carmo and Jacobi 2013) (Fig. 1.2).

This old growth grassland comprises one of the most speciose and endemic floras of the tropics. In the Espinhaço, the Rupestrian Grassland is home to more than 6000 plant species with some families reaching up to 80–90 % endemism (e.g., Giulietti et al. 1997; Rapini et al. 2008; Echternacht et al. 2011; Silveira et al. 2016).

Fig. 1.1 The mosaic of habitats most often found in quartzitic rupestrian grasslands in Serra do Cipó, Brazil. **a** Rock outcrops habitats immersed in a matrix of sandy and stony grasslands. **b** Aerial view showing gallery forests, rock outrcrops and peat bogs dispersed in the matrix of habitats. In this photo there is a large proportion of rock outcrops. **c** A close view of the diversity of plant species inhabiting crevices where soil accumulates on the rock outcrops. **d** The sandy grasslands which allow some larger species to develop. **e** Aerial view of a stony grassland, here dominated by *Vellozia* sp. **f** Seasonally dry forests are found at lower elevations on very rare sites where karstic soils occur. **g** Aerial view of a transition zone of quartztic and ironstone rupestrian grasslands and Atlantic rain forest. Habitats are separated by a few centimeters. **h** Aerial view of natural islands of Atlantic rain forest in a sandy grassland

Fig. 1.2 Some of the most common habitats found in ironstone rupestrian grasslands in the Amazonian region. **a** A ironstone rupestrian grasslands with many woody species and some species of cactus. **b** In moister and deeper soil sites a more woody vegetation develops. **c** The general appearance of bare rocks with its highly adapted vegetation (hard ferruginous laterite, known as *canga couraçada*). **d** The *reddish/orange colored termite mound* illustrates the ironstone soil in the region. **e** A lake environment found in a ironstone rupestrian grassland in the Carajás region. **f** In ironstone grasslands of Carajás the "veredas" are dominated by a few species of palms including *Mauritia flexuosa* (Buritis). **g** Even being in the Amazonian region where the precipitation is great, many species of the ironstone grasslands are deciduous. **h** A common species in the region is the native species *Norantea guianensis* (Marcgraviaceae)

The high biodiversity and endemism of the Rupestrian Grassland is influenced by latitudinal and altitudinal variations, land slope orientation, insularity and antiquity, which in synergy results in the creation of a great variety of soil types and micro-climates where temperature, light, and humidity vary enormously. In the Espinhaço Range, the role of humidity in the generation of more humid zones or microsites may represents another important source of variation to which many species respond positively (e.g., Barbosa et al. 2015, and references therein).

The uniqueness of this ecosystem given its diversity, endemism and exotic aspect, has long attracted the attention of people. Many early XIX Century European naturalists that traveled to Brazil, and whose importance is recognized around the world, explored the region. These include the German botanist Carl Friedrich Philipp von Martius, the German theologist and botanist Johann Baptist von Spix, the Russian botanist Georg Heinrich von Langsdorff, the French Augustin Saint-Hilaire, the Danish paleontologist Peter Wilhelm Lund and the Danish botanist Johannes Eugenius Bülow Warming (known as Eugen Warming and as the father of plant ecology, see Goodland 1975). The work by Warming in the cerrado-rupestrian grassland-seasonally dry forest transition region (1863 −1866) at the base of the Espinhaço chain provided the scenario on which the first books on plant ecology were constructed (Warming 1892, 1895).

1.2 Synergism Among Drivers and Environmental Filters Leading to Biodiversity

The variety of soil types, having originated from the decomposition of the nutrient-poor rock matrix of the mountain landscape, creates a diversity of habitats and microsites (heretofore referred to only as habitats) throughout the rupestrian land (see Benites et al. 2007; Le Stradic et al. 2015). In these different habitats communities of mutualistic mycorrhizae develop which may lead to new interactions with plants and the colonization by mountain plant species (Carvalho et al. 2012, 2014; Coutinho et al. 2015). The strong selective pressure imposed by the nutritional poverty of quartzite derived soils also drove the evolution of protocarnivory in *Paepalanthus bromelioides* (Eriocaulaceae), a large phytotelmata rosette plant that shares their habitats with small carnivorous *Drosera* spp. (Jolivet and Vasconcellos-Neto 1993; Figueira et al. 1994; Nishi et al. 2013). On the other hand, tolerance to heavy metals evolved independently many times in different plant lineages and metallophytes or metal-tolerant plant species possibly also evolved in the Amazon and the Iron Quadrangle canga's crusts (see Jacobi et al. 2007).

The ancient age of the geological formation and a certain climatic stability, at least in the last few thousand years (see Barbosa et al. 2015), may have facilitated many evolutionary pathways leading to fine-tuned adaptations of the flora and fauna. It is likely that climatic stability may have promoted habitat predictability and allowed the synchronization of biotic and abiotic phenomena; therefore creating

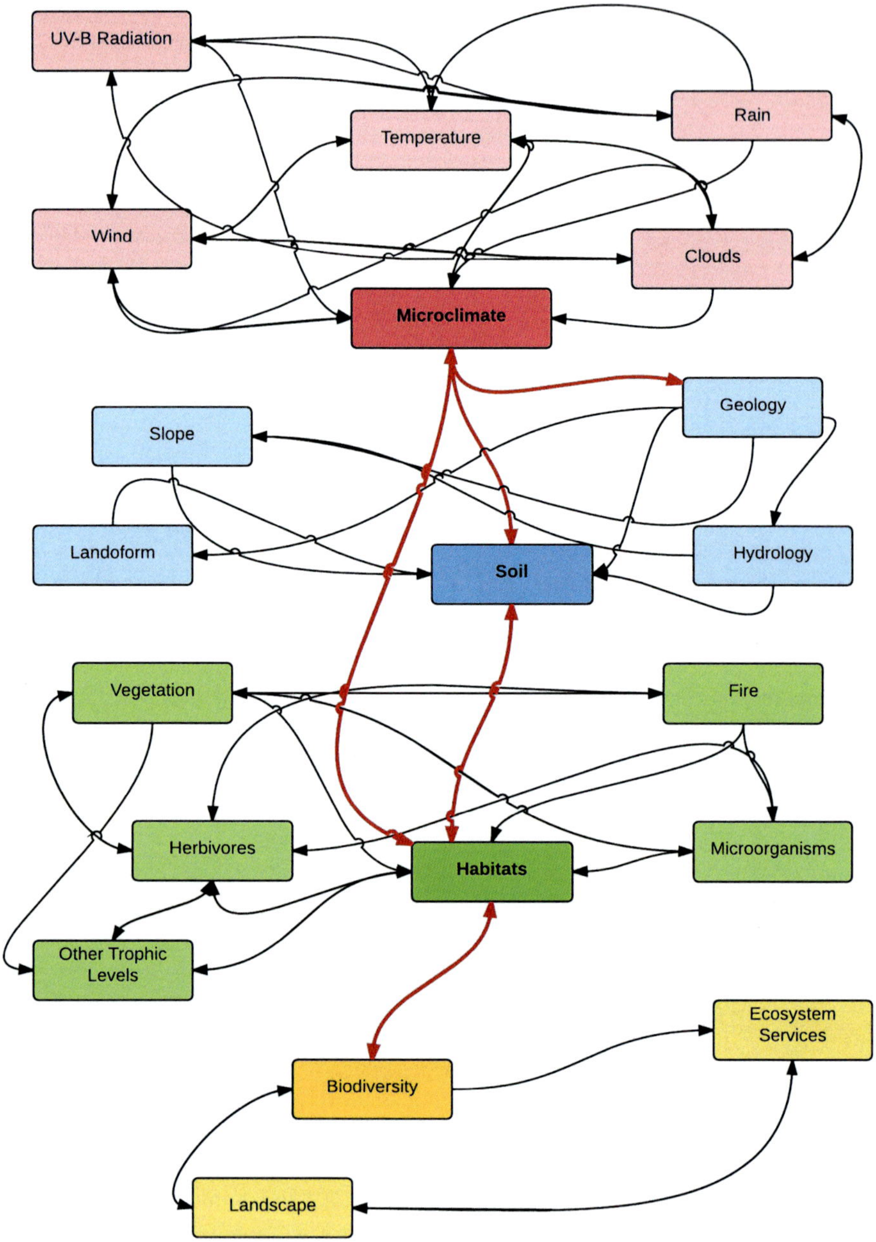
UV-B Radiation
Temperature
Rain
Wind
Clouds
Microclimate
Geology
Slope
Landoform
Soil
Hydrology
Vegetation
Fire
Herbivores
Habitats
Microorganisms
Other Trophic Levels
Ecosystem Services
Biodiversity
Landscape

◀ **Fig. 1.3** Factors influencing the biodiversity of the Rupestrian Grassland and, ultimately, its ecosystem services and landscape. The influence of climatic factors (*pinkish squares*) on biodiversity is very large and mostly driven by temperature and precipitation. Other factors of importance include wind, cloud cover, and UV-B radiation. The interaction of all factors creates a multitude of microclimates (*darker pink*). The geological origin of the terrain is also of major relevance to biodiversity (*bluish squares*). Under the influence of climate, mountains are shredded and eroded to form various landscapes and soils (*darker blue square*). Climatic variables act on the mother rock resulting in substrates of many textures and variable nutritional quality where plants colonize and persist (habitats; *greenish squares*). Such variables include the slope, landform, hydrology, site stability, and soils, which influence the availability of heat, light, water, nutrients and the photosynthetic output of plants. Plants colonizing the different habitats are strongly influenced by climatic and soil factors, but also by fire, herbivores, microorganisms, and other trophic levels (*greenish squares*). The combination of the interaction of climate, geology, soils, vegetation and associated fauna in the variable landscape promote biodiversity and ecosystem services (*yellowish squares*). Feedbacks exist among the many driving forces that shape biodiversity while its intensity changes both in space and time. *Red lines* and *arrows* connect the major compartments (*different colors*), while connections among influencing factors are connected by *black lines* and *arrows*

an optimum environment for the processes of speciation to take place, ultimately leading to increased biodiversity and to the specific arrangement of species in geographical space seen today. The complex topography of mountain environments may be responsible for the creation of a range of different microclimates that are of critical importance for species distribution (see Brown and Lomolino 1998) as well as for the existence of true microrefugia, hence increasing biodiversity and strong site to site variation in species composition (e.g. Barbosa et al. 2015).

The environmental filters and biotic interactions highlighted here may also be found in other ecosystems in the world. However, it is not only their intensity but also their interactive affect and arrangement in the geographical space provided by the mountain environment that contributed to the evolution of this immense biodiversity. Furthermore, the location of the evolutionary origin of rupestrian grasslands is significant. These mountain ranges separate the most diverse savanna of the world (the Cerrado) from the megadiverse Atlantic rain forest, that is they are inserted between the 6th and the 4th biodiversity hotspots of the world, respectively (Myers et al. 2000). The vegetation of these ecosystems has contributed enormously to the origin of the flora of the rupestrian grasslands of the Espinhaço range, although most of the influence comes from the Cerrado. The vegetation is mostly evergreen, sclerophyllous, and with many fire-adapted species (Fig. 1.1). The ironstone grasslands of northern Brazil share many species with the Cerrado and Amazonian rain forest. A summary of the likely ecological mechanisms and processes that most influence the Rupestrian Grassland is presented in Fig. 1.3, albeit a more complete figure shall be constructed in the next few years ahead as more studies are being developed.

1.3 The Human Dimension in the Rupestrian Grasslands

Owing to its singularity in terms of biodiversity, endemism and threats, the southern portion of the Espinhaço mountain range (in the state of Minas Gerais) was recognized as a Biosphere Reserve by UNESCO in 2005. The northern portion of the Espinhaço mountains and the other rupestrian grassland areas found in other isolated mountains were not honored with the UNESCO title, and it is very important that they become part of the biosphere reserve. In general, we need to see more and more specific policies generated towards better conservation practices for such a rare and vanishing ecosystem (see Domingues et al. 2012; Fernandes et al. 2014). Although the ironstone grasslands of northern Brazil are within a federal conservation unit, the Floresta Nacional de Carajás, special attention should be given to the region as well, given its unique and isolated species (see Skirycz et al. 2014; Chap. 2).

Traditionally, conservation research in the tropics has focused on forest ecosystems (e.g., Sanchez-Azofeifa et al. 2013; Veldman et al. 2015a), thereby neglecting the many other diverse ecosystems in the tropics. The tropics are not solely represented by the forest environment; grasslands and savannas, among others, play a major role. The scientific literature on the Rupestrian Grassland is underrepresented in comparison to the far more popular forest ecosystems (see Silveira et al. 2016). But nonetheless, we have seen an increasing number of ecological studies in rupestrian grasslands in recent years, and this shall contribute to bringing these other ecosystems, rarely known by the international community, to attention. This is important, as conservation efforts are largely influenced by public feelings and interest about a given environment.

Basic and descriptive botanical and zoological studies on the Rupestrian Grassland must be encouraged due to the great, but still unknown, number of species and their biology. This knowledge is crucial for conservation actions, as well as for the development of restoration and management practices. Unfortunately, we have witnessed increasing pressure on the Rupestrian Grassland ecosystem. The Rupestrian Grassland has been threatened by a multitude of factors, but mostly the lack of ecological knowledge, mismanagement and lack of governance of the environment (e.g. poorly-done environmental assessment plans, erroneous conservation policies, causing land conflicts, etc.) (Fernandes et al. 2014). These factors have resulted in increased fire frequency, introduction of exotic invasive species, uncontrollable ecotourism, leakage and irreversible environmental impacts caused by surface mining, road construction, real estate development, misinformed afforestation projects, and ongoing climate change (Fernandes et al. 2014; Veldman et al. 2015a, b, c; Silveira et al. 2016; Chap. 18). Undervaluation of the Rupestrian Grassland is reflected in environmental policies that inadvertently exacerbate conversion of rupestrian grasslands to agriculture and mining, degradation caused by inappropriate management and increasingly ill-placed tree plantings (Veldman et al. 2015a, b; see also Putz and Redford 2009; Parr et al. 2014; Gibbs et al. 2015), and even repeated criminal fire.

This book is the first attempt to bring forward a synthesis on the ecosystem of the Rupestrian Grassland. It brings to light state of the art information on its geological genesis, soil formation, environmental filters, biodiversity, species interactions and fine-tuned adaptations for survival and success in the harsh environment, as well as the human dimensions of the Rupestrian Grassland such as anthropogenic threats, ecosystem services and conservation needs in a time of climate change.

1.4 Book Sections

The book is divided into four sections, albeit the subjects are often connected to each other. Section 1.1 is a presentation of the physical environment, although a separation of such factors from biological information is difficult. Hence, a strict line between abiotic and biotic factors is not followed at all times. Chapters 2 and 3, led by Schaeffer et al., focus on the geological history of the Rupestrian Grassland and on the formation of soils that resulted in the many geo-environments found therein, respectively. In Chap. 4, Schaeffer et al. address, for the first time, other key factors that support life in the Rupestrian Grassland, such as air and soil temperature and hydrology. The pioneer work of Callisto et al. (Chap. 5) addresses water quality and the health of headwaters in a mountaintop rupestrian region in Serra do Cipó.

Section 1.2 is mostly centered on plant ecology and the interactions of plants and soil microrganisms. Conceição et al. (Chap. 6) review the biodiversity and endemism of the flora, highlighting principal trends among taxa and regions. Chapter 7 by Coelho et al. provide, for the first time, a synthesis of plant structure and composition of the natural islands of relict Atlantic rain forests in the Rupestrian Grassland matrix and provide some enlightening ideas on the functioning of such unique habitats. Chapter 8 by Oki et al. reviews the state of the art of recent studies on the interaction of plants and endophytes and mycorrhizae fungi in the Rupestrian Grassland. A novel proposition on succession on bare rocks is provided by Conceição & Pirani in Chap. 9, while in Chap. 10, Nunes et al. review the available information on germination strategies of plant species of the Rupestrian Grassland. Finally, Oliveira et al., in Chap. 11, present the unique adaptations of plants to survive in the rupestrian environment and describe many new types of strategies.

Section 1.3 of the book is devoted mostly to the interactions among species and with the harsh Rupestrian Grassland environment. Chapter 12, led by Morellato et al., provides, for the first time, a synthesis regarding the seasonal environment, bringing to light many new and important interactions. Positive interactions of organisms are addressed by Guerra et al. in Chap. 13. These authors bring to light numerous examples of new and amazing positive interactions among organisms and with their environment, as well as the resultant evolutionary outcomes. Antagonistic interactions among species are described by Neves et al. (Chap. 14), highlighting the importance of herbivory and parasitism in the Rupestrian Grassland. The influence of elevation, temperature, and precipitation on the distribution of some taxa of the Rupestrian Grassland have been studied since 1990 and a first synthesis on the topic is provided in Chap. 15 by Fernandes et al. Ending this section, Schaeffer et al. (Chap. 16) highlights the relevance of termites to the ironstone rupestrian grasslands in northern Brazil, by reporting their trends in richness and the nutritional value of termite mounds for the flora.

Section 1.4 of the book has a more trans-disciplinary perspective and presents new information on the human dimension of the Rupestrian Grassland. The issues found in this section are on land use change, cultural heritage, ecosystem services, conservation and threats to the Rupestrian Grassland. Chapter 17, led by Ribas et al. , addresses the growing pressure of afforestation on the Espinhaço mountains due to the introduction of plantations of exotic species of *Eucalyptus*. Chapter 18 by Figueira et al. brings to the reader a state of the art analysis of the effect of fire on rupestrian grassland species and ecosystem functioning, as well as possible fire management strategies. Chapter 19, by Fernandes et al. attempts to synthesize, for the first time, knowledge regarding restoration efforts in the Rupestrian Grasslands. Almada et al. (Chap. 20) bring to light cultural heritage associated with the Rupestrian Grassland by addressing the history of the rupestrian grassland people and their link to the harsh nature of the environment, their fate given the past and current land use pressures and conservation practices. Neves et al. (Chap. 21) focus on the ecosystem services provided by the Rupestrian Grassland. Albeit these studies are still in their infancy, they are of major importance and point to an increased need for conservation and sound ecological management. Chapter 22, lead by Barbosa and Fernandes, address the past, present, and future distribution of rupestrian grasslands given the increasing pressures imposed on them recently. Finally, Chap. 23 by Fernandes focuses on the threats to the Rupestrian Grassland, ways to mitigate impacts and conserve a megadiverse tropical old growth grassland.

Acknowledgements I thank FAPEMIG, CAPES and CNPq for the continuous support of my research that has led to this synthesis. I especially thank Gerhard Overback, José E.C. Figueira and Daniel Negreiros for reading the manuscript and for various suggestions.

References

Barbosa ALM, Sad JHG (1973) Tectonic control of sedimentation and trace-element distribution in iron ores in central Minas Gerais (Brazil). In: Proceedings of kiev symposium 1970, Genesis of precambrian iron and manganese deposits, UNESCO, pp. 125–131

Barbosa BC, Cappi VS, Ribeiro SP, Fernandes GW (2014) Avaliação da capacidade de rebrotamento pós-distúrbio das plantas lenhosas típicas dos campos rupestres. Ecol Aust 24:350–355

Barbosa NPU, Fernandes GW, Sanchez-Azofeifa A (2015) A relict species restricted to a quartzitic mountain in tropical America: An example of microrefugium? Acta Bot Bras 29:299–309

Belo MB, Negreiros D, Fernandes GW, Silveira FAO, Ranieri BD, Morellato PC (2013) Fenologia reprodutiva e vegetativa de arbustos endêmicos de campo rupestre na Serra do Cipó, Sudeste do Brasil. Rodriguesia 64:817–828

Benites VM, Schaefer CER, Simas FNB, Santos HG (2007) Soil associated with rock outcrops in the Brazilian mountain ranges Mantiqueira and Espinhaço. Rev Bras Bot 30:569–577

Brown JH, Lomolino MV (1998) Biogeography. Sinauer, Sunderland
Carmo FF, Jacobi CM (2013) A vegetação de canga no Quadrilátero Ferrífero, Minas Gerais: Caracterização e contexto fitogeográfico. Rodriguésia 64:527–541
Carvalho F, Souza FA, Carrenho R, Moreira FMS, Jesus EC, Fernandes GW (2012) The mosaic of habitats in the high-altitude Brazilian rupestrian fields is a hotspot for arbuscular mycorrhizal fungi. Appl Soil Ecol 52:9–19
Carvalho F, Godoy EL, Lisboa FJG, Moreira FMS, Souza FA, Berbara RLL, Fernandes GW (2014) Relationship between physical and chemical soil attributes and plant species diversity in tropical mountain ecosystems from Brazil. J Mt Sci 11:875–883
Coutinho ES, Fernandes GW, Berbara RLL, Valério HM, Goto BT (2015) Variation of arbuscular mycorrhizal fungal communities along an altitudinal gradient in rupestrian grasslands in Brazil. Mycorrhiza 25:627–638.
Domingues SA, Karez CS, Biondini IVF, Andrade MA, Fernandes GW (2012) Economic environmental management tools in the Serra do Espinhaço Biosphere Reserve. J Sust Develop 5:180–191
Dorr JVN (1969) Physiographic, stratigraphic, and structural development of the Quadrilátero Ferrífero Minas Gerais Brazil. Geological survey professional paper 641-a, United States Printing Office, Washington, DC
Echternacht L, Trovó M, Oliveira CT, Pirani JR (2011) Areas of endemism in the Espinhaço range in Minas Gerais, Brazil. Flora 206:782–791
Fernandes GW, Barbosa NPU, Negreiros D, Paglia AP (2014) Challenges for the conservation of vanishing megadiverse rupestrian grasslands. Nat Cons 12:162–165
Figueira JEC, Vasconcellos-Neto J, Jolivet P (1994) Une nouvelle plante protocarnivore *Paepalanthus bromelioides* Silv. (Eriocaulaceae) du Brésil. Revue d'Écologie 49:3–9
Gibbs HK, Rausch L, Munger J, Schelly I, Morton DC, Noojipady P, Soares-Filho B, Barreto P, Micol L, Walker NF (2015) Brazil's soy moratorium. Science 347:377–378
Giulietti AM, Pirani JR, Harley RM (1997) Espinhaço range region eastern Brazil. In: Davis SD et al. (eds). Centres of plant diversity: A guide and strategy for their conservation, WWF/IUCN, Cambridge, vol 3. pp. 397–404
Goodland RJ (1975) The tropical origin of ecology: Eugen warming's jubilee. Oikos 26:240–245
Jacobi CM, Carmo FF, Vincent RC, Stehmann JR (2007) Plant communities on ironstone outcrops: A diverse and endangered Brazilian ecosystem. Biodiv Cons 16:2185–2200
Jolivet P, Vasconcellos-Neto J (1993) Convergence chez les plantes carnivores. La Recherche 24:456–458
Le Stradic S, Buisson E, Fernandes GW (2015) Vegetation composition and structure of some Neotropical mountain grasslands in Brazil. J Mt Sci 12:864–877
Myers N, Mittermeier RA, Mittermeier CG, Fonseca GAB, Kent J (2000) Biodiversity hotspots for conservation priorities. Nature 403:853–858
Negreiros D, Le Stradic S, Fernandes GW, Rennó HC (2014) CSR analysis of plant functional types in highly diverse tropical grasslands of harsh environments. Plant Ecol 215:379–388
Nishi AH, Vasconcellos-Neto J, Romero GQ (2013) The role of multiple partners in a digestive mutualism with a protocarnivorous plant. Ann Bot 111:143–150
Parr CL, Lehmann CER, Bond WJ, Hoffmann WA, Andersen AN (2014) Tropical grassy biomes: Misunderstood, neglected, and under threat. Trend Ecol Evol 29:205–213
Putz FE, Redford KH (2009) Dangers of carbon-based conservation. Glob Environ Chan 19:400–401
Rapini A, Ribeiro PL, Lambert S, Pirani JR (2008) A flora dos campos rupestres da Cadeia do Espinhaço. Megadiv 4:16–24
Ribeiro KT, Fernandes GW (2000) Patterns of abundance of a narrow endemic species in a tropical and infertile montane habitat. Plant Ecol 147:205–218
Rizzini CT (1979) Tratado de fitogeografia do Brasil. 2nd ed. São Paulo/EDUSP
Sanchez-Azofeifa A, Power J, Fernandes GW, Quesada M (2013) Tropical dry forest in the Americas: Ecology, conservation and management. CRC Press, Boca Raton

Silveira FAO, Negreiros D, Barbosa NPU, Buisson E, Carmo F, Carstensen DW, Conceição AA, Cornelissen TG, Echternacht L, Fernandes GW, Garcia QS, Guerra TJ, Jacobi CM, Lemos-Filho JP, Le Stradic S, Morellato LPC, Neves FS, Oliveira RS, Schaefer CE, Viana PL, Lambers H (2016) Ecology and evolution of plant diversity in the endangered *campo rupestre*: A neglected conservation priority. Plant Soil doi: 10.1007/s11104-015-2637-8

Skirycz A, Castilho A, Chaparro C, Carvalho N, Tzotzos, Siqueira JO (2014) Canga biodiversity, a matter of mining. Front Plant Sci 5:653. doi:10.3389/fpls.2014.00653

Tolbert GE, Tremaine JW, Melcher GC, Gomes CB (1973). Geology and iron ore deposits of Serra dos Carajás, Pará, Brazil. In: Proceedings of kiev symposium 1970, Genesis of Precambrian iron and manganese deposits, UNESCO, pp. 271–280

Veldman JW, Buisson E, Durigan G, Fernandes GW, Le Stradic S, Mahy G, Negreiros D, Overbeck GE, Veldman RG, Zaloumis NP, Putz FE, Bond WJ (2015a) Toward an old-growth concept for grasslands, savannas, and woodlands. Front Ecol Environ 13:154–162

Veldman JW, Overbeck GE, Negreiros D, Mahy G, Le Stradic S, Fernandes GW, Durigan G, Buisson E, Putz FE, Bond WJ (2015b) Tyranny of trees in grassy biomes. Science 347:484–485

Veldman JW, Overbeck GE, Negreiros D, Mahy G, Le Stradic S, Fernandes GW, Durigan G, Buisson E, Putz FE, Bond WJ (2015c) Where tree planting and forest expansion are bad for biodiversity and ecosystem services. BioScience doi: 10.1093/biosci/biv118

Warming E (1892) Lagoa Santa: Et Bidragtil den Biologiske Plantegeografi med en Fortegnelse over Lagoa Santas Hvirveldyr—Naturvidenskabeligog Mathematisk Afdeling, 6. Rk. vol. 6, pp. 153−488

Warming E (1895) Plantesamfund—Grundtrækaf den økologiske Plantegeografi. P.G Philipsens Forlag, Kjøbenhavn. p. 335

Chapter 2
The Physical Environment of Rupestrian Grasslands (Campos Rupestres) in Brazil: Geological, Geomorphological and Pedological Characteristics, and Interplays

Carlos E.G.R. Schaefer, Guilherme R. Corrêa, Hugo G. Candido, Daniel M. Arruda, Jaquelina A. Nunes, Raphael W. Araujo, Priscyla M.S. Rodrigues, Elpídio I. Fernandes Filho, Aianã F. S. Pereira, Pedro Christo Brandão and Andreza V. Neri

Abstract The Rupestrian Grassland is a vegetational complex with grassy to shrubby formations that occur throughout the high mountains of Brazil, usually formed by structurally resistant rocks, little affected by late tectonics, and strongly eroded and weathered under long term geological stability. RGC is closely associated with high altitude landsurfaces, in which several factors have a determinant role: (1) extreme oligotrophy, and acid, nutrient-depleted parent materials; (2) resistance to weathering and erosion (chemical and physical); (3) constant wind exposure; (4) intense fire regime. Variations of RG phytophysiognomies are basically due to soil depth (edaphic

C.E.G.R. Schaefer (✉) · R.W. Araujo · E.I. Fernandes Filho · P.C. Brandão
Departamento de Solos E Nutrição de Plantas, Universidade Federal de Viçosa, Viçosa, MG 36570 900, Brazil
e-mail: carlos.schaefer@ufv.br

R.W. Araujo
e-mail: rapwa79@gmail.com

E.I. Fernandes Filho
e-mail: elpidio@ufv.br

P.C. Brandão
e-mail: pedrocbrandao@gmail.com

G.R. Corrêa
Instituto de Geografia, Universidade Federal de Uberlândia, Uberlândia, MG 38400902, Brazil
e-mail: guilhermeudi@yahoo.com.br

H.G. Candido · D.M. Arruda · P.M.S. Rodrigues · A.F.S. Pereira · A.V. Neri
Departamento de Biologia Vegetal, Universidade Federal de Viçosa, Viçosa, MG 36570 900, Brazil
e-mail: hugogalvaocandido@gmail.com

© Springer International Publishing Switzerland 2016
G.W. Fernandes (ed.), *Ecology and Conservation of Mountaintop Grasslands in Brazil*, DOI 10.1007/978-3-319-29808-5_2

factor), drainage and landscape stability and evolution. Landforms (geomorphological attributes) affect the RGC at continental scales (high landsurfaces), regional scales (regional landforms, such as escarpments, valleys, slopes) and local scales (soil depth, stoniness, rockiness, drainage). The most common occurrence of RGC in Brazil is on Quartzite and metarenites, followed by canga and other Fe-rich substrates, igneous rocks and metamorphics, hence displaying a high diversity of substrates, with a major trait of extreme soil oligotrophy and acidity, and crucial variations in soil depth. The occurrence of well-documented areas of RG on Granitic and gneissic terrains imply that even richer rocks, submitted to long term weathering and erosion, can lead to similar soils on Highlands, where rock outcrops are also common (e.g. Caparaó, Itatiaia, Brigadeiro, Serra dos Órgãos). RGC can occur immersed in different domains (Atlantic Forest, Caatinga, Cerrado), regardless of present day climates, since it represents an edaphic climax of long-term development. The widespread distribution of comparable RGC, from Amazonia (Carajás, Roraima, Pacaás Novos, Cachimbo) to the Central Plateau (Santa Barbara, Ricardo Franco, Pirineus) and Northeastern and Southeastern Brazil (Caparaó, Espinhaço, Sincorá, Jacobina, Itatiaia, Serra dos Órgãos, Brigadeiro, Carangola) raises the unresolved question of phylogenetic ancestry, age and similarities (floristic, structural) between those isolated islands of Rupestrian vegetation. These aspects are central to the evolution of Brazilian Biomes, representing key issues to resolve the late Quaternary Refuge Theory controversy, and test the validity of island biogeographical isolation theories.

Contents

D.M. Arruda
e-mail: arruda.dm@hotmail.com

P.M.S. Rodrigues
e-mail: riscylamsr@gmail.com

A.F.S. Pereira
e-mail: aiana.santos@yahoo.com.br

A.V. Neri
e-mail: andreza.neri@ufv.br

J.A. Nunes
Unidade Carangola, Praça dos Estudantes, Universidade do Estado de Minas Gerais (UEMG), 23-Santa Emília, Carangola, MG CEP36800-000, Brazil
e-mail: jaquelinabiologa@yahoo.com.br

2.1 The Geological Control

The rupestrian grassland constitute a singular kingdom of plants that form an ancient vegetation complex (rupestrian grassland complex—RGC). In Brazil, rupestrian grassland refers to rupicolous or rupestrian (rocky habitat) vegetation that grows on a range of geologies, including quartzite–sandstone (Harley and Simmons 1986), granite-gneiss (Queiroz et al. 1996) and itabirite—an iron-rich metal sedimentary rock (Jacobi and Carmo 2011; Jacobi et al. 2007; Vincent and Meguro 2008; Schaefer et al. 2009). Semir (1991) considered rupestrian grassland to include any open vegetation type associated with rocky outcrops. Veloso et al. (1991) classified it as "vegetation refuges or relic vegetation types" floristically different from the dominant surrounding flora. Bucher et al. (2007) proposed that such areas have acted as refuges during dry climate phases associated with major glaciations, as well as centers of recent speciation.

Rupestrian grassland is widely distributed in Brazil (Fig. 2.1), and forms a complex of sclerophyllous vegetation ranging from open rupestrian grassland *strictu* sensu, to denser physiognomies, varying from shrubs to low forests and thickets. In terms of vegetation life forms, RGC is rich in geophytes (bulbous plants), ericoids (heath-like shrubs), proteoids (protea shrubby with large leaves)

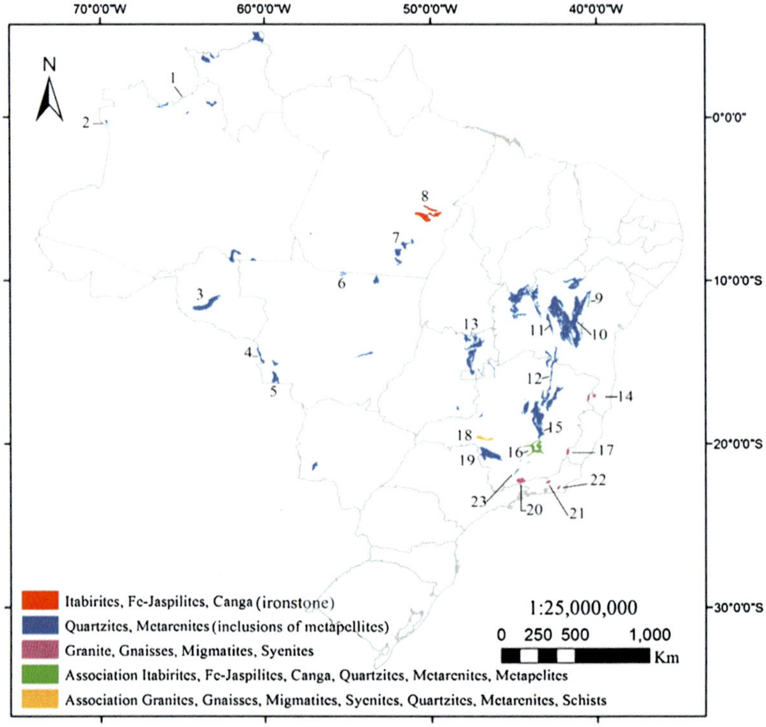

Fig. 2.1 The distribution of rupestrian grassland complex in relation to the geological substrate and Brazilian structural provinces. The main rocks are grouped for facilitating the comprehension of the lithological control. Areas of mixed and variable rock types form lithological associations. The numbers represent: *1* Tepuis, *2* Tunuí, *3* Serra dos Pacaás Novos, *4* Serra Ricardo Franco, *5* Santa Bárbara, *6* Serra do Cachimbo, *7* Serra da Seringa, *8* Serra dos Carajás, *9* Morro do Chapéu, *10* Chapada Diamantina (Serra do Sincorá), *11* Northern Espinhaço, *12* Southern Espinhaço, *13* Highlands of Central Plateau, *14* Pontões de Santo Antônio do Jacinto, *15* Serra do Cipó, *16* Iron Quadrangle, *17* Caparaó, *18* Araxá, *19* Canastra, *20* Itatiaia, *21* Serra dos Órgãos, *22* Macaé, *23* Ibitipoca. In the Appendix I, all pictures are related to the different areas of RGC throughout the Brazilian territory, as indicated by the number

and restiods (reed-like plants), acaulescent rossettes, all forming a grassy to shrubby physiognomy. The extreme oligothrophy, fire regime and wind exposure induce a great number of r-strategists, adapted to multiple environmental pressures (Grime 2001; Alves et al. 2014). The RGC are mainly associated with pre-Cambrian quartzites and metarenites, but important disjunct areas are also found on many contrasting lithologies (granites, gneisses, schists, syenites, metapelites, itabirites), all of which representing the most resistant rocks in terms of chemical and physical conditions. A general map of such distribution in relation to parent material (geology) is outlined in Fig. 2.2. There, the majority of RGC is developed on quartzite, and is clearly associated with the old Mobile Belts of Proterozoic age, as well as minor areas of younger plutonic rocks. Though tectonically stable today, these areas

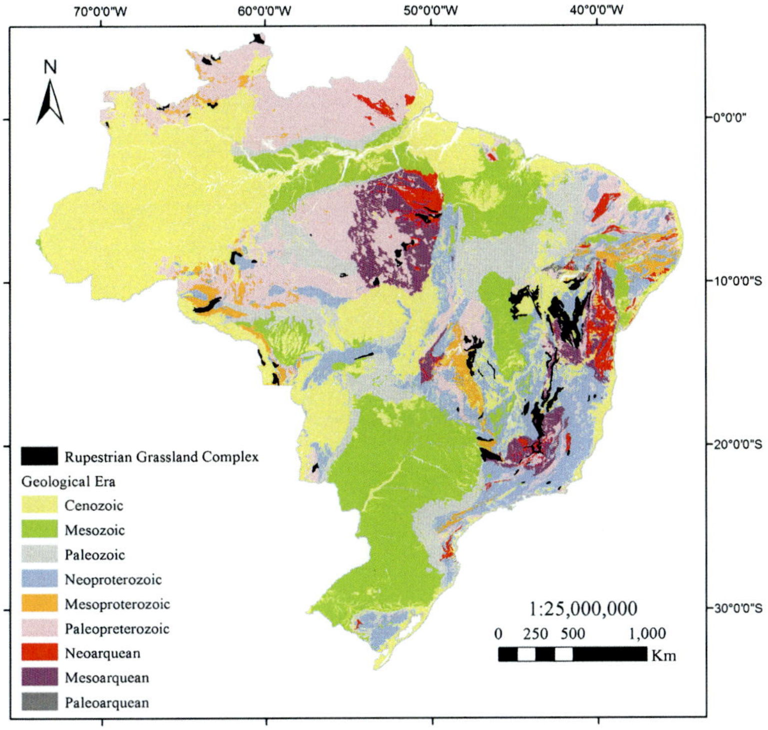

Fig. 2.2 The overlay between the distribution of rupestrian grassland complex (RGC) in function of the geological era on which it developed. The main areas mapped are associated with Pre-cambrian rocks, ranging from neoarchean (Carajás, for example) to meso- and neoproterozoic. All these areas have a common attribute of very poor, acid quartzite/camglomerate/metapelite substrates, upon which the soils formed are, by extension, acid, nutrient depleted, shallow and extremely weathered/eroded. It is remarkable that the central parts of the amazon and São Francisco cratons are practically devoid of significant RGC

are the "roots" of ancient mountain ranges former of greater development, during epochs of tectonic collision between South American and African Plates. Hence, the mountains associated with RGC are very old, residual cores of strongly weathered, folded, faulted and eroded landmasses, later subjected to regional uplift during renewed tectonic events of lower magnitude. Therefore, the diversified landforms of Proterozoic rock terrains are the result of former tectonic activity and mountain-building processes, creating strong regional heterogeneity at landscape scale that remains to this time.

The predominance of quartzite in RGC is based on the fact that these rocks combine all prerequisites for RGC development: (1) extreme oligotrophy; (2) chemical resistance to weathering; (3) Physical resistance; (4) High mountain occurrence due to differential erosion; and (5) widespread distribution across the Brazilian Mobile Belts (Schaefer 2013), or the Ancient Brasiliano Orogenic

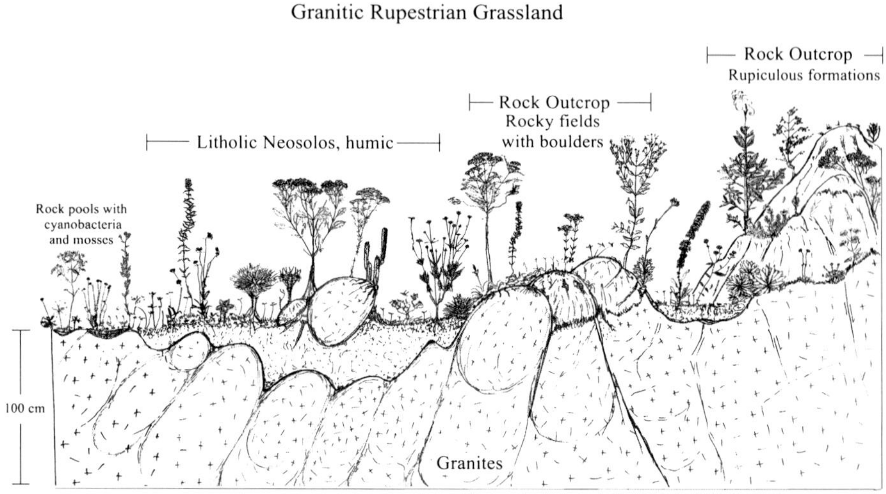

Fig. 2.3 The rupestrian grassland complex (RGC) with its physiognomic variations on granitic rocks, as illustrated by the high surfaces at o caparaó massif, the highest mountains of southeastern Brazil. Traditionally viewed as "Campos de Altitude" these RGC are, in fact, true rupestrian in many parts, where rocky surfaces predominate. There, subtle variations in soils depth, drainage and organic matter accumulation lead to varied niches, where a set of well-adapted, R-strategists, coexist. In some areas with rocky fields, low lying rocky terrains allow the adaptations of shrubby with xeromorphic adaptations, whereas rocky pools are associated with water accumulations and cyanobacteria mats and mosses growth, followed by abundant cyperaceae, xyridaceae and eriocaulaceae. rupiculous formation are found on rock outcrops with large blocks forming inselbergs, where lichens, mosses, bromeliads and shrubs coexist, with dissolution pathways (canelures) formed by the constant input of water saturated by organic acids. Biodiversity and endemism are characteristic of these isolated mountains, surrounded by a matrix of rainforest, where most rupestrian species are completely absent. (Drawing by C. Schaefer)

Systems (Alkmim 2015), of Neoproterozoic age. In addition, these highlands are very susceptible to frequent, severe fire regimes, so that most plants are fire-adapted and tolerant.

The main attribute for controlling the RGC occurrence is rock resistance and chemical impoverishment. Whenever a given rock is weathered and form deeper soil, other types of vegetation will develop, so that the RGC is, in fact, an edaphic climax related to resistant lithologies under a certain set of environmental conditions. They always occur in rocky terrains, which form resistant masses in the mid of deeply weathered saprolites. In the case of granites, magmatites or syenites, these massive rocks are more resistant to weathering and erosion than the surrounding gneiss, so that they stand out as protruding masses known as sugar loaves (type of inselbergs), with very steep topography (Fig. 2.3) (Martinelli 1989). In this case, the massivity of rock combined with the style and degree os fracturing, are additional factors controlling RGC formation.

The rugged mountainous relief of RGC provides a complex and varied combination of substrates, slope, aspect, altitude and edaphic condition, jointly promoting species diversification through niche specialization. The island-like distribution of RGC in Brazil points to a gene flow cut-off, from surrounding forested areas, keeping its singular floristic identity, by long-term biogeographic isolation (Fig. 2.1).

2.2 Landforms and Geomorphological Control

Looking at a general digital elevation model (topographic relief) of Brazil, it is clearly seen a close relation between the highlands above 1000 meters and the distribution of RGC (Fig. 2.4). In these highest landscape segments, rock outcrops are common, related to stripping of the weathering mantle and exhumation of resistant rock cores, exposed by differential erosion (Schaefer 2013). This process is variable in scale and depends on climate (and paleoclimate) conditions responsible for the balance between weathering, soil formation and erosion. At a local scale,

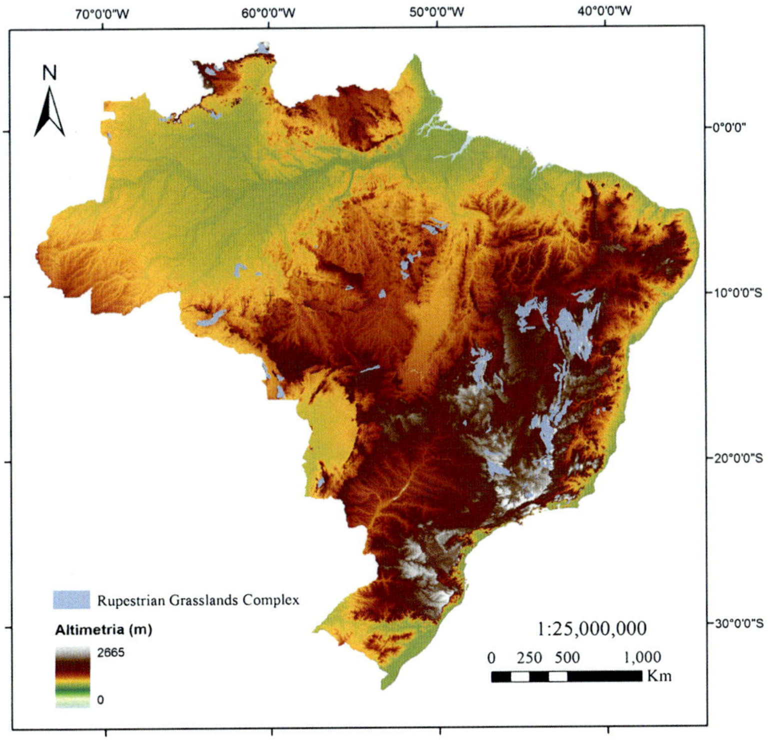

Fig. 2.4 The overlay of rupestrian grassland complex (RGC) distribution in Brazil in relation to topography. The close coincidence of highlands above 1000 m and RGC is striking. Most highlands are formed by residual, resistant tocks, rather than tectonically displaced mountains

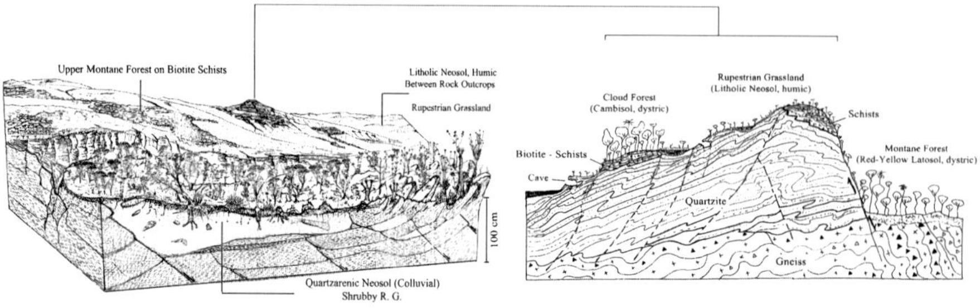

Fig. 2.5 A cross-section of the geology, landforms and soils on the rupestrian grassland complex of the Ibitipoca state park. In the second picture, a view from the Ibitipoca peak to the Pião peak (Background), showing the rupestrian complex as influenced by soil properties (depth, drainage, sandy texture)

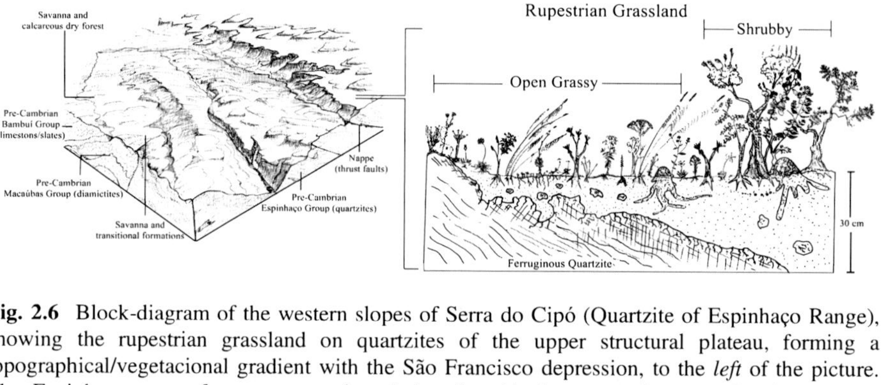

Fig. 2.6 Block-diagram of the western slopes of Serra do Cipó (Quartzite of Espinhaço Range), showing the rupestrian grassland on quartzites of the upper structural plateau, forming a topographical/vegetacional gradient with the São Francisco depression, to the *left* of the picture. The Espinhaço range form a mountains chain of residual, structural nature, by the extreme resistance of its quartzitic substrate. In the *right* picture, a sequence of soils at local scale is shown, as related to depth and drainage, affecting the distribution of open and shrubby physiognomies

three examples are offered here, to illustrate the variability of landforms in the RGC at local scale. In the first example, the Ibitipoca Mountains of Minas Gerais, RGC is found on quartzite, and minor variation on shallow soils on schists (Fig. 2.5), which favor forest vegetation in deeper soils (Dias et al. 2002). In another example, at the Serra do Cipó, western border of the Espinhaço Range, quartzite on the high plateau is closely associated with RGC with may different physiognomies (Fig. 2.6). Soils are invariably nutrient–poor, even though shallow and poorly developed. This apparent paradox is due to the extreme chemical poverty of the substrates, which have little nutrient to offer following exposure, weathering and soil formation, since these substrates are virtually depleted in most macronutrients (Ca, P, Mg, K and S) and micronutrients (Cu, Mn, Zn, B and Mo) (Schaefer 2013), as well as pre-weathered by long term evolution under tropical conditions (Silveira et al. 2016). A closer look on the soil aspect of RGC will be found in the next section.

Fire is an integral part of RGC ecosystems. It plays a key role in RGC development and survival of plants since it controls the cycles of destruction, regeneration, maturation and reproduction, hence having a major influence in the selective pressure, that allowed the enormous diversity to emerge in rupestrian grassland.

2.3 Soils, Landforms and Rupestrian Grassland Vegetation on Different Lithologies in Brazil

Together with geology and landforms, the soil factor plays a key role in controlling the physiognomies of rupestrian grasslands, as further described and discussed in Chap. 3. In this section, we describe the combined landscape attributes (soils, landforms, vegetation, geology) and their relationships, in four case studies representing the main types of rupestrian grasslands, and associated plant communities: (1) the canga vegetation on ironstone from Carajás, Amazonia; (2) The Ferruginous Rupestrian Complex on itabirites from the Iron Quadrangle (Moeda plateau) (3) the rupestrian metapellite to metarenite from the semiarid Serra Central State Park; and (4) the rupestrian grasslands on quartzite from the Espinhaço Mountains in Sempre-Vivas National Park.

2.3.1 Ferruginous Rupestrian Grasslands

The ferruginous rupestrian grasslands are herbaceous and shrubby physiognomy, associated with outcrops of Fe-rich rocks (itabirites/jaspelites), usually formed from the decomposition of ferruginous substrates (Schaefer et al. 2015). The ferruginous rupestrian grasslands are included in the open vegetation types associated with rocky outcrops (Rizzini 1979; Jacobi and Carmo 2011), called Rocky Complex by Semir (1991). Veloso et al. (1991) classifies such vegetation as refuges or relics, floristically different from the surrounding dominant flora. The vegetation associated with ferruginous rock outcrops was also called Rocky Fields over canga (Morellato and Rosa 1991; Silva 1992). In Brazil, they occur mainly as open vegetation islands in the region of the Iron Quadrangle (Minas Gerais) and in the Serra dos Carajás (Pará State) (Figs. 2.6, 2.7). In both cases, the occurrence of peculiar soils reveal an extraordinary adaptation of plants to an extremely restrictive edaphic scenario, both physically and chemically. From a climate and landscape viewpoint, the two regions are very different. Hence, the Amazonian Serra dos Carajás represent a very useful soil system for comparison to the canga soils of Iron Quadrangle, all extremely weathered rocky substrates.

The Iron Quadrangle is a mountainous region located in the southern portion of the Espinhaço Range. The Iron Quadrangle possesses a structural landform, heavily controlled by the resistance of quartzite and itabirite rocks that form the "frame" of

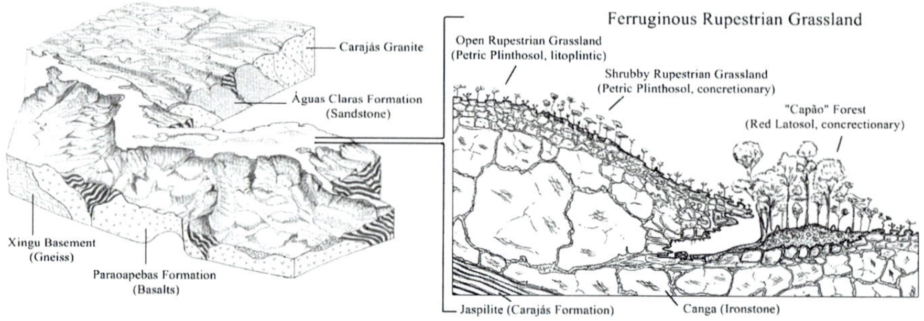

Fig. 2.7 Block-diagram of the Carajás plateau (Serra Sul) associated with ferrugineous rupestrian grasslands on the top surfaces, where ironstone (canga) is developed on Petric plinthosols. In the *right* picture, a sequence of soils at local scale is shown, as related to depth and drainage, affecting the distribution of open and shrubby physiognomies in these high landsurfaces according to topography

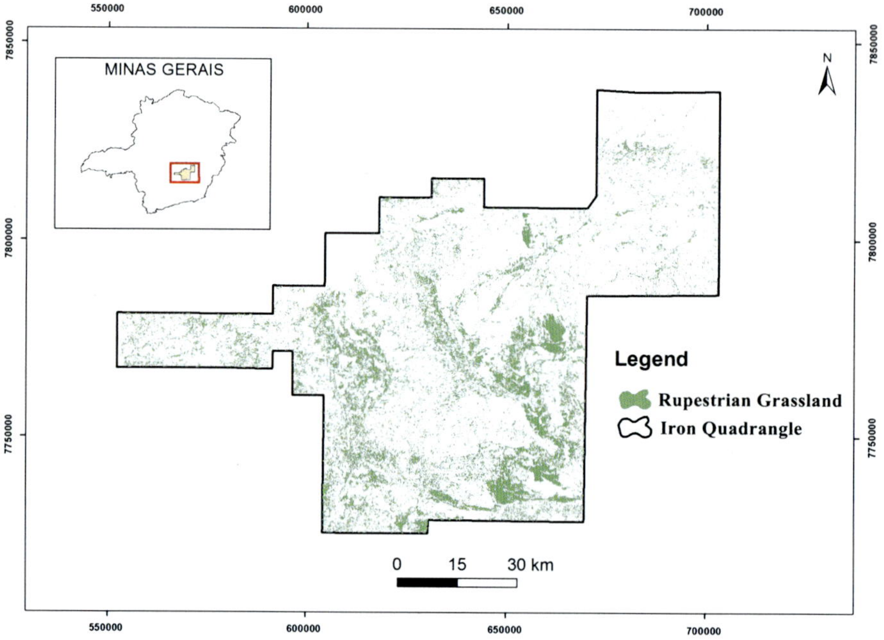

Fig. 2.8 Distribution of rupestrian formations in the iron quadrangle (Quadrilatero Ferrífero) as predicted from automatic classification using high resolution RapidEye and IKONOS imaging

the quadrangular highland region (Fig. 2.8). Because of its mineral wealth, extraction of mineral assets has taken place since the colonial period, when towns and cities emerged, as Mariana, Ouro Preto, Congonhas, Sabara, Nova Lima, among others. Currently, some studies have hypothesized an increase in mining activity and urbanization, exerting different pressures on remaining canga ecosystems (Jacobi et al. 2007; Carmo and Jacobi 2012; Diniz et al. 2014; Nunes et al. 2015). However, no quantification of these pressures (Jacobi et al. 2011), nor a realistic picture of the extent of current remaining areas exist.

In this mountainous region, the substrates rich in iron show the development of Rocky Fields, also called Ferruginous Rocky Fields, or "canga" vegetation (Vincent 2004; Viana and Lombardi 2007). Such ecosystems are distributed in the higher plateaus and in the mountain crests, mainly between 900–1800 m, on rocky substrates or ferruginous laterite. They are colonized by plants well adapted to oligotrophic environments, able to tolerate a many limiting environmental factors such as shallow depth soils, drought, low fertility, high concentration of oxidized iron, low water retention, marked daily temperature ranges, incidence frequent fire, high sun exposure and constant winds (Santos and Varajão 2004; Vincent 2004; Jacobi et al. 2007). Hence, the vegetation reveals numerous anatomical, morphological, physiological and reproductive adaptations that enable their survival in this harsh environment (Silveira et al. 2016; Chap. 11).

Based on the geological mapping available (Schobbenhaus and Brito-Neves 2003), an approximate total of 700,000 ha of Iron Quadrangle, approximately 90,800 ha (13 % of the area) are made up of iron formation. These formations have generally shallow soils, such as Litholic Neosols, Cambisols and Petric Plinthosols, with a complex of rupestrian grasslands vegetation, with savanna to montane forest formations, the latter in the pockets of Oxisols.

Although there is a relative knowledge about the flora found in the iron formations of the Iron Quadrangle, revealing its rich diversity, presence of endemic and endangered species (Porto and Silva 1989; Mendonça and Lins 2000; Vincent 2004; Jacobi et al. 2007; Mourão and Stehmann 2007; Viana and Lombardi 2007), there is still a lack of information about associated soils, due to its unsuitability to cultivation and extreme poverty. However, soil studies in the Iron Quadrangle can be very useful to developing tailored techniques for regeneration of native vegetation in mining, degraded areas by mining activities.

In the following sections, we present some of the main soils found in the canga environments of the Carajás and Iron Quadrangle, emphasizing its morphological chemical, physical and mineralogical attributes, and ecological relationships with the ferruginous rupestrian grasslands.

2.4 Canga: A Definition

The crusts of lateritic canga, or ferruginous "cuirasse" (ferricrete), comprise the surface formations resulting from processes associated with changes in landscape and relief, directly linked to polycyclic weathering and pedogenesis (Schaefer et al. 2015). Superficial formations on ironstone, in most cases, can be classified as soils originated through sub-aerial changes in banded–iron formations, and exerting control in the distribution of vegetation.

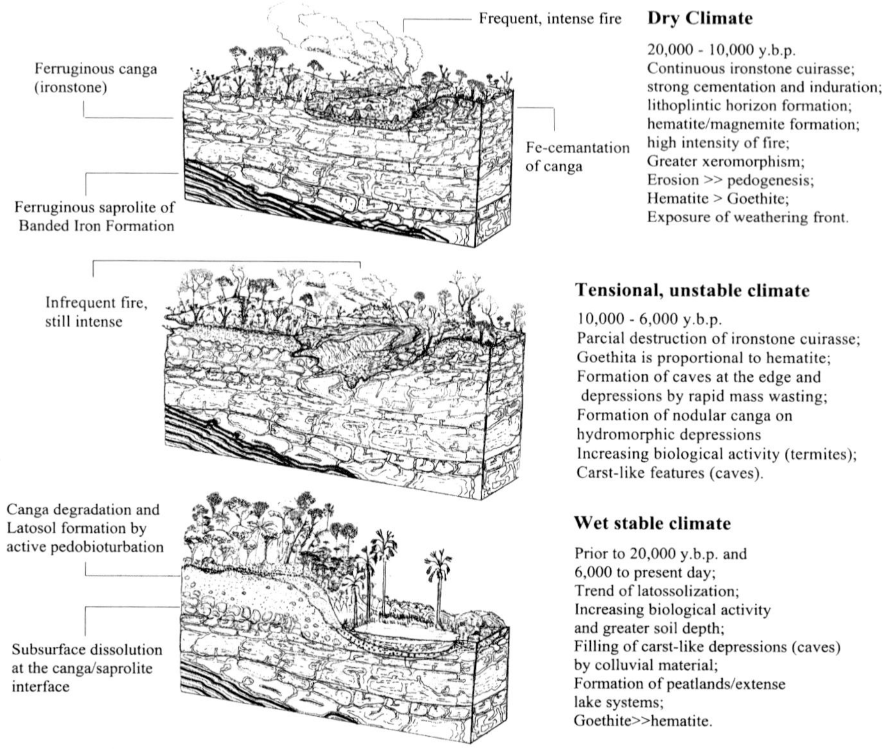

Fig. 2.9 Hypothetical cycle of canga formation and association with ferruginous rupestrian grassland evolution during dry and wet phases in the late quaternary (based on Schaefer et al. 2015). In the boxes, the most characteristic features of each phase are presented

The dynamics of formation of the ferricrete or ferruginous “cuirasse” at landscape level is an ongoing process, and strongly conditioned by Quaternary climatic variations, as illustrated in a hypothetical cycle in Fig. 2.9, starting with a dry phase and culminating in a wet phase. During the wet stages the canga degrades to produce a concretionary oxisol, with greater or lesser depth depending on the intensity of pedobioturbation, promoted by soil micro and mesofauna, especially termites and ants. This phenomenon promotes the selection and uptake of finer particles (clays and silts), which accumulates on the surface by biological turnover (Schaefer 2001). In the wet-dry climatic transitions, or vice versa, seasonality increases, and fire starts to rule the vegetation dynamics selecting resistant and/or tolerant species. Erosion losses increase, and the friable soil produced in the wet phase is markedly lost. In drier climate phases, there are outcrops of cemented canga (petroplinthite), where shallow soils prevail, with a litoplintic contact. This cycle can occur several times over the geological time, with variable duration of each phase without reaching a steady-state situation (equilibrium).

The canga formation is revealed by vertical and lateral movements of soluble and suspended products of weathering, responsible for the formation of soil horizons (Tardy and Nahon 1985). The alteration mantle on ferriferous itabiritic consists of a thick saprolite (30–300 m), wich forms the commercial interest ore. It is composed of iron oxy-hydroxides, kaolinite, gibbsite and eventually quartz, all alteration products of itabiritic rocks. In the upper part of the canga (petroplinthite), the degradation of the laterite is marked by the presence of pisolitic canga, with underground cavities, filled by small spherical aggregates (pisolites), composed by Fe and Al oxides, and several generations of coatings forming a typical coalesced hardened structure, with cracks, biological channels and tubules, with small laterite fragments and grains cemented by hematite/goethite (Schaefer et al. 2004).

These features are formed in acidic conditions and seasonal reduction, probably due to the accumulation of organic matter and water at the surface, which can cause dissolution of kaolinite, gibbsite and some of goethite, with remobilization of Si, Fe, and Al (Schaefer et al. 2004, 2008). While most of the Si is leached out of the profile, Al and Fe are repreciptated, after downward migration of colloidal solutions, followed by changing chemical conditions with depth and degradation of Al–Fe unstable organic complexes. Therefore, there is a cyclical succession of formation and destruction of the canga, with limited lowering of the landscape (Schaefer et al. 2008).

Subsurface degradation of the laterite crust occurs with the progressive lowering of the water table level, caused by regional uplift, which intensifies the vertical and lateral migration of aqueous solutions, reactivating the change process. For their insolubility, Al and Fe remobilized from the surface canga degradation is deposited in the form of gibbsite and iron oxyhydroxide along fractures and cracks on the top of the saprolite, forming a dense system of veins of gibbsite or goethite/hematite on the regolith.

2.5 Case Study 1—Island of Ferruginous Rupestrian Physiognomy in the Midst of the Amazonia Forest: A Sequence of Carajás Plateau, Brazilian Eastern Amazonia

Banded iron formations and associated ironstone soils and landscapes are hotspots of plant diversity around the world (Gibson et al. 2012; Porembski et al. 1994, 1997; Valentin et al. 1999; Vincent et al. 2002; Yates et al. 2007), supporting plant communities distinctive both structurally and floristically from surrounding vegetation (Gibson et al. 2010, 2012; Jacobi et al. 2008; Yates et al. 2011). These ironstone substrates present a series of restrictions to plant establishment, including shallow and patchy soils (resulting in low water-retention capacities and low nutrient availability), high insolation levels, elevated temperatures, and the presence of potentially toxic metal concentrations (Silva and Rosa 1984; Meirelles et al.

1997). The topographic and edaphic characteristics of rupestrian grassland rock outcrops likely influence the floristic composition of their vegetation cover, as described for many rocky outcrop environments elsewhere in the world (Gibson 2004a, b; Markey and Dillon 2008; Meissner and Caruso 2008; Porembski et al. 1994; Schaefer et al. 2009).

In this section, we present three distinct vegetation types that make up the RGC on the ferruginous canga, ranging from open rupestrian grasslands and shrubby rupestrian grasslands on rocky sites of the upper plateau (746–762 m asl), to low forest at lower elevations. Open rupestrian grasslands is an open formation of small shrubs, with *Vellozia glochidea* (Velloziaceae) dominant over a ground layer of grasses; shrubby rupestrian grassland is supported on more fragmented canga soil that allow greater root penetration, with a mix of tall shrubs and herbaceous plants; "capão" forest is a semi-deciduous forest formation isolated as small islands within the matrix of herbaceous and shrubby rupestrian grassland (see Chap. 7).

2.5.1 Study Site

The study was located on a plateau of ironstone canga (lateritic capping) known as Serra Sul (738–762 m a.s.l.) in the Carajás National Forest (FLONA) conservation unit, southeastern Pará State, Brazil (Fig. 2.10a). Ferruginous rupestrian grassland islands cover approximately 9,031 ha (2.28 %) of the total FLONA area of 395,827 ha.

The climate in the region is tropical, hot and humid (type "Aw" in the Köppen system; Ab'Saber 1986), with a dry season between May and October (average precipitation < 60 mm in the driest months), and a rainy season between November and April. Rainfall increases with altitude, lowland areas receiving annual averages of 1500 mm and more elevated sites receiving up to 1900 mm/year (IBAMA 2003). Average monthly temperatures vary between 19 and 31°C.

The RGC on ferruginous canga in Carajás ranges from open rupestrian grassland and shrubby rupestrian grassland on rocky sites to "capão" forest and fringing montane forest where the ironstone is fragmented and weathered to greater depths (see Fig. 2.7).

2.5.2 Open Rupestrian Grassland

The open rupestrian grassland (ORG) represents areas where rocky Fe-rich canga outcrops, with soil pockets rarely exceeding 10 cm, usually less than 5 cm of organic-rich soil. This organic surface layer is basically derived from termite nest surface erosion, spreading a friable soil that fills fractures and micro depressions on canga, where roots of *Vellozia glochidea* and tussocks of *Sobralia liliastrum* are dominating. At wetter areas usually shaded, these species display a taller development, where water deficit is reduced. The frequent fire finds difficult propagation

Fig. 2.10 National forest of Carajás, Pará state, Brazil. **a** Serra Sul plateau (06 24' 36,4" S and 50 21' 29,4" W), with the sequence of physiognomies found on ironstone. **b** Open rupestrian grassland. **c** Shrubby rupestrian grassland. **d** "capão forest

due to the existence of natural barriers of canga (lithoplintic) outcrops where no vegetation grows (Figs. 2.7, 2.10b).

The grassy, herbaceous physiognomy is dominating, but shrubby are common, side by side with herbs *Ipomoea carajaensis, Norantea goyasensis, Ipomoea marabaensis, Sobralia liliastrum.* According to Schaefer et al. (2008), soils in the ORG are Petric Plinthosols, Lithoplintic, dystric, a very shallow soil.

2.5.3 Shrubby Rupestrian Grassland

The shrubby Rupestrian grassland formation (SRG) is the most common physiognomy on the Carajás Plateau, with a marked homogeneity of the shrubby stratum, with variations in dominance and abundance. Soils are slightly deeper than in the ORG previously described, with depth reaching 30 cm. The canga (ironstone) is much more fragmented, giving way to a loose soil with easy root penetration. The alternation of soils with varying depths indicate a severe morphogenesis in the canga plateau and great erosion losses. Termite activity is also a keystone, with similar abundance compared with ORG (Fig. 2.10c). Trees are mainly *Callistene minor, Byrsonima coriacea, Bauhinia pulchella, Eugenia punicifolia, Alibertia cf. longiflora* e *Mimosa acutistipula* var. *ferrea.*

The Petric Plinthosols are usually concretionary or lithic, but always deeper (10–35 cm) than ORG (Schaefer et al. 2008).

2.5.4 "Capão" Forest

With soil development and accumulation, trees are commonly found, forming small islands of forest with ovoidal or circular shapes, usually associated with internal depressions or cave systems, where fine particles and sediments sink and accumulate (Fig. 2.10d). One such forest has been studied. According to Schaefer et al. (2008), a collapse of loose saprolite underneath the cemented canga (litoplintic) layer led to colluvial accumulation of deep loose soil material, well-aggregated and with good water retention depths are typically within 30–120 cm, with Latosols, Cambisols and Petric Plinthosols (deep, concrectionary) dispersed, fragmented cores of canga forming concrections. Termite mounds are abundant, formed by hematite-rich soil, rather than organic-rich sapric, black material, as in the case of ORG. The main forest species are *Pouteria ramiflora, Sacoglottis mattogrossensis* and *Myrcia splendens.*

2.6 Case Study 2—Canga Soils of the Iron Quadrangle, Serra Da Moeda, Minas Gerais, Brazil

The soil sequence in the Iron Quadrangle is rather similar to that of Carajás, and similar soil depths control the same physiognomy, in Carajás and Iron Quadrangle. The soils developed on Fe-rich substrates from Moeda are perferrics (>36 % Fe_2O_3, by sulfuric acid digestion, EMBRAPA 1997), and have clay textures, but are strongly affected by the presence of nodules and pisolithic concretions responsible for high levels of silt present (as microaggregates difficult to disperse) and sand (micronodules of hematite), in addition to gravels (bigger concretions, nodules, etc.). The water-dispersible clay content in water is low, especially in the Bw and Bi horizons.

The soils on canga (Table 2.1) show up a very poor nutrient status, with extremely low exchange capacity, reflecting its oxidic nature. The pH values are slightly acid, as a result of high weathering degree, which practically left an exhausted cation exchange capacity, which depends almost entirely of organic matter. This organic matter, in turn, is very poor and acidic, and mostly made up of black-carbon, resulting from incomplete burning (Benites et al. 2005, 2007). Fire appears to be a fundamental factor in the genesis of canga landscapes.

Table 2.1 Chemical characteristics of soils from Serra da Moeda

Hor.	Depth (cm)	pH		Ca^{2+}	Mg^{2+}	K^+	Na^+	SB[1]	Al^{3+}	H^+	T[2]	V[3]	COT	N[4]	m[5]	SiO_2	Al_2O_3	Fe_2O_3	TiO_2
		H_2O	KCl	$cmol_c\ kg^{-1}$								%							
P1—Red latosol, perferric typic (petroplintic), humic—Savanna grassland																			
AP	0–12	5.4	4.6	0.5	0.5	0.08	0.02	0.6	0.2	7.0	6.1	7	2.25	0.19	25	0.9	13.1	65.5	2.10
A1	12–35	5.7	5.1	0.3	0.3	0.04	0.02	0.4	0	5.2	5.6	7	1.82	0.14	0	0.9	12.7	65.0	1.97
A2	35–55	5.9	5.5	0.1	0.1	0.01	0.01	0.1	0	3.9	4.0	3	1.4	0.11	0	0.9	13.4	67.7	2.01
A3	55–80	5.9	5.7	0.1	0.1	0.01	0.01	0.1	0	3.0	3.1	3	1.36	0.10	0	0.8	12.5	69.3	1.95
AB	80–115	6.1	6.2	0.1	0.1	0.01	0.01	0.1	0	2.1	2.2	5	0.93	0.09	0	0.9	13.0	69.2	1.97
Bwc1	115–150	6.2	6.4	0.1	0.1	0.01	0.01	0.1	0	1.1	1.2	8	0.55	0.06	0	0.9	15.4	67.5	1.77
Bwc2	150–190	6.3	6.5	0.1	0.1	0.01	0.03	0.1	0	0.6	0.7	14	0.32	0.04	0	0.7	20.3	61.5	2.04
Bwc3	190–205+	6.2	6.5	0.1	0.1	0.02	0.03	0.2	0	1.1	1.3	15	0.4	0.05	0	0.7	20.2	59.2	2.03
P2—Red latosol, perferric typic—Rupestrian grassland																			
A	0–50	5.5	5.1	0.8	0.8	0.05	0.02	0.9	0	5.0	5.9	15	2.19	–	0	1.1	13.8	64.6	1.89
Bwc	85–115	5.9	6.2	0.2	0.2	0.01	0.01	0.2	0	0.6	0.8	25	1.16	–	0	1.3	15.9	64.2	1.99
P3—Haplic cambisol, dystric.perferric—Rupestrian grassland																			
A	0–40	5.8	5.2	0.8	0.8	0.03	0.02	0.9	0	1.4	2.3	–	0.62	–	0	0.8	0.6	66.4	0.18
Bic	40–60	6.1	6.7	0.6	0.4	0.01	0.02	1.0	0	0.2	1.2	–	0.18	–	0	1.3	2.4	63.9	0.20
P4—Haplic cambisol, dystric, perferric—Rupestrian grassland																			
AB	15–85	5.4	5.7	0.4	0.4	0.01	0.02	0.4	0	1.8	2.2	–	0.51	–	0	1.5	9.2	23.5	0.65
Bic	100–160	5.5	6.6	0.2	0.2	0.02	0.04	0.3	0	0	0.3	–	0.09	–	0	1.2	9.4	38.5	0.69
P5—Haplic cambisol, perferric—Dwarf forest																			
A1	0–20	4.57	–	0.21	0.09	0.13	0	0.43	2.60	24.4	24.83	1.7	4.70	–	85.8	–	–	–	–
A2	20–40	4.53	–	0.28	0.10	0.13	0	0.51	2.80	21.8	22.31	2.3	3.89	–	84.6	–	–	–	–
Bi	40–60	4.59	–	0.46	0.13	0.15	0	0.74	2.80	27.7	28.44	2.6	4.06	–	79.1	–	–	–	–
Bi	60–80	4.59	–	0.39	0.14	0.14	0	0.67	2.60	25.1	25.77	2.6	3.83	–	79.5	–	–	–	–

(continued)

Table 2.1 (continued)

Hor.	Depth (cm)	pH		Ca^{2+}	Mg^{2+}	K^+	Na^+	SB[1]	Al^{3+}	H^+	T[2]	V[3]	COT	N[4]	m[5]	SiO_2	Al_2O_3	Fe_2O_3	TiO_2
		H_2O	KCl	$cmol_c\ kg^{-1}$								%							
P—Haplic Cambisol, perferric—Dwarf forest																			
A1	0–20	4.80	–	0.25	0.14	0.20	0	0.59	1.60	16.2	16.79	3.5	2.73	–	73.1	–	–	–	–
A2	20–40	4.82	–	0.22	0.12	0.16	0	0.50	1.40	14.9	15.40	3.2	2.44	–	73.7	–	–	–	–
Bi	40–60	4.84	–	0.19	0.11	0.16	0	0.46	1.20	14.9	14.36	3.0	2.32	–	72.3	–	–	–	–

Ca, Mg and K levels are extremely low, consistent with results from elsewhere. The total organic carbon shows little significant values for horizons identified as humic, where organic matter coats aggregates and micronodules. The total N levels are too low and reflect the general poverty of the soil environment.

2.7 General Aspects of Canga Soils—A Discussion

Diverse terminologies have been used previously to describe the (mostly shrubby) vegetation growing on ferruginous laterites (canga) in eastern Amazonia. They have been classified as canga vegetation (Morellato and Rosa 1991; Secco and Mesquita 1983; Silva 1992), banded ironstone formations, ferruginous rupestrian grassland (Jacobi et al. 2007, 2008; Rizzini 1979; Vincent and Meguro 2008), and even metallophile savanna (Porto and Silva 1989). The plant species found growing there often show dwarfing (Porto and Silva 1989) and some may retain high concentrations of metals in their tissues (Teixeira and Lemos-Filho 1998; Vincent 2004). These high soil-metal concentrations may lead to the selection of tolerant individuals in these environments (Porto and Silva 1989), resulting in vegetation that is structurally and floristically distinct from the surrounding vegetation matrix. Regardless of differences between soil variables for the four areas sampled here, all were acidic and had low phosphorus levels; a resulted so found in other studies in ironstone (Markey and Dillon 2008, 2009, 2010; Meissner et al. 2009; Meissner and Wright 2010; Messias et al. 2013; Teixeira and Lemos-Filho 2002; Vincent and Meguro 2008). According to Jacobi and Carmo (2011), the comparison of open areas with forest islands, subjected to the same climatic conditions, shows the important influence of edaphic and microclimatic conditions in determining life-form spectra and species composition in these plant communities. The isolation of the edaphic environments of inselbergs produce endemic species, or species with distributions restricted largely to these conditions (Porembski et al. 1994), resulting in vegetation that is structurally and floristically distinct from the surrounding forest matrix, as reported by Viana and Lombardi (2007).

Soil bioavailable Al were markedly higher in open rupestrian grassland than in shrubby rupestrian grassland. High concentrations of Al can be toxic to plants (Table 2.2), but high levels of soil carbon can reduce the potentially toxic effects of high concentrations of Al (Barros 1979), and this may provide amelioration of potential toxicity effects in open rupestrian grassland. Similarly, although Al concentrations are high for surface soil in “Capão” forest, most nutrient cycling here is associated with the litter layer rather than with mineral soil. Melich-extractable Fe concentrations were high in all sites, with highest levels found in the rupestrian grassland sites (open and shrubby). The values for extractable Fe found here are higher than those reported by Silva (1992) for hematite-rich canga in Serra Norte, Carajás, and by Vincent and Meguro (2008) in the Iron Quadrangle of Minas Gerais, but there is no evidence to suggest that these high levels for extractable Fe (as Fe^{2+}) are driving vegetation differences among sites, since most Fe is present in

Table 2.2 Chemical characteristics of soils of the Carajás plateau (Serra Sul), Pará state, Brazil

Hor.	Depth (cm)	pH (1:2, 5)		ΔpH[1]											V[5]	m[6]	COT[7]	N[8]	C/N	Prem	Zn	Fe	Mn	Cu
					P	K	Na	Ca^{2+}	Mg^{2+}	Al^{3+}	H + Al	SB[2]	t[3]	T[4]										
		H_2O	KCl		mg dm^{-3}			$cmol_c$ dm^{-3}							%		dag kg^{-1}			mg L^{-1}	mg dm^{-3}			
Forest—Petric Plinthosol, concretionary, latosolic—"Capão" forest on deep soil of degraded canga																								
A	0–10	3.55	3.16	−0.39	16.9	21	1,5	0	0	2.89	21.9	0.07	2.96	21.97	0.3	97.6	9.8	0.12	80,0	21.7	0.77	236.3	0.8	0.37
AB	Oct-15	4.39	3.99	−0.4	18.9	8	0	0	0	1.06	10.8	0.02	1.08	10.82	0.2	98.1	2.1	–	–	13.8	0.79	124	2.3	0.67
Bwc1	15–80	4.2	4.09	−0.11	15.7	6	0	0	0	0.67	10.2	0.02	0.69	10.22	0.2	97.1	2.4	0.21	11,4	12.5	0.72	199.7	1.2	1.3
Bwc2	80–110⁺	4.2	4.08	−0.12	13.4	7	0,5	0	0	0.58	10.3	0.02	0.6	10.32	0.2	96.7	1.8	–	–	12.5	1.56	110.9	0.9	0.88
ORG—Petric plinthosol, typic, lithoplintic—Slopes with rupestrian grassland of ferriginous canga with Vellozia																								
A	0–5	4.89	3.39	−1,5	3.4	35	5.5	0.47	0.17	0.87	19.1	0.75	1.62	19.85	3.8	53.7	20.5	1.02	20,2	14.9	1.14	372.5	2.7	0.36
SRG—Petric plinthosol, typic, concrectionary—Slopes with shruby rupestrian grassland of ferruginous canga																								
A	0–25	5.51	4.02	−1.49	5.1	56	9.6	1.02	0.21	0.87	15.6	1.41	2.28	17.01	8.3	38.2	13.04	0.82	16,0	21.9	0.89	459	18,7	0.78
Af	25–30	6.01	5.12	−0.89	3.3	25	5.5	0.16	0.04	0.29	6.2	0.28	0.57	6.48	4.3	50.9	2.59	0.14	18,6	10.3	0.46	131.4	4.5	0.37

soil as Fe^{3+}, being insoluble and non-available. This indicates that seasonal reducing conditions occur in ferruginous rupestrian grasslands, increasing the amount of Fe^{2+} released from Fe-oxide dissolution. Further work on the effective role of high metal concentrations is recommended.

Botrel et al. (2002) reported that areas with dry soils and low chemical fertility tend to have fewer species, but in our study observed species richness was generally high throughout. Nevertheless, based on species richness estimators, total species richness was much higher in the extensive fringing montane forest, and also higher in "capão" forest, than in rupestrian grassland on canga (Nunes et al. 2015). In addition, the four vegetation types held few species in common, suggesting strong edaphic filtering from the available plant species pool acting most strongly in the rupestrian vegetation types. The composition of the isolated "capão" forest and rupestrian grassland sites may represent floras more specialized to the environmental filters associated with these ironstone habitat islands (Viana and Lombardi 2007).

According to Lista de Espécies da Flora do Brasil (2015), most of the species found in montane forest are widely distributed in Amazonian rainforest, including *Allophyllus floribundus, Cupania hispida, Dipterix odorata, Eugenia brachypoda, Eugenia patrisii, Guatteria foliosa, Inga stipulacea, Inga falcistipula, Nealchornea yapurensis, Vismia latifolia* and *Myrcia paivae*. In contrast, while some species in "capão" forest are also found in other Brazilian biomes (e.g. *Sacoglottis mattogrossensis, Myrcia splendens, Sclerolobium paniculatum, Matayba arborescens* and *Tapirira guianensis*), many are unique to this vegetation type in the Amazon area, including *Licania impressa, Alchornea fluviatilis, Connarus perrottetii, Cupania macrophylla, Emmotum fagifolium, Mouriri angulicosta, Ocotea tomentela,* and *Virola callophylla*. "Capão" forest also holds some species in common with the rupestrian grassland (*Callisthene minor, Eugenia punicifolia, Pouteria ramiflora* and *Myrcia cuprea*), as found also by Viana and Lombardi (2007).

The rupestrian complex (open and shrubby rupestrian grassland) of Carajás comprises a mix of widely distributed and locally endemic plant species. Several species are shared with the Brazilian Cerrado biome (e.g. *Chamaecrista desvauxi, Anemopaegma scabriusculum, Lippia alba, Norantea goyazensis, Guapira ferruginea, Pouteria ramiflora, Callisthene minor*) and Caatinga (e.g. *Pilosocereus* sp., *Mimosa acutistipula, Periandra mediterranea, Eugenia punicifolia, Polygala* sp.), while a few (including *Erythroxylum nelson-rosae, Ipomoea marabaensis, Ipomoea carajasensis, Ipomoea cavalcantei* plus two varieties of *Mimosa* and one of *Erythroxylum*) are restricted solely to ferruginous habitats in Carajás. However, ongoing studies (Schaefer et al. 2009; Nunes et al. 2015) indicate that many species considered ironstone lovers (e.g. *Vellozia glochidea*) are widely distributed across open vegetation supported by different lithologies in this region, not just ironstone outcrops. Overall, the species composition of vegetation types of the banded-iron canga islands is unique and shows very high beta diversity, reflecting a high diversity of habitats defined by soil physical and chemical properties. The high conservation value of these habitat islands, and their potential loss through anthropogenic impacts, requires urgent conservation management action (Nunes et al. 2015).

Table 2.3 Physical characteristics of soils of the Carajás plateau (Serra Sul), Pará state, Brazil

Horiz.	Depth (cm)	Color (Munsell)	cs[1]	fs	Silt	Clay	Textural class
			dag kg^{-1}				
Forest—Petric plinthosol, concretionary, latosolic—"Capão" forest on deep soil of degraded canga							
A	0–10	2.5YR 2.5/3	23	7	40	30	Sandy loam
AB	out/15	10R3/6	25	6	46	23	Loamy
Bwc1	15–80	10R3/6	24	6	47	23	Loamy
Bwc2	80–110+	10R3/6	25	7	44	24	Loamy
ORG—Petric plinthosol, typic, lithoplintic—Slopes with rupestrian grassland of ferriginous canga with Vellozia							
A	0–5	2.5Y 5/3	20	7	44	29	Sandy loam
SRG—Petric plinthosol, typic, lithoplintic—Slopes with shruby rupestrian grassland of ferruginous canga							
A	0–25	2.5 YR 3/4	45	5	21	29	Sandy clay loam
AF	25–30	2.5 YR 3/6	44	7	18	31	Sandy clay loam

Nunes et al. (2015) showed that chemical and physical soil attributes (Tables 2.2, 2.3) have a role beyond the definition of different phytophysiognomies, hence affecting species composition. For the latter, the environmental partitioning model explained 43 % of the observed variance. Soil variables, including pH and soil depth provided the greatest contribution explaining species composition, while combined soil chemical and physical properties also contributed in explaining the species distribution patterns observed. Differences in soil chemical attributes allowed clear separation of the rupestrian grassland sites, whereas soil depth appears to be a key factor for establishment and development of the forest sites.

Schaefer et al. (2009) noted that soil depth seemed to play a fundamental role in the definition of the phytophysiognomies seen in the Carajás Plateaux, which is supported by the relationships observed for soil depth in differentiating forest (deeper soils) and non-forest areas (shallower soils) on the one hand, and of SRG (deeper) and ORG (shallower) on the other. There is a clear vegetation gradient from ORG to "capão" forest accompanying the gradient of soil depth: ORG < SRG < "capão" forest. Yates et al. (2011) similarly identified species habitat preferences associated with soil depth for banded-iron plant communities in Western Australia.

Despite their shallowness, non-forest (open and shrubby rupestrian grassland) soils have higher surface (0–10 cm) organic matter levels than forest soils, probably due to the presence of charred organic matter produced by frequent fire. Thus, fire maybe an important environmental filter contributing to the floristic and structural differences in the vegetation. Furthermore, higher termite activity in the rupestrian grassland sites may also contribute to higher levels of surface soil organic carbon through redistribution of organic-rich materials (Schaefer et al. 2009). In their studies of the influence of termite activity on soil characteristics in Minas Gerais, Sarcinelli et al. (2009) concluded that high concentrations of organic matter, P and nutrients in termite mounds (in relation to the surrounding soils) were associated with the active incorporation of organic material during their construction. The higher organic matter in open rupestrian grassland may be also attributed to lower water availability and slower organic matter decomposition, and to differences in the soil fauna available for organic matter processing.

Thus, we can infer that edaphic (pedological) factors are crucial in explaining the types and distributions of rupestrian grasslands vegetation associated with ferruginous ironstone uplands (canga) in Carajás, eastern Amazon. Three different ferruginous vegetation types were identified, each with distinct soil physical and chemical, and floristic and structural characteristics. Vegetation differences were associated with variations in soil chemistry, depth and acidity, with forest soils showing greater depths, and higher concentrations of P and Al, but much lower levels of organic matter and exchangeable cations at the surface, reflecting major differences in nutrient cycles between forest and non-forest sites. The higher nutrient and organic carbon contents in rupestrian grasslands may be indirectly associated with frequent fire, filtering those species able to persist through recurrent fire events. Soil organic matter content is also likely a key variable for the development of the formations studied, especially in the ORG, where most of the inorganic nutrients were correlated with organic matter (carbon) amounts in the topsoil. High soluble Fe concentrations do not appear to restrict the establishment of species in the different formations, as this element was found in similar amounts throughout the study area. As such, the use of the term metal-loving savanna (or metallophile savanna) does not appear appropriate for the ironstone canga vegetation or rupestrian grassland at Carajás, and we suggest the use of the term ferruginous rupestrian grasslands complex for the open and shrubby rupestrian grasslands formations.

2.8 Case Study 3—Rupestrian Grassland on Metarenites and Metapellites from Serra Geral State Park, North Minas Gerais

Within the Brazilian semi-arid, there are many outliers of Proterozoic rocks (metarenites, quartzites, metacherts, metapelites) that form imposing Mountains in the ecotonal zone between the Atlantic Forest, Caatinga and Cerrado domains. On the rocky tops and escarpments of Serra Geral State Park, rupestrian formations are observed on very shallow soils (Fig. 2.11) (Rodrigues et al. 2014). In the following section, we discuss typical geoenvironments of rupestrian grasslands of this State Park, emphasizing and illustrating the most important landscape relationships. The landscapes associated with the soils described are illustrated in Fig. 2.11a–d. The Serra Central is bordered by the Limestone of Bambuí Group, to the west face (Fig. 2.11a), or basement rocks at the east facing slope (Fig. 2.11b), whereas dry forest occur on the depression lowlands, both east and west, downslope the escarpment.

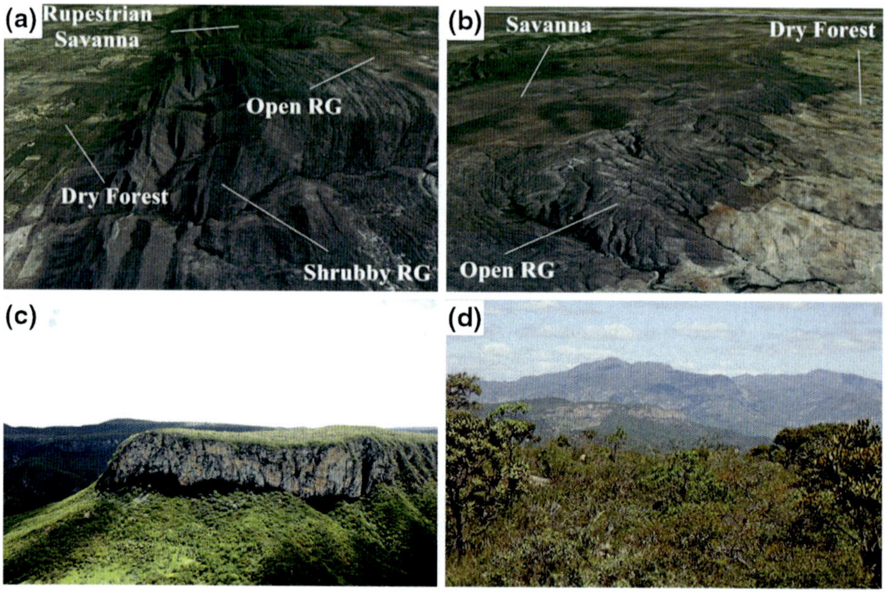

Fig. 2.11 A three-dimensional picture of the landforms at the Serra Geral state park, Minas Gerais states, illustrating the typical geoenvironments of rupestrian grasslands at the park, emphasizing and illustrating the most important landscape relationships. To the western slopes, the serra central is bordered by the limestone depression of bambuí group (São Francisco Depression) (**a**); to the east face, the plateau escarpment connects with the Espinosa depression, under semi-arid climates (**b**). In both depressions, dry forest occur on the lowlands, both east and west, downslope the escarpment. There is an abrupt change in vegetation as one approach the escarpment

2.8.1 High Top with Open Rupestrian Grassland on Metapellites

At the highest landscape position, shallow to moderately developed soils resting on rocky surfaces are gravelly and concretionary, being classified as Cambisols on metapelites/metacherts. Soil profile PSG1 (Table 2.4; Fig. 2.11d) is a Dystric Haplic Cambisol, clayey and gravelly, with very low fertility, very high toxic Al, low cation exchange capacity and sum of bases at the surface, as well at subsurface—2Bi horizon.

These unusual clayey soils are not typical of the RGC, but occur rather ocasionally, possessing a savanna-like physionomy. In these formations, high abundance of *Vochysia rufa* is diagnostic, besides *Tachigali aurea*, *Vellozia* sp., *Byrsonima pachyphylla*, *Vochysia elliptica*, *Chamaecrista orbiculata*, *Davilla rugosa* and *Miconia albicans*. The character of these formations is the marked abundance of Al- Hyperaccumulating plants, indicating extreme acidity and Al release from clays (Campos et al. 2014).

Table 2.4 Chemical characteristics of soils of Serra Geral state park, northern Minas Gerais, Brazil

Soil	Hor.	pH	pH	P	K	Na	Ca	Mg	Al	H-AL	SB	t	T	V	M	ISNa	OM	Prem	Zn	Fe	Mn	Cu
		H^2O	KCL	mg/dm^3			cmol/dm^3							%			dag/kg	mg/L	mg/dm^3			
	A	4.32	3.92	1.6	26	0	0	0.01	1.37	5.3	0.08	1.45	5.38	1.5	94.5	0	1.61	32.2	0.53	119.9	0.8	0.19
PSG1	2Bi	4.67	4	2	22	0	0	0.01	1.27	4.5	0.07	1.34	4.57	1.5	94.8	0	1.61	32.3	0.54	20	0.6	0.03
	2Cr	4.61	3.87	0.6	6	0	0	0	3.8	6	0.02	3.82	6.02	0.3	99.5	0	0.27	33.4	1.44	8.1	0.4	0.29
	A	4.62	3.63	1.5	24	3.3	0.07	0.04	1.56	7.4	0.18	1.74	7.58	2.4	89.7	0.82	4.16	41	0.72	75.1	1.2	0.14
PSG2	AC	4.56	3.76	1.6	14	5.3	0	0.01	1.66	6.6	0.07	1.73	6.67	1	96	1.33	3.22	43.3	0.99	167.6	0.6	0.35
	Cr	4.75	3.86	1.2	10	1.3	0.03	0.02	0.88	3.9	0.09	0.97	3.99	2.3	90.7	0.58	1.48	41.6	0.94	155.8	1.4	0.45
PSG3	A	4.8	3.85	2.3	36	4.7	0.06	0	1.07	4	0.17	1.24	4.17	4.1	86.3	1.65	1.88	45.9	0.88	42	0.8	0.63
	Cr	4.83	4.02	0.9	10	2.8	0	0	0.68	1.9	0.04	0.72	1.94	2.1	94.4	1.69	0.4	60	0.8	12.2	0.2	0.6

2.8.2 Slope and Crests with Open Rupestrian Grassland with **Vellozia** on Rocky Outcrops

In the steepy slopes at the fringes of Serra Central Plateau, or within the interior valleys where deep dissection took place, the metarenites and quartzite substrates outcrop as resistant rocks. There, typical rupestrian formations develop, following the structural alignments with the main geotectonic trend (NE/SW). Due to isolation within the semiarid domain, these Rupestrian Grasslands are characterized by high endemism, and presence of rare species. On rocky surfaces, lichen and mosses are very common, along crevices and shadowed faces. Soil occur as discontinuous pockets, rich in organic matter, but with very low fertility.

A representative soil is a Litholic Neosol, dystric, fragmentary (PSG2, Table 2.4). The sandy texture with a sequence of A-AC-Cr horizons, with unaltered rock and a lithic contact within 50 cm depth. Acid pH, very low contents of available P and CEC are typical.

The rupestrian vegetation is characterized by dominance of tall *Vellozia* sp, with common low trees and shrubs dominated by *Vochysia rufa*, *Lychnophora* e *Chamaecrista*, *Vochysia elliptica*, *Tachigali aurea* and *Byrsonima pachyphylla*.

2.8.3 Rupestrian Savanna (Cerrado Rupestre) on Escarpments

This environment represent an open to shrubby rupestrian grassland, in which many species of savanna (Cerrado) are found, on deeper soils. In the steep slopes and escarpments along the Plateau borders, shallow soils under strong erosion are covered by unusual savanna formation, with a rupestrian character. Soils are usually Regolitic Neosols (e.g. PSG3), dystric, typic, with a sequence of A-Cr horizon, and saprolite down to 50 cm, rather than fresh rock (Table 2.4). Texture is sandy loam with a predominance of fine sand (0,2–0,05 mm). Roots and biological channels are common, and many termite mounds highlight the importance of pedobioturbation in nutrient cycling in these soils. Values of cation exchange capacity, available P, organic matter and base saturation are very low, with a very acid pH (4.83 in the saprolite Cr), indicating very low nutrient reserves in the parent rock.

The vegetation of rupestrian savanna (cerrado rupestre) is characterized by trees and shrubs, mainly *Vochysia rufa*, *Byrsonima pachyphylla*, *Myrsine guianensis* (Aubl.) Kuntze, *Eremanthus erythropappus* (DC.) MacLeish, *Tibouchina* sp. and *Tachigali aurea*.

2.9 Case Study 4—A Soil Sequence of Quartzite and Micaceous Quartzite from the Espinhaço Range in the Sempre-Vivas State Park, Minas Gerais States

2.9.1 Study Area

The Espinhaço Meridional Range represent an ancient orogenic belt bordered by the São Francisco Craton, to the west, extending for 300 km in a N-S direction, from the Iron Quadrangle to the northern Minas Gerais. This is essentially built by the Espinhaço Supergroup (mainly quartzites rocks) with minor phyllites, conglomerates and acid volcanics (Almeida-Abreu 1995) (Fig. 2.12).

As part of Espinhaço Range, the National Park of Sempre-Vivas covers an 124,555 ha (IBAMA 2006) with an altitudinal range of 650–1,350 m (Almeida et al. 2007). The climate is wet subtropical, mesic (Cwb, according to Köppen's type) with mean annual temperature of 20° C, and rainfall of 1,500 mm (Almeida et al. 2007).

Approximately 80 % of the Park is on quartzites of Galho do Miguel Formation (Espinhaço Supergroup), whereas the remaining are minor metapelitic to conglomerate rocks, from the Macaúbas and Bambuí Groups.

The Galho do Miguel Formation is characterized by orthoquartzites, pure and fine-grained in more than 10 % of the unit. The remaining 10 % is basically

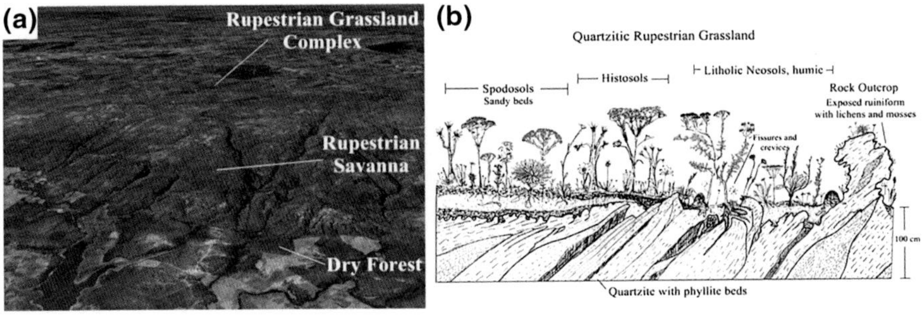

Fig. 2.12 a Three dimensional diagram of the west-facing slope fault escarpment of the Sempre-Vivas national park (Espinhaço Range) under rupestrian grassland complex; the São Francisco depression in the foreground, down the escarpment front (under dry forest). In the detailed scheme. **b** The most typical rupestrian grassland complex (RGC) on quartzites from the Sempre-Vivas park is illustrated. The RGC show a variety of landforms at microscales, from Karst-like rocky pinnacles of exposed cavernous (ruiniform) quartzite, to low lying depression where sand accumulated by aeolian or colluvial deposition form sandy plains, similar to Heathlands. Termite activity is key for RG development, since it concentrates most nutrients available for plants at niche, local scale. The presence of shallow spodosols is on account of intense organic matter illuviation down the soil profile, forming well-developed spodic Bh horions of dark colours. The dissolution forms indicates that, albeit resistant to weathering, quartzites are being slowly attacked by acid solutions, leading to the present-day ruiniform grounds where erosion exhumed the former subsurface rock substrate. (Drawing by C. Schaefer)

micaceous quartzites, with occasional shales. The original stratification is well preserved, and an aeolian origin is postulated (Dossin et al. 1987).

The soil sequence studied ranged from open rupestrian grassland with *Vellozia*, to rupestrian savanna, savanna and dwarf forest; the latter two formations are influenced by micaceaus quartzite (Fig. 2.12).

2.9.2 Open Rupestrian Grassland with Vellozia on Litholic Neosols

The open rupestrian physiognomy is invariably associated with very shallow soils (Litholic Neosol, BSV1) on quartzite outcrop, in which soils occur as discontinuous pockets amid the rocky surface. Abundant Velloziaceae dominates the formation, with a dense herbaceous cover of Poaceae, Cyperaceae and Xyridadeae (Fig. 2.13a). Dwarf trees and shrubby are common on fractured rocks, with their roots penetrating in fissures and weakness lives where organic matter accumulates.

Fig. 2.13 Sempre-Vivas national park, Minas Gerais state, Brazil. **a** Open rupestrian grassland with vellozia on litholic neosols. **b** Open and shrubby rupestrian savanna on cambisols. **c** Savanna (cerrado sensu stricto) on latosols and cambisols. **d** Dwarf forest in foreground and rupestrian grassland on the rock outcrop in background

The rugged, mountainous landform account for steep slopes where erosion is severe with intense soil loses! The structural control on rupestrian vegetation is evident with massive rocky covers on the top land surface. The main genera are *Lagenocarpus sp, Trachypogon sp, Vellozia sp, Lychnophora sp, Symplocos sp, Byrsonima sp* and *Chamaecrista sp.* Fire is intense and frequent, but its effect is reduced by the presence of little biomass and rocky surfaces.

2.9.3 Open and Shrubby Rupestrian Savanna on Cambisols

The rupestrian savanna is a typical open formation on metapelitic to metarenites rocks, with greater silt contents, and deeper soil, and part of the RGC. Rocky outcrops are not common, only where resistant rocky cores stand out (Fig. 2.13b). The physiognomy varies from open grassy savanna to typical savanna (cerrado sensu stricto). The herbaceous stratum is graminoid/cyperoid, and fire propagation is facilitated by a continuous grassy cover, compared with open rupestrian grassland. Soil have a cambic horizon, but are nutrient-poor and acid, as well as shallow and silt-rich. Colours are redder, resulting form higher Fe-content. Typical genera are *Qualea sp, Vellozia sp, Baccharis sp, Trachypogon sp* and *Vochysia sp.*

2.9.4 Savanna (Cerrado Sensu Stricto) on Latosols and Cambisols

Typical savannas occurs on deeper soils on slopes connecting the high plateau, under the influence of metapelitic rocks, savanna *s.s.* develops on deep Latosols or Cambisols, all very dystric and acid, with very low cation exchange capacity and virtually no nutrient reserve (acric, gibbsitic) (Fig. 2.13c). Plants are well adapted to extreme nutrient depletion, but require clayey and deeper soils. Tree typical species are *Caryocar brasiliense Cambess., Xylopia aromatica (Lam.) Mart., Qualea grandiflora Mart* and many siliceous Poaceae, forming a grassy cover. The main difference is the absence of rock outcrops and rupestrian character.

2.9.5 Dwarf Forest on Quartzarenic Neosols or Medium-Texture Latosols

The Dwarf Forests are cloud forest (rich in epiphytes) at altitudes higher than 1,000 m, usually on deep soils (PS) of medium to sandy textures (high fine sand content) (Fig. 2.13d). These dwarf forests are transitional with candeias shrubby, but with soils with better water retention. Soils are rich in organic matter at the surfaces although very poor in nutrients. Typical trees are *Protium heptaphyllum (Aubl.) Marchand, Guatteria sellowiana Schltdl.* and *Miconia cabucu Hoehne.*

2.10 Final Remarks

1. The Rupestrian Grassland complex is widely distributed across the Brazilian Territory, and closely follows the occurrence of highland resistant rocks, in the decreasing order of importance: Quartzites ≫ Fe-rich canga and banded iron formations > Igneous rocks (Granites, Syenites) or high-grade metamorphics (Gneiss, schists) > metapelitic rocks > other minor types (conglomerate, limestone). Quartzites comprise the vast majority of RG occurrences.
2. RGC is closely associated with high altitudes, in which a number of factors have a prominent role: (i) Extreme oligotrophy; (ii) chemical resistance to weathering; physical resistance; (iii) wind-prone environment at high landsurfaces; (iv) frequent fire, and harsh climatic regime.
3. Variations of RG physiognomies are basically related to soil depth and drainage (edaphic factor) and landscape stability and evolution.
4. The proposed map of RGC distribution included areas not well investigated by ecologists, such as Tepuis (Roraima and Amazonas), Serras de Santa Barbara and Ricardo Franco (Mato Grosso), Cachimbo (Pará). On the other hand, also included well-known areas, formerly classified as Campos de Altitude, such as Serra dos Órgãos (Rio de Janeiro), Itatiaia (Rio/Minas) and Caparaó (Minas/Esp. Santo). In all areas, true Rupestrian vegetation formation exists, so that there is no reason to constrain the limits of RGC to the quartzitic domains, as traditionally envisaged.
5. There is an urgent need for comparative studies (floristic, phylogenetics, pedological, structural and phytosociological) for investigating the similarities between these postulated occurrences of RGC in Brazil, so widely separated by surrounding dominant vegetation.
6. Case studies selected and presented here highlight the importance of parent materials and soil attributes (including pedobiological aspects) for determining the phytophysiognomies of RGC. In canga (Fe-rich substrates), for example, the degree of degradation of ironstone controls soils formation and, consequently, vegetation development. In granite, rockiness and soil attributes are the main drivers.
7. In synthesis, we present two basic diagrams of RGC in Brazil (Figs. 2.3, 2.12) to illustrate typical situations on (1) Granite, (2) Quartzite. In the first case, the granitic parent material is a massive rock body not so limiting in terms of chemistry, but with very poor soils resulting from extreme weathering and leaching of nutrients at these old, high mountain landscapes. In the second case, the chemical and physical resistance of quartz-rich substrates account for very shallow soils, in general. Variations of soil drainage, depth, organic-matter

accumulation, aspect, slope, stoniness and rockiness and termite activity are evident, and help to explain the different niches, in which resources (limiting) are differentially offered. Hence, all high landsurfaces of these resistant mountain tops are comparable, in terms of nutrient depletion, acidity, fire regime and climate, leading to a convergent evolution of RGCs.

8. Traditional Brazilian vegetation classification have emphasized the differences between Campos de Altitude (High Altitude Fields) (Martinelli 1989) on granite/gneiss and Campos Rupestres (Rupestrian Grasslands) on Canga or quartzite. However, we see no reason to treat these isolated hot-spots of true Rupestrian domains on Granite as separate phytogeographical entities. By the same rationale, Tepuis and Carajás should be treated as part of this single and complex Campo Rupestre domain. They do share many species in common, have a clear rupestrian physiognomy in most places, and are subjected to similar fire and wind regimes by high ground exposure. In addition, traits related to resource conservation are very similar, since soils are generally poor, acid, shallow and water deficient.
9. The importance of RGC as biodiversity pumps (sensu Silveira et al. 2016) or hot spots are really large, and unique. They occur on different rocks, are widely distributed across Brazil, within contrasting biomes, and are relatively well-preserved by mountain top isolation, under limited agricultural pressures. Other anthropic pressures, however, are mounting: mining, pasture expansion, tourism, water needs, so that urgent measures must be taken for its long term conservation. Sound scientific research is the necessary framework for any sustainable knowledge on RG, possibly one of the oldest kingdoms of the Brazilian Neotropics.

Acknowledgments Carlos Schaefer thanks Vale, CNPq and FAPEMIG for financing long-term studies on Carajás and Iron Quadrangle; thanks are due to Dr. Lilia Santos (Bioma), James Bockheim, and João Ker for many field trips and discussions on RGC ecology. Help for field work Sr. Alexander Castilho (Vale) is greatly appreciated.

Appendix 1

Picture illustrations of phytophysiognomies and landscape features of Rupestrian Grassland Complex (RGC) on different lithologies in Brazil. The striking similarities between widely separate geographical sits, very contrasting parent-materials, indicate a notable convergence of edaphic/landscape features at different localities. (a) Itatiaia highlands at Campo Belo Valley (2400 m) on syenite (alkaline rocks) (mark 20 in Fig. 2.14). (b) Caparaó highlands at Upper Crystal Valley (above Terreirão Plateau), with *Chusquea* sp. on granitoid rocks (mark 17 in Fig. 2.14). (c) Serra do Brigadeiro Plateau Pedra do Pato, migmatites/granites Paraíba do Sul headwaters. (d) Shrubby rupestrian grassland in Serra do Espinhaço, Diamantina, Minas Gerais States, with *Syagrus* sp. (Palmeira da Serra), on micaceous quartzites

Fig. 2.14 Phytophysiognomies and landscape features of Rupestrian Grassland Complex (RGC) on different lithologies in Brazil

Fig. 2.14 (continued)

Fig. 2.14 (continued)

(mark 12 in Fig. 2.14). (e) Sempre-Vivas National Park, shrubby rupestrian grassland on quartzite (mark 12 in Fig. 2.14). (f) Canastra National Park, open rupestrian grassland with *Vellozia* sp. on micaceaus quartzites/conglomerates (mark 19 in Fig. 2.14). (g) Ibitipoca State Park, with open rupestrian grassland on micaceous quartzites and metapelits (mark 23 in Fig. 2.14). (h) São Francisco River headwaters at Canastra National Park, with shrubby rupestrian grassland a quartzite plateau (mark 19 in Fig. 2.14). (i) Shrubby ferruginous rupestrian grassland on Fe-rich canga of Carajás Plateau, Santa Rita Durão, (see Caraça quartzite at background) (1000 m) (mark 8 in Fig. 2.14). (j) General view of ferruginous rupestrian grassland on the Fe-ich itabirite/canga on Serra Sul Plateau of Carajás (900 m) (mark 8 in Fig. 2.14). (k) Ferrugineous rupestrian grassland on the Fe-rich Itabirite/canga of Alto do Conta História (1600 m), Iron Quadrangle, Minas Gerais States (mark 16 in Fig. 2.14). (l) Exposed itabirite saprolite with overlying canga on the Moeda Plateau, with ferruginous rupestrian grassland (shrubby/open) (1380 m) (mark 16 in Fig. 2.14). (m) Open rupestrian savanna on metapelites from Serra de Grão Mogol (1200 m) (mark 12 in Fig. 2.14). (n) Open rupestrian grassland on conglomeratic quartzites of Diamantina Plateau, Minas Gerais States (mark 12 in Fig. 2.14) at 1420 m. (o) Sempre-Vivas National Park with open rupestrian grassland on quartzite with *Lychnophora* sp. (mark 12 in Fig. 2.14) at 1130 m. (p) Serra de São José Environmental Protected Area, Tiradentes, Minas Gerais States, with open rupestrian grassland on micaceous quartzite and schists at 1200 m (q) Pico da Neblina, the highest RG site in Brazil (close to 3000 m) on metarenites of Roraima Group, Amazonas State (mark 1 in Fig. 2.14). (r) Serra do Sol, at 2250 m altitude, on quartzites of the Roraima Group, Roraima State (mark 1 in Fig. 2.14). (s) Serra dos Pacaás Novos, with an altitude of 1050 m, an isolated spot of RG in the middle of the Amazon Forest domain, with metarenites to metapelitic rocks (mark 3 in Fig. 2.14). (t) (mark 1 in Fig. 2.14). (u) Serra Ricardo Franco, in the Brazil-Bolivia border, at Mato Grosso State, where RG occurs on metarenites (mark 4 in Fig. 2.14) at 1030 m. (v) Monte Roraima highest surface (mark 1 in Fig. 2.14), the largest Rupestrian Plateau in the Brazilian Amazonia (Roraima State), at 2850 m. (w) Monte Roraima top surface (2800 m), on Quartzite of the Roraima Group (mark 1 in Fig. 2.14). (x) Serra da Seringa, southeastern Pará State, an isolated RGC on Felsic Granites in the middle of the Caiapó Indigenous Land (mark 7 in Fig. 2.14).

References

Ab'Saber AN (1986) Geomorfologia da Região. In: Almeida JR (ed). Carajás: desafio político, ecologia e desenvolvimento. Brasiliense, Brasília, pp. 88–124

Alkmim FF (2015) Geological Background: A Tectonic Panorama of Brazil. In: Vieira BC, Salgado AAR, Santos LJC (eds) Landscapes and landforms of Brazil. Springer, Netherlands, pp 9–17

Almeida-Abreu PA (1995) O Supergrupo Espinhaço da Serra do Espinhaço Meridional (Minas Gerais): o Rifte, a Bacia e o Orógeno. Geonomos 3:1–18

Almeida TE, Souza DT, Salino A, Teles AM (2007) Levantamento florístico e caracterização das formações vegetacionais, Parque Nacional das Sempre-Vivas. Unpub. Technical report

Alves RJV, Silva NG, Oliveira JA, Medeiros D (2014) Circumscribing campo rupestre—megadiverse Brazilian rocky montane savanas. Braz J Biol 74:355–362

Barros NF (1979) Growth and foliar nutrient concentrations of *Eucalyptus grandis* in relation to spodasol properties in South Florida. Ph.D. Dissertation, University of Florida

Benites VM, Sá E, Schaefer CER, Novotny EH, Reis E, Ker JK (2005) Properties of black soil humic acids from high altitude rock complexes in Brazil. Geoderma 127:104–113

Benites VM, Schaefer CEGR, Simas FNB, Santos HG (2007) Soils associated with rock outcrops in the Brazilian mountain ranges Mantiqueira and Espinhaço. Rev Bras Bot 30:569–577

Botrel RT, Oliveira-Filho AT, Rodrigues LA, Curi N (2002) Influencia do solo e topografia sobre as variações da composição florística e estrutura da comunidade arbóreo–arbustiva de uma floresta estacional semidecidual em Ingaí, Mg. Rev Bras Bot 25:195–213

Butcher R, Byrne M, Crayn DM (2007) Evidence for convergent evolution among phylogenetically distant rare species of *Tetratheca* (Elaeocarpaceae formerly Tremandraceae) from Western Australia. Aust Syst Bot 20:126–138

Campos NV, Pereira TAR, Machado MF, Guerra MBB, Tolentino GS, Araújo JS, Rezende MQ, Silva MCNA, Schaefer CEGR (2014) Evaluation of micro-energy dispersive X-ray fluorescence and histochemical tests for aluminium detection in plants from high altitude rocky complexes, Southeast Brazil. An Acad Bras Ciênc 86:285–296

Carmo FF, Jacobi CM (2012) As cangas do Quadrilátero Ferrífero. In: Jacobi CM, Carmo FF (eds) Diversidade florística nas cangas do Quadrilátero Ferrífero. IDM, Belo Horizonte, pp 1–13

Dias HCT, Fernandes EIF, Schaefer CEGR, Fontes LEF, Ventorim LB (2002) Geoambientes do Parque Estadual do Ibitipoca, Município de Lima Duarte-MG. R Árv 26:777–786

Diniz JMFDS, Dos Reis AA, Junior FWA, Gomide LR (2014) Detecção da expansão da área minerada no Quadrilátero Ferrífero, Minas Gerais, no período de 1985 a 2011 através de técnicas de sensoriamento remoto. Bol Cien Geod 20:683–700

Dossin IA, Garcia AJV, Uhlein A, Dossin TM (1987) Facies eólico na Formação Galho do Miguel, Supergrupo Espinhaço-MG. Bol Soc Bras Geol 6:85–96

EMBRAPA—Empresa Brasileira De Pesquisa Agropecuária (1997) Manual de métodos de análise de solo. Centro Nacional de Pesquisas de Solos, Rio de Janeiro

Gibson N (2004a) Flora and vegetation of the eastern Goldfields ranges: Part 6. mt manning range. J R Soc West Aust 87:35–47

Gibson N (2004b) Flora and vegetation of the eastern Goldfields ranges: Part 7. middle and south Ironcap, Digger Rock and Hatter Hill. J R Soc West Aust 87:49–62

Gibson N, Yates CJ, Dillon R (2010) Plant communities of the ironstone ranges of South Western Australia: Hotspots for plant diversity and mineral deposits. Biodiv Cons 19:3951–3962

Gibson N, Meissner R, Markey AS, Thompson WA (2012) Patterns of plant diversity in ironstone ranges in arid south Western Australia. J Arid Environ 77:25–31

Grime JP (2001) Plant strategies, vegetation processes, and ecosystem properties, 2nd edn. John Wiley, Sons Ltd, Chichester

Harley RM, Simmons NA (1986) Florula of Mucugê, Chapada Diamantina—Bahia, Brazil. Royal Botanical Garden, Kew

IBAMA—Instituto Brasileiro do Meio Ambiente e dos Recursos Naturais Renováveis (2003) Plano de manejo para uso múltiplo da Floresta Nacional de Carajás. IBAMA, Brasília

IBAMA—Instituto Brasileiro do Meio Ambiente e dos Recursos Naturais Renováveis (2006) Plano Operativo de prevenção e combate aos incêndios do Parque Nacional das Sempre-Vivas. IBAMA, Diamantina

Jacobi CM, Carmo FF (2011) Life–forms, pollination and seed dispersal syndromes in plant communities on ironstone outcrops, SE, Brazil. Act Bot Bras 25:395–412

Jacobi CM, Carmo FF, Vincent RC, Stehmann JR (2007) Plant communities on ironstone outcrops: A diverse and endangered Brazilian ecosystem. Biodiv Cons 16:2185–2200

Jacobi CM, Carmo FF, Vincent RC (2008) Estudo Fitossociológico de uma comunidade vegetal sobre canga com subsídio para a reabilitação de áreas mineradas no Quadrilátero Ferrífero, MG. Rev Arv 32:345–353

Jacobi CM, Carmo FF, Campos IC (2011) Soaring extinction threats to endemic plants in Brazilian metal-rich regions. Ambio 40:540–543

Lista de Espécies da Flora do Brasil (2015) Jardim Botânico do Rio de Janeiro. Available in. http://floradobrasil.jbrj.gov.br/. Accessed 29 Jan 2016

Markey AS, Dillon SJ (2008) Flora and vegetation of the banded iron formations of the Yilgarn Craton: The central tallering land system. Cons Sci West Aust 7:121–149

Markey AS, Dillon SJ (2009) Flora and vegetation of the banded iron formations of the Yilgarn Craton: Herbert Lukin Ridge (Wiluna). Conserv Sci West Aust 7:391–412

Markey AS, Dillon SJ (2010) Flora and vegetation of the banded iron formations of the Yilgarn Craton: Gullewa. Conserv Sci West Aust 7:531–556

Martinelli G (1989) Campos de, altitude edn. Index, Rio de Janeiro

Meirelles ST, Mattos EA, Silva AC (1997) Potential desiccation tolerant vascular plants from southeastern Brazil. Pol J Environ Stud 6:17–21

Meissner R, Caruso Y (2008) Flora and vegetation of banded iron formations of the Yilgam Craton: Koolanooka and Pernjori hills. Cons Sci West Aust 7:73–88

Meissner R, Wright J (2010) Flora and vegetation of banded iron formations on the Yilgarn Craton: South Illaara Greenstone Belt. Conse Sci West Aust 7:605–616

Meissner R, Owen G, Bayliss B (2009) Flora and vegetation of the banded iron formation of the Yilgarn Craton: Robinson Ranges and Mount Gould. Cons Sci West Aust 7:363–376

Mendonça MP, Lins LV (2000) Lista Vermelha das Espécies Ameaçadas de Extinção da Flora de Minas Gerais. Biodiversitas e Fundação ZooBotânica de Belo Horizonte, Belo Horizonte

Messias MCTB, Leite MGP, Meira-Neto JA, Kozovits AR, Tavares R (2013) Soil–vegetation relationship in quartzitic and ferruginous Brazilian rocky outcrops. Folia Geobot 48:509–521

Morellato PC, Rosa NA (1991) Caracterização de alguns tipos de vegetação na região amazônica, Serra dos Carajás. Pará Brasil Rev Bras Bot 14:1–14

Mourão A, Stehmann JR (2007) Levantamento da flora do campo rupestre sobre canga hematítica couraçada remanescente na Mina do Brucutu, Barão de Cocais, Minas Gerais, Brasil. Rodriguésia 58:775–785

Nunes JA, Schaefer CEGR, Ferreira-Júnior WG, NERI AV, Correa, GR, Enright NJ (2015) Soil-vegetation relationships on a banded ironstone island, Carajás Plateau, Brazilian Eastern Amazonia. An Acad Bras Cien 87:2097–2110

Porembski S, Barthlott W, Dorrstock S, Biedinger N (1994) Vegetation of rock outcrops in Guinea: Granite inselbergs, sandstone table mountains and ferricretes—remarks on species numbers and endemism. Flora 189:315–326

Porembski S, Seine R, Barthlott W (1997) Inselberg vegetation and biodiversity of granite outcrops. J Roy Soc West Austral 80:193–197

Porto ML, Silva MFF (1989) Tipos de vegetação metalófila em áreas da Serra de Carajás e de Minas Gerais, Brasil. Act Bot Bras 3:13–21

Queiroz LP, Sena TSN, Costa MJSL (1996) Flora vascular da Serra da Jibóia, Santa Terezinha. Bahia. I: O campo rupestre. Sitientibus 15:27–40

Rizzini CT (1979) Tratado de fitogeografia do Brasil: Aspectos sociológicos e florísticos. Editora da Universidade de São Paulo, São Paulo

Rodrigues PMS, Schaefer CEGR, Corrêa GR, Campos PV, Neri AV (2014) Soils, landform and vegetation determine the geoenvironments at a conservation unit in northern Minas Gerais, Brazil. Neotrop Biol Cons 10:31–42

Santos MDC, Varajão AF (2004) Sedimentation and pedogenic features in a clay deposit in Quadrilátero Ferrífero, Minas Gerais, Brazil. An Acad Bras Ciênc 76:147–159

Sarcinelli TS, Schaeffer CEGR, Lynch LS, Arato HD, Viana JHM, Albuquerque Filho MR Gonçalves TT (2009) Chemical, physical and micromorphological properties of termite mounds and adjacent soils along a toposequence in Zona da Mata, Minas Gerais State, Brazil. Catena 76:107–113

Schaefer CEGR (2001) Brazilian latosols and their B horizon microstructure as long-term biotic constructs. Soil Res 39:909–926

Schaefer CEGR (2013) Bases físicas da paisagem brasileira: estrutura geológica, relevo e solos. In: Araújo AP, Alves BJR (eds) Tópicos em ciência do solo. Sociedade Brasileira de Ciência do Solo, Viçosa, pp 1–69

Schaefer CEGR, Gilkes RJ, Fernandes RBA (2004) EDS/SEM study on microaggregates of Brazilian Latosols, in relation to P adsorption and clay fraction attributes. Geoderma 123:69–81

Schaefer CEGR, MendonçA BAF, Ribeiro ASS (2008) Solos desenvolvidos sobre canga ferruginosa no quadrilaÏtero ferriÏfero, Minas Gerais. Biodiversidade, Conservaçafio e Perspectivas de Sustentabilidade. UFMF/ ICB, Belo Horizonte, SimpoÏsio Afloramentos Ferruginosos no QuadrilaÏtero FerriÏfero

Schaefer CEGR, Simas FNB, Mendonça BAF, Saboya AS, Ferreira Júnior WG, Nunes JA, Correa GR (2009) Geodiversidade dos ambientes de canga na região de Carajás– Pará. Technical Report. part 2. Vale Brazilian Company

Schaefer CEGR, Cândido HG, Corrêa GR, Pereira A, Nunes JA, Souza OF, Marins A, Fernandes-Filho E, Ker JC (2015) Solos desenvolvidos sobre canga ferruginosa no Brasil: Uma revisão crítica e papel ecológico de termiteiros. In: Carmo FF, Kamino LHY (eds) Geossistemas ferruginosos do Brasil: Áreas prioritárias para conservação da diversidade geológica e biológica. Patrimônio cultural e serviços ambientais. 3i, Belo Horizonte, pp. 77–102

Schobbenhaus C, Brito-Neves BB (2003) A geologia do Brasil no contexto da Plataforma Sul-Americana. In: Schobbenhaus C, Vidotti RM, Gonçalves JH (eds) Geologia, tectônica e recursos minerais do Brasil. CPRM, Brasília, pp 5–25

Secco RS, Mesquita AL (1983) Nota sobre a vegetação de canga da Serra Norte. I. Boletim Mus Par Em Goeldi 59:1–13

Semir J (1991) Revisão taxonômica de *Lychnophora* Mart. Vernoniaceae: Compositae. Ph.D. Thesis, Universidade Estadual de Campinas, Campinas

Silva MFF (1992) Distribuição de metais pesados na vegetação metalófila de Carajás. Act Bot Bras 6:107–22

Silva MFF, Rosa N (1984) estudos botânicos na área do projeto ferro de Carajás, Serra Norte. I— Aspectos ecológicos e vegetacionais dos campos rupestres. In: XXXV Congresso Nacional de Botânica, Sociedade Brasileira Botânica, Manaus, pp. 367–379

Silveira FA, Negreiros D, Barbosa NP, Buisson E, Carmo FF, Carstensen DW, Lambers H (2016) Ecology and evolution of plant diversity in the endangered campo rupestre: a neglected conservation priority. Plant and Soil. doi: 10.1007/s11104-015-2637-8

Tardy Y, Nahon D (1985) Geochemistry of laterites, stability of Al-goethite, Al-hematite, and Fe3 + -kaolinite in bauxites and ferricretes: An approach to the mechanism of concretion formation. Am J Sci 285:865–903

Teixeira WA, Lemos-Filho JP (1998) Metais pesados em folhas de espécies lenhosas colonizadoras de uma área de mineração de ferro em Itabirito, Minas Gerais. Rev Árv 22:381–388

Teixeira WA, Lemos-Filho JP (2002) Fatores edáficos e a colonização de espécies lenhosas em uma cava de mineração de ferro em Itabirito, Minas Gerais. Rev Árv 26:25–33

Valentin C, Herbes JM, Poesen J (1999) Soil and water components of banded vegetation patterns. Catena 37:1–24

Veloso HP, Rangel Filho ALR, Lima JCA (1991) Classificação da vegetação brasileira, adaptada a um sistema universal. Instituto Brasileiro de Geografia e Estatística, Rio de Janeiro

Viana PL, Lombardi JA (2007) Florística e caracterização dos campos rupestres sobre canga na Serra da Calçada, Minas Gerais, Brasil. Rodriguesia 58:159–177

Vincent RC (2004) Florística, fitossociologia e relações entre a vegetação e o solo em área de campos ferruginosos no Quadrilátero Ferrífero, Minas Gerais. Ph.D. Dissertation, Universidade de São Paulo

Vincent RC, Meguro M (2008) Influence of soil properties on the abundance of plant species in ferruginous rocky soils vegetation, southeastern Brazil. Rev Bras Bot 31:377–388

Vincent RC, Jacobi CM, Antonini Y (2002) Diversidade na adversidade. Cien Hoje 31:64–67

Yates CJ, Ladd PG, Coates DJ, Mcarthur S (2007) Hierarchies of cause: Understanding rarity in an endemic shrub *Verticordia staminosa* (Myrtaceae) with a highly restricted distribution. Aust J Bot 55:194–205

Yates CJ, Gibson N, Pettit NE, Dillon R, Palmer R (2011) The ecological relationships and demography of restricted ironstone endemic plant species: Implications for conservation. Aust J Bot 59:692–700

Chapter 3
Soils Associated with Rupestrian Grasslands

Carlos E. Schaefer, Hugo G. Cândido, Guilherme Resende Corrêa, Jaquelina A. Nunes and Daniel M. Arruda

Abstract Rupestrian Grassland (RG) is a peculiar type of vegetation characterized by a combination of shrubs, herbs, grasses and geophytes, growing on nutrient-poor rocky outcrops on the highest parts of the Mountain Landscape in Brazil. A broad review of existing soil data from RG revealed a trend of very poor nutrient status, shallow profiles, acidity and severe P depletion. Vegetation biomass generally increases with soil depth, accordingly. The RG can be found throughout Brazil, on different Lithologies (Quartzite, Itabirite, Phyllites, Granites, Gneisses, Syenites, Schists). Clustering of all RG soils allowed to distinguish two basic groups: (1) one related to higher organic matter accumulation and clayey/silty textures (Canga and Itabirite); (2) sandy soils with greater Al^{3+} exchangeable levels (Quartize and

C.E. Schaefer (✉)
Departamento de Solos, Universidade Federal de Viçosa (UFV),
Av. Peter Henry Rolfs, s/n – Campus Universitário, Viçosa,
Minas Gerais CEP 36570-900, Brazil
e-mail: carlos.schaefer@ufv.br

H.G. Cândido · D.M. Arruda
Centro de Ciências Biológicas 2, Departamento de Biologia Vegetal,
Universidade Federal de Viçosa (UFV), Secretaria de Pós-Graduação em
Botânica - Campus Universitário, Viçosa, Minas Gerais CEP 36570-900, Brazil
e-mail: hugogalvaocandido@gmail.com

D.M. Arruda
e-mail: arruda.dm@hotmail.com

G.R. Corrêa
Faculdade de Artes, Filosofia e Ciências Sociais, Insittuto de
Geografia – Santa Mônica, Universidade Federal de Uberlândia (UFU),
Uberlândia, Minas Gerais CEP 38400-902, Brazil
e-mail: guilhermeudi@yahoo.com.br

J.A. Nunes
Unidade Carangola, Praça dos Estudantes, Universidade do Estado de Minas Gerais
(UEMG), 23 – Santa Emília, Carangola, Minas Gerais CEP36800-000, Brazil
e-mail: jaquelinabiologa@yahoo.com.br

© Springer International Publishing Switzerland 2016

G.W. Fernandes (ed.), *Ecology and Conservation of Mountaintop Grasslands in Brazil*, DOI 10.1007/978-3-319-29808-5_3

granites, mainly). Physical, rather than chemical differences, exist amongst the RG soils, influencing water availability at short distances. The patchynature of RG is highly influenced by fire and soil depth, combined.

Contents

3.1 Introduction

3.1.1 Landscape Overview

The Brazilian highlands are part of the most ancient landscapes on earth, evolved under tectonic stability and extreme weathering. Landforms are structurally controlled, forming plateaux or Mountain ranges as part of the Mobile Belts zones: (i) Serra do Espinhaço–Chapada Diamantina, composed mainly by quartzites; (ii) Serra da Mantiqueira, with predominance of plutonic rocks (granite) and high grade metamorphics (migmatites, gneiss) and a few sparse quartzite areas (e.g. Serras do Ibitipoca, São José, Lenheiro); (iii) Tepuis of Roraima, basically metarenites and quartzites; (iv) Serra dos Pirineus e Mountain Ranges of the Central Plateau; (v) Serras de Ricardo Franco and Santa Barbara in Mato Grosso State, metarenites; (vi) Serra de Carajás with jaspilites/itabirites; (vii) Quadrilátero Ferrífero, with itabirites and quartzites (Fig. 3.1). These mountain ranges are extremely important from a geo-environmental point of view. There occur several watersheds that feed important urban centers of southeastern Brazil, as well as sheltering important vegetation fragments known as Rupestrian Grasslands (RG), characterizing "Hot Spots" for biodiversity. Hence, several conservation units have been created in order to protect these fragile environments. In the uppermost parts of these mountain ranges, a distinct ecosystem named High Altitude Rocky Complexes (HARC) (Semir 1991; Benites et al. 2003a, 2005) occurs, with peculiar soil and vegetation characteristics, as illustrated in Fig. 3.2. Although apparently monotonous, there can be observed a considerable diversity of pedoenvironments and associated vegetation mosaics, greatly determined by local topography and microenvironmental aspects.

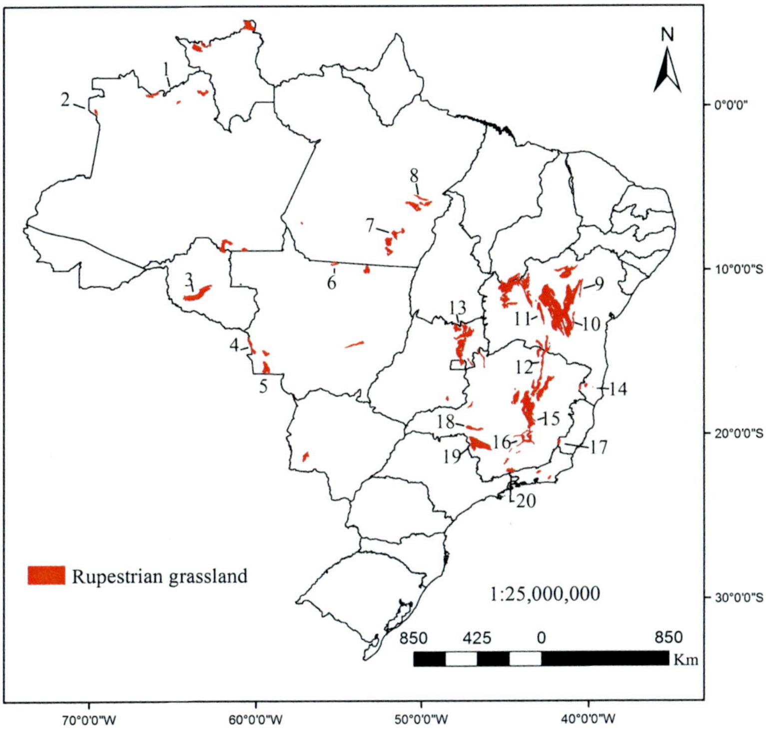

Fig. 3.1 General distribution of Campos Rupestres and associated complexes in the Brazilian Territory. *1* Tepuis, *2* Tunuí, *3* Serra dos Pacaás Novos, *4* Serra Ricardo Franco, *5* Santa Bárbara, *6* Serra do Cachimbo, *7* Serra da Seringa, *8* Serra dos Carajás, *9* Morro do Chapéu, *10* Chapada Diamantina (Serra do Sincorá), *11* Northern Espinhaço, *12* Southern Espinhaço, *13* Highlands of Central Plateau, *14* Pontões Santo Antônio do Jacinto, *15* Serra do Cipó, *16* Quadrilátero Ferrífero, *17* Caparaó, *18* Araxá, *19* Serra da Canastra, *20* Itatiaia

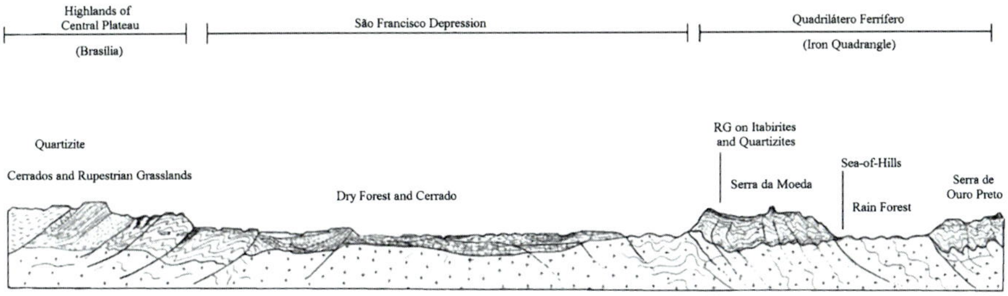

Fig. 3.2 General distribution of HARC complex and Campos Rupestres in relation to topography and landscape features in a west-east section across Minas Gerais State, ranging from Ouro Preto to Belo Horizonte (São Francisco Depression) and the Brasilia Central Plateau. Campos Rupestres occur on resistant rocks, which forms prominent mountains ranges of old age

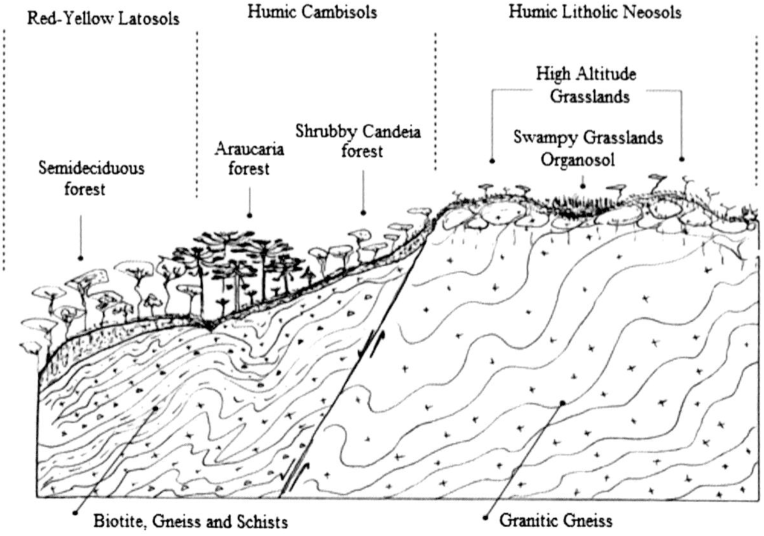

Fig. 3.3 Soil vegetation distribution on granitic gneiss of the Mitra do Bispo, Serra da Mantiqueira, MG. Humic litholic Neosols (Neossolos Litólicos Húmicos) occur on the top of the Moutain, associated with a complex of high altitude grasslands and scattered Rupestrian Grassland on very shallow soils, or rocky outcrops. Down the sequence, we find Shrubby Candeia Forest (transition) and Araucaria and semideciduous forest on deeper soils (Latossolos) at the footslopes. *Source* Author's illustration Schaefer (2013)

Markedly, rock outcrops are widespread, regardless of the predominant lithology (Benites et al. 2005). Therefore, contrary to common beliefs, rupestrian grasslands and associated formation form a Complex that can be found on different rocks (Granites, Syenites, Gneisses, Quartzites, Phyllites and itabirites), providing certain geomorphological and pedological conditions. Figures 3.3, 3.4, 3.5, 3.6, 3.7 and 3.8 show the different aspects of CR on different lithologies: Figs. 3.3 and 3.4—Granite; Figs. 3.5 and 3.6—Quarzite; Fig. 3.7—Itabirite; Fig. 3.8—Syenites.

A great number of endangered and endemic species are found in these highland refuges, highlighting their great importance for biodiversity preservation and scientific studies. However, few studies on soil identification and characterization has been undertaken (Volkoff et al. 1984; Benites et al. 2001; Dias et al. 2002; Benites et al. 2003a; Schaefer et al. 2002; Simas et al. 2004; Valente 2009; Ribeiro 2009; Carvalho Filho et al. 2010; Pereira 2010; Ferrari 2013), but they provide a broad understanding on these impoverished landscapes, on different lithologies.

3.1.2 Scope

Commonly, at small scale pedological surveys, soils associated to rock outcrops are mapped in units that do not identify the great variety of soils that can be observed at

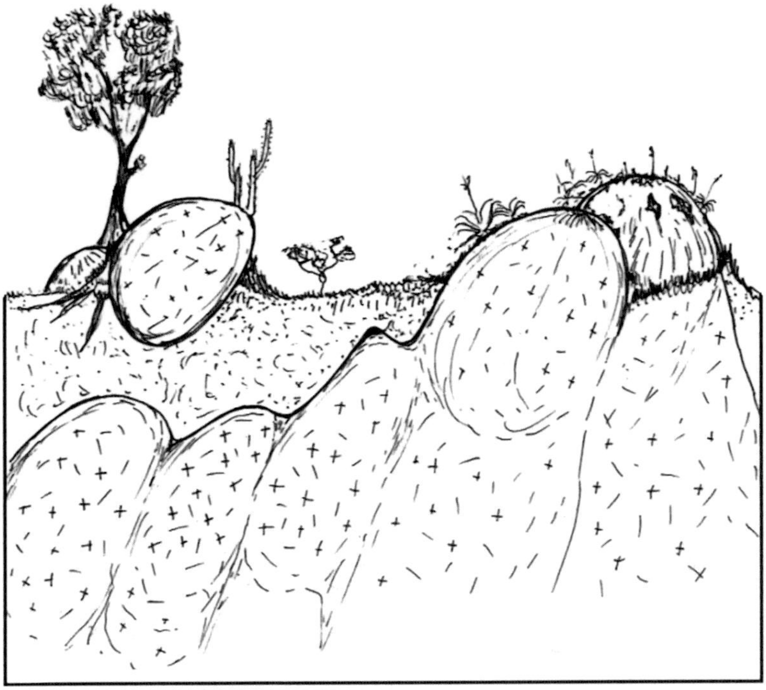

Fig. 3.4 A typical soil (Litholic Neosol) on Granite of Serra do Brigadeiro, MG. The massivity of the igneous rocks turn the weathering difficult, so that fractures and faults are normally explored by water to produce a landscape of rounded boulders and rock outcrops, where a typical rock-loving flora develops

a closer look. Here we summarized the existing information about soils associated to rock outcrops and rupestrian grasslands from all Brazilian highlands where soil data are available, illustrating the pedological diversity and the ecological importance of these ecosystems.

3.2 Soil Sampling and Characterization

We reviewed all available data published on HARC soils from all Brazilian regions, selecting 68 soil profiles data for statistical analysis. Most studies were carried out by the author, so that recognition of HARC was accomplished in areas surrounded by Caatinga, Cerrado, and Forest biomes, in which HARC is locally present at the highest positions, on resistant rock outcrops.

All analysis were carried out using the same methodology. Soils were air dried and sieved to 2 mm. The textural fractions (coarse sand, fine sand, silt and clay) was determined in the <2 mm fraction according Embrapa (1997). pH and exchangeable

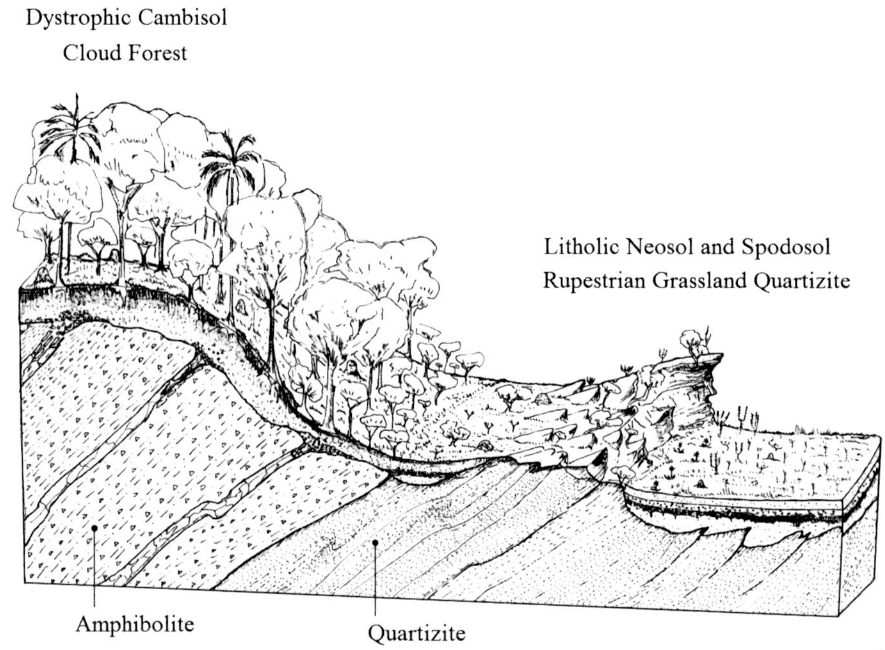

Fig. 3.5 Block diagram illustrating a transition between a Cloud semideciduous forest to Rupestrian Grassland (RG) following a soil gradient at the Serra do Cipó National Park, highlighting the close association between forest on mafic intrusions (amphibolite) and CR on quartzite

Ca + Mg and Al^{3+} were also determined according Embrapa (1997). Total soil organic carbon was determined by the wet combustion method (Yeomans and Bremner 1988).

3.3 Soils Associated with Rock Outcrops

3.3.1 Soils Overview

Paradoxically, soils associated with rock outcrops are generally weakly developed but extremely leached and impoverished, making their properties strongly influenced by the acid, chemically poor, parent materials. *Solum* depth is extremely variable, as a function of local topography and faulting/fracturing, with very shallow soils on steep slopes and deeper weathered soils on more stable areas. Normally, rock outcrops occur scattered amongst soil patches (Fig. 3.3), or very small soils spots are formed directly on bare rock (Fig. 3.4). In places, the presence of a deep pre-weathered saprolite, with depths exceeding 50 m, is also important, as

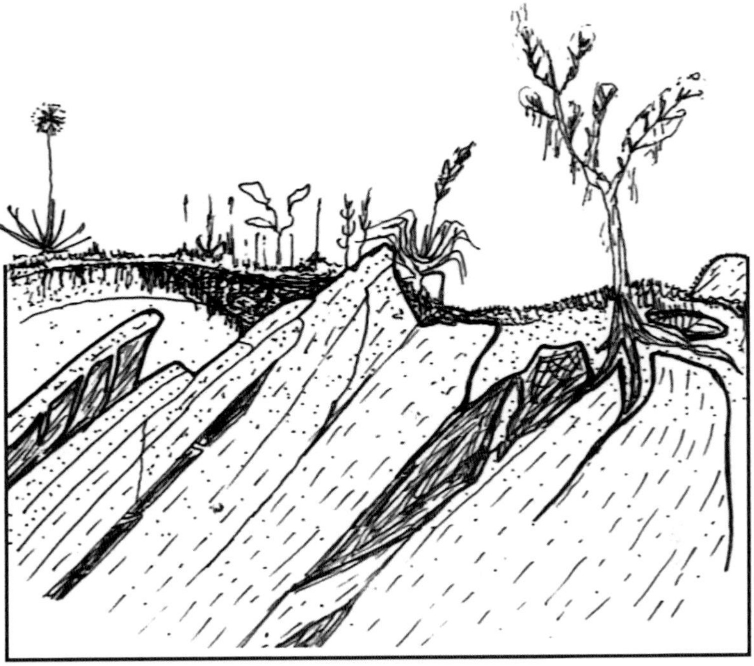

Fig. 3.6 A typical soil (Litolic Neosol) with humic A horizon on the C.R. of Serra do Cipó, MG (Quartzite)

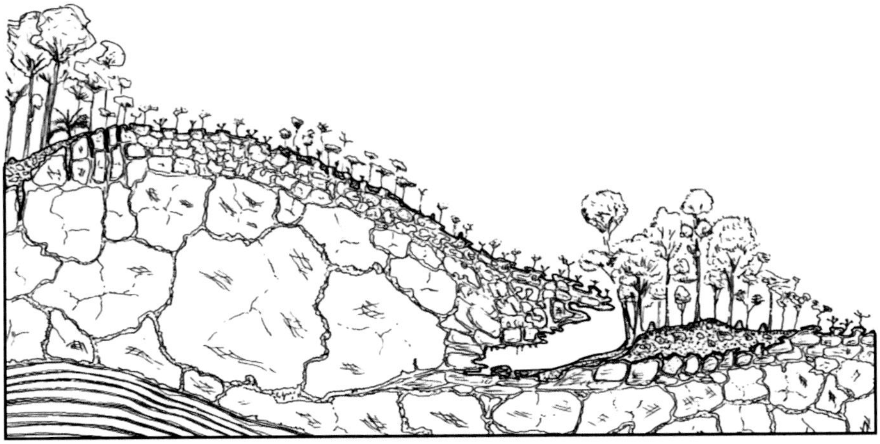

Fig. 3.7 Canga Quadrilátero. Soil—vegetation sequence on Ironstone from the Quadrilátero Ferrífero (Iron Quadrangle)/Carajás, in wich open Vellozia RG is associated with the shallower soils on canga outcrops

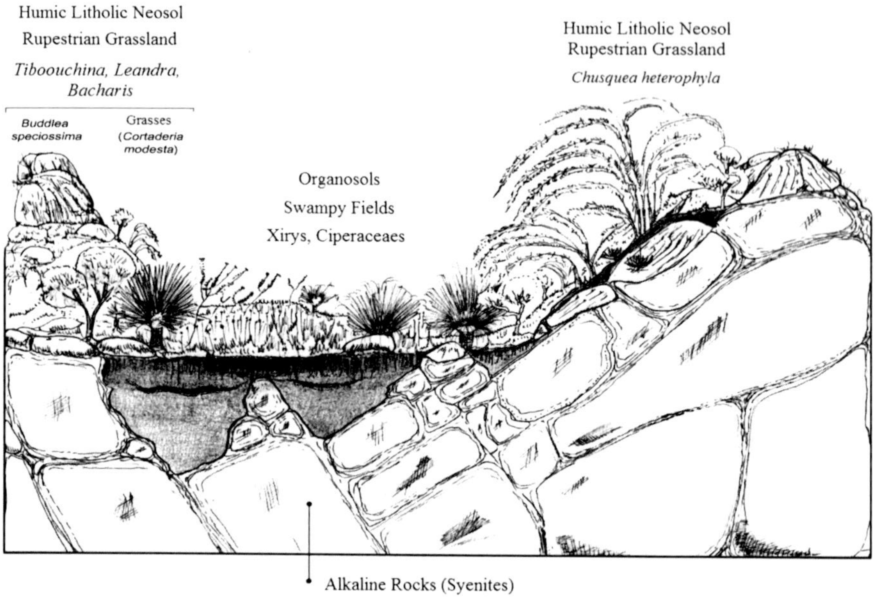

Fig. 3.8 Soil-landform-vegetation in the Rupestrian Grassland-Highland Grassland complex at the Itatiaia Mountain Range, on alkaline rocks (Syenite), at altitudes higher than 2000 m. The presence of well drained soils with organic horizons indicate a Late Quaternary paleo-landscape with colder and wetter climate than the present one (Carlos Schaefer illustration, Schaefer 2013)

soils developed from these mantles are generally extremely poor, even though they can be shallow. According to Benites et al. (2003a, b), there are marked differences between shallow soils on either granitic or quartzitic rocks (Fig. 3.3). In the first, boulders associated with deeper weathered mantles are observed and scrubby formation prevail between large rock outcrops. In the latter, fracturing, schistosity and faulting are prominent features controlling soils/vegetation development.

Soils at the HARC are young and shallow, and show little structure development. Thus, contrary to Lambers et al. (2008) who considered most P-impoverished soils as being aged and old soils, HARC soils are not old, and nutrient depletion is basically inherited from the nutrient-poor parent rock. Most soils associated to rock outcrops are classified as "Neossolos Litólicos" in the Brazilian Soils Classification System (Embrapa 1999), which correspond the Orthents suborder of the Soil Taxonomy (USDA 1998) and to the Leptosols of the FAO soil classification system. The "Neossolos Litólicos" class comprises little developed soils, up to 50 cm deep, without a clear diagnostic B horizon. These soils are normally characterized by a surface horizon (A) resting directly on the underlying rock or on a C horizon or saprolite (decomposed parent rock).

3.3.2 Chemical Status of Rupestrian Grasslands Soils

With reference to soil chemistry, all soil and regions studied are basically very acid, extremely oligotrophic, with remarkably low levels of P, K, Ca and Mg, as well as basis sum, and high (but varying) levels of exchangeable Al (Table 3.1). The only exception is Chapada Diamantina soils developed from Quartzite, with showed relatively higher amounts of P and BS, atypical for these soils. This may be attributed to analytical errors.

Hence, there is a general trend of low fertility (comparable with the lowest soil fertility status range for global soils) for all RG soils, so that all areas share a similar vegetation (in terms of physiognomy) with a vast repertoire of traits for overcoming these severe nutrient limitations. This, in turn, leads to a particular syndrome: a plant community that evolved under a combination of all possible environmental pressures: (1) high-frequency fire regime, (2) chemical poverty and (3) low water availability by shallow soils. This situation is in marked contrast with the postulate by Lambers et al. (2008) who emphasized soil age as a driver for nutrient depletion and plant adaptation. In the case of RG, soils are young, although rocks are ancient and either weathered or physically resistant. Thus, RG soils are nutrient-poor because rocks are strongly leached and weathered under well-drained tropical conditions (Itabirites, Gneiss, Granites), or alternatively acid, nutrient poor parent rock (Quartzite, Phyllites). In this sense, Schaefer (2013) emphasized the need to take into account the residual, structural nature of Brazilian Mountain ranges and high altitude landforms to RG development. Ecollogically, RG is also a residual, relict vegetation that persists on stable, high mountains due to its long term tectonic stability, since tectonic uplift is negligible.

Soil texture is variable according to parent material, being richer in clay and silt for RG soils on Itabirite and Canga (Carajás, QF) and much richer in sand for Quartzite.

Most RG soils are not only shallow but covered by a detritic pavement (stony lag deposit), composed of quartz's gravels and cobbles and, occasionally, Fe concretions of varying sizes (petroplinthite). Normally poor in nutrients and with low organic carbon content, these soils are classified as "Neossolos Litólicos distróficos" or "Neossolos Litólicos psamíticos" (high sand content). They are characterized by a moderate A horizon overlying a coarse textured mineral layer of up to 50 cm, which rests on rock (quartzite) or deep saprolite (schists/pellitic rocks). When the solum is deeper than 50 cm, composed basically of quartzous sand and gravels, it is classified as "Neossolo Quartzarênico Órtico", as observed in Serra do Ibitipoca (Dias et al. 2002; Benites et al. 2003b) and in some areas of Serra do Cipó and the Diamantina Plateau (Benites et al. 2003a; Schaefer et al. 2002) and Quadrilátero Ferrífero (Ker and Schaefer 1995; Schaefer et al. 2005), where quartzite is the main lithology (Fig. 3.6).

Soils with an incipient B horizon underlying a humic A horizon, classified as "Cambissolos Húmicos" (USDA 1998; Embrapa 1999) are also found associated with rock outcrops and saprolites (Dias et al. 2002; Simas et al. 2004). Despite less common in occurrence, these soils are found on both quartzite and igneous/metamorphic

Table 3.1 General chemical characteristic o Rupestrian Grasslands from different locations in Brazil

		Roraima (Tepuis)	Espinhaço (C. Diamantina)	Carajás	Mantiqueira	Espinhaço (Cipó)	Ibitipoca	Quadrilátero Ferrífero		Ricardo Franco	Brigadeiro/Caparaó
	Numbers of pedon	6	4	5	5	7	6	9	16	3	7
	Lithology	Metarenite	Quartzite	Canga	Granite/Gneiss	Quartzite	Schistose quartzite	Canga	Quartzite	Metarenite	Granite/Gneiss
A (non humic)	pH (H_2O)	5.2 (4.1–5.4)	3.7 (3.6–3.9)	5.0 (4.9–5.1)	4.7 (4.2–5.2)	4.3 (4.0–4.6)	3.9 (3.7–4.1)	4.9 (4.7–5.1)	4.8 (4.6–5.0)	4.1 (3.9–4.2)	4.7 (4.3–5.0)
	P (mg/dm^3)	1 (0.6–1.5)	12.9 (6.4–19.5)	3.5 (3.1–3.8)	1.9 (1.1–2.8)	2.9 (1.4–4.4)	7.3 (2.1–12.5)	2.6 (2.1–3.1)	1.1 (0.8–1.4)	0.4 (0.1–0.7)	3.9 (2.5–5.4)
	K (cmol$_c$/dm^3)	0.05 (0.01–0.08)	0.4 (0.2–0.6)	0.1 (0.1–0.1)	0.07 (0.01–0.12)	0.0 (0.0–0.1)	0.1 (0.1–0.2)	0.1 (0.1–0.2)	0.1 (0.0–0.1)	0.1 (0.1–0.1)	0.1 (0.1–0.1)
	Ca (cmol$_c$/dm^3)	0.3 (0.1–0.8)	1.0 (0.0–2.0)	0.6 (0.5–0.7)	0.2 (0.1–0.3)	0.0 (0.0–0.1)	0.4 (-0.1–0.9)	0.7 (0.3–1.0)	0.3 (0.1–0.5)	0.1 (0.1–0.2)	0.1 (0.0–0.1)
	Mg (cmol$_c$/dm^3)	0.3 (0.1–0.3)	1.2 (0.3–2.0)	0.2 (0.1–0.2)	0.05 (0.01–0.1)	0.0 (0.0–0.1)	0.1 (0.0–0.2)	0.2 (0.1–0.3)	0.1 (0.0–0.1)	0.1 (0.1–0.1)	0.1 (0.1–0.1)
	Al (cmol$_c$/dm^3)	1.6 (1–1.9)	2.5 (1.7–3.4)	0.9 (0.7–1.1)	2.0 (1.0–3.0)	1.6 (1.0–2.2)	1.9 (1.4–2.3)	0.8 (0.6–1.0)	1.4 (0.8–2.1)	2.4 (1.9–2.9)	2.0 (1.2–2.9)
	H + Al (cmol$_c$/dm^3)	8.3 (6.1–12.8)	20.8 (16.9–24.8)	15.9 (13.5–18.3)	13.1 (6.5–19.8)	11.8 (6.5–17.1)	11.7 (7.0–16.4)	12.3 (10.7–14.0)	7.5 (4.6–10.3)	11.7 (10.2–13.3)	15.8 (12.0–19.6)
	BS (%)	2.0 (1–3.5)	11.7 (3.1–20.2)	5.7 (4.7–6.7)	2.5 (1.6–3.4)	1.9 (0.5–3.2)	5.7 (3.5–7.9)	7.3 (4.3–10.2)	6.3 (3.7–8.9)	3.2 (1.8–4.7)	1.8 (1.45–2.3)
	COT (dag/kg)	1.9 (1.6–2.3)	8.5 (7.8–9.2)	13.2 (11.6–14.8)	5.3 (0.1–10.53)	4.1 (2.6–5.5)	4.8 (2.8–6.8)	3.7 (2.5–4.8)	2.0 (1.3–2.7)	3.7 (3.2–4.2)	6.6 (5.1–8.1)
	CEC(T) (cmol$_c$/dm^3)	9.2 (7.1–14.5)	23.5 (20.6–26.5)	16.8 (14.3–19.4)	13.6 (6.8–20.4)	12.0 (6.7–17.2)	13.4 (8.9–17.9)	28.7 (16.4–41.0)	7.6 (4.6–10.7)	12.1 (10.7–13.6)	17.1 (12.3–22.0)

(continued)

Table 3.1 (continued)

		Roraima (Tepuis)	Espinhaço (C. Diamantina)	Carajás	Mantiqueira	Espinhaço (Cipó)	Ibitipoca	Quadrilátero Ferrífero		Ricardo Franco	Brigadeiro/Caparaó
	Sand (%)	61.0 (55–67)	77.8 (71.1–84.4)	41.8 (36.7–46.9)	76.0 (71.4–80.6)	56.0 (51.9–60.1)	84.2 (81.4–86.9)	55.1 (43.6–66.6)	62.9 (54.9–70.9)	62.0 (57.9–66.1)	71.1 (63.6–78.7)
	Silt (%)	19.0 (17–22)	13.0 (7.3–18.6)	29.8 (26.1–33.5)	12.2 (10.5–13.9)	28.9 (24.0–33.7)	5.7 (3.4–7.9)	25.7 (17.0–34.4)	21.2 (15.7–26.6)	21.0 (16.1–25.9)	15.3 (11.1–19.5)
	Clay (%)	20.0 (15–22)	9.3 (8.2–10.5)	28.4 (26.4–30.4)	11.8 (8.7–14.9)	15.1 (13.2–17.1)	10.1 (8.1–12.2)	19.2 (13.6–24.8)	15.3 (11.4–19.1)	17.0 (15.9–18.1)	13.6 (9.7–17.4)

Table 3.2 Chemical characteristics of selected Humic Litholic Neosols, Dystrophics and Psamitics, associated to rocky outcrops in Mantiqueira and Espinhaço (Benites et al. 2003a)

Horiz	Depth	Color	cs[a]	fs	Silt	Clay	Fe^{b}_{DCB}	pH	Ca + Mg	Al[c]	C^{d}_{org}
	Cm		%				g kg^{-1}		$cmol_c$ kg^{-1}		g kg^{-1}
A_h	0–20	2.5/0	7	58	25	10	0.7	4.6	0.2	2	60
A	0–15	2/1 2.5Y	51	23	14	12	1.8	4.5	0.3	5.1	71
A	0–8	3/1 7.5YR	80	11	5	4	0.3	4.3	0.2	0.1	14
C	Ago/45	5/2 7.5YR	65	24	7	4	0	5	0	0	3
A	0–16	2/1 10YR	12	44	18	26	40.7	4.7	0.1	5	57
A	0–15	3/1 2.5Y	23	50	16	11	0.5	4.8	0.2	1.9	12
C	15–45	5/3 2.5Y	58	27	10	5	1.2	5	0	0.6	3

[a]Coarse sand, fine sand, silt and clay in the <2 mm fraction determined according to Embrapa (1997)
[b]Free Fe extracted with ditionite-citrate-bicarbonate according to Embrapa (1997)
[c]Exchangeable Al^{3+} (Embrapa 1997)
[d]Total organic carbon (Yeomans and Bremner 1988)

lithologies (see Table 3.2). They usually occupy more stable, lower areas which are stable enough for allowing colluvial accumulation, in situ pedogenesis and B horizon differentiation. However, on quartzite, due to the normally sandy texture of the soils, the occurrence of the "Neossolo Quartzarênico" class is more common than the "Cambissolo Húmico".

Organic substances accumulate in soils associated to rock outcrops due unfavorable conditions for microbial decomposition. The main factors are: lack of nutrients, high Al^{3+} levels and lower temperatures, all of which reduce microbial activity and decomposition rates, promoting the accumulation of organic C. This, in turn, favors the establishment of vegetation by feedback mechanisms, because the accumulated organic layer ultimately enhances nutrients and water retention, serving as a substratum for plant development. In addition, mesothermic climates in these highlands, with cool winters can also account for a relative decrease of carbon mineralization, leading to high soil organic matter content (Table 3.2).

3.3.3 Soil and Vegetation Relationships

The extremely low nutritional status of these soils required the development of survival strategies by the vegetation, involving physiological and morphological adaptations. Some nutrients, notably P, which is extremely limiting for plant development, show negligible amounts in some soils (Oliveira et al. 2015). In igneous rock outcrops, despite the generalized lack of P in the soil, the soils still maintains some reserve of this element in primary apatite minerals. In this case, plants release root exudates capable of solubilizing this element directly from the rock. In quartzitic outcrops, where rock apatite is absent, P uptake mechanisms are even more notable. Insectivorous plants are frequently observed in these environments and the organic P assimilated from insects may represent a greater part of available P for plants. P amounts are particularly critical for ferruginous RG developed on ironstone (Itabirites) from Fe-mining areas of Carajás and Quadrilátero Ferrífero in Minas Gerais (Fig. 3.7).

3.3.4 Similarity and Differences Between RG Soils

The PCA analysis (Fig. 3.9) enable the clustering of all RG soils into two basic groups: (1) one related to higher organic matter accumulation and clayey/silty textures (Carajás and QF, both on Canga and Itabirite); and (2) another group of sandt soils with greater Al^{3+} exchangeable levels, interestingly grouping both Granite/Gneiss and Quartzite/Phyllites. An outlier formed by CR soil from the Chapada Diamantina-BA was formed by anomously high available P levels (data from Abel Conceição), without any possible explanation, since soil data from soil surveys show a similar chemical pattern with Espinhaço CR soils from MG. This was attributed to analytical error.

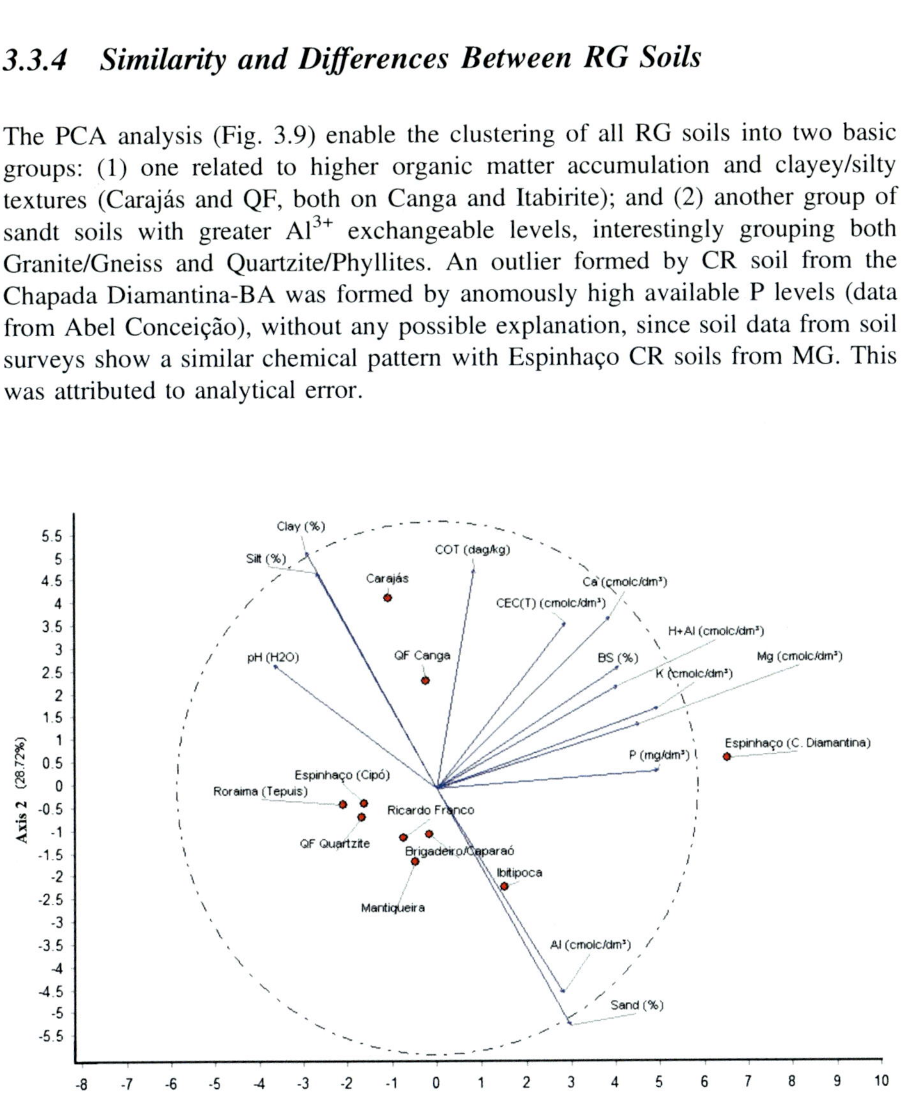

Fig. 3.9 PCA analysis of CR soils from Brazil

In conclusion, the range of soil fertility for CR soils are close to the lower detection limit for most major nutrients, and cluster analysis indicate that physical, rather than chemical differences, exist amongst the CR soils. The low biomass status of this vegetation is closely linked to a very low supply of nutrients (particularly P), rather than Al toxidity, since high biomass forest occurs in soils with even greater Al^{3+} levels in Brazil.

References

Benites VM, Schaefer CERG, Mendonça ES, Martin Neto L (2001) Caracterização da matéria orgânica e micromorfologia de solos sob Campos de Altitude no Parque Estadual da Serra do Brigadeiro. Rev Bras Cien Solo 25:661–674

Benites VM, Caiafa AN, Mendonça ES, Schaefer CER, Ker JC (2003a) Solos e vegetação nos complexos rupestres de altitude da Mantiqueira e do Espinhaço. Rev Flor Amb 10:1

Benites VM, Cunha TJF, Ferraz RPD, Caiafa AN, Simas FNB, Mendonça ES (2003b) Caracterização de solos em duas topossequências em diferentes litologias em áreas altimontanas na Serra da Mantiqueira. Bol Pesq 57, Embrapa Solos, Rio de Janeiro

Benites VM, Sá E, Schaefer CER, Novotny EH, Reis E, Ker JK (2005) Properties of black soil humic acids from high altitude rock complexes in Brazil. Geoderma 127:104–113

Carvalho Filho A, Curi N, Shinzato E (2010) Relações solo-paisagem no Quadrilátero Ferrífero em Minas Gerais. Pesq Agropec Bras 45:903–916

Dias HCT, Fernandes EIF, Schaefer CEGR, Fontes LEF, Ventorim LB (2002) Geoambientes do Parque Estadual do Ibitipoca, Município de Lima Duarte-MG. Rev Arv 26:777–786

Embrapa S (1997) Manual de métodos e análises de solo. Embrapa Solos, Rio deJaneiro

Embrapa S (1999) Sistema brasileiro de classificação de solos. Embrapa Solos, Rio de Janeiro

Ferrari LT (2013) Geoambientes na RPPN Capivari II e monitoramento térmico e hídrico em solos de canga ferrífera e quartzito, quadrilátero ferrífero—Minas Gerais. MSc Dissertation, Universidade Federal de Viçosa

Ker JC, Schaefer CE (1995) Roteiro da excursão pedológica Viçosa-Sete Lagoas. Cong Brasl de Ciência do Solo, 25. Viçosa, SBCS/DPS—UFV/EMBRAPA—Centro Nacional de Pesquisas de Solos

Lambers H, Raven JA, Shaver GR, Smith SE (2008) Plant nutrient acquisition strategies change with soil age. Trends Ecol Evol 23:95–103

Oliveira RS, Galvão HC, De Campos MCR, Eller CB, Pearse SJ, Lambers H (2015) Mineral nutrition of campos rupestres plant species on contrasting nutrient-impoverished soil types. New Phytol 205:1183–1194

Pereira AFS (2010) Florística, fitossociologia e relação solo vegetação em campo rupestre ferruginoso do quadrilátero ferrífero, MG. MSc Dissertation, Universidade Federal de Viçosa

Ribeiro ASS (2009) Pedologia Edogênese nos geoambientes de canga na serra sul, floresta nacional de Carajás, Para—MG. MSc Dissertation, Universidade Federal de Viçosa

Schaefer CEGR (2013) Bases físicas da paisagem brasileira: estrutura geológica, relevo e solos. Top Cien Solo 8:1–69

Schaefer CEGR, Ker JC, Gilkes RJ, Campos JC, Da Costa LM, Saadi A (2002) Pedogenesis on the uplands of the Diamantina Plateau, Minas Gerais, Brazil: A chemical and micropedological study. Geoderma 107:243–269

Schaefer CEGR, Albuquerque-Filho MR, Oliveira DA (2005) Solos da Serra de Ouro Branco In: De Paula CC, Oliveira DA, Da Silva RR (Org.) A Serra de Ouro Branco. Viçosa, DI Gra e Edi 1:17–29

Semir J (1991) Revisão taxonômica de *Lychnophora* Mart. (Vernoniaceae: Compositae). PhD Thesis, Universidade Estadual de Campinas

Simas FNB, Schaefer CEGR, Fernandes Filho EI, Chagas AC, Brandão PC (2004) Chemistry, mineralogy and micropedology of highland soils on crystalline rocks of the Serra da Mantiqueira, southeastern Brazil. Geoderma 125:187–201

USDA—United States Department of Agriculture (1998) Keys to soil taxonomy. Cornell University, New York

Valente EL (2009) Relações solo-vegetação do Parque Nacional da Serra do Cipó, Espinhaço meridional, Minas Gerais. PhD thesis, Universidade Federal de Viçosa

Volkoff B, Cerri CC, Melfi AJ (1984) Húmus e a mineralogia dos horizontes superficiais de três solos de campos de altitude dos estados de Minas Gerais, Paraná e Santa Catarina. R Bras Cien Solo 8:277–283

Yeomans JC, Bremner JM (1988) A rapid and precise method for routine determination of organic carbon in soil. Commun Soil Sci Plant Anal 19:1467–1476

Chapter 4
Thermic and Hydric Dynamics of Ironstone (Canga) and Quartzite Rupestrian Grasslands in the Quadrilátero Ferrífero: The Ecological Importance of Water

Lucas Teixeira Ferrari, Carlos Ernesto Gonçalves Reynaud Schaefer, Raphael Bragança Alves Fernandes, Bruno Araújo Furtado Mendonça, Davi Feital Gjorup, Guilherme Resende Corrêa and Eduardo Osório Senra

Abstract Most Rupestrian Grasslands are associated with the more resistant and prominent rocks in the landscape (quartzites and itabirites). However, there is little knowledge on the ecological importance of water and near-surface temperature regime to this peculiar vegetation, as well as their relationship with the soil and

L.T. Ferrari (✉)
Campus Teófilo Otoni, Departamento de Meio Ambiente, Instituto Federal Norte de Minas Gerais (IFNMG), Teófilo Otoni, MG, Brazil
e-mail: lucas.ferrari@ifnmg.edu.br

C.E.G.R. Schaefer · R.B.A. Fernandes
Departamento de Solos, Universidade Federal de Viçosa, Viçosa, MG, Brazil
e-mail: carlos.schaefer@ufv.br

R.B.A. Fernandes
e-mail: raphael@ufv.br

B.A.F. Mendonça
Departamento de Silvicultura, Instituto de Florestas, Universidade Federal Rural do Rio de Janeiro, Rio de Janeiro, Brazil
e-mail: brunoafmendonca@gmail.com

D.F. Gjorup · E.O. Senra
Doutorando em Solos e Nutrição de Plantas, Departamento de Solos, Universidade Federal de Viçosa, Viçosa, MG, Brazil
e-mail: davifeitalgjorup@gmail.com

E.O. Senra
e-mail: senra.eduardo@gmail.com

G.R. Corrêa
Instituto de Geografia, Universidade Federal de Uberlândia, Uberlândia, MG, Brazil
e-mail: guilhermeudi@yahoo.com.br

© Springer International Publishing Switzerland 2016
G.W. Fernandes (ed.), *Ecology and Conservation of Mountaintop Grasslands in Brazil*, DOI 10.1007/978-3-319-29808-5_4

environmental importance. We characterized and compared the thermic and hydric dynamics of soils on ironstone and quartzite in the Iron Quadrangle (IQ). In situ monitoring of moisture and soil temperature and air temperature in four different environments, namely: Rupestrian Grassland on Ironstone, Forest on Ironstone, Rupestrian Grassland on Quartzite, and Forest on Quartzite are reported. The soil moisture monitoring method was based on the TDR (Time Domain Reflectometer), by means of sensors calibration in a laboratory, in which the propagation time of electromagnetic pulse detected by sensors installed in the soil were converted into gravimetric moisture values. The pioneer data indicated that temperature, and soil moisture and depth have strong influence on vegetation physiognomies in both rupestrian grassland environments. The magnitude of soil temperature and soil moisture variations in the rupestrian grasslands is very high; hence highlighting the extreme environmental conditions to which they are subjected. Results allowed a deeper understanding of the pedoclimatic dynamics on these environments.

Contents

4.1 Introduction

Rupestrian Grasslands are associated with resistant and prominent rocks in the landscape (quartzites and itabirites), and soil and hidrological properties play a prominent role in defining the plant communities evolved and adapted to such extreme environments. To this day, there is little knowledge on the ecological importance of water and near-surface temperature regime to this peculiar vegetation, as well as its relationship with the soil and environmental importance for its conservation. Against this background, we characterized and compared the thermic and hydric dynamics of soils on ironstone and quartzite in the Iron Quadrangule (Quadrilátero Ferrífero) in Minas Gerais state, Brazil. For this purpose, we performed the in situ monitoring of moisture and soil temperature and air temperature in four different environments, namely: Rupestrian Grassland on Ironstone (RC), Forest on Ironstone (FC), Rupestrian Grassland on Quartzite (RQ) and Forest on Quartzite (FQ). The soil moisture monitoring method was based on the TDR (Time Domain Reflectometer), by means of sensor calibration in a laboratory, in which the propagation time of electromagnetic pulse detected by sensors installed in the soil were converted into gravimetric moisture values (Gong et al. 2003; Campbell Scientific 2006; Alfaro Soto et al. 2007).

4.2 Monitoring Sites Location

Two contrasting vegetation types were selected for comparative studies. A forest and a neighbouring rupestrian grassland (rocky outcrop) on different substrates (Itabira Group Canga derived from itabirite and Quartzites of the Moeda Formation), with a total of four soil sites monitored, identified by the codes RC, FC, RQ and FQ. The location was close to the highlands of the Capanema district, at the vicinity of Ouro Preto, in the coordinates according Table 4.1.

4.3 Soil Characterization

The soil and vegetation gradient in the selected environments represent a continuum between extreme herbaceous and woody formations. However, the abrupt change between the two made possible the simple division used, based on the two extremes: Rupestrian Grassland and well-drained, upper montane Forest, on ironstone and quartzite. Figure 4.1 illustrates the two soil-vegetation systems, as well as representative profiles of the four monitoring sites (RC, FC, RQ e FQ).

Soil depth is key for controlling the vegetation gradient. In the forest environments (FC and FQ) soils are much deeper than in the Rupestrian Grasslands (RC and RQ). Similarly, at the Amazonian "Serra de Carajás" the degree of fracturing

Table 4.1 Monitoring sites with associated vegetations, lithology, GPS position and depths, in Capanema district

Sites	Vegetation	Lithology	Latitude Longitude	Altitude	Layers
Rupestrian grassland on ironstone (RC)	Ironstone rupestrian grassland	Ironstone (Canga)	−20° 11′ 15″ −43° 37′ 39″	1650 m	10 cm—RC10 30 cm—RC30
Forest on ironstone (FC)	Cloud forest				10 cm—FC10 30 cm—FC30 100 cm—FC100
Rupestrian grassland on quartzite (CQ)	Quartzite rupestrian grassland	Quartzite	−20° 9′ 1″ −43° 37′ 58″	1390 m	10 cm—RQ10 30 cm—RQ30
Forest on quartzite (FQ)	Seasonaly dry forest				10 cm—FQ10 30 cm—FQ30 100 cm—FQ100

Fig. 4.1 Soils and vegetation at the monitored sites: Forest and Rupestrian Grassland on Ironstone (ferruginous canga) (**a**) and on Quartzite (**b**); and soils P1 (**c**), P2 (**d**), P3(**e**) and P4(**f**), respectively monitoring sites RC, FC, RQ and FQ

and canga degradation, with resulting deepening of soils, the hardened laterite (lithoplintic horizon) change into a loose concretionary horizon, with greater development of vegetation biomass (Ribeiro 2009).

Table 4.2 presents the general results of chemical and physical characteristics of the studied soils. Except for FQ, all soils are skeletic, with large amounts of coarse materials (Gravels) in the form of concretions. In FC, the amount of concretions throughout the profile led to its classification as Plinthosol at order level, rather than Latosol, despite the deep weathering degree.

The amounts of available P, CEC, bases sum (SB) and bases saturation (V%) are very low in all soils, indicating severe dystrophy, combined with high Al saturation (high Al^{3+}), typical of such soils from the IQ. At the surface horizons of soils P2 and P4 the CEC is slightly greater, due to higher Organic matter related to higher biomass and vegetation cycling.

At Soil P3, the chemical characteristics indicate the influence of a phyllite layer (yellowish color) normally richer in nutrients. The Organic carbon increase may be attributed to fire regime, very common in such Rupestrian Grasslands (Scarano 2007; Chaps. 22 and 17).

Table 4.2 Chemical and physical analysis of soils P1, P2, P3 e P4, on the rupestrian grassland (RC) and forest (FC) on ironstone and rupestrian grassland (RQ) and forest (FQ) on quartzite, respectively, from Capanema district

Horizon	pH		P	SB	T	T	Al^{3+}	H + Al	V	m	TOC	P-rem	CSand	Fine sand	Clay	Silt	Textural class
Cm	H_2O	KCl	mg/dm^3	$cmol_c/dm^3$					%		dag/kg	mg/L	dag/kg				
P1—Petric Plinthosol, lLithoplintic, Typic—RC																	
Ac (0.15; 23)	5.03	4.11	2.0	0.85	1.24	11.2	0.39	10.3	7.6	31.5	2.45	21.1	34	18	29	19	Loamy
A/F (15; 23–30; 40)	4.99	3.90	2.4	0.86	1.45	10.4	0.59	9.5	8.3	40.7	2.91	26.9	37	33	17	13	Sandy loam
P2—Petric Plinthosol, Concretionary, Latosolic—FC																	
Ac/F (0–14)	4.4	3.66	5.6	1.68	3.05	19.2	1.37	17.5	8.8	44.9	7.05	23.9	30	04	37	29	Sandy loam
ABc/F (14–33)	4.71	4.20	3.8	0.41	2.07	17.9	1.66	17.5	2.3	80.2	3.14	7.7	41	4	37	29	Sandy loam
Bwc/F (33–58)	4.95	4.45	2.2	0.30	0.98	15.1	0.68	14.8	2	69.4	2.76	5.3	37	3	18	42	Clay
Cc/F (58–93+)	5.05	4.84	0.9	0.10	0.20	7.2	0.10	7.1	1.4	50.0	1.38	5.3	26	4	27	43	Clay
P3—Litholic Neosol, Distrophic, Typic—RQ																	
AR (0–8; 27)	5.05	4.84	0.9	0.10	0.20	7.20	0.10	7.1	1.4	50.0	1.38	5.3	35	33	15	17	Sandy loam
CR (8; 27–38+)	4.62	3.88	3.5	0.49	2.05	7.69	1.56	7.2	6.4	76.1	2.45	33.7	37	32	14	17	Sandy loam

(continued)

Table 4.2 (continued)

Horizon	pH		P	SB	T	T	Al^{3+}	H + Al	V	m	TOC	P-rem	CSand	Fine sand	Clay	Silt	Textural class
Cm	H_2O	KCl	mg/dm^3	$cmol_c/dm^3$					%		dag/kg	mg/L	dag/kg				
P4—Quartzarenic Neosol, Ortic, Typic—FQ																	
A (0–22)	4.09	3.61	6.5	0.48	3.8	18.0	3.32	17.5	2.7	87.4	4.78	17.1	40	22	16	22	Sandy clay loam
C1 (22–43)	4.20	3.81	2.1	0.17	0.76	16.1	0.59	15.9	1.1	77.6	2.30	11.7	42	23	13	22	Sandy clay loam
C2 (43–86)	4.51	4.23	1.1	0.11	2.35	8.61	2.24	8.5	1.3	95.3	0.99	14.2	48	21	26	5	Sandy loam
C3 (86–108+)	4.63	4.37	0.9	0.10	1.95	6.7	1.85	6.6	1.5	94.9	0.84	20.1	47	22	14	17	Sandy loam

The soil texture on canga soils (P1 and P2) indicates the dominance of sand fraction, even in soils with very high amounts of oxides. This is due to the difficulties in dismantling the hardened, cemented nodules and concretions using the dispersing agents. The high amount of iron also contributes for this behaviour, for the cementing effect promoted by hematite (Silveira et al. 2016). The sandy behavior of canga soils is crucial for determining the thermal and hydric dynamics of these soils.

4.4 Soil Climate Aspects and Vegetation Implications

4.4.1 Soil Moisture

Figure 4.2 illustrates the behavior of the gravimetric soil moisture (U) during the monitoring period, compared with rainfall data form a nearby station (Represa das Codornas), as well as the coefficient of variation (CV) obtained from three replicates of soil moisture sensors. The correspondence of rainfall events and soil humidity is clear, with marked peaks of moisture increase during the rainy season (summer), whereas practically no peaks in the dry season (winter).

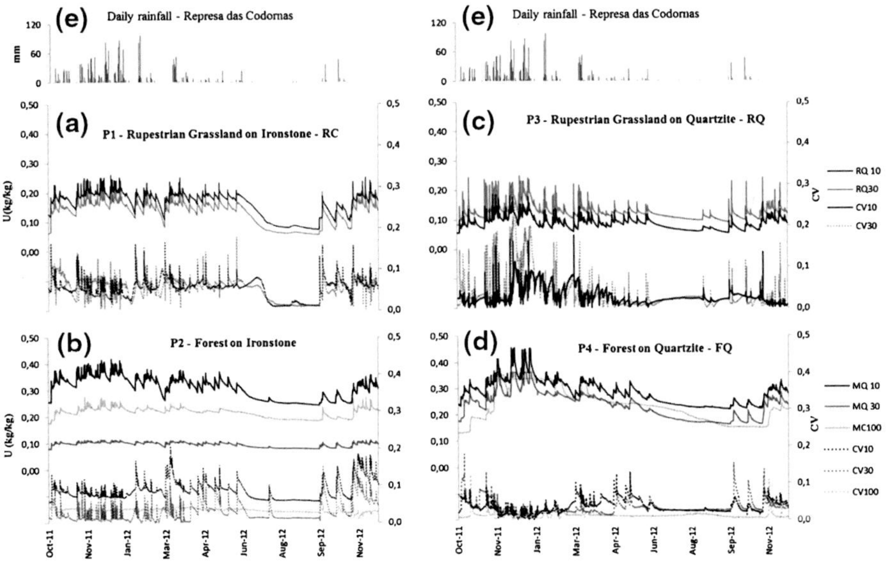

Fig. 4.2 Gravimetric moisture (kg/kg) and coefficient of variation (%) of the signal obtained by the three sensors at different depths, for each monitoring site: RC (**a**), FC (**b**), RQ (**c**) and FQ (**d**). *Continuous lines* represent values of gravimetric moisture at 10, 30 and 100 cm depths (U10 and U30 for RC e RQ; U10, U30 and U100 for FC and FQ) and *broken lines*, the CV values at the same depths (CV10 and CV30 for RC and RQ; CV10, CV30 and CV100 for FC and FQ); recent climatic data range to 30 Sept 2012 (**e**), whereas logger data of monitoring sites to 6 Dec 2012

The seasonal variation of moisture is evident, and even for the coefficients of variation, with a similar trend to the moisture regime. During the rainy season, the CV values are oscillating, with peaks of very high amounts, such as those of the RQ 30 cm layer, where CV reaches 25 %. However, at the dry season, the CV is close to zero, and stable. This can be explained by wetting and drying cycles, leading to the development of heterogeneity and sites of greater moisture, redistributed by vertical or lateral movements. For this reason, the deep soil layers at 100 cm depth (FC100 e FQ100), are less susceptible to climatic variations, having greater stability in the CV values along the year, as well as buffering effect. Table 4.3 shows the values for CV at each layer, suggesting a possible influence of substrates and organic matter content. Greater CV values were observed at RC and FC, especially at the surface (FC10 cm).

The gravimetric moisture in Rupestrian Grasslands (RC and RQ) is much more variable than values for forest sites, with a trend of smoothed values, especially at the 30 and 100 cm layers (Fig. 4.2). The soil moisture in RC and RQ reaches values lower than 0.10 kg/kg, indicating a severe water deficit in these environments. Table 4.3 shows the number and duration of water deficit peaks, with marked values for both Rupestrian Grasslands (RC and RQ). In the quartzite (RQ) the greater number of very low values of available water points to a greater susceptibility of such Rupestrian Grasslands on quartzite to water stress, compared with canga. Greater clay and silt (fine) particles in the canga environment accounts for greater water retention, and water availability and this has a profound influence on the vegetation and other organisms (e.g., Carvalho et al. 2012; Negreiros et al. 2014).

The extremely low values of moisture in both RC and RQ suggest that the Permanent Wilting Point (PWP) for plants adapted to these soils are extremely low, in view of the low water field capacity of these soils, so that the total amount of available water is very low. According to Rizzini (1997), the PWP of a given plant is highly dependent on soil texture, so that in Sandy soils under open grassy formations, the PWP can reach down to less than 4 %, while in clayey soil, exceeds 30 % (Lüttge 2008; Lüttge et al. 2007, see also Chap. 11).

In RC, the surface layer stayed wetter compared to the subsurface during most monitoring period (Fig. 4.2), although we would expect a water saturation down to 30 cm, as that found in RQ. In both Rupestrian Grasslands, layer with less water retention are those with greater coarse fraction (RC30 and RQ10). Hence, higher moisture values of 10 cm in RC suggest an input of humidity from dew (clouds), as nebular condensation, due to high altitude of the canga site (1650 m). The pattern of epiphytism of vegetation in RC and FC suggest a nebular character, corroborating the external input from clouds, as described by Valente (2009). This indicate a great importance of nocturnal dew for the water budget on these high mountain landscapes, at the dry season, as pointed out by Rizzini (1997), Baêta (2012), and Barbosa et al. (2015) (see also Chap. 7). This, in turn, mitigates the water deficit character of Rupestrian Grasslands.

The moisture behavior for the forest environments (FC and FQ) was very contrasting (Fig. 4.2). In the ironstone forest, all layers presented different values throughout the year, while at the quartzite forest (FQ), moisture values had

Table 4.3 Coarse fraction (CF), Coefficient of variation between three replicates of measured soil moisture (CV), particle density (Dp), Organic Matter amount (OM), gravimetric moisture (U) maximum, medium e minimun, average day amplitude of U, number of events with <10 % U, and maximum time in which U stayed <10 %, for the four monitoring sites

Layer depth	CF	CV mean	D_p	OM	$U_{máx}$	U_{mean}	$U_{mín}$	Mean day amplitude	Number of events U < 10 %	Maximum time < 10 %
	%		kg/dm^3	dag/kg	kg/kg					Hours
Rupestrian grasslands on ironstone (RC)										
RC10	60.5	5	3.89	4.21	0.26	0.16	0.07	0.011	1646	1636
RC30	78.9	6	4.04	5.00	0.32	0.13	0.06	0.013	2588	2067
Forest on ironstone (FC)										
FC10	48.7	9	2.54	12.12	0.42	0.31	0.24	0.011	0	0
FC30	65.3	3	3.44	5.07	0.11	0.10	0.08	0.003	5987	2656
FC100	47.7	3	2.96	2.37	0.28	0.22	0.18	0.004	0	0
Rupestrian grasslands on quartzite (RQ)										
RQ10	60.8	3	2.52	2.37	0.19	0.09	0.05	0.010	7546	2563
RQ30	45.8	3	2.57	4.21	0.25	0.13	0.08	0.014	736	321
Forest on quartzite (FQ)										
FQ10	0.0	4	1.94	8.23	0.46	0.28	0.22	0.010	0	0
FQ30	0.0	4	2.94	2.83	0.36	0.23	0.16	0.006	0	0
FQ100	0.0	1	2.96	1.45	0.35	0.23	0.13	0.003	0	0

alternating highs and lows at 10, 30 and 100 cm, throughout the year. The ironstone cangas are spatially heterogeneous in both surface and subsurface, making it distinct which are distinguishable from typical rocky systems (Pifano et al. 2010; Carmo and Jacobi 2013). The heterogeneity at soil profile 2, on Ironstone (FC), is the reason for contrasting curves of moisture, since high organic matter at FC10, compared with the dominance of high density minerals in FC30 influence the U values (Table 4.3). In the later, the high density can mask the effects of moisture variations, since it is based on gravimetric, rather than volumetric correction, which would be the most adequate in this case.

At the deepest layer (FC100), little variation in moisture can help dissolve Fe-forms, favoring canga degradation. Hence, Dp and CF values are reduced in FC100, compared with FC30 (Table 4.3). In other words, friable, loose soil produced and accumulated by weathering of canga can enhance water retention and storage. Conversely, in FQ, where soil is much more homogeneous, the moisture curves are closer, with a trend of increasing variability and alternations at the surface, compared with 100 cm, where moisture is somewhat buffered. A noticeable fact is the dryer condition of the 30 cm depth in FQ, compared with 100 cm. The explanation for less moisture at the 30 cm in Forest (FQ) is the high evapotranspiration at the dry season, since water-taping tree roots are concentrated in subsurface. Similar results were described by Carvalho (2011) in a moisture monitoring in soils under coffee in Minas Gerais.

4.4.2 Soil Temperature

The variation in soil and air temperature in terms of mean monthly values are reported in Fig. 4.3. Air temperatures in Rupestrian Grassland and forest environments are similar, but the surface soil temperature in the forest closely follows the air temperature (more buffered), while in Rupestrian Grassland a great seasonal variation exists. Also, annual variability is much greater for Rupestrian Grassland. In Rupestrian Grassland on Ironstone (RC), the range (amplitude) is much greater than in Rupestrian Grassland on Quartzite (RQ); also soil temperature in RC is higher than air temperature during most monitoring period, with closer values in the winter time. In RQ the soil temperature is lower than the air temperature during the winter. Thus, soil temperature tends to follow the vegetation gradient, while the air temperature is independent of the local vegetation.

Differences between soil and air temperatures along the monitoring period are shown in Fig. 4.3 (b, c, e, f). In this case, we can discount the altitude effect between the Ironstone and Quartzite sites. Negative values in the ordinates indicate higher air temperatures in relation to soil. Once again, the buffering effect of the forest is clear, with delta temperatures values closer to zero, in comparison with CR values, in both Ironstone and Quartzite. However, differences in the thermal behavior of Ironstone and quartzite are marked. In RC, the mean daily soil temperature is nearly 7 °C higher than air temperatures, whereas in RQ this difference

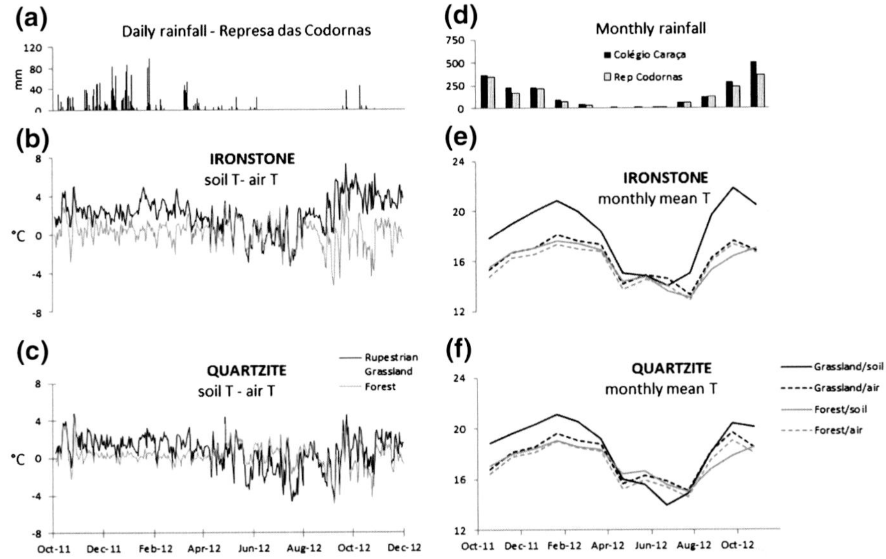

Fig. 4.3 Air and soil temperature data at monitoring sites. Differences between air and soil temperatures, for ironstone (**b**) and quartzite (**c**), forest and rupestrian grassland; monthly mean data of soil and air temperature on ironstone (**e**) and quartzite (**f**), also at both vegetation type; rainfall data from a nearby station (Represa das Codornas) is shown at the top (**a**). All data was based on historic mean data from Represa das Codornas and Colégio Caraça (1985–2011) (**d**)

does not exceed 5 °C. This can be explained by albedo differences between the two substrates: a dark, reddish, dense canga, is able to absorb much more energy than the light, whitish quartzite. Hence, heating is greater in canga substrates. In the dry winter season we observed an inversion in the termic behavior between Rupestrian Grasslands and Forest. In the Rupestrian Grasslands on quartzite (RQ) the delta values reached negative figures for most time, indicating rapid cooling of surface soils in comparison with forest on the same lithology (FQ). A similar trend was observed for Ironstone (RC and FC), but to a lesser intensity, since it heats up more rapidly and easily.

Table 4.4 shows the events of maximum and minimum for air and soil temperatures, thermic amplitude (daily and monthly), number of events with temperatures higher than 35 °C, and maximum time under temperatures higher than 35 °C. According to Lambers et al. (2006), when plants reach foliar temperatures above 35–40 °C, they emit isoprene as a mechanism for protecting against damages to photosynthesis. The results corroborate the greater heating capacity of Ironstone by comparing RC and RQ figures. Curiously, soil in FQ reached greater values and variability compared with RQ soil, even though the mean indicate a buffering effect of forest soil temperatures. Although soil temperature in FQ reached the highest

Table 4.4 Temperature (soil and air) maximum and minimum; thermic amplitude (daily and monthly), number of events >35 °C and maximum time with T > 35 °C, for each monitored site

		Temperature (°C)		Daily amplitude			Monthly amplitude			n T > 35 °C	Max time (h) T > 35 °C
		Max	Mín	máx	méd	mín	máx	méd	mín		
RC	Soil	38	9	18	8	1	26	15	8	14	4
	Air	32	3	18	9	1	26	19	13	0	0
FC	Soil	21	11	4	2	0	9	5	4	0	0
	Air	31	4	14	7	1	25	15	12	0	0
RQ	Soil	31	5	19	7	1	25	14	7	0	0
	Air	33	5	16	9	1	26	17	13	0	0
FQ	Soil	42	5	26	9	1	28	20	13	8	1
	Air	31	5	15	7	1	25	15	11	0	0

Fig. (42 °C), the number of events and time under temperatures >35 °C are bigger in Ironstone (RC).

Figure 4.4 shows the soil temperature and moisture variations at the surface (10 cm) along a week, in three different periods: end of rainy season (1–7 april 2012), peak of dry season (1–7 Aug 2012) and beginning of the rainy season (1–7 Nov 2012). The mean values of temperatures are continuous lines (T) and moisture by broken lines (U). Graphs indicate that greater moisture can lead to lower temperatures. In ironstone environments, soil temperature variation is much greater in CR than in forest, with closer values during the night.

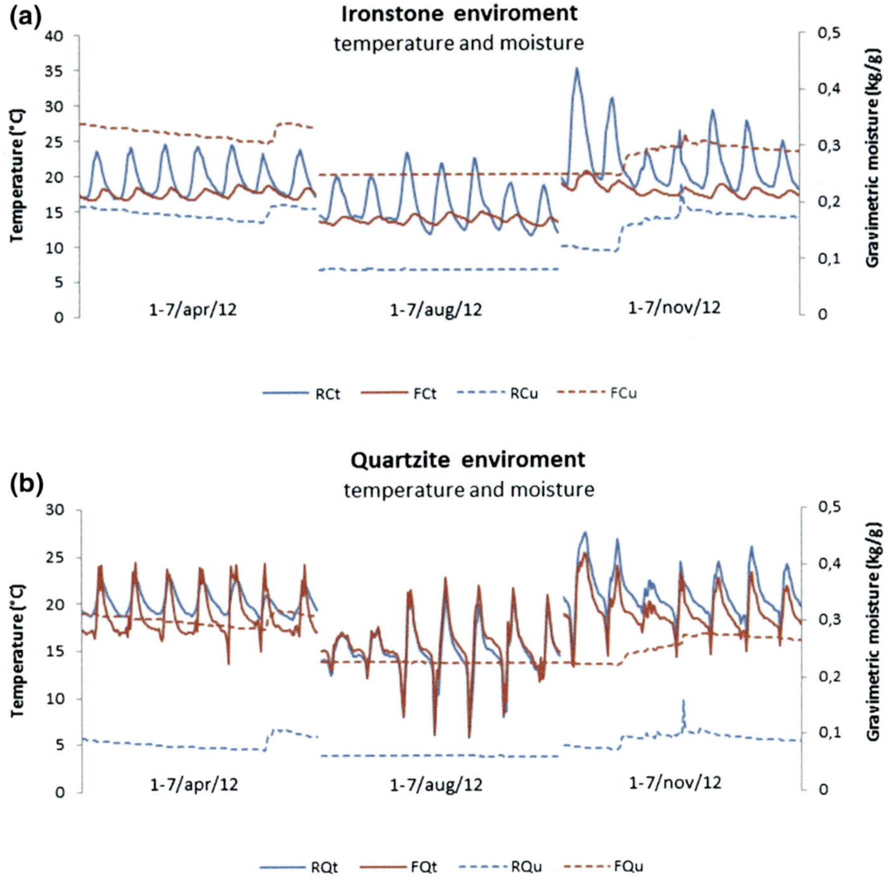

Fig. 4.4 Variation in soil temperature (t), at 5 cm depth, and moisture (u) at 10 cm depth, during a week, in three selected periods of time, on ironstone (**a**) and quartzite environment (**b**), at both vegetation type: rupestrian grassland (*blue line*) and forest (*red line*)

4.5 Conclusions

There are notable differences in soil temperature and moisture between the Rupestrian Grassland and Forest vegetation developed on Ironstone or Quartzite. For Ironstone, besides the pronounced water deficit, there are many diurnal events of high temperatures (>35 °C), limiting for plant growth. For Quartzite, the water deficits are less severe, and soil texture is sandy.

The monitoring of soil water and temperature revealed the importance of soil depth and granulometry, besides the presence of roots and microclimatic aspects, such as dew condensation during the night. The in situ monitoring of water and temperature allowed a much better knowledge of Ironstone and Quartzite ecology of these singular, multiple-stressed environments.

The use of TDR devices for moisture monitoring soils dominated by coarse and dense materials presents methodological challenges, in view of the need for calibrating the sensors for volumetric data.

Acknowledgements We thank Vale and CNPq for grants and scholarships that made possible this study, as well as for all logistic support.

References

Alfaro Soto MA, Kumayama DM, Chang HK (2007) Calibração de um reflectômetro para estudos do fluxo de água em solo não saturado. Geocien 26:357–368

Baêta HE (2012) Contribuição da deposição úmida (chuva e neblina) nas relações hídricas e nutricional de fisionomias de campos ferruginosos na Serra da Brígida, Ouro Preto, MG. Dissertation, Universidade Federal de Ouro Preto, Ouro Preto

Barbosa NPU, Fernandes GW, Sanchez-Azofeifa A (2015) A relict species restricted to a quartzitic mountain in tropical America: an example of microrefugium? Acta Bot Bras 29:299–309

Campbell Scientific (2006) CS616 and CS625 water content reflectometers. Instruction manual revision. Campbell Scientific Inc., North Logan

Carmo CM, Jacobi FF (2013) A vegetação de canga no Quadrilátero Ferrífero, Minas Gerais: caracterização e contexto fitogeográfico. Rodriguésia 64:527–541

Carvalho AF (2011) Água e radiação em sistemas agroflorestais com café no Território da Serra do Brigadeiro—MG. PhD Thesis Universidade Federal de Viçosa, Viçosa

Carvalho F, Souza FA, Carrenho R, Moreira FMS, Jesus EC, Fernandes GW (2012) The mosaic of habitats in the high–altitude Brazilian rupestrian fields is a hotspot for arbuscular mycorrhizal fungi. Appl Soil Ecol 52:9–19

Gong Y, Cao Q, Sun Z (2003) The effects of soil bulk density, clay content and temperature on soil water content measurement using time-domain reflectometry. Hydrol Proc 17:3601–3614

Lambers H, Chapin FS, Pons TL (2006) Plant physiological ecology. Springer, New York, 540 p

Lüttge U (2008) Savannas II. The environmental factors water, mineral nutrients and fire. In: Lüttge U (ed) Physiological ecology of tropical plants. Springer, Heidelberg, pp 313–377

Lüttge U, Duarte HM, Scarano FR, Attos EA, Cavalin PO, Franco AC, Fernandes GW (2007) Physiological ecology of photosynthesis of five sympatric species of Velloziaceae in the rupestrian fields of Serra do Cipó, Minas Gerais, Brazil. Flora 202:637–646

Negreiros D, Le Stradic S, Fernandes GW, Rennó HC (2014) CSR analysis of plant functional types in highly diverse tropical grasslands of harsh environments. Plant Ecol 215:379–388

Pifano DS, Valente ASM, Almeida HS, Melo PHA, Castro RM, Van Den Berg E (2010). Caracterização florística e fitofisionômica da Serra do Condado, Minas Gerais, Brasil. Biota Neotropica, vol. 10, no. 1. http://www.biotaneotropica.org.br/v10n1/en/abstract?article+bn01010012010

Ribeiro AS de S (2009). Caracterização física, química e micromorfológica dos solos da Serra Sul, Floresta Nacional de Carajás, Pará. Dissertation, Universidade Federal de Viçosa, Viçosa

Rizzini CT (1997) Tratado de fitogeografia do Brasil. V.2. Ed. Hucitec, São Paulo 374p

Scarano FR (2007) Rock outcrop vegetation in Brazil: a brief overview. Rev Bras Bot 30:561–568

Silveira FAO, Negreiros D, Barbosa NPU, Buisson E, Carmo FF, Carstensen D, Conceição AA, Cornelissen TG, Echternacht L, Fernandes GW, Garcia QS, Guerra TJ, Jacobi CM, Lemos-Filho JP, Le Stradic S, Morellato LPC, Neves FS, Oliveira RS, Schaefer CE, Viana PL, Lambers H (2016) Ecology and evolution of plant diversity in the endangered campo rupestre: a neglected conservation priority. Plant and Soil (in press)

Valente EL (2009) Relações Solo-Vegetação no Parque Nacional da Serra do Cipó, Espinhaço Meridional, Minas Gerais. PhD Thesis, Universidade Federal de Viçosa, Viçosa

Chapter 5
Water Resources in the Rupestrian Grasslands of the Espinhaço Mountains

Marcos Callisto, José Francisco Gonçalves and Raphael Ligeiro

Abstract The rupestrian grasslands in Minas Gerais State comprise headwaters of important watersheds that drainages for millions citizens in over 400 cities in São Francisco and Doce river basins. The human activities in the rupestrian grasslands include domestic supply, agriculture, forestry, cattle raising, industry, and mineral extraction. This chapter addresses the ecological conditions of streams in terms of water quality (physical and chemical characteristics, nutrient availability), habitat quality and structure (diversity of benthic macroinvertebrates, structure of the riparian vegetation, riparian food webs, invertebrate drift), and ecosystem functioning (allochthonous and autochthonous production, dynamics of coarse and fine particulate organic matter, leaf litter breakdown of native and alien species). A synthesis of 20 years of ongoing research on the headwaters in the rupestrian grasslands is included, together with perspectives for future conservation and management of water resources.

M. Callisto (✉)
Departamento de Biologia Geral, Instituto de Ciências Biológicas, Universidade Federal de Minas Gerais, Minas Gerais, Brazil
e-mail: callistom@ufmg.br

J.F. Gonçalves
Departamento de Ecologia, Instituto de Ciências Biológicas, Universidade de Brasília, Brasília, Distrito Federal, Brazil
e-mail: jfjunior@unb.br

R. Ligeiro
Instituto de Ciências Biológicas, Universidade Federal do Pará, Belém, Pará, Brazil
e-mail: ligeirobio@gmail.com

© Springer International Publishing Switzerland 2016

G.W. Fernandes (ed.), *Ecology and Conservation of Mountaintop Grasslands in Brazil*, DOI 10.1007/978-3-319-29808-5_5

Contents

5.1 The Headwater Streams of the Espinhaço Mountains

Freshwaters harbor a rich diversity of species and habitats. Although they cover only 0.8 % of the Earth's surface, 10 % of all animal species live in freshwaters (at least 126,000 plant and animal species), including 35 % of all vertebrate species. These species combine to provide a wide range of critical services for humans, such as flood protection, food, water filtration and carbon sequestration (Collen et al. 2013). Despite the high number of species described in freshwaters and their importance for biodiversity, anthropogenic disturbances have been increasing dramatically since the last Century. As a result, the decline in freshwater biodiversity is outpacing that of other ecosystems (Dudgeon et al. 2006). Freshwaters are hotspots of biodiversity that are being threatened by fragmentation and habitat destruction, introduction of invasive species, pollution, human population growth, as well as overharvesting (Stendera et al. 2012; Macedo et al. 2014).

The Espinhaço Mountain Range shows headwaters for a large number of watercourses which house a high number of aquatic and semi-aquatic plant and animal species. The water quality in the headwaters in this region is excellent and achieves the highest level of the Brazilian Environmental Legislation ("special class" according to CONAMA 357/2005). These headwaters form three important river basins in the Cerrado biome: São Francisco, Doce and Jequitinhonha. The freshwater biodiversity in the Espinhaço Mountain Range includes 27 endemic fish species, 162 fish species that inhabit small creeks and low-order streams, and 12 fish species threatened of extinction (Alves et al. 2008); 105 anuran species, of which 28 are endemics species (Leite et al. 2008).

Human activities in the Espinhaço Mountain Range, including agriculture, mining, and dam construction, have increased the environmental impacts on the freshwaters, which change physical habitats, chemical water quality and negatively influence native plant and animal species. The anthropogenic disturbances by industrial, urban, agro-cattle and mining are responsible not only for local changes that threaten aquatic biodiversity but also influence basin water quality due to the bioaccumulation and biomagnification of heavy metals, Polycyclic Aromatic Hydrocarbons (PAHs), insecticides, herbicides, fungicides and other chemical compounds along the food webs. Experiments in Parque Nacional da Serra do Cipó streams have been conducted to assess the habitat-species composition associations and to test global hypotheses in international collaboration networks. In this chapter, we describe the features of the land and the scientific experience of the last 20 years of studies conducted in Espinhaço Mountain Range region.

5.2 River Basin Beta Diversity and Trends in the Distribution of Aquatic Invertebrates in the Headwater Streams of the Parque Nacional da Serra do Cipó

5.2.1 Beta Diversity in River Basins

Headwaters comprise the majority (c. 70 %) of the total length of any river basin (Benda et al. 2005). Due to the longitudinal linkage, the water quality upstream affects the water quality downstream. Headwater streams are also important for the maintenance of the biodiversity of the whole river basin. Many invertebrates and vertebrates reproduce in headwaters, from which they can potentially colonize downstream reaches (Clarke et al. 2008).

Whittaker (1960) coined the term "alpha diversity" (α) to account to the variety of species found locally and the term "beta diversity" (β) to account to the composition dissimilarity between local sites. Because headwaters generally have narrow and shallow channels, each stream reach usually harbor less species than medium size river reaches do. Authors argue that medium-size river reaches are likely to present the highest taxonomic diversity of the river network, small streams and large rivers presenting lower diversity (Allan and Castillo 2007). Thus, headwater streams usually present lower alpha diversity when compared to wider and deeper medium size downstream reaches. On the other hand, because headwater reaches are usually highly isolated from each other, they typically present higher beta diversity, especially when landscape features like hilly terrain and dense vegetation hamper aerial dispersion (Finn and Poff 2005). Consequently, the total species diversity found in all headwater reaches accounts for the majority of the total species diversity found in the whole basin. Whittaker named this regional diversity as "gamma diversity" (γ).

For a long time, stream ecologists have focused on the factors driving the diversity of species at the site scale (i.e., α diversity). More recently, a large number of studies started to investigate the dissimilarity of species composition among sites (i.e., β diversity) (Legendre et al. 2005; Tuomisto and Ruokolainen 2006; Melo et al. 2011). Both aspects of diversity must be considered because it is the interaction between alpha and beta diversities that defines the total diversity (gamma) found in any region.

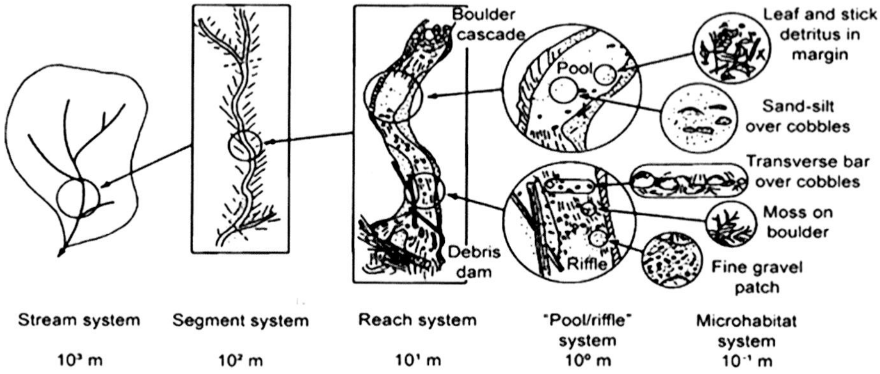

Fig. 5.1 Hierarchical approaches in stream ecology, after Frissel et al. (1986)

5.2.2 Assessing Key Spatial Scales

Stream ecosystems can be investigated at many spatial scales, from centimeters to kilometers. Frissel et al. (1986) recognized several discrete spatial levels, ranging from microhabitats to river basins (Fig. 5.1). The biological assemblages at any site are structured by environmental factors acting at all those spatial levels (Allan 2004). For aquatic invertebrates, at the microhabitat level assemblages are constrained by the type of bottom substrate, microflow regimes near the streambed and, possibly, by interactions among species. At the stream reach level, riparian zone characteristics can play a stronger role; while at the basin level, assemblages are affected by the vegetation type and human land uses. Identifying the spatial level that presents the highest beta diversity of biological assemblages is critical to determine which spatial scale should be mostly considered in conservation and ecosystem management actions. For instance, if we find that most of the assemblage dissimilarity is found between stream reaches, it is not worthwhile to apply all the available resources for conserving a single or few stream reaches. In this scenario, even if one stream reach is superbly well protected, much of the regional biodiversity can be lost because other stream reaches were not considered.

A long term study was performed in the Parque Nacional da Serra do Cipó to identify the "key" spatial scales responsible for the highest β diversity of aquatic invertebrates (Ligeiro et al. 2010a). We used a hierarchical nested sampling design across multiple spatial levels at the Mascates River basin. The Mascates basin belongs to the São Francisco River Basin, one of the most important basins for the Brazilian economy. We selected three headwater streams in this basin, representing the "stream segment" spatial level of Frissel et al. (1986): the Pedras stream, the Farofa stream, and the Taioba stream (Fig. 5.2a–c). They are all narrow (~4 meters wide) and shallow (~0.5 meters deep) streams, with sandy/rocky substrate, presenting riffle/pool sequences and ecological preservation but without dense riparian vegetation. In all aspects, they well represent the typology of the headwater streams

Fig. 5.2 Representation of typical stream reaches of three headwater streams in Parque Nacional da Serra do Cipó (Brazil). **a** Pedras stream. **b** Farofa stream. **c** Taioba stream

found in the Serra do Cipó region. In each stream segment, we sampled in two stream reaches, longitudinal stretches of the streams of approximately 50 meters each. In each reach, we sampled in three riffles, representing the "habitat" spatial level of Frissel et al. (1986). Finally, in each riffle we took three samples of the stream benthos, the "microhabitat" spatial level of Frissel et al. (1986). We performed this sampling design for each of the three types of substrates, representing the heterogeneity at the microhabitat level: stones, gravel, and leaf litter. In total, 162 samples were taken from the three types of substrates in the Mascates River basin (Fig. 5.3). The detailed procedures can be found in Ligeiro et al. (2010a).

For assessing the spatial levels of the greater aquatic invertebrate beta diversity ("key spatial scales"), we used an additive partitioning diversity approach (Veech et al. 2002). First, the average species richness is calculated for the sampling units of each spatial level. Then, the β diversity of each spatial level (L) is defined by subtracting the average species richness of that level from the average species richness of the next higher spatial level (L + 1). For instance, the β diversity at the stream reach level is equal to the average species richness of the stream segment level minus the average species richness of the stream reach level, and so on for all the spatial levels considered. Consequently, the γ diversity coincides with the sum of the α diversity (the average species richness found in the smallest spatial level) with the beta diversity calculated for each spatial level considered ($\gamma = \alpha + \beta\ 1 + \beta\ 2 + \beta\ 3 + \beta\ 4$).

For the three substrate types selected we determined that most of the variation in aquatic invertebrates composition was found at the stream segment spatial level, indicating that the biological dissimilarity between the three headwater streams studied (β 4) was much greater than the dissimilarities found between stream reaches (β 3), habitats (β 2), and microhabitats (β 1) (Fig. 5.4). This high similarity found within each headwater stream can be partially explained by the unidirectional water flow from upstream to downstream, which promotes the passive and active dispersion of invertebrate adults and juveniles through water drift, homogenizing the many sites along the same stream (Allan and Castillo 2007).

In parallel to the partitioning of the aquatic invertebrate diversity across multiple spatial levels, we found great dissimilarities between the samples of the three substrate types considered, mainly between the organic substrate (leaf litter) and the inorganic substrates (stones and gravel) (Ligeiro et al. 2010a). The substrate types

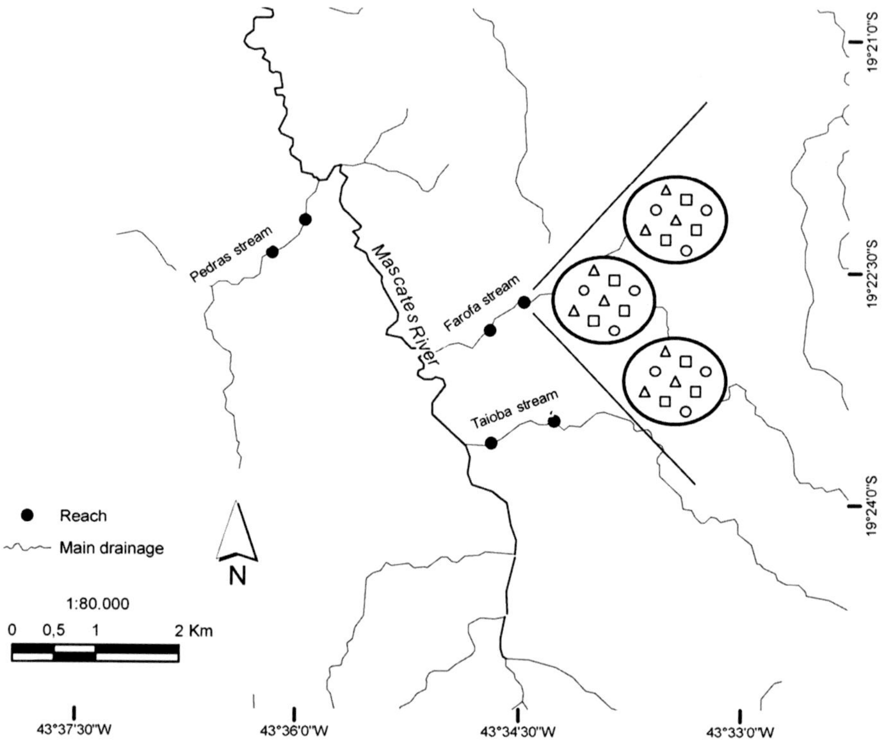

Fig. 5.3 Sampling design for assessing the beta diversity of macroinvertebrate assemblages across multiple spatial scales in a small river basin in Parque Nacional da Serra do Cipó (Brazil). Adapted from Ligeiro et al. (2010a)

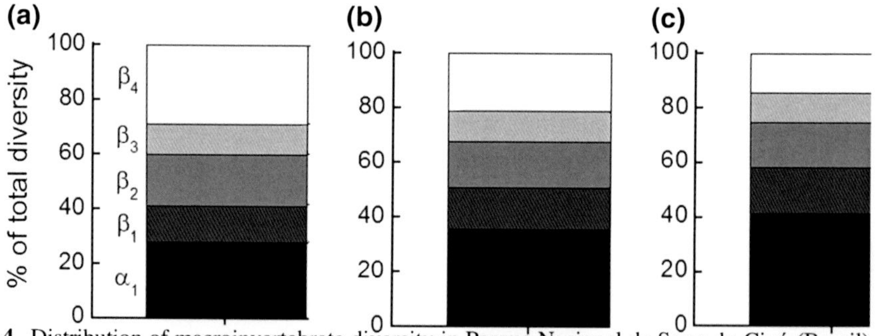

Fig. 5.4 Distribution of macroinvertebrate diversity in Parque Nacional da Serra do Cipó (Brazil) across multiple spatial scales according to the additive diversity partitioning approach (more details in the text), for each of the main substrate types found in the streambeds. **a** Leaf litter. **b** Stones. **c** Gravel

are very important for aquatic invertebrates, defining the availability of shelter from predators and natural disturbances (e.g., flashfloods), and also the availability and quality of food resources (e.g., particulate organic matter, prey).

5.2.3 Implications of β Diversity for Ecosystem Management and Conservation

The fact that we identified the stream segment (represented in our study by the three headwater streams sampled) as the key spatial level for the biodiversity of aquatic invertebrates has important implications for management and conservation thinking of Serra do Cipó streams and Cerrado streams in general. First, this result indicates that sampling many headwater streams is essential to properly represent the biodiversity of headwater basins. As resources (money, time, qualified personal) are often limited for any scientific research or conservation program (Hughes and Peck 2008), it is ineffective to concentrate many samples in one single stream segment, rather the sampling effort should be spread through the highest possible number of headwater streams. The substrate type was an important factor differentiating invertebrate assemblages; therefore, it is also important to consider microhabitat heterogeneity when conducing biodiversity surveys.

Because most of the diversity of the aquatic invertebrates of Parque Nacional da Serra do Cipó was generated by the dissimilarities between the headwater streams, conservation and management actions should invest in protecting the highest number possible of headwater streams in the region. Each stream impaired by human activities can mean the loss of aquatic species in the whole region. Because each stream is relatively internally homogeneous, occasional alterations conducted in some parts of the streams are likely not to greatly impair the biodiversity at a larger spatial scale. This scenario suggests that the best management approach is not to isolate a few streams from human contact, but to wisely promote sustainable human use (recreational, human and familiar agriculture supply) of the existing streams, in this way allowing each one to harbor and conserve its unique diversity.

5.3 Leaf Breakdown and Organic Matter Dynamics

In low order streams, tree canopy cover commonly limits light availability to streambed. Thus, the reduced stream's primary production and allochthonous organic matter inputs are the main energysource for the aquatic ecosystems. The Espinhaço Range is rich in headwater streams, particularly between the altitudes of 500 and 2000 meters a.s.l. In this region different habitats of the rupestrian grasslands co-occur, including rock outcrops, open grasslands and forests (Gonçalves et al. 2006a, b, c; Chaps. 1, 6 and 7). Headwaters at higher altitudes (1st and 2nd order streams) are bordered by woody grasslands, while at the lower sections rivers flow through forest habitats.

The energetic base is maintained by the production of algae and bacteria (Callisto et al. 2004). Bacteria use the dissolved organic carbon, which can be observed by the tea color of the water (Janzen 1974). The waters are generally nutrient-poor due to the low nutrient content in the soil (Chap. 3), which also limits the aquatic productivity. We hypothesize that this could lead to the dependency of the productivity on the biofilm (e.g., algae and periphyton) that colonizes rocks, gravel, and trunks in the streambed. Along the longitudinal gradient, larger streams also have low productivity due to lack of light (due to water turbidity), despite possessing higher concentration of nutrients in the water. In those segments of the watercourse, energy depends on the organic matter derived from the plants on its banks (Coarse Particulate Organic Matter, CPOM, and Fine Particulate Organic Matter, FPOM), while their productivity depends on the dynamics of the allochthonous organic matter. On the other hand, studies on the primary production in these ecosystems are still rare. Furthermore, little is known about the abundance and dynamics of dissolved organic matter (DOM) and fine particulate organic matter (FPOM).

In general, the abundance of gathering-collectors is negatively related with the FPOM; where the FPOM concentration is low, gathering-collectors drift away (Castro et al. 2013a). When the FPOM is available in high concentration, the filtering collectors are abundant. This suggests that invertebrates' drift is related to food availability while high drift movement by the gathering-collectors is related to their searching for deposited FPOM as food. During the rainy seasons and flash-floods, bottom instability may lead to FPOM drift, the main food resource in tropical headwater streams (Castro et al. 2013b). For instance, the drift of invertebrate assemblages along a longitudinal gradient in a headwater stream were mostly composed by aquatic insects. Among the 91 taxa, Chironomidae (Diptera: 33 genera), Trichoptera (18 genera), and Ephemeroptera (13 genera) showed the highest taxonomic richness values (Callisto and Goulart 2005).

We conducted three studies on the dynamics of organic matter in the Espinhaço Range (Gonçalves et al. 2006a; França et al. 2009; Gonçalves and Callisto 2013) (Fig. 5.5). The diversity of plant species in the riparian zone ranged from 15 (stream reaches surrounded by rocky outcrops) to 192 in a reach 1300 m a.s.l. (along with 128 species contributed for inputs of CPOM). In the three sites, 13 species were common to at least two sites, and the genus *Ocotea* was common in all sites. The CPOM productivity from the riparian vegetation (measured by vertical and terrestrial inputs) ranged from 1.8–4.6 t ha^{-1} $year^{-1}$. This productivity was seasonal being higher during the transition between periods of drought and rain, with maximum values occurring in October. Most of the CPOM falling in the soil from the riparian zone goes to the stream (between 82 and 100 % of the total CPOM that reaches the ground). This occurs because headwater streams in the Espinhaço Range have a steep slope and narrow channel morphology. Thus, the decomposition of CPOM (especially leaves, representing more than 60 % of the total CPOM) is fundamental to understand the functioning of the stream ecosystems in the Espinhaço Range. We found 9 studies

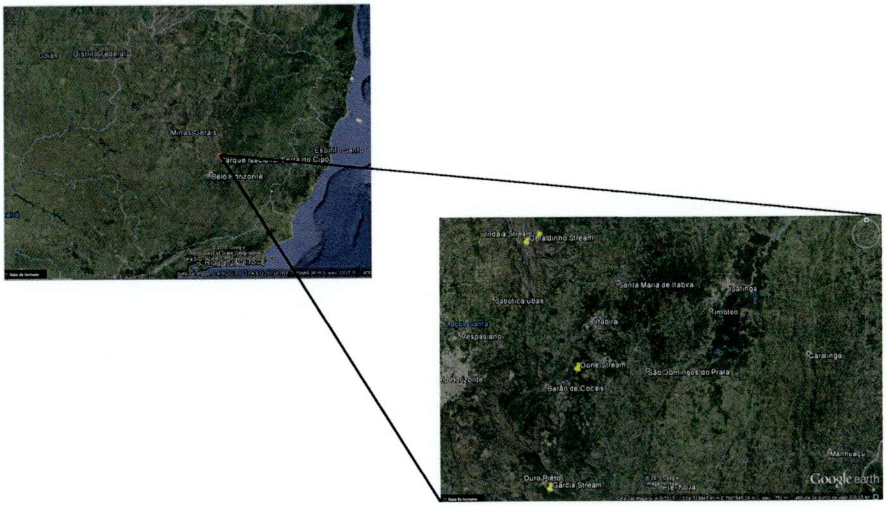

Fig. 5.5 Localization of sites studied in the Espinhaço range showing the distance among streams

on this topic in the rupestrian grasslands of the Espinhaço Range (Fig. 5.5), where nine native species were studied (*Myrcia guyanensis*, *Miconia chartacea*, *Protium brasiliense*, *Protium heptaphyllum*, *Coccoloba cereifera*, *Ocotea* sp., *Baccharis platypoda*, *Baccharis concinna*, *Baccharis dracunculifolia*), two exotic species (*Eucalyptus grandis* and *Alnus glutinosa*) and two studies of "mixed leaves" (Gonçalves et al. 2006b; Moretti et al. 2007a, b). Most studies were conducted between April and September during the years 2001–2009.

The leaves of the native plant species had low nitrogen (1.15 %) and phosphorus (0.0914 %), and high levels of lignin (31 %) and polyphenols (15 %). The leaf breakdown rate of the native species was -0.0039 day^{-1}, mixed leaves -0.0111 day^{-1}, *Eucalyptus grandis* -0.005 day^{-1} and *Alnus glutinosa* -0.014 day^{-1}. The resulting k values indicate that the native leaves have slow decomposition, while the mixed and exotic species are characterized by intermediate decomposition (Ligeiro et al. 2010b; Gonçalves et al. 2014). There are two factors that could influence these results: the low quality of the leaf litter of the rupestrian grassland plants and the low concentration of the nutrients in the headwater streams in the Espinhaço Range, as demonstrated by Medeiros et al. (2015).

Rupestrian grassland leaves decomposing in streams had low fungal biomass, measured as ergosterol concentration. Here, ergosterol values ranged from 62 to 278 µg g^{-1}. On the other hand, for the small leaves of *Baccharis dracunculifolia* and *B. concinna* (Alvim et al. 2015), the ergosterol values were high (908 µg g^{-1} in both) compared to the concentrations in the leaves of an exotic species widely used in decomposition experiments (*A. glutinosa*). This last species was also studied in the headwater streams at Espinhaço Range region, and presented ergosterol max value of 573 µg g^{-1} (Gonçalves et al. 2006c).

The Total Microbial Biomass estimations (fungi, bacteria and other microorganisms through the concentration of ATP) also indicate low participation of the microbial decomposers (concentration in rupestrian grassland leaves between 42 and 194 nmoles ATP g^{-1}, 4278–4023 nmoles ATP g^{-1} *B. dracunculifolia* and *B. concinna*, respectively, and 531 nmoles ATP g^{-1} in *A. glutinosa*). These results indicate that, besides fungi, other microorganisms, such as bacteria, proportionally may play a major role in litter decomposition. We believe that the lack of nutrients in leaf litter and in the waters negatively affect fungal activity. Before that, bacteria could be more effective in the decomposition of leaf detritus than fungi, especially in the first phase of the decomposition process (*sensu* Degradative Ecological Succession, Begon et al. 2007; Gonçalves et al. 2006b, c, 2007).

Detritivorous invertebrates are largely represented by the larvae of Chironomidae (Diptera) (over 60 % of all assemblages associated with organic detritus, Callisto et al. 2007). Among the trophic guilds, we found the gathering-collectors (using fine particles of particulate organic matter—FPOM) were dominant, representing an average of 62 % of the individuals (mainly chironomids), while the shredders (which feed directly from particles greater than 1 mm—CPOM) represented 0–20 % of invertebrates. These results indicate that the influence of invertebrates in leaf fragmentation is low, which is likely due to the low nutritional value of the leaves associated with the reduced fungal biomass. However, the aquatic invertebrates preferentially use FPOM with a high biomass of bacteria and easy digestibility (Callisto and Graça 2013).

Due to the low quality of the leaves from the riparian zone (e.g., low nitrogen and phosphorus, high cellulose and ligninins), the decomposition is slow, leading to the accumulation of CPOM on the streambeds. We postulate that this CPOM is basically processed by physical fragmentation because the biological effect is reduced. Thus, the energy could be dependent on the FPOM (originated by aquatic or terrestrial origin by physical and biological factors) when the processing of the CPOM is inefficient. On the other hand, the FPOM can have a proportionally greater role as energy and nutrient sources to the trophic web, despite its biomass being lower when compared to the fraction of the CPOM (Callisto and Graça 2013). These ecosystems have complex food webs, which are still poorly known. Another factor to be considered in ecological studies could be the low redundancy in the components of the trophic web and a dependency on allochthonous organic matter (CPOM and FPOM) indicating a high ecological fragility of these ecosystems. This is due to the dependence of the dynamic and complex processes that are connected by many environmental variables (vegetation composition, climate, roughness, slope the banks and the stream channel, flow, etc.), which may lead to the decreased resilience of these ecosystems to anthropic activities or global climate change.

Due to the ecological fragility of the aquatic ecosystems of the Espinhaço Range, the need for a biodiversity conservation policy is urgently needed. This could be implemented through policies from the Federal Government that are extended for Minas Gerais and Bahia states and their watersheds, which involve many more states. Another important aspect is a network of systematic studies to raise the level of basic information, allowing the development of an ecological model for sustainable growth of the entire mountain range complex. From this, the ecological and economic zoning of the region should be established, considering the use and occupation of the soil and water resources.

5.4 The Conservation of Benthic Invertebrates and Their Use as Bioindicators in a Global Perspective

In addition to its importance as a potential area for water supply and aquatic resources, Espinhaço Range headwater streams have been used for agriculture and mine companies, and also represent important aquatic resources for other uses, such as recreation and wildlife maintenance. The stream ecology studies in this region have been developed during the last 20 years. The first years (1995–2000) were dedicated to surveys of the distribution and structure of the aquatic communities, assessments of the existing habitats/micro-habitats, the availability of trophic resources, and of the distribution patterns of the aquatic communities (Galdean et al. 1999). These studies adopted the utilization of higher taxonomic levels and/or functional groups of organisms, together with the basic ecological features of the environments (Barbosa and Galdean 1997). The basic information provided the basis for defining conservation priorities and sustainable uses for the area in the Parque Nacional da Serra do Cipó. Some regions located within the Parque Nacional da Serra do Cipó are in "nearly" pristine conditions, while some other regions have been affected by human activities to varying degrees (Galdean et al. 2000). The most important impacts are related to past deforestation, the disposal of untreated organic sewage from the riverine human populations, cattle ranching, and agriculture. In recent years, some parts of the Espinhaço Range have been severely degraded by iron mining, which caused heavy siltation of the headwaters and deforestation of the riparian zone, resulting in the local extinctions of native aquatic species.

In general, the waters are dark in color, relatively well oxygenated (>80 % saturation), nutrient poor (soluble reactive phosphorus <20 $\mu g\ L^{-1}$, total phosphorus <50 $\mu g\ L^{-1}$) and of low conductivity (<20 uS cm^{-1}), with a pH ranging between 5 and 6 (Galdean et al. 1999, 2000; Fernandes et al. 2014). The benthic macroinvertebrate communities are dominated by larvae of the aquatic insects Ephemeroptera, Plecoptera, Trichoptera, and Diptera-Chironomidae. Crustacea, Bivalvia and Oligochaeta may also be found in low densities in some areas (Galdean et al. 1999). These taxa colonize riverbed rocks, clay and stony substrates, aquatic macrophytes and riparian vegetation. Filamentous algae, mosses and lichens, and coarse and fine organic detritus, on or in between these habitats, form the major microhabitats in the headwater streams. These results have been used as an instrument to assess the present biodiversity, and thus represent an aid in conserving the existing rivers' benthic fauna.

A global survey of the stream fauna (146 sites from 16 areas located in six continents) included data from the streams at the Espinhaço Range to test the hypothesis on the scarcity of shredders in tropical sites. Both the evolutionary adaptation of shredders to cool waters, which might cause their scarcity in the tropics for physiological reasons, as well as more plant defenses against herbivores in tropical areas than those in the temperate zone, were assessed by Boyero et al. (2011a, b). In fact, tropical streams possess few shredders compared to temperate streams, but their distribution varies at multiple spatial scales and the great majority of the variation occurred among areas within zones, particularly when the richness was quantified as the number of species per site. However, shredder assemblage composition differs between the temperate and tropical zones, which suggest that the distribution of several taxa strongly depends on temperature (Boyero et al. 2012). The hypothesis of plant defenses against herbivores was not corroborated, because leaf toughness was not related to shredder densities or species richness. Perhaps, this finding is supported by the fact that the richness of litter entering into and retained in streams is greater in the tropics. Thus, shredders have a lower chance of re-encountering a leaf of any particular species, which makes it difficult to find palatable leaves and/or develop anti-herbivore strategies (as proposed by Wantzen et al. 2002). Additional global biodiversity aspects were discussed in a series of articles by Boyero et al. (2011a, b, 2012, 2015a, b).

5.5 Anthropogenic Disturbances, Threats to Headwaters, Future Perspectives and Global Climate Change

Freshwater species are consistently under a greater level of threat than those resident in terrestrial ecosystems (Collen and Boehm 2012; Collen et al. 2013). These patterns of threat are mediated by the high rates of habitat loss and degradation, pollution and overexploitation, which are particularly problematic for species inhabiting flowing waters. Land-use change driving habitat loss and degradation affects the majority of threatened freshwater species. As stated by Kominoski and Rosemond (2012), fluxes and pools of organic matter are critical food and habitat resources at relatively small scales. On larger scales, alterations in the availability, retention, and processing rates of detritus can affect the delivery of organic matter to downstream organisms and fluxes of C to the atmosphere and oceans.

The Espinhaço Range headwater streams are impacted by different human uses. Climate change is expected to affect aquatic ecosystems strongly by altering the quantity, characteristics, processing, and retention of inputs of terrestrial detritus and sediments (www.ipcc.ch). The principal climate drivers affecting detrital dynamics in aquatic ecosystems are increases in atmospheric CO_2 and its effects on terrestrial organic matter, changes in precipitation and associated hydrologic impacts, and increases in temperature and siltation.

Altered patterns of precipitation will make extreme events (e.g., droughts and floods) part of flow regimes, which directly influence water quantity and indirectly influence the availability and biological processing of detrital resources and sediment deposits. Drought reduces physical breakdown rates and slows biological processing. Higher discharge will accelerate the transport of sediments and physical breakdown of organic matter. High-flow events can scour detritus from stream reaches.

According to our results the future research projects on headwater streams at the Espinhaço Range should focus on strategic issues and perspectives for future conservation and the management of water resources, including:

(i) Incorporate measures of organic matter processing as indices of stream ecosystem function to complement structural measures (water quality, taxonomic composition) in stream health monitoring (Piggott et al. 2015);
(ii) Investigate how deposited fine sediments and elevated nutrient concentrations interact to affect stream water quality and freshwater biodiversity in a climate-change future scenario.

The resulting information will allow the testing of individual stressor hypotheses, such as: (i) nutrient enrichment generally increase the measures of decomposition rate; (ii) the positive effect of nutrient enrichment are weaker or suppressed when fine sediment is also present.

Acknowledgments The authors are grateful to Geraldo W. Fernandes for the invitation to contribute with this chapter and for his constructive comments on early versions. Manuel Graça and Irineu Bianchini offered carefull reading and insights to previous version of this chapter. Our studies have been supported by Conselho Nacional de Pesquisa e Desenvolvimento (CNPq), Fundação de Amparo à Pesquisa do Estado de Minas Gerais (FAPEMIG), Fundação Coordenação de Aperfeiçoamento de Pessoal de Nível Superior (CAPES), P&D ANEEL (GT-487) and CEMIG-Programa Peixe Vivo. MC was awarded research productivity grants CNPq (No. 303380/2015-2) and CNPq research grant (446155/2014-4) and Minas Gerais researcher grant FAPEMIG PPM-77/13 and PPM IX-02/2015. JFG-Jr was supported by CNPq (Proc. 472.328/01–8), FAPEMIG (Proc. 1085/03, APQ-2051-5.03/07, and FAPEMIG/PRONEX 20/2006, 465/07) and PADI-Project AWARE Foundation. The authors are grateful for the logistical support provided by IBAMA, IEF-MG and the US Fish and Wildlife Service.

References

Allan JD (2004) Landscape and riverscapes: The influence of land use on river ecosystems. Ann Rev Ecol Evol Syst 35:257–284

Allan JD, Castillo MM (2007) Stream ecology: Structure and function of running waters. Springer, Netherlands

Alves CBM, Lean CG, Brito MFG, Santos ACA (2008) Biodiversidade e conservação de peixes do complexo do Espinhaço. Megadiv 4:145–164

Alvim EACC, Medeiros AO, Rezende RS, Gonçalves JF (2015) Small leaf breakdown in a Savannah headwater stream. Limnol 51:131–138

Barbosa FAR, Galdean N (1997) Ecological taxonomy: A basic tool for biodiversity conservation. Tren Ecol Evol 12:359–360

Begon M, Townsend CR, Harper JL (2007) Ecologia: de indivíduos a ecossistemas. Artmed, São Paulo

Benda L, Hassan MA, Church M, May CL (2005) Geomorphology of steepland headwaters: The transition from hill slopes to channels. J Am Water Res Assoc 41:835–851

Boyero L, Pearson R, Dudgeon D, Graça MAS, Gessner MO, Albarino R, Ferreira V, Mathuriau C, Boulton A, Arunachalam M, Callisto M, Chauvet E, Ramirez A, Chara J, Moretti MS, Gonçalves JFJr (2011a) Global distribution of a key trophic guild contrasts with common latitudinal diversity patterns. Ecol 92:1839–1848

Boyero L, Pearson RG, Gessner MO, Barmuta LA, Ferreira V, Graça MAS, Dudgeon D, Boulton AJ, Callisto M, Chauvet E, Helson JE, Bruder A, Albariño RJ, Yule CM, Arunachalam M, Davies JN, Figueroa R, Flecker AS, Ramírez A, Death RG, Iwata T, Mathooko JM, Mathuriau C, Gonçalves, JF, Moretti MS, Jinggut T, Lamothe S, M Erimba C, Ratnarajah L, Schindler MH, Castela J, Buria LM, Cornejo A, Villanueva V, West DC (2011b) A global experiment suggests climate warming will not accelerate litter decomposition in streams but might reduce carbon sequestration. Ecol Lett 14:289–294

Boyero L, Pearson RG, Dudgeon D, Ferreira V, Graça MAS, Gessner MO, Boulton AJ, Chauvet E, Yule CM, Albariño RJ, Ramírez A, Helson JE, Callisto M, Arunachalam M., Chará J, Figueroa R, Mathooko JM, Gonçalves JF, Moretti MS, Chará-Serna AM, Davies JN, Encalada A, Lamothe S, Buria LM, Castela J, Cornejo A, Li AOY, M'Erimba C, Villanueva VD, Zúñiga C, Swan CM, Barmuta LA (2012) Global patterns of stream detritivore distribution: Implications for biodiversity loss in changing climates. Global Ecol Biogeog 21:134–141

Boyero L, Pearson R, Gessner MO, Dudgeon D, Ramirez A, Yule C, Callisto M, Pringle C, Encalada A, Arunachalam M, Mathooko J, Helson J, Rincon J, Bruder A, Cornejo A, Flecker AS, Mathuriau C, Merimba C, Moretti MS, Gonçalves JF, Jinggut T (2015a) Leaf-litter breakdown in tropical streams: Is variability the norm? Freshwat Sci. doi:10.1086/681093

Boyero L, Pearson R, Swan CM, Hui C, Albarino R, Arunachalan R, Callisto M, Chara J, Chara-Serna AM, Chauvet E, Cornejo A, Dudgeon D, Encalada A, Ferreira V, Gessner MO, Gonçalves JF, Graça MAS, Helson J, MathookoJ, McKie B, Moretti MS, Yule C (2015b) Latitudinal gradient of nestedness and its potential drivers in stream detritivores. Ecography 38:1–7

Callisto M, Goulart MD (2005) Invertebrate drift along a longitudinal gradient in a Neotropical stream in Serra do Cipó National Park, Brazil. Hydrobiol 539:47–56

Callisto M, Graça MAS (2013) The quality and availability of fine particulate organic matter for collector species in headwater streams. Int Rev Hydrobiol. doi:10.1002/iroh.201301524

Callisto M, Goulart M, Medeiros AO, Moreno P, Rosa CA (2004) Diversity assessment of benthic macroinvertebrates, yeasts, and microbiological indicators along a longitudinal gradient in Serra do Cipó, Brazil. Braz J Biol 64:743–755

Callisto M, Gonçalves JF, Graca MAS (2007) Leaf litter as a possible food source for chironomids in headwater streams. Rev Brasil Zool 24:442–448

Castro DMP, Hughes RM, Callisto M (2013a) Effects of flow fluctuations on the daily and seasonal drift of invertebrates in a tropical river. Ann Limnol Int J Limnol. doi:10.1051/limn/2013051

Castro DMP, Hughes RM, Callisto M (2013b) Influence of peak flow changes on the macroinvertebrate drift downstream of a Brazilian hydroelectric dam. Braz J Biol 73:775–782

Clarke A, MacNally R, Bond N, Lake PS (2008) Macroinvertebrate diversity in headwater streams: A review. Freshwat Biol 53:1707–1721

Collen B, Boehm M (2012) The growing availability of invertebrate extinction risk assessments—A response to Cardoso et al. (October 2011): Adapting the IUCN red list criteria for invertebrates. Biol Conserv 149:145–146

Collen B, Whitton F, Dyer EE, Baillie JEM, Cumberlidge N, Darwall WRT, Richman NI, Pollock C, Soulsby A-M, Bohm M (2013) Global patterns of freshwater species diversity, threat and endemism. Global Ecol Biogeog. doi:10.1111/geb.12096

Dudgeon D, Arthington AH, Gessner MO, Kawabata Z-I, Knowler DJ, Lévêque C, Naiman RJ, Prieur-Richard A-H, Soto D, Stiassny MLJ, Sullivan CA (2006) Freshwater biodiversity: Importance, threats, status and conservation challenges. Biol Rev 81:163–182

Fernandes GW, Barbosa NPU, Negreiros D, Paglia AP (2014) Challenges for the conservation of vanishing megadiverse rupestrian grasslands. Nat Cons 12:162–165

Finn DS, Poff NL (2005) Variability and convergence in benthic communities along the longitudinal gradients of four physically similar rocky mountain streams. Freshwat Biol 50:243–261

França JS, Gregório RS, D'Arc JD, Gonçalves JF, Ferreira FA, Callisto M (2009) Composition and dynamics of allochthonous organic matter inputs and benthic stock in a Brazilian stream. Mar Fresh Res 60:990–998

Frissel CA, Liss WJ, Warren CE, Hurley MD (1986) A hierarchical framework for stream habitat classification—viewing streams in a watershed context. Environ Manag 10:199–214

Galdean N, Barbosa FAR, Callisto M, Rocha LA, Margarida MGSMM (1999) A proposed typology for the rivers of Serra do Cipó (Minas Gerais, Brazil) based on the diversity of benthic macroinvertebrates and the existing habitats. Trav Mus Natl Hist Nat Grigore Antipa 41:445–453

Galdean N, Callisto M, Barbosa FAR (2000) Lotic ecosystems of Serra do Cipó, Southeast Brazil: Water quality and a tentative classification based on the benthic macroinvertebrate community. Aq Ecosyst Heal Manag 3:545–552

Gonçalves JF, Callisto M (2013) Organic-matter dynamics in the riparian zone of a tropical headwater stream in Southern Brazil. Aquat Bot 109:8–13

Gonçalves JF, França JS, Callisto M (2006a) Dynamics of allochthonous organic matter in a tropical Brazilian headstream. Braz Arch Biol Technol 49:967–973

Gonçalves JF, França JS, Medeiros AO, Rosa CA, Callisto M (2006b) Leaf breakdown in a tropical stream. Int Rev Hydrobiol 91:164–177

Gonçalves JF, Graça MAS, Callisto M (2006c) Leaf-litter breakdown in 3 streams in temperate, mediterranean, and tropical Cerrado climates. J N Am Benthol Soc 25:344–355

Gonçalves JF, Graça MAS, Callisto M (2007) Litter decomposition in a cerrado savannah stream is retarded by leaf toughness, low dissolved nutrients and a low density of shredders. Freshwat Biol 52:1440–1451

Gonçalves JF, Martins RT, Ottoni BMP, Couceiro SRM (2014) Uma visão sobre a decomposição foliar em sistemas aquáticos brasileiros. In: Hamada N, Nessimian J.L, Querino R.B (eds). Insetos aquáticos na Amazônia brasileira: taxonomia, biologia e ecologia. 1ed. Manaus: Editora do INPA, vol. I, pp. 89–116

Hughes RM, Peck DV (2008) Acquiring data for large aquatic resource surveys: The art of compromise among science, logistics, and reality. J N Am Benthol Soc 27:837–859

Janzen DH (1974) Tropical blackwater rivers, animals, and mast fruiting by the dipterocarpaceae. Biotropica 6:69–103

Kominoski JS, Rosemond AD (2012) Conservation from the bottom up: Forecasting effects of global change on dynamics of organic matter and management needs for river networks. Freshwat Sci 31:51–68

Legendre P, Borcard D, Peres-Neto PR (2005) Analyzing beta diversity: Partitioning the spatial variation of community composition data. Ecol Monogr 75:435–450

Leite FSF, Juncá FA, Eterovick PC (2008) Status do conhecimento, endemismo e conservação de anfíbios anuros da Cadeia do Espinhaço, Brasil. Megadiv 4:182–200

Ligeiro R, Melo AS, Callisto M (2010a) Spatial scale and the diversity of macroinvertebrates in a neotropical catchment. Freshw Biol 55:424–435

Ligeiro R, Moretti MS, Gonçalves JF, Callisto M (2010b) What is more important for invertebrate colonization in a stream with low-quality litter inputs: Exposure time or leaf species? Hydrobiol 235:1–2

Macedo DR, Hughes RM, Ligeiro R, Ferreira W, Castro MA, Junqueira NT, Oliveira DR, Firmiano KR, Kaufmann PR, Pompeu PS, Callisto M (2014) The relative influence of catchment and site variables on fish and macroinvertebrate richness in cerrado biome streams. Landsc Ecol 29:1001–1016

Medeiros AO, Callisto M, Graça MAS, Ferreira V, Rosa CA, França JS, Eller AP, Rezende RS, Gonçalves JF (2015) Microbial colonization and litter decomposition in a cerrado stream is limited by low dissolved nutrient concentration: Evidence from a manipulative experiment. Limnetica (in press) 34:283–292

Melo AS, Shneck F, Hepp LU, Simões NR, Siqueira T, Bini LM (2011) Focusing on variation: Methods and applications of the concept of beta diversity in aquatic ecosystems. Acta Limnol Bras 23:318–331

Moretti M, Gonçalves JF, Callisto M (2007a) Leaf breakdown in two tropical streams: Differences between single and mixed species packs. Limnol 37:250–258

Moretti MS, Gonçalves JF, Ligeiro R, Callisto M (2007b) Invertebrates colonization on native tree leaves in a neotropical stream (Brazil). Int Rev Hydrobiol 92:199–210

Piggott JJ, Nyougi DK, Townsend CR, Mathaei CD (2015) Multiple stressors and stream ecosystem functioning: Climate warming and agricultural stressors interact to affect processing of organic matter. J Appl Ecol. doi:10.1111/1365-2664.12480

Stendera S, Adrian R, Bonada B, Cañedo-Argüelles M, Hungueny B, Januschke K, Pletterbauer F, Hering D (2012) Drivers and stressors of freshwater biodiversity patterns across different ecosystems and scales: A review. Hydrobiol 696:1–28

Tuomisto H, Ruokolainen K (2006) Analyzing or explaining beta diversity? Understanding the targets of different methods of analysis. Ecol 87:2697–2708

Veech JA, Summerville KS, Crist TO, Gering JC (2002) The additive partitioning of species diversity: recent revival of an old idea. Oikos 99:3–9

Wantzen KM, Wagner R, Suetfeld R, Junk WJ (2002) How do plant–herbivore interactions of trees influence coarse detritus processing by shredders in aquatic ecosystems of different latitudes? Verh Int Ver Theor Angew Limnol 28:815–821

Whittaker RH (1960) Vegetation of the Siskiyou mountains, Oregon and California. Ecol Monog 26:1–80

Part II
Foreword 2 + Photo

While the geologist has an eye on the Earth's shape and rocks, its minerals and the climatic factors that shape, bend, and shred the magnificent mountains and valleys transforming them into soils of different colors and quality, the very first attraction of biologists is that of an ecosystem, plants, animals, and the surrounding ecological forces. In this part, some of the forces that shape the vegetation, and the interactions among the physical and biological worlds, above and below ground, in rupestrian grasslands are revealed.

The rupestrian grasslands are ancient ecosystems where the land was colonized by a rich flora of herbaceous and shrubby species. Different combinations of plants and animals inhabit the different mountains with valleys acting as insurmountable oceans separating them, isolating them from one another. The diversity of plants is highly variable at every scale studied, with a high level of endemism and unique species compositions. Strong environmental and biological filters have shaped it all by limiting colonization by species adapted to soils high in nutrition, by eliminating those not adapted to the rigors of climate, fire and, possibly, herbivory. To survive, plants have evolved particular adaptations, be it phenological behavior or traits such as specialized underground organs that enable them to resprout repeatedly after damage, leaf orientation and trichomes to avoid excessive light and heat, or even new avenues such as the amazing capacity to feed on nematodes.

In spite of the importance that has come to be associated with remnant Atlantic forest patches because of the astonishing level of fragmentation and destruction they suffered during the last 50 years in Brazil, the existence of natural patches of relict Atlantic rain forests in the rupestrian grassland matrix is of major importance as well because many of them are still pristine, yet the knowledge about them is anecdotal. This part unveils some trends in the biology and natural history of these Atlantic rain forest relicts. Recent studies on the association of microorganisms and rupestrian plants are unraveling many new interactions, hence opening possibilities for an infinity of new studies. These might even include the very first events of colonization of bare rocks by microorganisms and rock-dwelling plant species. To survive in such harsh environments, disconnected sometimes by few centimeters, an important strategy is seed germination. In the rupestrian grasslands, these studies have revealed many germination syndromes, each adapted to small singularities of the habitats. But these fine-tuned adaptations do not end there, as the plant species of the rupestrian grassland have evolved a wide spectrum of possibilities and strategies.

Chapter 6
Rupestrian Grassland Vegetation, Diversity, and Origin

Abel A. Conceição, Alessandro Rapini, Flávio F. do Carmo, Juliana C. Brito, Gabriela A. Silva, Sâmia P.S. Neves and Claudia M. Jacobi

Abstract Rupestrian grasslands (*campos rupestres*) are tropical landscapes consisting of a mosaic of herbaceous and shrubby physiognomies on quartzite or ironstone that occur in highlands, usually above 900 m. These landscapes encompass a high diversity of habitats, under different environmental conditions, such as on rock outcrops dominated by desiccation-tolerant species, as well as grasslands and shrublands dominated by resprouter species. Therefore, water availability is one of the most significant selective pressures on rock outcrops, while fire is more important on grasslands and shrublands. Poaceae and Velloziaceae are the two dominant plant families in terms of cover area. There is predominance of hemicryptophytes, chamaephytes, and phanerophytes. Autochory is the main dispersal syndrome and contributes to high plant endemism. Floristic and vegetation structural patterns are affected by geology, geography, habitat, and disturbance. Rupestrian grasslands are Old Stable Landscapes and high specialization, phylo-

---

A.A. Conceição (✉) · A. Rapini · S.P.S. Neves
Departamento de Ciências Biológicas, Universidade Estadual de Feira de Santana (UEFS), Av. Transnordestina S/N, 44036-900 Feira de Santana, Bahia, Brazil
e-mail: abel18@gmail.com

A. Rapini
e-mail: rapinibot@yahoo.com.br

S.P.S. Neves
e-mail: samia_neves@yahoo.com.br

F.F. do Carmo
Instituto Prístino, Minas Gerais, Brazil
e-mail: flavio@institutopristino.org.br

J.C. Brito · G.A. Silva
Programa de Pós-Graduação em Botânica, UEFS, Bahia, Brazil
e-mail: julicarvalhais@gmail.com

G.A. Silva
e-mail: gabrielabiologa@gmail.com

C.M. Jacobi
Departamento de Biologia Geral, Universidade Federal de Minas Gerais, Minas Gerais, Brazil
e-mail: jacobi@icb.ufmg.br

© Springer International Publishing Switzerland 2016

G.W. Fernandes (ed.), *Ecology and Conservation of Mountaintop Grasslands in Brazil*, DOI 10.1007/978-3-319-29808-5_6

genetic conservatism, and low dispersal ability characterize most lineages in these landscapes. Rather than plant refuges during the warmer and moister Pleistocene interglacial periods, highlands have probably worked as refuges for fire-sensitive lineages since the expansion of fire-prone savannas (*cerrados*) in the late Tertiary. Most lineages from these mountain ranges then diversified during the Quaternary as rupestrian grasslands were finely fragmented by fire-prone landscapes. The fragmented and rich biodiversity in rupestrian grasslands is naturally vulnerable to anthropogenic disturbances. Therefore, protected areas even if small, scattered along the whole landscape, will help to protect them for a while. However, areas suitable for rupestrian grasslands will probably be greatly reduced in the next few decades as seasonality increases. This scenario calls for immediate ex situ conservation measures.

Contents

6.1 General Characterization

6.1.1 Vegetation Overview

Rupestrian grasslands (*campos rupestres*) comprise herbaceous and shrubby vegetation on nutrient-poor and toxic soils of quartzite or ironstone highlands, displaying a physiognomy similar to other tropical mountain vegetation around the word, with a mixture of grasslands, shrublands and rock outcrops (Fig. 6.1). Despite their physiognomic resemblance, rupestrian grasslands are characterized by a high number of endemic plant species and lineages, distributed along a great diversity of habitats on account of topography, edaphic conditions, and substrates (Harley 1995; Giulietti et al. 1997; Conceição and Pirani 2005, 2007; Jacobi et al. 2007; Rapini et al. 2008; Ribeiro et al. 2012).

(a)
(b)
(c)
(d)
(e)
(f)
(g)
(h)
(i)

◀ **Fig. 6.1** Rupestrian grasslands habitats in northern (**a**, **c**, **d**) and southern (**b**) Espinhaço Range: cleft on rock outcrop (**a**); rock outcrop (**b**); *Vellozia* aff. *sincorana*'s shrublands (**c**); grasslands with cryptophytic orchid (**d**). Epilithic chamaephytic orchid *Hadrolaelia sincorana* (Schltr.) Chiron & V.P.Castro (**e**). Terophytic *Dactylaena microphylla* Eichler (**f**). Phanerophytic *Sauvagesia paniculata* D.Cardoso & A.A.Conc. (**g**). Desiccation-tolerant chamaephyte (*Vellozia punctulata* Seub.) in wet (**h**) and dry (**i**) period. Species endemic to rupestrian grasslands of Bahia (**c**, **e**, **g**–**i**). Photographs by F.F. Carmo (**b**) and A.A. Conceição (**a**, **c**–**i**)

Distinct habitats in rupestrian grasslands result in a wide range of environmental conditions and allow the coexistence of different species and functional groups (Conceição and Pirani 2005; Jacobi et al. 2007). For instance, *grassland* habitat on sandy soils is dominated by graminoid plants of Poaceae (mainly hemicryptophytes), with Cyperaceae, Xyridaceae, and Eriocaulaceae being conspicuous especially in areas that are periodically wet. On the other hand, *rock outcrops* habitat is dominated by desiccation-tolerant chamaephytes and phanerophytes (Velloziaceae and Cyperaceae), plus epilythic Orchidaceae and basal rosettes (hemicryptophytes, mainly Bromeliaceae), growing on islands of vegetation within a rocky matrix. *Cleft* habitat occurs on rock outcrops with shrubs and treelets in rifts, and *shrubland* habitat occurs most usually over a mosaic of substrates, as boulders, rocky surfaces and sandy soils, with shrubs scattered among graminoids (Harley 1995; Giulietti et al. 1997; Conceição and Giulietti 2002; Conceição and Pirani 2005; Conceição et al. 2007a, b; Jacobi et al. 2007; Neves and Conceição 2010; Jacobi and Carmo 2011; Messias et al. 2011).

The Espinhaço Range in Brazil is the core area of rupestrian grasslands. In its southern portion (Minas Gerais state), mountaintops are surrounded by Cerrado (Brazilian savanna) and Atlantic forest, whereas in its northern portion (Bahia state), the matrix vegetation is the Caatinga (Brazilian semiarid vegetation) (Harley 1995; Giulietti et al. 1997). These isolated highlands encompass a very distinct flora and are important for biodiversity conservation (Echternacht et al. 2011; Ribeiro et al. 2012, 2014; Bitencourt and Rapini 2013). Floristic surveys in southern Espinhaço Range (e.g., Giulietti et al. 1987; Alves and Kolbek 2009; Jacobi and Carmo 2012) and in Chapada Diamantina (Stannard 1995; Zappi et al. 2003) have shown a high number of species and endemism, as well as a high beta diversity, in rupestrian grasslands (Zappi et al. 2003; Alves and Kolbek 2009). The high concentration of narrow endemics (see examples of endemic species in Fig. 6.1) contributes to this high beta diversity, since many species are restricted to small areas of the Espinhaço Range.

6.1.2 Scope

In the first part of this chapter, we describe the rupestrian grassland vegetation patterns based on quantitative data sampled by a standard methodology, encompassing different habitats (rock outcrops, grasslands, shrublands, and clefts), disturbances (fire and mining), substrates (ironstone and quartzite), and regions (southern and northern Espinhaço Range). In the second part, we discuss the diversification and conservation of rupestrian grasslands based on phylogenetic data and on the Old Stable Landscape framework.

6.1.3 Methodological Summary

We used 39 communities from phytosociological studies in rupestrian grasslands of the Espinhaço Range (Fig. 6.2). Thirty-four communities were sampled in Chapada

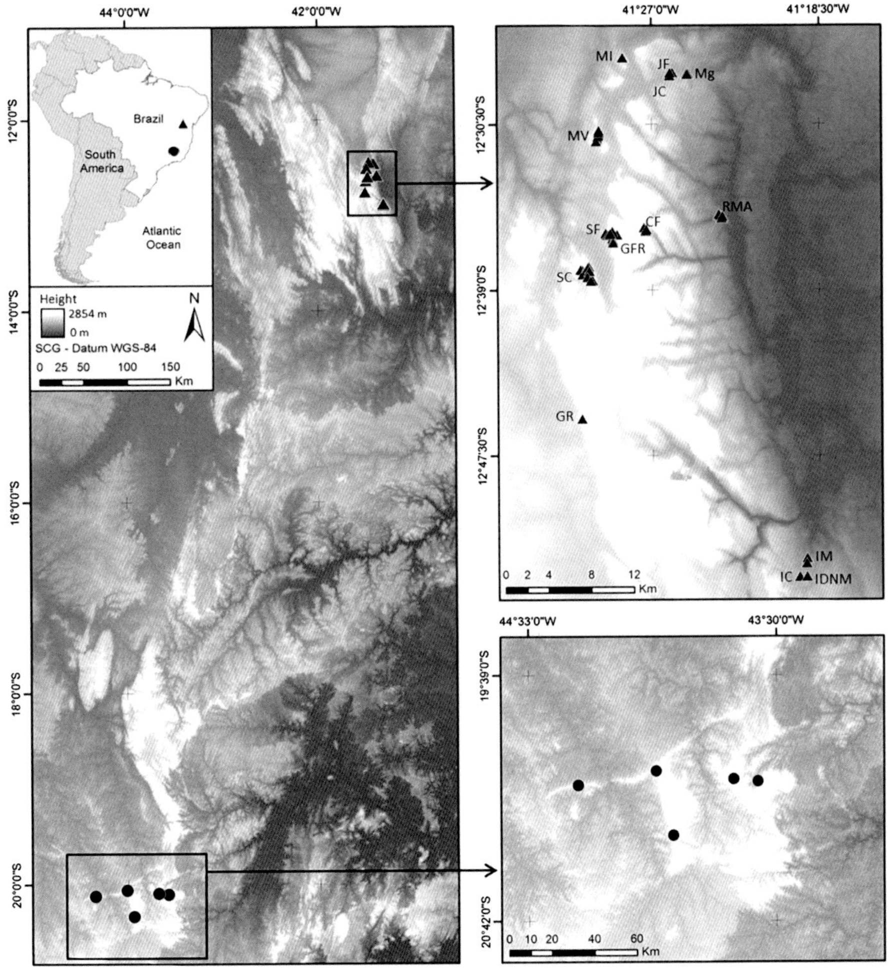

Fig. 6.2 Location of the 39 plant communities from the northern and southern (Quadrilátero Ferrífero) Espinhaço Range. Sites on *quartzites* are represented by *triangles* and on ironstones by *circles*

Diamantina (northern Espinhaço Range), using 2 m × 2 m plots (Conceição and Pirani 2005; Conceição et al. 2005, 2015; Neves and Conceição 2007, 2010; Brito 2011; Silva 2013), and five communities on ironstone in the Quadrilátero Ferrífero (southern Espinhaço Range), using 1 m × 2 m plots (C.M. Jacobi and F.F. Carmo, unpublished data). The area sampled in each community varied between 60 and 80 m^2. Among these communities, 23 were on rock outcrops (18 on quartzites and five on ironstones), 10 on shrublands, four on grasslands, and two on clefts. To obtain information on restoration, rock outcrops near or on abandoned mining sites —six on quartzite (diamond) and one on ironstone (iron)—were analyzed separately, since mining disturbance is common on rock outcrops and important for biodiversity conservation (Rumble 1989; Jacobi et al. 2007, 2015). Cover percentage of vascular plant species were used to construct a matrix of abundances of species, family (APG III 2009), and life-form (Ellenberg and Muller-Dombois 1967). The life-forms are categorized according to the position of buds to plant grow after the unfavorable season (dry season in the rupestrian grasslands). The main life-forms studied were hemicryptophyte (buds at or near the sediment surface), chamaephyte (aerial buds, but no more than 50 cm above the sediment surface), phanerophyte (buds more than 50 cm above the sediment surface), cryptophyte (below ground buds), and therophyte (annual plants, remaining in the area as seeds). Species geographic distribution (according to the *Lista de espécies da flora do Brasil* 2014), dispersal syndrome (Pijl 1982), and functional group were used to describe the vegetation and analyze local endemism, physiognomy, and diversity.

6.2 Chapada Diamantina and Quadrilátero Ferrífero Plant Communities

6.2.1 Habitats, Plant Families, Life-Forms, and Dispersal Syndromes

The taxonomic distribution of the rupestrian grasslands vegetation is represented by 51 families of Eudicots that cover 41.2 % of the rupestrian grassland vegetation, 15 Monocot families (56.5 % of plant cover area), ten families of Lycophytes and Monilophytes (2.1 %), and four Magnoliid families (0.2 %). The two most abundant families are Monocots, and together they cover almost half of the rupestrian grassland (42.3 %): Poaceae (25.3 %) and Velloziaceae (17.0 %). Fabaceae is the third most abundant family (8.3 %). The high abundance of some taxa is remarkable, but there are many inconspicuous families, as 34 families combined covering only 1 % of the vegetation.

There is evidence of habitat preference among the most abundant families (Fig. 6.3). Poaceae predominates in grasslands, where the Velloziaceae is not abundant. The Asteraceae, Bromeliaceae, Clusiaceae, and Velloziaceae thrive on rock outcrops and clefts. The Lamiaceae and Selaginellaceae are abundant on clefts, where the Fabaceae is not abundant. The Cactaceae and Orchidaceae are abundant

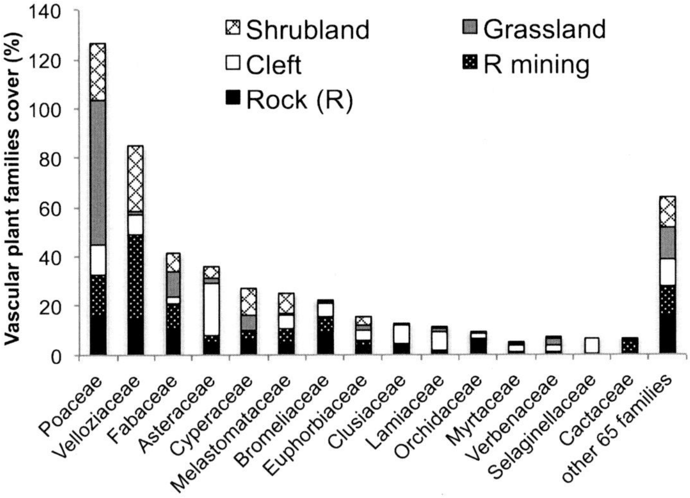

Fig. 6.3 Percentage of area covered by plant families in 39 rupestrian grassland communities according to the habitat

on rock outcrops, but the former family is abundant on mining areas and the latter is not. The abundance of Cactaceae on ironstones was related to only one species, *Arthocereus glaziovii* (K. Schum.) N.P.Taylor & D.C.Zappi, which is endemic to this lithotype (Zappi and Taylor 2012).

Rupestrian grasslands have a predominance of chamaephytes, hemicryptophytes, and phanerophytes, whereas the abundance of therophytes, cryptophytes, lianas, and epiphytes is low. In grasslands, chamaephytes and phanerophytes are not so abundant, and cryptophytes and therophytes tend to increase. Phanerophytes are more abundant in clefts (Fig. 6.4). The abundance of therophytes increases in disturbed areas. On average, therophyte proportion increases after fires (4.9 % of the vegetation), while in rupestrian grasslands without recent fire the therophyte percentage is much lower (0.7 %). The proportion of therophytes in rock mining sites is intermediate (2.9 %). Fire is important for the vegetation dynamics in rupestrian grasslands (Kolbek and Alves 2008; Neves and Conceição 2010; Conceição et al. 2013; Chap. 18), strongly affecting therophyte abundance, particularly at the onset of vegetation recovery, when more space and light boost seed germination and plant establishment.

Autochory is the main dispersal syndrome in rupestrian grasslands (49.3 %), followed by anemochory (19.9 %) and endozoochory (21.2 %) (Fig. 6.5). This pattern is similar to rock outcrops of Pai Inácio hill's rupestrian grasslands (Conceição et al. 2007c). The cover area of epizoochoric species is low (3.9 %), and the syndrome of 5.7 % of the species cover area was indeterminate. Epizoochory tends to be high on grasslands and endozoochory low on shrublands (Fig. 6.5). Epizoochory is more effective in grasslands than in vegetation islands because the continuous graminoid vegetation supports a richer mammalian fauna. On the other

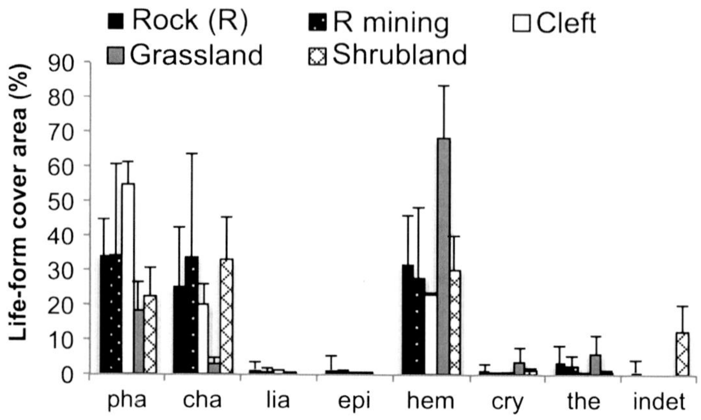

Fig. 6.4 Abundance of life-forms according to rupestrian grassland habitats. *Pha* phanerophyte; *cha* chamaephyte; *lia* liana; *epi* epiphyte; *hem* hemicryptophyte; *cry* Cryptophyte; *the* therophyte; *indet* indeterminate life-form

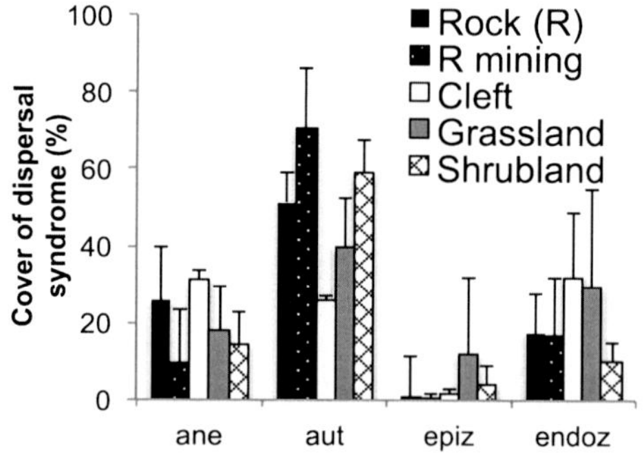

Fig. 6.5 Percentage of area covered by dispersal syndromes per habitat in rupestrian grasslands. *Aut* autochory; *ane* anemochory; *epiz* epizoochory; *endoz* endozoochory

hand, dispersal between vegetation islands seems to be more effective via endozoochory, especially by birds (Conceição et al. 2007c).

6.2.2 *Diversity and Endemism*

A total of 507 species of vascular plant was inventoried in the 785 rupestrian grassland plots (2840 m^2) from the 39 communities of the Espinhaço Range analyzed here, an average of 0.18 species/m^2. These rupestrian grasslands are dominated by species endemic to Brazil. Overall, 165 species are endemic either to southern or northern Espinhaço Range (32.7 %), particularly on rock outcrops and shrublands, comprising 38.1 % of the plant cover area (Fig. 6.6). The percentage of species endemic to rupestrian grasslands, however, is higher than values presented

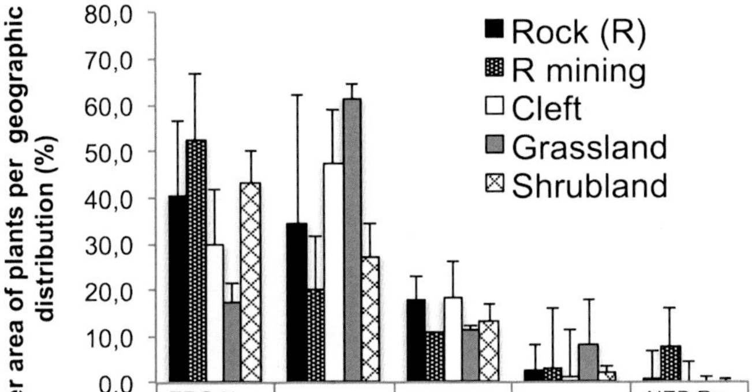

	ERG N or S of ER	OEB	NEB	NEB Ru.	NEB Ru. Ex.Inv.
Rock (R)	40,4	34,6	18,0	2,7	0,0
R mining	52,6	20,1	10,6	3,1	7,7
Cleft	29,8	47,2	18,2	1,0	0,0
Grassland	17,3	61,3	11,5	8,1	0,0
Shrubland	43,2	27,1	13,1	2,2	0,0

Fig. 6.6 Percentage of area covered by the pool of vascular plant species in the same geographic distribution category per rupestrian grassland habitat. ERG N or S of *ER* species endemic to rupestrian grasslands either in northerner or southern Espinhaço Range; *OEB* other endemics to Brazil; *NEB* not endemic to Brazil; *Ru* ruderal species; *Ex.Inv.* exotic invasive species

here because species endemic to rupestrian grasslands from more than one state, mainly Minas Gerais, Bahia, and Goiás, were classified as 'other endemics to Brazil'. Species endemic to Brazil are mainly autochoric (Mann-Whitney U test, $p = 0.0001$), whereas epizoochory is more common in non-endemics (Fig. 6.7). This pattern suggests that local dispersal (autochory) may affect species distribution in rupestrian grasslands. Exotic invasive species are more abundant on rock outcrops mining, while non-endemic ruderal species are more abundant in grasslands (Fig. 6.6).

6.2.3 *Floristic and Structural Patterns*

The cluster of the 39 communities shows groups according to floristic composition and vegetation structure (Fig. 6.8). Geology and geography determine the most distinctive groups, one on ironstone in southern Espinhaço Range and another on quartzite in northern Espinhaço Range. In the ironstone group, the only area under mining influence is isolated, probably as result of floristic change caused by local disturbance. In the quartzite group, the influence of geography, habitat, and disturbance is evident. Shrublands dominated by *Vellozia* aff. *sincorana* (MV, SC, and SF) are segregated by habitat, whereas their internal division is mostly affected by

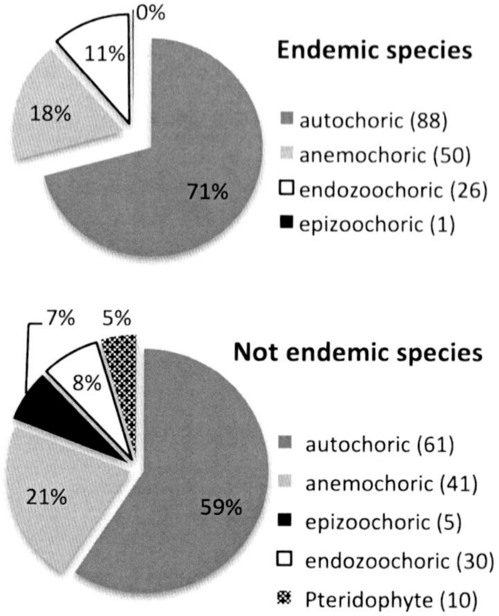

Fig. 6.7 Percentage of area covered by species pooled by dispersal syndrome. *Above* endemic species to rupestrian grasslands in the north or south of the Espinhaço Range. *Below* species not endemic to Brazil. The number of species is indicated in *brackets*

geography (relief and distance), with all areas in MV separated of SC and SF (see Fig. 6.2). Geographic influence is also visible in the Mãe Inácia rock outcrop and cleft (MI R-Cl) and Cachoeira da Fumaça shrublands and cleft (CF Shr-Cl). Fire probably affects the clustering between communities on rock outcrops and on grasslands (PF6m-PF18m). Using presence/absence data (Fig. 6.9), geography is the most influential factor contributing to floristic composition patterns, seen as local pools of species assembly, regardless of habitats differences.

Carmo and Jacobi (2013) distinguished ironstone vegetation from quartzitic systems in southeastearn Brazil based on the larger phytogeographic influence of floristic elements from the Atlantic forest domain, the higher frequency of sinusiae formed by treelets and shrubs, and species-rich genera, such as *Andropogon*, *Cattleya, Chromolaena* and *Ocotea.* This difference is also correlated with the geomorphology and mineralogy of ironstones. Disturbance is another important factor influencing rupestrian grassland diversity. Recurrent fires result in patches with different times of vegetation recovery, increasing the diversity of therophitic species and affecting the functioning of rupestrian grasslands (Chap. 18). On the other hand, perennial resprouter species may display massive post-fire flowering, offering a great amount of resources to animals (e.g., Conceição and Orr 2012; Conceição et al. 2013). The variety of habitats side-by-side harbors distinct plant communities, which in turn provide for a regional wealth of plant diversity.

The top five dominant species in percent cover summarize the main ecological features in each habitat (Table 6.1). Overall, half of these 20 species in the Table 6.1 is endemic to rupestrian grasslands, and the fire resprouter *Renvoizea trinii* is among the top five in all habitats, highlighting the importance of fire disturbance in rupestrian grasslands (Kolbek and Alves 2008; Neves and Conceição

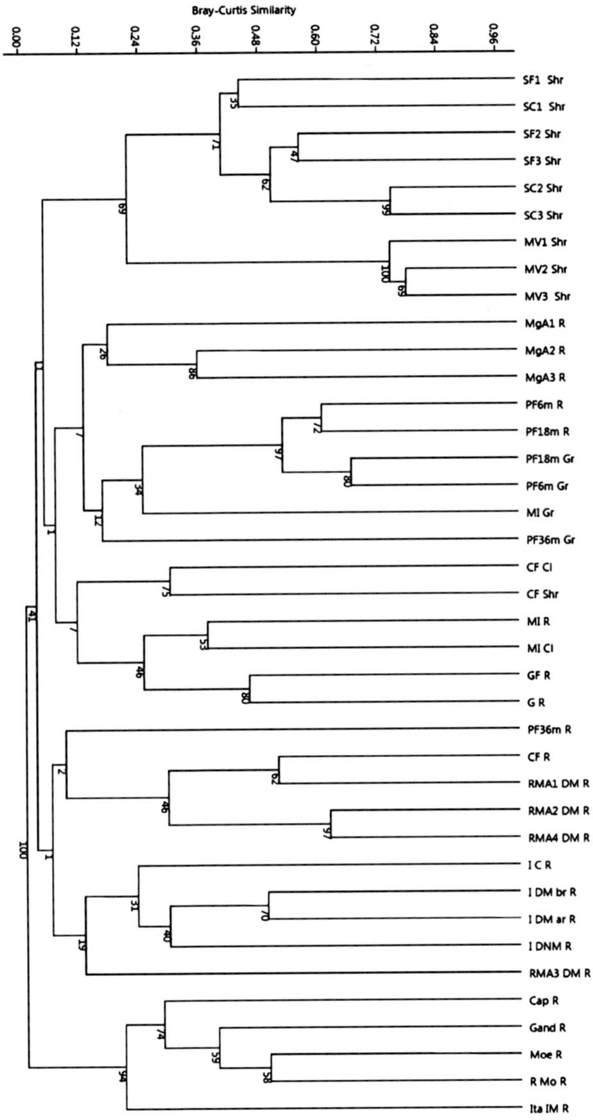

Fig. 6.8 Cluster grouping 39 rupestrian grasslands communities based on quantitative data (Bray-Curtis, UPGMA; values at nodes = bootstrap support; cophenetic correlation coefficient = 0.92). Sites: Serra da Fumaça (SF), Serra do Candombá (SC), Morro dos Ventos (MV), Mucugezinho (Mg), Pelé (P), Mãe Inácia (MI), Cacheira da Fumaça (CF), Gerais da Fumaça (GF), Guiné (G), Ribeirão do Meio (RM), Igatu (I), Serra da Capanema (Cap), Serra da Gandarela (Gand), Serra da Moeda (Moe), Rola Moça (R Mo), and Itatiaiuçu (Ita). Habitats: Shrubland (Shr), Rock outcrop (R), Grassland (Gr), Cleft (Cl). Disturbances: diamond mining (DM), iron mining (IM), fire (F) 18 and 36 months (m), disturbance not mining (DNM). Before river (br). After river (ar)

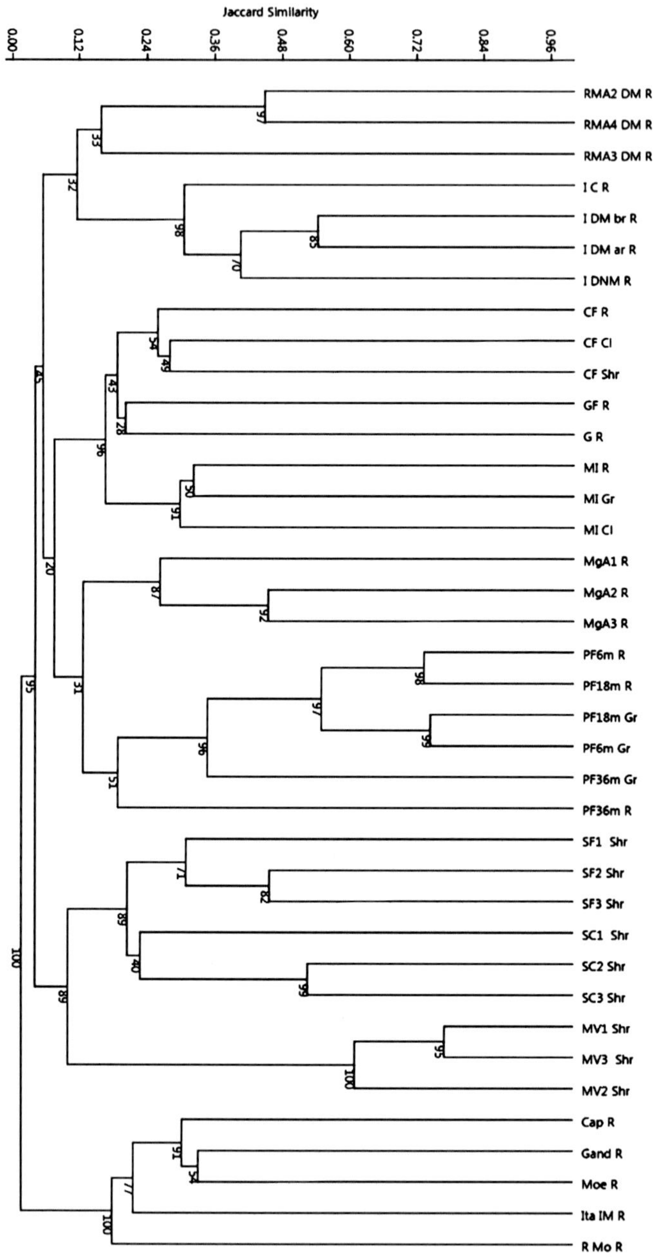

Fig. 6.9 Cluster grouping 38 rupestrian grasslands communities based on qualitative data (Jaccard, UPGMA; values at nodes = bootstrap support; Cophenetic correlation coefficient = 0.96). An outlier mining area with *Vellozia punctulata* totally dominant (99.9 % of cover area) was removed. For site codes, see Fig. 6.8

Table 6.1 Top five abundant species in percent cover in each habitat in rupestrian grasslands (R: rock outcrop; R m: rock outcrop near or on mining sites

Vascular plant species	Plant family	GD	Functional group	R	R m	Cl	Sh	Gr
Renvoizea trinii (Kunth) Zuloaga & Morrone	Poaceae	OEB	Graminoid resprouter	4.6	0.0	6.4	5.0	28.0
Vellozia punctulata Seub.	Velloziaceae	End RG	Stem rosette desic-tolerant	3.8	21.2	2.9	0.0	0.0
Vellozia aff. *sincorana* L.B.Sm. & Ayensu	Velloziaceae	End RG	Stem rosette resprouter	0.0	0.0	0.0	26.3	0.0
Trachypogon macroglossus Trin.	Poaceae	OEB	Graminoid resprouter	0.4	0.0	1.2	0.1	12.0
Baccharis reticularia DC.	Asteraceae	OEB	Shrub	0.2	0.0	12.1	1.1	0.1
Clusia obdeltifolia Bittrich	Clusiaceae	OEB	Tree	3.9	0.0	7.8	0.0	0.0
Calliandra calycina Benth.	Fabaceae	End RG	Shrub resprouter	1.6	0.0	0.0	0.0	9.2
Axonopus pellitus (Ness ex Trin.) Hitchc. & Chase	Poaceae	NEB	Graminoid resprouter	2.2	1.4	0.0	0.0	7.2
Tibouchina pereirae Brade & Markgr.	Melastomataceae	End RG	Shrub resprouter	2.2	5.1	2.1	0.1	0.3
Orthophytum burle-marxii L.B.Sm. & Read	Bromeliaceae	End RG	Basal rosette	2.1	5.2	0.7	0.0	0.0
Barbacenia blanchetii Goethart & Henrard	Velloziaceae	End RG	Stem rosette desic-tolerant	3.5	3.0	1.1	0.0	0.1
Lagenocarpus rigidus Nees	Cyperaceae	End RG	Graminoid resprouter	0.3	0.5	0.0	3.8	3.0
Melinis minutiflora P.Beauv.	Poaceae	Ex i. Ru	Graminoid resprouter	0.0	7.6	0.0	0.0	0.0
Selaginella marginata (Humb. & Bonpl. ex Willd.) Spring	Selaginellaceae	NEB	Desiccation-tolerant herb	0.2	0.0	6.4	0.0	0.0

(continued)

Table 6.1 (continued)

Vascular plant species	Plant family	GD	Functional group	R	R m	Cl	Sh	Gr
Eriope exaltata Harley	Lamiaceae	End RG	Tree	0.3	0.0	6.1	0.0	0.0
Micranthocereus purpureus (Gürke) F.Ritter	Cactaceae	End RG	Aphyllous succulent	0.3	5.1	0.3	0.0	0.0
Calliandra asplenioides (Nees) Renvoize	Fabaceae	OEB	Shrub resprouter	0.0	0.0	0.1	4.5	0.0
Trichanthecium cyanescens (Nees ex Trin.) Zuloaga & Morrone	Poaceae	NEB Ru	Graminoid resprouter	0.0	0.0	0.0	0.0	4.4
Poaceae sp.5	Poaceae	–	Graminoid	0.0	0.0	0.0	4.2	0.0
Vellozia jolyi L.B.Sm.	Velloziaceae	End RG	Stem rosette desic-tolerant	4.1	0.0	0.0	0.0	0.0
				29.8	**49.0**	**47.3**	**45.3**	**64.4**

Cl cleft; *Sh* shrubland; *Gr* grassland). Geographic distribution (GD): endemic to rupestrian grasslands (End RG); other endemics to Brazil (OEB); not endemic to Brazil (NEB); ruderal species (Ru); exotic invasive (Ex i.). Desiccation-tolerant functional group (desic-tolerant)

2010; Chap. 18). On rock outcrops, there are three endemic, stem rosette desiccation-tolerant species of Velloziaceae, plus one tree and one graminoid species, and the desiccation-tolerant *Vellozia punctulata* dominates under mining influence. This dominance of desiccation-tolerant species confirms water as one of the main constraints on rock outcrops. The top five species in mining areas are classified in different functional groups, including a shrub, a basal rosette, a graminoid, and an aphyllous succulent plant. This diversity of functional groups is probably related to secondary succession and rock outcrop colonization, with plants resistant to mechanical impacts (Velloziaceae, Bromeliaceae, and Cactaceae) and able to propagate by vegetative reproduction, even in the case of the shrub *Tibouchina pereirae* (Melastomataceae), which is a versatile and wide niche breath species (Conceição and Pirani 2007; Chap. 9). The graminoid species in these areas is the exotic invasive *Melinis minutiflora*, the only species that is not endemic to rupestrian grasslands. This suggests that vegetation recovery in areas under mining influence occurs mainly by clonal regrowth of endemic species and exotic species invasion (Conceição et al. 2015), which can be devastating for rupestrian grasslands (Barbosa et al. 2010; Hilário et al. 2011; Fernandes et al. 2014). Therefore, mining activities threat native species because of habitat loss (Jacobi et al. 2007, 2011, 2015; Jacobi and Carmo 2012; Fernandes et al. 2014) and because they facilitate biological invasions (Barbosa et al. 2010; Hilário et al. 2011; Fernandes et al. 2014).

Three of top five species on cleft are trees and shrubs, since in this habitat the soil is deeper, shaded, and not so nutrient-poor as in other habitats (Conceição and Pirani 2005). The desiccation-tolerant herb *Selaginella marginata* occurs on clefts borders, directly on rocky surfaces, and the Poaceae *R. trinii* occurs in grassland or shrubland contact zone. All top five species in shrublands and in grasslands are resprouters, confirming the importance of fire in continuous habitats (Neves and Conceição 2010). The stem rosette *Vellozia* aff. *sincorana* is dominant in the fire-prone shrublands, which is endemic to rupestrian grasslands. Three of the top five species on shrublands are graminoids and one is a shrub (*Calliandra asplenioides*). In grasslands, there is a predominance of graminoids (four of top five), besides the shrub *Calliandra calycina*, which is endemic to rupestrian grasslands. Thus, *Calliandra* (Fabaceae) is also an important genus in these continuous fire-prone habitats.

6.3 Origins and Diversification of Rupestrian Grasslands

6.3.1 The Traditional Pleistocene "Pump" Hypothesis

Rupestrian grasslands consist of mosaics of open vegetation physiognomies with high plant diversity, mainly shaped by edaphic, topographic and climatic conditions. They are rich in narrow endemics, but widespread species are common, and species disjunctly distributed between rupestrian grassland nuclei at different scales have been reported (e.g., Giulietti and Pirani 1988; Rando and Pirani 2011).

The traditional hypothesis for diversification in rupestrian grasslands postulates successive retraction-expansion events driven by Pleistocene climatic fluctuations, working as an evolutionary pump (Harley 1988, 1995). During the moister and warmer interglacial periods, forests would expand their distribution to higher altitudes and rupestrian grasslands would become restricted to mountaintops, forming archipelagic distributions. According to this hypothesis, mountain chains serve as long-term refuges for rupestrian grasslands. Their diverse topography would favor migrations to suitable places, maintaining lineages in the region despite of climatic changes. Small populations, disjunctly distributed in highlands, are more susceptible to inbreeding depression and genetic drift, tending to become extinct or to diverge from each other to eventually be recognized as different species (Rapini et al. 2008). In contrast, during dryer and cooler glacial periods, forests would retract to lower altitudes and rupestrian grasslands would expand their distribution to lowlands, allowing gene flow between populations that were previously isolated in highlands. These hybridizations may homogenize differences between populations or result in lineages that evolve independently from parental lineages, being recognized as distinct species.

Although the origin and distribution of rupestrian grassland have intrigued botanists for a long time, little data are available to investigate their diversification. Phylogenetic studies with groups predominantly distributed in rupestrian grasslands are few and only two provided time-calibrated phylogenies so far. Antonelli et al. (2010) suggested that *Hoffmannseggella*, a genus of orchids, diversified mainly by hybridization. According to them, the cooling climate since the Late Miocene Optimum (15 million years ago) promoted the expansion of rupestrian grasslands and the connectivity between previously disjunct populations would result in a rapid pre-Pleistocene radiation. In contrast, Ribeiro et al. (2014) suggested that *Minaria*, a genus of asclepiads, diversified mainly by geographic isolation and non-adaptive radiation. They postulated a general trend toward retraction of rupestrian grasslands, interrupted by few Pleistocene episodes of expansion that would boost diversification by providing source for new retractions. Both studies, however, agree that diversification in rupestrian grassland lineages preceded the expansion of fire-prone savannas (*cerrados*), a pattern also recovered in other plant groups (e.g., Loeuille et al. 2015).

6.3.2 Old Stable Landscapes

Rupestrian grasslands occupy old, geologically stable areas, dominated by ancient erosion (Saadi 1995; Pedreira 1997), climatically buffered by topographic diversity, with oligotrophic, severely phosphorous-impoverished soils (Oliveira et al. 2015). The exposed landsurface in the Quadrilátero Ferrífero, for instance, is among the most resistant to weathering and most ancient on Earth (Monteiro et al. 2014) and the extremely oligotrophic soil in these ironstones are the result of millions of years of weathering (Carvalho Filho et al. 2010; Salgado and Carmo 2015). Areas such as the Pantepuis in the Guayana Shields, northern South America, and the Espinhaço Range in Brazilian Shield, eastern South America, can be considered typical old, climatically buffered, infertile landscapes (OCBILs; Hopper 2009) and probably share ecological and evolutionary patterns with other highly diverse and endemic-rich landscapes, such as the Southwest Australia and the Greater Cape. The OCBIL theory integrates evolution and ecology and provides a conservation framework for highly diverse and naturally vulnerable biotas. Mucina and Wardell-Johnson (2011) refined the OCBIL concept and proposed the idea of Old Stable Landscapes (OSLs) by reducing soil infertility as a function of landscape age and adding fire regime predictability as a new dimension of the theory.

Lineages in OSLs (Hopper 2009; Mucina and Wardell-Johnson 2011) are usually highly specialized and present low dispersal capability. Anemochory, for instance, is a common syndrome in rupestrian grasslands, as shown above. Short-distance dispersal minimizes propagule loss and usually guarantees occupation of suitable places, but prevents migrations between disjunct nuclei of OSLs. Consequently, OSLs are rich in narrow endemics and their populations tend to be strongly differentiated. As result, alpha and beta diversities are high in OSLs, as shown for rupestrian grasslands in the Chapada Diamantina (e.g., Conceição et al. 2005). Due to the long-term isolation caused by natural fragmentation, mechanisms that prevent variability loss caused by inbreeding are selected (James Effect); pollination seems to be one of them, contributing to population variability (Hopper 2009).

Lineages in OSLs also present high phylogenetic conservatism. Due to climatic stability, however, they can persist for long periods in these landscapes, which worked as refuges, acting as cradles and/or museums of biodiversity. Bromeliads, currently with more than 3000 species, for instance, probably arose on the tepuis, approximately 100 million years ago (Givinish et al. 2007, 2011). Asclepiads present different refuge patterns along the Espinhaço Range (Bitencourt and Rapini 2013). The northern Espinhaço Range seems to represent a museum for asclepiads, sheltering palaeoendemic species, while the southern Espinhaço Range houses several microneoendemic species (but possibly palaeoendemics too) and probably represents an asclepiad cradle.

As in other OSLs, plants from rupestrian grasslands are also adapted to impoverished soils and present strategies that enhance the acquisition and conservation of nutrients, such as root specializations (Oliveira et al. 2015), underground system and sclerophyllous leaves (Giulietti et al. 1987, 1997). Some ironstone outcrop species, for instance, have a specialized root system that is able to penetrate through fissures and microcrevices in the rock and may form, when reaching a cavity, massive columnar structures known as rhyzothems (Ferreira 2005; Carmo and Jacobi 2012). Perennial plants are dominant and deciduousness is uncommon, probably because development and regeneration on nutrient-impoverished soils are slow. Several characteristics in rupestrian grasslands, such as resprouting ability, look like adaptations to fire regimes. However, as addressed by Hopper (2009), they may have evolved in response to different stresses and later were recruited under fire regime conditions, representing exaptations.

Fire affects the composition and function of biotas and is responsible for shaping several biomes (Mucina and Wardell-Johnson 2011). The expansion of fire-prone savanna (cerrados) during the Late Miocene-Pliocene was boosted by aridity increase and dominance of C4 flammable grasses (Beerling and Osborne 2006) and lineages from fire-free biomes then colonized the fire-prone cerrado (e.g. Simon et al. 2009; Simon and Pennington 2012). However, several fire-sensitive and phylogenetic conservative lineages from rupestrian grasslands did not. They became restricted to highlands, where fire range tend to be smaller and fire frequency lower than in lowlands because of lower proportion of C4 grasses, higher concentration of rocky outcrops, and milder weather, with lower seasonality because of orographic moisture. According to this hypothesis, therefore, highlands are not cooler and dryer refuges for rupestrian grasslands during Pleistocene interglacial periods, but refuges for fire-sensitive lineages since the Tertiary (Ribeiro et al. 2012). Ribeiro et al. (2014) postulated that lineages endemic from rupestrian grasslands possibly derived from a type of proto-savanna, a general open vegetation that preceded the full establishment of fire-regimes. Rupestrian grasslands started to diversify in the Late Miocene and most extant species would have emerged during the Quaternary as fire-sensitive lineages were fragmented not only by geological erosion, but also by fire-prone landscapes.

Rupestrian grasslands and cerrados occur in two different Pyro-landscapes (sensu Mucina and Wardell-Johnson 2011): cerrados are fire-dependent and highly resilient to fire regimes (Pre-Pleistocene-pyro-landscapes) whereas rupestrian grasslands are fire-refuges within a matrix of fire-dependent biome (Pleistocene Pyro-landscapes). Although physiognomically similar, they are historically, ecologically, and floristically distinct. Diversity in cerrados derived mainly from recruiting unrelated lineages from other biomes (Simon et al. 2009; Simon and Pennington 2012), while a great proportion of diversity in rupestrian grasslands derived from in situ radiations, resulting in high concentration of closely related microendemic species. Rupestrian grasslands, however, are neither static nor hermetically closed and their diversification cannot be reduced to in situ diversification only; they provide lineages to other biomes and also receive lineages from them.

6.4 Conservation

OSLs shelter high concentrations of exclusive diversity. In rupestrian grasslands, plant diversity are higher than average. The Espinhaço Range, for instance, correspond to approximately 1 % of the Brazilian territory, but houses more than 10 % of plant species reported to Brazil and, in average, 30 % of them are endemic (Ribeiro et al. 2012). Due to high phylogenetic conservatism, lineages restricted to OSLs do not respond quickly to new conditions and are usually unable to colonize new habitats. Furthermore, most of them cannot reach distant suitable areas because of low dispersal capability. These lineages, therefore, are highly vulnerable to disturbances and their conservation deserves special attention.

Based on *Minaria*, Ribeiro et al. (2012) pointed three priority areas for conservation in the Espinhaço Range, each based on different aspects of biodiversity: (1) Serra do Cipó and Diamantina Plateau in the southern Espinhaço Range, because of the highest phylogenetic diversity and endemism; (2) the Chapada Diamantina, in the northern portion, because of the high levels of exclusive evolutionary history, represented by endemic remnants of ancient lineages; and (3) the Quadrilátero Ferrífero, in the southernmost Espinhaço Range, because of its endemic species are the most threatened ones. However, the high frequency of narrowly endemic species dispersed across the entire landscape result in many areas with exclusive biodiversity. In these cases, selection of priority areas for conservation seems less effective and the protection of the entire Espinhaço Range has been invoked (Rapini et al. 2002; Echternacht et al. 2011). Since OSLs are resilient to fragmentation and their populations are adapted to long persistence in small areas, conservation of several fragments, even if small, scattered across the entire landscape can be an effective strategy (Hopper 2009). These small-scattered areas should comprise different habitats devoid of mining influence.

OSLs have been buffered for a long time from climatic changes and rapid changes may greatly reduce their biodiversity. By the end of this century, the Pantepui is estimated to lose 80 % of its current habitats, and several patches will disappear (Vegas-Vilarrúbia et al. 2012). The Espinhaço Range is estimated to lose approximately 50 % of the area currently suitable for rupestrian grasslands, which will almost disappear in the more drier and seasonal northern portion. The reduction of rupestrian grasslands will extinguish many endemic species and current conservation units will not be effective to protect several of them from extinction (Fernandes et al. 2014; Bitencourt et al. 2016). Furthermore, communities in rupestrian grasslands are clustered phylogenetically and therefore they are more vulnerable to similar disturbances (Pugliese and Rapini 2015). The areas of rupestrian grasslands in the Espinhaço Range are probably decreasing due to seasonality increase (Bitencourt et al. 2016). This process was probably gradual and continuous since the Tertiary, but has been greatly accelerated because of anthropogenic changes and soon several highlands will not be suitable for lineages currently endemic to rupestrian grasslands. With this perspective, an ex situ conservation of their biodiversity needs to start right now (Vegas-Vilarrúbia et al. 2012; Pugliese and Rapini 2015).

Acknowledgements The authors would like to thank Instituto Chico Mendes de Conservação da Biodiversidade for the license to work in Chapada Diamantina National Park. This study received financial support from the Brazilian *Conselho Nacional de Desenvolvimento Científico e Tecnológico* (CNPq, National Council for Scientific and Technological Development; research grants no. 311301/2011-8 and 479834/2008-3, general research grant no. 556820/2011-0 and *Pesquisa Ecológica de Longa Duração* [PELD, Long-term Ecological Research] grant no. 59/2009), the Minas Gerais State Research Foundation (FAPEMIG, grants APQ-00851-11 and PPM-00224-13), and the *Fundação de Amparo à Pesquisa do Estado da Bahia* (FAPESB, Foundation for the Support of Research in the State of Bahia; grant no. TO 26/2011).

References

Alves RJV, Kolbek J (2009) Summit vascular flora of Serra de São José, Minas Gerais, Brazil. Check List 5:35–73

Antonelli A, Verola CF, Parisod C, Gustafsson ALS (2010) Climate cooling promoted the expansion and radiation of a threatened group of South American orchids (Epidendroideae: Laeliinae). Biol J Linn Soc 100:597–607

Apg III (2009) An update of the Angiosperm Phylogeny Group classification for the orders and families of flowering plants: APG III. Biol J Linn Soc 161:105–121

Barbosa NPU, Fernandes GW, Carneiro MAA, Junior LAC (2010) Distribution of non-native invasive species and soil properties in proximity to paved roads and unpaved roads in a quartzitic mountainous grassland of southeastern Brazil (rupestrian fields). Biol Inv 12:3745–3755

Beerling DJ, Osborne CP (2006) The origin of the savanna biome. Glob Chan Biol 12:2023–2031

Bitencourt C, Rapini A (2013) Centres of endemism in the Espinhaço Range: identifying cradles and museums of Asclepiadoideae (Apocynaceae). Syst Biodiv 11:525–536

Bitencourt C, Rapini A, Damascena LS, De Marco Junior P (2016) The worrying future of the endemic flora of a tropical mountain climate change. Flora 218:1–10

Brito JC (2011) Efeitos do fogo sobre a vegetação em duas áreas de campo rupestre na Chapada Diamantina, Bahia, Brasil. MSc Dissertation, Universidade Estadual de Feira de Santana

Carmo FF, Jacobi CM (2012) The cangas of the Iron Quadrangle. In: Jacobi CM, Carmo FF (eds) Floristic diversity of the quadrilátero ferrífero cangas. Ed. IDM, Belo Horizonte, pp 14–30

Carmo FF, Jacobi CM (2013) Canga vegetation in the Iron Quadrangle, Minas Gerais: characterization and phytogeographical context. Rodriguésia 64:527–541

Carvalho Filho A, Curi N, Shinzato E (2010) Relações solo-paisagem no Quadrilátero Ferrífero em Minas Gerais. Pesq Agropec Bras 45:903–916

Conceição AA, Giulietti AM (2002) Composição florística e aspectos estruturais de campo rupestre em dois platôs do Morro do Pai Inácio, Chapada Diamantina, Bahia, Brasil. Hoehnea 29:37–48

Conceição AA, Orr BJ (2012) Post-fire flowering and fruiting in the caulescent rosette *Vellozia sincorana*, an endemic plant to the Northeast of Brazil. Acta Bot Bras 26:94–100

Conceição AA, Pirani JR (2005) Delimitação de habitats em campos rupestres na Chapada Diamantina: substratos, composição florística e aspectos estruturais. Bol Bot Univ São Paulo 23:85–111

Conceição AA, Pirani JR (2007) Diversidade em quatro áreas de campos rupestres na Chapada Diamantina, Bahia, Brasil: espécies distintas, mas riquezas similares. Rodriguésia 58:193–206

Conceição AA, Rapini A, Pirani JR, Giulietti AM, Harley R, Silva TRS, Funch R, Santos AKA, Correia C, Andrade IM, Costa JAS, Souza LRS, Andrade MJG, Freitas TA, Freitas AMM, Oliveira AA (2005) Campos rupestres. In: Juncá FA, Funch L, Franca-Rocha W (eds) Biodiversidade e conservação da Chapada Diamantina. Biodiversidade 13. Ministério do Meio Ambiente, Brasília, pp 153–180

Conceição AA, Funch LS, Pirani JR (2007a) Reproductive phenology, pollination and seed dispersal syndromes on sandstone outcrop vegetation in the "Chapada Diamantina", northeast Brazil: population and community analyses. Rev Bras Bot 30:475–485

Conceição AA, Giulietti AM, Meirelles ST (2007b) Ilhas de vegetação em afloramentos de quartzito-arenito no Morro do Pai Inácio, Chapada Diamantina, Bahia, Brasil. Acta Bot Bras 21:335–347

Conceição AA, Pirani JR, Meirelles ST (2007c) Floristics, sctructure and soil of insular vegetation in four quartzite-sandstone outcrops of "Chapada Diamantina", northeast Brazil. Rev Bras Bot 30:641–655

Conceição AA, Alencar TG, Souza JM, Moura ADC, Silva GA (2013) Massive post-fire flowering events in a tropical mountain region of Brazil: high episodic supply of floral resources. Acta Bot Bras 27:847–850

Conceição AA, Cristo FH, Santos AA, Santos JB, Freitas EL, Borges BPS, Macêdo LSS, Oliveira RCS (2015) Vegetação endêmica e espécie invasora em campos rupestres de áreas garimpadas. Rodriguésia 66:675–683

Echternacht L, Trovó M, Oliveira CT, Pirani JR (2011) Areas of endemism in the Espinhaço Range in Minas Gerais, Brazil. Flora 206:782–791

Ellenberg H, Müller-Dombois D (1967) Tentative physiognomic-ecological classification of plant formations of the earth. Ber Geob Inst Rubel Zurich 37:21–55

Fernandes GW, Barbosa NPU, Negreiros D, Paglia AP (2014) Challenges for the conservation of vanishing megadiverse rupestrian grasslands. Nat Cons 12:162–165

Ferreira RL (2005) A vida subterrânea nos campos ferruginosos. Carste 3:106–115

Giulietti AM, Pirani JR (1988) Patterns of geographic distribution of some plant species from the Espinhaço Range, Minas Gerais and Bahia, Brazil. In: Vanzolini PE, Heyer WR (eds) Proceedings of a workshop on Neotropical distribution patterns. Academia Brasileira de Ciências, Rio de Janeiro, pp 39–69

Giulietti AM, Menezes NL, Pirani JR, Meguro M, Wanderley MGL (1987) Flora da Serra do Cipó, Minas Gerais: caracterização e lista das espécies. Bol Bot Univ São Paulo 9:1–151

Giulietti AM, Pirani JR, Harley RM (1997) Espinhaço Range Region, Eastern Brazil. In: Davis SD, Heywood VH, Villa-Lobos J, Hamilton AC (eds) Centres of plant diversity. A guide and strategy for their conservation, vol 3. The Americas. IUCN Publication Unity, Cambridge, pp 397–404

Givnish TJ, Millam KC, Berry PE, Sytsma KJ (2007) Phylogeny, adaptive radiation, and historical biogeography of Bromeliaceae inferred from *Ndhf* sequence data. In: Columbus JT, Friar EA, Porter JM, Prince LM, Simpson MG (eds) Monocots: comparative biology and evolution—Poales. Rancho Santa An Botanic Garden, Claremont, California, pp 3–26

Givnish, TJ Barfuss MHJ, Ee BV, Riina R, Schulte K, Horres R, Gonsiska PA, Jabaily RS, Crayn DM, Smith JAC, Winter K, Brown GK, Evans TM, Holst BK, Luther H, Till W, Zizca G, Berry PE, Sytsma KJ (2011) Phylogeny, adaptive radiation, and historical biogeography in Bromeliaceae: insights from an eight-locus plastid phylogeny. Am J Bot 98:872–895

Harley RM (1988) Evolution and distribution of *Eriope* (Labiatae), and its relatives, in Brazil. In: Vanzolini PE, Heyer WR (eds) Proceedings of a workshop on neotropical distribution patterns. Academia Brasileira de Ciências, Rio de Janeiro, pp 71–120

Harley RM (1995) Introduction. In: Stannard BL (ed) Flora of the Pico das Almas, Chapada Diamantina, Brazil. Royal Botanic Gardens, Kew, pp 1–42

Hilário RR, Castro SAB, Ker FTO, Fernandes GW (2011) Unexpected effects of pigeon-peas (*Cajanus cajan*) in the restoration of rupestrian grasslands. Planta Daninha 29:717–723

Hopper SD (2009) OCBIL theory: towards an integrated understanding of the evolution, ecology and conservation of biodiversity on old, climatically buffered, infertile landscape. Plant Soil 322:49–86

Jacobi CM, Carmo FF (2011) Life-forms, pollination and seed dispersal syndromes in plant communities on ironstone outcrops, SE Brazil. Acta Bot Bras 25:395–412

Jacobi CM, Carmo FF (eds) (2012) Floristic Diversity of the Quadrilátero Ferrífero Cangas. Editora IDM, Belo Horizonte

Jacobi CM, Carmo FF, Vincent RC, Stehmann JR (2007) Plant communities on ironstone outcrops: a diverse and endangered Brazilian ecosystem. Biodiv Cons 16:2185–2200

Jacobi CM, Carmo FF, Campos IC (2011) Soaring extinction threats to endemic plants in Brazilian metal-rich regions. Ambio 40:540–543

Jacobi CM, Carmo FF, Campos IC (2015) Iron geosystems: priority areas for conservation in Brazil. In: Tibbett M (ed) Mining in ecologically sensitive landscapes. CRC Press, Leiden, pp 55–77

Kolbek J, Alves RJV (2008) Impacts of cattle, fire and wind in rocky savannas, southeastern Brazil. Acta Univ Carol Biol 22:111–130

Lista de Espécies da Flora do Brasil (2014) Jardim Botânico do Rio de Janeiro. http://www.floradobrasil.jbrj.gov.br. Accessed 29 Oct 2014

Loeuille B, Semir J, Lohmann LG, Pirani JR (2015) A phylogenetic analysis of Lychnophorinae (Asteraceae: Vernonieae) based on molecular and morphological data. Syst Bot 40:299–315
Messias MCTB, Leite MGP, Meira-Neto JAA, Kozovits AR (2011) Life-form spectra of quartzite and itabirite rocky outcrops sites, Minas Gerais, Brazil. Biota Neotrop 11:255–268
Monteiro HS, Vasconcelos PM, Farley KA, Spier CA, Mello CL (2014) (U–Th)/He geochronology of goethite and the origin and evolution of cangas. Geochim Cosmochim Acta 131:267–289
Mucina L, Wardell-Johnson GW (2011) Landscape age and soil fertility, climatic stability, and fire regime predictability: beyond the OCBIL framework. Plant Soil 341:1–23
Neves SPS, Conceição AA (2007) Vegetação em afloramentos rochosos na Serra do Sincorá, Chapada Diamantina, Bahia Brasil. Sitientibus 7:36–45
Neves SPS, Conceição AA (2010) Campo rupestre recém-queimado na Chapada Diamantina, Bahia, Brasil: plantas de rebrota e sementes, com espécies endêmicas na rocha. Acta Bot Bras 24:697–707
Oliveira RS, Galvão HC, Campos MCR, Eller CB, Pearse SJ, Lambers H (2015) Mineral nutrition of campos rupestres plant species on contrasting nutrient-impoverished soil types. New Phytol 205:1183–1194
Pedreira AJ (1997) Sistemas deposicionais da Chapada Diamantina centro-oriental, Bahia. Rev Bras Geoc 27:229–240
Pugliese L, Rapini A (2015) Tropical refuges with high phylogenetic diversity reveal contrasting phylogenetic structures. Int J Biodiv. doi:10.1155/2015/758019
Rando JG, Pirani JR (2011) Padrões de distribuição geográfica das species de *Chamaecrista* sect. *Chamaecrista* ser. *Coriaceae* (Benth.) H.S. Irwin & Barneby, Leguminosae – Cesalpinoideae. Rev Bras Bot 34:499–513
Rapini A, Mello-Silva R, Kawasaki ML (2002) Richness and endemism in Asclepiadoideae (Apocynaceae) from the Espinhaço Range of Minas Gerais, Brazil – a conservationist view. Biodiv Cons 11:1733–1746
Rapini A, Ribeiro PL, Lambert S, Pirani JR (2008) A flora dos campos rupestres da Cadeia do Espinhaço. Megadiv 4:15–23
Ribeiro PL, Rapini A, Silva UCS, Konno TUP, Damascena LS, Berg C (2012) Spatial analyses of the phylogenetic diversity of *Minaria* (Apocynaceae): assessing priority areas for conservation in the Espinhaço Range, Brazil. Syst Biodiv 10:317–331
Ribeiro PL, Rapini A, Damascena LS, Berg C (2014) Plant diversification in the Espinhaço Range: insights from the biogeography of *Minaria* (Apocynaceae). Taxon 63:1253–1264
Rumble MA (1989) Wildlife associated with scoria outcrops: implications for reclamation of surface-mined lands. USDA Forest Service 285:1–6
Saadi A (1995) A geomorfologia da Serra do Espinhaço em Minas Gerais e de suas margens. Geonomos 3:41–63
Salgado AAR, Carmo FF (2015) 'Quadrilátero Ferrífero': a beautiful and neglected landscape between the gold and iron ore reservoirs. In: Vieira BC, Salgado AAR, Santos LJC (eds) Landscapes and landforms of Brazil, world geomorphological landscapes. Springer, Dordrecht, pp 319–330
Silva GA (2013) Florística e estrutura de campos rupestres com diferentes tempos desde o último incêndio, Chapada Diamantina, Brasil. MSc Dissertation, Universidade Estadual de Feira de Santana
Simon MF, Pennington T (2012) Evidence for adaptation to fire regimes in the tropical savanna of the Brazilian cerrado. Int J Plant Sci 173:711–723
Simon MF, Grether R, Queiroz LP, Skema C, Pennington RT, Hughes CE (2009) Recent assembly of the Cerrado, a neotropical plant diversity hotspot, by in situ evolution of adaptations of fire. Proc Natl Acad Sci USA 48:20359–20364
Stannard BL (ed) (1995) Flora of the Pico das Almas, Chapada Diamantina. Brazil, Royal Botanic Gardens, Kew
Van der Pijl L (1982) Principles of dispersal in higher plants. Springer, Berlin
Vegas-Vilarrúbia T, Nogué S, Rull V (2012) Global warming, habitat shifts and potential refugia for biodiversity conservation in the neotropical Guayana Highlands. Biol Cons 152:159–168

Zappi DC, Taylor NP (2012) Cactaceae. In: Jacobi CM, Carmo FF (eds) Floristic diversity of Quadrilátero Ferrífero Cangas. Código Editora, BR, pp 98–100

Zappi DC, Lucas E, Stannard B, Lughadha EN, Pirani JR, Queiroz LP, Atkins S, Hind DJN, Giulietti AM, Harley RM (2003) Lista das plantas vasculares de Catolés, Chapada Diamantina, Bahia, Brasil. Bol Bot Univ São Paulo 21:345–398

Chapter 7
Archipelago of Montane Forests Surrounded by Rupestrian Grasslands: New Insights and Perspectives

Marcel S. Coelho, G. Wilson Fernandes, Priscila Pacheco, Victor Diniz, Alline Meireles, Rubens M. dos Santos, Felipe A. Carvalho and Daniel Negreiros

Abstract In this chapter, we describe, for the first time, the natural islands of Atlantic rainforests, regionally known as *capões* (heretofore called Atlantic forest islands, or simply forest islands), associated to the Espinhaço Range under the perspective of the landscape context in which they are immersed. The structure, composition and reproductive aspects of the vegetation are analyzed, as are their soil properties. Forest islands of the Espinhaço Range have a similar floristic composition to the semi-deciduous forests of southeastern Brazil, that are associated to the Atlantic rainforest domain, despite having been classified in its climate

M.S. Coelho (✉) · G.W. Fernandes · P. Pacheco · V. Diniz · A. Meireles
F.A. Carvalho · D. Negreiros
Ecologia Evolutiva & Biodiversidade/DBG, ICB, Universidade Federal de Minas Gerais, 30161-970 Belo Horizonte, MG, Brazil
e-mail: marcel.s.coelho@gmail.com

P. Pacheco
e-mail: priscila.p.carlos@gmail.com

V. Diniz
e-mail: vdinizpinto@gmail.com

A. Meireles
e-mail: aline.meirelles01@hotmail.com

F.A. Carvalho
e-mail: felipecarvalho.email@gmail.com

D. Negreiros
e-mail: negreiros.eco@gmail.com

G.W. Fernandes
Department of Biology, Stanford University, Stanford, CA 94395, USA
e-mail: gw.fernandes@gmail.com

R.M. dos Santos
Departamento de Ciências Florestais, Campus Universitario Jardim Eldorado, Universidade Federal de Lavras, 37200000 Lavras, MG, Brazil
e-mail: rubensmanoel@dcf.ufla.br

© Springer International Publishing Switzerland 2016

G.W. Fernandes (ed.), *Ecology and Conservation of Mountaintop Grasslands in Brazil*, DOI 10.1007/978-3-319-29808-5_7

regime as ombrophilous vegetation. The forest islands of the Espinhaço Range are classified as disjunctions of broadleaved, evergreen, cloud, montane tropical forest located on mountain ridges, rocky slopes or swamps. Forest islands are always associated to mountains, which have concentrated rainfall and cloudiness. The forest islands are located in an ecological transition zone and are strongly influenced by elements of the Atlantic rainforest biome, despite the elements of the Cerrado biome (in the present case, in a matrix of rupestrian grasslands), which are present in minor proportions. The influence of the rupestrian grassland ecosystem increases according to the levels of disturbance suffered by the vegetation. The forest islands are an edaphoclimatic formation, hence dependent on specific climate and soil properties for their development. Soil physical properties, such as drainage, are more relevant than their chemical properties, like nutrient input, to the establishment and development of the forest islands. Among the several threats, fire stands out, causing large decreases in the dimensions of the forest and changes in the successional dynamics of the vegetation. These natural islands, which are mostly immersed in matrices of rupestrian grasslands, have a high biogeographical importance, especially as a refuge for the seasonal surrounding environment. However, they may be under a fast-paced threat due to recurrent anthropogenic fires and the potential effects of climate and land use changes.

Contents

7.1 Atlantic Forest Islands Associated with the Espinhaço Range

The Atlantic Rainforest, although dominated by montane rainforest, is a complex of vegetations which encompasses distinct formations. Among these, the following stand out: montane rainforest, Araucaria forest, riparian and isolated rainforests. Rainfall generally exceeds 2000 mm/year. The forest is located along a lengthy coastal mountain range following the Atlantic Ocean from Rio Grande do Sul to the northeast region. Its main or central area lies in the Mar and Mantiqueira Ranges, encompassing the states of São Paulo, Minas Gerais, Rio de Janeiro, and Espírito

Santo. Aside from these regions, the Atlantic Rainforest extends out through *capões* (heretofore called Atlantic forest islands, or simply forest islands) and galleries into mountain ranges (to the South, central region, and Northeast). On the coast and inland areas of the Northeast, there are remnants of the Atlantic rainforest in every state, whether in inland pockets or on dune and plains formations along the coast (Rizzini 1997). Dense, moist, high-mountain forests are mesophanerophytic formations approximately 20 m in height, found at the tops of high mountains on litholic soils and peat accumulations in the depressions where the forest is located. These forests are comprised of phanerophytes with thin branches and trunks, small and leathery leaves, and thick, fissured bark. The species composition includes families with wide geographical distributions, although many species are largely endemic and reveal the former isolation of the refuge, which can be called a cloud forest.

As a result of climate change, the montane forests are home to particular characteristics arising from their biological and geographic history, which together with local factors, such as topography, form a gradient of unique vegetative formations with distinct ecological processes characterized by a high degree of endemism (Gentry 1995; Webster 1995). Compared with lowland forests, montane forests have a lower canopy, a smaller richness of angiosperms, and a greater abundance of epiphytes, with a significant representation of Araceae, Orchidaceae, Bromeliaceae, Pteridophyta, and bryophytes. This elevated representativeness of epiphytes is a consequence of the high humidity in these formations (Whitmore 1990; Van der Hammen 1995; Webster 1995).

In his treatise on the phytogeography of Brazil, Rizzini (1997) gave more attention to the gallery forests, but described the forest islands as related to savannah vegetation. The forest islands can be connected to the rainforests of the northeastern mountain ranges, which are isolated among the Caatinga (seasonally dry forests). In these regions, despite the climate favoring other types of vegetation, the altitude and local conditions (especially the substrate) are the elements which facilitate the establishment of forest islands. They are found in water-soaked soils when a formation similar to the montane rainforest develops, especially with the presence of mesophilic and macrophilic herbaceous plants and epiphytes, as well as the palm *Euterpe edulis*. To date, the classification of Brazilian vegetation by Veloso et al. (1991) has not addressed forest islands as a characteristic vegetation type. However, these authors cite the existence of enclaves, or disjointed areas that connect to form mosaics. These mosaics are comprised of areas situated between two isolated ecological regions, with them being exclusively defined by sketched out maps and continually dependent on scale, because on a smaller scale they can always be separated. Therefore, it is not difficult to demarcate this type of vegetation in transitional soils, either by similar vegetation type or by structurally distinct types, such as moist forests/savannah (Cerrado), for example, where forest islands can be found. The classification proposed by Oliveira-Filho (2009) neither explicitly deals with such a separation, nor specifically includes forest islands.

There are few studies that provide information about forest islands associated with the Espinhaço Range; most are dissertations and theses that might not be ever published (Campos 1995; Meguro et al. 1996a, b; Santos 2009; Souza 2009;

Valente 2009; Coelho 2014). All of these works focus on the study of structure and composition (Campos 1995; Meguro et al. 1996a, b; Souza 2009; Valente 2009; Coelho 2014), while only three discuss the edaphic aspects of this formation (Meguro et al. 1996a, b; Valente 2009; Coelho 2014). Because almost all of the studies (Campos 1995; Meguro et al. 1996a, b; Souza 2009; Valente 2009; Coelho 2014) of forest islands environments in the Espinhaço Range were carried out in the Serra do Cipó, and since this region is very similar to other locations in the Espinhaço Range, henceforth in this work we will use it as our model for analysis. Nevertheless, relevant conclusions obtained by Souza (2009) from sampling conducted in forest islands associated with Rio Preto State Park (PERP) and by Meguro et al. (1996b) in the municipality of Datas, on the Diamantina plateau, will be discussed when appropriate.

7.2 The Cipó Mountains in a Landscape Context

The mountains of Serra do Cipó occupy a zone of transition, or ecological tension (ecotone), between the Cerrado and the Atlantic Rainforest, which is one of the reasons for its high level of biological diversity (e.g., Giulietti et al. 1987; Ribeiro et al. 2009; Fernandes et al. 2014; Chap. 6). The distribution of vegetation forms true mosaics of predominantly grassland vegetation with patches of shrubby Cerrado together with forest formations (along its western side), while the eastern part hosts more homogeneous forest vegetation. The grassland regions of the western portion are also not uniform. The high levels of topographic and edaphic complexity generate at least four distinct vegetation types. These were classified by Carvalho et al. (2012) as: (1) rupestrian grasslands, (2) sandy bogs, (3) peat bogs, (4) rocky outcrops (5) and quartz gravel grasslands (see review in Chap. 6). In all its complexity, grassland vegetation covers more than 40 % of the Serra do Cipó National Park (a representative conservation unit in the region), while rupestrian grasslands covers approximately 20 %, Cerrado vegetation (sensu stricto, drained, and typical Cerrado) covers approximately 18 %, and forests (semi-deciduous seasonal forests, dense Cerrado, and *Eucalyptus* and pine plantations) cover another 18.8 % (Peloso 2009; Peloso and Shimabukuro 2010). However, Ribeiro et al. (2009) recalculated the size of the Atlantic Rainforest in the Serra do Cipó National Park to 25 % because of the following indicators: (1) past reports and contemporary statements, mapping of indicator species (e.g., *Euterpe edulis*, *Cecropia hololeuca*, *Attalea oleifera*), (2) proportion of forest remnants, (3) geographic distribution of *Vellozia gigantea*, and (4) the front of stationary nebulosity. The main criteria for the demarcation of the areas of Cerrado and Atlantic Rainforest was the division of drainage lines defined by the mountain ridges, as they drain into the Rio Doce river basin (on the eastern side) and towards the Rio São Francisco river (western side). Furthermore, the zone of influence of the Atlantic tropical mass plays an important role in forming an area of stationary nebulosity in the highest regions of Serra do Cipó. These are concentrated on the eastern faces, but their influence extends beyond the ridges to cover higher altitudinal zones on the western side. Cloud cover is proportional to the action of this air mass; in other words, the greater its intensity, the more clouds are present (Peloso 2009; Ribeiro et al. 2009; Peloso and Shimabukuro 2010). An interesting pattern was detected by Peloso (2009) through monochromatic views of TM images during two periods of drought in 1990 and 2000, and TM imaging from the end of the dry season and beginning of the rainy season in 2008, as well as a color composition of R(5)G(4)B(3) for better observation of the phenomenon. In all of the images, the pattern is clear: even in the dry season, the region exhibits high humidity, with more on the east side than on the west (Peloso 2009; Ribeiro et al. 2009). Therefore, from the strong influence of the nebulosity generated by the intensity of the tropical Atlantic mass, it is possible to detect a concentration of clouds on the eastern side of the ridge advancing into the highest areas of the western side, as well as low temperatures directly related to a higher humidity (Figs. 7.1 and 7.2). This phenomenon occurs on a daily basis, with humid oceanic masses concentrating throughout the early morning and dispersing at around 10:00 am (Valente 2009; unpublished data—*Long-Term Ecological Research Program Project—PELD-Cipó Site*). Similar orographic effects occur in the differentiation of plant composition in geographically close areas. Along these lines, in the Mantiqueira Range (Ab`Saber 2003) and Serra do Mar Range (Durigan et al. 2008), differences were detected in the composition of species; these differences were clearly related to the humidity and temperature gradients caused by the orographic effects.

The forest islands in the mountains of Serra do Cipó investigated by Coelho (2014) are located in an altitudinal stratum above 1240 m. These forest islands (Fig. 7.3) were used as a model to analyze the matrix of their surroundings. This

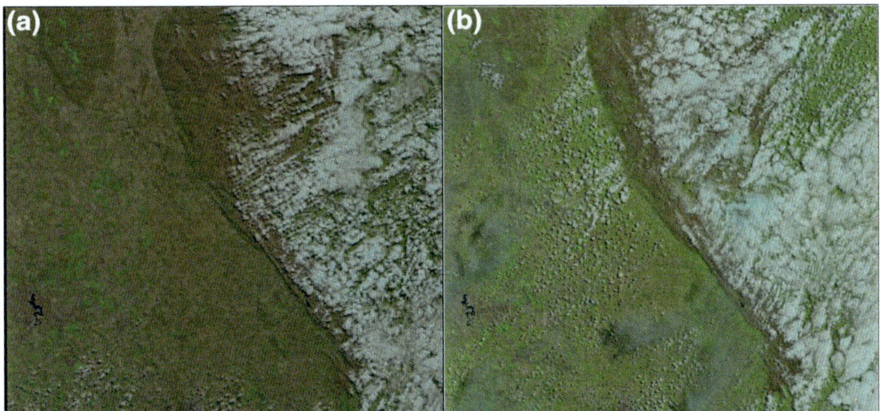

Fig. 7.1 LANDSAT images (resolution = 30 m) of the windward and leeward sides, with accumulation of stationary nebulosity from the Atlantic Tropical Mass in the region of Serra do Cipó, Minas Gerais: **a** Date: July 24, 2013, ID LC82180732013205LGN00, **b** Date: March 5, 2014, ID LC82180732014064LGN00. *Source* http://glovis.usgs.gov

Fig. 7.2 Image of a *Atlantic forest islands* system in Serra do Cipó, immersed in fog. Date of photograph: June 23, 2013. Geographic coordinates of the *atlantic forest islands*: 0656853–7874458 UTM. Elevation: 1322 m. *Source* Private collection—Marcel S. Coelho

was done through classification using land use and coverage data with the aid of SPRING software (http://www.dpi.inpe.br/spring/); this program processed Rapid Eye images from 2010, acquired by the State Institute of Forests (IEF). The image has a resolution of 5 m and is composed of five spectral bands. For classification purposes, the composition of spectral bands was 5, 4, and 3. The location used in the classification has the following UTM coordinates: 643148.57/7879094.83 and 666213.57/7863134.83. The classification was done after rules of contrast were applied. The contrast permitted the emphasis of land use typologies. Seven classes were defined, for which polygon samples were collected and subsequently validated during five field visits. The results of the classification can infer that the forest islands are located within a setting of predominantly rupestrian grasslands dominated by herbaceous and schlerophyllous vegetative cover, elements which are associated with the Cerrado (Fig. 7.4). Around the forest islands, it is also possible to observe mono-dominant vegetation made up of Candeia (*Eremanthus erythropappus* (DC.) Mac). By assessing the location of these forest islands associated with the basins and sub-basins of the Rio São Francisco and Rio Doce rivers (source: Minas Gerais State Water Management Institute), it can also be stated that they are closely associated with headwaters, rivers, creeks, and small streams, and it can be inferred that there are no water restrictions in these plant formations (Figs. 7.5 and 7.6).

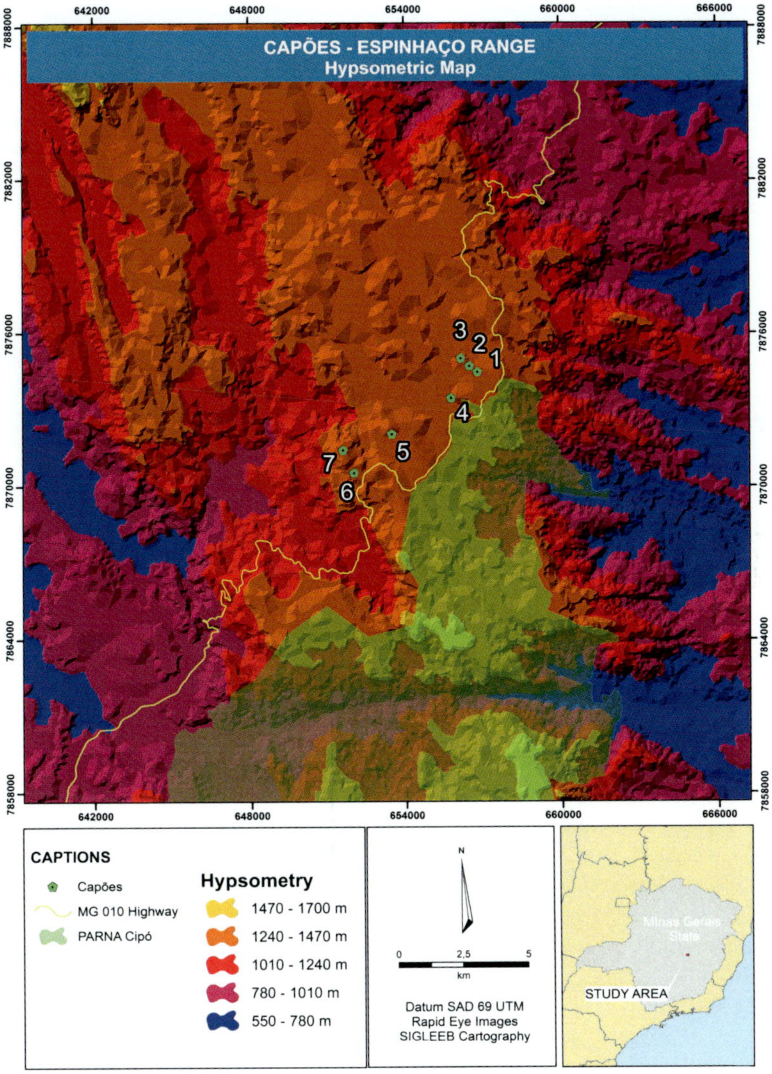

Fig. 7.3 Altitudinal strata in the region of Serra do Cipó, highlighting the location of seven atlantic forest islands sampled by Coelho (2014) in the higher altitudinal strata (1240–1470 m)

7.3 Structure and Floristic Composition

The structure and composition of the vegetation comprise the goals of pioneering studies, in attempts to classify vegetative formations using systematically collected data. Very few studies have been conducted in forest islands associated with the Espinhaço Range, compared with those carried out involving other plant formations

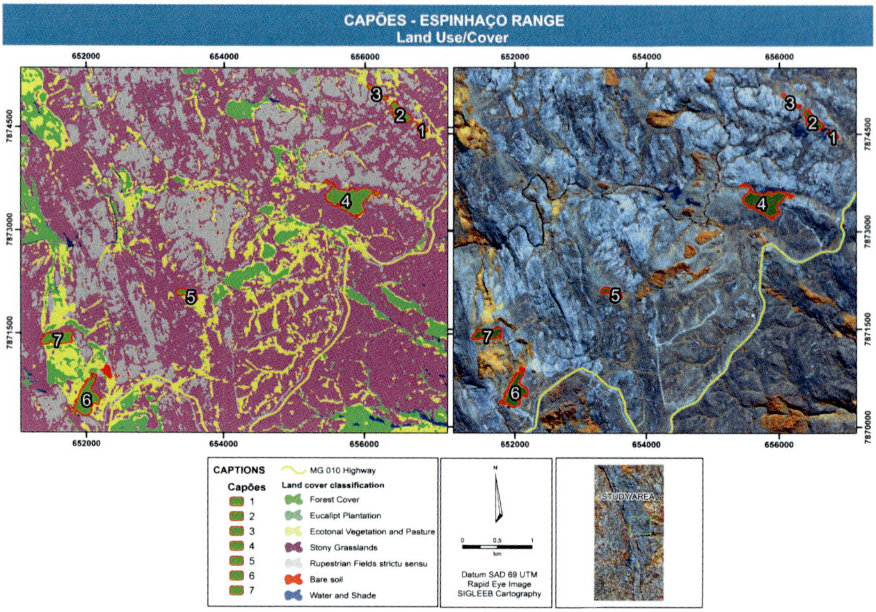

Fig. 7.4 Map of land use and soil cover in Serra do Cipó, showing the seven Atlantic forest islands sampled by Coelho (2014)

(e.g., gallery forests, seasonal semi-deciduous forests, moist forests, seasonal deciduous forests). All of these utilize an approach that focuses on identifying patterns of structure and flora composition, revealing the pioneering nature that still exists and which is also associated with the lack of systematic studies on this type of vegetation (Campos 1995; Meguro et al. 1996a, b; Souza 2009; Valente 2009; Coelho 2014). It is important to highlight the differences in sampling the arboreal components within these studies. While Campos (1995), Valente (2009), and Souza (2009) sampled individuals with diameter at breast height (DBH) greater than or equal to 5 cm, Meguro et al. (1996a, b) and Coelho (2014) sampled individuals with DBH greater than 2.5 cm.

According to Meguro et al. (1996a), the canopy height of an forest island of approximately 2 ha in size, analyzed in the Serra do Cipó region, was approximately 12–16 m in the central region, decreasing to 6–8 m in border areas, with an average height of 7.08 m, and showed a variation in the secondary stratum between 4 and 8 m. Campos (1995) sampled the structure of three small forest islands also in the Serra do Cipó, and recorded heights of 10–12 m in the first forest island, with an area of 0.8 ha, while in the second (area of 2.0 ha) the dominant tree heights were 12–15 m. Also in Serra do Cipó, Valente (2009) sampled two forest islands with areas of 14.5 ha and 35.3 ha, respectively. The first featured a canopy 20 m tall, with some areas reaching 30 m; the second had a canopy of 25 m, with emergent trees up to 30 m tall. By sampling seven forest islands of various sizes and shapes

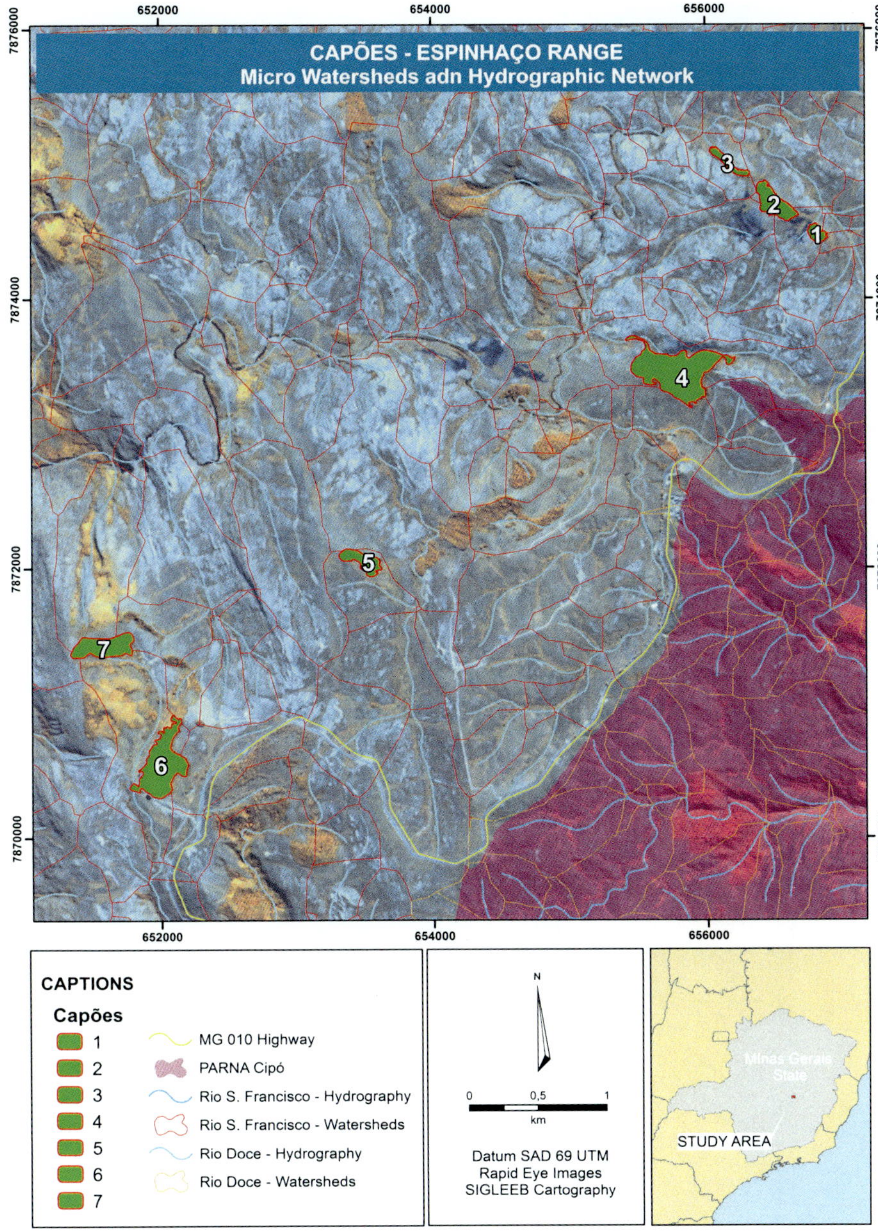

Fig. 7.5 Location of the seven atlantic forest islands sampled by Coelho (2014) in Serra do Cipó associated with the basins and sub-basins of the Rio São Francisco and Rio Doce rivers (*Source* Minas Gerais State Water Management Institute)

Fig. 7.6 External and internal views of the Atlantic forest islands in Serra do Cipó

that ranged from 1.6 to 16.9 ha, Coelho (2014) recorded a variation in mean heights from 6.21 to 8.20 m, and emergent individuals approximately 18.3 m tall. Additionally, Souza (2009) stated that the forest islands in PERP are characterized by canopies with many clearings and very irregular heights of approximately 10 m, with some emergent trees reaching 15 m. In this same study, Souza (2009) surveyed several sampling plots in central areas and on the edges of the forest islands. The mean height values were 6.82 and 4.83 m, respectively.

The basal area of the forest island sampled by Meguro et al. (1996a) was 31 m^2/ha. Meanwhile, Coelho (2014) recorded distinct values ranging from 18 to 41.2 m^2/ha. And Souza (2009) recorded very different mean basal areas between the edges and the central areas in the PERP forest islands, 15 m^2/ha and 51 m^2/ha, respectively.

The broad structural variations (in height and basal area) that are observed in forest islands may be related to the wide variety of environmental conditions associated with the mountains, including edaphic and topographical conditions. Additionally, the history of land use and fire may be yet another influence on the forest structure, especially in areas of ecological tension, such as the drainage zones of the Cipó mountains. Fire frequently plays a distinct role in Cerrado ecosystems, compared to formations associated with the Atlantic Rainforest. Various studies have demonstrated the influence of topographical and edaphic characteristics (e.g., Oliveira-Filho et al. 1994a, b, 1997, 2001, 2004; Carvalho et al. 2000a, b, 2005; Botrel et al. 2002; Toniato and Oliveira-Filho 2004; Pinto et al. 2005; Budke et al. 2007; Pereira et al. 2007; Higuchi et al. 2008; Budke et al. 2010; Machado and Oliveira-Filho 2010; Sanchez et al. 2013), the effect of fire (Cochrane 2003; Ivanauskas et al. 2003; Silva et al. 2005), and the impact of historical land use (Pereira et al. 2007) in the structure and vegetative composition of formations associated with the Atlantic Rainforest (see also Ross et al. 2002). Fire can also play an important role in maintaining the vegetative formations at intermediate or even initial stages of succession (Castellani and Stubblebine 1993).

Even with the different methods of analysis and sampling efforts, comparison of floral richness is also valid, along with the species abundance among the different forest islands associated with the Espinhaço Range. Meguro et al. (1996a) recorded 32 tree species and a density of 4629 ind/ha. The densities recorded in the forest islands sampled by Coelho (2014) ranged from 1990 to 4620 ind/ha. The richness of tree species ranged from 39 to 69, with total richness of 143 species. The overall species richness, obtained by compiling all the studies conducted on this type of forest in the Serra do Cipó, was 243 tree species. In a survey of individuals with DBH greater than or equal to 5 cm, Souza (2009) recorded a density lower than the values sampled by Meguro et al. (1996a), which were 1407.2 ind/ha on the edges and 2889.7 ind/ha in the central areas of the forest islands associated with PERP. The richness of tree species by forest island ranged from 20 to 34 species, with a total richness of 93 tree species. Of these, 79 were recorded within the forest islands and 53 on their edges.

Because of the different sampling efforts and collection methodologies used, from this point onward, the data related to the floristic composition of the Cipó mountains (Campos 1995; Meguro et al. 1996a, b; Souza 2009; Valente 2009;

Coelho 2014) and those generated by Souza (2009) will be analyzed separately. Furthermore, the data from projects conducted in the Serra do Cipó region will be compiled (since this region is the study model) and compared to the results obtained by Souza (2009), when appropriate.

The families with the greatest species richness in the forest islands from Serra do Cipó were Myrtaceae, Lauraceae, and Melastomataceae, followed by Fabaceae, Asteraceae, and Rubiaceae (Fig. 7.7). In the PERP region, the families with the greatest species richness in the central region of the forest islands were Myrtaceae, Lauraceae, and Myrsinaceae, followed by Asteraceae, Melastomataceae, and Cyatheaceae. The families with the highest importance value (IVI) were Myrtaceae, Anacardiacae, and Phyllanthaceae, followed by Clusiaceae, Magnoliaceae, and Monimiaceae. The families richest in species on the edges were Asteraceae, Myrtaceae, and Melastomataceae, followed by Cyatheaceae, Myrsinaceae, and Burseraceae, while those with the highest IVI were Myrsinaceae, Anacardiaceae, and Myrtaceae (Souza 2009).

The genera most rich in species in the Cipó mountains were: *Ocotea* (Lauraceae), *Miconia* (Melastomataceae), and *Myrcia* (Myrtaceae), followed by *Nectandra*

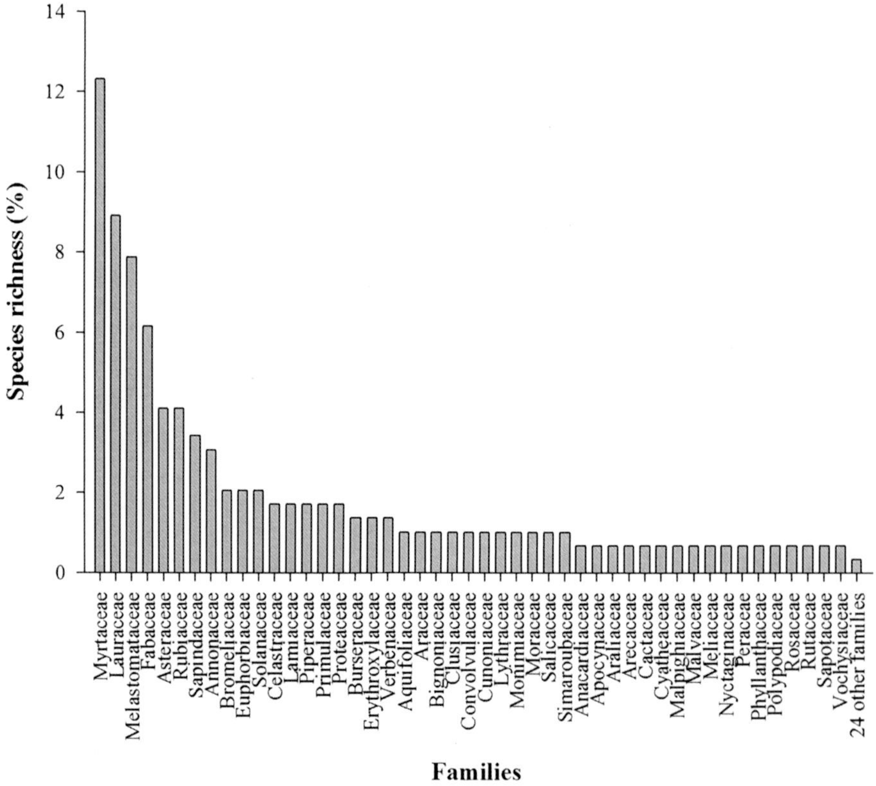

Fig. 7.7 Species richness per family in Atlantic forest islands in Serra do Cipó

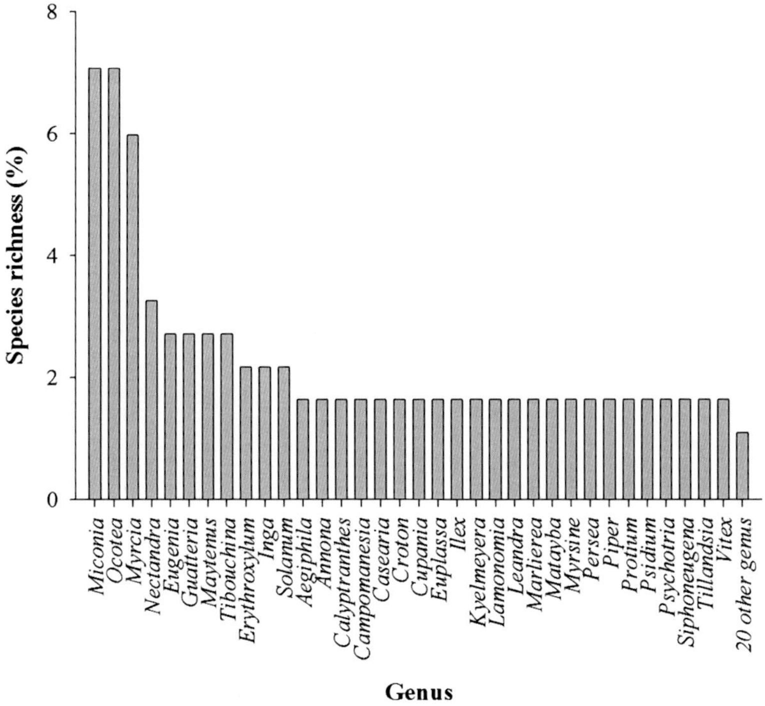

Fig. 7.8 Species richness per genus in Atlantic forest islands in Serra do Cipó

(Lauraceae), *Eugenia* (Myrtaceae), and *Guatteria* (Annonaceae) (Fig. 7.8). Meanwhile, in PERP the most significant genera were *Myrcia* (Myrtaceae), *Miconia* (Melastomataceae), *Ocotea* (Lauraceae), *Cyathea* (Cyatheaceae), and *Myrsine* (Myrsinaceae). In a robust review, where botanical inventories of 125 forest areas fed a database of over 2532 species, Oliveira-Filho and Fontes (2000) identified the variations in floristic composition in moist and semi-deciduous forests throughout their geographical and climatic variations. This work identified a high floristic similarity between seasonal semi-deciduous forests and moist forests. Many families and genera are species-rich in both formations, such as Myrtaceae, Melastomataceae, Lauraceae and Rubiaceae. According to Oliveira-Filho and Fontes (2000), the composition of families in the Cipó mountains and PERP are very similar in both formations (Oliveira-Filho and Fontes 2000; Souza 2009).

The same occurs for certain genera in the forest islands, which host species that are common in both moist and semi-deciduous forests; an example is the high representativeness of the genera *Miconia*, *Myrcia*, and *Ocotea.* However, the high representativeness of the genera *Campomanesia* (Myrtaceae), *Guatteria* (Annonaceae), *Ilex* (Aquifoliaceae), and *Nectandra* (Lauraceae), which are very representative only in seasonal semi-deciduous forests, indicate greater floristic similarity between these formations and the forest islands associated with the Cipó mountains. Due to the floristic composition of seasonal semi-deciduous forests, these can be considered a subset or part of the moist forests. Distance from the ocean appears to be an important driver in the floristic differentiation of the forests associated with the Atlantic Rainforest domain. As moist forests are located farther from the coast, some species do not seem to tolerate the decreased moisture and availability of water. Some species acclimatize, and their leaves are seen to become deciduous. Therefore, inland forests have similar floristic composition as a result of this stress, which is characterized by the seasonal impact of the climate (Oliveira-Filho and Fontes 2000). Despite the importance of altitude in the floristic composition of the semi-deciduous and moist forests, and the cloudy nature of the landscape in which the forest islands are associated with the Espinhaço Range (Peloso 2009; Ribeiro et al. 2009; Peloso and Shimabukuro 2010), the composition of the arboreal flora of the forest islands systems in the Cipó mountains appears to depend on the plants' dispersion sources (Oliveira-Filho et al. 1994a, b, 2004; Carvalho et al. 2000a, b, 2005; Oliveira-Filho and Fontes 2000; França and Stehmann 2004). Since the eastern region of the Cipó mountains is covered by seasonal semi-deciduous forests (Santos et al. 2011, 2012; Coelho 2012), it would be logical to think that the population source (where the seeds come from) for forest islands are such semi-decidious forests, as well as the vegetation associated with gallery forests that often connect forest islands. Because the landscape in PERP is very similar to the landscape of the Cipó mountains, dispersal and colonization processes seem to be similar (Souza 2009).

Although sampling efforts are still incipient, 31 % of the species recorded in forest islands in the Serra do Cipó mountains were comprised of shrubs, shrublet, vines, and herbs (Fig. 7.9). Furthermore, the forest islands in the Cipó mountains and PERP exhibit some elements of humid montane and high-montane moist

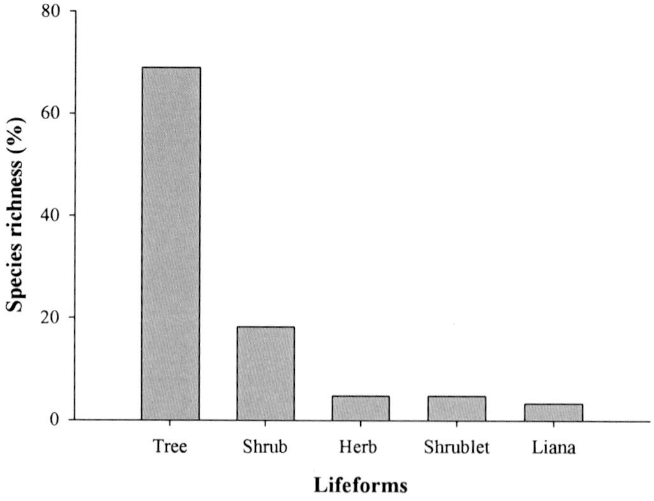

Fig. 7.9 Species richness of plants by life form in Atlantic forest islands in Serra do Cipó

forests, such as diverse epiphytic flora with the occurrence of the families Araceae, Bromeliaceae, and Orchidaceae, as well as bryophytes, ferns, and lichens. The presence of leafy and filamentous lichens, especially those pertaining to the genera *Usnea* and *Cladonia*, are very common (Campos 1995; Meguro et al. 1996a, b; Valente 2009). In the forest islands associated with PERP, Souza (2009) also highlighted the presence of herbaceous angiosperms, recording 37 species, 29 genera, and 11 families, with the families Orchidaceae, Araceae, Bromeliaceae, Piperaceae, and Poaceae being the most common. The same author also recorded 38 species of pteridophytes, with the families Polypodiaceae, Dropteridaceae, Hymenophyllaceae, and Aspleniaceae exhibiting the greatest species richness. The high diversity of both pteridophytes and herbaceous angiosperms is a strong indicator of an environment without water restrictions.

By analyzing the phytogeographic areas where all the species of the forest islands in the Cipó mountains occur (LEFB 2014), it can be stated that most species are characteristic of both the Atlantic Rainforest and the Cerrado, with a greater number of species unique to the Atlantic Rainforest compared to those exclusive to the Cerrado

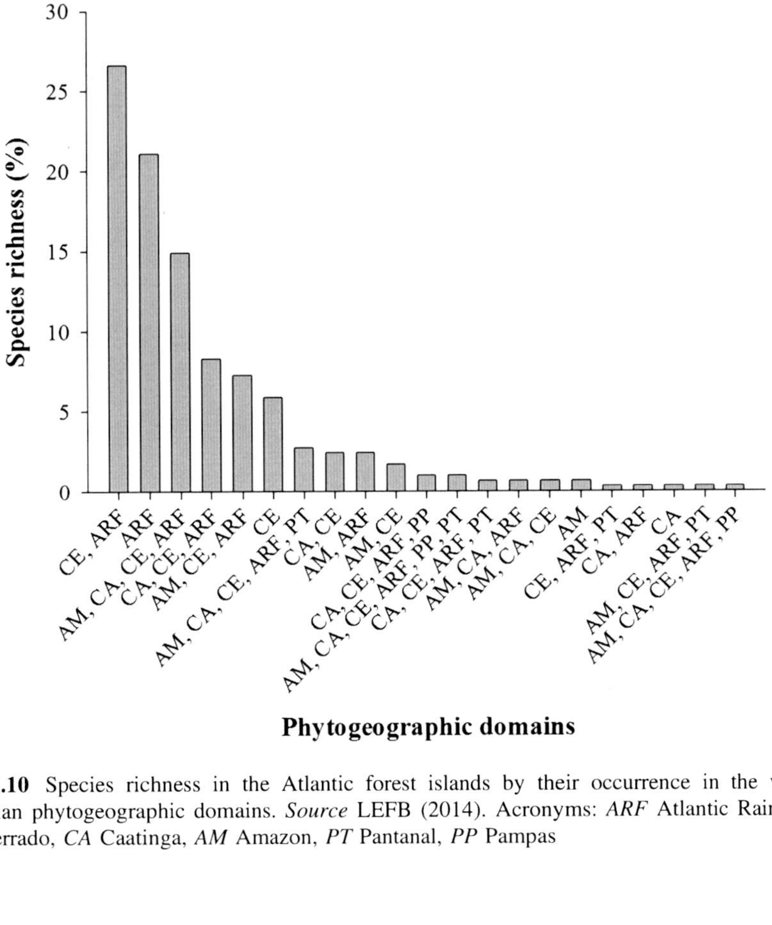

Fig. 7.10 Species richness in the Atlantic forest islands by their occurrence in the various Brazilian phytogeographic domains. *Source* LEFB (2014). Acronyms: *ARF* Atlantic Rainforest, *CE* Cerrado, *CA* Caatinga, *AM* Amazon, *PT* Pantanal, *PP* Pampas

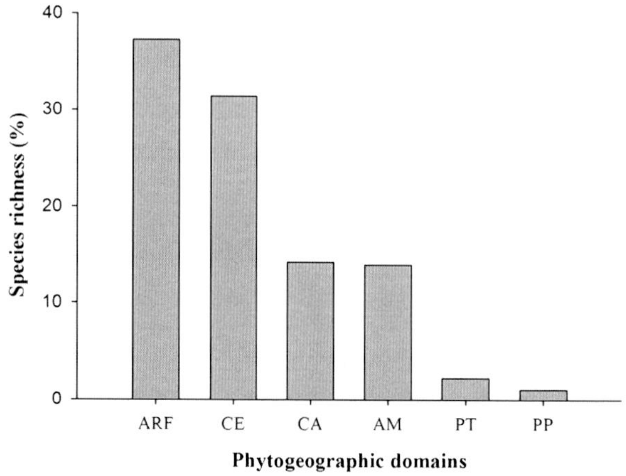

Fig. 7.11 Species richness in the Atlantic forest islands by their total occurrence in the various Brazilian phytogeographic domains. *Source* LEFB (2014). Acronyms: *ARF* Atlantic Rainforest, *CE* Cerrado, *CA* Caatinga, *AM* Amazon, *PT* Pantanal, *PP* Pampas

(Fig. 7.10). When the occurrence (presence/absence) of the species in each phytogeographic domain is considered, the Atlantic Rainforest predominates (Fig. 7.11).

Species richness alone is not sufficient to classify a type of vegetation as being associated with a specific phytogeographic domain. There are cases where one formation hosts greater species richness characteristic of a given domain, but the species with the highest phytosociological indexes (e.g., Importance Value Index, Value Coverage Index) are characteristic of a different phytogeographical domain. Therefore, this possibility was tested in the forest islands of the Cipó mountains, with three phytosociological parameters: (1) species richness, (2) abundance of individuals, and (3) coverage value index (%) (See Coelho 2014 for details). These parameters were taken from seven forest islands stands of the Cipó mountains where data were collected from individual trees with DBH greater than or equal to 2.5 cm (Fig. 7.4). The means of these three parameters for the Atlantic Rainforest were compared with the means of the Cerrado. Paired analyses (*t* test) were conducted to test for significant differences. The test was conducted on two scales. In the first, each *island* stand was analyzed separately, replicating the parameter values for the 10 sub-plots (10 × 10 m) in each of the seven forest islands ($n = 10$; Fig. 7.12). In the second, the forest islands were grouped and the replicates were the mean values of the parameters in each island ($n = 7$; Fig. 7.13).

On both scales, and for almost all of the tested parameters, the means were significantly higher for the Atlantic Rainforest. The only exceptions were the parameters of richness and of abundance for forest islands 5, 6, and 7, in which the means in the Cerrado did not vary significantly compared to the Atlantic Rainforest (Figs. 7.12 and 7.13).

Because they are located in transitional zones, commonly in ridge regions between the east and west sides of the Espinhaço Range, and are also ecotones between the Atlantic Rainforest and Cerrado environments, there is a need to understand the similarities or dissimilarities between the different vegetation physiognomies in this landscape. To this end, the Cipó mountains serves as an

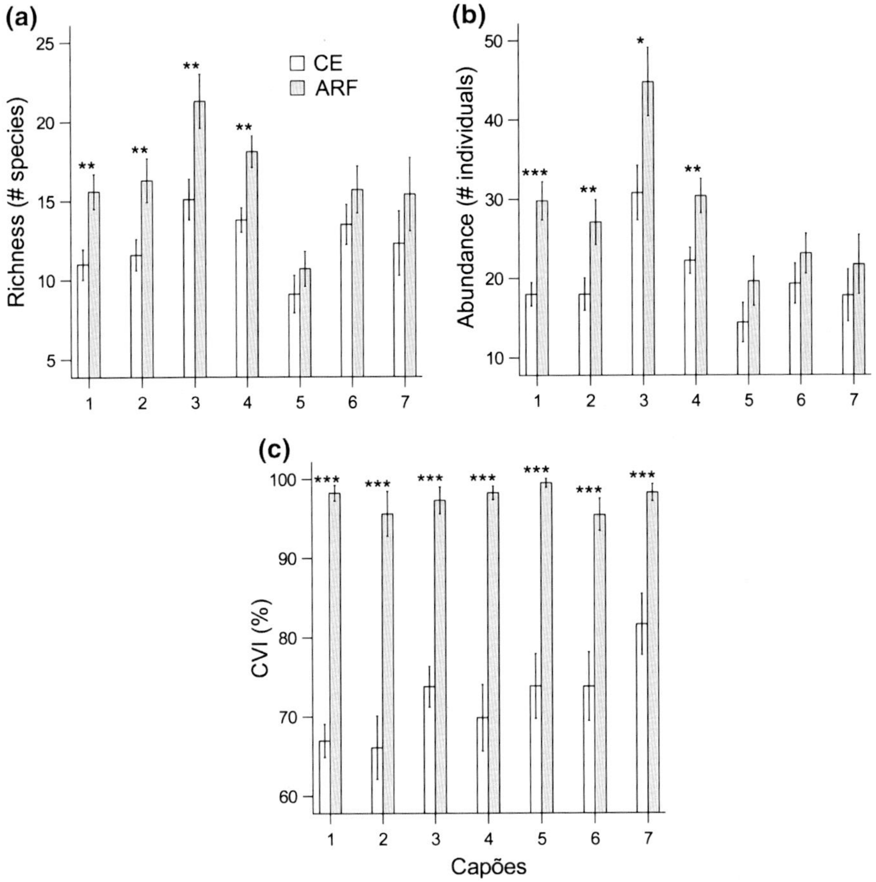

Fig. 7.12 Means and standard errors for three phytosociological parameters in the Cerrado (CE) and Atlantic Rain forest (AFR), for the seven Atlantic forest islands (**a** richness, **b** abundance, **c** cover value index) in the Cipó mountains. *Asterisks* indicate the level of significance (*$P < 0.05$; **$P < 0.01$; ***$P < 0.001$)

excellent model combining floristic studies on different vegetation types: seasonal semi-deciduous forests in its eastern area (Santos 2009; Santos et al. 2011, 2012), forest islands in regions above 1300 m in its western slopes and in regions close to the ridge line (Campos 1995; Meguro et al. 1996a, b; Souza 2009; Valente 2009; Coelho 2014), gallery forests (Meguro et al. 1996a, b) in its western slopes, and dry forests on limestone outcroppings at low altitudes also in its western region (Coelho et al. 2012). Furthermore, it is important to detect patterns of floristic similarity among the physiognomies, using the only floristic study conducted on forest islands associated with the Espinhaço Range outside the Cipó mountains, which was conducted in PERP by Souza (2009). This was achieved by building a grid showing the presence and absence of tree species in the following environments: (1) *Capões*

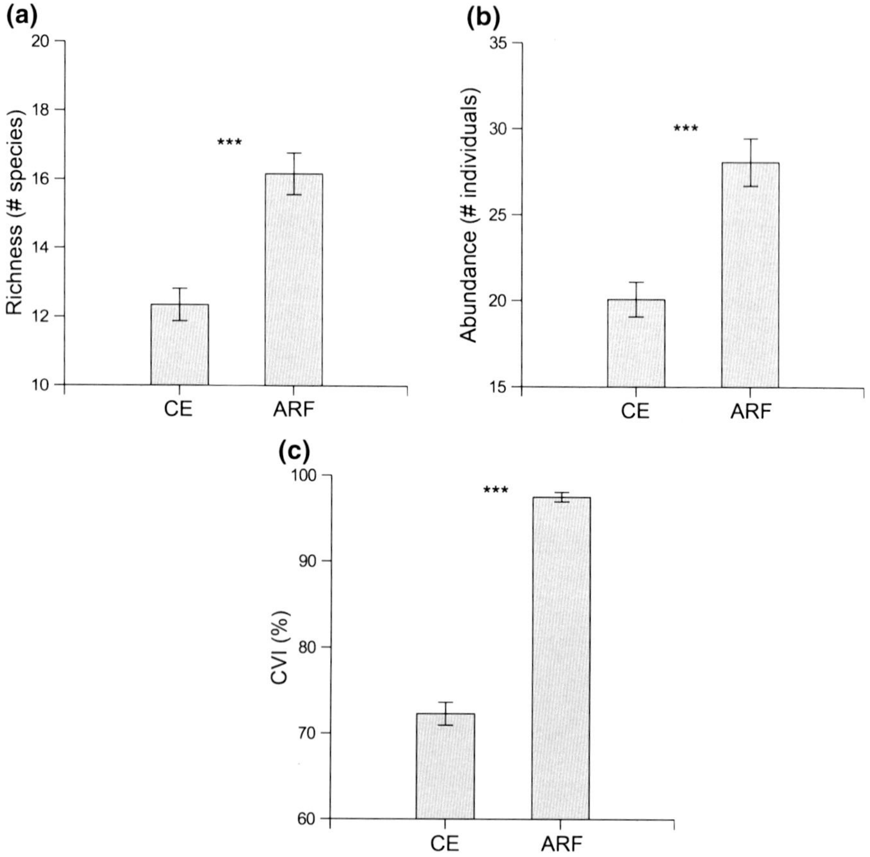

Fig. 7.13 Means and standard errors for three phytosociological parameters in the Cerrado (CE) and Atlantic Rainforest (AFR), for the compiled data from the Atlantic forest islands (**a** richness, **b** abundance, **c** cover value index) for the Cerrado and the Atlantic rain forest domains in the Cipó mountains. *Asterisks* indicate the level of significance (***$P < 0.001$)

in the Cipó mountains, (2) *Capões* associated with PERP, (3) Gallery forests, (4) Atlantic Rainforest (seasonal semi-deciduous forests), (5) Dry forests on limestone outcroppings. With the exception of the *capões* associated with PERP, all of the environments are part of the complex landscape comprising the Serra do Cipó. Dissimilarity analyses were used to compare the environments, according to Legendre and Legendre (1998). Jaccard dissimilarity (1-Jaccard similarity) was obtained using the "vegan" package (Oksanen 2009), and two-dimensional projection of this dissimilarity matrix was created by analyzing the principal coordinates (PCoA) using the "ade4" package (Dray and Dufour 2007), both in the R environment (R Development Core Team 2013).

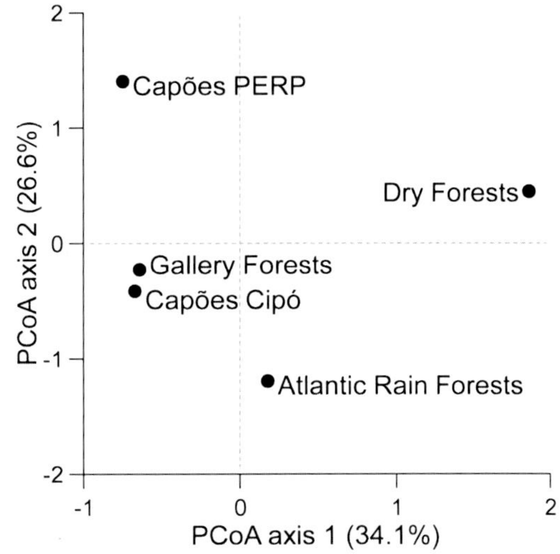

Fig. 7.14 Two-dimensional projection of the dissimilarity matrix analyzing the principal coordinates (PCoA). The vegetation formations are the *capões* of the Cipó mountains, *capões* associated with PERP, gallery forests, Atlantic Rainforest (seasonal semi-deciduous forests), and dry forests on limestone outcroppings. Eigenvalues are indicated in *parentheses*

Table 7.1 Jaccard dissimilarity matrix showing the relationship between vegetative formations

	Capões Cipó	*Capões* PERP	Gallery forests	Atlantic rainforest	Dry forests
Capões Cipó	0				
Capões PERP	0.852	0			
Gallery forests	0.746	0.847	0		
Atlantic rainforest	0.830	0.930	0.848	0	
Dry forests	0.967	0.994	0.959	0.929	0

The results indicated greatest similarity between the forest islands associated with the Cipó mountains and the gallery forests in its western area. Second highest values in the similarity ranking were obtained between the forest islands of the Cipó mountains and the Atlantic Rainforest (Fig. 7.14). Highest dissimilarity was always obtained when comparing any vegetation type with dry forests located on limestone outcroppings (Table 7.1; Fig. 7.14).

The results show that, although they are located between two important phytogeographic domains, and present elements of both Cerrado and Atlantic Rainforest, the floristic, composition, structural and phytosociological characteristics of the forest islands of the Cipó mountains are more associated with the Atlantic Rainforest domain. However, because forest islands are mostly immersed in an array of sandy, and rocky grasslands (see Fig. 7.4), it is possible that ruderal species already established in the surroundings and common to the Cerrado have favoured the colonization of those typical Cerrado species in the forest islands. This might have been possibly facilitated by the increasing intensity of disturbances, either natural (strong gusts of wind associated with concentrated rain, which are very frequent events in these areas) or anthropogenic (clearing, fire, or the presence of livestock).

7.4 Reproductive and Phenological Characteristics

The first work that sought to identify the reproductive characteristics of the species in the forest islands of the Espinhaço Range was carried out by Campos (1995) in three small forest islands stands where 66 % of species utilized zoochory as their main dispersal mode. Only 29 % of the species relied on anemochory as their main dispersal mechanism, and these plants were characteristically located on the border regions, or were vines, tall trees, or emerging species. Also in the Cipó mountains and the Diamantina plateau, species were predominantly zoochoric, followed by barychoric species and a few that utilized wind (Meguro et al. 1996b). Corroborating the previous work, Coelho (2014) recorded the tree species in the forest islands associated with the Cipó mountains as predominantly zoochoric (Fig. 7.15).

As for pollination in the tree species recorded by Coelho (2014), there is a clear predominance of pollination mechanisms mediated by animals (zoophily), compared to those mediated by wind (anemophily). Among the various pollination syndromes, the following were most notable, in descending order of species richness: melittophily (Hymenoptera, especially bees), undetermined, cantharophily (Coleoptera), psychophily (Lepidoptera, butterflies), phalaenophily (Lepidoptera, moths), myophily (Diptera), chiropterophily (bats), and ornithophily (birds) (Fig. 7.16).

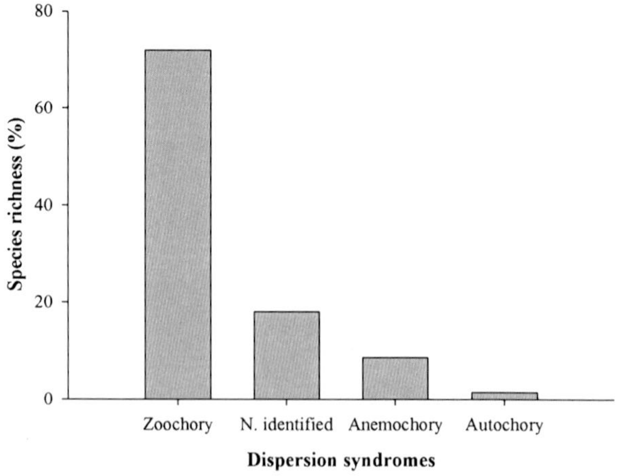

Fig. 7.15 Percentage of zoochoric, anemochoric, and autochoric tree species in Atlantic forest islands associated with the Cipó mountains, sampled by Coelho (2014)

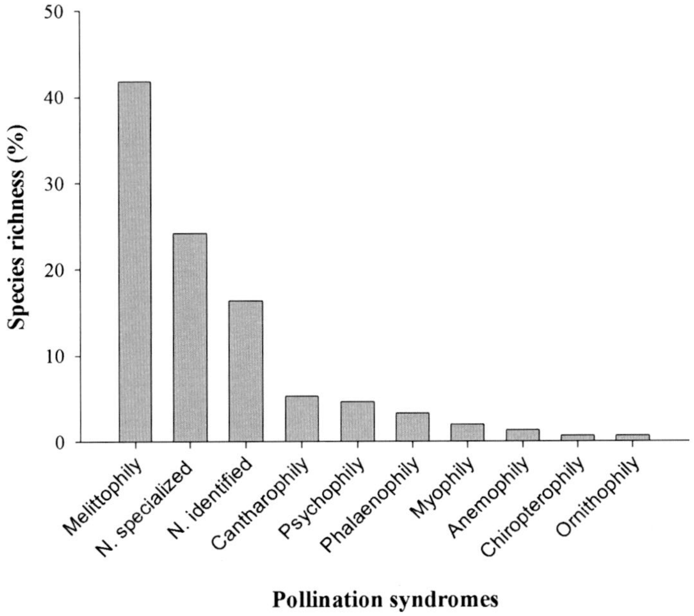

Fig. 7.16 Percentage of species by pollination syndrome category for tree species in Atlantic forest islands associated with the Cipó mountains, sampled by Coelho (2014)

It is possible that the population source for establishing forests in grassland formations on mountaintops is the drainage pathways that cut through the surfaces of the mountain ridges and slopes (Rizzini 1997). However, it would also be appropriate to consider that the species which are dispersed and pollinated by animals would have greater success in moving from the low slopes to mountaintops than those colonized by wind. Even so, only reaching the mountaintop areas is not sufficient for successful establishment. If this were the case, mountaintops (especially those associated with the Espinhaço Range) would be covered by forest formations and not by rupestrian grasslands (Fig. 7.4). These species must find adequate soils that allow trees to become established and develop (see Sect. 7.5).

Although there are as yet no published studies related to the phenology of the communities of forest islands associated with the Espinhaço Range, there is ongoing discussion specifically related to the level of deciduousness. In the most recent attempt to classify South American cis-Andean tropical and subtropical vegetation types, Oliveira-Filho (2009) highlighted climate among the classification attributes. Consequently, the climatic regimes specific to each vegetation type are particularly defined by: (1) duration of the dry season, (2) annual rainfall, (3) annual thermal amplitude, and (4) horizontal precipitation (the proportion of precipitation resulting from water trapped in the clouds coming into contact with vegetation). These climatic attributes are useful to define the next vegetation types: coastal, semi-arid, seasonal, rain, and cloud. The forest islands of the Espinhaço Range are often associated with rivers, streams, and springs, and are also located at high altitudes with strong orographic effects, subject to high rates of relative humidity, as well as rainfall (e.g., Campos 1995; Meguro et al. 1996a, b; Rizzini 1997; Peloso 2009; Ribeiro et al. 2009; Souza 2009; Valente 2009; Peloso and Shimabukuro 2010; Coelho 2014; See Figs. 7.2, 7.3, 7.4 and 7.5). Because of this association, this type of vegetation is not subject to water restriction during any period of the year. According to Oliveira-Filho and Fontes (2000), the arboreal flora of semi-deciduous forests can be considered a subset of moist forest flora, probably a subset of species which can survive a longer dry season. Therefore, as the forests associated with the Atlantic Rainforest move inland, some species cannot move with them because they cannot tolerate the climatic seasonality, a fact which modifies the floristic composition between moist forests near the coast and semi-deciduous moist forests located inland. With this distancing from coastal areas, not only the floristic composition changes, but also the phenological behavior of some species is modified. In environments with marked seasonality, some species become deciduous as they acclimatize to water restrictions. However, islands of forests located at high altitudes far from the coast do not suffer from water restrictions. In addition to distance from the coast, one of the main variables influencing the composition of vegetation formations associated with the Atlantic Rainforest is geographical distance from sources of colonizers (i.e., other forests) (Oliveira-Filho et al. 1994a, b; Van den Berg and Oliveira-Filho 2000; Gonzaga et al. 2008). The forest islands associated with the Espinhaço Range have diverse epiphyte flora, with greater representation of the Araceae, Orchidaceae, and Bromeliaceae families, as well as bryophytes, and pteridophytes, with greater representation of the Polypodaceae, Dropteridaceae, Hymenophyllaceae, and Aspleniaceae families (Campos 1995; Meguro et al. 1996a, b; Souza 2009; Valente 2009). Furthermore, in a study of forest communities, the phenology of forest islands in the Cipó mountains was recorded for one year (Coelho 2014). Although some species were deciduous, the community did not lose more than 30 % of foliar coverage during the winter (dry and cold season), which would classify this phytophysiognomy as semi-deciduous (Oliveira-Filho 2009). This fact indicates that the vegetation in the forest islands of the Cipó mountains is mostly evergreen or moist based on its foliar renewal regime, despite the fact that the vegetation shows greater floristic similarity with semi-deciduous forests. This most likely results from the short geographical distance separating them, compared to the moist forests near the coast.

7.5 Soils of the *Atlantic Forest Islands*

In a study on the soil gradient in different vegetative physiognomies, Valente (2009) classified the soils of the forest islands associated with the Cipó mountains into two classes: red-yellow latosols derived from metapelitic rock (phyllite), and spodosols derived from the weathering of quartzite. In general, the soils are chemically acid, have low natural fertility, and have elevated levels of aluminum saturation associated with low rates of base saturation. Soils ranged in texture from clayish to sandy clay loam, sandy loam, and very clayish (Meguro et al. 1996a; Valente 2009; Coelho 2014). Despite low fertility, Valente (2009) stated that, regardless of the array of geological and chemical conditions, the soil exercises an incredible influence over the plant formations that cover it. Phyto-physiognomic conditioning is related to physical properties of the soil, such as depth and texture. The depth and texture of the soil, which permit moisture retention, would compensate for the lack of fertility associated with greater efficiency in nutrient cycling. Therefore, considering the chemical conditions of the soil in the forest islands, the vegetation depends on the presence of moisture in the system. This is an important mechanism, which causes forest islands to always be associated with rivers, streams, or wetlands and headwater regions (Fig. 7.5). Since the establishment of the forest islands depends on both weather and soil conditions, these can be classified as edaphoclimatic formations (Rizzini 1997) (see Sect. 7.1).

7.6 Ecological Dynamics

The processes of establishment and genesis for the forest islands are related to the potential expansion, regression, or stability of these plant formations in the Espinhaço Range. According to Meguro et al. (1996b), the forest islands are successional stages of colonization by forest components in the grassland areas of the Espinhaço Range, in increasing dimensions. At the same time, Meguro et al. (1996b) cite the availability of water as a factor limiting the expansion of forest islands, stating that further expansion of the treeline can only occur as far as the substrate and groundwater can guarantee proper water supply to the arboreal vegetation. However, the results obtained by other authors (Valente 2009; Coelho et al. 2014), who have conducted broader sampling efforts, suggest the opposite. Forest islands would be subjected to soil and climate controls, and are necessarily limited to zones with soils and climate appropriate for the development of the vegetation that comprises these formations (Valente 2009). The controlling factors are both the soil and the climate, creating some areas of environmental suitability for the development of forest islands. Due to the fact that these processes are dynamic, when they are subjected to changes in land use (livestock, fire, cutting), their balance can be altered. Expansion is limited by both soil and climatic conditions, characteristics of an edaphoclimatic vegetational formation (see Sect. 7.1). Some studies have shown strong adaptation to specific climatic conditions, not only herbaceous angiosperms and epiphytes, but also by arboreal vegetation. Foliar absorption of water by arboreal vegetation is an essential mechanism for establishment of these vegetation types (Eller et al. 2013). In this way, expansion or retraction of forest islands is associated with climatic changes, which occur on distinct time scales, while the natural dimensions of the forest islands are naturally stable, with no processes of expansion or retraction (Silva et al. 2008).

7.7 Threats to Conservation

In addition to natural phenomena, such as wind and harsh weather action, which is very intense on mountaintops and can topple trees and open clearings, the forest islands associated with the Espinhaço mountains have been subjected to intense anthropogenic interference (Souza 2009). This interference can range from the pasturing of livestock to the cutting of individual trees for firewood. Because the forest islands vegetative formations are home to a variety of pteridophytes and angiosperms, these ornamental species are often collected for use in landscaping, which is a result of insufficient monitoring. Orchids, bromeliads, and Araceae species have been collected for illegal trade for decades.

However, fire might stands as one of the most serious threats. As a result of both climatic conditions and land use, fire is recurrent in the region. While the vegetation of the matrix (rupestrian grassland) has adapted in reponse to fire (e.g., Coutinho 1982, 1990; Howe 1994; Whelan 1997; Bond et al. 2005; Bond and Keeley 2005; Veldman et al. 2015), the rain forest vegetation of the forest islands is totally vulnerable to it. Fire is common in the rupestrian grasslands (Barbosa et al. 2015; see review in Chap. 16). During the dry months of the year these islands of forests are frequently seen immersed into huge bushfires, that ends up penetrating from the drier borders.

In the forest islands of the Espinhaço Range, fire has resulted in tree mortality, causing changes in structure and successional dynamics (see reports in Campos 1995; Meguro et al. 1996a, b; Souza 2009; Valente 2009; Coelho 2014). Because they are high-humidity environments, and the highest levels of humidity are encountered in the central regions of the forest islands, fire has a more intense impact on the border areas. In a study conducted on forest islands associated with PERP, Souza (2009) detected significant floristic and structural differences between the border and central regions. Measures such as density, diameter, average individual height, richness, and species diversity varied visibly between the border and central regions of the evaluated forest islands, showing that, despite their small size compared to contiguous forests, forest islands systems exhibit a clear edge effect. In a study of soils in a grassland-forest vegetation gradient in the Serra do Cipó, Valente (2009) also identified differences in the physical and chemical properties of soils collected at different points in the forest islands; that is, soil conditions in the central areas were more favorable to the establishment of species characteristic of late stages of succession. In this context, fire plays a predominant role in intensifying the impacts associated with the edge effect. Another consequence, in addition to those already mentioned, is facilitated biological invasion (Laurance and Cochrane 2001; Tabarelli et al. 2008; Souza 2009; Barbosa et al. 2010; Coelho 2014). For instance, ruderal species, such as *Eremanthus erythropappus* (DC.) Macleish (Asteraceae), are already found in abundance in forest islands with high levels of disturbance.

7.8 Conclusion

The forest islands of the Espinhaço Range feature a floristic composition similar to semi-deciduous forests in southeastern Brazil that are associated with the Atlantic Rainforest. Following the most recent classification proposed by Oliveira-Filho (2009), the forest islands in the Espinhaço Range would be classified as Disjunctions of Tropical Broadleaf Evergreen Upper Montane Cloud Forest, which can be found on Crassifolium substrates, rocky slopes, ridges, or *Arenaria Paludicola*, which are associated with mountains where rain and cloudiness are concentrated. The forest islands are an edaphoclimatic formation since they are dependent on specific climate and soil factors for their development. Physical characteristics of the soil, such as drainage, are more relevant than chemical characteristics, such as nutrient supply. These natural islands are strongly influenced by elements of the Atlantic rainforest, despite those of the Cerrado being present in smaller proportions. On the other hand, the influence of the Cerrado on these islands is likely to increase due to increasing levels of disturbances caused by land use and climate change. Among various threats, fire is the most significant because it decreases the dimensions of the forest islands and alters successional dynamics.

Acknowledgements We thank FAPEMIG, CNPq and Reserva Vellozia for finantial and logistical support. To Ramón Perea (Stanford University) for reading the manuscript and for suggestions.

References

Ab`Saber AN (2003) Os domínios de natureza no Brasil: potencialidades paisagísticas. Ateliê Editorial, São Paulo

Barbosa NPU, Fernandes GW, Carneiro MAA, Júnior LAC (2010) Distribution of non-native invasive species and soil properties in proximity to paved roads and unpaved in a quartzitic mountainous grassland of southeastern Brazil (rupestrian fields). Biol Inv 12:3745–3755

Barbosa NPU, Fernandes GW, Sanchez-Azofeifa A (2015) A relict species restricted to a quartzitic mountain in tropical America: an exemple of microrefugium? Acta Bot Bras 29:299–309

Bond WJ, Keeley JE (2005) Fire as a global 'herbivore': the ecology and evolution of flammable ecosystems. Trends Ecol Evol 20:387–394

Bond WJ, Woodward FI, Midgley GF (2005) The global distribution of ecosystems in a world without fire. New Phytol 165:525–538

Botrel RT, Oliveira-Filho AT, Rodrigues AL, Curi N (2002) Influência do solo e topografia sobre as variações da composição florística e estrutura da comunidade arbóreo-arbustiva de uma floresta estacional semidecidual em Ingaí, MG. Rev Bras Bot 25:195–213

Budke JC, Jarenkow JA, Oliveira-Filho AT (2007) Relationships between tree component structure, topography and soils of a riverside forest, Rio Botucaraí, southern Brazil. Plant Ecol 189:187–200

Budke JC, Jarenkow JA, Oliveira-Filho AT (2010) Intermediary disturbance increases tree diversity in riverine forest of southern Brazil. Biodiv Cons 19:2371–2387

Campos MTVA (1995) Composição Florística e aspectos da estrutura e da dinâmica de três capões na Serra do Cipó, Minhas Gerais, Brasil. MSc Dissertation. Universidade de São Paulo
Carvalho LMT, Fontes MAL, Oliveira-Filho AT (2000a) Tree species distribution in canopy gaps and mature forest in an area of cloud forest of the Ibitipoca Range, south-eastern Brazil. Plant Ecol 149:9–22
Carvalho DA, Oliveira-Filho AT, Vitela EA, Curi N (2000b) Florística e estrutura da vegetação arbórea de um fragmento de floresta semidecidual às margens do reservatório da Usina Hidrelétrica Dona Rita (Itambé do Mato Dentro, MG). Acta Bot Bras 14:37–55
Carvalho DA, Oliveira-Filho AT, Van Den Berg E, Fontes MAL, Vilela EA, Marques JJGSM, Carvalho WAC (2005) Variações florísticas e estruturais do componente arbóreo de uma floresta ombrófila alto-montana às margens do Rio Grande, Bocaina de Minas, MG, Brasil. Acta Bot Bras 19:91–109
Carvalho F, Souza FA, Carrenho R, Moreira FMS, Jesus EC, Fernandes GW (2012) The mosaic of habitats in the high-altitude Brazilian rupestrian fields is a hotspot for arbuscular mycorrhizal fungi. Appl Soil Ecol 52:9–19
Castellani TT, Stubblebine WH (1993) Sucessão secundária inicial em uma mata tropical mesófila, após perturbação por fogo. Rev Bras Bot 16:181–203
Cochrane MA (2003) Fire science for rainforests. Nature 421:913–919
Coelho MS (2014) Padrões e processos ecológicos em capões de mata da cadeia da Serra do Espinhaço. PhD Thesis. Universidade Federal de Minas Gerais
Coelho MS, Almada ED, Quintino AV, Fernandes GW, Santos RM, Sanchez-Azofeifa A, Espítito-Santo MM (2012) Floristic composition and structure of a tropical dry forest at different successional stages in the Espinhaço Mountains, southeastern Brazil. Interciencia 37:190–196
Coutinho LM (1982) Ecological effects of fire in Brazilian Cerrado. In: Huntley BJ, Walker BH (eds) Ecological studies: ecology of tropical savannas. Springer, New York, pp 273–291
Coutinho LM (1990) Fire in the ecology of Brazilian Cerrado. In: Goldammer JG (ed) Ecological studies: fire in the tropical biota. Springer, New York, pp 82–105
Dray S, Dufour AB (2007) The ade4 package: implementing the duality diagram for ecologists. J Stat Softw 22:1–20
Durigan G, Bernacci LC, Franco GADC, Arbocz GDF, Metzger JP, Catharino ELM (2008) Estádio sucessional e fatores geográficos como determinantes da similaridade florística entre comunidades florestais no Planalto Atlântico, estado de São Paulo, Brasil. Acta Bot Bras 22:51–62
Eller CB, Lima AL, Oliveira RS (2013) Foliar uptake of fog water and transport belowground alleviates drought effects in the cloud forest tree species, *Drimys brasiliensis* (Winteraceae). New Phytol 199:151–162
Fernandes GW, Barbosa NPU, Negreiros D, Paglia AP (2014) Challenges for the conservation of vanishing megadiverse rupestrian grasslands. Nat Cons 12:162–165
França GS, Stehmann JR (2004) Composição florística e estrutura do component arbóreo de uma floresta altimontana do sul de Minas Gerais, Brasil. Rev Bras Bot 27:19–30
Gentry AH (1995) Patterns of diversity and floristics composition in a neotropical montane forests. In: Churchill SP, Balslev H, Forero E, Luteyn JL (eds) Neotropical montane forests biodiversity and conservation symposium. The New York Botanical Garden, New york, pp 103–126
Giulietti AM, Menezes NL, Pirani JR, Meguro M, Wanderley MGL (1987) Flora da Serra do Cipó, Minas Gerais: caracterização e lista de espécies. Bol Bot Univ São Paulo 9:1–159
Gonzaga APD, Oliveira-Filho AT, Machado ELM, Hargreaves P, Machado JNN (2008) Diagnóstico florístico-estrutural do component arbóreo da floresta da Serra de São José, Tiradentes, MG, Brasil. Acta Bot Bras 22:505–520
Higuchi P, Oliveira-Filho AT, Bebber DP, Brown ND, Silva AC, Machado ELM (2008) Spatio-temporal patterns of tree community dynamics in a tropical forest fragment in South-east Brazil. Plant Ecol 1999:125–135

Howe HF (1994) Managing tall-grass species diversity: assumptions and implications. Cons Biol 8:691–704

Ivanauskas NM, Monteiro R, Rodrigues RR (2003) Alterations following a fire in a Forest community of Alto Rio Xingu. For Ecol Manag 184:239–250

Laurance WF, Cochrane MA (2001) Synergistic effects in fragmented landscapes. Special section. Cons Biol 15:1488–1535

Legendre P, Legendre L (1998) Numerical ecology. Elsevier Science, Amsterdam

Lista de Espécies da Flora do Brasil Jardim Botânico do Rio de Janeiro (LEFB). http://floradobrasil.jbrj.gov.br/. Accessed 25 May 2014

Machado ELM, Oliveira-Filho AT (2010) Spatial patterns of tree community dynamics are detectable in a small (4 ha) and disturbed fragment of the Brazilian Atlantic forest. Acta Bot Bras 24:250–261

Meguro M, Pirani JR, Mello Silva R, Giulietti AM (1996a) Caracterização florística e estrutural de matas ripárias e capões de altitude da Serra do Cipó, Minas Gerais. Bol Bot Univ São Paulo 15:13–29

Meguro M, Pirani JR, Mello-Silva R, Giullietti AM (1996b) Estabelecimento de matas ripárias e capões nos ecossistemas campestres da Cadeia do Espinhaço. Bol Bot Univ São Paulo 15:1–11

Oksanen J (2009) Multivariate analysis of ecological communities in R: Vegan Tutorial. url:http://cran.r-project.org

Oliveira-Filho AT (2009) Classificação das fitofisionomias da América do Sul Cisandina Tropical e Subtropical: proposta de um novo sistema – prático e flexível – ou uma injeção a mais de caos? Rodriguésia 60:237–258

Oliveira-Filho AT, Fontes MAL (2000) Patterns of floristic differentiation among atlantic forest in southeastern Brazil and the influence of climate. Biotropica 32:793–810

Oliveira-Filho AT, Vilela EA, Carvalho DA, Gavilanes ML (1994a) Effects of soil and topography on the distribution of tree species in a tropical riverine forest in South-eastern Brazil. J Trop Ecol 10:483–508

Oliveira-Filho AT, Vilela EA, Gavilanes ML, Carvalho DA (1994b) Comparison of the woody flora and soils of six montane semi-deciduous Forest in Southern Minas Gerais, Brazil. Edinb J Bot 51:355–389

Oliveira-Filho AT, Mello JM, Scolforo RS (1997) Effects of past disturbance and edges on tree community structure and dynamics within a fragment of tropical semideciduous forest in south-eastern Brazil over a five-year period (1987–1992). Plant Ecol 131:45–66

Oliveira-Filho AT, Curi N, Vilela AE, Carvalho DA (2001) Variation in tree community composition and structure with changes in soil properties within a fragment of semideciduous forest in south-eastern Brazil. Edinb J Bot 58:139–158

Oliveira-Filho AT, Carvalho DA, Fontes MAL, Van Den Berg E, Curi N, Carvalho WAC (2004) Variações estruturais do compartimento arbóreo de uma floresta semidecídua altimontana na chapada das Perdizes, Carrancas, MG. Rev Bras Bot 27:291–309

Peloso BDA (2009) Classificação da vegetação do Parque Nacional da Serra do Cipó (MG) e sua relação com variáveis morfométricas. MSc Dissertation, Universidade Federal de Minas Gerais

Peloso BDA, Shimabukuro YE (2010) Caracterização de unidades geo-botânicas do Parque Nacional da Serra do Cipó (MG) através da integração de imagens ópticas e modelo digital de elevação. Rev Bras Cart 62:103–118

Pereira JAA, Oliveira-Filho AT, Lemos-Filho JP (2007) Environmental heterogeneity and disturbance by humans control much of the tree species diversity of Atlantic montane forest fragments in SE Brazil. Biodiv Cons 16:1761–1784

Pinto JRR, Oliveira-Filho AT, Hay JDV (2005) Influence of soil and topography on the composition of a tree community in a central Brazilian valley forest. Edinb J Bot 62:69–90

R Development Core team. 2013. R: a language and environment for statistical computing. Version 3.0.2. R Foundation for Statistical Computing

Ribeiro KT, Nascimento JS, Madeira JA, Ribeiro LC (2009) Aferição dos limites da Mata Atlântica na Serra do Cipó, MG, Brasil, visando maior compreenão e proteção de um mosaico vegetacional fortemente ameaçado. Nat Cons 7:30–49

Rizzini CT (1997) Tratado de fitogeografia do Brasil: aspectos ecológicos, sociológicos e florísticos. Âmbito, Rio de Janeiro
Ross KA, Fox BJ, Fox MD (2002) Changes to plant species richness in forest fragments: fragmente age, disturbance and fire history may be as importante as area. J Biogeogr 29:749–765
Sanchez M, Pedroni F, Eisenlohr PV, Oliveira-Filho AT (2013) Changes in tree community composition and structure of Atlantic rain forest on a slope of the Serra do Mar range, southeartern Brazil, from near sea level to 1000 m of altitude. Flora 208:184–196
Santos MF (2009) Análise florística em floresta estacional semidecidual na encosta leste da Serra do Cipó, MG. MSc Dissertation, Universidade de São Paulo
Santos MF, Serafim H, Sano PT (2011) Fisionomia e composição da vegetação florestal na Serra do Cipó, MG, Brasil. Acta Bot Bras 25:793–814
Santos MF, Serafim H, Sano PT (2012) Composição e estrutura arbórea em floresta estacional semidecidual no Espinhaço meridional (Serra do Cipó, MG). Rodriguésia 63:985–997
Silva VF, Oliveira-Filho AT, Venturin N, Carvalho WAC, Gomes JBV (2005) Impacto do fogo no componente arbóreo de uma floresta estacional semidecídua no município de Ibituruna, MG, Brasil. Acta Bot Bras 19:701–716
Silva LCR, Sternberg L, Haridasans M, Hoffmann WA, Miralles-Wilhelm F, Franco AC (2008) Expansion of gallery forests into central Brazilian savannas. Glob Change Biol 14:2108–2118
Souza DT (2009) Composição florística e estrutura dos capões de altitude no parque estadual do Rio Preto, Minas Gerais, Brasil. MSc Dissertation, Universidade Federal de Minas Gerais
Tabarelli M, Lopes AV, Peres CA (2008) Edge-effects drive tropical forest fragments towards an early-successional system. Biotropica 40:657–661
Toniato MTZ, Oliveira-Filho AT (2004) Variations in tree community composition and structure in a fragment of tropical semideciduous forest in southeastern Brazil related to different human disturbance histories. Forest Ecol Manag 198:319–339
Valente EL (2009) Relações solo-vegetação no Parque Nacional da Serra do Cipó, Espinhaço Meridional, Minas Gerais. PhD Thesis, Universidade Federal de Viçosa
Van den Berg EVD, Oliveira-Filho AT (2000) Composição florística e estrutura fitossociológica de uma floresta ripária em Itutinga, Minas Gerais, e comparação com outras áreas. Rev Bras Bot 23:231–253
Van der Hammen T (1995) Global change, biodiversity, and conservation of neotropical montane forests. In: Churchill SP, Balslev H, Forero E, Luteyn JL (eds) Biodiversity and conservation of neotropical montane forests. The New York botanical Garden, New York, pp 603–607
Veldman JW, Buisson E, Durigan G, Fernandes GW, Le Stradic S, Mahy G, Negreiros D, Overbeck GE, Veldman RG, Zaloumis NP, Putz FE, Bond WJ (2015) Toward an old-growth concept for grasslands, savannas, and woodlands. Front Ecol Environ 13:154–162
Veloso HP, Ranel Filho ALR, Lima JCA (1991) Classificação da vegetação brasileira adaptada a um sistema universal. Instituto Brasileiro de Geografia e Estatística, Rio de Janeiro
Webster GL (1995) The panorama of Neotropical Cloud Forests. In: Churchill SP, Balslev H, Forero E, Luteyn JL (eds) Biodiersity and conservation of neotropical montane forests: proceedings of neotropical montane forest biodiversity and conservation symposium. The New York Botanical Garden, New York, pp 57–77
Whelan RJ (1997) Ecology of fire. Cambridge University, United kingdon
Whitmore TC (1990) An introduction to tropical rain forests. Clarendon Press, Oxford

Chapter 8
Arbuscular Mycorrhiza and Endophytic Fungi in Ruspestrian Grasslands

Yumi Oki, Bruno Tomio Goto, Khadija Jobim, Luiz Henrique Rosa, Mariana Costa Ferreira, Etiene Silva Coutinho, João Henrique de Azevedo Xavier, Fernanda Carvalho, Fatima Maria de Souza Moreira, Francisco Adriano de Souza, Ricardo Luís Louro Berbara and G. Wilson Fernandes

Abstract Rupestrian grasslands are a biodiversity hotspot for arbuscular mycorrhizal fungi (AMF) and endophytic fungi. These groups have major influence on the structure and dynamics of the plant community in this ecosystem. AMF are generalists, associating with a wide variety of host plants, while endophytic fungi tend to establish more specific relations with host plants. AMF distribution is mainly influenced by soil attributes, while the distribution of endophytic fungi is mostly driven by the host plant. Both fungal groups influence the host plant metabolism and ecology, enhancing the host plant survival in this harsh ecosystem. There are no studies relating both fungal groups in rupestrian grasslands. A strong relation may exist between them, and exploring this gap will contribute to the understanding of this unique ecosystem and the relations in it.

Y. Oki (✉) · G.W. Fernandes
Ecologia Evolutiva and Biodiversidade/DBG, CP 486, ICB, Universidade Federal de Minas Gerais, Belo Horizonte, MG 31270-901, Brazil
e-mail: yumioki1@gmail.com

G.W. Fernandes
e-mail: gw.fernandes@gmail.com

B.T. Goto
Departamento de Botânica e Zoologia, CB, Universidade Federal do Rio Grande do Norte, Natal, RN, Brazil
e-mail: brunogoto@hotmail.com

K. Jobim
PPG em Sistemática e Evolução, CB, Universidade Federal do Rio Grande do Norte, Natal, RN, Brazil
e-mail: khadija_jobim@hotmail.com

L.H. Rosa · M.C. Ferreira
Departamento de Microbiologia, ICB, Universidade Federal de Minas Gerais, Belo Horizonte, MG, Brazil
e-mail: lhrosa@icb.ufmg.br

© Springer International Publishing Switzerland 2016

G.W. Fernandes (ed.), *Ecology and Conservation of Mountaintop Grasslands in Brazil*, DOI 10.1007/978-3-319-29808-5_8

Contents

8.1 Introduction

Endophytic fungi and arbuscular mycorrhizal fungi (AMF) play important roles in the functioning of ecosystems and influence the structure and dynamics of plant communities. Their role in fragile ecosystems with low regenerative capacity has been argued to be even more relevant. The relationships among these two

M.C. Ferreira
e-mail: marianacostaferreira@yahoo.com.br

E.S. Coutinho · J.H. de Azevedo Xavier
PPG em Ecologia, Conservação da Vida Silvestre, ICB, Universidade Federal de Minas Gerais, Belo Horizonte, MG, Brazil
e-mail: etienescoutinho@yahoo.com.br

J.H. de Azevedo Xavier
e-mail: eco.joaoxavier@gmail.com

F. Carvalho · F.M. de Souza Moreira
Departamento de Ciência do Solo, Universidade Federal de Lavras, Lavras, MG, Brazil
e-mail: fernandacarva@hotmail.com

F.M. de Souza Moreira
e-mail: fmoreira@dcs.ufla.br

F.A. de Souza
Núcleo de Biologia Aplicada, Embrapa Milho e Sorgo, Sete Lagoas, MG, Brazil
e-mail: francisco.adriano@embrapa.br

R.L.L. Berbara
Departamento de Solos, Instituto de Agronomia, Universidade Federal Rural Do Rio de Janeiro, Seropédica, RJ, Brazil
e-mail: rberbara@yahoo.com.br

G.W. Fernandes
Department of Biology, Stanford University, Stanford, CA 94395, USA

distinctive fungal groups and their associated plants may have had a crucial role in plant adaptation to environmental harshness. Endophytic fungi and AMF are considered ubiquitous and important components of fungal diversity [estimated at 1.5 million species, Hawksworth (2001)], although the knowledge about their diversity, host range, and geographical distribution remains mostly unclear.

Endophytic fungi are a diverse group composed of Basidiomycota and Ascomycota. They colonize host plant tissues without causing damage (Petrini et al. 1992; Wilson 1993; Saikkonen et al. 2004; Arnold and Lutzoni 2007; Tejesvi et al. 2007) and their diversity varies enormously among plant species. Their presence can alter host plant physiology. These fungi have been linked to increased plant biomass accumulation (Breen 1994; Preszler et al. 1996), and have often been related to increased host plant resistance against herbivores (Fernandes and Price 1992; Gaylord et al. 1996), and pathogens (Clay 2004). They also improve host plant survival rates under adverse abiotic conditions (Redman et al. 2002; Rodriguez et al. 2004). Some leaf endophytes are latent saprotrophs such as *Xylaria*, found in the shrub *Baccharis dracunculifolia*, with high cellulase and ligninaseproduction, which contributes to the degradation of leaves.

Arbuscular mycorrhizal fungi (AMF) belong to the phylum Glomeromycota and are the most abundant and relevant group of mycorrhizal fungi in grasslands, savannas and tropical forests (Read and Perez-Moreno 2003). In this symbiosis, the plant supplies the fungus with energy to grow and reproduce, while the fungus a range of services to the host plant (Gupta et al. 2000; Souza et al. 2007; Smith and Read 2008), including nutrient uptake in nutrient-poor soils. Therefore, the community of arbuscular mycorrhizal fungi has been considered an indicator of the quality and sustainability of soils (Moreira et al. 2007; Miranda 2008; Leal et al. 2009; Gianinazzi et al. 2010).

This chapter focuses on the diversity and distribution of endophytic and AM fungi in rupestrian grasslands, the environmental factors controlling their occurrence, their relevance and use in plant-soil ecology studies and the economic implications of these interactions.

8.2 Distribution and Diversity of Endophytic Fungi and Arbuscular Mycorrhizal Fungi in Rupestrian Grasslands

Studies on endophytic and arbuscular mycorrhizal fungi are concentrated in plants of agronomic interest (e.g., Clay 1990; Marks and Clay 1996; Rodriguez et al. 2004; Kivlin et al. 2011; Lee et al. 2013). In the rupestrian grasslands, the research on them is still beginning. Most publications about these fungi are restricted to a few localities (e.g. Oki et al. 2008, 2009; Pagano and Scotti 2009; Vaz et al. 2009; Carvalho et al. 2012; Vieira et al. 2014; Coutinho et al. 2015). In rupestrian grasslands, the major sample methods are soil samples for AMF and host plant leaves for endophytic fungi.

8.2.1 Endophytic Fungi

Plant species included in the study on the diversity of endophytic fungi in rupestrian grasslands are: *Acianthera hamosa*, *Acianthera teres*, *Anathallis sclerophylla*, *Baccharis dracunculifolia*, *Baccharis trimera*, *Bulbophyllum involutum*, *Coccoloba cereifera*, *Epidendrum filicaule*, *Epidendrum martianum*, *Epidendrum secundum*, *Habenaria setacea*, *Isochilus linearis*, *Isochilus linearis*, *Maxillaria brasiliensis*, *Maxillaria rigida*, *Octomeria crassifolia*, *Oncidium warmingii*, *Scuticaria irwiniana*, *Sophronitis fournieri*, *Sophronitis longipes*, *Vellozia compacta*, *Vellozia nanuzae*, and *Zygopetalum maxillare* (see Table 8.1). Only the culturable endophytic fungi were evaluated in these plants, since evaluating the non-culturable was prohibitively expensive.

Despite the few rupestrian grassland plant species studied so far, a high number of culturable endophytic fungi per plant species has been recorded. In the rupestrian grasslands, 282 endophytes were found in *Vellozia graminea* (Santiago 2008; Fig. 8.1), 104 in *C. cerefeira* (Fernandes et al. 2011), 60 in *B. dracunculifolia* (Oki et al. 2008, 2009), 25 in *Baccharis trimera* (Vieira et al. 2014), and 21 in *Vellozia compacta* (Rodrigues 2010).

Endophyte diversity increases with the age of host tissue. Mature leaves of *B. dracunculifolia* (Oki et al. 2008) and *C. cereifera* (Fernandes et al. 2011) presented a greater diversity of endophytic fungi compared to young leaves. In *C. cereifera*, the endophyte richness found in old leaves (87 species) was 15 times greater than in young leaves (6 species). Arnold and Herre (2003) speculate that the higher diversity of endophytes in mature leaves is due to longer exposure time, which would favor the infection by endophytes. Sanchez-Azofeifa et al. (2011) analyzed the structural differences and chemical diversity of endophytic fungi at different leaf ages of *C. cereifera* and highlighted a correlation of endophyte diversity and leaf development. They argued that endophyte diversity was related to structural changes (wax, fiber, biomass) and secondary chemistry of leaves during development.

Despite the scarcity of studies on endophytic fungi in rupestrian grassland plants, a trend of specificity in endophyte-plant interactions has emerged. Studies have shown that the diversity of endophytic fungi changes with plant genus, plant species and/or ontogenetic plant characteristics. Some common generalist genus are *Phomopsis*, *Acremonium*, and *Phoma* (Fig. 8.2). The most abundant endophytic fungal species in *B. trimera* were closely related to *Diaporthe phaseolorum*, *Pestalotiopsis* sp. 1 and *Preussia pseudominima*, while in *B. dracunculifolia* the most abundant endophytic fungi were *Biscogniauxia* sp. 1, *Biscogniauxia* sp. 2, *Cladosporium cladosporioides*, *Penicillium* sp. 1, and *Aureobasidium pullulans* (Table 8.1). *Trichoderma asperellum* was recorded in two species of Orchidaceae (*Epidendrum martianum*, *Zygopetalum maxillare*). *Fusarium* was commonly found in Orchidaceae species (*Acianthera teres*, *Epidendrum secundum*, *Scuticaria irwiniana*, *Bulbophyllum involutum*, and *Isochilus linearis*). The composition of endophytic fungal species also differs with host plant sex, as in *B. dracunculifolia* (Oki et al. 2009). In this plant, 80 % of endophytic fungal species found in female plants did not occur in male plant individuals.

Table 8.1 Endophytic fungi taxa found in plants species from rupestrian grasslands in the Espinhaço mountains

Endophytic fungal taxa	Host plant	Family plant	Reference
Acarospora sp. 1.	*Vellozia compacta* Mart. ex Schult. f.	Velloziaceae	Rodrigues (2010)
Acremonium strictum	*Isochilus linearis* (Jacq.) R. Br.	Orchidaceae	Vaz et al. (2009)
Alternaria sp. 1	*Baccharis trimera* (Less.)	Asteraceae	Vieira et al. (2014)
Alternaria sp. 2	*Oncidium warmingii* Rchb. f.	Orchidaceae	Vaz et al. (2009)
Aureobasidium pullulans	*Baccharis dracunculifolia* D. C.	Asteraceae	Oki et al. (2008, 2009)
Bensingtonia sp. 1	*Octomeria crassifolia* Lindl.	Orchidaceae	Vaz et al. (2009)
Biscogniauxia sp. 1	*Baccharis dracunculifolia* D. C.	Asteraceae	Belmiro (2010)
Biscogniauxia sp. 2	*Baccharis dracunculifolia* D. C.	Asteraceae	Belmiro (2010)
Candida parapsilosis	*Anathallis sclerophylla* (Lindl.) Pridgeon and M.W. Chase	Orchidaceae	Vaz et al. (2009)
Chaetomium sp. 1	*Baccharis trimera*	Asteraceae	Vieira et al. (2014)
Cladosporium cladosporioides	*Baccharis dracunculifolia* D.C.	Asteraceae	Oki et al. (2008, 2009)
Cladosporium cladosporioides	*Vellozia compacta* Mart. ex Schult. f.	Velloziaceae	Rodrigues (2010)
Cochliobolus lunatus	*Baccharis trimera* (Less.)	Asteraceae	Vieira et al. (2014)
Coniophora sp. 1	*Epidendrum filicaule* Lindl.	Orchidaceae	Vaz et al. (2009)
Cylindrocarpon sp. 1	*Sophronitis fournieri* (Cogn.) Van den Berg and M.W. Chase	Orchidaceae	Vaz et al. (2009)
Diaporthe phaseolorum	*Baccharis trimera* (Less.)	Asteraceae	Vieira et al. (2014)
Diaporthe sp. 1	*Baccharis trimera* (Less.)	Asteraceae	Vieira et al. (2014)
Dothideomycetes sp. 1	*Vellozia compacta* Mart. ex Schult. f.	Velloziaceae	Rodrigues (2010)
Epicoccum nigrum	*Baccharis trimera* (Less.)	Asteraceae	Vieira et al. (2014)
	Maxillaria rigida Barb. Rodr.	Orchidaceae	Vaz e. al. (2009)
Epicoccum sp. 1	*Baccharis trimera* (Less.)	Asteraceae	Vieira et al. (2014)

(continued)

Table 8.1 (continued)

Endophytic fungal taxa	Host plant	Family plant	Reference
Fusarium oxysporum	*Acianthera teres* (Lindl.) Luer	Orchidaceae	Vaz et al. (2009)
	Epidendrum secundum Jacq.	Orchidaceae	Vaz et al. (2009)
	Scuticaria irwiniana Pabst	Orchidaceae	Vaz et al. (2009)
	Vellozia compacta Mart. ex Schult. f.	Velloziaceae	Rodrigues (2010)
Fusarium sp. 1	*Bulbophyllum involutum* Borba, Semir and F. Barros	Orchidaceae	Vaz et al. (2009)
Fusarium sp. 2	*Isochilus linearis* (Jacq.) R. Br.	Orchidaceae	Vaz et al. (2009)
Gelasinospora seminuda	*Vellozia compacta* Mart. ex Schult. f.	Velloziaceae	Rodrigues (2010)
Gibberella moniliformis	*Maxillaria brasiliensis* Brieger and Bicalho	Orchidaceae	Vaz et al. (2009)
Guignardia sp. 1	*Baccharis trimera* (Less.)	Asteraceae	Vieira et al. (2014)
Hypocrea sp. 1	*Sophronitis longipes* (Rchb.f.) Van den Berg and M.W. Chase	Orchidaceae	Vaz et al. (2009)
Nigrospora sp. 1	*Baccharis trimera* (Less.)	Asteraceae	Vieira et al. (2014)
Nigrospora sp. 2	*Baccharis dracunculifolia* D. C.	Asteraceae	Oki et al. (2008, 2009)
Penicillium sp. 1	*Vellozia compacta* Mart. ex Schult. f.	Velloziaceae	Rodrigues (2010)
Penicillium sp. 2	*Baccharis dracunculifolia* D. C.	Asteraceae	Oki et al. (2008, 2009)
Pestalotiopsis microspora	*Vellozia compacta* Mart. ex Schult. f.	Velloziaceae	Rodrigues (2010)
Pestalotiopsis sp. 1	*Baccharis trimera* (Less.)	Asteraceae	Vieira et al. (2014)
Pestalotiopsis sp. 2	*Vellozia compacta* Mart. ex Schult. f.	Velloziaceae	Rodrigues (2010)
Phaeosphaeria avenaria	*Habenaria setacea* Lindl.	Orchidaceae	Vaz et al. (2009)
Phoma herbarum	*Vellozia compacta* Mart. ex Schult. f.	Velloziaceae	Rodrigues (2010)
Phoma sp. 1	*Baccharis trimera* (Less.)	Asteraceae	Vieira et al. (2014)
Phomopsis sp. 1	*Baccharis dracunculifolia* D. C.	Asteraceae	Oki et al. (2008, 2009)

(continued)

Table 8.1 (continued)

Endophytic fungal taxa	Host plant	Family plant	Reference
Phomopsis sp. 2	*Baccharis trimera* (Less.)	Asteraceae	Vieira et al. (2014)
Phomopsis sp. 3	*Vellozia compacta* Mart. ex Schult. f.	Velloziaceae	Rodrigues (2010)
Podospora sp. 1	*Baccharis trimera* (Less.)	Asteraceae	Vieira et al. (2014)
Preussia africana	*Baccharis dracunculifolia* D. C.	Asteraceae	Oki et al. (2008, 2009)
Preussia africana	*Baccharis trimera* (Less.)	Asteraceae	Vieira et al. (2014)
Preussia pseudominima	*Baccharis trimera* (Less.)	Asteraceae	Vieira et al. (2014)
Preussia sp. 1	*Baccharis trimera* (Less.)	Asteraceae	Vieira et al. (2014)
Rhodotorula mucilaginosa	*Acianthera hamosa* (Barb. Rodr.) Pridgeon and M.W. Chase	Orchidaceae	Vaz et al. (2009)
Sclerostagonospora opuntiae	*Epidendrum filicaule* Lindl.	Orchidaceae	Vaz et al. (2009)
Sporormiella sp.	*Baccharis trimera* (Less.)	Asteraceae	Vieira et al. (2014)
Trichoderma asperellum	*Epidendrum martianum* Lindl.	Orchidaceae	Vaz et al. (2009)
Trichoderma asperellum	*Zygopetalum maxillare* Lodd.	Orchidaceae	Vaz et al. (2009)
Xylaria sp. 1	*Baccharis dracunculifolia* D. C.	Asteraceae	Oki et al. (2008, 2009)
Xylaria sp. 2	*Baccharis trimera* (Less.)	Asteraceae	Vieira et al. (2014)
Xylaria venulosa	*Baccharis dracunculifolia* D. C.	Asteraceae	Oki et al. (2008, 2009)

Endophytic fungi may produce the same secondary compounds as the host plant. The endophytic fungus *A. pullulans* produced the same triterpenes group found in host plant, *B. dracunculifolia* (Nascimento 2010). These specific relationships are likely due to horizontal gene transfer between fungi and their host plants (Azevedo 1998).

Endophytes can strongly influence interactions between plants and herbivores and therefore the dynamics of the ecosystem (Oki et al. 2009). Individuals of *B. dracunculifolia* with more endophyte species presented a lower abundance of herbivores (Oki et al. 2008). Additional studies showed that some endophytic

Fig. 8.1 **a** *Vellozia graminea* Pohl; **b** flowers of *V. gaminea*; **c** diversity of morphospecies colonies of endophytic fungal isolates recovered from leaves of *V. graminea*

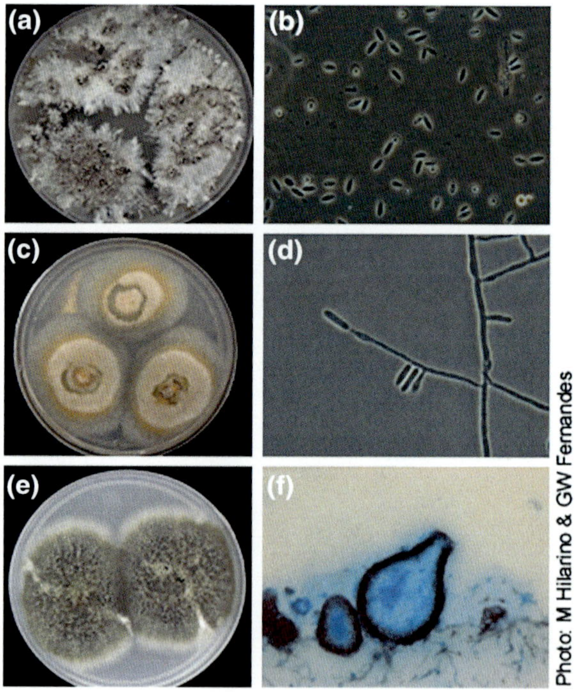

Fig. 8.2 Macroscopic (*left column*) and microscopic (*right column*) aspects of dominant species of endophytic fungi. **a** *Phomopsis* sp. 1, **b** *Phomopsis* sp. 1 conidia, **c** *Acremonium* sp., **d** *Acremonium* sp. conidia, **e** *Phoma* sp., **f** *Phoma* sp. pycnidia

fungal species (such as *A. pullulans*, *P. africana* and *X. venulosa*) found in this host plant produce secondary metabolites that minimize herbivory (Nascimento 2010) and are lethal to its main herbivore, the aphid *Uroleucon erigeronensis*. Thus, the diversity of endophytic fungi directly impacts a plant's performance and survival.

8.2.2 Arbuscular Mycorrhizal Fungi

In ecosystems where environmental conditions are harsh, the occurrence of symbiotic relationship between plants and AMF is important for native plants survival (Smith and Read 2008). At the rupestrian grasslands, plants must deal with extreme soil nutrient stress (Carvalho et al. 2012; Coutinho et al. 2015).

Pagano and Scotti (2009) led the first initiative to assess AMF occurrence and diversity in rupestrian grasslands, but the study was restricted to the examination of two endemic plants species. Only three AMF genera were identified, two at the genus level, *Acaulospora* and *Glomus*, and one at specie level, *Dentiscutata biornata*. In contrast, a high spore density was obtained (77-139 glomerospores per 100 g of soil-1), possibly due to the typical environmental conditions such as the diversity of plant species and soil properties that influence the sporulation (Smith and Read 2008).

Carvalho et al. (2012) conducted an important study on AMF diversity in rupestrian grassland habitats, which countered results obtained by Pagano and Scotti (2009). This pioneer study showed that this complex ecosystem represents a biodiversity hotspot for AMF. A total of 49 AMF species were identified, a high number when compared to the total number found in the Cerrado (90 species) (Bononi and Trufem 1983; Schenck et al. 1989; Siqueira et al. 1989; Pagano and Scotti 2009). Among these species, *Acaulospora colossica* and *Pacispora dominikii* were recorded for the first time in Brazil. The genus *Glomus* showed the highest number of recorded species (19) followed by *Acaulospora* (13) and *Scutellospora* (3). The high AMF diversity found in rupestrian grasslands was related to the variety of habitats found in the studied area, notably, the differences in physical and chemical properties of soil and the high dominance of plant species found in these habitats showed a strong influence on the diversity of AMF (Carvalho et al. 2012). The influence of soil conditions on AMF diversity was also observed by Coutinho et al. (2015) who found a larger number of species (53) compared to the inventory conducted by Carvalho et al. (2012). AMF species richness recorded in these studies exceeds that found in any other taxonomic survey from the tropics thus far (e.g., Trufem et al. 1989, 1994; Trufem 1990, 1995; Santos et al. 1995; Cordoba et al. 2001; Cordazo and Stürmer 2007; Oliveira et al. 2009; Silva et al. 2012; Stürmer et al. 2013). Coutinho et al. (2015) recorded three species for the first time

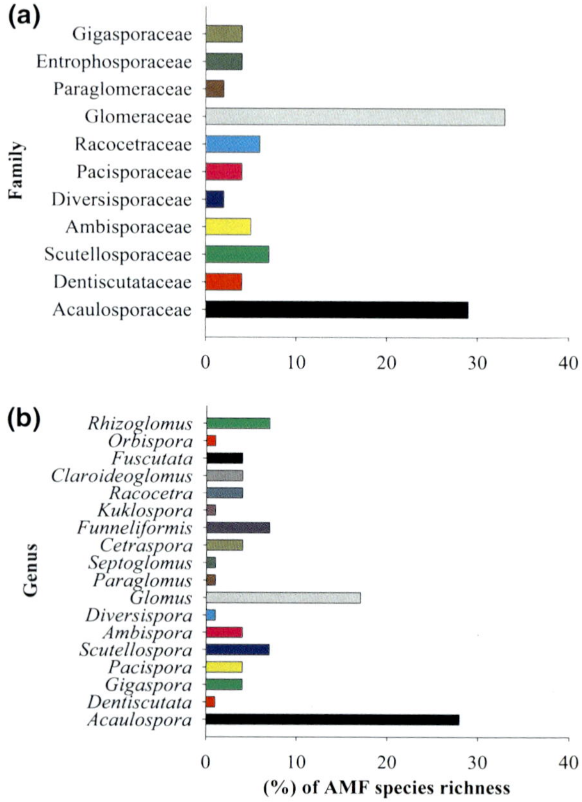

Fig. 8.3 **a** Representation (%) of AMF species richness per family in rupestrian grasslands (Modified from Carvalho et al. 2012; Lima et al. 2014; Coutinho et al. 2015); **b** Representation (%) of the AMF fungi species richness by genus in rupestrian grasslands (Modified from Carvalho et al. 2012; Lima et al. 2014; Coutinho et al. 2015)

in rupestrian grasslands (*Acaulospora denticulata*, *Ambispora callosa,* and *Gigaspora gigantea*) and 14 potentially new species. All taxonomic inventories conducted in the rupestrian grassland ecosystem totaled 71 AMF species (Fig. 8.3; Table 8.2), which represents 26 % of all known AMF species, and 46 % of the species recorded in Brazil. These results show that the rupestrian grasslands are a potential AMF diversity source. The species cataloged for rupestrian grasslands are distributed across 11 families and 18 genera (Fig. 8.3a, b). The most represented families are Acaulosporaceae (34 %) and Glomeraceae (28 %), with the most represented genera being *Acaulospora* (27 %) and *Glomus* (15 %). This pattern is similar to those recorded for other Brazilian biomes (Amazon, Caatinga, Atlantic Forest, and Pantanal), where these taxonomic groups are also highly represented (e.g., Prance 1994; Goto et al. 2010, 2012; Souza et al. 2010; Mello et al. 2012; Silva et al. 2012; Bonfim et al. 2013; Leal et al. 2013; Stürmer et al. 2013; Gomide et al. 2014; Novais et al. 2014; Pereira et al. 2014).

Table 8.2 AMF species reported in the Cerrado, Brazil

Species	Reference
[a]*Acaulospora* aff. *bireticulata* F.M. Rothwell and Trappe	Carvalho et al. (2012), Coutinho et al. (2015)
[a, b]*Acaulospora cavernata* Błaszk.	Carvalho et al. (2012)
[a]*Acaulospora colossica* P.A. Schultz, Bever and J.B. Morton	Carvalho et al. (2012), Coutinho et al. (2015)
[a]*Acaulospora delicata* C. Walker, C.M. Pfeiffer and Bloss	Carvalho et al. (2012), Coutinho et al. (2015)
[a, b]*Acaulospora denticulata* Sieverd. and S. Toro	Coutinho et al. (2015)
[a, b]*Acaulospora koskei* Błaszk.	Carvalho et al. (2012), Coutinho et al. (2015)
[a]*Acaulospora longula* Spain and N.C. Schenck	Souza et al. (2010), Carvalho et al. (2012)
[a, b]*Acaulospora mellea* Spain and N.C. Schenck	Carvalho et al. (2012), Coutinho et al. (2015)
[a]*Acaulospora morrowiae* Spain and N.C. Schenck	Souza et al. (2010), Carvalho et al. (2012); Coutinho et al. (2015)
[a]*Acaulospora* aff. *rhemii* Sieverd. and S. Toro	Coutinho et al. (2015)
[a]*Acaulospora rugosa* J.B. Morton	Carvalho et al. (2012), Coutinho et al. (2015)
[a]*Acaulospora scrobiculata* Trappe	Souza et al. (2010), Carvalho et al. (2012), Coutinho et al. (2015)
[a]*Acaulospora* sp. 1	Carvalho et al. (2012)
[a]*Acaulospora* sp. 2	Carvalho et al. (2012)
[a]*Acaulospora* sp. 3	Coutinho et al. (2015)
[a]*Acaulospora* sp. 4	Coutinho et al. (2015)
[a]*Acaulospora* sp. 5	Coutinho et al. (2015)
[a]*Acaulospora* sp. 6	Coutinho et al. (2015)
[a]*Acaulospora spinosa* C. Walker and Trappe	Souza et al. (2010), Carvalho et al. (2012), Coutinho et al. (2015)
Acaulospora tuberculata Janos and Trappe	Souza et al. (2010)
[a]*Ambispora appendicula* (Spain, Sieverd., N.C. Schenck) C. Walker	Souza et al. (2010), Carvalho et al. (2012), Coutinho et al. (2015)
[a, b]*Ambispora brasilensis* B.T. Goto, L.C. Maia and Oehl	Souza et al. (2010), Carvalho et al. (2012), Coutinho et al. (2015)
[a, b]*Ambispora callosa* (Sieverd.) C. Walker, Vestberg and A. Schüssler	Coutinho et al. (2015)
Ambispora fecundispora (N.C. Schenck and G.S. Sm.) C. Walker, Vestberg and A. Schüssler	Souza et al. (2010)
Ambispora gerdemannii (S.L. Rose, B.A. Daniels and Trappe) C. Walker, Vestberg and A. Schüssler	Souza et al. (2010)
[a, b]*Cetraspora auronigra* Oehl, L.L. Lima, Kozovits, Magna and G.A. Silva	Lima et al. (2014)

(continued)

Table 8.2 (continued)

Species	Reference
[a]*Cetraspora gilmorei* (Trappe and Gerd.) Oehl, F.A. de Souza and Sieverd.	Souza et al. (2010); Carvalho et al. (2012)
[a, b]*Cetraspora spinosissima* (C. Walker and Cuenca) Oehl, F.A. de Souza and Sieverd.	Lima et al. (2014)
Cetraspora pellucida (T.H. Nicolson and N.C. Schenck) Oehl, F.A. de Souza and Sieverd.	Souza et al. (2010)
[a, b]*Claroideoglomus claroideum* (N.C. Schenck and G.S. Sm.) C. Walker and A. Schüssler	Carvalho et al. (2012)
[a]*Claroideoglomus etunicatum* (W.N. Becker and Gerd.) C. Walker and A. Schüssler	Souza et al. (2010), Carvalho et al. (2012), Coutinho et al. (2015)
[a]*Claroideoglomus lamellosum* (Dalpé, Koske and Tews) C. Walker and A. Schüssler	Carvalho et al. (2012), Coutinho et al. (2015)
[a]*Dentiscutata biornata* (Spain, Sieverd. and S. Toro) Sieverd., F.A. de Souza and Oehl	Souza et al. (2010), Carvalho et al. (2012), Coutinho et al. (2015)
Dentiscutata cerradensis (Spain and J. Miranda) Sieverd., F.A. de Souza and Oehl	Souza et al. (2010)
Dentiscutata heterogama (T.H. Nicolson and Gerd.) Sieverd., F.A. de Souza and Oehl	Souza et al. (2010)
Dentiscutata nigra (J.F. Readhead) Sieverd., F.A. de Souza and Oehl	Souza et al. (2010)
Dentiscutata reticulata (Koske, D.D. Miller and C. Walker) Sieverd., F.A. de Souza and Oehl	Souza et al. (2010)
[a]*Diversispora* aff. *inscultpta* (Błaszk.) Oehl, G.A. Silva and Sieverd.	Carvalho et al. (2012)
[a, b]*Diversispora* sp.	Coutinho et al. (2015)
[a]*Funneliformis geosporus* (T.H. Nicolson and Gerd.) C. Walker and A. Schüssler	Souza et al. (2010), Carvalho et al. (2012), Coutinho et al. (2015)
Funneliformis monosporus (Gerd. and Trappe) Oehl, G.A. Silva and Sieverd.	Souza et al. (2010)
[a]*Funneliformis mosseae* (T.H. Nicolson and Gerd.) C. Walker and A. Schüssler	Souza et al. (2010), Carvalho et al. (2012), Coutinho et al. (2015)
[a, b]*Funneliformis multiforus* (Tadych and Błaszk.) Oehl, G.A. Silva and Sieverd.	Carvalho et al. (2012)
[a]*Funneliformis* aff. *verruculosum* (Błaszk.) C. Walker and A. Schüssler	Carvalho et al. (2012)
[a]*Funneliformis* sp.	Coutinho et al. (2015)
[a, b]*Fuscutata heterogama* (T.H. Nicolson and Gerd.) Sieverd., F.A. de Souza and Oehl	Carvalho et al. (2012), Coutinho et al. (2015)
[a]*Fuscutata rubra* (Stürmer and J.B. Morton) Oehl, F. A. de Souza and Sieverd.	Carvalho et al. (2012), Coutinho et al. (2015)
Gigaspora albida N.C. Schenck and G.S. Sm.	Souza et al. (2010)
[a]*Gigaspora decipiens* I.R. Hall and L.K. Abbott	Carvalho et al. (2012), Coutinho et al. (2015)

(continued)

Table 8.2 (continued)

Species	Reference
[a]*Gigaspora gigantea* (T.H. Nicolson and Gerd.) Gerd. and Trappe	Souza et al. (2010), Coutinho et al. (2015)
[a]*Gigaspora margarita* W.N. Becker and I.R. Hall	Souza et al. (2010), Carvalho et al. (2012), Coutinho et al. (2015)
Gigaspora rosea T.H. Nicolson and N.C. Schenck	Souza et al. (2010)
Glomus clavisporum (Trappe) R.T. Almeida and N.C. Schenck	Souza et al. (2010)
[a]*Glomus diaphanum* J.B. Morton and C. Walker	Carvalho et al. (2012), Coutinho et al. (2015)
[a, b]*Glomus glomerulatum* Sieverd.	Carvalho et al. (2012), Coutinho et al. (2015)
[a]*Glomus macrocarpum* Tul. and C. Tul.	Souza et al. (2010), Carvalho et al. (2012), Coutinho et al. (2015)
[a]*Glomus microcarpum* Tul. and C. Tul.	Souza et al. (2010), Carvalho et al. (2012), Coutinho et al. (2015)
[a]*Glomus* aff. *pellucidum* McGee and Pattinson	Carvalho et al. (2012)
[a]*Glomus* aff. *rubiforme* (Gerd. and Trappe) R.T. Almeida and N.C. Schenck	Souza et al. (2010), Coutinho et al. (2015)
[a]*Glomus* sp. 1	Carvalho et al. (2012)
[a]*Glomus* sp. 2	Coutinho et al. (2015)
[a]*Glomus* sp. 3	Coutinho et al. (2015)
[a]*Glomus* sp. 4	Coutinho et al. (2015)
[a]*Glomus* sp. 5	Coutinho et al. (2015)
[a]*Glomus* sp. 6	Coutinho et al. (2015)
[a]*Glomus* sp. 7	Coutinho et al. (2015)
[a]*Kuklospora colombiana* (Spain and N.C. 68 Schenck) Oehl and Sieverd.	Souza et al. (2010), Carvalho et al. (2012)
[a]*Orbispora pernambucana* (Oehl, D.K. Silva, N. Freitas, L.C. Maia) Oehl, G.A.Silva and D.K. Silva	Carvalho et al. (2012), Coutinho et al. (2015)
[a, b]*Pacispora dominikii* (Blaszk.) Sieverd. and Oehl	Carvalho et al. (2012)
[a](cf) *Pacispora robigina* Sieverd. and Oehl	Carvalho et al. (2012)
Pacispora scintilans (S.L. Rose and Trappe) Sieverd. and Oehl ex C Walker, Vestberg and Schuessler	Souza et al. (2010)
[a]*Pacispora* sp.	Coutinho et al. (2015)
Paraglomus albidum (C. Walker and L.H. Rhodes) Oehl, F.A. Souza, G.A. Silva and Sieverd.	Souza et al. (2010)
[a]*Paraglomus occultum* (C. Walker) J.B. Morton and D. Redecker	Souza et al. (2010), Carvalho et al. (2012), Coutinho et al. (2015)
Racocetra coralloidea (Trappe, Gerd. and I. Ho) Oehl, F.A. de Souza and Sieverd.	Souza et al. (2010)
[a]*Racocetra fulgida* (Koske and C. Walker) Oehl, F.A. de Souza and Sieverd.	Carvalho et al. (2012), Coutinho et al. (2015)

(continued)

Table 8.2 (continued)

Species	Reference
Racocetra persica (Koske and C. Walker) Oehl, F.A. de Souza and Sieverd.	Souza et al. (2010)
Redeckera fulva (Berk. and Broome) C. Walker and A. Schüssler	Souza et al. (2010)
[a]*Rhizoglomus clarum* (T.H. Nicolson and N.C. Schenck) Sieverd., G.A. Silva and Oehl	Souza et al. (2010), Carvalho et al. (2012), Coutinho et al. (2015)
[a]*Rhizoglomus fasciculatum* (Thaxt.) Sieverd., G.A. Silva and Oehl	Souza et al. (2010), Carvalho et al. (2012), Coutinho et al. (2015)
[a]*Rhizoglomus invermaium* (I.R. Hall) Sieverd., G.A. Silva and Oehl	Carvalho et al. (2012), Coutinho et al. (2015)
[a]*Rhizoglomus microaggregatum* (Koske, Gemma and P.D. Olexia) Sieverd., G.A. Silva and Oehl	Souza et al. (2010), Carvalho et al. (2012), Coutinho et al. (2015)
Sclerocystis coremioides, Berk. and Broome	Souza et al. (2010)
[a, b]*Sclerocystis sinuosa* Gerd. and B.K. Bakshi	Coutinho et al. (2015)
Scutellospora auriglоba (I.R. Hall) C.Walker and F. E. Sanders	Souza et al. (2010)
[a]*Scutellospora calospora* (T.H. Nicolson and Gerd.) C. Walker and F.E. Sanders	Souza et al. (2010), Carvalho et al. (2012), Coutinho et al. (2015)
[a, b]*Scutellospora dipurpurescens* J.B. Morton and Koske	Carvalho et al. (2012), Coutinho et al. (2015)
[a]*Scutellospora* sp. 1	Carvalho et al. (2012)
[a]*Scutellospora* sp. 2	Coutinho et al. (2015)
[a]*Septoglomus constrictum* (Trappe) Sieverd G.A. Silva and Oehl	Carvalho et al. (2012), Coutinho et al. (2015)

[a]AMF species reported in the rupestrian grasslands
[b]AMF species reported in the Cerrado sensu lato collected exclusively in rupestrian grasslands

8.3 The Influence of Environmental Factors on Endophytic Fungi and Arbuscular Mycorrhizal Fungi

8.3.1 Endophytic Fungi

A factor that greatly influences the diversity of endophytic fungi is the season. Numerous studies on oaks (*Quercus* spp.) have shown a greater diversity of endophytes in the rainy season, when the relative humidity and water-mediated spore dispersal is higher. In rupestrian grasslands, the influence of seasonality on endophytic fungi is relatively unknown. However, one study showed that the number of endophytic fungal species in *B. dracunculifolia* was higher during the dry season (42 species) than in the wet season (18 species) (Oki et al. 2008).

Endophytes may be strongly affected by natural disturbances that severely impact their host plants. After being damaged by a strong hailstorm, *Coccoloba cereifera* showed lower performance, decreased photosynthetic rates and increased production of secondary metabolites, which negatively influenced the number of endophytic fungi (Fernandes et al. 2011). These authors reported that endophyte species richness decreased by two thirds.

Increased atmospheric CO_2 concentrations also significantly affect the frequency and survival of endophytes in *B. dracunculifolia*. Under higher concentrations of CO_2, plants presented great structural, nutritional, and chemical modifications that changed endophyte community composition (Oki and Fernandes in preparation). About 50 % more endophytic fungal species were found in *B. dracunculifolia* under a CO_2 enriched environment. Curiously, some endophytic species with the potential to minimize infection by pathogens were not found in the treatment with elevated concentration of CO_2.

8.3.2 Arbuscular Mycorrhizal Fungi

Soil characteristics, plant diversity, altitude and season are often evaluated as factors affecting the occurrence and diversity of AMF.

The structure of rupestrian grassland communities varies across short distances. The substrate often differs within a few meters or even centimeters. Carvalho et al. (2012) showed that the different habitats (cerrado, rocky outcrops, sandy grasslands, peat bogs, stony grasslands) presented different AMF communities and probably influences plant communities. However, the authors did not find a significant correlation between AMF diversity and plant diversity, although they reported distinct patterns of organization according to habitat type. The lowest AMF species richness was found in sandy grasslands and rocky outcrops, while the greatest AMF species richness was reported in the stony grasslands (see Carvalho et al. 2012). These differences can be related to dominant plant species found in these habitats. In sandy grasslands the habitat was dominated by *Lagenocarpus rigidus* (Cyperaceae), a specie recognized as non-mycorrhizal or with lower incidence of mycorrhizal colonization. In areas of rocky outcrops, the dominant species was *Trachypogon spicatus*, a slow-growing climax Poaceae that is slightly dependent on AMF. In AMF rich stony grasslands, the dominant species was a *Vellozia* sp., a monocotyledon that ceases photosynthetic activity under drought deficit conditions (Carvalho et al. 2012).

The influence of soil characteristics on AMF richness was also reported by Coutinho et al. (2015) in a study on AMF occurrence and diversity along an altitudinal gradient (800–1400 m). The chemical attributes most strongly correlated with AMF occurrence were the sum of bases, organic matter content and cation exchange capacity and soil structure. This study revealed higher spore density at 1100 m, where fine sand content and aluminum saturation were higher, and phosphorus levels were lower. These soil properties might have led AMF to produce a greater number of propagules (Coutinho et al. 2015).

Although Stürmer and Siqueira (2008) emphasized the difficulty in establishing a clear link between soil and environmental variables and the diversity and occurrence of AMF, it is clear that the intrinsic characteristics, such as soil attributes and vegetation, found in the rupestrian grassland habitat mosaic, ought to influence the occurrence and diversity of AMF.

8.4 Bioprospecting for Endophytic Fungi and AMF from Rupestrian Grasslands

Endophytic fungi have been described as a promising source of compounds with potential medicinal applications (Strobel 2002). In recent years, this fungal group has become an attractive target for the discovery of new classes of natural products, and many studies on tropical plants have documented remarkable endophyte richness (e.g., Vieira et al. 2014). The possibilities of finding new and potential compounds from these fungi are higher in endemic plants and in those with medicinal importance (Strobel 2002). Among Brazilian ecosystems, the rupestrian grasslands have a rich biodiversity of potential source plants, including endemic species and species with established medicinal applications (Dantas et al. 2005; Jacobi et al. 2007; Ventura et al. 2007).

Despite their potential, only a few endophytic fungal taxa found in rupestrian grasslands have been reported as producers of antimicrobial compounds (Table 8.3). Vieira et al. (2014) obtained different endophytic fungi associated with *B. trimera* (Less) D.C., a medicinal plant, commonly known as "carqueja", that is used to treat skin diseases, ulcers, anemia, diabetes, malaria, inflammatory reactions, and hepatic diseases (Soicke et al. 1986; Souza et al. 1991; Gene et al. 1996). Among the fungi isolated from this species, 23 isolates displayed antimicrobial activities, of which *Epicoccum* sp., *Pestalotiopsis* sp. 1, *Cochliobolus lunatus,* and *Nigrospora* sp. presented the best Minimum Inhibitory Concentration (MIC) values.

Three endophytic fungal species in *B. dracunculifolia* (*P. africana*, *X. venulosa, A. pullulans*) showed a significant effect on phyto and human pathogens (Oki et al. data not shown). The extract of *P. africana* had the greatest inhibitory activity against *Penicillium digitatum* and prevented the growth of three pathogens. The efficiency of *A. pullulans* and *X. venulosa* against pathogens may be associated with the presence of triterpenes and phenols, detected in the extracts. Triterpenes have a variety of functions in plants, with emphasis on protection against herbivory, induction of seed germination and antimitotic agents (Novaes et al. 2007). Investigations of endophytic fungi belonging to the genus *Xylaria* demonstrated that they produce substances with potential pharmaceutical and agricultural application (Sette et al. 2006), highlighting the importance of this genus.

From leaves, stems, and roots of orchids, Vaz et al. (2009) isolated 395 endophytic fungi that were examined for the production of crude extracts with

Table 8.3 Biological activity of endophytic fungi associated with plant species from rupestrian grasslands

Endophytic fungi	Biological activity	Reference
Acremonium strictum	Antibacterial—against *Escherichia coli, Staphylococcus aureus, Bacillus cereus, Salmonella thyphimurium*	Vaz et al. (2009)
Alternaria sp.	Antibacterial—*E. coli, S. aureu, B. cereus, S. thyphimurium* Antifungal—*Paracoccidiodes brasiliensis* (Pb18)	Vaz et al. (2009), Vieira et al. (2014)
Aureobasidium pullulans (de Bary) G. Arnaud (1918)	Antifungal—*Penicillium digitatum, Colletotricum acutatum* and *Botrytis cinerea*	Nascimento (2010)
Bensingtonia sp.	Antibacterial—*E. coli, S. aureu, B. cereus, S. thyphimurium*	Vaz et al. (2009)
Candida parapsilosis	Antibacterial—*E. coli, S. aureu, B. cereus, S. thyphimurium*	Vaz et al. (2009)
Chaetomium sp.	Antifungal—*Cryptococcus neoformans, P. brasiliensis* (Pb18)	Vieira et al. (2014)
Cochliobolus lunatus	Antifungal—*P. brasiliensis* (Pb18)	Vieira et al. (2014)
Coniophora sp.	Antifungal—*Candida albicans, Candida krusei, Candida glabrata, C. neoformans*	Vaz et al. (2009)
Cylindrocarpon sp.	Antibacterial—*E. coli, S. aureus, B. cereus, S. thyphimurium* Antifungal—*C. albicans, C. krusei, C. glabrata, C. neoformans*	Vaz et al. (2009)
Diaporthe phaseolorum	Antifungal—*C. neoformans, P. brasiliensis* (Pb18)	Vieira et al. (2014)
Diaporthe sp.	Antifungal—*C. neoformans, P. brasiliensis* (Pb18)	Vieira et al. (2014)
Epicoccum nigrum	Antibacterial—*E. coli, S. aureus, B. cereus, S. thyphimurium* Antifungal—*C. albicans, C. krusei, C. glabrata, C. neoformans*	Vaz et al. (2009) Vieira et al. (2014)
Epicoccum sp.	Antifungal—*C. tropicalis, P. brasiliensis* (Pb18)	Vieira et al. (2014)
Fusarium oxysporum	Antifungal—*C. albicans, C. krusei, C. glabrata, C. neoformans*	Vaz et al. (2009)
Fusarium sp.	Antifungal—*C. albicans, C. krusei, C. glabrata, C. neoformans*	Vaz et al. (2009)
Gibberella moniliformis	Antifungal—*C. albicans, C. krusei, C. glabrata, C. neoformans*	Vaz et al. (2009)
Guignardia sp.	Antifungal—*P. brasiliensis* (Pb18)	Vieira et al. (2014)
Hypocrea sp.	Antifungal—*C. albicans, C. krusei, C. glabrata, C. neoformans*	Vaz et al. (2009)
Nigrospora sp.	Antifungal—*C. tropicalis, P. brasiliensis* (Pb18)	Vieira et al. (2014)

(continued)

Table 8.3 (continued)

Endophytic fungi	Biological activity	Reference
Pestalotiopsis sp.	Antifungal—*C. albicans, C. krusei, C. glabrata, C. neoformans, P. brasiliensis* (Pb18)	Vieira et al. (2014)
Phaeosphaeria avenaria	Antifungal—*C. albicans, C. krusei, C. glabrata, C. neoformans* Antibacterial—*E. coli, S. aureus, B. cereus, S. thyphimurium*	Vaz et al. (2009)
Phoma sp.	Antifungal—*C. albicans, C. tropicalis, C. neoformans, C. gatti, P. brasiliensis* (Pb18), *C. parapsilosis*	Vieira et al. (2014)
Phomopsis sp.	Antibacterial—*E. coli* Antifungal—*C. neoformans, P. brasiliensis* (Pb18)	Vieira et al. (2014)
Podospora sp.	Antifungal—*P. brasiliensis* (Pb18)	Vieira et al. (2014)
Preussia africana	Antifungal—*P. brasiliensis* (Pb18)	Vieira et al. (2014)
	Antifungal—*P. digitatum, B. cinerea, Pestalotiopsis longisetula* and *C. albicans*	Nascimento (2010)
Preussia pseudominima	Antifungal—*C. tropicalis, P. brasiliensis* (Pb18), *C. neoformans*	Vieira et al. (2014)
Preussia sp.	Antifungal—*P. brasiliensis* (Pb18)	Vieira et al. (2014)
Rhodotorula mucilaginosa	Antibacterial—*E. coli, S. aureus, B. cereus, S. thyphimurium*	Vaz et al. (2009)
Sclerostagonospora opuntiae	Antifungal—*C. albicans, C. krusei, C. glabrata, C. neoformans*	Vaz et al. (2009)
Sporormiella sp.	Antifungal—*P. brasiliensis* (Pb18)	Vieira et al. (2014)
Trichoderma asperellum	Antifungal—*C. albicans, C. krusei, C. glabrata, C. neoformans* Antibacterial—*E. coli, S. aureus, B. cereus, S. thyphimurium*	Vaz et al. (2009)
Xylaria sp.	Antifungal—*P. brasiliensis* (Pb18)	Vieira et al. (2014)
Xylaria venulosa	Antifungal—*P. digitatum, B. cinerea*	Nascimento (2010)

antimicrobial activity. Among these fungi, 33 % displayed antimicrobial activity against at least one target microorganism. *Alternaria* sp. and *Fusarium oxysporum* presented the strongest antibacterial activity. Three *Fusarium* isolates, *Epicoccum nigrum,* and *Sclerostagonospora opuntiae* showed the greatest Minimum Inibitory Concentration (MIC) values against the pathogenic yeasts.

At present, only one compound has been isolated from an endophytic fungus associated with a rupestrian plant species. Altenusin is a metabolite obtained from the organic extract of a broth culture of *Alternaria* sp. UFMGCB 55 recovered from the plant *Trixis vauthieri*. Altenusin inhibited TryR enzymatic activity with an IC_{50} value of 4.3 mM (Cota et al. 2008).

Preliminary studies indicate that endophytes associated with rupestrian plant taxa may represent an excellent source of bioactive compounds with potential applications in the pharmaceutical industry. With the application of new genomic, proteomic and drug design techniques, endophytic fungal communities associated with rupestrian grassland plants may emerge as important sources of novel bioactive compounds.

While AMF may reduce the negative effects of nutrient limitation stress (Sylvia and Williams 1992), infection by some endophytic fungal species can increase plant tolerance of acidic soils with high aluminum concentrations and increase the plant growth (Liu et al. 1996; Zaurov et al. 2001). AMF species perform multiple environmental services related to soil structure, plant nutrition, nutrient cycling, pathogen control, agroforestry production, recovery of degraded areas, and environmental restoration (e.g., Carneiro et al. 1998; Dodd 2000; Berbara et al. 2006). However, studies on the relationship between ecosystem dynamics and AMF are still needed in the rupestrian grasslands.

In summary, both endophytic fungi and AMF are of major importance in rupestrian grasslands. The significance of their role in plant metabolism and ecology is undeniable, allowing their hosts to persist in this harsh ecosystem. Studies on their diversity and distribution may contribute to a better understanding of how this ecosystem functions. A strong relationship may exist between both, but there are no studies relating these fungal groups in rupestrian grasslands, probably due to the differences in sampling methodologies required for AMF and endophytes. Exploring this gap will contribute to our understanding of this unique ecosystem and the relations in it.

Acknowledgements We thank for CNPq, FAPEMIG and CAPES for grants and scholarships to the authors. We also thank for John Schroeder for careful reading the manuscript and for his helpful comments and suggestions.

References

Arnold AE, Herre EA (2003) Canopy cover and leaf age affect colonization by tropical fungal endophytes: Ecological pattern and process in *Theobroma cacao* (Malvaceae). Mycol 95:388–398

Arnold AE, Lutzoni F (2007) Diversity and host range of foliar fungal endophytes: are tropical leaves biodiversity hostpots? Ecology 88:541–549

Azevedo JL (1998) Microorganismos endofíticos. In: de Melo IS, Azevedo JL (eds) Ecologia microbiana. Embrapa, Jaguariúna, pp 117–137

Belmiro MS (2010) Herbivoria e resinas - fatores mediadores da composição de fungos endofíticos em *Baccharis dracunculifolia* DC (Asteraceae). 2010. Monograph, Pontifícia Universidade Católica de Minas Gerais

Berbara RLL, De Sousa FA, Fonseca HMA (2006) Fungos micorrízicos arbusculares: muito além da nutrição. In: Fernandes MS (ed) Nutrição mineral de plantas. SBCS, Viçosa, pp 53–88

Bonfim JA, Vasconcellos RLF, Stürmer SL, Cardoso EJBN (2013) Arbuscular mycorrhizal fungi in the Brazilian Atlantic forest: a gradient of environmental restoration. Appl Soil Ecol 71:7–14

Bononi VLR, Trufem SFB (1983) Endomicorrizas vesículo-arbusculares do cerrado da Reserva Biológica de Moji-Guaçú, SP, Brasil. Rickia 10:55–84

Breen JP (1994) Acremonium endophyte interactions with enhanced plant resistance to insect. Annu Ver Entomol 86:401–423

Carneiro MAC, Siqueira JO, Moreira FMS, Carvalho D, Botelho S (1998) Micorriza arbusculares em espécies arbóreas e arbustivas nativas de ocorrência no Sudeste do Brasil. Rev Cerne 4:129–145

Carvalho F, Souza FA, Carrenho R, Moreira FMS, Jesus EC, Fernandes GW (2012) The mosaic of habitats in the high-altitude Brazilian rupestrian fields is a hotspot for arbuscular mycorrhizal fungi. Appl Soil Ecol 52:9–19

Clay K (1990) Fungal endophytes of grasses. Annu Rev Ecol Evol Syst 21:275–297

Clay K (2004) Fungi and the food of the gods. Nature 427:401–402

Cordazo CV, Stürmer SL (2007) Ocorrência de fungos micorrízicos arbusculares em *Panicum Racemosum* (P. Beauv.) Spreng (*Poaceae*) em dunas costeiras do extremo sul do Brasil. Atlântica 29:65–68

Cordoba AS, Mendonça MM, Stürmer SL, Rygiewicz PT (2001) Diversity of arbuscular mycorrhizal fungi along a sand dune stabilization gradient: a case study at Praia da Joaquina, Ilha de Santa Catarina, South Brasil. Mycoscience 42:379–387

Cota B, Rosa L, Caligiorne R, Rabello T, Rosa C, Zani C (2008) Altenusin, a biphenyl isolated from the endophytic fungus *Alternaria* sp. inhibits trypanothione reductase from *Trypanosoma cruzi*. FEMS Microbiol Lett 285:177–182

Coutinho ES, Fernandes GW, Berbara RLL, Valerio HM, Goto BT (2015) Variation of arbuscular mycorrhizal fungal communities along an altitudinal gradient in rupestrian fields in Brazil. Mycorrhiza 25:627–638

Dantas ALA, Valente LMM, Morais LASD, Feliciano G, Bernardo-Filho M (2005) Vitro study of *Vellozia pusilla* pohl (Velloziaceae), a Brazilian plant species: antitumoral activity and labeling of blood elements. Braz Arch Bio Tech 48:63–67

Dodd JC (2000) The role of arbuscular mycorrizal fungi in agro and natural ecosystems. Outlook Agri 29:55–62

Fernandes GW, Price PW (1992) The adaptive significance of insect gall distribution: survivorship of species in xeric and mesic habitats. Oecol 90:14–20

Fernandes GW, Oki Y, Sanchez-Azofeifa A, Faccion G, Amaro-Arruda HC (2011) Hail impact on leaves and endophytes of the endemic threatened *Coccoloba cereifera* (Polygonaceae). Plant Ecol 212:1687–1697

Gaylord ES, Preszler RW, Boecklen WJ (1996) Interaction between host plants, endophytic fungi and phytophagous insect in an oak (*Quercus grisea* × *Quercus gambelli*) hybrid zone. Oecol 105:336–342

Gene RM, Panizza S, Correa PL (1996) Anti-inflamatory and analgesic activity of *Baccharis trimera*: identification of its constituents. Planta Med 62:232–235

Gianinazzi S, Gollote A, Binet MN, Van Tuinen T, Redecker D, Wipf D (2010) Agroecology: the key role of arbuscular mycorrhizas in ecosystem services. Mycorrhiza 20:519–530

Gomide PHO, Silva MLN, Soares CRFS, Cardoso EL, Carvalho F, Leal PL, Marques RM, Stürmer SL (2014) Fungos micorrízicos arbusculares em fitofisionomias do pantanal da Nhecolândia, Mato Grosso do Sul. Rev Bras Ci Solo 38:1114–1127

Goto BT, Silva GA, Yano-Melo AM, Maia LC (2010) Checklist of the arbuscular mycorrhizal fungi (*Glomeromycota*) in the Brazilian semiarid. Mycotaxon 113:251–254

Goto BT, Silva GA, Assis DM, Silva DKA, Souza RG, Ferreira ACA, Jobim K, Mello CMA, Vieira HEE, Maia LC, Oehl F (2012) Intraornatosporaceae: (Gigasporales), a new family with two new genera and two new species. Mycotaxon 119:117–132

Gupta V, Satyanarayana T, Garg S (2000) General aspects of mycorrhiza. In: Mukerji KG, Chamola BP, Singh J (eds) Mycorrhizal biology. Kluwer Academic/Plenum Publishers, New York, pp 27–44

Hawksworth DL (2001) The magnitude of fungal diversity: the 1.5 million species estimate revised. Mycol Res 105:1422–1432

Jacobi CM, Do Carmo FF, Vincent RC, Stehmann JR (2007) Plant communities on ironstone outcrops: a diverse and endangered Brazilian ecosystem. Biodiv Cons 16:2185–2200

Kivlin SN, Hawkes CV, Treseder KK (2011) Global diversity and distribution of arbuscular mycorrhizal fungi. Soil Biol Biochem 43:2294–2303

Leal PL, Stürmer SL, Siqueira JO (2009) Occurrence and diversity of arbuscular mycorrhizal fungi in trap cultures from soils under different land use systems in the Amazon, Brazil. Braz J Microb 40:111–121

Leal PL, Siqueira JO, Stürmer SL (2013) Switch of tropical Amazon forest to pasture affects taxonomic composition but not species abundance and diversity of arbuscular mycorrhizal fungal community. Appl Soil Ecol 71:72–80

Lee EH, Eo JK, Ka KH, Eom AH (2013) Diversity of arbuscular mycorrhizal fungi and their roles in ecosystems. Mycobiology 41:121–125

Lima LL, Kozovits AR, Assis DMA, Silva GA, Oehl F (2014) *Cetraspora auronigra*, a new glomermycetes species from Ouro Preto, Minas Gerais (Brazil). Sydowia 66:299–308

Liu H, Heckman JR, Murphy JA (1996) Screening fine fescues for aluminum tolerance. J Plant Nutr 19:677–688

Marks S, Clay K (1996) Physiological responses of *Festuca arundinacea* to fungal endophyte infection. New Phytol 133:727–733

Mello CMA, Silva GA, Vieira HEE, Silva IR, Maia LC, Oehl F (2012) *Fuscutata aurea*, a new species in the Glomeromycetes from cassava and maize fields in the Atlantic rainforest zone in Northeastern Brazil. Nova Hedwigia 95:267–275

Miranda JCC (2008) Cerrado: micorriza arbuscular: ocorrência e manejo. Embrapa Cerrados, Planaltina 169p

Moreira M, Baretta D, Tsai SM, Gomes-da-Costa SM, Cardoso EJBN (2007) Biodiversity and distribution of arbuscular mycorrhizal fungi in *Araucaria angustifolia*. Forest Sci Agric 64:393–399

Nascimento IM (2010) Avaliação da atividade antipatogênica e enzimática de três fungos endofiticos de *Baccharis dracunculifolia* (Asteraceae). Monograph, Pontifícia Universidade Católica de Minas Gerais

Novaes MRCG, Novaes LCG, Taveira VC (2007) Efeitos farmacológicos dos fungos Agaricales: uma revisão de evidência. Rev Cienc Med 16:87–95

Novais CB, Borges WL, Jesus EC, Saggin Junior OJ, Siqueira JO (2014) Inter and intraspecific functional variability of tropical arbuscular mycorrhizal fungi isolates colonizing corn plants. Appl Soil Ecol 76:78–86

Oki Y, Fernandes GW, Correa Junior A (2008) Fungos: Amigos ou Inimigos? Cien Hoje 42:64–66

Oki Y, Soares NR, Belmiro MS, Correa-Junior A, Fernandes GW (2009) The influence of the endophytic fungi on the herbivores from *Baccharis dracunculifolia* (Asteraceae). Neotrop Biol Cons 4:83–88

Oliveira JRG, Souza RG, Silva FSB, Mendes ASM, Yano-Melo AM (2009) O papel da comunidade de fungos micorrízicos arbusculares (FMA) autóctones no desenvolvimento de espécies vegetais nativas em área de dunas de restinga revegetadas no litoral do Estado da Paraíba. Rev Bras Bot 32:663–670

Pagano MC, Scotti MR (2009) A survey of the mycorrhiza occurrence in *Paepalanthus bromelioids* and *Bulbostylis* sp. in rupestrian fields, Brazil. Micol Appl Int 21:1–10

Pereira CMR, Silva DKA, Ferreira ACA, Goto BT, Maia LC (2014) Diversity of arbuscular mycorrhizal fungi in Atlantic forest areas under different land uses. Agr Ecosyst Environ 185:245–252

Petrini O, Sieber TH, Toti L, Viret O (1992) Ecology, metabolite production, and substrate utilization in endophytic fungi. Nat Tox 1:185–196

Prance GT (1994) A comparison of the efficacy of higher taxa and species numbers in the assessment of biodiversity in the Neotropics. Philos Trans R Soc Lond B Biol Sci 345:89–99

Preszler RW, Gaylord ES, Boecklen WJ (1996) Reduced parasitism of a leaf-mining moth on trees with high infection frequencies of an endophytic fungus. Oecol 108:159–166

Read DJ, Perez-Moreno J (2003) Mycorrhizas and nutrient cycling in ecosystems–a journey towards relevance? New Phytol 157:475–492

Redman RS, Sheehan KB, Stout RG, Rodriguez JR, Henson JM (2002) Thermotolerance generated by plant/fungal symbiosis. Science 298:1581p

Rodrigues RL (2010) Fungos endofíticos associados à *Vellozia compacta* Mart. ex Schult. F. (Velloziaceae) presente em afloramentos rochosos nos estados de Minas Gerais e Tocantins. Dissertacion, Universidade Federal de Ouro Preto

Rodriguez JR, Redman RS, Henson JM (2004) The role of fungal symbioses in the adaptation of plants to high stress environments. Mitig Adap Strat Glob Chan 9:261–272

Saikkonen K, Wäli P, Helander M, Faeth SH (2004) Evolution of endophyte–plant symbioses. Trends Plant Sci 9:275–280

Sanchez-Azofeifa A, Oki Y, Fernandes GW, Ball RA, Gamon J (2011) Relationships between endophyte diversity and leaf optical properties. Trees 26:291–299

Santiago IF (2008) Bioprospecção de fungos endofíticos associados à *Vellozia graninea* Pohl. presente em afloramentos rochosos da Cadeia do Espinhaço, Minas Gerais. Monograph, Universidade Federal de Ouro Preto

Santos OM, Oliveira NC, Novais RF (1995) Observações preliminares sobre fungos micorrízicos vesículo-arbusculares em plantas crescendo em dunas na Bahia. Rev Ceres 42:191–202

Schenck NCL, Siqueira JO, Oliveira E (1989) Changes in the incidence of VA mycorrhizal fungi with changes in ecosystems. Dev Soil Sci 18:125–129

Sette LD, Passarini MRZ, Delarmelina C, Salati F, Duarte MCT (2006) Molecular characterization and antimicrobial activity of endophytic fungi from coffee plants. World J Microb Biot 22:1185–1195

Silva DKA, Pereira CMR, Souza RG, Silva GA, Oehl F, Maia LC (2012) Diversity of arbuscular mycorrhizal fungi in resting and dunes areas in Brazilian Northeast. Biodiv Cons 21:2361–2373

Siqueira JO, Colozzi-Filho A, Oliveira E (1989) Ocorrência de micorrizas vesicular-arbusculares em agro e ecossistemas do estado de Minas Gerais. Pesq Agropec Bras 24:1499–1506

Smith SE, Read D (2008) Mycorrhizal symbiosis, 3ª edn. Academic Press, San Diego

Soicke H, Almeida ER, Costa AF (1986) Characterisation of flavonoids from *Baccharis trimera* and their properties. Planta Med 52:37–39

Sousa MP, Matos MO, Matos FJ (1991) Constituintes químicos de plantas medicinais brasileiras. UFC Edições, Fortaleza

Souza FA, Silva ICL, Berbara RLL (2007) Fungos micorrízicos arbusculares: muito mais diversos do que se imaginava. In: Moreira FMS, Siqueira JO, Brussard L (eds) Biodiversidade do solo em ecossistemas brasileiros. UFLA, Lavras, pp 483–536

Souza FA, Stürmer SL, Carrenho R, Trufem SFB (2010) Classificação e taxonomia de fungos micorrízicos arbusculares e sua diversidade e ocorrência no Brasil. In: Siqueira JO, Souza FA, Cardoso EJBN, Tsai SM (eds) Micorrizas: 30 anos de pesquisas no Brasil. Editora UFLA, Lavras p, pp 15–73

Strobel G (2002) Endophytes as sources of bioactive products. Microb Infect 5:535–544

Stürmer SL, Siqueira JO (2008) Diversidade de fungos micorrízicos arbusculares em ecossistemas brasileiros. In: Moreira FMS, Siqueira JO, Brussard L (eds) Biodiversidade do solo em ecossistemas brasileiros. UFLA, Lavras, pp 537–584

Stürmer SL, Stürmer R, Pasqualini D (2013) Taxonomic diversity and community structure of arbuscular mycorrhizal fungi (Phylum Glomeromycota) in three maritime sand dunes in Santa Catarina State, South Brazil. Fungal Ecol 6:27–36

Sylvia DM, Williams SE (1992) Vesicular-arbuscular mycorrhizae and environmental stress. In: Bethlenfalvay GJ, Linderman RG (eds) Mycorrhizae in sustainable agriculture. Madison, J AM Soc Agron, pp 101–124

Tejesvi MV, Kini KR, Prakash HS, Subbiah V, Shetty HS (2007) Genetic diversity and antifungal activity of species of *Pestalotiopsis* isolated as endophytes from medicinal plants. Fungal Divers 24:37–54

Trufem SFB (1990) Aspectos ecológicos de fungos micorrízicos vesículo-arbusculares da mata tropica tropical úmida da Ilha do Cardoso, SP, Brasil. Acta Bot Bras 4:31–45

Trufem SFB (1995) Aspectos ecológicos de fungos micorrízicos arbusculares na rizosfera de plantas de restinga da Ilha do Cardoso, SP, Brasil. Rev Bras Bot 18:51–60

Trufem SFB, Otomo HS, Malatinszky SMM (1989) Fungos micorrízicos vesículos arbusculares em rizosferas de plantas em dunas do Parque Estadual da Ilha do Cardoso, São Paulo, Brasil. Acta Bot Bras 3:141–152

Trufem SFB, Malatinszky SMM, Otomo HS (1994) Fungos micorrízicos arbusculares em rizosferas de plantas do litoral arenoso do Parque Estadual da Ilha do Cardoso, SP, Brasil. Acta Bot Bras 8:219–229

Vaz ABM, Mota RC, Bomfim MRQ, Vieira MLA, Zani CL, Rosa CA, Rosa LH (2009) Antimicrobial activity of endophytic fungi associated with *Orchidaceae* in Brazil. Can J Microbiol 55:1381–1391

Ventura CP, Oliveira ABD, Braga FC (2007) Antimicrobial activity of *Trembleya laniflora*, *Xyris platystachia* and *Xyris pterygoblephara*. Rev Bras Farm 17:17–22

Vieira ML, Johann S, Hughes FM, Rosa CA, Rosa LH (2014) The diversity and antimicrobial activity of endophytic fungi associated with medicinal plant *Baccharis trimera* (Asteraceae) from the Brazilian savannah. Microbiol 60:847–856

Wilson D (1993) Fungal endophytes: out of sight but should not be out of mind. Oikos 68:379–384

Zaurov DE, Bonos S, Murphy JA, Richardson M, Belanger FC (2001) Endophyte infection can contribute to aluminum tolerance in fine fescues. Crop Sci 41:1981–1984

Chapter 9
Succession on the Rocky Outcrop Vegetation: A Rupestrian Grassland Scheme

Abel A. Conceição and José R. Pirani

Abstract We investigated the nature of the temporal processes involved in the formation of vegetation on the heterogeneous landscape of rocky outcrops on ridge tops in rupestrian grasslands (*campos rupestres*), Northeastern Brazil. Relationships were established between spatial patterns and gradual processes of succession and their respective species life-forms. The first plants to occupy the isolated rock are mainly chamaephyte monocotyledons (tolerant to desiccation or epilithic) and hemicryptophytes (with or without rosettes). With time, species with other life-forms and eudicotyledons join them, gradually to go from a solitary spatial distribution on the rock to a more continuous distribution. In the initial steps, droughts and storms are the most intense disturbances. In subsequent steps, due to the continuity of the vegetation and high frequency of fires, fire is the predominant factor. The scheme presented here follows the general pattern described for vegetation in other rocky outcrops, with similar taxonomic and/or functional groups. The simultaneous occurrence of several successional steps in the same area increases the probability that part of the vegetation will remain when confronted with disturbances, and constitutes an important diversity factor in rupestrian grasslands.

A.A. Conceição (✉)
Departamento de Ciências Biológicas, Universidade Estadual de Feira de Santana, Feira de Santana, BA, Brazil
e-mail: abel18@gmail.com

J.R. Pirani
Instituto de Biociências, Departamento de Botânica, Universidade de São Paulo, São Paulo, SP, Brazil
e-mail: pirani@usp.br

© Springer International Publishing Switzerland 2016
G.W. Fernandes (ed.), *Ecology and Conservation of Mountaintop Grasslands in Brazil*, DOI 10.1007/978-3-319-29808-5_9

Contents

9.1 Succession on Rocky Outcrops

In rocky outcrops, the vegetation isolated by the exposed rock facilitates the physical delimitation of natural units, where migration is restricted, constituting useful systems for the preparation of models (McCormick et al. 1974; Wiser et al. 1996). Severe environmental conditions, such as soil deficiencies, high daily temperature oscillations and intense insolation restrict the occupation of the exposed rock by many organisms, limiting their species composition (Shure and Ragsdale 1977).

Due to the great heterogeneity verified on the surface of rocky outcrops, several habitats have been described and related to succession (e.g. Oosting and Anderson 1937; Burbanck and Platt 1964; Hambler 1964; Oliveira et al. 1975; Shure and Ragsdale 1977; Houle 1990; Ibisch et al. 1995; Alves and Kolbek 2000; Ribeiro and Medina 2002). Characterized by a sequence of temporal changes in the species composition of a community, the succession is supposed to be associated with a sequence of changes in its structural and functional properties (Drury and Nisbet 1973), driven towards greater diversity of species, biomass and vertical stratification, with each taxonomic group increasing at different rates and to different degrees depending on particular causal factors (Shure and Ragsdale 1977).

The vegetation on rocky outcrops is special in terms of succession, since it includes primary succession organisms colonizing the rock surface (*litosere*), essentially cyanobacteria and cyanolichens, contributing to the formation of the soil required for the accommodation of larger terrestrial plants (Büdel et al. 2000). This will further lead to a dynamic process of alterations, capable to include all successional seres in a small rocky outcropping area (Weaver and Clements 1938). Because the spatial patterns on the rock are analogous to temporal patterns, the processes responsible for structural alterations in the community are also likely to explain temporal changes (Houle and Phillips 1989).

The rupestrian grasslands (*campos rupestres*) harbor specific plant communities on rock outcrops in the mountainous regions, especially diverse in the Espinhaço Range, in Minas Gerais and Bahia states in Brazil (Harley 1995; Giulietti et al. 1997).

Successional processes in rupestrian grasslands can occur directly on the rock surface (Alves and Kolbek 2000; Conceição and Pirani 2005, 2007; Jacobi et al. 2007; Conceição et al. 2007a, b; Chap. 6) or in the fire-disturbed areas (Kolbek and Alves 2008; Neves and Conceição 2010; Chap. 18).

In the rupestrian grasslands, local communities encompassing endemic species and lineages are dispersed across the mountains in a wide geographic range (Ribeiro et al. 2012, 2014), and as the floristic composition is highly inconstant throughout the different areas (Zappi et al. 2003; Conceição and Pirani 2007; Conceição et al. 2007a; Neves and Conceição 2010; Chap. 6), many communities in the rupestrian grasslands are still unknown.

Although these communities possess different species composition, the plant families and structure are similar, probably because the environmental conditions affecting the local processes are strong (Conceição and Pirani 2005, 2007; Conceição et al. 2007a; Le Stradic et al. 2015). In this way, the life-forms keep the same general patterns, but depend on the habitats and substrate properties (Conceição and Pirani 2005, 2007; Conceição et al. 2007a; Jacobi et al. 2007; Jacobi and Carmo 2011; Messias et al. 2011, 2013; Chap. 6).

The rupestrian grasslands are characterized by shrubs and herbs associated with sandy soils and rocky surfaces on the high-elevation places in the mountain ranges (Harley 1995; Giulietti et al. 1997; Conceição and Pirani 2005; Jacobi et al. 2007; Le Stradic et al. 2015). At ridge tops, the high diversity of habitats intercalated with the rocky surface results in a heterogeneous system in relation to the composition and structure of the vegetation, in a constrained area (Conceição and Pirani 2005, 2007; Jacobi et al. 2007; Neves and Conceição 2010). Hence, the nature of the temporal processes involved in the formation of vegetation in different habitats on the heterogeneous landscape of quartzite outcrops was investigated at the Chapada Diamantina, Bahia, Northeastern Brazil. To achieve this, we intend to detect vegetation structural patterns and to prepare a hypothetical scheme of succession on rocky outcrops in rupestrian grasslands.

9.2 Habitats

In spite of the generalized floristic unity found in the plant families of rupestrian grasslands on different locations at the Espinhaço Range (Harley 1995), there is an evident heterogeneity of habitats, with differentiations in vegetation composition and structure (Giulietti et al. 1997; Conceição and Giulietti 2002; Conceição and Pirani 2005, 2007; Conceição et al. 2005, 2007a; Messias et al. 2013; Le Stradic et al. 2015). In extensive and flat areas of sandy soils on the ridge tops there is a predominance of *campos gerais* (general grasslands), marked by a rich carpet of Cyperaceae, Eriocaulaceae, and Poaceae species. In places where drainage is difficult because of the topography or substrate, there can be the formation of patches of humic soil, occupied by herbaceous species of Xyridaceae, Eriocaulaceae, Cyperaceae, Lentibulariaceae, Orchidaceae, and Gentianaceae (Giulietti et al. 1996).

Fig. 9.1 Rocky outcrop vegetation in the rupestrian grasslands, Northeast Brazil. **a**, **b** Insular habitat with soil islands dominated by *Vellozia hemisphaerica* Seub. (Velloziaceae) (**a**), and by *Vellozia punctulata* Seub. (Velloziaceae) (**b**), **c** Cleft habitat surrounded by an insular habitat. *Photographs* by Abel A. Conceição

Fig. 9.2 Rocky outcrop vegetation in the rupestrian grasslands, Northeast Brazil. **a**, **b** Continuous habitat dominated by graminoid plants surrounding the island habitat, **c** Detail of continuous habitat with high shrub and dead biomass proportions. *Photographs* by Abel A. Conceição

(a)
(b)
(c)
(d)

◀ **Fig. 9.3** Disturbance evidences on rocky outcrop vegetation in the rupestrian grasslands, Northeast Brazil. **a** Drastic fire impact on islands of soil killing almost all individuals, **b**, **c** Drought effetcs on island of soil showing desiccation-tolerant species (*Trilepis lhotzkiana* Ness and *Barbacenia blanchetii* Goethart & Henrard) (**b**) and *Tibouchina pereirae* Brade & Markgr. sprouting by the stem base after aerial branches died (**c**), **d** Recent exposed rock surface after island rupture. *Photographs* by Abel A. Conceição

In rocky outcrops, many of the dominant species belong to the families Velloziaceae, Clusiaceae, Orchidaceae, Bromeliaceae, Melastomataceae, and Asteraceae, composing a more shrubby physiognomy than in the grasslands. Sandy sediment areas between rocky outcrops are mostly covered by Poaceae, Cyperaceae, and Leguminosae species, among several other families (Conceição and Giulietti 2002; Conceição and Pirani 2005; Conceição et al. 2005, 2007a, b; Chap. 6).

In this study, three habitats were specified, according to rock surface characteristics and vegetation limits (Conceição and Pirani 2005; Figs. 9.1, 9.2, 9.3 and 9.4).

Insular habitat: areas largely composed by exposed rock, with a remarkably discontinuous shrubby-herbaceous vegetation. It comprises two microhabitats: **crevice**—fissures on the rock surface, variable in size, containing one or more species; **soil island**—small patches of vegetation composed by two or more species of vascular plants, limited by the rock surface around the entire edge (Figs. 9.1a, b, 9.3b–d and 9.4);

Continuous habitat: areas with more extensive soils, usually extending from tens to hundreds of meters, covered by a shrubby-herbaceous vegetation, lacking a defined limit between itself and the rock surface (Figs. 9.2 and 9.4d);

Cleft habitat: a more restricted habitat, represented by arboreal-shrubby vegetation installed on clefts limited by rocky walls, usually (sub)parallel, with a height greater than 25 cm and smaller than 2 m (Fig. 9.1c). This habitat receives the smallest incidence of sun and wind, which makes it the most humid among the habitat types described here.

The above-mentioned habitats are similar to some types described for inselbergs (Ibisch et al. 1995; Porembski et al. 2000) and ironstones (Jacobi et al. 2007). Generally, members of the family Velloziaceae prevail in the insular habitat, especially on soil islands (Conceição and Pirani 2005; Conceição et al. 2007a, b, c). The same is observed on crevices, but in this type of habitat other families are also important, such as Orchidaceae, Melastomataceae and Poaceae. The continuous habitat is dominated by Poaceae. In clefts, the Asteraceae are quite abundant. The continuous and cleft habitats show greater proportions of eudicotyledon species; clefts have higher numbers of families, species, and life-forms, in addition to being more vulnerable due to their restricted occurrence and higher numbers of unique species (Conceição and Pirani 2005, 2007). The locations with higher species richness per unit area are the mixed habitat areas, and the continuous habitat shows a smaller number of species per vegetation area than the insular habitats (Conceição 2003).

(a)
(b)
(c)
(d)
(e)
(f)
(g)
(h)

◀ **Fig. 9.4** Rocky outcrop vegetation in the rupestrian grasslands, Northeast Brazil. **a–c** Pioneer vascular plants. **a** *Cattleya elongata* Barb.Rodr. (Orchidaceae) fixed on the rock after shedding the scape, **b**, **c** Fissures on the rock occupied by *Tibouchina pereirae* Brade & Markgr. (Melastomataceae) (**b**) and by *Schizachyrium sanguineum* (Retz.) Alst. (Poaceae) and *T. pereirae* (**c**), **d** Continuous habitat fragmented resulting in two soil islands, **e** Soil island almost united to another island, **f** Soil islands dispersed on rock outcrop, **g** Basal rosette of *Orthophytum albopictum* Philcox (Bromeliaceae), **h** Soil island with epilithic chamaephyte *Hadrolaelia sincorana* (Schltr.) Chiron & V.P. Castro (Orchidaceae) flowering. *Photographs* by Abel A. Conceição

9.3 Methodological Summary

The study of temporal processes was carried out at four rupestrian grasslands sites with a great proportion of exposed rock on hilltops and ridge tops at Serra do Sincorá (Sincorá Range), Chapada Diamantina, Bahia state, Brazil. **Morro da Mãe Inácia** (12° 27′S–41° 28′W) is located in the municipality of Palmeiras, and is the only site outside the limits of the Chapada Diamantina National Park, with elevations between 1100 and 1140 m. The **Cachoeira da Fumaça** (12° 35′S–41° 27′W) and **Gerais da Fumaça** (12° 36′S–41° 28′W) outcrops are located at Serra da Larguinha, in the municipality of Palmeiras, between 1310 and 1360 m elevation. The **Guiné** outcrops, in the municipality of Mucugê (12° 45′S–41° 30′W), are located at Serra do Esbarrancado, at about 1400 m elevation (Conceição et al. 2005, 2007a).

The climate at the study locations is sub-warm, semi-humid, with a rainy season usually from January to April and dry season from August to November (Nimer 1989). Sandstones and quartzites formed in the Pre-Cambrian make up the rugged topography of the Chapada Diamantina (Moreira and Camelier 1977). The rocky outcrops studied are included in the Chapada Diamantina Group, and are part of the Tombador Formation (Torquato and Fogaça 1981). The sediments on rocky outcrops studied are shallow, sandy, and acidic, but on the insular habitat the organic matter and clay contents are higher than on the continuous habitat (Conceição and Giulietti 2002; Conceição and Pirani 2005; Conceição et al. 2007a).

During the rainy season, in the years from 2000 to 2002, 48 10 × 10 m units were randomly located in the four study sites, which were selected based on their high proportions of rock surface above 1000 m elevation. In each of the four sites, four 10 × 10 m units stratified in the insular habitat were drawn, in addition to another four non-stratified (simple random) units, totaling 32 units (Conceição et al. 2005, 2007a). In addition, 16 units were drawn at Mãe Inácia and at Cachoeira da Fumaça, four stratified in the cleft habitat and four in the continuous habitat, corresponding to eight units per location (Conceição and Pirani 2005). Each of the 48 10 × 10 m units were divided into 25 2 × 2 m subunits, of which five were drawn, totaling 240 2 × 2 m subunits; in these, all occurring species had their percentage cover values estimated by vertical projections of their above-ground parts (Westhoff and van der Maarel 1978), expressed over the total found in the unit. Raunkiaer's life-forms were determined according to Ellenberg and Müeller-Dombois (1967) (further details in Conceição and Pirani 2005). Species that occurred isolated on the rock were distinguished for each 10 × 10 m unit, and the total by species was divided by the total number of occurrences of the species in all 48 units. Low values of these proportions indicate sporadic occurrence on the rock, while high values indicate preference for the rocky surfaces, especially when the total number of occurrences is high.

The ordination of communities was accomplished by correspondence analysis (CA), using the reciprocal average (RA), with the Multivariate Statistics Package (MVSP) software program, version 2.2, L. Kovach 1986–1995, Anglesey, Wales. Three matrices were used, including the cover area percentages for the species, summed by family. The analysis at the family level is more comprehensive than the one performed at the specific level, since many distinct species, however belonging to the same families, occur in outcrops from different regions and substrates (Ibisch et al. 1995; Meirelles et al. 1999; Alves and Kolbek 2000; Conceição and Pirani 2005, 2007; Conceição et al. 2007a). In the two analyses that focused on the three habitats in each site, the matrices included families with more than two occurrences at the 2 × 2 m subunits (24 families in 59 subunits at Mãe Inácia and 24 families in 56 subunits at Cachoeira da Fumaça). The joint analysis of the non-stratified subunits and subunits stratified in the insular habitat, in the four sites, included families occurring in more than 10 subunits (20 families in 150 subunits). The analyses displaying the areas in biplots were performed but not showing here.

9.4 Spatial Patterns

9.4.1 Isolated Species on the Rock

Among the 40 species that occurred isolated on the rock (Table 9.1), 14 occurred only once, while *Schizachyrium sanguineum* and *Tibouchina pereirae* occurred in more than 20 10 × 10 m units. Many species had their occurrences restricted to only one of the sites studied. The families with the higher number of isolated occurrences on the rock were Orchidaceae (32), Poaceae (29), Melastomataceae (27), Velloziaceae (24), Asteraceae (12), Bromeliaceae (12), Amaryllidaceae (nine), and Cyperaceae (eight).

9.4.2 Habitat Gradient

The ordination of families in different habitats at Morro da Mãe Inácia showed that in the axis 1, the families Cyperaceae and Velloziaceae are opposite to Poaceae. In the axis two Euphorbiaceae and Primulaceae are opposite to Cyperaceae and

Table 9.1 Vascular plant species with isolated occurrence on rocky surfaces, with respective families and life-forms (mp: microphanerophyte, np: nanophanerophyte, ec: epilithic chamaephyte, dc: desiccation-tolerant chamaephyte, ch: chamaephyte not epilithic and not desiccation-tolerant, hem: hemicryptophyte, cryp: cryptophyte, ther: therophyte, succ: succulent), distinguished on four study sites: Morro da Mãe Inácia (M), Cachoeira da Fumaça (F), Guiné (G), and Gerais da Fumaça (GF), Chapada Diamantina, Bahia, Brazil. T: sum of isolate occurrences. % 48 un: percentage of isolate occurrences over total occurrences in 48 10 × 10 m units, regardless of whether they are alone or not

Family	Species	Life-form	Isolated occurrence					% 48
			F	M	G	GF	T	un
Poaceae	*Schizachyrium sanguineum* (Retz.) Alst.	hem	10	6	4	4	24	63
Melastomataceae	*Tibouchina pereirae* Brade & Markgr.	np/mp/ch	9	7	1	5	22	48
Orchidaceae	*Acianthera ochreata* (Lindl.) Pridgeon & M.W. Chase	ec	1	8	0	0	9	90
Amaryllidaceae	*Hippeastrum solandriflorum* (Lindl.) Herb.	cryp	2	0	1	6	9	50
Velloziaceae	*Vellozia jolyi* L.B.Sm.	dc	1	0	4	3	8	50
Bromeliaceae	*Vriesea chapadensis* Leme	hem	0	3	5	0	8	35
Asteraceae	*Lychnophora bishopii* H. Rob.	np	4	1	3	0	8	30
Orchidaceae	*Cattleya elongata* Barb.Rodr.	ec	1	5	1	0	7	78
Velloziaceae	*Vellozia dasypus* Seub.	dc	7	0	0	0	7	30
Orchidaceae	*Hadrolaelia sincorana* (Schltr.) Chiron & V.P. Castro	ec	0	0	6	0	6	75
Orchidaceae	*Epidendrum secundum* Jacq.	ec	4	0	0	1	5	17
Orchidaceae	*Hoffmannseggella pfiesteri* (Pabst & Senghas) Chiron & V.P. Castro	ec	0	0	4	0	4	80
Velloziaceae	*Barbacenia blanchetii* Goethart & Henrard	dc	0	4	0	0	4	40
Velloziaceae	*Vellozia punctulata* Seub.	dc	4	0	0	0	4	29
Cyperaceae	*Trilepis lhotzkiana* Nees	dc	0	3	0	0	3	60
Asteraceae	*Lasiolaena morii* R.M. King & H. Rob.	np	0	3	0	0	3	50
Lythraceae	*Cuphea ericoides* Cham. & Schltdl.	np	0	0	0	3	3	30
Melastomataceae	*Marcetia velutina* Markgr.	np	0	1	0	2	3	27
Loganiaceae	*Spigelia cremnophila* Zappi & E. Lucas	ch	0	3	0	0	3	25
Eriocaulaceae	*Paepalanthus pulchellus* Herzog	ther	1	0	2	0	3	12
Cyperaceae	*Abildgaardia baeothryon* A. St-Hil.	ther	1	0	2	0	3	9
Leguminosae	*Calliandra viscidula* Benth.	np	0	0	0	2	2	40

(continued)

Table 9.1 (continued)

Family	Species	Life-form	Isolated occurrence					% 48
			F	M	G	GF	T	un
Bromeliaceae	*Orthophytum albopictum* Philcox	hem	0	0	2	0	2	33
Cactaceae	*Micranthocereus purpureus* (Gürke) F. Ritter	succ	0	2	0	0	2	25
Cyperaceae	*Lagenocarpus rigidus* (Kunth) Nees	hem	0	0	0	2	2	14
Poaceae	*Trachypogon macroglossus* Trin.	hem	2	0	0	0	2	14
Asteraceae	*Richterago discoidea* (Less.) Kuntze	cryp	0	0	1	0	1	50
Bromeliaceae	*Neoregelia bahiana* (Ule) L.B. Sm.	hem	0	0	1	0	1	50
Melastomataceae	*Microlicia* sp.1	np	0	0	0	1	1	50
Orchidaceae	*Octomeria alexandri* Schltr.	ec	0	1	0	0	1	50
Poaceae	*Paspalum pumilum* Nees	hem	0	0	0	1	1	50
Piperaceae	*Peperomia galioides* Humb, Bonpl. & Kunth	ec	0	1	0	0	1	33
Melastomataceae	*Microlicia* sp.2	np	0	0	0	1	1	17
Bromeliaceae	*Orthophytum burle-marxii* L.B. Sm. & Read	hem	0	0	0	1	1	14
Lamiaceae	*Eriope exaltata* Harley	np/mp	0	1	0	0	1	14
Selaginellaceae	*Selaginella marginata* (Humb. & Bonpl.) Spring	dc	1	0	0	0	1	9
Leguminosae	*Calliandra* sp.1	ch	0	0	0	1	1	6
Poaceae	*Dichanthelium cumbucana* (Renvoize) Zuloaga	hem	0	0	0	1	1	6
Poaceae	*Paspalum minarum* Hack.	hem	1	0	0	0	1	5
Verbenaceae	*Lantana caatingensis* Moldenke	ch	1	0	0	0	1	5

Eriocaulaceae. The families Euphorbiaceae and Primulaceae are located near one another on the three axes (Fig. 9.5).

The ordination of families in different habitats at Cachoeira da Fumaça showed that the families Bromeliaceae and Velloziaceae polarize one side of axis one, while Aquifoliaceae, Convolvulaceae, and Lythraceae another side. Dryopteridaceae and Verbenaceae polarize one side of axis two, while Lythraceae another side. Families located near each other on the three axes are: Bromeliaceae-Velloziaceae-Amaryllidaceae, Begoniaceae-Gesneriaceae, Malvaceae-Asteraceae, and Euphorbiaceae-Aquifoliaceae-Convolvulaceae-Lamiaceae-Poaceae-Eriocaulaceae-Gentianaceae-Leguminosae (Fig. 9.6).

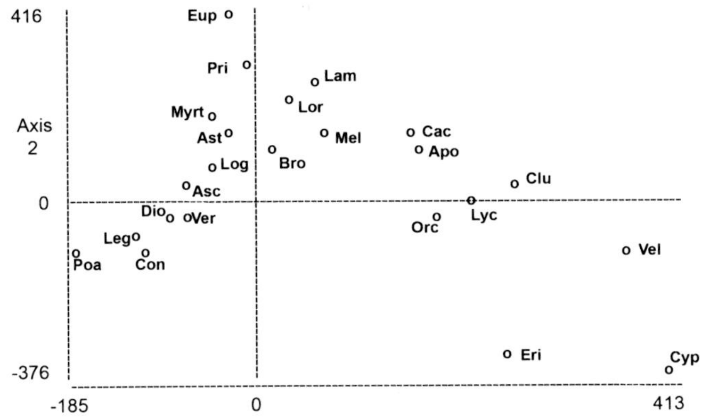

Fig. 9.5 Biplot showing the ordination in the first two axes of the correspondence analysis (CA) of 24 families in 59 subunits of three habitats at Morro da Mãe Inácia, Chapada Diamantina, Northeast Brazil (the cumulative variance explained by the first three axes was 18, 32 and 44 %). *Apo* Apocynaceae; *Asc* Asclepiadoideae; *Ast* Asteraceae; *Bro* Bromeliaceae; *Cac* Cactaceae; *Clu* Clusiaceae; *Con* Convolvulaceae; *Cyp* Cyperaceae; *Dio* Dioscoreaceae; *Eri* Eriocaulaceae; *Eup* Euphorbiaceae; *Lam* Lamiaceae; *Leg* Leguminosae; *Log* Loganiaceae; *Lor* Loranthaceae; *Lyc* Lycopodiaceae; *Mel* Melastomataceae; *Myrt* Myrtaceae; *Orc* Orchidaceae; *Pri* Primulaceae; *Poa* Poaceae; *Pol* Polygalaceae; *Vel* Velloziaceae; *Ver* Verbenaceae. Families that remained close to each other at the intersections of axis one with axes two and three were encompassed by ellipses

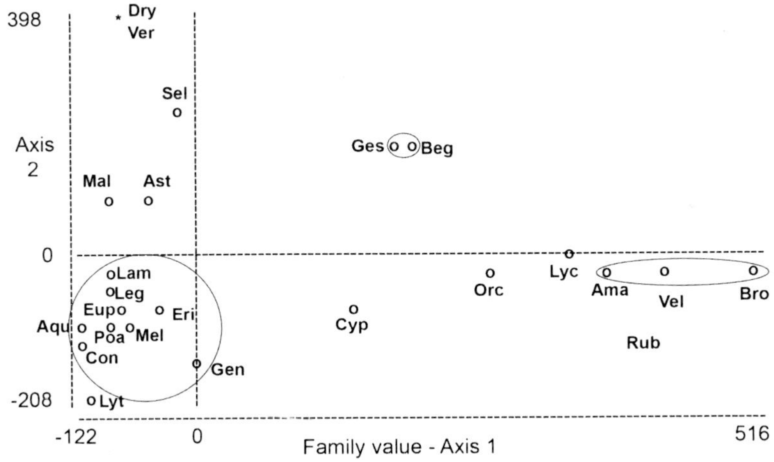

Fig. 9.6 Biplot showing the ordination in the first two axes of correspondence analysis (CA) of 24 families in 56 subunits of three habitats at Cachoeira da Fumaça, Chapada Diamantina, Northeast Brazil (the cumulative variance explained by the first three axes was 24, 39 and 48 %). *Ama* Amaryllidaceae; *Aqu* Aquifoliaceae; *Ast* Asteraceae; *Beg* Begoniaceae; *Bro* Bromeliaceae; *Con* Convolvulaceae; *Cyp* Cyperaceae; *Dry* Dryopteridaceae; *Eri* Eriocaulaceae; *Eup* Euphorbiaceae; *Gen* Gentianaceae; *Ges* Gesneriaceae; *Lam* Lamiaceae; *Leg* Leguminosae; *Lyc* Lycopodiaceae; *Lyt* Lythraceae; *Mal* Malvaceae; *Mel* Melastomataceae; *Orc* Orchidaceae; *Poa* Poaceae; *Rub* Rubiaceae; *Sel* Selaginellaceae; *Vel* Velloziaceae; *Ver* Verbenaceae. Families that remained close to each other at the intersections of axis one with axes two and three were encompassed by ellipses

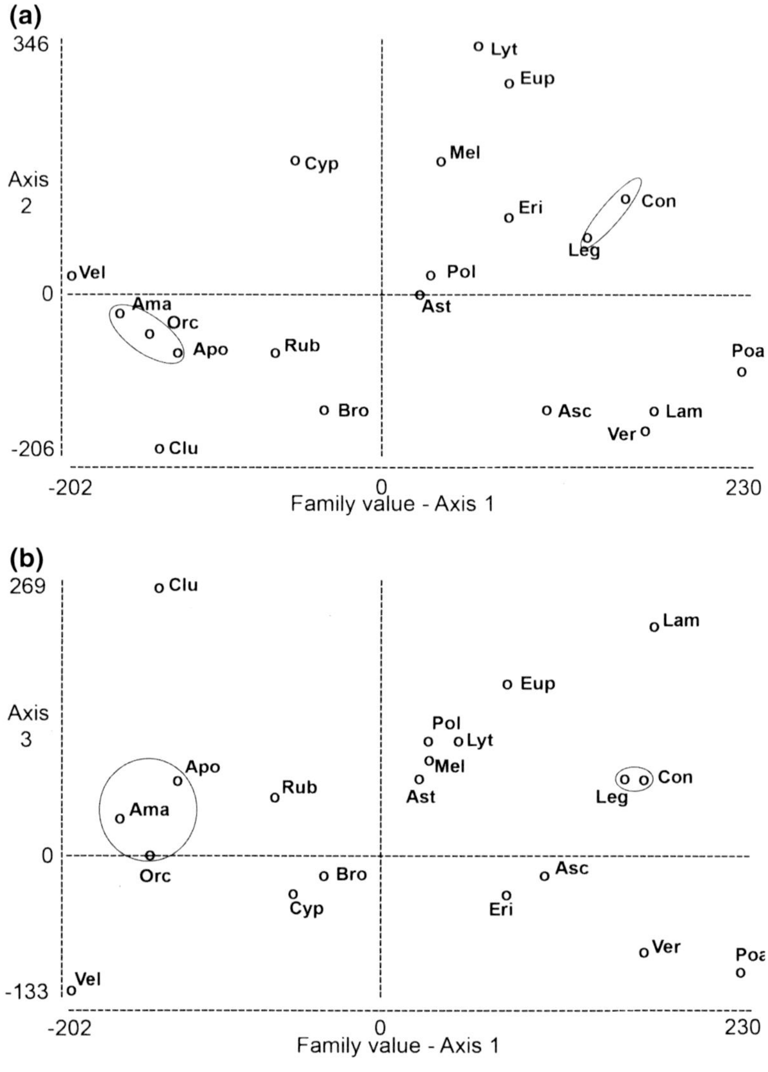

Fig. 9.7 Biplot showing the ordination in the first three axes of correspondence analysis (CA) of 20 families in 150 simple random and stratified random subunits at the insular habitat, at four sites. Chapada Diamantina, Northeast Brazil (the cumulative variance explained by the first three axes was 17, 30 and 41 %). **a** axes 1 and 2, **b** axes 1 and 3. *Ama* Amaryllidaceae; *Apo* Apocynaceae; *Asc* Asclepiadoideae; *Ast* Asteraceae; *Bro* Bromeliaceae; *Clu* Clusiaceae; *Con* Convolvulaceae; *Cyp* Cyperaceae; *Eri* Eriocaulaceae; *Eup* Euphorbiaceae; *Lam* Lamiaceae; *Leg* Leguminosae; *Lyt* Lythraceae; *Mel* Melastomataceae; *Orc* Orchidaceae; *Poa* Poaceae; *Pol* Polygalaceae; *Rub* Rubiaceae; *Vel* Velloziaceae; *Ver* Verbenaceae. Families that remained close to each other at the intersections of axis one with axes two and three were encompassed by ellipses

The joint analysis of the units stratified in the insular habitat together with non-stratified units showed that the family Poaceae polarizes one side of axis one, while Velloziaceae another side. Axis two is polarized by Lythraceae, whith Clusiaceae in the other side (Fig. 9.7a). Axis three is polarized by Clusiaceae in one sector and by Velloziaceae in another one (Fig. 9.7b). The families clustered by the three axes are Amaryllidaceae-Orchidaceae-Apocynaceae and Convolvulaceae-Leguminosae (Fig. 9.7).

9.5 Spatial Patterns Explaining Temporal Changes

9.5.1 Overview of Successional Scheme

The isolated occurrence of a plant species on the rock surface may happen by the direct occupation of the bare rock or because the species remains after a disturbance. Only in the first case the colonizers are dispersed across the space, while in the second they characterize an essentially temporal process (Grubb 1987), remaining alive because they possess physiological and/or morphological mechanisms specialized to cope with the actions of fire, storm, or drought, or yet to remain by seeds, which constitutes a primary and secondary succession complex (Grubb 1987; Peet 1992).

Vegetative reproduction is important in physically unstable habitats (Grubb 1987), like those studied here, responsible for part of the isolated occurrence on the rock, verified in Orchidaceae and Bromeliaceae, but probably common in other recurrent monocotyledons in the study, since it includes advantages, such as the formation of large populations from an individual established at isolated locations, perpetuation during a long time, reduction of extinction risks, and agility in horizontal expansion, effecting a quick occupation of the more favorable locations (Biedinger et al. 2000).

Out of the total 247 species sampled in the 48 units studied (Conceição and Pirani 2005; Conceição et al. 2005), only 16.2 % (40) occurred isolated on the rock, a fact that indicates how peculiar this type of occupation is. The four species with more than 70 % of the isolated occurrences on the rock are all epilithic Orchidaceae, confirming the importance of this family and its functional characteristic in primary succession on rocky substrates (Alves and Kolbek 2000). Species such as *Neoregelia bahiana* and *Octomeria alexandrii*, although occurring in few units, must also possess ability to colonize and to inhabit the exposed rocky substrate, since a great proportion of their occurrences are isolated. Conversely, species like *Lantana caatingensis* and *Paspalum minarum* occur in many units, rarely in an isolated manner, suggesting greater affinity with other substrates (see Table 9.1).

The species found isolated in higher numbers, *Schizachyrium sanguineum* and *Tibouchina pereirae*, were found in all habitats described (Conceição and Pirani 2005, 2007; Conceição et al. 2007a, b) and at the four sites, demonstrating great ecological amplitudes. The wide distribution of *S. sanguineum* across the tropics (Renvoize 1984) demonstrates a high dispersal capacity, an important factor for pioneer species (Grubb 1987). Although treated as endemic in the Chapada Diamantina, *Tibouchina pereirae* shows a wide morphological variation, raising questions with regard to its taxonomic circumscription and therefore to its geographic distribution and dispersal capacity (Baumgratz and Souza 1995). This is a fast-growing shrubby species, with an underground and aerial system at ground level which regrows quickly after an intense drought or fire (Fig. 9.3c), making it very suitable to undergo secondary succession processes (Grubb 1987). The other species with higher number of isolated occurrences on the rock (four to nine) possesses special structural and functional characters that enable them to live on xeric environments, such as the presence of organs that accumulate water and reserve substances (bulbs in Amaryllidaceae, pseudobulbs and fleshy leaves in Orchidaceae, rosettes in Bromeliaceae), branches with dense and imbricate microphyllous leaves in *Lychnophora* (Asteraceae) and *Cuphea* (Lythraceae), roots with a velamen (Orchidaceae), desiccation tolerance in Velloziaceae and Cyperaceae (Meirelles et al. 1997; Conceição 2003), and CAM metabolism in Orchidaceae and Bromeliaceae (Lüttge 1997).

The vegetation structure is determined by recurrent families in all sites studied, as well as families that are characteristic in certain sites and/or habitats. At Morro da Mãe Inácia, a composition and abundance gradient is observed between cleft and continuous and between cleft and insular habitats (Fig. 9.5). Families Cyperaceae and Velloziaceae include the most abundant species in the insular. Families Euphorbiaceae and Primulaceae refer to cleft trees, forming the vertex of a triangle across the gradient, probably resulting from the mixture of habitats in the cleft, which shows insular vegetation on the rock domain that surrounds it and continuous vegetation on the limit with the inner part of the cleft.

At Cachoeira da Fumaça (Fig. 9.6), axis one separates the most typical families in the insular habitat. On axis two, the most typical cleft families are separated from continuous families. The gradient between insular and cleft provides evidence that Begoniaceae and Gesneriaceae are typical of the insular and cleft habitats. Amaryllidaceae and Velloziaceae form a group with little variation, together with Bromeliaceae, Lycopodiaceae, Orchidaceae, and Rubiaceae. The most typical cleft families are Dryopteridaceae and Verbenaceae, but Selaginellaceae, Melastomataceae, Malvaceae, and Asteraceae are also listed. The continuous habitat includes the highest number of families, such as Poaceae, Lythraceae, Aquifoliaceae, Leguminosae, Convolvulaceae, Lamiaceae, Eriocaulaceae, Melastomataceae, Asteraceae, Gentianaceae, and Euphorbiaceae. In the figure, a more isolated arrangement of families in the insular in relation to those of the continuous is also observed in the sites studied. A family located in the typical group of a habitat may also have one or more species occurring in another habitat. This is particularly valid in the case of the richest families, like Poaceae, Orchidaceae, Asteraceae, and Melastomataceae, usually with species present in all habitats.

When non-stratified and insular units are viewed together (Fig. 9.7), the typical insular families are separated on axis one from those in the continuous, revealing probable aspects related to early successional seres from the rock, since the sample includes insular habitats typical of the *litosere*, such as crevices and soil islands, characterized by one to a few species surrounded by the rock surface, in addition to continuous habitat, which would be the result of gradual changes in the composition and abundance of species with time (Berrío et al. 2003). In the insular habitat, the soil richest in organic matter and least sandy in relation to the continuous (Conceição and Pirani 2005) may also indicate succession from the insular to the continuous habitat, where lichen and moss residues initially result in a high organic matter percentage, with an increase in sand percentage caused by disaggregation of the sandstone (Pidgeon 1940).

In the biplot of axes one and three (Fig. 9.7b), insular-related families form a gradient from families containing desiccation tolerant members (Velloziaceae and Cyperaceae), rosette-bearing plants (Bromeliaceae), or plants provided with roots that adhere directly onto the rock surface (Orchidaceae), or with well-developed subterraneous systems (Orchidaceae, Amaryllidaceae, Apocynaceae, Rubiaceae), and even treelets (Clusiaceae). In the case of continuous vegetation, two gradients were observed on the biplot of axes one and two (Fig. 9.7a), which may indicate two succession trends, one from Cyperaceae and the other from Bromeliaceae and Rubiaceae. The first could be related to the continuous vegetation in flat places, more similar to *campos gerais*, while the second could be more related to the continuous vegetation interspersed with portions of exposed rock at more hilly places. At Chapada Diamantina, Cyperaceae and Bromeliaceae are important families resprouting after fire (Neves and Conceição 2010). In general, the families studied in this chapter have representatives adapted to fire (Safford 2001; Neves and Conceição 2010; Chap. 17), allowing a great composition variety resulting from fires, which must vary, among other factors, according to the degree of disturbance and composition prior to the fire. Plant cover regeneration, with many individuals regrowing and blooming within a short time interval (one to three months), was observed at burned rupestrian grassland locations (Neves and Conceição 2010; Conceição and Orr 2012; Conceição et al. 2013; Chap. 17). Fire-resistant plants are also widely known on the Cerrado (Brazilian savanna) flora, which share several species with the rupestrian grasslands flora, and adaptation to fire-regimes is usually considered as a fundamental step that allowed occupation of the open habitats in the Neotropics (e.g. Simon and Pennington 2012).

9.5.2 Successional Scheme

Based on this and other studies on insular, continuous, and cleft habitats in elevations at Chapada Diamantina (Conceição 2003; Conceição and Pirani 2005, 2007; Conceição et al. 2005, 2007a, b, c; Neves and Conceição 2010; Chap. 6), a hypothetical scheme of the succession on the rock outcrops summits was prepared

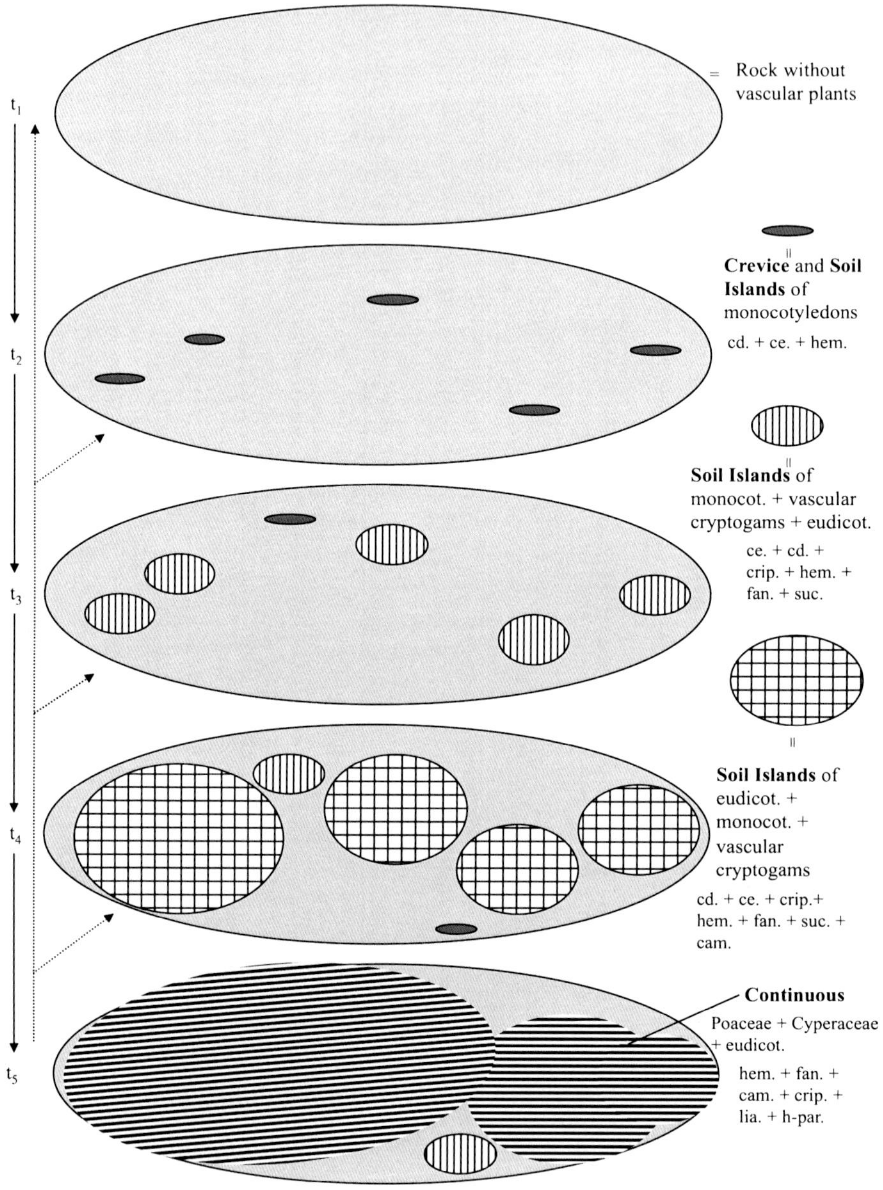

Fig. 9.8 Schematic representation of probable successional seres in rocky outcrops of rupestrian grasslands. The *continuous arrows* indicate the species turnover direction in time. The *broken arrows* refer to retrogression. Monocot. = monocotyledons. Eudicot. = eudicotyledons. Life-forms: desiccation tolerant chamaephyte (cd), epilithic chamaephyte (ce), hemicryptophyte (hem), phanerophyte (fan), succulent (suc), cryptophyte (crip), chamaephyte not epilithic and not desiccation tolerant (cam), therophyte (tero) and liana (lia)

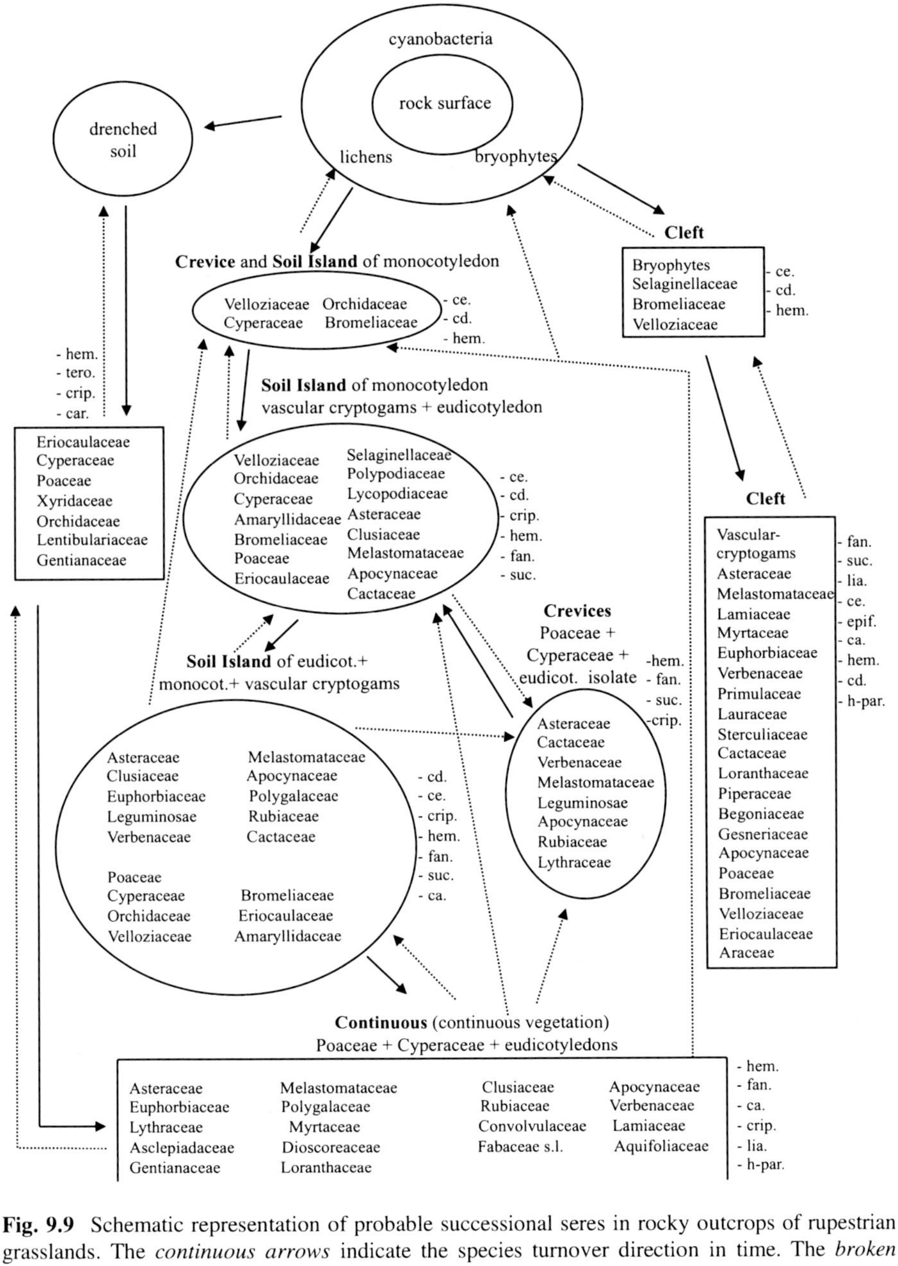

Fig. 9.9 Schematic representation of probable successional seres in rocky outcrops of rupestrian grasslands. The *continuous arrows* indicate the species turnover direction in time. The *broken arrows* refer to retrogression. Monocot = monocotyledons. Eudicot = eudicotyledons. Life-forms: desiccation tolerant chamaephyte (cd), epilithic chamaephyte (ce), hemicryptophyte (hem), phanerophyte (fan), succulent (suc), cryptophyte (crip), chamaephyte not epilithic and not desiccation tolerant (cam), therophyte (tero), epiphyte (epif), hemiparasite (h-par), carnivorous (car) and liana (lia)

(Figs. 9.8 and 9.9). In rocky outcrop sites fully exposed to the sun and where no plants exist, lichens are the most abundant organisms: crustose lichens impart a colorful aspect to the rock and foliose lichens form extensive 'mats'. Portions of rock showing evidence of cyanobacteria are not very frequent, generally associated with water-flowing places, showing a dark reddish-brown color. Bryophytes also occur, but in much smaller quantities than lichens.

Isolated monocotyledon species occur on the rock (crevices) at restricted sites usually associated with irregularities on the rock surface that allow some sediment accumulation. These species possess special adaptations to withstand extreme limiting conditions, and are the first vascular plants to become established on the rock. They include desiccation-tolerant chamaephytic species (Velloziaceae: *Barbacenia blanchetii* and *Vellozia* spp.; Cyperaceae: *Trilepis lhotzkiana*); chamaephytic epilithic species (Orchidaceae) in the genera *Hoffmannseggella* (*Sophronitis* or *Laelia*), *Hadrolaelia* (*Sophronitis* or *Laelia*), *Epidendrum*, *Acianthera* (*Pleurothallis*), and *Cattleya*, and hemicryptophytic species without (*Schizachyrium sanguineum*, Poaceae) or with rosettes (Bromeliaceae: *Vriesea chapadensis*, *Orthophytum* spp., *Neoregelia bahiana*). As time goes on, increasing sediment accumulation and new monocotyledon species are added to the crevices, forming soil islands that increase in size, gradually lessen the severity of the environment. Vascular cryptogams and shrubby eudicotyledons may become installed on the islands, like *Huperzia mooreana* (Lycopodiaceae), *Tibouchina pereirae* (Melastomataceae), and *Lychnophora bishopii* (Asteraceae), imparting variety to the microhabitats, with shady locations or places exposed to the sun, and differentiation in the quality of the sediment.

As the islands area expands, the diversity of life-forms increases, as well as the diversity and abundance of eudicotyledons, mainly *Clusia obdeltifolia* (Clusiaceae), *Mandevilla* spp. (Apocynaceae), and *Polygala* spp. (Polygalaceae). Little by little, adjacent islands may connected to each other, losing insularity and becoming a more continuous vegetation, gradually dominated by Poaceae, Cyperaceae, and eudicotyledon species. This last step of succession (t4 → t5) may be restricting just to side-by-side habitats. The likely taxa included in the successional stages, starting from the rock, are indicated by the arrows (Fig. 9.9). The dotted lines represent retrogressions, induced by the actions of fire, droughts, and storms. Contrary to the gradual succession process, such disturbances can be dramatic, reducing the continuous vegetation cover directly to crevices and monocotyledon islands (Figs. 9.8 and 9.9), from which the succession process will start over again.

The presence of thickened stems in low individuals of *Tibouchina pereirae*, isolated on the rock, indicates that these shrubs may be also a remnant of a more extensive patch of vegetation which was disturbed, giving rise to a crevice (Fig. 9.4b). In visits made after the phenological monitoring initiated in 2000 (Conceição et al. 2007c), the union of nearby islands was observed, as well as the occurrence of vegetative propagation, with the rosette of *Vriesea chapadensis* (Bromeliaceae) detaching from an island and becoming attached, alone, on the rock. In the same way, fallen *Cattleya elongata* (Orchidaceae) scapes became attached to the rock surface (Fig. 9.4a). During the study, one of the islands marked at Guiné disappeared, probably as a result of the joint action of drought and storm. Also islands resulting from fragmentation of continuous habitat were observed during the study (Fig. 9.4d). Disturbance by fire was verified in the year 2000 in soil islands marked in 1997 at Morro do Pai Inácio (Conceição et al. 2007b), when most islands were not affected by the fire due to areas without vegetation. But after the fire that occurred in 2012, the effects on islands near of continuous habitat were drastic, killing all species on many soil islands (Fig. 9.3a). During the driest period of the study, in the continuous habitat and larger soil islands, shrubs like *Cuphea ericoides* (Lythraceae), *Clusia obdeltifolia*, *Lychnophora triflora*, and Melastomataceae species, in addition to *Renvoizea trinii* (Poaceae), *Rumohra adiantiformis* (Dryopteridaceae), and *Abildgaardia baeothryon* (Cyperaceae), dried up completely and died.

Some taxa involved in succession at perpendicular-walled clefts or other similar habitats are different from those involved in the succession from the rock more exposed to the sun and to the wind. In clefts, species of bryophytes, Selaginellaceae, Bromeliaceae, Cactaceae, and Velloziaceae occupy the rocky portions, including walls and edges. The volume of sediment inside the cleft increases with time, and eudicotyledon (such as Asteraceae, Melastomataceae, Primulaceae, Euphorbiaceae, and Myrtaceae), monocotyledon (such as Poaceae and Eriocaulaceae), and vascular cryptogam species (such as Dryopteridaceae and Cyatheaceae) are added. The limits of these locations enlarge with time, meeting the continuous vegetation, but remaining distinct from the continuous habitat because their walls continue parallel-shaped, interfering with quality in the cleft habitat.

Depending on the slope and occurrence of concavity on the rock surface, some locations that accumulate periodically-saturated soil may be colonized by small herbs, especially in the families Eriocaulaceae, Cyperaceae, Xyridaceae, Poaceae, Lentibulariaceae, and Gentianaceae. With time, these locations will increase in size and, similarly as occurs in the succession starting from rock to crevice to island to continuous, they may start joining other drenched soil patches, forming continuous habitat.

The model presented here follows the general pattern described for vegetation in other rocky outcrops, with similar taxonomic and/or functional groups (Oosting and Anderson 1937; Pidgeon 1940; Burbanck and Platt 1964; Hambler 1964; Shure and Ragsdale 1977; Houle 1990; Ibisch et al. 1995; Porembski et al. 2000). In the Neotropical region, similarities occur at an even greater scale (Oliveira et al. 1975; Ibisch et al. 1995; Sarthou and Villiers 1998; Alves and Kolbek 2000; Michelangeli 2000; Ribeiro and Medina 2002; Berrío et al. 2003; Arbeláez and Duivenvoorden 2004), especially in the case of the other study carried out on sandstone-quartzitic in Brazil (Alves and Kolbek 2000), where many genera and some species are common, like *Acianthera* (*Pleurothallis*), *Polypodium*, *Sophronitis* (*Hoffmannseggella*, *Hadrolaelia* or *Laelia*), *Clusia*, *Myrcia*, *Baccharis*, *Lychnophora*, *Encyclia*, *Vriesea*, *Vellozia*, *Hippeastrum*, *Barbacenia*, and *Anthurium*, among others, in addition to the species *Doryopteris ornithopus*, *Cuphea ericoides*, *Coppensia blanchetii* (*Oncidium*), *Lagenocarpus rigidus*, and *Marcetia taxifolia*. Attention is drawn to the smaller representativeness of *Selaginella* as a pioneer element on quartzite outcrops in relation to granite gneiss ones (Oliveira et al. 1975; Ibisch et al. 1995; Meirelles et al. 1997).

Changes in species composition and abundance are gradual, and are governed by many factors, from the genetic makeup of organisms, biological interactions, history, and chance (Connell and Slatyer 1977; Shure and Ragsdale 1977; Peet 1992), and it is impossible to establish actual groups, even if observations lasted decades. The same is valid for exposed habitats in the hypothetical model, since blocks (boulders) and rock stairs variable in size are other examples of variations in the substrate, characteristic in rocky outcrops. Depending on the objective of the research, it is possible to refine the discernment between habitats into several scales. In each of the four study sites there are particular species, and the hypothetical model proposed uses the most recurring and abundant taxa. The fact that a taxon is not listed as characteristic in a given sere does not exclude the possibility that it will occur in other seres. In continuous habitat, for example, members belonging to several monocotyledon families other than Poaceae and Cyperaceae also occur, such as Orchidaceae, Bromeliaceae, Eriocaulaceae, and Velloziaceae. This model is also based on correspondence analyses processed with soil islands from each location studied, since they can be under different successional processes, such as the compositional and structural differences observed in islands from Gerais da Fumaça after fire disturbance (Conceição et al. 2007a).

In the case of succession in drenched soils, the continuous vegetation resulting from the succession process must be more similar to what is locally called *campos gerais*. These are extensive and continuous grasslands, sampled at small stretches, especially at Gerais da Fumaça and at Guiné, common in more plain places on the ranges, at the lowlands of summits. The most important physiognomic difference between *campos gerais* and the continuous vegetation predominant in this study is the smaller proportion of shrubs in the *campos gerais*, generally represented by species of Lythraceae, Melastomataceae, Asteraceae, Convolvulaceae, and Lamiaceae.

Organisms react differently when faced with the physical and biotic environment (Peet 1992), resulting in high diversity, since in each stage and at each location there are differentiations in species composition and in functional characteristics. This reinforces the dynamic and complex character of succession (Peet 1992). The occurrence of all successional seres at the same time (Fig. 9.8), as also observed by Weaver and Clements (1938), ensures that, in the case of fire, at least monocotyledon islands and crevices are left, protected by the discontinuity of the vegetation. Then, after fire the insular habitat can be an important source of propagules for the colonization of continuous disturbed areas, while the continuous habitat is a source of propagules to colonize the rock surface, to enrich the islands, and to start the secondary succession on islands impacted by fire, since in the continuous border occur typical species from the soil islands. According to Porembski (2000) and Porembski and Barthlott (1997), the occurrence of similar habitats in adjacent rocky outcrops cause local populations to not go extinct due to integrated dispersal.

Only islands closer to continuous vegetation may suffer the action of fire, suggesting that intense droughts and storms are the main responsible by species mortality in crevices and soil islands, acting on retrogression, consisting in markedly seasonal factors, important in the dynamics of this ecosystem. The proposed model takes into consideration autogenic and allogenic processes (Van der Maarel 1988; Glenn-Lewin and van der Maarel 1992), the first being more evident at the early stages of colonization of rigorously extreme environments, such as facilitation, probably followed by inhibition or tolerance (Connell and Slatyer 1977; Houle and Phillips 1989). Competitive relationships probably increase at later successional stages (Peet 1992), such as continuous and cleft, which involve greater diversity of organisms. The existence of the same general pattern in distinct outcrops indicates certain stability in the system as a whole.

Isolated eudicotyledon species on the rock are treated in the present model as remnants from disturbances, but could also be related to direct succession from the rock in locations with greater irregularities, which would be occupied mainly by Melastomataceae and Asteraceae species. Studies for longer periods of time must be carried out in order to refine the model, considering processes of soil formation, dispersal, establishment, mortality, biological interactions, and environmental disturbances.

Acknowledgements We thank FAPESP for a scholarship granted to the first author, to CNPq for support to the first and second authors, and to IBAMA for issuing a license for collection at Chapada Diamantina National Park. To F.R. Scarano, J.A.N. Conceição, G.W. Fernandes, and to Elise Buisson for reading the manuscript and for suggestions, and to the botanists that helped with the identification of the plants.

References

Alves RJV, Kolbek J (2000) Primary succession on quartzite cliffs in Minas Gerais, Brazil. Biol Brat 55:69–83

Arbeláez MV, Duivenvoorden JF (2004) Patterns of plant species composition on Amazonian sandstone outcrops in Colombia. J Veg Sci 15:181–188

Baumgratz JFA, Souza MLR (1995) *Tibouchina* Aubl. In: Stannard BL (ed) Flora of the Pico das Almas, Chapada Diamantina, Bahia, Brazil. Royal Botanic Gardens, Kew, pp 480–482

Berrío JC, Arbeláez MV, Duivenvoorden JF, Cleef AM, Hooghiemstra H (2003) Pollen representation and successional vegetation change on the sandstone plateau of Araracuara, Colombian Amazonia. Rev Palaeobot Palynol 2538:1–19

Biedinger N, Porembski S, Barthlott W (2000) Vascular plants on inselbergs: vegetative and reproductive strategies. In: Porembski S, Barthlott W (eds) Inselbergs. Ecological studies, vol 146. Springer, Heidelberg, pp 117–142

Büdel B, Becker U, Follmann G, Sterflinger K (2000) Algae, fungi, and lichens on inselbergs. In: Porembski S, Barthlott W (eds) Inselbergs. Ecological studies, vol 146. Springer, Heidelberg, pp 69–90

Burbanck MP, Platt RB (1964) Granite outcrop communities of the piedmont plateau in Georgia. Ecol 45:292–306

Conceição AA (2003) Ecologia da vegetação em afloramentos rochosos na Chapada Diamantina, Bahia, Brasil. PhD thesis, Universidade de São Paulo

Conceição AA, Giulietti AM (2002) Composição florística e aspectos estruturais de campo rupestre em dois platôs do Morro do Pai Inácio, Chapada Diamantina, Bahia, Brasil. Hoehnea 29:37–48

Conceição AA, Orr BJ (2012) Post-fire flowering and fruiting in the caulescent rosette *Vellozia sincorana*, an endemic plant to the Northeast of Brazil. Acta Bot Bras 26:94–100

Conceição AA, Pirani JR (2005) Delimitação de habitats em campos rupestres na Chapada Diamantina: substratos, composição florística e aspectos estruturais. Bol Bot Univ São Paulo 23:85–111

Conceição AA, Pirani JR (2007) Diversidade em quatro áreas de campos rupestres na Chapada Diamantina, Bahia, Brasil: espécies distintas, mas riquezas similares. Rodriguésia 58:193–206

Conceição AA, Rapini A, Pirani JR, Giulietti AM, Harley R, Silva TRS, Funch R, Santos AKA, Correia C, Andrade IM, Costa JAS, Souza LRS, Andrade MJG, Freitas TA, Freitas AMM, Oliveira AA (2005) Campos rupestres. In: Juncá FA, Funch L, Franca-Rocha W (eds) Biodiversidade e conservação da Chapada Diamantina. Biodiversidade, vol 13. Ministério do Meio Ambiente, Brasília, pp 153–180

Conceição AA, Funch LS, Pirani JR (2007a) Reproductive phenology, pollination and seed dispersal syndromes on sandstone outcrop vegetation in the "Chapada Diamantina", northeast Brazil: population and community analyses. Braz J Bot 30:475–485

Conceição AA, Giulietti AM, Meirelles ST (2007b) Ilhas de vegetação em afloramentos de quartzito-arenito no Morro do Pai Inácio, Chapada Diamantina, Bahia, Brasil. Acta Bot Bras 21:335–347

Conceição AA, Pirani JR, Meirelles ST (2007c) Floristics, sctructure and soil of insular vegetation in four quartzite-sandstone outcrops of "Chapada Diamantina", northeast Brazil. Braz J Bot 30:641–655

Conceição AA, Alencar TG, Souza JM, Moura ADC, Silva GA (2013) Massive post-fire flowering events in a tropical mountain region of Brazil: high episodic supply of floral resources. Acta Bot Bras 27:847–850

Connell JH, Slatyer RO (1977) Mechanisms of succession in natural communities and their role in community stability and organization. Amer Nat 111:1119–1144

Drury WH, Nisbet CT (1973) Succession. J Arnold Arb 54:331–368

Ellenberg H, Müller-Dombois D (1967) Tentative physiognomic-ecological classification of plant formations of the earth. Ber Geob Inst Rubel Zurich 37:21–55

Torquato JR, Fogaça, ACC (1981) Correlação entre o Supergrupo Espinhaço no Brasil, o Grupo Chela em Angola e as Formações Nasib e Khoabendus da Namíbia. Anais do simpósio sobre o Craton do São Francisco e suas faixas marginais. Sociedade Brasileira de Geologia - Núcleo da Bahia, coordenação da produção mineral, pp 87–99

Giulietti AM, Queiroz LP, Harley RM (1996) Vegetação e flora da Chapada Diamantina, Bahia. Anais 4ª reunião especial da SBPC, Feira de Santana, Bahia, p 144–156

Giulietti AM, Pirani JR, Harley RM (1997) Espinhaço Range Region, Eastern Brazil. In: Davis SD, Heywood VH, Herrera-Macbryde O, Villa-Lobos J, Hamilton AC (eds) Centres of plant diversity. A guide and strategy for their conservation, vol 3. The Americas. IUCN Publication Unity, Cambridge, pp 397–404

Glenn-Lewin DC, van der Maarel E (1992) Patterns and processes of vegetation dynamics. In: Glenn-Lewin DC, Peet RK, Veblen TT (eds) Plant Succession: theory and prediction. Chapman & Hall, Cambridge, pp 11–59

Grubb PJ (1987) Some generalizing ideas about colonization and succession in green plants and fungi. In: Gray AJ, Crawley MJ, Edwards PJ (eds) Colonization, succession and stability. Blackwell, London, pp 81–102

Hambler DJ (1964) The vegetation of granitic outcrops in western Nigeria. J Ecol 52:573–594

Harley RM (1995) Introduction. In: Stannard BL (ed) Flora of the Pico das Almas, Chapada Diamantina, Bahia, Brazil. Royal Botanic Gardens, Kew, pp 1–42

Houle G (1990) Species-area relationship during primary succession in granite outcrop plant communities. Amer J Bot 77:1433–1439

Houle G, Phillips DL (1989) Seed availability and biotic interaction in granite outcrop plant communities. Ecol 70:1307–1316
Ibisch PL, Rauer G, Rudolph D, Barthlott W (1995) Floristic, biogeographical, and vegetational aspects of Pre-Cambrian rock outcrops (inselbergs) in eastern Bolivia. Flora 190:299–314
Jacobi CM, Carmo FF (2011) Life-forms, pollination and seed dispersal syndromes in plant communities on ironstone outcrops, SE Brazil. Acta Bot Bras 25:395–412
Jacobi CM, Carmo FF, Vincent RC, Stehmann JR (2007) Plant communities on ironstone outcrops: a diverse and endangered Brazilian ecosystem. Biodiv Cons 16:2185–2200
Kolbek J, Alves RJV (2008) Impacts of cattle, fire and wind in rocky savannas, southeastern Brazil. Acta Univ Carol Biol 22:111–130
Le Stradic S, Buisson E, Fernandes GW (2015) Vegetation composition and structure of some neotropical mountain grasslands in Brazil. J Mt Sci 12:864–877
Lüttge U (1997) Physiological ecology of tropical plants. Springer, Heidelberg
McCormick JF, Ariel EL, Sharitz RR (1974) Experimental analysis of ecosystems. In: Strain BR, Billings WD (eds) Vegetation and environment. The Hague, Netherlands, pp 151–179
Meirelles ST, Mattos EA, Silva AC (1997) Potential desiccation tolerant vascular plants from Southeastern Brazil. Pol J Environ Stud 6:17–21
Meirelles ST, Pivello VR, Joly CA (1999) The vegetation of granite rock outcrops in Rio de Janeiro, Brazil, and the need for its protection. Environ Cons 26:10–20
Messias MCTB, Leite MGP, Meira-Neto JAA, Kozovits AR (2011) Life-form spectra of quartzite and itabirite rocky outcrops sites, Minas Gerais, Brazil. Biota Neotrop 11:255–268
Messias MCTB, Leite MGP, Meira-Neto JAA, Kozovits AR, Tavares R (2013) Soil-vegetation relationship in quartzite and ferruginous Brazilian rocky outcrops. Folia Geobot 48:509–521
Michelangeli FA (2000) Species composition and species-area relationships in vegetation isolates on the summit of sandstone mountain in southern Venezuela. J Trop Ecol 16:69–82
Moreira AAN, Camelier C (1977) Relevo. Geografia do Brasil, IBGE, Rio de Janeiro
Neves SPS, Conceição AA (2010) Campo rupestre recém-queimado na Chapada Diamantina, Bahia, Brasil: plantas de rebrota e sementes, com espécies endêmicas na rocha. Acta Bot Bras 24:697–707
Nimer N (1989) Climatologia do Brasil. IBGE, Rio de Janeiro
Oliveira RF, Coimbra AFF, Silva ZL (1975) Sobre litosere: algumas espécies para revestimento de encostas rochosas. Bras Florest 6:3–18
Oosting HJ, Anderson LE (1937) The vegetation of a barefaced cliff in western North Carolina. Ecol 18:280–292
Peet KR (1992) Community structure and ecosystem function. In: Glenn-Lewin DC, Peet RK, Veblen TT (eds) Plant Succession: theory and prediction. Chapman & Hall, Cambridge, pp 103–151
Pidgeon IM (1940) The ecology of the central coastal area of New South Wales. III. Types of primary succession. Proc Linn Soc NSW 65:221–249
Porembski S (2000) Biodiversity of terrestrial habitat islands—the inselberg evidence. In: Porembski S, Barthlott W (eds) Inselbergs. Ecological studies, vol 146. Springer, Heidelberg, pp 507–514
Porembski S, Barthlott W (1997) Seasonal dynamics of plant diversity on inselbergs in the Ivory Coast (West Africa). Bot Acta 110:466–472
Porembski S, Becker U, Seine R (2000) Islands on islands: habitats on inselbergs. In: Porembski S, Barthlott W (eds) Inselbergs. Ecological studies, vol 146. Springer, Heidelberg, pp 49–67
Renvoize SA (1984) The grasses of Bahia. Royal Botanic Gardens, Kew
Ribeiro KT, Medina BMO (2002) Estrutura, dinâmica e biogeografia das ilhas de vegetação sobre rocha do Planalto do Itatiaia, RJ. Boletim do Parque Nacional do Itatiaia 10:11–82
Ribeiro PL, Rapini A, Silva UCS, Konno TUP, Damascena LS, Berg C (2012) Spatial analyses of the phylogenetic diversity of *Minaria* (Apocynaceae): assessing priority areas for conservation in the Espinhaço Range, Brazil. Syst Biodivers 10:317–331
Ribeiro PL, Rapini A, Damascena LS, Berg C (2014) Plant diversification in the Espinhaço Range: insights from the biogeography of *Minaria* (Apocynaceae). Taxon 63:1253–1264

Safford HD (2001) Brazilian Páramos. III. Patterns and rates of postfire regeneration in the campos de altitude. Biotropica 33:282–302
Sarthou C, Villiers J (1998) Epilithic plant communities on inselbergs in French Guiana. J Veg Sci 9:847–860
Shure DJ, Ragsdale L (1977) Patterns of primary succession on granite outcrop surfaces. Ecol 58:993–1006
Simon MF, Pennington T (2012) Evidence for adaptation to fire regimes in the tropical savannas of the Brazilian Cerrado. Int J Plant Sci 173:711–723
Van der Maarel E (1988) Vegetation dynamics: patterns in time and space. Vegetatio 77:7–19
Weaver J, Clements FE (1938) Plant ecology. McGraw-Hill, New York
Westhoff V, van der Maarel E (1978) The Braun-Branquet approach. In: Whittaker RH (ed) Classification of plant communities. Dr W, Junk, Boston, pp 287–399
Wiser SK, Peet RK, White PS (1996) High-elevation rock outcrop vegetation of the southern Appalachian Mountains. J Veg Sci 7:703–722
Zappi DC, Lucas E, Stannard B, Lughadha EN, Pirani JR, Queiroz LP, Atkins S, Hind DJN, Giulietti AM, Harley RM (2003) Lista das plantas vasculares de Catolés, Chapada Diamantina, Bahia, Brasil. Bol Bot Univ São Paulo 21:345–398

Chapter 10
Seed Germination Ecology in Rupestrian Grasslands

Flávia Peres Nunes, Roberta L.C. Dayrell, Fernando A.O. Silveira, Daniel Negreiros, Denise Garcia de Santana, Fabio Janoni Carvalho, Queila Souza Garcia and G. Wilson Fernandes

Abstract Germination behavior is a phenotype that results from the interaction between seed properties (intrinsic factors) and environmental (extrinsic) factors. Since germination is an irreversible process and seedling stage is particularly susceptible to harsh conditions, selection should favor strategies that prevent seeds from germinating under unfavorable conditions for plant establishment. Plant species inhabiting rupestrian grasslands of Espinhaço Mountain Range in Brazil have been considered to have particular germination behaviors that allow plant community to persist in an environment with marked seasonal and spatial heterogeneity. Here, we aimed to review data on

F.P. Nunes (✉) · D. Negreiros · G.W. Fernandes
Ecologia Evolutiva & Biodiversidade/DBG, CP 486, ICB, Universidade Federal de Minas Gerais (UFMG), Belo Horizonte, MG 30161-901, Brazil
e-mail: flaviaperesnunes@yahoo.com.br

D. Negreiros
e-mail: negreiros.eco@gmail.com

G.W. Fernandes
e-mail: gw.fernandes@gmail.com

R.L.C. Dayrell · F.A.O. Silveira · Q.S. Garcia
Instituto de Ciências Biológicas, Depto. Botânica, Universidade Federal de Minas Gerais (UFMG), Belo Horizonte, MG 30161-901, Brazil
e-mail: dayrell.rlc@gmail.com

F.A.O. Silveira
e-mail: faosilveira@icb.ufmg.br

Q.S. Garcia
e-mail: queilagarcia@gmail.com

D.G. de Santana · F.J. Carvalho
Instituto de Ciências Agrárias, Universidade Federal de Uberlândia (UFU), Uberlândia, MG 38408-100, Brazil
e-mail: dgsantana@umuarama.ufu.br

F.J. Carvalho
e-mail: fabiojanoni@ufu.br

G.W. Fernandes
Department of Biology, Stanford University, Stanford, CA 94395, USA

© Springer International Publishing Switzerland 2016

G.W. Fernandes (ed.), *Ecology and Conservation of Mountaintop Grasslands in Brazil*, DOI 10.1007/978-3-319-29808-5_10

germination ecology of rupestrian grasslands species and to describe how environmental and intrinsic factors control seedling emergence, trying to disentangle the contribution of each factor and emphasize how they may interact. We further discuss how the breadth of germination niches may influence the ecological distribution of adult plants at both microhabitat and biogeographic scale and how knowledge on seed biology can contribute to restoration of degraded areas in the rupestrian grasslands.

Contents

10.1 Introduction

The propagation and the colonization of plant species are widely influenced by the patterns of seed germination, which indirectly determine the structure of a plant community (Donohue et al. 2010). Seed germination is a key phase modeled by a group of conditions named “regeneration niche” (Grubb 1977) or “safe-site” (Harper 1977). Therefore, the period in which seed germination occurs is crucial for the establishment of seedlings and, consequently, for the spatial distribution of the species, as this is normally the most vulnerable phase in the life cycle of plants (Harper 1977; Leck et al. 2008). Here, we review the literature on the germination ecology of species inhabiting the quartzitic and ironstone rupestrian grasslands of the Espinhaço mountains (locally known as campo rupestre, canga, campo de altitude, among other terms; Chap. 1). There are few studies published to date, but the available evidence allowed us to: first, describe how environmental factors and intrinsic factors control seedling emergence in natural conditions; to discuss how knowledge on seed germination ecology can be useful to explain the distribution of adult plants at both local (microhabitat) and regional (biogeographic scale); and finally, to discuss how knowledge on seed biology can contribute to restoration of degraded areas in the rupestrian grasslands (see also Chap. 19).

Seed germination ecology is a complex subject that aims to understand what controls the timing of seed germination in nature. Addressing this question requires knowledge on seed biology, environmental conditions in the habitat and how these two interact from time of seed dispersal to germination (Baskin and Baskin 2014). For that matter, the problem needs to be divided into a series of specific subjects revolving around germination, e.g. timing of seed dispersal, dormancy classification at the time of dispersal, environmental conditions in the habitat between dispersal and germination, environmental requirements for germination, and dormancy break.

Data from seed germination studies both in the laboratory and greenhouse are useful to explain timing and control of seed germination of species in nature (Baskin and Baskin 2014). Seeds are usually tested for germination under both light and darkness conditions at regular intervals over a range of temperatures. How much and how fast seeds germinate can be then further correlated with environmental conditions in different microhabitats and seasons. Complementarily, field studies are extremely important to reveal patterns and test laboratory findings. Several germination studies have aimed at better understanding seed germination in the rupestrian grasslands and the main results are discussed below, though most of them have been carried out under laboratory conditions due to the typical small-sized seeds of the species of this ecosystem.

10.2 Environmental Control of Seed Germination

10.2.1 Water

Water is essential for seed germination. During embryo maturation, seeds usually undergo a desiccation process that results in the inactivation of the embryo's metabolism. Seeds remain quiescent until favorable and moist conditions allow water uptake by the embryos (imbibition), which leads to the reactivation of their metabolism and allows germination to occur (Castro et al. 2004). Imbibition largely depends on environmental conditions such as temperature and osmotic potential, as well as on water-permeability of seed or fruit coats (Castro and Hilhorst 2004; Baskin and Baskin 2014).

Rupestrian grasslands have well-drained soils that present very low levels of moisture during a 5-month dry season (Madeira and Fernandes 1999; Ribeiro and Fernandes 2000; Coelho et al. 2006; Medina and Fernandes 2007; Chap. 4). During this period, seed germination is unlikely to take place due to severe water limitation that prevents seed imbibition (Silveira et al. 2012a). However, timing of seed dispersal and germination has other important implications for plant reproductive success in seasonal environments, since it also influences seedling survival by determining the environment to which this vulnerable life stage is exposed (Donohue et al. 2010). Soil moisture is still high during the rainy-to-dry transition in this ecosystem, but it significantly drops as the dry season approaches (Coelho et al. 2006; Silveira 2011).

Thus, seeds are able to germinate during rainy-to-dry transitional period but most seedlings cannot survive the following dry season (Coelho et al. 2008).

Theoretically, natural selection should favor plants that either avoid dispersing seeds during rainy-to-dry transition or delay germination until favorable conditions are met. In fact, seed dormancy was shown to have evolved multiple times in seeds of Melastomataceae of rupestrian grasslands dispersed in late-rainy season (Silveira et al. 2012a). Unfortunately, studies on the phenology of the rupestrian grassland species are still scarce, considering the great diversity of phenological patterns already reported (Belo et al. 2013), therefore preventing the recognition and delineation of more general patterns (but see Chap. 12).

10.2.2 Light

The light effect on seed germination is mediated by phytochrome (chromoproteins), and varies from species to species within a range of possible outcomes (Casal et al. 1998). The majority of studies up to this date have reported positive photoblastic seeds —that present light requirement for germination or increase in germination in light— for the rupestrian grassland species (Table 10.1). Nonphotoblastic seeds of rupetrian grassland have only been demonstrated for species of a few families, such as Fabaceae, Bromeliaceae and Verbenaceae (Table 10.1), and, to our knowledge, *Lippia filifolia* (Verbenaceae) is the only report of negative photoblastic seeds (Pimenta et al. 2007).

Small-sized seeds are more likely to require light for germination than larger seeds (Milberg et al. 2000). Small seeds have limited resources that restrict seedling emerge from more than superficial depths (Bewley and Black 1994; Hewitt 1998). These seeds probably use presence or absence of light as a cue to perceive their location, i.e. if buried at some depth or near the soil surface, and germinate where they can easily find light. Indeed, many species of rupestrian grasslands with positive photoblastic seeds are small-sized, e.g. Eriocaulaceae, Xyridaceae, Velloziaceae (Garcia and Oliveira 2007; Oliveira and Garcia 2011; Soares da Mota and Garcia 2013) and most Melastomataceae (Silveira 2011), and many nonphotoblastic seeds, such as those of Fabaceae species, are considerably larger. However, the relationship between light requirement and seed mass in this environment remains to be formally tested.

Moreover, light sensitivity of seeds may be affected by temperature (Pons 1992). For instance, seeds of several *Vellozia* species are able to germinate in the dark only at high temperatures (Garcia and Diniz 2003; Garcia et al. 2007; Soares da Mota and Garcia 2013). This response to temperature can be mediated by phytochromes (Heschel et al. 2007), but it is unclear whether it has any ecological meaning.

10.2.3 Temperature

Temperature is a major environmental factor that controls emergence either by determining the capacity and rate of germination, or by removing or inducing

Table 10.1 Examples of rupestrian grassland species in which seeds are positively photoblastic (positive), negatively photoblastic (negative), or light insensitive

Family/species	Seeds photoblastism	References
Asteraceae		
Baccharis retusa	Positive	Garcia et al. (2006)
Lychnophora pinaster	Positive	Melo et al. (2014)
Bromeliaceae		
Aechmea nudicaulis	Light insensitive	Marques et al. (2014)
Alcantarea imperialis	Positive	Pereira et al. (2009)
Cryptanthus schwackeanus	Positive	Marques et al. (2014)
Dyckia spp.	Positive	Tarré et al. (2007), Marques et al. (2014)
Encholirium spp.	Positive	Tarré et al. (2007)
Neoregelia bahiana	Light insensitive	Marques et al. (2014)
Pitcairnia flammea	Positive	Pereira et al. (2009)
Tillandsia stricta	Light insensitive	Marques et al. (2014)
Vriesea spp.	Light insensitive	Marques et al. (2014)
Cactaceae		
Arthrocereus spp.	Positive	Cheib and Garcia (2012)
Eriocaulaceae		
Actinocephalus spp.	Positive	Kraus et al. (1996)
Comanthera spp.	Positive	Oliveira and Garcia (2011)
Paepalanthus spp.	Positive	Kraus et al. (1996)
Syngonanthus spp.	Positive	Oliveira and Garcia (2011)
Fabaceae		
Calliandra fasciculata	Light insensitive	Silveira et al. (2005)
Chamaecrista dentata	Light insensitive	Maia et al. (2010)
Mimosa spp.	Light insensitive	Silveira and Fernandes (2006), Dayrell et al. (2015)
Lythraceae		
Diplusodon spp.	Positive	Silveira et al. (2012a, b)
Melastomataceae		
Acisanthera spp.	Positive	Silveira (2011)
Chaetostoma armatum	Positive	Silveira (2011)
Clidemia spp.	Positive	Silveira (2011)
Comolia spp.	Positive	Silveira (2011)
Lavoisiera spp.	Positive	Silveira (2011)
Leandra spp.	Positive	Silveira (2011)
Macairea spp.	Positive	Silveira (2011)
Marcetia taxifolia	Positive	Silveira et al. (2004), Silveira (2011)
Miconia spp.	Positive	Silveira (2011)
Microlicia spp.	Positive	Silveira (2011)

(continued)

Table 10.1 (continued)

Family/species	Seeds photoblastism	References
Tibouchina heteromalla	Positive	Garcia et al. (2006), Silveira (2011)
Trembleya laniflora	Positive	Silveira (2011), Rodrigues and Silveira (2013)
Velloziaceae		
Vellozia spp.	Positive	Garcia and Diniz (2003), Soares da Mota and Garcia (2013)
Verbenaceae		
Lippia filifolia	Negative	Pimenta et al. (2007)
Lippia origanoides	Light insensitive	Pimenta et al. (2007)
Lippia rotundifolia	Positive	Pimenta et al. (2007)
Stachytarpheta spp.	Positive	Santos (2008)
Xyridaceae		
Xyris spp.	Positive	Abreu and Garcia (2005)

dormancy (Bewley and Black 1994). It is important to stress that dormancy break and germination are two different processes and may have different requirements (Vleeshouwers et al. 1995; Thompson and Ooi 2010). In other words, seeds may require one set of environmental conditions for dormancy break and another one for germination (Baskin and Baskin 2014). Unfortunately, it is not always possible to distinguish from the available experimental work which of these is involved, but the distinction will be made here whenever possible.

Most experimental studies with seeds done so far with rupestrian grassland species have investigated germination in temperatures between 15 and 40 °C at 5 °C intervals. The choice of tested temperatures is mostly based on daily and seasonal temperature ranges reported for the rupestrian ecosystem, which varies approximately from 18 to 30 °C in the summer and from 8 to 25 °C in the winter (Madeira and Fernandes 1999; Chap. 15).

Low temperatures usually decrease germination and germination rate of seeds. Exceptionally, few species, such as *Comanthera bisulcata* and *Sinningia rupicola*, produce seeds that present the highest germination percentages in low temperatures (e.g. 10 and 15 °C). Still, the germination process of these species is slower at lower temperatures than at higher ones (Oliveira and Garcia 2011; Ranieri et al. 2012a). Conditional (or relative) dormancy is likely to play a role in shaping this behavior (see Baskin and Baskin 2014), as evidenced by the lower germination percentage of these seeds at other temperatures.

High temperatures (i.e. 30 and 35 °C) are able to increase germination of *Mimosa* species, but this effect is rather a consequence of tegument disruption by heat, which allows subsequent imbibition of the physically dormant seeds. *Mimosa* seeds germinate over a broad thermal gradient when a dormancy breaking treatment (scarification) is given prior to incubation (Silveira and Fernandes 2006; Dayrell et al. 2015). This is a good example of temperature control of germination by removing dormancy.

Many studies in the rupestrian grasslands have found that optimum temperature for germination ranges between 20 and 30 °C (e.g. Ranieri et al. 2003; Abreu and Garcia 2005; Oliveira and Garcia 2011; Cheib and Garcia 2012; Ranieri et al. 2012b; Silveira et al. 2012b; Soares da Mota and Garcia 2013; Marques et al. 2014), which is the same pattern reported for Cerrado seeds (Zaidan and Carreira 2008). Despite typically germinating to higher percentages at intermediate temperatures, seeds of rupestrian grasslands are very distinct regarding their temperature ranges for germination. Some species are able to maintain high germination percentages over a relatively wide range of temperatures, which is the case of several Velloziaceae and Bromeliaceae species that germinate from 15 to 40 °C and from 15 to 35 °C respectively (Soares da Mota and Garcia 2013; Marques et al. 2014). On the other extreme, *Vellozia maxillarioides* (Velloziaceae) and *Pitcairnia flammea* (Bromeliaceae) have considerably narrower germination ranges, 30–40 and 15–25 °C respectively (Pereira et al. 2009; Soares da Mota and Garcia 2013).

Furthermore, much can be learned about germination in the field by incubating seeds at alternating temperatures simulating those in the natural habitat (Baskin and Baskin 2014), but few studies of this kind have been conducted in this ecosystem. Available data suggests that the effect of temperature fluctuation is variable, ranging from no effect on *Xyris* (Abreu and Garcia 2005), *Vellozia* (Le Stradic et al. 2015) and *Syngonanthus* (Oliveira and Garcia 2005) to considerable increases in germination of many *Stachytarpheta* species (Santos 2008). This calls for further studies aiming to compare the effects of constant vs fluctuating temperatures on germination of rupestrian grassland species.

10.3 Intrinsic Factors of Seed Germination

There are several reasons why seeds may not germinate under a set of favorable environmental conditions, including adequate levels of water, temperature, light and gases (Hilhorst 2011). First, many species have high proportions of embryoless seeds or seeds with unviable embryos. Second, a large number of plant populations produce dormant seeds, i.e. viable seeds that are prevented from germinating under favorable conditions due to some intrinsic seed characteristic (Baskin and Baskin 2014). Despite being completely different causes, these factors converge to produce exactly the same response: the absence of germination (see Table 10.2; Silveira 2013).

10.3.1 Seed Viability

Embryo presence and viability are often overlooked in germination studies, and very few have investigated the importance of these factors in rupestrian grassland seeds. Available data shows that Velloziaceae and Xyridaceae seeds presented relatively

Table 10.2 Viable and dormant seeds (mean percentage) for rupestrian grassland species

Family/species	Viable seeds (%)	Dormant seeds (%)	References
Asteraceae			
Eremanthus incanus (2001)	50.8	0	Velten and Garcia (2005)
Eremanthus incanus (2002)	16.7	0	Velten and Garcia (2005)
Lessingianthus linearifolius	41	0	Le Stradic et al. (2015)
Richterago arenaria	1	*NC*	Le Stradic et al. (2015)
Cyperaceae			
Lagenocarpus alboniger	7	*NC*	Le Stradic et al. (2015)
Lagenocarpus tenuifolius	38	100	Le Stradic et al. (2015)
Rhynchospora ciliolata	5	*NC*	Le Stradic et al. (2015)
Rhynchospora consanguinea	39.5	100	Le Stradic et al. (2015)
Rhynchospora riedeliana	30	100	Le Stradic et al. (2015)
Melastomataceae			
Chaetostoma armatum	20.2	93.6	Silveira et al. (2012a)
Comolia sertularia	36.7	87.7	Silveira et al. (2012a)
Miconia corallina	17.0	94.1	Silveira et al. (2012a)
Tibouchina cardinalis	20.4	53.4	Silveira et al. (2012a)
Poaceae			
Aristida torta	28	67.9	Le Stradic et al. (2015)
Echinolaena inflexa	0	*NC*	Le Stradic et al. (2015)
Velloziceae			
Vellozia caruncularis	86	2.3	Le Stradic et al. (2015)
Vellozia epidendroides	89.5	7.3	Le Stradic et al. (2015)
Vellozia resinosa	93.5	2.7	Le Stradic et al. (2015)
Vellozia variabilis	89	1.1	Le Stradic et al. (2015)
Xyridaceae			
Xyris obtusiuscula	67	56.7	Le Stradic et al. (2015)
Xyris pilosa	89.5	11.2	Le Stradic et al. (2015)

All seeds have been recently collected kept at natural conditions for the tests. Viable seeds are in respect to total number of seeds. Dormant seeds were calculated as the final germination percentage over the total number of viable seeds. *NC* Non-conclusive data

high percentages of embryo viability, but species from Asteraceae, Cyperaceae, Melastomataceae and Poaceae studied to date produce a large amount (at least 50 %) of unviable seeds (Table; Velten and Garcia 2005; Silveira et al. 2012a; Le Stradic et al. 2015). Another important aspect worth mentioning is the inter-population and inter-annual variation in seed viability. Seed viability of *Eremanthus incanus* (Asteraceae) varied significantly at different times of collection (see Table 10.2; Velten and Garcia 2005) and different populations of *Eremanthus erythropappus* produced remarkably different percentages of empty seeds (Feitosa et al. 2009). Therefore, low seed viability might be a major cause of low germination percentages in rupestrian grassland species and deserves attention in future works.

It is still unclear what causes the production of high percentages of empty or unviable seeds in this ecosystem, and some possible explanations reside on: (1) the phosphorus-impoverished soils that limits investment in sexual reproduction (Fujita et al. 2014); (2) breeding system limitation, such as pollen limitation; (3) genetic load, which occurs in populations that rely on clonal reproduction to ensure reproductive success; (4) investment in a resprouting strategy, rather than on reproduction through seeds, to persist in the environment after fires (Hoffmann 1998; Lamont and Wiens 2003; Montoro and Santos 2007; Holmes et al. 2008; Fidelis et al. 2010; Vos et al. 2012; Barbosa et al. 2014; Chaps. 6 and 11). Further experimental studies are needed to disentangle the role of each factor in causing such remarkable high levels of unviable seeds.

10.3.2 Seed Dormancy

Seed dormancy is an innate constraint on germination under conditions that would otherwise promote germination in nondormant seeds (Baskin and Baskin 2014). This condition enables seeds to have control over the timing of their own germination, avoiding it to occur during periods that are only ephemerally favorable. By producing seeds with various degrees of dormancy, plants may also use it as a strategy to distribute their offspring across time when environmental conditions are unpredictable (Venable 2007; Poisot et al. 2011).

Dormancy can be expressed by diverse mechanisms originated from physiological, morphological, and/or anatomical characteristics of seeds, what has lead to various systems for classifying seed dormancy (Baskin and Baskin 2014). According to Baskin and Baskin (2004) classification system, physiological dormancy is the most frequent dormancy class and allows seeds to perceive seasonal cues, ensuring that germination occurs at the appropriate time (Baskin and Baskin 2014). Besides having a non-negligible amount of empty and/or unviable seeds, many species of Cyperaceae, Melastomataceae, Poaceae, and Xyridaceae produce a large percentage of dormant seeds (Table 10.2), which have physiological dormancy (Silveira et al. 2012a; Baskin and Baskin 2014). There are a few other reports of physiological dormancy occurrence in other rupestrian grassland populations, e.g. *Lychnophora pinaster* (Melo et al. 2009), *Comanthera elegans* (Oriani et al. 2009), *Lippia* spp. (Pimenta et al. 2007) and *Stachytarpheta* spp. (Santos 2008). Silveira et al. (2012a) have shown an association between physiological dormancy occurrence in Melastomataceae and timing of seed dispersal, but further studies are necessary to confirm whether the patterns found can be extrapolated to the community-level in this ecosystem.

Physical dormancy is caused by water-impermeable seed or fruit coats that prevent germination. Physical dormancy is present in at least 18 families of angiosperms and reaches its greater geographical importance precisely in tropical savannas, where nearly 34 % of the species have this type of dormancy (Baskin and Baskin 2014), but in rupestrian grasslands it was only documented in Fabaceae, i.e. *Mimosa* spp. (Silveira and Fernandes 2006; Dayrell et al. 2015) and *Chamaecrista* spp. (Gomes et al. 2001; Maia et al. 2010). It should be noted, however, that not all members of Fabaceae have seeds with physical dormancy. In fact, *Calliandra fasciculata*, a sympatric legumes, has nondormant seeds (Silveira et al. 2005). Since *C. fasciculata* habitats are mesic contrasting with the xeric habitats colonized by *Mimosa* and *Chamaecrista*, low levels of soil moisture might be a major selective pressure towards physical dormancy in Fabaceae, but this hypothesis remains to be tested.

Furthermore, Ribeiro et al. (2015) demonstrated for the first time the occurrence of water-impermeable seed coat in Myrtales. The seeds of *Chaetostoma armatum* (Melastomataceae) have physiophysical dormancy (physical dormancy + physiological dormancy), a rare dormancy type, only reported for nine families up to this date. The phylogenetic distribution of the seed dormancy types suggests that physical dormancy + physiological dormancy in *C. armatum* is likely derived from physiological dormancy, because this type of dormancy is common in melastomes (Silveira et al. 2013) and physical dormancy had never been previously reported for Myrtales (Baskin and Baskin 2014; Willis et al. 2014). Since physical dormancy appears to have originated during periods of increased aridity (Baskin and Baskin 2014), this study strengthens the hypothesis of selective pressure towards physical dormancy in xeric environments.

10.4 Soil Seed Bank and Seasonal Cyclical Dormancy

Soil seed banks constitute an important strategy for plants that experience adverse environmental conditions (Moles et al. 2003), increasing the probability of successful seedling establishment (Meyer and Kitchen 1994). However, there is still little information on the persistence of seeds in the soil of rupestrian grasslands, where most native species are endemic and of outstanding interest for conservation. In situ experiments have shown that buried seeds of some typical species of rupestrian grasslands maintain the viability for periods exceeding one year and have potential to form persistent soil seed bank (sensu Thompson 1993). Species of Cactaceae (Cheib and Garcia 2012), Eriocaulaceae (Garcia et al. 2014), Velloziaceae (Garcia and Oliveira 2007; Munné-Bosch et al. 2011) and Xyridaceae (Garcia et al. 2012) are examples of typical and endemic plants of this environment that are able to form persistent soil seed banks.

Seeds can experience changes in their germinability under natural conditions synchronizing their germination/dormancy cycles with the seasons of the year

(Footitt et al. 2011). Seasonal cycles of dormancy are an endogenous mechanism that regulates seed germination preventing its occurrence under unfavorable environmental conditions for seedling establishment (Vleeshouwers et al. 1995). Whereas burial did not alter the seed germination in response to seasonal changes of Velloziaceae (Munné-Bosch et al. 2011) and Cactaceae species (Cheib and Garcia 2012), some Eriocaulaceae and Xyridaceae species showed germination/dormancy cycles synchronized with the different seasons of the year.

Temperature and humidity are the main factors influencing seed dormancy and longevity. The involvement of these environmental factors (especially temperature) in dormancy cycling is well known in temperate annual species (Baskin and Baskin 1998). However, little is known about such processes in perennial species of tropical regions, where temperature does not show marked seasonal variations. Although seeds are exposed to lower temperature variations in tropical environments, they may suffer from more extreme degrees of seasonal water restrictions. Some studies have shown that the humidity cycles together with temperature can regulate the degrees of seed dormancy (Kruk and Benech-Arnold 1998).

Seeds of Eriocaulaceae (*Comanthera bisulcata*, *C. elegans*, *Syngonanthus anthemidiflorus* and *S. verticillatus*) and Xyridaceae (*Xyris bialata* and *X. peregrina*) show an absolute requirement of light to germinate, but they are non-dormant when recently collected, i.e. do not have primary dormancy (Oliveira and Garcia 2011; Abreu and Garcia 2005; Garcia et al. 2012). Recent studies have demonstrated that seeds buried at their natural habitat acquire secondary dormancy and change their dormancy level in response to environmental factors throughout the year (Garcia et al. 2012, 2014).

Seeds respond to environmental signals, adjusting their dormancy/non-dormancy cycles to the seasons and thereby determining the optimum time for plant establishment (Footitt et al. 2011). Under field conditions, seeds are exposed to fluctuating temperature and humidity. Rupestrian grassland environments are characterized by great variation in rainfall during the year, with well-defined dry (winter) and rainy (summer) seasons. The results of these studies show that buried seeds of Eriocaulaceae and Xyridaceae acquired secondary dormancy during the rainy season (spring/summer), which was alleviated at the end of the subsequent dry season. Furthermore, this acquired dormancy can be induced, lost and re-induced repetitively in response to seasonal changes. Annual cycles of dormancy are observed for at least two consecutive years for *S. anthemidiflorus* and *S. verticillatus* seeds (Garcia et al. 2014). Hence, the strategy of these species is to prevent seed germination when rainy season becomes established and to break dormancy at the end of dry season or beginning of the wet season, adjusting the seasonal emergence to guarantee their reproductive success.

In addition, Duarte and Garcia (2015) demonstrated interaction between temperature and soil moisture variation in signaling cycles of dormancy in buried seeds of *C. bisulcata* and *S. verticillatus*. This study showed that low temperatures favor dormancy break, while soil moisture regulates both induction and breaking of dormancy, thus demonstrating a fine tuning with the local seasonality. In a scenario of global change, further studies are required to better understand how typical and endemic species of rupestrian grasslands will respond to changes in climatic conditions, which provides for significant changes in temperature and rainfall distribution.

10.5 Ecological and Evolutionary Aspects

10.5.1 Ecological Breadth and Geographical Range

The quartzitic rupestrian grassland consists of a mosaic of environments conditioned primarily by the particle-size of the substrate (Carvalho et al. 2012, 2014; Negreiros et al. 2014) and proximity to watercourses (Conceição and Pirani 2005; Medina and Fernandes 2007). Distinct microhabitats in the rupestrian grasslands show different environmental conditions, including light intensity, temperature ranges and soil moisture (Silveira 2011; Marques et al. 2014; Chap. 6). Given that each species has a restricted set of conditions in which germination takes place (the niche breadth), we expect that germination requirements are related to species ecological breadth and geographic range—i.e. the regeneration niche hypothesis (Grubb 1977).

The combination of the remarkable habitat heterogeneity of rupestrian grassland with the high diversity and varying degrees of endemism of congeneric species (Silveira et al. 2016) provides a great opportunity for testing the regeneration niche hypothesis. The prediction that geographic distribution is associated with regeneration niche was tested in three species of plants in Gesneriaceae. Species with the smallest geographic range also had the narrowest germination breadth, corroborating the hypothesis (Ranieri et al. 2012a). Similarly, the prediction that ecological breadth is associated with regeneration niche was supported by a study carried out with 12 bromeliads species (Marques et al. 2014). There was a significant association between germination traits, and the characteristics of sites where adult plants occur. Habitat-generalist species had broader germination niches compared to habitat-specialist species and germination traits were associated with environmental conditions, mainly with substrate moisture and light environment (Marques et al. 2014). However, Silveira et al. (2012b) found an opposite relationship for two sympatric *Diplusodon* species differing both in their geographical range and microhabitat. Germinability of the more widespread species was lower than that of the endemic congener, indicating that other ecological and historical factors have a major influence in this case.

These few available studies do not yet provide a general conclusion about the influence of germination traits in determining distribution of phylogenetically related species, but highlight the need to expand the studies to species of more representative families of the rupestrian grasslands, such as Velloziaceae, Xyridaceae, Eriocaulaceae and endemic genera of Asteraceae, Melastomataceae and Fabaceae. Comparative studies with species from other vegetation can also provide valuable information about the evolution of the germination behavior of rupestrian grassland species.

10.5.2 Environmental Filters of Seedling Recruitment

The rupestrian grasslands occurs in shallow, acidic, nutrient-poor and excessively drained quartzite-derived or ironstone soils (Oliveira et al. 2015; Chap. 3), where plants often experience strong winds, high irradiance exposure, frequent fires, high daily thermal amplitudes and water shortage during dry season (Giulietti et al. 1997; Madeira and Fernandes 1999; Chap. 1). The combined effects of all these environmental conditions act as a major environmental filters leading to a clear predominance of a stress-tolerant strategy in plant communities (Negreiros et al. 2014).

Environmental filtering is a very important driver of plant community assembly. The biotic and abiotic environment selectively filters out certain species for their physiological, morphological and life-history traits, while favoring other species to establish, grow and reproduce (Harper 1977; Bazzaz 1991; Keddy 1992; Grime 2006). Since both plant traits and requirements may change throughout plants' life stages, for example seedlings are normally the most vulnerable stage (Harper 1977; Leck et al. 2008), the filter of the regenerative phase (i.e. their regeneration niche; Grubb 1977) may differ considerably from the filter of the established phase.

The regenerative phase of rupestrian grassland has been poorly studied up to this date. However, growing evidences suggest that environmental conditions in those landscapes may hinder germination and establishment of plant species, requiring mechanisms and strategies to ensure reproductive success. For instance, seed dormancy seem to prevent germination from occurring during dry season or at the end of rainy season (Garcia et al. 2012, 2014; Silveira et al. 2012a), possibly to guarantee that seedlings—which are more susceptible to desiccation (Coelho et al. 2008)—have time to settle enough to endure the dry period.

Although plants on pristine areas seem to find their way around the filtering of regenerative stages and to maintain their populations, degraded areas of rupestrian grassland do not share the same fate (Le Stradic et al. 2014b). Areas resulting from processes of historical use remain devoid of vegetation, even after been abandoned for decades (e.g. Giulietti et al. 1997; Negreiros et al. 2011; Chap. 19). Several non-conflicting hypotheses may explain the lack of spontaneous recruits in degraded areas, such as absence of seed rain and seed bank, low seed viability, lack of appropriate conditions for germination, substrate toxicity and seedling death due to the extreme harshness of the degraded site (Bradshaw 2000; Baskin and Baskin 2014).

Indeed, despite the typically high species richness of the mature vegetation, seed bank of campo rupestre was shown to be poorer in species and less dense than seed bank of neighboring cerrado, forests, and gallery forest (Medina and Fernandes 2007). Moreover, many typical families from these environments exhibit no obvious mechanisms for seed dispersal (Silveira et al. 2016) and several native species produce high percentages of empty or unviable seeds (Dayrell 2015). Taken together, these studies suggest that seed availability is at least one of the limiting steps for spontaneous regeneration of degraded sites.

It is far less clear whether seed germination of rupestrian grassland species is a limiting step for the restoration of degraded sites. Several germination studies for some of the most representative families of these environments have been conducted, but no obvious and general pattern has emerged. Seed germination trials carried out under controlled environmental conditions show a wide range of germinative profile for rupestrian grassland species, as we discussed throughout this chapter. For example, Bromeliaceae, Velloziaceae and Xyridaceae species usually present high seed germination percentages (Garcia and Oliveira 2007), while germination percentages of Poaceae and Cyperaceae are usually low (Figueiredo et al. 2012; Le Stradic et al. 2015). Most Fabaceae species require tegument break to germinate (Gomes et al. 2001; Silveira and Fernandes 2006; Dayrell et al. 2015), and Melastomataceae (Silveira 2011) and Asteraceae (Velten and Garcia 2005; Melo et al. 2014; Le Stradic et al. 2015) germination percentages are highly variable. This points out to a high functional diversity of germination strategies, which may be partially explained by remarkable habitat heterogeneity, and partially by phylogenetic niche conservatism.

There is no evidence, however, that higher germination percentages and wide range of germination temperature in laboratory trials are translated into higher ability of native species to colonize degraded areas in the field. Hay transfer from natural to degraded areas has been tested, but in spite of the presence of seeds in hay, only a few seedlings emerged within 14 months of the treatment application and they were mostly ruderal species (Le Stradic et al. 2014a). Negreiros et al. (2011) have also highlighted the prevalence of ruderal species in established vegetation of abandoned degraded sites. It still remains unclear if seeds of most native species were unable to germinate and/or environmental conditions were unsuitable for their establishment. There is huge knowledge gap on the establishment of plants in rupestrian grassland, but available data suggest that this phase is a major bottleneck for sexual reproduction (Coelho et al. 2008). Further studies of seed germination and seedling establishment in the field are necessary to shed a light on this matter (Chap. 19).

10.6 Conclusions

Germination ecophysiology was studied in a relatively low number of species of the megadiverse rupestrian grassland. The discussion of the studies presented in this chapter is only a first approximation to describe and interpret the seed germination patterns and is part of an effort to objectively estimate the regenerative capacity of rupestrian grassland plant communities. The data summarized here will be useful to a greater understanding of the plant ecology of this ecosystem and are vital for the development of conservation and management plans. Given the high sensitivity of rupestrian grassland to soil removal and increasing threats from opencast mining (Fernandes et al. 2014; Silveira et al. 2016), knowledge on seed ecology will be vital to assist development of more effective ecological restoration programs (Chap. 19).

Acknowledgements Several grants by CNPq, Fapemig, and logistic support by Reserva Vellozia are acknowledged by F.P. Nunes and G.W. Fernandes. R.L.C. Dayrell acknowledges a Ph.D. scholarship from CAPES. Q.S. Garcia, G.W. Fernandes and F.A.O. Silveira receive a research productivity grant from CNPq.

References

Abreu MEP, Garcia QS (2005) Efeito da luz e da temperatura na germinação de sementes de quatro espécies de *Xyris* ocorrentes na Serra do Cipó (MG). Act Bot Bras 19:149–154

Barbosa BC, Cappi VS, Ribeiro SP, Fernandes GW (2014) Avaliação da capacidade de rebrotamento pós-distúrbio das plantas lenhosas típicas dos campos rupestres. Ecol Austral 24:350–355

Baskin CC, Baskin JM (1998) Seeds: ecology, biogeography, and evolution of dormancy and germination. Academic Press, San Diego

Baskin CC, Baskin JM (2004) A classification system for seed dormancy. Seed Sci Res 14:1–16

Baskin CC, Baskin JM (2014) Seeds: ecology, biogeography, and evolution of dormancy and germination. Academic Press, San Diego

Bazzaz FA (1991) Habitat selection in plants. Am Nat 137:S116–S130

Belo RM, Negreiros D, Fernandes GW, Silveira FAO, Ranieri BD, Morellato PC (2013) Fenologia reprodutiva e vegetativa de arbustos endêmicos de campo rupestre na Serra do Cipó, Sudeste do Brasil. Rodriguésia 64:817–828

Bewley JD, Black M (1994) Seeds: physiology of development and germination. Plenum Press, New York

Bradshaw AD (2000) The use of natural processes in reclamation—advantages and difficulties. Landscape Urban Plans 51:89–100

Carvalho F, Souza FA, Carrenho R, Moreira FMS, Jesus EC, Fernandes GW (2012) The mosaic of habitats in the high altitude Brazilian rupestrian fields is a hotspot for arbuscular mycorrhizal fungi. Appl Soil Ecol 52:9–19

Carvalho F, Godoy EL, Lisboa FJ, Moreira FMS, Souza FA, Berbara RLL, Fernandes GW (2014) Relationship between physical and chemical soil attributes and plant species diversity in tropical mountain ecosystems from Brazil. J Mt Sci 11:875–883

Casal JJ, Sánchez RA, Botto JF (1998) Modes of action of phytochromes. J Exp Bot 49:127–138

Castro RD, Hilhorst HWM (2004) Embebição e reativação do metabolismo. In: Ferreira AG, Borghetti F (eds) Germinação: do básico ao aplicado. Artmed, Porto Alegre, pp 149–162

Castro RD, Bradford KJ, Hilhorst HWM (2004) Desenvolvimento de sementes e conteúdo de água. In: Ferreira AG, Borghetti F (eds) Germinação: do básico ao aplicado. Artmed, Porto Alegre, pp 51–67

Cheib AL, Garcia QS (2012) Longevity and germination ecology of seeds of endemic Cactaceae species from high-altitude sites in southeastern Brazil. Seed Sci Res 22:45–53

Coelho FF, Capelo C, Neves ACO, Martins RP, Figueira JEC (2006) Seasonal timing of pseudoviviparous reproduction of *Leiotrix* (Eriocaulaceae) rupestrian species in South-eastern Brazil. Ann Bot 98:1189–1195

Coelho FF, Capelo C, Figueira JEC (2008) Seedlings and ramets recruitment in two rhizomatous species of Campo rupestre: *Leiothrix curvifolia* var. *lanuginosa* and *Leiothrix crassifolia* (Eriocaulaceae). Flora 203:152–161

Conceição AA, Pirani JR (2005) Delimitação de habitats em campos rupestres na Chapada Diamantina, Bahia: substratos, composição florística e aspectos estruturais. Bol Bot Univ São Paulo 23:85–111

Dayrell RLC (2015) Ecology and evolution of seed dormancy in campos rupestres. Mater's thesis, Universidade Federal de Minas Gerais

Dayrell RLC, Gonçalves-Alvim SJ, Negreiros D, Fernandes GW, Silveira FAO (2015) Environmental control of seed dormancy and germination of *Mimosa calodendron* (Fabaceae): implications for ecological restoration of a highly threatened environment. Braz J Bot 38:395–399

Donohue K, de Casas RR, Burghardt L, Kovach K, Willis CG (2010) Germination, postgermination adaptation, and species ecological ranges. Annu Rev Ecol Evol Syst 41:293–319

Duarte DM, Garcia QS (2015) Interactions between substrate temperature and humidity in signaling cyclical dormancy in seeds of two perennial tropical species. Seed Sci Res 25: 170–178

Feitosa SS, Davide AC, Tonetti OAO, Fabricante JR, Lui JJ (2009) Estudos de viabilidade de sementes de candeia *Eremanthus erythropappus* (DC). MacLeish por meio de testes de germinação e raios X. Floresta 39:393–399

Fernandes GW, Barbosa NPU, Negreiros D, Paglia AP (2014) Challenges for the conservation of vanishing megadiverse campo rupestre. Nat Cons 12:162–165

Fidelis A, Müller SC, Pillar VD, Pfadenhauer J (2010) Population biology and regeneration of forbs and shrubs after fire in Brazilian campos grasslands. Plant Ecol 211:107–117

Figueiredo MA, Baêta HE, Kozovits AR (2012) Germination of native grasses with potential application in the recovery of degraded areas in Quadrilátero Ferrífero, Brazil. Biota Neotrop 12:118–123

Footitt S, Douterelo-Soler I, Clay H, Finch-Savage WE (2011) Dormancy cycling in *Arabidopsis* seeds is controlled by seasonally distinct hormone-signaling pathways. Proc Nat Acad Sci USA 108:20236–20241

Fujita Y, Venterink HO, Bodegom PM, Douma JC, Heil GW, Hölzel N, Jabłońska E, Kotowski W, Okruszko T, Pawlikowski P, Ruiter PC, Wassen MJ (2014) Low investment in sexual reproduction threatens plants adapted to phosphorus limitation. Nature 505:82–86

Garcia QS, Diniz ISS (2003) Comportamento germinativo de três espécies de *Vellozia* da Serra do Cipó, MG. Acta Bot Bras 17:487–494

Garcia QS, Oliveira PG (2007) Germination patterns and seed longevity of Monocotyledons from the Brazilian campos rupestres. Seed Sci Biotech 1:35–41

Garcia LC, Barros FV, Lemos-Filho JP (2006) Comportamento germinativo de duas espécies de canga ferrífera: *Baccharis retusa* DC (Asteraceae) e *Tibouchina multiflora* Cogn. (Melastomataceae). Acta Bot Bras 20:443–448

Garcia QS, Jacobi CM, Ribeiro BA (2007) Resposta germinativa de duas espécies de *Vellozia* (Velloziaceae) dos campos rupestres de Minas Gerais, Brasil. Acta Bot Bras 21:451–456

Garcia QS, Giorni VT, Müller M, Munné-Bosch S (2012) Common and distinct responses in phytohormone and vitamin E changes during seed burial and dormancy in *Xyris bialata* and *X. peregrina*. Plant Biol 14:347–353

Garcia QS, Oliveira PG, Duarte DM (2014) Seasonal changes in germination and dormancy of buried seeds of endemic Brazilian Eriocaulaceae. Seed Sci Res 24:113–117

Giulietti AM, Pirani JR, Harley RM (1997) Espinhaço range region: eastern Brazil. In: Davis SD, Heywood VH, Herrera-MacBryde O, Villa-Lobos J, Hamilton AC (eds) Centre of plants diversity: a guide and strategy for their conservation. World Wildlife Fund/World Conservation Union, Cambridge, pp 397–404

Gomes V, Madeira JA, Fernandes GW, Lemos-Filho JP (2001) Seed dormancy and germination of sympatric species of *Chamaecrista* (Leguminosae) in a rupestrian field. Int J Ecol Environ Sci 27:191–197

Grime JP (2006) Trait convergence and trait divergence in herbaceous plant communities: mechanisms and consequences. J Veg Sci 17:255–260

Grubb PJ (1977) The maintenance of species-richness in plant communities: the importance of the regeneration niche. Biol Rev 52:107–145

Harper JL (1977) Population biology of plants. Academic Press, London

Heschel MS, Selby J, Whitelam GC, Sharrock RA, Donohue K (2007) A new role for phytochromes in temperature-dependent germination. New Phytol 174:735–741

Hewitt N (1998) Seed size and shade-tolerance: a comparative analysis of North American temperate trees. Oecol 114:432–440

Hilhorst HWM (2011) Standardizing seed dormancy research. In: Kermode AR (ed) Seed dormancy: methods and protocols. Springer, New York, pp 43–52

Hoffmann WA (1998) Post-burn reproduction of woody plants in a neotropical savanna: the relative importance of sexual and vegetative reproduction. J Appl Ecol 35:422–433

Holmes GD, James EA, Hoffmann AA (2008) Limitations to reproductive output and genetic rescue in populations of the rare shrub *Grevillea repens* (Proteaceae). Ann Bot 102:1031–1041

Keddy PA (1992) Assembly and response rules: two goals for predictive community ecology. J Veg Sci 3:157–164

Kraus JE, Scatena VL, Lewringer ME, Trench KUS (1996) Morfologia externa e interna de *Paepalanthus* Kunth. (Eriocaulaceae) em desenvolvimento pós-seminal. Bol Bot Univ São Paulo 15:45–53

Kruk BC, Benech-Arnold RL (1998) Functional and quantitative analysis of seed thermal responses in prostrate knotweed (*Polygonum aviculare*) and common purslane (*Portulaca oleracea*). Weed Sci 46:83–90

Lamont BB, Wiens D (2003) Are seed set and speciation rates always low among species that resprout after fire, and why? Evol Ecol 17:277–292

Le Stradic S, Buisson E, Fernandes GW (2014a) Restoration of neotropical grasslands degraded by quarrying using hay transfer. Appl Veg Sci 17:482–492

Le Stradic S, Buisson E, Negreiros D, Campagne P, Fernandes GW (2014b) The role of native woody species in the restoration of *campos rupestres* in quarries. Appl Veg Sci 17:109–120

Le Stradic S, Silveira FAO, Buisson E, Cazelles K, Carvalho V, Fernandes GW (2015) Diversity of germination strategies and seed dormancy in herbaceous species of campo rupestre grasslands. Aust Ecol 40:537–546

Leck MA, Parker VT, Simpson RL (2008) Why seedlings? In: Leck MA, Parker VT, Simpson RL (eds) Seedling ecology and evolution. Cambridge University Press, Cambridge, pp 3–13

Madeira JA, Fernandes GW (1999) Reproductive phenology of sympatric taxa of *Chamaecrista* (Leguminosae) in Serra do Cipó, Brazil. J Trop Ecol 15:463–479

Maia FV, Messias MCTB, Moraes MG (2010) Efeitos de tratamentos pré-germinativos na germinação de *Chamaecrista dentata* (Vogel) H.S. Irwin and Barneby. Flor Amb 17:44–50

Marques AR, Atman APF, Silveira FAO, Lemos-Filho JP (2014) Are seed germination and ecological breadth associated? Testing the regeneration niche hypothesis with bromeliads in a heterogeneous neotropical montane vegetation. Plant Ecol 215:517–529

Medina BMO, Fernandes GW (2007) The potential of natural regeneration of rocky outcrop vegetation on rupestrian field soils in "Serra do Cipó", Brazil. Rev Bras Bot 30:665–678

Melo PRB, Oliveira JA, Carvalho MLM, Guimaraes RM, Carvalho BO (2009) Aplicação do teste de raios X no estudo da morfologia interna e da qualidade fisiológica de aquênios de arnica (*Lychnophorapinaster* Mart.). Rev Bras Sementes 31:146–154

Melo PRB, Oliveira JA, Guimarães RM, Pereira CE, Pinto JEBP (2014) Germinação de aquênios de *Lychnophora pinaster* em função de estádios de maturação, temperatura e luz. Científica 42:404–410

Meyer SE, Kitchen SG (1994) Life history variation in blue flax (*Linum perenne*: Linaceae): seed germination phenology. Am J Bot 81:528–535

Milberg P, Andersson L, Thompson K (2000) Large-seeded species are less dependent on light for germination than small-seeded ones. Seed Sci Res 10:99–104

Moles AT, Warton DI, Westoby M (2003) Seed size and survival in the soil in arid Australia. Austral Ecol 28:575–585

Montoro GR, Santos ML (2007) Fenologia e biologia reprodutiva de *Tibouchina papyrus* (Pohl) Toledo (Melastomataceae) no Parque Estadual da Serra dos Pireneus, Goiás. Rev Biol Neotrop 4:21–29

Munné-Bosch S, Oñate M, Oliveira PG, Garcia QS (2011) Changes in phytohormones and oxidative stress markers in buried seeds of *Vellozia alata*. Flora 206:704–711

Negreiros D, Fernandes GW, Berbara RLL, Rodarte LHO, Barbosa NPU (2011) Caracterização físico-química de solos quartzíticos degradados e áreas adjacentes de campo rupestre na Serra do Cipó, MG, Brasil. Neotrop Biol Conserv 6:156–161

Negreiros D, Le Stradic S, Fernandes GW, Rennó HC (2014) CSR analysis of plant functional types in highly diverse tropical grasslands of harsh environments. Plant Ecol 215:379–388

Oliveira PG, Garcia QS (2005) Efeitos da luz e da temperatura na germinação de sementes de *Syngonanthus elegantulus*, *S. elegans* e *S. venustus* (Eriocaulaceae) Acta Bot Bras 19:627–633

Oliveira PG, Garcia QS (2011) Germination characteristics of *Syngonanthus* seeds (Eriocaulaceae) in campos rupestres vegetation in south-eastern Brazil. Seed Sci Res 21:39–45

Oliveira RS, Galvão HC, de Campos MC, Eller CB, Pearse SJ, Lambers H (2015) Mineral nutrition of *campos rupestres* plant species on contrasting nutrient-impoverished soil types. New Phytol 205:1183–1194

Oriani A, Sano PT, Scatena VL (2009) Pollination biology of *Syngonanthus elegans* (Eriocaulaceae—Poales). Aust J Bot 57:94–105

Pereira AR, Andrade ACS, Pereira TS, Forzza RC, Rodrigues AS (2009) Comportamento germinativo de espécies epífitas e rupícolas de Bromeliaceae do Parque Estadual do Ibitipoca, Minas Gerais, Brasil. Rev Bras Bot 32:827–838

Pimenta MR, Fernandes LS, Pereira UJ, Garcia LS, Leal SR, Leitão SG, Salimena FRG, Viccini LF, Peixoto PHP (2007) Floração, germinação e estaquia em espécies de *Lippia* L. (Verbenaceae). Rev Bras Bot 30:211–220

Poisot T, Bever JD, Nemri A, Thrall PH, Hochberg ME (2011) A conceptual framework for the evolution of ecological specialization. Ecol Lett 14:841–851

Pons TL (1992) Seed responses to light. In: Fenner M (ed) Seeds: the ecology of regeneration in plant communities. CAB International, Wallingford, pp 259–284

Ranieri BD, Lana TC, Negreiros D, Araújo LM, Fernandes GW (2003) Germinação de sementes de *Lavoisiera cordata* e *Lavoisiera francavillana* (Melastomataceae), espécies simpátricas da Serra do Cipó, Brasil. Acta Bot Bras 17:523–530

Ranieri BD, Pezzini FF, Garcia QS, Chautems A, França MGC (2012a) Testing the regeneration niche hypothesis with Gesneriaceae (Tribe Sinningiae) in Brazil: implications for the conservation of rare species. Austral Ecol 37:125–133

Ranieri BD, Negreiros D, Lana TC, Pezzini FF, Fernandes GW (2012b) Fenologia reprodutiva, sazonalidade e germinação de *Kielmeyera regalis* Saddi (Clusiaceae), espécie endêmica dos campos rupestres da Cadeia do Espinhaço, Brasil. Acta Bot Bras 26:632–641

Ribeiro KT, Fernandes GW (2000) Patterns of abundance of a narrow endemic species in a tropical and infertile montane habitat. Plant Ecol 147:205–218

Ribeiro RC, Oliveira DM, Silveira FAO (2015) A new seed coat water-impermeability mechanism in *Chaetostoma armatum* (Melastomataceae): evolutionary and biogeographical implications of physiophysical dormancy. Seed Sci Res 25:179–193

Rodrigues ERS, Silveira FAO (2013) Seed germination requirements of *Trembleya laniflora* (Melastomataceae), an endemic species from neotropical montane rocky savannas. Plant Spec Biol 28:165–168

Santos FMG (2008) Comportamento germinativo de espécies do gênero *Stachytarpheta* Vahl. (Verbenaceae) ocorrentes nos campos rupestres da Cadeia do Espinhaço em Minas Gerais. MSc dissertation, Universidade Federal de Minas Gerais

Silveira FAO (2011) Evolutionary ecophysiology of seed dormancy and germination of Melastomataceae from rupestrian fields. PhD thesis, Universidade Federal de Minas Gerais

Silveira FAO (2013) Sowing seeds for the future: the need for establishing protocols for the study of seed dormancy. Acta Bot Bras 27:264–269

Silveira FAO, Fernandes GW (2006) Effect of light, temperature and scarification on the germination of *Mimosa foliolosa* (Leguminosae) seeds. Seed Sci Technol 34:607–614

Silveira FAO, Negreiros D, Fernandes GW (2004) Influência da luz e da temperatura na germinação de sementes de *Marcetia taxifolia* (A. St.-Hil.) DC. (Melastomataceae). Acta Bot Bras 18:847–851

Silveira FAO, Fernandes F, Fernandes GW (2005) Light and temperature influence on seed germination of *Calliandra fasciculata* Benth. (Leguminosae). Lundiana 6:95–97

Silveira FAO, Ribeiro RC, Oliveira DMT, Fernandes GW, Lemos-Filho JP (2012a) Evolution of physiological dormancy multiple times in Melastomataceae from neotropical montane vegetation. Seed Sci Res 22:37–44

Silveira FAO, Negreiros D, Araújo LM, Fernandes GW (2012b) Does seed germination contribute to ecological breadth and geographic range? A test with sympatric *Diplusodon* (Lythraceae) species from rupestrian fields. Plant Spec Biol 27:170–173

Silveira FAO, Fernandes GW, Lemos JP (2013) Seed and seedling ecophysiology of neotropical melastomataceae: implications for conservation and restoration of savannas and rainforests. Ann Missouri Bot Gard 99:82–99

Silveira FAO, Negreiros D, Barbosa NPU, Buisson E, Carmo FF, Carstensen DW, Conceição AA, Cornelissen TG, Echternacht L, Fernandes GW, Garcia QS, Guerra TJ, Jacobi CM, Lemos-Filho JP, Le Stradic S, Morellato LPC, Neves FS, Oliveira RS, Schaefer CE, Viana PL, Lambers H (2016) Ecology and evolution of the endangered campo rupestre: a neglected biodiversity conservation priority. Plant Soil (in press). doi:10.1007/s11104-015-2637-8

Soares da Mota LA, Garcia QS (2013) Germination patterns and ecological characteristics of Vellozia seeds from high-altitude sites in southeastern Brazil. Seed Sci Res 23:67–74

Tarré E, Pires BBM, Guimarães APM, Carneiro LA, Forzza RC, Mansur E (2007) Germinability after desiccation, storage and cryopreservation of seeds from endemic *Encholirium* Mart. Ex Schult. and Schult. f. and *Dyckia* Schult. and Schult.f. species (Bromeliaceae). Acta Bot Bras 21:777–783

Thompson K (1993) Persistence in soil. In: Hendry GAF, Grime JP (eds) Methods in comparative plant ecology. Chapman & Hall, London, pp 176–205

Thompson K, Ooi MKJ (2010) To germinate or not to germinate: more than just a question of dormancy. Seed Sci Res 20:209–211

Velten SB, Garcia QS (2005) Efeitos da luz e da temperatura na germinação de *Eremanthus* (Asteraceae), ocorrentes na Serra do Cipó, MG, Brasil. Acta Bot Bras 19:755–763

Venable DL (2007) Bet hedging in a guild of desert annuals. Ecol 88:1086–1090

Vleeshouwers LM, Bouwmeester HJ, Karssen CM (1995) Redefining seed dormancy: an attempt to integrate physiology and ecology. J Ecol 83:1031–1037

Vos JM, Keller B, Isham ST, Kelso S, Conti E (2012) Reproductive implications of herkogamy in homostylous primroses: variation during anthesis and reproductive assurance in alpine environments. Funct Ecol 26:854–865

Willis CG, Baskin CC, Baskin JM, Auld JR, Venable DL, Cavender-Bares J, Donohue K, Casas RR, The NESCent Germination Working Group (2014) The evolution of seed dormancy: environmental cues, evolutionary hubs, and diversification of the seed plants. New Phytol 203:300–309

Zaidan LBP, Carreira RC (2008) Seed germination in Cerrado species. Braz J Plant Physiol 20:167–181

Chapter 11
Ecophysiology of *Campos Rupestres* Plants

Rafael S. Oliveira, Anna Abrahão, Caio Pereira, Grazielle S. Teodoro, Mauro Brum, Suzana Alcantara and Hans Lambers

Abstract *Campos rupestres* are rocky seasonally-dry environments that occur over mountaintops in central Brazil. Plant growth is limited and prone to fire during the dry winter, and soils are severely nutrient-impoverished. Plants in these habitats exhibit a wide range of strategies to cope with these limitations. *Campos rupestres* plants show different rooting depths, according to plant habit and substrate. Water status in plants varies between stable water potentials in isohydric species to water potentials changing according to air humidity in anisohydric plants, some of them being dessication-tolerant and dormant during the dry season (resurrection plants). Carbon assimilation in the dry season is therefore limited by water availability, especially in dessication-tolerant species. A wide variety of mineral nutrition strategies allow effective nutrient acquisition in *campos rupestres*: mycorrhizas, non-mycorrhizal sand-binding root specializations, symbiotic nitrogen fixation, carnivory, parasitism. The incidence of natural fires may have played a role in the ecological and evolutionary processes that have molded the current flora of *campos rupestres*, as well as the occurrence of specific functional traits in these habitats. Unveiling the relative importance and prevalence of in situ adaptation and adaptive divergence in the lineages that diversified in *campos rupestres* will allow us to further discuss mechanisms related to trait evolution and adaptive radiation in *campos rupestres*.

R.S. Oliveira (✉) · A. Abrahão · C. Pereira · G.S. Teodoro · M. Brum
Departamento de Biologia Vegetal, Universidade Estadual de Campinas, Campinas, Brazil
e-mail: rafaelsoliv@gmail.com

R.S. Oliveira · A. Abrahão · C. Pereira · H. Lambers
School of Plant Biology, University of Western Australia, Perth, Australia

G.S. Teodoro
Departamento de Engenharia Florestal, Universidade Federal de Lavras, Lavras, Brazil

S. Alcantara
Departamento de Botânica, Universidade Federal de Santa Catarina, Florianópolis, Brazil

© Springer International Publishing Switzerland 2016

G.W. Fernandes (ed.), *Ecology and Conservation of Mountaintop Grasslands in Brazil*, DOI 10.1007/978-3-319-29808-5_11

Contents

11.1 Introduction

Campos rupestres occur over mountain tops in central Brazil and are dominated by small sclerophyllous trees and shrubs interspersed in a matrix of grasses and sedges (Giulietti et al. 1997). Climatically, the *campos rupestres* are characterized by wet summers and dry winters, high vapor pressure deficit (VPD) and high total radiation. Soils are usually shallow in a matrix of decomposed quartzite rocks with fissures, sand and a mixture of small portions of organic matter with low water-holding capacity (Fernandes et al. 2014; Negreiros et al. 2014; Chap. 3). The severely nutrient-impoverished soils impose a high cost in terms of nutrient acquisition relative to biomass accumulation, and, consequently, most *campos rupestres* plants have a perennial-leaf phenology (Le Stradic 2012; Oliveira et al. 2015). The majority of *campos rupestres* plants show reduced stature and longevity associated with a large investment in durable and well-defended structures (Negreiros et al. 2014). The low productivity of *campos rupestres*, as shown by the low leaf area per unit ground area, partially results from the low nutrient availability, but also from the long period of low physiological activity during the dry season (Lambers et al. 2015; Oliveira et al. 2015). In this chapter, we provide an overview of the ecophysiology of *campos rupestres* plants. First, we focus on strategies related to carbon acquisition and water relations. Then, we discuss the different strategies related to nutrient acquisition and use, and how these relate to soil properties. Finally, we discuss the importance of fire, trait evolution and adaptive radiation in *campos rupestres*.

11.2 Carbon and Water Relations of *Campos Rupestres* Plants

Water is a major limiting factor for gas exchange for species in *campos rupestres*, as in other water-limited environments, such as the *kwongkan* (south-western Australia) and *fynbos* (South Africa) (Lambers et al. 2014; West et al. 2012). *Campos rupestres* soils are shallow with low moisture retention (Ribeiro and Walter 2008). During the dry season, soils dry out and VPD increases, imposing even drier conditions for plants.

Plants from water-limited environments have evolved several morphological and physiological adaptations to improve water acquisition/use during periods of low water availability (Porembski and Barthlott 2000; Oliveira et al. 2005; Jacobi et al. 2007; Peñuelas et al. 2011). The strategies to cope with the seasonal water availability vary from drought avoidance to desiccation tolerance. Drought-tolerating species maintain their metabolism under conditions that are suboptimal for growth, while desiccation-tolerant species enter an apparent anabiosis during periods unfavorable for growth (Tuba et al. 1998; Porembski and Barthlott 2000). The *campos rupestres* are considered a center of diversity for desiccation-tolerant vascular plants (also known as resurrection plants or poikilohydric plants) (Porembski and Barthlott 2000; Oliveira et al. 2005; Gaff and Oliver 2013). Desiccation tolerance is the ability of plants under water-limiting conditions to equilibrate their water content with that of dry air (Alpert 2000; Proctor and Tuba 2002). In their desiccated state, these plants remain functionally inactive and appear to be dead; when water is supplied, they rehydrate and resume their normal metabolism within hours or days (Alpert 2000). Drought-avoiding species are also found in the *campos rupestres*, and include species with deep or dimorphic root systems with access to deep groundwater (i.e. *Eremanthus seidelii* MacLeish and Schumacher (Asteraceae);. *Mimosa clausenii* Benth. (Fabaceae); Brum 2013) or, more rarely, drought-deciduous species.

11.2.1 Water-Acquisition Mechanisms

Root water uptake is the most common water-acquisition mechanism in terrestrial plants and plant survival in *campos rupestres* might depend directly on positioning roots in pockets of moist soil deposited on outcrops and sandy soil. In fact, the water niche partitioning by roots in the soil profile has been used to explain the coexistence between different plant functional types, as a function of soil water availability (Walter 1971; Jackson et al. 1999; Rossatto et al. 2013). Hydrological niche segregation is defined here as "partitioning of space on fine-scale soil moisture gradients, or partitioning of water through different strategies of water acquisition and use such as different phenologies and/or root depth" (Araya et al. 2011; Silvertown et al. 2015). Hydrological niche segregation is likely also important in *campos rupestres*.

Campos rupestres plants are expected to show a large investment of biomass in root systems. Belowground structures are required for water uptake, as well as for storage of nonstructural carbohydrates that allow resprouting after drought (Bell et al. 1996). However, the shallow and gravel soil hamper root growth, and the below-ground root investment can vary among species and the site where they grow. For example, the Fabaceae species *Collaea cipoensis* Fortunato, *Calliandra fasciculata* Benth., *Chamaecrista ramosa* (Vog.) H.S. Irwin and Barneby, and *Mimosa foliolosa* Benth. in Serra do Cipó, Minas Gerais, show a low root/shoot ratio ranging from 0.2 to 0.8 (Negreiros et al. 2009). Species adapted to an arid and nutrient-poor environment are expected to have a root/shoot ratio greater than one, as also found in *cerrado* trees with most species having root/shoot ratios ranging from one to six (Negreiros et al. 2009). The low root biomass allocation can be explained by the shallow soil that can constrain root growth. On the other hand, the tap-rooted species *Euphorbia sarcodes* Boiss. (Euphorbiaceae) from Chapada Diamantina shows six times greater investment in belowground structures than in aerial parts which was attributed to the deeper white sands in its natural habitat (Ferreira and Conceição 2012).

The coarse texture of the stony grassland is associated with a low water-retention capacity compared with that of sandy grasslands in *campos rupestres* (Negreiros et al. 2014). Only plants able to make morphological adjustments in roots or possessing specialized flat root systems can explore deep water sources in rocky substrate (Schwinning 2010). For example, *Leandra aurea* (Cham.) Cogn., a multi-stemmed abundant shrub with a dimorphic root system, shows a flattened roots morphology in the tap-root growing into rock fissures (Brum 2013) (Fig. 11.1). Root foraging for water maximizes chances to reach cracks in the quartzite rock that might be essential for survival. Once the roots occupy the fissure, the plant can explore, free of competition, a water-storage pocket in rocks (Schwimning 2010; Poot et al. 2012).

The prevalence of different rooting depths may differ across communities that occur in a mosaic of substrates conditioned by particle sizes that vary from finer texture, such as in peat bogs near river courses and in crevices of rocky outcrops, to soils where sand and gravel predominate (Le Stradic 2012). A survey on the diversity of root system morphologies of 15 co-occurring species in a quartzite gravel soil at Serra da Canastra, in the southeast of Minas Gerais, showed that one-third of the species had dimorphic root systems (i.e. a tap-root with prominent lateral roots close to the soil surface, Fig. 11.2a) (Brum 2013). Dimorphic roots in shrubs and small trees allow plants to switch water sources between seasons, buffering the lack of water in superficial soil. In fact, the two dimorphic-rooted species *Eremanthus seidelli* and *Mimosa clausenii* are able to change water sources between soil layers between seasons. This is evidenced by changes in the isotopic composition of xylem water (δD and $\delta^{18}O$) from shallow to deeper water sources (Brum 2013). The species that have access to deep water can maintain higher pre-dawn shoot water potential values and faster transpiration rates during the dry season compared with those of shallow-rooted plants (Fig. 11.3). Dimorphic root systems may also allow hydraulic redistribution, i.e. the passive movement of water between soil layers with contrasting water potential via roots during the night

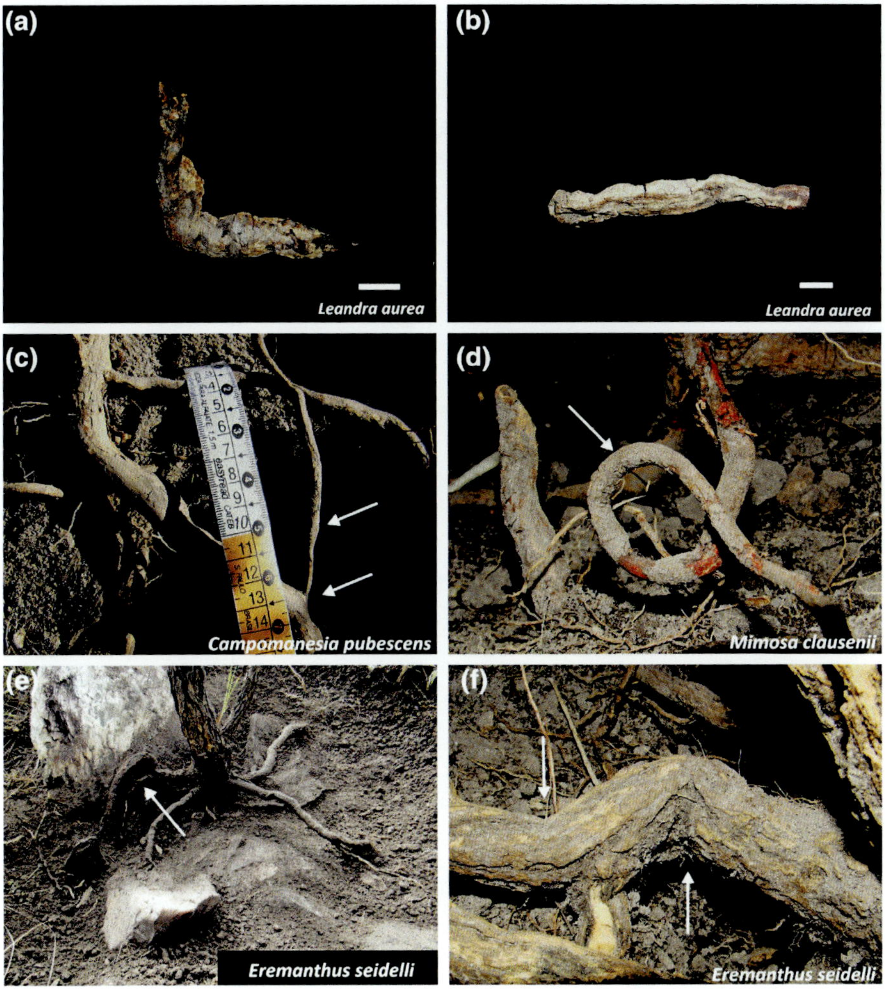

Fig. 11.1 Examples of underground systems with morphological variations common in rocky habitats. **a**, **b** A piece of tap-root of *Leandra aurea* showing the morphological flattening. Scale bars = 5 cm. **c** Flattening of *Campomanesia pubescens* (Myrtaceae) to cross rock fissures on horizontal growth. **d** Roots of *Mimosa clausseni* (Fabaceae) growing in a loop around a stone. **e** Rock deviation and **f** zig-zag growth in *Eremanthus seidelli* (Asteraceae). These adaptation allow plants to cross the fissures in the rock matrix in *campos rupestres* in Serra da Canastra, Minas Gerais. Photos from Brum (2013)

(Caldwell et al. 1998). However, hydraulic redistribution has never been investigated in *campos rupestres* plants.

As in many savanna ecosystems (Walter 1971), grasses in *campos rupestres* are predominantly rooted in the shallowest soil (Fig. 11.2b) (Rossatto et al. 2013). *Echinolaena inflexa* (Poir.) Chase and *Hypolytrum pulchrum* (Rudge) H. Pfeiff., abundant grasses in Chapada Diamantina, show five and six times greater

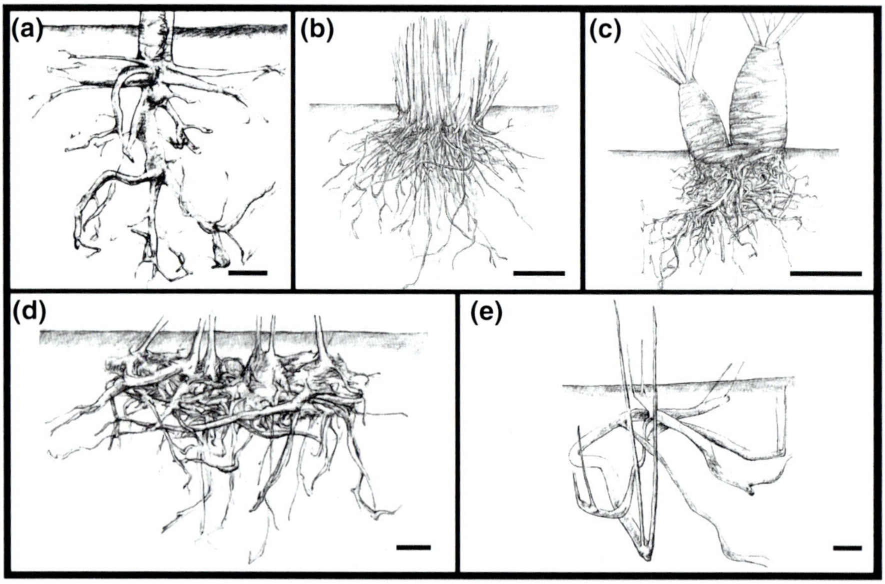

Fig. 11.2 Schematic view of general root-types in *campos rupestres* in Serra da Canastra in Minas Gerais. **a** Dimorphic root system **b** Fibrous roots of grasses. **c** Thick fibrous roots of Velloziaceae. **d** Lignotuber of *Lessingianthus warmingianus* (Asteraceae). **e** Stem tuber of *Campomanesia pubescens* (Myrtaceae). Scale bars: **a** = 10 cm, **b–e** = 5 cm. Illustrations of D.A. Figueiredo

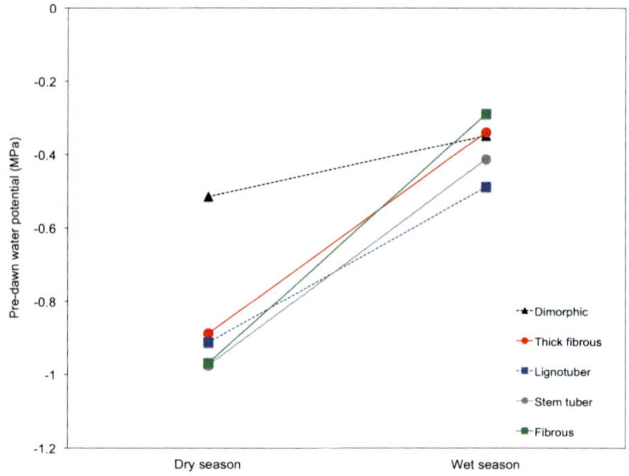

Fig. 11.3 Seasonal pre-dawn water potential of 15 species grouped in five functional root types that occur in Serra da Canastra, Minas Gerais. Data from Brum (2013)

investment in aboveground than in belowground biomass, respectively (Ferreira and Conceição 2012). Shallow soil layers can also be exploited by shrubs and rosettes with contrasting underground systems (Fig. 11.2c–e) *Lessingianthus warmingianus* (Baker) H. Rob. (=*Vernonia warmingiana* Baker, Asteraceae) and *Campomanesia pubescens* (DC.) O. Berg. (Myrtaceae) are two common shrub species in Serra da Canastra with contrasting root architectures that use water exclusively from shallow soil (Fig. 11.2d, e). *Campomanesia pubescens*, *H. pulchrum*, *Andira humilis* Mart. ex Benth. (Fabaceae), *Jacaranda decurrens* Cham. (Bignoniaceae) and *Marcetia taxifolia* (A. St.-Hil.) DC. (Melastomataceae), have underground stems that grow close to the soil surface and allow for foraging of resources (Appezzato-da-Glória 2003; Alves et al. 2013; Brum 2013; Ferreira and Conceição 2012). *Lessingianthus warmingianus* has lignotubers/xylopodia, which are underground organs that store carbohydrates that might be involved in osmotic adjustment. These organs also allow resprouting after prolonged drought periods (Lopes-Mattos et al. 2013; Dietrich et al. 2012).

Many epiphytic and epilithic plants from *campos rupestres* show velamen radicum (Pita and Menezes 2002), which is a structure that has been associated with root specializations for water acquisition from the atmosphere (Porembsky and Barthlott 1995). The velamen is a multiseriate epidermis that recovers the roots. This root structure is present in Velloziaceae, Orchidaceae, Araceae, Cyperaceae and Bromeliaceae (Porembski and Barthlott 1995; Pita and Menezes 2002). For example, *Philodendron adamantinum* Mart. ex Schott shows these structures and grows directly on bare rocks (Fig. 11.9e). It could allow rapid water absorption, an important trait for plants that depend on intermittent water supply. Direct water absorption by shoots has been observed in the desiccation-tolerant plant *Vellozia squamata* Pohl (=*V. flavicans* Mart.), a widely-distributed species in *campos rupestres*. The species has a distinct morphology, the trunks (or pseudostem) consist of persistent and fibrous leaf sheaths under which many adventitious roots with velamen radicum can be found. Reverse flow in xylem water movement found in *V. squamata* indicates that this species can take up water from the canopy and redistribute it to stems or roots (Oliveira et al. 2005). This water uptake by aerial parts contributes to maintain a favorable plant water status in long periods of drought. Moreover, this water uptake can be associated with the absorption of nutrients deposited by dew or fog. The nutrient deposition by fog on leaves is 90 times greater than that in rain events (Baêta 2012). When leaf water potential is too low in the dry season, water deposited on leaves, dew or fog might help the translocation of nutrients from leaves to stem or roots.

Some species in *campos rupestres* have leaf structures that secrete hygroscopic mucilaginous or resinous polysaccharides that possibly facilitate moisture absorption of fog and light rain water (Lusa et al. 2014a). For example, *Lychnophora diamantinana* Coile and S.B. Jones (Asteraceae), has nonglandular trichomes and Apocynaceae species have colleters (i.e. multicellular secretory structures attached to the differents plants organs) with hydrophilic substances (Lusa et al. 2014a; Morokawa et al. 2013). Furrows and stomatal crypts are found on leaves of species of the subtribe Lychnophorinae (Asteraceae) and in many species of *Vellozia* (Velloziaceae) (Mello-Silva et al. 2011; Lusa et al. 2014b; Alcantara et al. 2015). These structures have been hypothesized to play a role in the diffusion of carbon dioxide from the abaxial surface to adaxial palisade cells (Jordan et al. 2008; Lambers et al. 2014) in some *campos rupestres* species (i.e. Lychnophorinae subtribe) (Lusa et al. 2014b), but their functional importance still remains to be assessed.

11.2.2 Photosynthesis and Water Relations

Grasses (Poaceae) with a C_4 photosynthetic pathway are widely distributed in *campos rupestres* (Garcia et al. 2009). In the Espinhaço Mountain Range this photosynthetic pathway represents 60–85 % of the total grass species. In other Brazilian grasslands, the two photosynthetic types (C_3 and C_4) are almost equally represented (Garcia et al. 2009). The C_4 photosynthetic pathway consists of a number of anatomical and biochemical modifications, in which the primary CO_2 acceptor is phospho*enol*pyruvate (PEP) producing oxaloacetate, catalyzed by the enzyme PEP carboxylase. These plants decrease their water loss relative to CO_2 assimilation. Under conditions of high temperatures and low water availability, C_4 species are generally competitively superior compared with C_3 species (de Souza et al. 2005; Lambers et al. 2008a).

Plants exhibiting the crassulacean acid metabolism (CAM) photosynthetic pathway are also found in *campos rupestres*. CAM plants show a temporal separation of carbon assimilation and photosynthesis: during the night they take up CO_2, storing it as C_4 acids, and during the day, they process CO_2 in photosynthesis (Lambers et al. 2008a). These plants open their stomata at night thus reducing water loss to the atmosphere. The orchid *Epidendrum secundum* Jacq. (which occurs in *campos rupestres* although not restricted to this vegetation type), shows diurnal acid fluctuations, a thick cuticle, large and vacuolated cells, indicating a CAM metabolism (Moreira et al. 2009).

In *campos rupestres*, plants are subjected to high irradiance; however, photoinhibition does not appear to be a major limitation for photosynthesis (Lüttge et al. 2007). In a study with five Velloziaceae species, Lüttge et al. (2007) measured the effective quantum yield of photosystem II and apparent electron transport rate in the morning, at mid-day and in the afternoon. The photosynthetic capacity was greater in the morning and mid-day than in the afternoon, and this was not a result of photoinhibition, but related to stomatal limitations. This is firm evidence that the carbon assimilation of some *campos rupestres* plants is determined by water availability.

Plants show different strategies to deal with the conflicting demand of water loss regulation and carbon uptake (Filella and Peñuelas 2003; Peñuelas et al. 2011). Stomatal regulation has a central role to control plant water use under varying conditions of water availability (Schwinning and Ehleringer 2001; Damour et al. 2010) and this seems to be the case for species in *campos rupestres*. Brum (2013) measured the gas exchange and water potential of 12 species in *campos rupestres* at Serra da Canastra during the transition between wet and dry season. In general, the species experienced lower pre-dawn and mid-day water potentials in the dry season and reduced stomatal conductance (Fig. 11.4). Shrubs in *campos rupestres* formed a continuum of hydraulic strategies related to stomatal control and water potential regulation, varying from isohydric to anisohydric behavior. Isohydric species have a strong regulation of stomatal conductance and maintain nearly constant water potential despite reductions in soil water potential (Fig. 11.5). Typically, isohydric plants exhibit a high stomatal sensitivity to evaporative demand. In contrast,

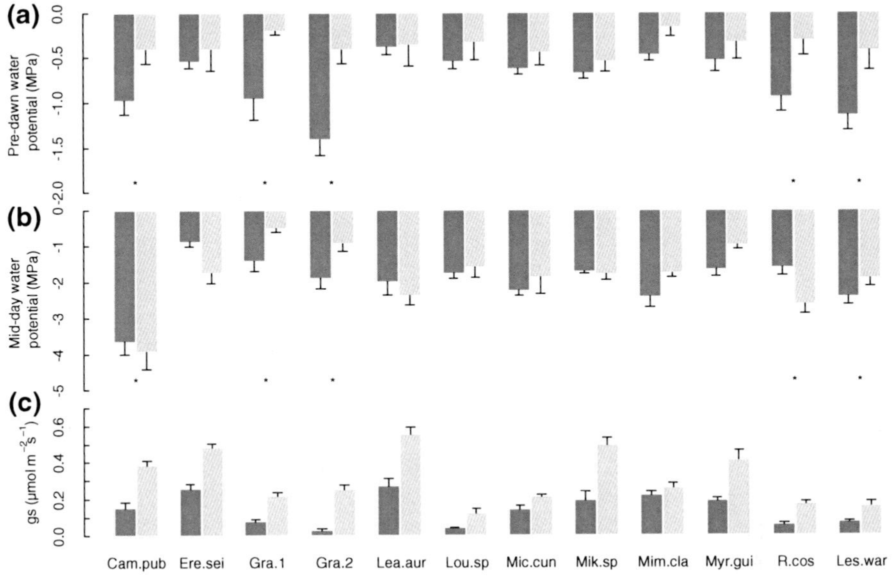

Fig. 11.4 Water potential and stomatal conductance in wet and dry season. *Light gray bars* represent the end of wet season and *dark gray bars* represents the peak of the dry season for 12 species with different rooting depth at Serra da Canastra National Park; **a** Pre-dawn water potential; **b** Mid-day water potential; **c** Stomatal conductance. The (*) represents species with a shallow root system. Cam. pub: *Campomanesia pubecens* (Myrtaceae), Ere.sei: *Eremanthus seidelli* (Asteraceae), Gra.1: Poaceae sp.1, Gra.2: Poaceae sp.2, Lea.aur: *Leandra aurea* (Melastomataceae), Lou.sp: *Loudetiopsis* sp. (Poaceae), Mic.cun: *Microlicea cuneata* (Melastomataceae), Mik.sp: *Mikania* sp. (Asteraceae), Mim.cla: *Mimosa clausennii* (Fabaceae), Myr.gui: *Myrsine guianensis* (Myrsinaceae), R.cos: *Rhyncospora cosanguinea* (Cyperaceae), Ver.war: *Lessingianthus warmingianus* (Asteraceae). Data from Brum (2013)

anisohydric plants allow mid-day water potential to decline and maintain gas exchange even under drought conditions (Tardieu and Simonneau 1998; McDowell et al. 2008; Plaut et al. 2012). These water-use strategies are related to a species' capacity to survive drought in many ecosystems (Meinzer et al. 2009; West et al. 2012; Plaut et al. 2012; McDowell et al. 2013). *Eremanthus seidelli* exhibits an isohydric behavior and maintains a nearly constant mid-day water potential (Fig. 11.5a). Stomatal conductance for this species is strongly related to vapor pressure deficit and soil water content, corroborating the prediction that stomatal regulation in isohydric species depends on water availability (Ogle and Reynolds 2002; Damour et al. 2010; Teodoro 2014). *Campomanesia pubecens* shows an anisohydric strategy, exhibiting a strong decrease in mid-day water potential (Fig. 11.5b). Among species of the monocot family Velloziaceae, the differences in gas exchange between seasons depend on the strategies to cope with drought, i.e. the desiccation-tolerance and the non-desiccation-tolerance strategy. The desiccation-tolerant (DT) plants show inhibition of carbon assimilation during the desiccated state; carbon acquisition is restricted to the periods that plants are

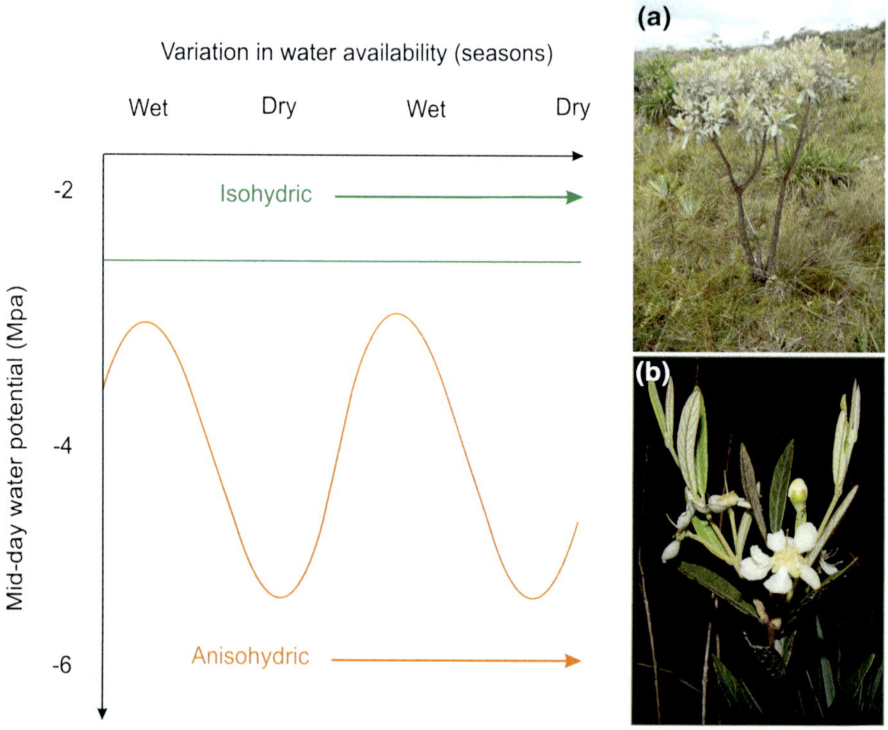

Fig. 11.5 Seasonal variation in mid-day water potential of two shrub species from *campos rupestres* that show contrasting hydraulic strategies: **a** *Eremanthus seidelli* MacLeish and Schumacher (Asteraceae) showing an isohydric strategy. **b** *Campomanesia pubecens* Mart. ex DC. (Myrtaceae) with an anisohydric strategy. Data from Teodoro (2014) (with permission). *Photos* G. S. Teodoro

hydrated and metabolically active (Alpert 2006). Alcantara et al. (2015) studied the carbon acquisition and habitat segregation of 20 species of Velloziaceae that differ in their ability to tolerate desiccation (Fig. 11.6). During the dry season, the non-DT species maintain carbon gain; however, they show small reductions in photosynthesis, stomatal conductance and water potential (Fig. 11.7a–c). On the other hand, the DT species reach negative values of carbon assimilation during the dry season (Fig. 11.7c). The DT species are less productive than the non-DT species, suggesting a trade-off between DT and productivity in these plants (Alcantara et al. 2015). In addition, Velloziaceae species with contrasting strategies to cope with drought occur in different soil patches reflecting habitat segregation (Alcantara et al. 2015). The differences in soil properties among patches reflect differences in water retention capacity and nutrient availability. Desiccation-tolerant species occur in rocky outcrops, in many cases, directly on the exposed rocks, while the non-DT species occur on deeper sandy soils. The DT species are largely restricted to areas that are not inhabited by desiccation-sensitive species. In these habitats, the effects

(a)
(b)
(c)
(d)
(e)
(f)
(g)
(h)
(i)

◀ **Fig. 11.6** Species of Velloziaceae with contrasting strategies to cope with drought: desiccation tolerance (DT) and non-DT. **a–f** DT species: **a** *Barbacenia macrantha* Lem. hydrated; **b** *B. macrantha* desiccated; **c** *Vellozia caruncularis* Mart. ex Seub. hydrated; **d** *V. caruncularis* desiccated; **e** *V. resinosa* Mart. ex Schult. and Schult. hydrated; **f** *V. resinosa* desiccated; **g–i** Non-DT species: **g** *Barbacenia* sp. (Alcantara 118); **h** *V. strangii* L.B.Sm. ex Mello-Silva; **i** *V. variabilis* Mart. ex Schult. & Schult. f. *Photos* **a, h, i** S. Alcantara. **b–g** G.S. Teodoro

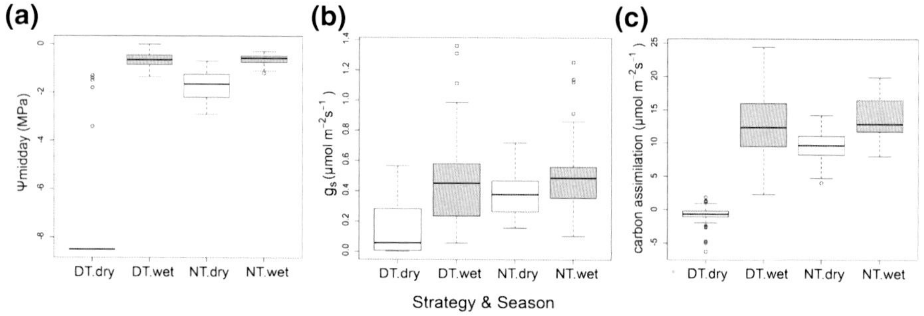

Fig. 11.7 Physiological attributes of desiccation—(DT) and non-desiccation—(NT) tolerant species of Velloziaceae in the dry (*white boxes*) and the wet (*grey boxes*) seasons measured in the field. **a** Ψleaf-MD: leaf water potential at mid-day (MPa). The minimal water potential indicated on the y-axes of the graphs (8.5 MPa) represents the measurement limit of the equipment used. **b** gs: stomatal conductance. **c** Carbon-assimilation rates. DT: N = 13 spp.; NT: N = 7 spp. Data modified from Alcantara et al. (2015), with data available from the Dryad Digital Repository: http://dx.doi.org/10.5061/dryad.51n8n

of dry seasons or persistent droughts can be magnified by the hard substrates impenetrable to roots (Gaff 1987; Porembski and Barthlott 2000; Alpert 2006; Alcantara et al. 2015). Coexistence of both DT and non-DT strategies in *campos rupestres* reflects the trade-off between the ability to tolerate desiccation and occupy rocky sites *versus* the ability to be more productive and competitively dominant in sandy soils.

A key aspect of the carbon metabolism of perennial plants in seasonally-dry environments is how they allocate the carbon fixed towards long-term storage (nonstructural carbohydrates) that can be drawn upon and used either during drought condition or for regrowth after drought (Chapin et al. 1990; Sala et al. 2010). In *campos rupestres*, some species produce lignotubers that can store large amounts of nonstructural carbohydrates (e.g., starch, fructans). *Vernonia* species (Asteraceae) accumulate fructans, which are an important reserve in species in seasonally-dry environments with impoverished soils (Tertuliano and Figueiredo-Ribeiro 1993; Appezzato-da-Glória et al. 2008). These nonstructural carbohydrates are related by osmoregulation, due to fast polymerization and depolymerization reactions (Tertuliano and Figueiredo-Ribeiro 1993; Appezzato-da-Glória et al. 2008).

Campos rupestres plants exhibit a wide range of morphological as well as phenological strategies to cope with water limitation in the seasonally-dry environments. Deep rooting allows drought avoidance, and dessication tolerance permits the survival on substrates where water supply is the most limiting. Water absorption by leaves and adventitious roots also comprise important mechanisms that allow water and nutrient acquisition deposited by fog. Alternative photosynthetic pathways as C_4 and CAM are found in several plant species in *campos rupestres* and increase water-use efficiency in hot and dry habitats. Carbon assimilation can be limited by water availability, and species vary between isohydric and anisohydric, exhibiting stable or variable water potentials. The storage of nonstructural carbohydrates in *campos rupestres* plants is related to osmoregulation and the carbohydrates can be used for regrowth after drought. Taken together, these strategies indicate major adaptations to survive seasonal droughts.

11.3 Plant Mineral Nutrition in *Campos Rupestres*

Nutrient availability is one of the main environmental filters influencing plant distribution (Lambers et al. 2008a). Plant productivity in old landscapes tends to be phosphorus-limited, rather than nitrogen-limited (Walker and Syers 1976) because phosphorus (P) is originally present in the soil parent material and then lost during soil development (pedogenesis), while nitrogen (N) is scarce in the parent material and is gradually accumulated by biological N_2 fixation and atmospheric deposition. Therefore, unless soils are continually rejuvenated, ecosystem development is accompanied by nutrient impoverishment (Turner and Condron 2013), especially P, and also by increasing species diversity (Pekin et al. 2012). This diversity is also reflected in plant nutritional strategies (Zemunik et al. 2015).

Hopper (2009) referred to ancient environments as OCBILs (Old, Climatically-Buffered Infertile Landscapes) and based his OCBIL theory on three floristic provinces: the succulent Karoo vegetation from South Africa, the Southwest Australian Floristic Region (SWAFR), and the Venezuelan Pantepuis, but he also suggested that other parts of South America might function as OCBILs, without specifically naming them. Since Hopper's work, several papers found similarities in plant functioning between *campos rupestres* and OCBILs (Pereira et al. 2012; Oliveira et al. 2015; Abrahão et al. 2014; Silveira et al. 2016). This section will focus on further similarities between these environments.

11.3.1 Pedogenesis, Sources of N and P

Campos rupestres present extremely nutrient-impoverished acidic soils (Benites et al. 2003, Table 11.1). The parental material is mainly quartzite (Fig. 11.8a, b), which is naturally nutrient poor (Porder and Ramachandran 2013). According to a recent review by Vasconcelos (2011), plant vegetation over ironstone, locally called *cangas* (Fig. 11.8c, d), can also be called *campos rupestres*. *Cangas* also present

Table 11.1 Properties of *campos rupestres* soils from quartzitic and ironstone outcrops under different vegetation types

Locality	Vegetation type	P ($mg\ kg^{-1}$)	N ($g\ kg^{-1}$)	pH	OM ($g\ kg^{-1}$)	Mn ($mg\ kg^{-1}$)	Fe ($mg\ kg^{-1}$)	References
Quartzitic outcrops								
Serra da Calçada—PESRM	Grassland	5.2		4.0	39.8	21.5	251.0	Vincent and Meguro (2008)
Several localities	Rock outcrop			3.7		6.0		Benites et al. (2003)
Several localities	Grassland			4.4		4.0		Benites et al. (2003)
Serra de Ouro Preto—Mariana	Quartzite outcrop slopes	4.1	0.4	4.9	26.2	7.9	155.0	Messias et al. (2013)
Serra de Ouro Preto—Mariana	Quartzite outcrop plateau	1.2	0.2	5.3	10.5	2.6	284.8	Messias et al. (2013)
Serra do Cipó	Outcrop grassland	1.6	2.0	4.4				Negreiros et al. (2011)
Serra do Cipó	Outcrop grassland	0.6			57.2	3.9	74.9	Negreiros et al. (2008)
Serra do Cipó	Sandy bog	2.0	0.7	4.2	13.5	0.3	79.7	de Carvalho et al. (2012)
Serra do Cipó	Rock outcrop	2.0	0.9	4.1	20.3	0.5	66.3	de Carvalho et al. (2012)
Serra do Cipó	Quartz gravel field	3.0	0.8	4.4	23.2	0.8	59.3	de Carvalho et al. (2012)
Serra do Cipó	Sandy grassland	2.0		4.2				Ribeiro and Fernandes (2000)
Serra do Cipó	Quartz gravel field	1.8		4.8				Ribeiro and Fernandes (2000)
Serra do Cipó	Rock outcrop	3.0		4.2				Ribeiro and Fernandes (2000)
Serra do Cabral	White sand	0–3						Oliveira et al. (2015)
Ironstone outcrops								
Serra da Mutuca—PESRM	Shruby grassland	14.8		3.4	136.0	5.1	508.0	Vincent and Meguro (2008)
Serra de Ouro Preto—Ouro Preto	Itabirite outcrop slopes	2.6	1.0	5.0	115.5	54.6	588.9	Messias et al. (2013)
Serra de Ouro Preto—Ouro Preto	Itabirite outcrop plateau	2.4	0.1	5.1	95.6	49.9	455.1	Messias et al. (2013)

P refers to available phosophorus an N to total soil nitrogen (Kjedahl)
For conversion purposes, soil density was considered to be 1 $kg\ dm^{-3}$ (Neves et al. 2005). *PESRM* Parque Estadual da Serra do Rola Moça

Fig. 11.8 *Campos rupestres* landscapes. **a** Overview of quartzitic campos rupestres in the Espinhaço range, Minas Gerais, Brazil; **b** Quartz gravel and graminoid matrix in quartzitic campos rupestres at Serra do Cipó, Minas Gerais, Brazil; **c** Overview of ironstone *campos rupestres*, Parque Estadual da Serra do Rola Moça, Minas Gerais, Brazil; **d** Ironstone *campos rupestres* soil covered with Velloziaceae *Photos* **a** R.S. Oliveira, **b**. A. Abrahão, **c**, **d**. R. Dayrell

acidic, nutrient-poor soils (Vincent and Meguro 2008), but contain higher concentrations of heavy metals (Messias et al. 2013).

The Espinhaço range is the largest range containing *campos rupestres* (Alves et al. 2014). Other smaller sites include Serra da Canastra, in Minas Gerais, and Chapada dos Veadeiros, in Goiás. However, since most of the studies were on the Espinhaço range, we will give some details on the geomorphology and stratigraphy of the Espinhaço range, a complex Precambrian range formed by three main lithologic groups. The Precambrian rock outcrops were formed during Archean and Paleoproterozoic, 3600–1600 million years ago (Saadi 1995). The Espinhaço Supergroup was a sandy sedimentary basin formed during the Paleoproterozoic (Saadi 1995). The compression from east to west during the Mesoproterozoic lifted the Espinhaço Supergroup and formed the mountain range. At the end of the Mesoproterozoic (1000 million years ago), sedimentation accompanied by glaciation formed the Macaúbas Group. During the Neoproterozoic, a large distention event caused the subsidence of the São Francisco craton, and allowed the deposition of the pelitic-carbonatic sediments of the Bambuí group. At the end of the Neoproterozoic, the junction of Gondwana pushed the range further west, and imposed the superposition of the Espinhaço Supergroup over the Macaúbas and

Bambuí groups (Saadi 1995). The final tectonic events that resulted in the current relief are inferred from Neoproterozoic (~900 Ma) (Almeida-Abreu and Pflug 1994). Erosion and weathering throughout the Paleozoic, Mesozoic and Tertiary were responsible for the current landscape. The quartzitic *campos rupestres* outcrops consist mostly of the Espinhaço Supergroup rocks.

In quartzitic *campos rupestres*, the soils are rather heterogeneous, ranging from thin soil layers between rock outcrops, to deep sandy depositions in valleys and depressions, depending on the local topography (Benites et al. 2007), with high porosity and permeability (Neves et al. 2005). The soils are usually formed by a sandy A horizon over a saprolithic C horizon, or directly over the rock parent material. When these soils are shallow, they are classified as Entisols in the US Soil Taxonomy (USDA 1998) and Leptosols in the FAO soil classification system (FAO 1998). However, depending on the topography, deeper soils can be formed. When the sandy A horizon is deeper than 50 cm, the soils are classified as Quartzipsamments (USDA 1998) or Arenosols (FAO 1998). When topography is smooth, the soil can present stony lag deposits made of gravels, quartz and sometimes iron concretions (Benites et al. 2007). Incomplete burning of plant tissues causes the formation of charcoal fragments that contain stable aromatic compounds (Benites et al. 2005). Some of the soils are seasonally water saturated, which contributes to the accumulation of nondecomposed organic matter. Low temperatures, nutrient-poor soils, and aluminum (Al) saturation also contributes to the low microbial activity (Benites et al. 2007). Organic matter accumulation promotes water and nutrient retention, which improves the conditions for plant establishment and growth.

Banded Ironstone Formations (BIFs) of itabirite, hematite and limonite that predominate in the southern Espinhaço range *campos rupestres* (i.e. *canga* vegetation) were produced as a result of metamorphosis during Proterozoic, up to 2500 million years ago (Pomerene 1964; Simmons 1968). The soils are composed of iron oxides, silica and quartz, and present much lower permeability than quartzitic *campos rupestres*. In addition to the harsh conditions experienced by quartzitic *campos rupestres*, ironstone *campos rupestres* are exposed to high heavy metal concentrations, and some of the species are metal-accumulators (metallophytes), with high heavy-metal concentrations in their leaves or roots (Teixeira and Lemos-Filho 1998).

Microorganisms contribute to the N input into *campos rupestres*. Rock outcrops of the ironstone and quartzitic *campos rupestres* show colonies of epilithic and endolithic N_2-fixing cyanobacteria (Jacobi et al. 2007; Alves et al. 2014).

11.3.2 Nutrient-Acquisition Strategies

Nutrient-acquisition strategies in *campos rupestres* range from mycorrhizal associations and N_2-fixing root nodules to root specializations that vary in morphology and anatomy (Araújo 2014; Abrahão et al. 2014; Oliveira et al. 2015). *Campos rupestres* present carnivorous (Pereira et al. 2012), as well as protocarnivorous plants (Nishi et al. 2013). In addition, *campos rupestres* exhibit hemiparasites (Arruda et al. 2012) as well as holoparasites (Gomes and Fernandes 1994).

11.3.2.1 Root Specializations for Nutrient Acquisition

Root specializations for nutrient acquisition involve different root morphologies, but usually a similar function (Miller 2005). Root specializations are capable of soil "mining", and intensively exploiting sources of soil inorganic as well as organic P that are not readily available for other roots (Lambers et al. 2008b). They are highly efficient at acquiring nutrients, because of their combination of morphology and physiology (Lambers et al. 2006). Cluster roots are a well-known root specialization, first described in Proteaceae, and now recognized in several other families (Lambers et al. 2006). They are composed of a dense set of lateral rootlets with determined growth, that can be simple, as in most species, or compound, as in *Banksia* (Lambers et al. 2003). Compound cluster roots can form mats at the soil surface, locally building up high concentrations of carboxylates and enzymes and allowing the acquisition of both poorly-available organic and inorganic P (Shane and Lambers 2005). This strategy allows intense exploitation of small soil patches, and has been called a "mining" strategy (Lambers et al. 2008b).

Cluster roots have not been described for *campos rupestres*, but we can find functionally-similar root specializations such as dauciform roots in Cyperaceae (Oliveira et al. 2015). Dauciform roots are carrot-shaped sections of lateral roots with long root hairs (Fig. 11.9a) that release carboxylates in the soil, solubilizing P similarly to cluster roots (Shane et al. 2006). As Cyperaceae is an abundant and widespread family in *campos rupestres*, dauciform roots likely play an important role in P acquisition in these soils. Not all Cyperaceae produce dauciform roots (Shane et al. 2005), but quite likely those that do not produce these specialized roots do release carboxylates also (Hayes et al. 2014).

Members of the Eriocaulaceae also present interesting root specializations (Fig. 11.9b). Eriocaulaceae roots usually show a dense sand cover around the root hairs (Oliveira et al. 2015) which increases the root-soil contact, providing more intimate exchange with the rhizosphere (McCully 1999). The sand sheath is usually formed by the release of mucilage by the roots, and helps prevent desiccation during the dry season, by reducing the gaps between the soil particles (North and Nobel 1997). This intimate contact between roots and soil particles, likely in combination with the release of P-mobilizing exudates, also allows more efficient P acquisition (Smith et al. 2011; Hayes et al. 2014). Sand sheaths have been observed within Eriocaulaceae in species of *Syngonanthus* and *Actinocephalus* (Scatena et al. 2005), but also in other families such as Xyridaceae, for example in *Xyris obcordata* Kral and Wand. (Oliveira et al. 2015), *Xyris nubigena* (Fig. 11.9c), and Cactaceae, as in *Discocactus placentiformis* (Lehm.) K. Schum. (Fig. 11.9d) (Abrahão et al. 2014).

Fig. 11.9 Plant mineral nutrition strategies in *campos rupestres*. **a** Dauciform roots in *Bulbostylis paradoxa*, Cyperaceae, in Serra do Cipó, Minas Gerais, Brazil; **b** Root of Eriocaulaceae, with long root hairs; Serra do Cipó, Minas Gerais, Brazil; **c** Root of *Xyris nubigena*, Xyridaceae, with long root hairs; Serra do Cipó, Minas Gerais, Brazil; **d** Sand-binding roots of *Discocactus placentiformis*, Cactaceae, in Serra do Cabral, Minas Gerais, Brazil; **e** Roots of *Philodendron adamantinum*, Araceae, Serra do Cabral, Minas Gerais, Brazil; **f** *Aechmea phanerophlebia*, Bromeliaceae, a tank-forming species in Parque Nacional das Sempre Vivas, Minas Gerais, Brazil. Scale bars 500 μm. *Photos* A. Abrahão

Discocactus placentiformis presents very long root hairs that release carboxylates into the rhizosphere at a low P supply (Abrahão et al. 2014). These carboxylates help release sparingly-available forms of P and micronutrients that are unavailable to plants without specialized traits (Ryan et al. 2001). The exudate-release activity can also supply neighboring plants with limiting nutrients and facilitate the establishment of other species in nutrient-limited environments (Muler et al. 2014). Although further carboxylate-exudation analyses have not yet been published for *campos rupestres* species, we have strong indications that carboxylate release is an important strategy in this habitat, where root specializations for nutrient acquisition are still largely unknown (Lambers et al. 2015).

Important evidence for the ability to acquire strongly bound forms of P is leaf manganese (Mn) concentration in nonmycorrhizal species. In Mn-poor acidic soils, the release of carboxylates into the soil solution mobilizes P previously bound to oxides and hydroxides of iron (Fe) or Al, but also desorbs Mn and zinc (Zn) that can be taken up by broad-specificity transporters (Shane and Lambers 2005). Therefore, under these circumstances, leaf Mn concentrations can be used as a proxy for carboxylate release (Lambers et al. 2015). In a dune chronosequence study in the *kwongkan* vegetation in south-western Australia, nonmycorrhizal species presented higher leaf Mn concentrations than mycorrhizal species (Hayes et al. 2014). These nonmycorrhizal species were more abundant on the oldest low-P soils of the chronosequence, and presented several carboxylate-releasing root specializations (Hayes et al. 2014; Zemunik et al. 2015). In several species from *campos rupestres*, leaf Mn concentrations are very high (Lambers et al. 2015; Oliveira et al. 2015). However, we still do not have studies on leaf Mn concentrations, root specializations and carboxylate release in the same species to allow a detailed assessment of the associations that have been shown in the *kwongkan*.

Plants in *campos rupestres* show several root morphologies, and many of them exhibit underground storage organs (Mamede 1993; Alves et al. 2014; Veldman et al. 2015). As *campos rupestres* are fire-prone environments within the *cerrado* (Simon et al. 2009), underground storage tissues allow resprouting after fire (Joaquim et al. 2014) as well as after prolonged drought (Garcia et al. 2011), although the development of these structures is usually not so marked as in *cerrado* soils (see Sect. 11.2.1). Despite the body of literature on the underground anatomy of *campos rupestres* families like Asteraceae, Xyridaceae and Eriocaulaceae (Pillai and Pillai 1961; Menezes et al. 1979; Scatena et al. 2005), little is known about the nutrient-acquisition strategies of these root structures and what exactly is stored in these structures (Oliveira et al. 2015).

11.3.2.2 Mycorrhizas

Mycorrhizas are symbiotic associations between roots and fungi that belong to a range of functional types (Smith and Read 2008). The most common type is arbuscular mycorrhiza (AM), estimated to colonize 74 % of Angiosperm species globally (Brundrett 2009). Mycorrhizal associations allow exchanges between the plant and the fungus where the plant transfers carbon fixed in photosynthesis to the fungus, while the fungus transfers highly-immobile nutrients such as P and Zn (Marschner 1995). AM fungi do not have access to a chemically different source of P (Yao et al. 2001), but their hyphae can extend beyond the depletion zone reached by the roots and root hairs. This foraging strategy is referred to as "scavenging" (Lambers et al. 2008b) as opposed to the 'mining' strategy of the specialized carboxylate-releasing roots. Mycorrhizal 'scavengers' exploit much larger areas than nonmycorrhizal roots (Lambers et al. 2008a). Ectomycorrhiza, orchid mycorrhiza and ericoid mycorrhiza release phosphatases and organic acids in the soil, therefore have access to additional pools of P (Williamson 1973; Landeweert et al. 2001; Buée et al. 2005).

Campos rupestres comprise a high diversity of arbuscular mycorrhizal fungi (de Carvalho et al. 2012; Coutinho et al. 2015, see Chap. 8). de Carvalho et al. (2012) studied a wide range of soil types in *campos rupestres* of Serra do Cipó, in the Espinhaço range. The soils investigated include sandy bogs, peat bogs, rocky outcrops, quartz gravel field, and surrounding areas of *cerrado*. They found 49 species of AM fungi in the five different soil types, including four new species. The most species-rich genera in *campos rupestres* are *Glomus*, *Acaulospora* and *Scutellospora*. Fungus identification was based on spores, and soil collection was carried out in the dry season. Although the dry season is expected to show most of the fungi sporulated, AM species richness could increase if the sampling were conducted in the wet season. Despite the sampling limitations, the authors noted that the species richness of a single collection in *campos rupestres* was higher than that in several tropical forests. Coutinho et al. (2015) collected soil samples for mycorrhizal spore analysis during the wet season along an altitudinal gradient of Serra do Cipó (800–1400 m). They found 54 species of fungi that differed in composition depending on soil conditions at each altitude.

Root colonization by AM fungi is known for the most common *campos rupestres* families (Oliveira et al. 2015). For example, Scatena et al. (2005) studied anatomical sections of 14 species of *Actinocephalus* (Eriocaulaceae) and reported root colonization by fungi in all of them. They reported the formation of intracellular papillae that were further investigated by Oriani and Scatena (2007). The authors interpret these papillae as structures to prevent further mycorrhizal colonization. Ultrastructural and histochemical analyses showed cellulose, pectine and callose deposition between the cell wall and the plasma membrane in response to fungal colonization. Most of the roots of *Actinocephalus* species studied by Oriani and Scatena showed fungal colonization, but none of the colonized cells developed the arbuscules, which allow exchange between the root and the fungus, typical in AM. In seven species of Microliceae (Melastomataceae) from Chapada dos Veadeiros, unpublished material by Rodrigues et al. (2013) shows high rates of colonization (80 %) and the presence of many arbuscules (30 % of the roots analyzed). Further evidence is needed to understand the participation of these AM in mineral nutrition of these species.

Mycorrhizal colonization was also recorded on eight species of orchids collected both in quartzitic and ironstone *campos rupestres* (Nogueira et al. 2005). The fungi were isolated and grown to allow identification. The authors identified *Mycelium radicis atrovirens* Melin, classified as dark septate fungi in field-collected roots. They also obtained isolates of *Rhizoctonia*, *Epulorhiza* and *Ceratorhiza* fungi. Approximately 70 or 80 % of all angiosperm species are mycorrhizal (Smith and Read 2008; Bundrett 2009). A model proposed by Lambers et al. (2008b) suggests that the proportion of scavenging mycorrhizal species decreases with severe P impoverishment, giving place to mining strategies. With increasing soil development, it is therefore expected that the scavenging strategy is partly replaced by the mining strategy. The carbon cost of maintaining a mycorrhizal symbiosis can be too high in extremely nutrient-impoverished soils (Lynch and Ho 2005); therefore, the carbon is expected to be invested in mining root specializations, towards their construction and functioning. This prediction was confirmed for the *kwongkan* vegetation (Hayes et al. 2014; Zemunik et al. 2015). They studied mineral nutrition along a chronosequence and found that the oldest, most severely P-impoverished soils bear the lowest proportion of mycorrhizal species, and the highest with mining root specializations. In *campos rupestres*, the proportion of known species bearing AM is relatively low (Oliveira et al. 2015). In the white sands in Serra do Cabral and Parque Estadual do Rio Preto, for example, ca. 18 and 25 % of the species show AM colonization, respectively. In more fertile peat bogs, however, only around 10 % of the species were not colonized, but almost 60 % of the mycorrhizal species presented less than 5 % root colonization (Oliveira et al. 2015). Cyperaceae, Cactaceae and Velloziaceae are predominantly nonmycorrhizal. In Eriocaulaceae, both mycorrhizal and nonmycorrhizal species occur. In Poaceae, the species sampled showed a very low percentage of root colonization, or they were nonmycorrhizal.

Colonization by mycorrhizal fungi may be important for plant establishment on bare rock outcrops (Nogueira et al. 2005), but in general, the percentage of root colonization of *campos rupestres* species is low (Oliveira et al. 2015). As discussed above, there appear to be mechanisms to resist colonization in Eriocaulaceae (Oriani and Scatena 2007), reinforcing the idea that mycorrhiza may not act as beneficial symbionts on the oldest most severely nutrient-impoverished soils.

11.3.2.3 Symbiotic Nitrogen Fixation

In *campos rupestres*, there are continuous losses of N from the system, predominantly due to fire. These losses need to be balanced, and symbiotic N_2 fixation is expected to play a major role in this habitat. Nitrogen is abundant in the Earth's atmosphere, but cannot be used by plants in the form of N_2. Gaseous nitrogen needs to be converted into plant-available N species, for example in microbial N_2 fixation. This process comprises the conversion of atmospheric N_2 into ammonia, catalyzed by nitrogenase. Nitrogen fixation can be carried out by some free-living microorganisms or in symbiotic associations, for example by rhizobia in legume root nodules. The ammonia produced by nitrogenase can be taken up and assimilated by plants or converted into other readily available compounds like nitrate or amino acids by soil microorganisms (Lambers et al. 2008a). Nitrogenase is extremely sensitive to oxygen, and, thus, an anaerobic environment must be maintained near the enzyme. In symbiotic associations with vascular plants, this anaerobic environment is the result of an oxygen barrier surrounding the symbiotic structure. Leghemoglobin ensures that sufficient oxygen arrives near the microorganisms to generate sufficient ATP in respiration to drive the energetically-expensive process catalyzed by nitrogenase (Lambers et al. 2008a).

Symbiotic N_2 fixation is not only costly in terms of carbon, but also with respect to P (Raven 2012). The very low soil P availability in *campos rupestres* might therefore restrict the N_2-fixing process. Some species are incapable of nodulating in P-impoverished soils, unless they are colonized by AM fungi, which increase the plant's P uptake (Sprent 1979; Vitousek et al. 2002). However, this has never been studied in *campos rupestres*. In fact, little is known about the quantitative significance of N_2 fixation in the *campos rupestres*, although Fabaceae species can be common in some areas (Oliveira and Godoy 2007; Ferreira and Forzza 2009; Silva and Martins 2013; Dutra and Garcia 2014). Fabaceae is an important family of angiosperms in *campos rupestres* in terms of species numbers (Table 11.2), and species in this family are known for their association with N_2-fixing rhizobia. However, the relative cover of these species is generally quite low (less than 3 % in areas at Serra do Cipó; Le Stradic 2012). Moreover, the expectation that N_2 fixation

Table 11.2 Abundance of Fabaceae within *campos rupestres* (habitat type, species abundance in percentage of total and family rank regarding the abundance of species; data on cover not available)

Locality	Habitat type	Species (%)	Family rank	References
Morro do Forno, SP	Rocky outcrops	8.9	1st	Oliveira and Godoy (2007)
Chapada Diamantina, BA	*Campos rupestres*	7.7	3rd–5th	Conceição and Pirani (2005)
Chapada Diamantina, BA	*Campos rupestres*	3.7	6th–7th	Conceição et al. (2007)
Chapada dos Veadeiros, GO	Rocky outcrops	16.0	1st	Santos et al. (2012)
Serra de Ouro Preto, MG	*Campos rupestres*	6.1	4th	Messias et al. (2012)
Serra do Itacolomi e Serra do Ouro Branco, MG	Rocky outcrops	2.7	8th	Lemes (2009)
Serra do Rola-Moça, MG	*Campos rupestres*	6.9	4th	Vincent and Meguro (2008)
Serra do Rola-Moça, MG	*Campos rupestres*	6.4	4th–9th	Jacobi et al. (2008)
Toca dos Urubus, MG	*Campos rupestres*	7.9	3rd	Ferreira and Forzza (2009)
Serra de Antônio Pereira, MG	*Campos rupestres*	6.7	4th–5th	Ataíde et al. (2011)
Serra do Cipó, MG	*Campos rupestres*	6.3	11th	Le Stradic (2012)

is widespread in areas of *campos rupestres* that show a relative abundance of legumes might be incorrect, considering the P-limited nature of the environment. For example, Pate et al. (1998) showed that the contribution of *Acacia* species (Fabaceae) to N_2 fixation in a P-limited arid environment (mulga) in Western Australia is rather low, in spite of the great abundance of acacias in such areas. Instead, in mulga, termites were shown to depend on N_2 fixation. Many termites have a symbiotic association with N_2 bacteria (French et al. 1976; Tayasu et al. 1998). Ecophysiological data on N_2 fixation in *campos rupestres* are limited to 38 species of the *Mimosa* genus (sub-family Mimosoideae) (Reis et al. 2010). The authors surveyed the frequency of nodulation within the *cerrado* and *caatinga* Biomes and tested the nodules for the presence of nitrogenase by immunolabeling with an antibody against a nitrogenase subunit and using the $\delta^{15}N$ technique to assess their N_2-fixation activity (Reis et al. 2010). The majority of the species in their survey were found above 1000 m, mainly in *campos rupestres* areas, and nodulation was observed in 67 out of the 70 species collected. The activity of nitrogenase was shown for 20 out of the 23 species tested. More quantitative data are required to understand the significance of symbiotic N_2 fixation by legumes in *campos rupestres* (see Chap. 8). The presence of N_2-fixing symbionts can have a positive effect on plant growth and soil N concentrations in *campos rupestres*. Matias et al. (2009), in an effort to ascertain suitable species to restore degraded areas of the *campos rupestres*, carried out an experiment with two plant species that are widespread in the *cerrado* domain. One of these species was *Centrosema coriaceum* Benth. (Leguminosae), which was grown with different levels of AM fungi and rhizobium inoculation. The inoculated individuals grew better, and showed greater survival, and higher leaf P and N concentrations.

The role of termites in other P-impoverished systems is well established (Pate et al. 1998; Tayasu et al. 1998). Symbiotic N_2 fixation is widespread amongst termites with a N-depleted diet (French et al. 1976; Bourguignon et al. 2011). Given the abundance of termite mounds in *campos rupestres*, the significance of termite activity to the N input into the system is worth further exploring.

11.3.2.4 Nutrition from Tanks (Phytotelmata)

Some species have leaves arranged in compact rosettes, which forms a tank that can retain water, known as phytotelmata, for example, in *Aechmea phanerophlebia* Baker (Fig. 11.9f). This tank can harbor a wide range of microorganisms, for example yeasts, which release extracellular enzymes. *Vriesea minarum* L.B.Sm., a bromeliad species from ironstone outcrops, bears 36 species of yeast, where 22 are basidiomycetous and 14 are ascomycetous species (Gomes et al. 2014). These yeast species release a wide range of extracellular enzymes, including proteases, xylanases, amylases, pectinases and cellulase (Gomes et al. 2014). This active community of yeasts is capable of rapid decomposition of plant litter inside the tank which can benefit the plant.

The formation of tanks also occurs in Eriocaulaceae, as in *Paepalanthus bromelioides* Silveira. In addition to plant detritus, the tanks receive animal faeces. Some spider species, for example *Alpaida quadrilorata* (Simon) 1897 (Araneidae) and *Latrodectus geometricus* CL Koch, 1841, (Theridiidae), are associated with *P. bromelioides*. They build their webs on the leaves, and as a result of predation, their faeces can fall in the tanks (Nishi et al. 2013). Leaf $\delta^{15}N$ can be used as an integrator of N sources for the plant. Experimental addition of ^{15}N-labeled spider faeces demonstrated that the contribution of the faeces to the N nutrition was almost 19 % of the total N in the leaves. The tanks also contain bacteria with proteolytic capacity, but the authors found no effect of the bacteria in N cycling and acquisition by the plant, since the addition of antibiotics to the phytotelmata did not affect leaf $\delta^{15}N$. When compared with other species from the same environment, *P. bromelioides* presented a leaf $\delta^{15}N$ closer to that of the carnivorous plant *Drosera hirtella* A.St.-Hil., than to noncarnivorous plants. It suggests that animal debris are absorbed by special trichomes on the leaves and contribute to the plant's N nutrition; hence, *P. bromelioides* can be considered a protocarnivorous plant (Nishi et al. 2013).

11.3.2.5 Carnivory

Carnivorous plants occupy a diverse array of nutrient-impoverished environments (Ellison 2006). Their habitats are usually characterized by being well lit, temporarily waterlogged and nutrient-poor, especially in terms of available N, P and potassium (Givnish et al. 1984). Therefore, evolutionary pressures towards carnivory are similar across the world, and, consequently evolutionary convergence has occurred (Benzing 1987). The carnivorous syndrome is relatively rare among flowering plants, with only approximately 0.2 % of all angiosperms presenting such habit. The use of prey by plants may not be a very economical way to secure nutrients, but when these are scarce, the strategy increases in importance (Benzing 1987; Givnish 1989; Lambers et al. 2014).

Carnivory is a common strategy in the *campos rupestres*, as is the case in *kwongkan* (Lambers et al. 2014). Nutrient-poor and wet patches are usually occupied by populations of *Drosera* (Fig. 11.10a), a genus comprising almost exclusively perennial carnivorous species (McPherson 2010; Silva et al. 2011). Differently, sites subjected to seasonal oscillations of water availability, especially areas where the soils are waterlogged during the summer months, are usually inhabited by annual species of *Lentibulariaceae* (Silva et al. 2011). Each year, with the beginning of the rainy season, the outcrops' surfaces and moss layers become moist and, where there used to be colonies of *Sphagnum* and *Campylopus*, seeds of Lentibulariaceae such as *Genlisea violacea* A.St.-Hil. and *Utricularia amethystina* Salzm. ex A.St.-Hilaire and F. Girard and *U. neottioides* A. St.-Hil. (Fig. 11.10b) germinate (Silva et al. 2011).

Fig. 11.10 Carnivorous plant strategies in *campos rupestres*. **a** Flypaper traps of *Drosera hirtella* (Droseraceae), with adhesive tentacles that trap and digest the preys, in Serra do Cipó, Minas Gerais, Brazil; **b** *Utricularia neottioides* (Lentiburaceae), Diamantina Plateau, Minas gerais, Brazil; *Utricularia* species produce bladders that suck in their prey; **c**, **d** *Philcoxia minensis* (Plantaginaceae), which produces underground leaves with adhesive glands that trap and digest nematodes; in Serra do Cabral, Minas Gerais, Brazil. Scale bar 1 cm. *Photos* **a** R.S. Oliveira, **b**. H. Lambers, **c** and **d**. A. Abrahão

Most families of carnivorous plants exhibit low species diversity, with the notable exception of two families: Droseraceae, a family with close to 200 species, most of them in the genus *Drosera* (Rivadavia et al. 2009; McPherson 2010); and Lentibulariaceae, a family with more than 330 species, approximately 220 in the genus *Utricularia* and 30 in the genus *Genlisea* (Muller et al. 2006). Plants from these three genera can be found throughout the Espinhaço Range and in other disjunct areas of *campos rupestres*. The species diversity of these families within *campos rupestres* is unknown, mainly due to the lack of studies in these environments. *Philcoxia* P. Taylor and V.C. Souza (Plantaginaceae) comprises five species, all rare endemic to seasonally-dry areas of deep white sand among *campos rupestres* in Minas Gerais, Bahia and Goiás (Fritsch et al. 2007; Taylor et al. 2000). The genus exhibits several characteristics, in common with other carnivorous species, especially a nonmycorrhizal and poorly-developed root system. A striking feature of *Philcoxia* is the placement of its minute leaves (0.5–1.5 mm blade diameter) under the surface of the white sand in which it grows. Numerous glands on the surface of its leaves produce mucilage that trap free-living nematodes that are the only known prey of these plants. Taylor et al. (2000) described the genus for the first time as well as three species (*P. minensis* Fig. 11.10c, d, *P. bahiensis* and *P. goiasensis*), each named after the State in which they were collected. Since then, three more species were discovered in Minas Gerais and Bahia, which have yet to be described (A. Scatigna, pers. comm.).

11.3.2.6 Parasitic Plants

Parasitic plants make a connection with the host through a haustorium that connects either with the host xylem (hemiparasites) or phloem (holoparasites) (Kuijt 1969; Press and Phoenix 2005; Press et al. 1999). These species can be either root or shoot parasites and can be facultative (not dependent on a host to complete their life cycle) or obligate (a host is essential to complete their life cycle) parasites. Hemiparasites in *campos rupestres* are represented mainly by mistletoes, which have leaves with functional chloroplasts and produce photoassimilates, although they may also import a significant fraction of their carbon from their host via the xylem (Lambers et al. 2008a).

To obtain water and nutrients from their hosts, hemiparasitic plants require a more negative water potential in their xylem that in the host xylem, so as to maintain a gradient for water to move. Generally, hemiparasites show faster rates of transpiration and lower water-use efficiency compared with their host. The nutrition of host-parasite systems is closely linked to their water relations. With the rapid transpiration rates, the parasites acquire the nutrients transported in the xylem sap (Glatzel and Geils 2009).

Parasitic plants account for 1 % of all angiosperm species ($\sim$4100) (Nickrent and Musselman 2004) and they occur in many ecosystems. In *cerrado*, there are many parasitic plants ($\sim$ 105 spp/source: http://floradobrasil.jbrj.gov.br). However, the number of species in *campos rupestres* is only about 15 (0.03 % of the flora);

that is, much less than the global percentage of 1 % (http://floradobrasil.jbrj.gov.br). A similar under-representation has been recorded for another nutrient-impoverished and seasonally-dry system, *kwongkan* in south-western Australia (Lambers et al. 2014).

In *campos rupestres*, there are stem holoparasites (*Cuscuta* - Convolvulaceae and *Pilostyles*—Acanthaceae), root holoparasites (*Langsdorffia*—Balanophoraceae), stem hemiparasites (*Psittacanthus* and *Struthanthus* (Fig. 11.11a–c)—Loranthaceae and *Phoradendron*—Santalaceae) and root hemiparasites (*Physocalyx*—Orobanchaceae, Fig. 11.11d) (http://floradobrasil.jbrj.gov.br).

Parasitic plants may exert significant negative effects on host growth, reproduction and physiology (Press and Phoenix 2005). *Struthanthus flexicaulis* (Mart.) Mart. (Loranthaceae, Fig. 11.11a–c) is an abundant mistletoe in *campos rupestres*. This is a

Fig. 11.11 Holoparasitic and hemiparasitic species in the campos rupestres. **a** Fruits of the mistletoe *Struthanthus flexicaulis* Mart. (Loranthaceae); **b** Seedling of *S. flexicaulis*; **c** Secondary haustorium of *S. flexicaulis*; **d** The root hemiparasite *Physocalyx* sp.(Orobanchaceae); **e** The holoparasite *Pilostyles ingae* (Acanthaceae); **f** Flowers of *Pilostyles ingae* and a recently planted seed from a mistletoe species, *Psittacanthus robustus* Mart. (Loranthaceae). *Photos* **a–c** H. Lambers; **d–f** R.S. Oliveira

generalist species, parasitizing at least 44 species (from an area with 85 vascular plant species), often multiple species at the same time, in an ironstone outcrop. It is most commonly found parasitizing on *Mimosa calodendron* Mart. (Fabaceae), an endemic species on ironstone outcrops (Mourão et al. 2006). The parasitism negatively impacts the reproductive success of *M. calodendron.* In the parasitized individuals, the seeds are smaller. The parasites can alter the population structure and dynamics of *M. calodendron*, due to decreased survival and fecundity (Mourão et al. 2009).

Another abundant species in *campos rupestres* is the holoparasite *Pilostyles ingae* Karst (Fig. 11.11e, f). This is a specialist species, parasitizing mainly on *Mimosa* species in *campos rupestres* (*M. maguirei* Barneby, *M. setosa* Benth, and *M. foliolosa* Benth. (Fabaceae)). In *Pilostyles*, the endophytic system is completely mixed within the host tissues (Gomes and Fernandes 1994). As a consequence of the parasitism, infested individuals of *Mimosa* change their wood anatomy, showing an increase in vessel frequency and with vessels that are narrower, thus increasing the hydraulic safety (Amaral and Ceccantini 2011).

Parasites appear to be abundant in *campos rupestres*; however, taking into account the total number of species, they are actually underrepresented. As suggested by Lambers et al. (2014), this nutrient-acquisition strategy may incur risks in a seasonally-dry environment. The high demand for water of hemiparasites, required to allow import of nutrients in the xylem stream, may lead to severe water stress in the host in this seasonally-dry environment. This may cause severe dehydration, xylem cavitation, and even death of the host, with fatal consequences for the hemiparasite, especially if that is a mistletoe. Another risk in these areas is the fire occurrence. Fire is considered the main natural suppressing agent of mistletoe species in many ecosystems (Shaw et al. 2004).

11.3.2.7 Nutrient Use and Economy

In order to assess mineral nutrition in *campos rupestres*, Oliveira et al. (2015) analyzed N and P concentrations of more than 50 species as dependent on soil P availability. Leaf nutrient concentrations in *campos rupestres* are much lower than the global average values. Nitrogen concentrations in *campos rupestres* mature leaves range from 5.6 to 27.9 mg N g^{-1} leaf dry weight (DW), and the average was 12 mg N g^{-1} leaf DW, while the global average is 17 mg N g^{-1} leaf DW (Vergutz et al. 2012). Oliveira et al. (2015) found a weak increase in leaf N concentrations with increasing soil P availability. The wide range of root specializations found in the most P-impoverished soils might contribute to the additional acquisition of N. Leaf P concentrations in mature leaves ranged from 0.07 to 1.2 mg P g^{-1} leaf DW, with an average of 0.3 mg P g^{-1} leaf DW, which is also lower than the global average of 1.0 mg P g^{-1} leaf DW (Vergutz et al. 2012). Leaf N to P ratios (N:P) are high, ranging from 16 to 116, with an average of 43, and indicate P limitation of plant productivity rather than N limitation (Güsewell 2004). The N:P ratios of these plants are among the highest reported, even higher than for well-studied P-limited environments as the *kwongkan* in south-western Australia, and the Cape floristic region in South Africa (Lambers et al. 2010; Stock and Verboom 2012).

The leaf nutrient concentrations, also in combination with the high N:P ratios reinforce the concept of plant productivity being limited by P, rather than N. The concept of nutrient limitation, usually applied in cultivated plants, must be used carefully in an old-growth vegetation (Veldman et al. 2015). Nutrient limitation implies biomass increase in response to nutrient addition (Haridasan 2008). However, nutrient enrichment in *campos rupestres* can allow the establishment of faster-growing species, frequently exotic invasive species. Native species can cope with low nutrient availability in a wide range of strategies (Gomes and Fernandes 1994; Pereira et al. 2012; Nishi et al. 2013; Abrahão et al. 2014; Veldman et al. 2015) and nutrient addition can be harmful for this vegetation type, either because the native species get outcompeted by faster-growing species, or even because of nutrient toxicity (Aerts and Chapin 2000; Lambers et al. 2013a).

In addition to low nutrient availability, some species in *campos rupestres* are exposed to high heavy-metal concentrations in the soil. Teixeira and Lemos-Filho (1998) analyzed leaf and soil concentration of Zn, Fe, Mn, copper (Cu), cadmium (Cd), chromium (Cr), lead (Pb), and nickel (Ni) in five species (three species of Melastomataceae and two species of Asteraceae) in both ironstone and quartzitic *campos rupestres*. Quartzitic *campos rupestres* presented lower soil Mn, Zn and Cu concentrations. In both quartzitic and ironstone *campos rupestres*, Cr and Ni were not detected in the soil, but soil Fe, Zn and Cu concentrations were especially high. Despite the low Ni concentration in the soils, some species presented relatively high concentrations in the leaves. The leaves of all species accumulated Zn, Ni, Cr and Zn. However, only Asteraceae accumulated Ni and Cu. This tolerance to metal, or even metal accumulation, is an important trait to survive in these soils, and can be important for restoration of mined areas.

11.3.3 Fluoroacetate-Bearing Plant Species

Globaly, very few plant species accumulate fluoroacetate (CH_2FCOO^-), a highly-toxic, acute-acting plant secondary compound as a chemically-mediated defense against herbivory (Twigg 2014). In *campos rupestres*, *Palicourea* (Fig. 11.12a, b) (Rubiaceae) species are known to produce fluoroacetate (de Oliveira 1963). In addition to the genus *Palicourea*, which also occurs outside *campos rupestres*, this compound is only known to accumulate in five other plant genera, in Brazil (*Amorimia*, Lee et al. 2012; and *Arrabidea*, Krebs et al. 1994), Australia (*Gastrolobium* and *Acacia*, Twigg 2014) and southern Africa (*Dichapetalum*, Vickery and Vickery 1972). *Palicourea marcgravii* A.St.-Hil., whose natural habitat is moist woodland, is a shrub that has long been known as one of the most toxic plants of local pastures (de Oliveira 1963). In south-western Australia, native animals are considerably less sensitive to fluoroacetate than introduced species (Twigg 2014). There is no evidence in the literature that native

Fig. 11.12 Fluoracetate-bearing *Palicourea rigida* Kunth (Rubiaceae) from Diamantina Plateau, Minas Gerais, Brazil. **a** *P. rigida* growing on quartzitic rock outcrops; **b** Detail of the flowers of *P. rigida*. *Photos* H. Lambers

animals in Brazil have similarly evolved mechanisms to cope with high fluoroacetate concentrations in *Palicourea* species, but this would appear highly likely and worthy of further investigation (Twigg 2014).

11.4 Fire

In addition to the limitations of water and carbon acquisition caused by the seasonally-dry climate in *campos rupestres*, the occurrence of fire is a strong factor affecting species distribution in this habitat (see Chap. 17). Soil analyses from several areas in *campos rupestres* indicate high levels of dark matter in the organic portion of the soil, as a consequence of long-term occurrence of natural fires (Benites et al. 2003). Charcoal and ash deposition is associated with an increase in soil water retention capacity, also affecting soil pH and consequently plant nutrient availability (Coutinho et al. 2002; Glaser 2007). The dynamics of natural fires in *campos rupestres* is closely related with the water and carbon balance of the system, since natural fires are mainly caused by lightning strikes that characterize the starting of rainy season. Moreover, the occurrence of rainfall episodes soon after these natural fires, the large areas of bare rock and the island vegetation, prevent large-scale effects by allowing fast burning and decreasing the severity of these natural fire episodes (Alves and Silva 2011; Conceição and Orr 2012). To the best of our knowledge, there are no studies assessing specifically the ecological relationships between water/carbon balance and fire in *campos rupestres*. However, the incidence of natural fires might have played a role in the ecological and evolutionary processes that have molded the current flora of *campos rupestres*, as well as the distribution of specific functional traits in these habitats.

Several morphological traits associated with resprouting in *campos rupestres* (like lignotubers and xylopodial roots, underground stems and terminal buds protected in the center of tussock grasses, see Fig. 11.2) might be associated with fire survival (Lüttge 2008; Simon et al. 2009). Most of our knowledge about strategies of post-fire regeneration has been established in areas of *cerrado*, the vegetation domain that borders most of the *campos rupestres* (Eiten 1990; Giulietti and Pirani 1988; Giulietti et al. 2005). Such near- and belowground protected structures characterize a re-sprouting strategy, allowing the species to re-grow from underground buds and organs that remain viable after a fire (Coutinho 1990; Fidelis et al. 2010).

In addition to the re-sprouting strategy conferred by underground organs, many *campos rupestres* species also show aboveground characteristics that prevent/reduce fire damage (Carmo et al. 2007; Alves and Silva 2011; Lusa et al. 2014a). Such traits are especially relevant for plants on shallow soils over bare rocks, where underground structures are limited. In this sense, the presence of coating/corky stems, persistent leaf sheaths, rosette forms of life and overlapping leaves (Fig. 11.13), have frequently been reported as "fire-protection adaptations" (Givnish et al. 1986; Alves and Silva 2011; Lusa et al. 2014a). However, such identification a priori of fire-related adaptive traits still remains to be carefully evaluated (Bradshaw et al. 2011). For instance, the stem rosette life-form is common in monocots occurring in fire-prone mountains across the world (Safford 2001; Givnish et al. 2006; Conceição and Orr 2012). However, congeneric fire-sensitive species occurring sympatrically with fire-resistant species, where both have a stem rosette life-form, has been reported for *campos rupestres* (Conceição and Orr 2012). This raises questions on the adaptive value of this habit and other hypothesized fire-adapted traits that have not been properly evaluated under complementary ecological and evolutionary approaches. This is an area for further study in *campos rupestres*.

Comparisons between *campos rupestres* and others fire-prone environments across the world are rare, and the few studies carried out have revealed interesting aspects. Most floristic studies report a smaller proportion of therophytes on rocky outcrops (Meirelles et al. 1999; Ribeiro and Medina 2002; Conceição and Pirani 2005; Jacobi et al. 2007; Ribeiro et al. 2007; Messias et al. 2011) than that found in other fire-prone habitats, like Mediterranean areas (Specht 1969; Litsios et al. 2013). Le Stradic et al. (2015) also demonstrated that fire-related cues do not significantly affect seed germination in herbs from *campos rupestres*, in spite of the variability in fire tolerance and seed germination found there. Thus, so far the available evidence indicates that resprouting is the dominant mechanism for fire regeneration in *campos rupestres*, instead of a reseeder strategy, similarly to the

Fig. 11.13 Aboveground traits associated with fire resistance/tolerance in *campos rupestres*. **a** General life-forms observed on top of the outcrops—note the abundance of rosette and stem-rosette life-forms. **b**, **c** Asteraceae: **b** Coating stems and reduced, scleromorphic leaves in *Lychnophora*; **c** Rosette life-form in *Minasia*. **d** Overlapping reduced leaves in *Lavoisiera* (Melastomataceae). **e–g** Velloziaceae: **e.** persistent leaves and stem rosettes in *Barbacenia markgrafii* Schulze-Menz; **f**, **g** Corky stems formed by persistent leaf sheaths in **f** *Vellozia declinans* Goethart and Henrard and **g** *Barbacenia macrantha* Lem. *Photos* **a**, **c** and **e–g**: S. Alcantara; **b**. R.S. Oliveira; **d** J. Lovo

situation in other Neotropical fire-prone communities (Carmo et al. 2007; Overbeck and Pfadenhauer 2007; Simon et al. 2009; Le Stradic et al. 2015).

Many re-sprouters and fire-tolerant species also show massive flowering after fire, and some of these seem to be dependent on fire for flowering. That is the case for *Bulbostylis paradoxa* (Spreng.) Lindm. (Cyperaceae), which can spend several years without flowering in the absence of a fire stimulus (Gottsberger and Silberbauer-Gottsberger 2006). Other examples of fire-dependent or increased flowering include the rosette or stem rosette species: *Actinocephalus polyanthus* (Bong.) Sano (Figueira 1998) and *Actinocephalus ramosus* (Wikstr.) Sano (Souza 2011) (Eriocaulaceae), *Dyckia dissitiflora* Schult. & Schult. f., *Cottendorfia florida* Schult. & Schult. f. (Souza 2011) (Bromeliaceae), *Vellozia alata* L.B. Sm. *V. peripherica* Mello-Silva (2004), and *V. sincorana* L.B. Sm. and Ayensu (Conceição and Orr 2012) (Velloziaceae). These species show long life cycles and massive flowering after fire which are traits that could mitigate the reproductive effects of putative long periods (=several years, see Conceição and Orr 2012) without flowering.

The occurrence of a fire-induced flowering in *campos rupestres* indicates that some species might benefit after fire (Lüttge 2008; Simon et al. 2009). However, fire effects depend on the timing (i.e. which season) and duration of the burning episodes, which is directly related to the dry matter accumulated between the fire occurrences (Neves and Conceição 2010; Alves and Silva 2011). Natural fires in *campos rupestres* are usually episodic, associated with the end of the dry season, and occur fast and at low intensity, since the bare rock and the islands of vegetation slow down spreading of fire (Neves and Conceição 2010; Conceição and Orr 2012). On the other hand, anthropogenic fires in these areas are usually started at the beginning of the dry season, especially in the grasslands occurring between outcrops, for pasture purposes (Giulietti et al. 2005). Such anthropogenic fires might have catastrophic effects, since in periods without enough availability of water, efficient re-sprouting after fire is compromised (Medina 1982). Consequently, areas of *campos rupestres* that have been exposed recently to high-frequency fires already show a shift of native and introduced pyrophilous species (i.e. plants that have superior competitive ability under frequent fire regimes), like the Poaceae species *Chusquea pinifolia* (Nees) Nees and *Melinis minutiflora* P. Beauv. Alves and Silva (2011) report the intrinsic association between pastures, the increase of the outcrops basal area coverage by tussocks of *Melinis minutiflora* ("capim-gordura") and the devastating effects of anthropogenic fires for native species of *campos rupestres*, like many Velloziaceae. Velloziaceae are usually fire tolerant or fire resistant in natural conditions, when fires occur sporadically and at low intensity, and many of them show increased metabolic rates (Teodoro 2014) and fire-induced flowering (Mello-Silva 2004; Conceição and Orr 2012). Based on effects of cumulative and frequents anthropogenic fires in *campos rupestres*, we urgently need studies that aim to fill the gap of knowledge concerning functional traits associated with fire on *campos rupestres* (Alves and Silva 2011; Souza 2011). Specifically, approaches that provide more evidence in order to understand (i) how fire frequency affects water and carbon balance dynamics in species and communities of *campos rupestres*, as well as (ii) how specific traits affect species performance during fires and regeneration afterwards, are needed.

11.5 Trait Evolution and Adaptive Radiation in *Campos Rupestres*

The *campos rupestres* flora has been described since it was first recognized as "highly adapted to these very specialized conditions" (Harley 1988). The evidence for such adaptations is the high levels of morphological convergence, which prevails in several plant families/genera occurring in a particular microhabitat, among the many environments observed in the *campos rupestres* mosaic (Giulietti and Pirani 1988). Moreover, many of these phenotypic convergences are also found in similar habitats in other parts of the world, like *páramos* and other seasonally-dry and nutrient-impoverished floras such as the *kwongkan* (south-western Australia) and other Mediterranean areas around the globe. Examples of these long-term hypothesized morphological "adaptations" are scleromorphic leaves, often involved in defense against herbivores (Lambers et al. 2008a), and specialized trichomes and roots.

The large floristic diversity and levels of endemism (Giulietti and Pirani 1988; Giulietti et al. 1997; Rapini et al. 2008) challenge any effort to establish simple scenarios of trait evolution for *campos rupestres*, at least while general and specific patterns of evolution of groups representative of this habitat remain unknown (Fiaschi and Pirani 2009). Some of the fire-associated traits in *campos rupestres* might be interpreted as indications for the putative *cerrado-campos rupestres* species turnover during the recent *cerrado* diversification. However, there is no firm evidence for what proportion of these traits is derived from floristic exchange with neighboring vegetation and/or resulting from in situ adaptation in *campos rupestres* (e.g., Souza 2011).

Convergent evolution cannot be considered indicative of natural selection operating on the focal phenotype, even in similar environments (Losos 2011). The occurrence of trait convergence requires investigation of underlying mechanisms (Gould and Vrba 1982; Nielsen 2009). In this sense, recent ecophysiological studies in *campos rupestres* have allowed evaluation of the functional traits in these habitats, as highlighted in this chapter (i.e. Oliveira et al. 2005; Abrahão et al. 2014; Alcantara et al. 2015; Oliveira et al. 2015).

Phylogenetic and population genetic studies have only been carried out in parallel with ecophysiological studies in the last few years, instigated by species diversity and endemisms observed in *campos rupestres*. These studies focus on taxonomic and historical biogeographic and phytogeographic approaches, without specific tests of hypotheses on adaptations. In general, the phylogenies are poorly resolved (e.g., Fritsch et al. 2004; Andrade et al. 2010; Antonelli et al. 2010; Mello-Silva et al. 2011; Lovo et al. 2012; Trovó et al. 2013; Ribeiro et al. 2014; Louille et al. 2015) and endemic species are usually subdivided genetically by small-scale disjunctions associated with microhabitat heterogeneity (Borba et al. 2001; Jesus et al. 2001, 2009; Lambert et al. 2006; Ribeiro et al. 2008; Barbosa et al. 2012). These patterns, coupled with a high indices of alpha diversity and the occurrence of highly diverse and endemic lineages suggest fast diversification. However, as highlighted previously (Echternacht et al. 2011), to date no study has associated species richness or micro-endemism with habitat complexity in *campos rupestres*. Thus, this putative pattern resulting from adaptive divergence still remains to be evaluated.

Few studies focusing on lineage diversification indicate contrasting evidence for adaptive radiation in *campos rupestres* (Drummond et al. 2012; Ribeiro et al. 2014). Drummond et al. (2012) were the first to characterize an increased diversification rate for a lineage occurring in *campos rupestres* (*Lupinus*, Fabaceae). The authors interpret this pattern as an example of adaptive radiation conferred by the perennial life cycles in "montane" habitats. However, Givnish (2015) pointed out that explosive diversification is not an obligatory consequence of adaptive radiation, and provided different explanations for the diversification of *Lupinus* in *campos rupestres*. A selective disadvantage of annual species in tropical mountains and on nutrient-poor soils would select for perennial and N_2-fixing lineages, like in *Lupinus*. In addition, *Lupinus* species typically use carboxylate-releasing P-mobilizing strategies, which we discussed above (Lambers et al. 2013b). Coupled with limited dispersal this would favor diversification, instead of functional divergence of sympatric conspecifics which is expected as a result of adaptive radiation (Givnish 2015). Ribeiro et al. (2014) also did not find evidence for adaptive differentiation among *Minaria* species (Apocynaceae) when assessing large-scale environmental variables. Instead, these authors detected important influences of geographic isolation driving diversification in this genus. The importance of vicariance and genetic drift during cycles of climatic fluctuations has been considered fundamental to the formation of the endemic floras of *campos rupestres* (Giulietti and Pirani 1988). Phylogenetic and phylogeographic studies indicate strong barriers to the dispersion between southern and northern areas of *campos rupestres* (Barbosa et al. 2012; Bitencourt and Rapini 2013), also evidencing the importance of geographical isolation. However, the evidence concerning the time and the dynamics of climatic fluctuations affecting diversification of lineages in *campos rupestres* is scarce and conflicting (see Antonelli et al. 2010; Ribeiro et al. 2014).

11.6 Summary

In this chapter we reviewed water and carbon relations, mineral nutrition strategies and trait evolution of *campos rupestres* plants.

Shallow rooting depth, morphological plasticity of roots and water absorption by shoots are common water-acquisition strategies on rocky substrates with seasonal rains, such as *campos rupestres*. *Campos rupestres* plants exhibit a continuum of variation between isohidry and anisohidry; some species tightly regulate their water potential, maintaining it within a narrow range, while others regulate their water potential as dependent on air humidity, drying and rehydrating accordingly. Desiccation tolerance is also a common strategy found in *campos rupestres* plants, mostly in Velloziaceae.

Campos rupestres plants show several mineral nutrition strategies, including sand-binding root specializations, mycorrhizal colonization, nitrogen fixation, carnivory, parasitism, both hemi- and holoparasitism. This diversity of strategies allows effective nutrient acquisition on the different substrates (or hosts) in *campos rupestres*. However, we still need evidence on the ecophysiological significance of these strategies to further understand their functioning, as well as the relative importance of each strategy in *campos rupestres*.

We still lack empirical evidence on the importance and prevalence of in situ adaptation and adaptive divergence in the lineages that diversified in *campos rupestres*. This will likely change in the near future, based on (i) the development of comparative methods allowing the evaluation of incomplete taxon sampling, (ii) the increasing accumulation of molecular phylogenies and phylogeographic studies for endemics and conspicuous lineages of *campos rupestres*, and (iii) the functional trait information gathered for several species/areas of *campos rupestres*. Only integrative studies that test specific predictions about adaptive evolution of traits and adaptive radiation in lineages will allow a test of hypothesized adaptations to several environmental conditions in different regions of *campos rupestres*. The perspectives of incorporation of ecological and evolutionary information in the studies of *campos rupestres* are particularly exciting, given its unique flora (Joly 1970; Giulietti and Pirani 1988; Oliveira et al. 2015; Silveira et al. 2016). This perspective would allow a comparison of the diversification of *campos rupestres* floras with the patterns of surrounding areas (i.e. *cerrado*, Simon et al. 2009), and consequently the putative floristic interchange between those areas. Moreover, comparisons with habitats subject to similar selective pressures, displaying similar morphological convergences to that observed in *campos rupestres* (i.e. *fynbos*: Linder 2003; Litsios et al. 2013; *kwongkan*: Lambers et al. 2014; *páramos*: Madriñán et al. 2013) is crucial to the identification of "replicated adaptive radiations" (sensu Schluter 1990) associated with tropical montane environments, nutrient-poor soils and soil heterogeneity, and natural fire regimes around the world.

Acknowledgements We would like to acknowledge Fundo de Apoio à Pesquisa do Estado de São Paulo for funding our research (project 2010/17204-0), Conselho Nacional de Desenvolvimento Científico e Tecnológico and Conselho Aperfeiçoamento de Pessoal de Nível superior for the scholarships, scholarship 2010/50327-8 to G.S. Teodoro, and project grant number CAPES PVE 88887.108541/2015-00. We thank Instituto Chico Mendes de Conservação da Biodiversidade (Sisbio 43511-2) and Instituto Estadual de Florestas de Minas Gerais (056/2014) for the licenses to work in campos rupestres. We also acknowledge the Graduate Programs of Ecology and Plant Biology at Universidade Estadual de Campinas. Finally, we would like to thank the editor for the invitation to write the chapter.

References

Abrahão A, Lambers H, Sawaya ACHF, Mazzafera P, Oliveira RS (2014) Convergence of a specialized root trait in plants from nutrient-impoverished soils: phosphorus-acquisition strategy in a nonmycorrhizal cactus. Oecologia 176:345–355

Aerts R, Chapin FS (2000) The mineral nutrition of wild plants revisited: a re-evaluation of processes and patterns. Adv Ecol Res 30:1–67

Alcantara S, Mello-Silva R, Teodoro GS, Drequeceler K, Ackerly DD, Oliveira RS (2015) Carbon assimilation and habitat segregation in resurrection plants: a comparison between desiccation- and non-desiccation-tolerant species of Neotropical Velloziaceae (Pandanales). Funct Ecol. doi:10.1111/1365-2435.12462

Almeida-Abreu PA, Pflug R (1994) The geodynamic evolution of the southern Serra do Espinhaço, Minas Gerais, Brazil. Zentralbl Geol Paläontol Teil 2 Hist Geol Paläontol 1:21–44

Alpert P (2000) The discovery, scope, and puzzle of desiccation tolerance in plants. Plant Ecol 151:5–17

Alpert P (2006) Constraints of tolerance: why are desiccation-tolerant organisms so small or rare? J Exp Biol 209:1575–1584

Alves RJ, Silva NG (2011) O fogo é sempre um vilão nos campos rupestres? Biodivers Bras 1:120–127

Alves RJV, da Silva NG, Fernandes Júnior AJ, Guimarães AR (2013) Longevity of the Brazilian underground tree *Jacaranda decurrens* Cham. An Acad Bras Cienc 85:671–678

Alves RJV, Silva NG, Oliveira JA, Medeiros D (2014) Circumscribing campo rupestre—megadiverse Brazilian rocky montane savanas. Braz J Biol 74:355–362

Amaral MM, Ceccantini G (2011) The endoparasite *Pilostyles ulei* (Apodanthaceae—Curcubitales) influences wood structure in three host species of *Mimosa*. IAWA J 32:1–13

Andrade MJG, Giulietti AM, Rapini A, de Queiroz LP, Conceição AS, de Almeida PRM, van den Berg C (2010) A comprehensive molecular phylogenetic analysis of Eriocaulaceae: evidence from nuclear ITS and plastid *psbA-trnH* and *trnL-trnF* DNA sequences. Taxon 59:379–388

Antonelli A, Verola CF, Parisod C, Gustafsson ALS (2010) Climate cooling promoted the expansion and radiation of a threatened group of South American orchids (Epidendroideae: Laeliinae). Biol J Linn Soc 100:597–607

Appezzato-da-Glória B (2003) Morfologia de sistemas subterrâneos. Histórico e evolução do conhecimento no Brasil. Alexandre Sene Pinto, Ribeirão Preto

Appezzato-da-Glória B, Cury G, Soares MKM, Rocha R, Hayashi AH (2008) Underground systems of Asteraceae species from the Brazilian Cerrado. J Torrey Bot Soc 135:103–113

Araújo KS (2014) Eficiência simbiotica e identificação de estirpes de *Burkholderia* oriundas de campos rupestres. Dissertation, Universidade Federal de Lavras

Araya YN, Silvertown J, Gowing DJ, McConway GJ, Linder HP, Midgley G (2011) A fundamental, eco-hydrological basis for niche segregation in plant communities. New Phytol 189:253–258

Arruda R, Fadini RF, Carvalho LN, Del-Claro K, Mourão FA, Jacobi CM, Teodoro GS, van den Berg C, Caires CS, Dettke GA (2012) Ecology of neotropical mistletoes: an important canopy-dwelling component of Brazilian ecosystems. Acta Bot Bras 26:264–274

Ataíde ES, Castro PTA, Fernandes GW (2011) Florística e caracterização de uma área de campo ferruginoso no Complexo Minerário Alegria, Serra de Antônio Pereira, Ouro Preto, Minas Gerais, Brasil. Rev Arv 35:1265–1275

Baêta HE (2012) Contribuição da deposição úmida (chuva e neblina) nas relações hídricas e nutricionais de fisionomias de campos ferruginosos na Serra da Brígida, Ouro Preto, MG. Dissertation, Universidade Federal de Ouro Preto

Barbosa AR, Fiorini CF, Silva-Pereira V, Mello-Silva R, Borba EL (2012) Geographical genetic structuring and phenotypic variation in the *Vellozia hirsuta* (Velloziaceae) ochlospecies complex. Am J Bot 99:1477–1488

Bell TL, Pate JS, Dixon KW (1996) Relationships between fire response, morphology, root anatomy and starch distribution in South-West Australian Epacridaceae. Ann Bot 77:357–364
Benites VDM, Caiafa AN, de Mendonça ES, Schaefer CE, Ker JC (2003) Solos e vegetação nos complexos rupestres de altitude da Mantiqueira e do Espinhaço. Flor Ambient 10:76–85
Benites VDM, de Sá Mendonça E, Schaefer CEGR, Schaefer CEGR, Novotny EH, Reis EL, Ker JC (2005) Properties of black soil humic acids from high altitude rocky complexes in Brazil. Geoderma 127:104–113
Benites VDM, Schaefer CEGR, Simas FNB, Santos HG (2007) Soils associated with rock outcrops in the Brazilian mountain ranges Mantiqueira and Espinhaço. Rev Bras Bot 30:569–577
Benzing DH (1987) The origin and rarity of botanical carnivory. Trends Ecol Evol 2:364–369
Bitencourt C, Rapini A (2013) Centres of endemism in the Espinhaço range: identifying cradles and museums of Asclepiadoideae (Apocynaceae). Syst Biodiv 11:525–536
Borba EL, Felix JM, Solferini VN, Semir J (2001) Fly-pollinated *Pleurothallis* (Orchidaceae) species have high genetic variability: evidence from isozyme markers. Am J Bot 88:419–428
Bourguignon T, ŠObotnÍK JAN, Lepoint G, Martin J-M, Hardy OJ, Dejean A, Roisin Y (2011) Feeding ecology and phylogenetic structure of a complex neotropical termite assemblage, revealed by nitrogen stable isotope ratios. Ecol Entomol 36:261–269
Bradshaw SD, Dixon KW, Hopper SD, Lambers H, Turner SR (2011) Little evidence for fire-adapted plant traits in Mediterranean climate regions. Trends Plant Sci 16:69–76
Brum M Jr (2013) Partição de recursos hídricos em comunidades vegetais de campo rupestre e campo de altitude no Sudeste brasileiro—Brasil. Dissertation, University of Campinas
Brundrett MC (2009) Mycorrhizal associations and other means of nutrition of vascular plants: understanding the global diversity of host plants by resolving conflicting information and developing reliable means of diagnosis. Plant Soil 320:37–77
Buée M, Vairelles D, Garbaye J (2005) Year-round monitoring of diversity and potential metabolic activity of the ectomycorrhizal community in a beech (*Fagus silvatica*) forest subjected to two thinning regimes. Mycorrhiza 15:235–245
Caldwell MM, Dawson TE, Richards JH (1998) Hydraulic lift: consequences of water efflux from the roots of plants. Oecol 113:151–161
Carmo FF, Sousa E, Fonseca FC, Ribeiro LC (2007) Jacobi CM (2007) Recrutamento pós-fogo em dois habitats de um campo rupestre ferruginoso (canga) na Serra da Moeda. Anais do VIII Congresso de Ecologia do Brasil, Caxambu, MG. In
Chapin FS, Schulze ED, Mooney HA (1990) The ecology and economics of storage in plants. Annu Rev Ecol Syst 21:423–447
Conceição AA, Orr BJ (2012) Post-fire flowering and fruiting in *Vellozia sincorana*, a caulescent rosette plant endemic to Northeast Brazil. Act Bot Bras 26:94–100
Conceição AA, Pirani JR (2005) Delimitação de habitats em campos rupestres na Chapada Diamantina, Bahia: substratos, composição florística e aspectos estruturais. Bol Bot USP 23:85–111
Conceição AA, Pirani JR, Meirelles ST (2007) Floristics, structure and soil of insular vegetation in four quartzite-sandstone outcrops of "Chapada Diamantina", Northeast Brazil. Braz J Bot 30:641–656
Coutinho LM (1990) Fire in the ecology of the Brazilian Cerrado. In: Goldammer JG (ed) Fire in the tropical biota: ecosystem, processes and global challenges. Springer-Verlag, Berlin, pp 82–105
Coutinho LM, Miranda HS, Morais HC (2002) O Bioma do Cerrado e o fogo I. Caracterização. Série Cien Amb 20:1–48
Coutinho ES, Fernandes GW, Berbara RLL, Valério HM, Goto BT (2015) Variation of arbuscular mycorrhizal fungal communities along an altitudinal gradient in rupestrian grasslands in Brazil. Mycorrhiza. doi:10.1007/s00572-015-0636-5
Damour G, Simonneau T, Cochard H, Urban L (2010) An overview of models of stomatal conductance at the leaf level. Plant Cell Environ 33:1419–1438

de Carvalho F, de Souza FA, Carrenho R, Moreira FMS, Jesus EC, Fernandes GW (2012) The mosaic of habitats in the high-altitude Brazilian rupestrian fields is a hotspot for arbuscular mycorrhizal fungi. Appl Soil Ecol 52:9–19

de Oliveira M (1963) Chromatographic isolation of monofluoroacetic acid from *Palicourea marcgravii* St. Hil. Experientia 19:586–587

de Souza A, de Moraes MG, Ribeiro RDCLF (2005) Gramíneas do cerrado: carboidratos não-estruturais e aspectos ecofisiológicos. Acta Bot Bras 19:81–90

Dietrich RC, Bengough AG, Jones HG, White PJ (2012) A new physical interpretation of plant root capacitance. J Exp Bot 63:6149–6159

Drummond CS, Eastwood RJ, Miotto STS, Hughes CE (2012) Multiple continental radiations and correlates of diversification in *Lupinus* (Leguminosae): testing for key innovations with incomplete taxon sampling. Syst Biol 61:443–460

Dutra VF, Garcia FCP (2014) Mimosa L. (Leguminosae-Mimosoideae) dos campos rupestres de Minas Gerais, Brasil. Iheringia 69:49–88

Echternacht L, Trovó M, Oliveira CT, Pirani JR (2011) Areas of endemism in the Espinhaço range in Minas Gerais, Brazil. Flora 206:782–791

Eiten G (1990) Vegetação. In: Pinto MN (ed) Cerrado: Caracterização, Ocupação e Perspectivas. Editora Universidade de Brasília, Brasília, pp 9–65

Ellison A (2006) Nutrient limitation and stoichiometry of carnivorous plants. Plant Biol 8:740–747

FAO (1998) World soil resources report 84. FAO/ISRIC/ISSS, Rome

Fernandes GW, Barbosa NP, Negreiros D, Paglia AP (2014) Challenges for the conservation of vanishing megadiverse rupestrian grasslands. Nat Conserv 12:162–165

Ferreira MMAA, Conceição AA (2012) Alocação preferencial de recursos e morfologia de órgãos subterrâneos em plantas resistentes ao fogo em vegetação campestre. Sitientibus 12:143–149

Ferreira FM, Forzza RC (2009) Florística e caracterização da vegetação da Toca dos Urubus, Baependi, Minas Gerais, Brasil. Biota Neotrop 9:131–148

Fiaschi P, Pirani JR (2009) Review of plant biogeographic studies in Brazil. J Syst Evol 47:477–496

Fidelis A, Müller SC, Pillar VDP, Pfadenhauer J (2010) Population biology and regeneration of forbs and shrubs after fire in Brazilian Campos grasslands. Plant Ecol 211:107–117

Figueira JEC (1998) Dinâmica de populações de *Paepalanthus polyanthus* (Eriocaulaceae) na Serra do Cipó, MG. PhD Thesis, Universidade Estadual de Campinas

Filella I, Peñuelas J (2003) Partitioning of water and nitrogen in co-occurring species of a mediterranean shrubland. Oecologia 137:51–61

French JRJ, Turner GL, Bradbury JF (1976) Nitrogen fixation by bacteria from the hindgut of termites. J Gen Microbiol 95:202–206

Fritsch PW, Almeda F, Renner SS, Martins AB, Cruz BC (2004) Phylogeny and circumscription of the near-endemic Brazilian tribe Microlicieae (Melastomataceae). Am J Bot 91:1105–1114

Fritsch PW, Almeda F, Martins AB, Cruz BC, Estes D (2007) Rediscovery and phylogenetic placement of *Philcoxia minensis* (Plantaginaceae), with a test of carnivory. Proc Calif Acad Sci 58:447–467

Gaff DF (1987) Desiccation tolerant plants in South America. Oecol 74:133–136

Gaff DF, Oliver M (2013) The evolution of desiccation tolerance in Angiosperm plants: a rare yet common phenomenon. Funct Plant Biol 40:315–328

Garcia RJF, Longhi-Wagner HM, Pirani JR, Meirelles ST (2009) A contribution to the phytogeography of Brazilian campos: an analysis based on Poaceae. Braz J Bot 32:703–713

Garcia PM, Asega AF, Silva E, Carvalho MA (2011) Effect of drought and re-watering on fructan metabolism in *Vernonia herbacea* (Vell.) Rusby. Plant Physiol Biochem 49:664–670

Giulietti AM, Pirani JR (1988) Patterns of geographic distribution of some plant species from the Espinhaço range, Minas Gerais and Bahia, Brazil. In: Vanzolini PE, Heyer WR (eds) Proceedings of a workshop on neotropical distribution patterns. Academia Brasileira de Ciências, Rio de Janeiro, pp 39–69

Giulietti AM, Pirani JR, Harley RM (1997) Espinhaço range region, eastern Brazil. In: Davis SD et al (eds) Centres of plant diversity: a guide and strategy for their conservation, vol 3. WWF/IUCN, Cambridge, pp 397–404
Giulietti AM, Harley RM, de Queiroz LP, Wanderley MGL, van den Berg C (2005) Biodiversity and conservation of plants in Brazil. Cons Biol 19:632–639
Givnish TJ (1989) Ecology and evolution of carnivorous plants. In: Abrahamson WG (ed) Plant-animal interactions. McGraw-Hill, New York, pp 243–290
Givnish TJ (2015) Adaptive radiation versus 'radiation' and 'explosive diversification': why conceptual distinctions are fundamental to understanding evolution. New Phytol 2:297–303
Givnish TJ, Burkhardt EL, Happel RE, Weintraub JD (1984) Carnivory in the bromeliad *Brocchinia reducta*, with a cost/benefit model for the general restriction of carnivorous plants to sunny, moist, nutrient-poor habitats. Am Nat 124:479–497
Givnish TJ, McDiarmid RW, Buck WR (1986) Fire adaptation in *Neblinaria* (Theaceae), a high-elevation rosette shrub endemic to a wet equatorial tepui. Oecol 70:481–485
Givnish TJ, Pires JC, Graham SW, McPherson MC, Prince LM, Rai HS, Patterson TB, Roalson EH, Evans TM, Hahn WJ, Millam KC, Meerow AW, Molvray M, Kores PJ, O'Brien HE, Hall JC, Kress WJ, Sytsma KJ (2006) Phylogenetic relationships of monocots based on the highly informative plastid gene *ndh*F: evidence for widespread concerted convergence. Aliso 22:28–51
Glaser B (2007) Prehistorically modified soils of central Amazonia: a model for sustainable agriculture in the twenty-first century. Phil Trans Roy Soc B 362:187–196
Glatzel G, Geils BW (2009) Mistletoe ecophysiology: host-parasite interactions. Bot 87:10–15
Gomes AL, Fernandes GW (1994) Influence of parasitism by *Pilostyles ingae* (Rafflesiaceae) on its host-plant, *Mimosa naguirei* (Leguminosae). Ann Bot 74:205–208
Gomes FCO, Safar SVB, Marques AR, Medeiros AO, Santos ARO, Carvalho C, Lachance MC, Sampaio JP, Rosa CA (2014) The diversity and extracellular enzymatic activities of yeasts isolated from water tanks of *Vriesea minarum*, an endangered bromeliad species in Brazil, and the description of *Occultifur brasiliensis* f.a., sp. nov. Antonie Van Leeuwenhoek 107:597–611
Gottsberger G, Silberbauer-Gottsberger I (2006) Life in the Cerrado, a South American tropical seasonal ecosystem, vol 1: Origin, structure, dynamics and plant use. Reta Verlag, Ulm
Gould SJ, Vrba ES (1982) Exaptation—a missing term in the science of form. Paleobiol 8:4–15
Güsewell S (2004) N: P Ratios in terrestrial plants: variation and functional significance. New Phytol 164:243–266
Haridasan M (2008) Nutritional adaptations of native plants of the cerrado biome in acid soils. Braz J Plant Physiol 20:183–195
Harley RM (1988) Evolution and distribution of *Eriope* (Labiatae), and its relatives, in Brazil. In: Vanzolini PE, Heyer WR (eds) Proceedings of a workshop on neotropical distribution patterns. Academia Brasileira de Ciências, Rio de Janeiro, pp 71–121
Hayes P, Turner BL, Lambers H, Laliberté E (2014) Foliar nutrient concentrations and resorption efficiency in plants of contrasting nutrient-acquisition strategies along a 2-million-year dune chronosequence. J Ecol 102:396–410
Hopper SD (2009) OCBIL theory: towards an integrated understanding of the evolution, ecology and conservation of biodiversity on old, climatically buffered, infertile landscapes. Plant Soil 322:49–86
Jackson PC, Meinzer FC, Bustamante M, Goldstein G, Franco A, Rundel PW, Caldas L, Igler E, Causin F (1999) Partitioning of soil water among tree species in a Brazilian Cerrado ecosystem. Tree Physiol 19:717–724
Jacobi CM, Do Carmo FF, Vincent RC, Stehmann JR (2007) Plant communities on ironstone outcrops: a diverse and endangered Brazilian ecosystem. Biodiv Cons 16:2185–2200
Jacobi CM, Carmo FF, Vincent RC (2008) Estudo fitossociológico de uma comunidade vegetal sobre canga como subsídio para a reabilitação de áreas mineradas no Quadrilátero Ferrífero, MG. Rev Arv 32:345–353
Jesus FF, Solferini VN, Semir J, Prado PI (2001) Local genetic differentiation in *Proteopsis argentea* (Asteraceae), a perennial herb endemic in Brazil. Plant Syst Evol 226:59–68

Jesus FF, Abreu AG, Semir J, Prado PI (2009) Low genetic diversity but local genetic differentiation in endemic *Minasia* (Asteraceae) species from Brazil. Plant Syst Evol 277:187–196

Joaquim EDO, de Figueirado-Ribeiro RCL, Hayashi AH, de Carvalho MAM (2014) Inulin contents and tissue distribution in storage underground organs of Asteraceae from the Brazilian rocky fields. Bot 92:827–836

Joly AB (1970) Conheça a vegetação brasileira. EDUSP e Polígono, São Paulo

Jordan GJ, Weston PH, Carpenter RJ, Dillon RA, Brodribb TJ (2008) The evolutionary relations of sunken, covered, and encrypted stomata to dry habitats in Proteaceae. Am J Bot 95:521–530

Krebs H, Kemmerling W, Habermehl G (1994) Qualitative and quantitative determination of fluoroacetic acid in *Arrabidea bilabiata* and *Palicourea marcgravii* by 19F-NMR spectroscopy. Toxicon 32:909–913

Kuijt J (1969) The biology of parasitic plants. University of California Press, Berkeley

Lambers H, Cramer MD, Shane MW, Wouterlood M, Veneklaas EJ (2003) Structure and functioning of cluster roots and plant responses to phosphate deficiency. Plant Soil 248:693–713

Lambers H, Shane MW, Cramer MD, Pearse SJ, Veneklaas EJ (2006) Root structure and functioning for efficient acquisition of phosphorus: matching morphological and physiological traits. Ann Bot 98:693–713

Lambers H, Chapin FS, Pons TL (2008a) Plant physiological ecology. Springer, New York

Lambers H, Raven JA, Shaver GR, Smith SE (2008b) Plant nutrient-acquisition strategies change with soil age. Trends Ecol Evol 23:95–103

Lambers H, Brundrett MC, Raven JA, Hopper SD (2010) Plant mineral nutrition in ancient landscapes: high plant species diversity on infertile soils is linked to functional diversity for nutritional strategies. Plant Soil 334:11–31

Lambers H, Ahmedi I, Berkowitz O, Lambers H, Ahmedi I, Berkowitz O, Dunne C, Finnegan PM, Hardy GEStJ, Jost R, Laliberté E, Pearse SJ, Teste FP (2013) Phosphorus nutrition of phosphorus-sensitive Australian native plants: threats to plant communities in a global biodiversity hotspot. Cons Physiol 1:cot010

Lambers H, Clements JC, Nelson MN (2013b) How a phosphorus-acquisition strategy based on carboxylate exudation powers the success and agronomic potential of lupines (*Lupinus*, Fabaceae). Am J Bot 100:263–288

Lambers H, Colmer TD, Hassiotou F, Mitchell PM, Poot P, Shane MW, Veneklaas EJ (2014) Carbon and water relations. In: Lambers H (ed) Plant life on the sandplains in southwest Australia, a global biodiversity hotspot. UWA Publishing, Crawley

Lambers H, Hayes PE, Laliberté E, Oliveira RS, Turner BL (2015) Leaf manganese accumulation and phosphorus-acquisition efficiency. Trends Plant Sci 20:83–90

Lambert SM, Borba EL, Machado MC (2006) Allozyme diversity and morphometrics of the endangered *Melocactus glaucescens* (Cactaceae), and investigation of the putative hybrid origin of *Melocactus* × *albicephalus* (*Melocactus ernestii* × *M. glaucescens*) in north-eastern Brazil. Plant Spec 21:93–108

Landeweert R, Hoffland E, Finlay RD, Kuyper TW, Van Breemen N (2001) Linking plants to rocks: ectomycorrhizal fungi mobilize nutrients from minerals. Trends Ecol Evol 16:248–254

Le Stradic S (2012) Composition, phenology and restoration of campo rupestre mountain grasslands-Brazil. PhD Thesis, Université d'Avignon et des Pays de Vaucluse, Avignon and Universidade Federal de Minas Gerais

Le Stradic S, Silveira FAO, Buisson E, Cazelles K, Carvalho V, Fernandes GW (2015) Diversity of germination strategies and seed dormancy in herbaceous species of *campo rupestre* grasslands. Aust Ecol 40:537–546

Lee ST, Cook D, Riet-Correa F, Pfister JA, Anderson WR, Lima FG, Gardner DR (2012) Detection of monofluoroacetate in *Palicourea* and *Amorimia* species. Toxicon 60:791–796

Lemes FOA (2009) Relações florísticas, fitossociológicas e aspectos edáficos de comunidades de campos rupestres da Serra do Itacolomi e Serra de Ouro Branco, Minas Gerais. Dissertation, Universidade Federal de Ouro Preto

Linder HP (2003) The radiation of the Cape flora, southern Africa. Biol Rev 78:597–638
Litsios G, Wüest RO, Kostikova A, Forest F, Lexer C, Linder HP, Pearman PB, Zimmermann NE, Salamin N (2013) Effects of a fire response trait on diversification in replicated radiations. Evol 68:453–465
Lopes-Mattos KLB, Azevedo AA, Soares AA, Meira RMSA (2013) Underground system of *Mandevilla atroviolacea* (Stadelm.) Woodson (Apocynaceae, Apocynoideae) from the Brazilian high-altitude grassland. South Afr J Bot 87:27–33
Losos JB (2011) Convergent, adaptation, and constraint. Evol 65:1827–1840
Louille B, Semir J, Lohmann LG, Pirani JR (2015) A phylogenetic analysis of Lychnophorinae (Asteraceae: Vernonieae) based on molecular and morphological data. Syst Bot 40:299–315
Lovo J, Winkworth RC, Mello-Silva R (2012) New insights into Trimezieae (Iridaceae) phylogeny: what do molecular data tell us? Ann Bot 110:689–702
Lusa MG, Cardoso EC, Machado SR, Appezzato-da-Glória B (2014a) Trichomes related to an unusual method of water retention and protection of the stem apex in an arid zone perennial species. AoB Plants plu088
Lusa MG, Appezzato-da-Glória B, Loeuille B, Bartoli G, Ciccarelli D (2014b) Functional groups in Lychnophorinae (Asteraceae: Vernonieae) based on morphological and anatomical traits. Aust J Bot 62:150–163
Lüttge U (2008) Savannas II. The environmental factors water, mineral nutrients and fire. In: Lüttge U (ed) Physiological ecology of tropical plants. Springer, Berlin, pp 313–377
Lüttge U, Duarte HM, Scarano FR, Attos EA, Cavalin PO, Franco AC, Fernandes GW (2007) Physiological ecology of photosynthesis of five sympatric species of Velloziaceae in the rupestrian fields of Serra do Cipó, Minas Gerais, Brazil. Flora 202:637–646
Lynch JP, Ho MD (2005) Rhizoeconomics: carbon costs of phosphorus acquisition. Plant Soil 269:45–56
Madriñán S, Cortés AJ, Richardson JE (2013) Páramo is the world's fastest evolving and coolest biodiversity hotspot. Front Gen 4:1–7
Mamede MCH (1993) Anatomia dos orgãos Vegetativos de *Camarea* (Malpighiaceae). Acta Bot Bras 7:3–19
Marschner P (1995) Mineral nutrition of higher plants, 2nd edn. Academic Press, London
Matias SR, Pagano MC, Muzzi FC, Oliveira C, Carneiro A, Horta N, Scotti M (2009) Effect of rhizobia, mycorrhizal fungi and phosphate-solubilizing microorganisms in the rhizosphere of native plants used to recover an iron ore area in Brazil. Eur J Soil Biol 45:259–266
McCully ME (1999) Roots in soil: unearthing the complexities of roots and their rhizospheres. Annu Rev Plant Physiol Plant Mol Biol 50:695–718
McDowell NG, Pockman WT, Allen CD, Breshears DD, Cobb N, Kolb T, Plaut J, Sperry J, West A, Williams DG, Yepez EA (2008) Mechanisms of plant survival and mortality during drought: why do some plants survive while others succumb to drought? New Phytol 178:719–739
McDowell NG, Fisher RA, Xu C, Domec JC, Höltta T, Mackay DS, Sperry JS, Boutz A, Dickman LT, Gehres N, Limousin M, Macalady A, Martınez-Vilalta J, Mencuccini M, Plaut JA, Ogée J, Pangle RE, Rasse DP, Ryan MG, Sevanto S, Waring RH, Williams AP, Yepez EA, Pockman WT (2013) Evaluating theories of drought-induced vegetation mortality using a multimodel—experiment framework. New Phytol 200:304–321
McPherson S (2010) Carnivorous plants and their habitats, 2nd edn. Red Fern Natural History, Poole
Medina E (1982) Physiological ecology of neotropical Savanna plants. In: Huntles BJ, Walker BH (eds) Ecological studies, vol 42., Ecology of tropical savannasSpringer, Berlin, pp 308–335
Meinzer FC, Johnson DM, Lachenbruch B, McCulloh KA, Woodruff DR (2009) Xylem hydraulic safety margins in woody plants: coordination of stomatal control of xylem tension with hydraulic capacitance. Funct Ecol 23:922–930
Meirelles ST, Pivello VR, Joly CA (1999) The vegetation of granite rock outcrops in Rio de Janeiro, Brazil, and the need for its protection. Environ Cons 26:10–20

Mello-Silva R (2004) Novitates Velloziacearum florae phanerogamicae Sancti Pauli. Rev Brasil Bot 27:453–462
Mello-Silva R, Santos DYAC, Salatino MLF, Motta LB, Cattai MB, Sasaki D, Lovo J, Pita PB, Rocini C, Rodrigues CDN, Zarrei M, Chase MW (2011) Five vicarious genera from Gondwana: the Velloziaceae as shown by molecules and morphology. Ann Bot 108:87–102
Menezes NL, Müller C, Sajo MDG (1979) Um novo e peculiar tipo de sistema subterrâneo em espécies de *Vernonia* da Serra do Cipó. Bol Bot USP 7:33–38
Messias MCTB, Leite MGP, Meira-Neto JAA, Kozovits AR (2011) Life-form spectra of quartzite and itabirite rocky outcrop sites, Minas Gerais, Brazil. Biota Neotrop 11:255–268
Messias MCTB, Leite MGP, Meira-Neto JAA, Kozovits AR (2012) Fitossociologia de campos rupestres quartzíticos e ferruginosos no Quadrilátero Ferrífero, Minas Gerais. Acta Bot Bras 26:230–242
Messias MCTB, Leite MGP, Meira Neto JAA, Kozovits AR, Tavares R (2013) Soil-vegetation relationship in quartzitic and ferruginous Brazilian rocky outcrops. Folia Geobot 48:509–521
Miller RM (2005) The nonmycorrhizal root: a strategy for survival in nutrient-impoverished soils. New Phytol 165:655–658
Moreira ASFP, Lemos-Filho JP, Zotz G, Isaias RMS (2009) Anatomy and photosynthetic parameters of roots and leaves of two shade-adapted orchids, *Dichaea cogniauxiana* Shltr. and *Epidendrum secundum* Jacq. Flora 204:604–611
Morokawa R, Simões AO, Kinoshita LS (2013) Apocynaceae s. str. do Parque Nacional da Serra da Canastra, Minas Gerais, Brasil. Rodriguésia 64:179–199
Mourão FA, Carmo FF, Ratton P, Jacobi CM (2006) Hospedeiras da hemiparasita *Struthanthus flexicaulis* (Mart.) Mart. (Loranthaceae) em campos rupestres ferruginosos do Quadrilátero Ferrífero, Minas Gerais. Lundiana 7:103–109
Mourão FA, Jacobi CM, Figueira JEC, Batista EKL (2009) Effects of the parasitism of *Struthanthus flexicaulis* (Mart.) Mart. (Loranthaceae) on the fitness of *Mimosa calodendron* Mart. (Fabaceae), an endemic shrub from rupestrian fields over ironstone outcrops, Minas Gerais State, Brazil. Acta Bot Bras 23:820–825. 2009
Muler AL, Oliveira RS, Lambers H, Veneklaas EJ (2014) Does cluster-root activity benefit nutrient uptake and growth of co-existing species? Oecologia 174:23–31
Muller KF, Borsch T, Legendre L, Porembski S, Barthlott W (2006) Recent progress in understanding the evolution of carnivorous Lentibulariaceae (Lamiales). Plant Biol 8:748–757
Negreiros D, Borges Moraes ML, Wilson Fernandes G (2008) Caracterização da fertilidade dos solos de quatro leguminosas de campos rupestres, Serra do Cipó, MG, Brasil. Rev Cien Suelo Nutr Veg 8:30–39
Negreiros D, Fernandes GW, Silveira FA, Chalub C (2009) Seedling growth and biomass allocation of endemic and threatened shrubs of rupestrian fields. Acta Oecol 35:301–310
Negreiros D, Fernandes GW, Berbara RLL, Rodarte LHO, Barbosa NPU (2011) Caracterização físico-química de solos quartzíticos degradados e áreas adjacentes de campo rupestre na Serra do Cipó, MG, Brasil. Neotrop Biol Cons 6:156–161
Negreiros D, Le Stradic S, Fernandes GW, Rennó HC (2014) CSR analysis of plant functional types in highly diverse tropical grasslands of harsh environments. Plant Ecol 215:379–388
Neves SPS, Conceição AB (2010) Campo rupestre recém-queimado na Chapada Diamantina, Bahia, Brasil: plantas de rebrota e sementes, com espécies endêmicas na rocha. Acta Bot Bras 24:697–707
Neves SC, Abreu PAAS, Fraga LM (2005) Fisiografia. In: Silva AC, Pedreira LCVSF, Abreu PAA (eds) Serra do Espinhaço meridional—paisagens e ambientes. O Lutador, Diamantina, pp 45–58
Nickrent DL, Musselman LJ (2004) Introduction to parasitic flowering plants. Plant Health Instructor. doi:10.1094/PHI-I-2004-0330-01
Nielsen R (2009) Adaptionism - 30 years after Gould and Lewontin. Evolution 63:2487–2490
Nishi AH, Vasconcellos-Neto J, Romero GQ (2013) The role of multiple partners in a digestive mutualism with a protocarnivorous plant. Ann Bot 111:143–150

Nogueira RE, Pereira OL, Kasuya MCM, Lanna MCS, Mendonça MP (2005) Fungos micorrízicos associados a orquídeas em campos rupestres na região do Quadrilátero Ferrífero, MG, Brasil. Acta Bot Bras 19:417–424

North GB, Nobel PS (1997) Drought-induced changes in soil contact and hydraulic conductivity for roots of *Opuntia ficus-indica* with and without rhizosheaths. Plant Soil 191:249–258

Ogle K, Reynolds JF (2002) Desert dogma revisited: coupling of stomatal conductance and photosynthesis in the desert shrub, *Larrea tridentata*. Plant, Cell Environ 25:909–921

Oliveira RB, Godoy SAP (2007) Floristic composition of the rock outcrops of the Morro do Forno, Altinópolis-SP. Biota Neotrop 7:37–47

Oliveira RS, Dawson TE, Burgess SS (2005) Evidence for direct water absorption by the shoot of the desiccation-tolerant plant *Vellozia flavicans* in the savannas of central Brazil. J Trop Ecol 21:585–588

Oliveira RS, Galvão HC, De Campos MCR, Eller CB, Pearse SJ, Lambers H (2015) Mineral nutrition of campos rupestres plant species on contrasting nutrient-impoverished soil types. New Phytol 205:1183–1194

Oriani A, Scatena VL (2007) Intracellular papillae of *Actinocephalus* (Eriocaulaceae-Poales) roots and their interaction with fungi: a light and transmission electron microscopy study. Micron 38:611–617

Overbeck GE, Pfadenhauer J (2007) Adaptive strategies in burned sub-tropical grassland in southern Brazil. Flora 202:27–49

Pate JS, Unkovich MJ, Erskine PD, Stewart GR (1998) Australian mulga ecosystems ^{13}C and ^{15}N natural abundances of biota components and their ecophysiological significance. Plant, Cell Environ 21:1231–1242

Pekin BK, Boer MM, Wittkuhn RS, Macfarlane C, Grierson PF (2012) Plant diversity is linked to nutrient limitation of dominant species in a world biodiversity hotspot. J Veg Sci 23:745–754

Peñuelas J, Terradas J, Lloret F (2011) Solving the conundrum of plant species coexistence: water in space and time matters most. New Phytol 189:5–8

Pereira CG, Almenara DP, Winter CE, Fritsch PW, Lambers H, Oliveira RS (2012) Underground leaves of *Philcoxia* trap and digest nematodes. Proc Natl Acad Sci 109:1154–1158

Pillai SK, Pillai A (1961) Root apical organization in Monocotyledons—Xyridaceae. Proc Indian Acad Sci Sect B 54:234–240

Pita PB, Menezes NL (2002) Anatomia da raiz de espécies de *Dyckia* Schult. f. e *Encholirium* Mart. ex Schult. & Schult. f. (Bromeliaceae, Pitcairnioideae) da Serra do Cipó (Minas Gerais, Brasil), com especial referência ao velame. Rev Bras Bot 25:25–34

Plaut JA, Yepez EA, Hill J, Pangle R, Pangle R, Sperry JS, Pockman WT, Mcdowell NT (2012) Hydraulic limits preceding mortality in piñon-juniper woodland under experimental drought. Plant Cell Environ 35:1601–1617

Pomerene JB (1964) Geology and ore deposits of the Belo Horizonte, Ibirite, and Macacos Quadrangles, Minas Gerais, Brazil. Geol Surv Prof Pap 341-D:1–84

Poot P, Hopper SD, van Diggelen JM (2012) Exploring rock fissures: does a specialized root morphology explain endemism on granite outcrops? Ann Bot 110:291–300

Porder S, Ramachandran S (2013) The phosphorus concentration of common rocks-a potential driver of ecosystem P status. Plant Soil 367:41–55

Porembski S, Barthlott W (1995) On the occurrence of a velamen radicum in Cyperaceae and Velloziaceae. Nord J Bot 15:625–629

Porembski S, Barthlott W (2000) Granitic and gneissic outcrops (inselbergs) as centers of diversity for desiccation-tolerant vascular plants. Plant Ecol 151:19–28

Press MC, Phoenix GK (2005) Impacts of parasitic plants on natural communities. New Phytol 166:737–751

Press MC, Scholes JD, Watling JR (1999) Parasitic plants: physiological and ecological interactions with their hosts. In: Press MC, Scholes JD, Barker MG (eds) Physiological plant ecology. Blackwell Science, Oxford, pp 175–197

Proctor MCF, Tuba Z (2002) Poikilohydry and homoiohydry: antithesis or spectrum of possibilities? New Phytol 156:327–349

Rapini A, Ribeiro PL, Lambert S, Pirani JR (2008) A flora dos campos rupestres da Cadeia do Espinhaço. Megadiv 4:15–23

Raven JA (2012) Protein turnover and plant RNA and phosphorus requirements in relation to nitrogen fixation. Plant Sci 188–189:25–35

Reis FB Jr, Simon MF, Gross E, Boddey RM, Elliott GN, Neto NE, Loureiro MF, Queiroz LP, Scotti MR, Chen WM, Noren A, Rubio MC, Faria SM, Bontemps C, Goi SR, Young PW, Sprent JI, James EK (2010) Nodulation and nitrogen fixation by Mimosa spp. in the Cerrado and Caatinga biomes of Brazil. New Phytol 186:934–946

Ribeiro KT, Fernandes GW (2000) Patterns of abundance of a narrow endemic species in a tropical and infertile montane habitat. Plant Ecol 147:205–218

Ribeiro KT, Medina BMO (2002) Estrutura, dinâmica e biogeografia das ilhas de vegetação sobre rocha do Planalto do Itatiaia, RJ. Bol Parque Nac Itatiaia 10:1–84

Ribeiro JF, Walter BMT (2008) As principais fitofisionomias do bioma cerrado. In: Sano SM, Almeida SP, Ribeiro JF (eds) Cerrado: ecologia e flora. Embrapa/CPAC, Planaltina

Ribeiro KT, Medina BMO, Scarano FR (2007) Composição de espécies e relações biogeográficas da flora sobre afloramentos rochosos no Planalto do Itatiaia, SE do Brasil. Rev Bras Bot 30:623–639

Ribeiro PL, Borba EL, Smidt EC, Lambert SM, Schnadelbach AS, van den Berg C (2008) Genetic and morphological variation in the *Bulbophyllum exaltatum* (Orchidaceae) complex occurring in the Brazilian campos rupestres: implications for taxonomy and biogeography. Plant Syst Evol 270:109–137

Ribeiro PL, Rapini A, Damascena LS, van den Berg C (2014) Plant diversification in the Espinhaço Range: insights from the biogeography of *Minaria* (Apocynaceae). Taxon 63:1253–1264

Rivadavia F, Vicentini A, Fleischmann A (2009) A new species of sundew (*Drosera*, Droseraceae), with water-dispersed seed, from the floodplains of the northern Amazon Basin, Brazil. Ecotropica 15:13–21

Rodrigues KF, Batista LG, Carrenho R (2013) Micorrizas arbusculares em espécies de Microlicieae (Melastomataceae) da Chapada dos Veadeiros, Goiás. In: 64 Congresso Nacional de Botânica. Belo Horizonte

Rossatto DR, da Silveira Lobo Sternberg L, Franco AC (2013) The partitioning of water uptake between growth forms in a Neotropical savanna: do herbs exploit a third water source niche? Plant Biol 15:84–92

Ryan PR, Delhaize E, Jones DL (2001) Function and mechanism of organic anion exudation from plant roots. Annu Rev Plant Physiol Plant Mol Biol 52:527–560

Saadi A (1995) A geomorfologia da Serra do Espinhaço em Minas Gerais e de suas margens. Geonomos 3:41–63

Safford HD (2001) Brazilian Páramos. III. Patterns and rates of postfire regeneration in Campos de Altitude. Biotrop 33:282–302

Sala A, Piper F, Hoch G (2010) Physiological mechanisms of drought-induced tree mortality are far from being resolved. New Phytol 186:274–281

Santos TRD, Pinto JRR, Lenza E, Mews HA (2012) The tree-shrub vegetation in rocky outcrop cerrado areas in Goiás State, Brazil. Braz J Bot 35:281–294

Scatena VL, Oriani A, Sano PT (2005) Anatomia de raízes de *Actinocephalus* (Koern.) Sano (Eriocaulaceae). Acta Bot Bras 19:835–841

Schluter D (1990) Species-for-species matching. Am Nat 136:560–568

Schwinning S (2010) The ecohydrology of roots in rocks. Ecohydrol 3:238–245

Schwinning S, Ehleringer JR (2001) Water use trade-offs and optimal adaptations to pulse-driven arid ecosystems. J Ecol 89:464–480

Shane MW, Lambers H (2005) Cluster roots: a curiosity in context. Plant Soil 274:101–125

Shane MW, Dixon KW, Lambers H (2005) The occurrence of dauciform roots amongst western Australian reeds, rushes and sedges, and the impact of phosphorus supply on dauciform-root development in *Schoenus unispiculatus* (Cyperaceae). New Phytol 165:887–898

Shane MW, Cawthray GR, Cramer MD, Kuo J, Lambers H (2006) Specialized "dauciform" roots of Cyperaceae are structurally distinct, but functionally analogous with "cluster" roots. Plant Cell Environ 29:1989–1999

Shaw DC, Watson DM, Mathiasen RL (2004) Comparison of dwarf mistletoe (*Arceuthobium* spp., Viscaceae) in the weastern United States with mistletoe (Amyema spp., Loranthaceae) in Australia: ecological analogs and reciprocal models for ecosystem management. Aust J Bot 52:481–498

Silva ED, Martins AB (2013) Leguminosae-Papilionoideae in the Serra do Cabral, Minas Gerais State, Brazil. Hoehnea 40:293–314

Silva NG, Alves RJV, Pereira JF, Rivadavia F (2011) Lentibulariaceae, Serra de São José, Minas Gerais, Brazil. Checklist 7:120–127

Silveira FAO, Negreiros D, Barbosa NPU, Buisson E, Carmo FF, Carstensen D, Conceição AA, Cornelissen TG, Echternacht L, Fernandes GW, Garcia QS, Guerra TJ, Jacobi CM, Lemos-Filho JP, Le Stradic S, Morellato LPC, Neves FS, Oliveira RS, Schaefer CE, Viana PL, Lambers H (2016) Ecology and evolution of plant diversity in the endangered campo rupestre: a neglected conservation priority. Plant and Soil (in press)

Silvertown J, Araya Y, Gowing D (2015) Hydrological niches in terrestrial plant communities: a review. J Ecol 103:93–108

Simmons GC (1968) Geology and iron deposits of the Western Serra do Curral, Minas Gerais, Brazil. Geol Surv Prof Pap 341-G:1–57

Simon MF, Grether R, de Queiroz LP, Skemae C, Pennington RT, Hughes CE (2009) Recent assembly of the Cerrado, a neotropical plant diversity hotspot, by in situ evolution of adaptations to fire. Proc Natl Acad Sci USA 106:20359–20364

Smith SE, Read DJ (2008) Mycorrhizal symbiosis, 3rd edn. Academic Press, London

Smith RJ, Hopper SD, Shane MW (2011) Sand-binding roots in Haemodoraceae: global survey and morphology in a phylogenetic context. Plant Soil 348:453–470

Souza JMS (2011) Efeitos do fogo na fenologia reprodutiva de angiospermas em vegetações campestres na Chapada Diamantina, Brasil. PhD Thesis, Universidade Estadual de Feira de Santana

Specht RL (1969) A comparison of the sclerophyllous vegetation characteristic of Mediterranean type climates in France, California, and Southern Australia. I. Structure, morphology, and succession. Aust J Bot 17:277–282

Sprent JI (1979) The biology of nitrogen-fixing organisms. McGraw-Hill, London

Stock WD, Verboom GA (2012) Phylogenetic ecology of foliar N and P concentrations and N: P ratios across mediterranean-type ecosystems. Glob Ecol Biogeogr 21:1147–1156

Tardieu F, Simonneau T (1998) Variability among species of stomatal control under fluctuating soil water status and evaporative demand: modelling isohydric and anisohydric behaviours. J Exp Bot 49:419–432

Tayasu I, Inoue T, Miller LR, Sugimoto A, Takeichi S, Abe T (1998) Confirmation of soil-feeding termites (Isoptera; Termitidae; Termitinae) in Australia using stable isotope ratios. Funct Ecol 12:536–542

Taylor P, Souza VC, Giulietti AM, Harley RM (2000) *Philcoxia*: a new genus of Scrophulariaceae with three new species from eastern Brazil. Kew Bull 55:155–163

Teixeira WA, Lemos-Filho JP (1998) Metais pesados em folhas de espécies lenhosas colonizadoras de uma área de mineração de ferro em Itabirito, Minas Gerais. Rev Árvore 22:381–388

Teodoro GS (2014) Extreme drought effects on the phenology, growth and ecophysiology performance of campos rupestres species—Brazil. PhD Thesis, Universidade Estadual de Campinas

Tertuliano MF, Figueiredo-Ribeiro RCL (1993) Distribution of fructose polymers in herbaceous species of Asteraceae from the cerrado. New Phytol 123:741–749

Trovó M, Andrade MJG, Sano PT, Ribeiro PL, Berg C (2013) Molecular phylogenetics and biogeography of Neotropical Paepalanthoideae with emphasis on Brazilian *Paepalanthus* (Eriocaulaceae). Bot J Linn Soc 171:225–243

Tuba Z, Protor CF, Csintalan Z (1998) Ecophysiological responses of homoiochlorophyllous and poikilochlorophyllous desiccation tolerant plants: a comparison and an ecological perspective. Plant Growth Reg 24:211–217

Turner BL, Condron LM (2013) Pedogenesis, nutrient dynamics, and ecosystem development: the legacy of T.W. Walker and J.K. Syers. Plant Soil 367:1–10

Twigg LE (2014) Fluoroacetate, plants, animals and a biological arms race. In: Lambers H (ed) Plant life on the sandplains in southwest Australia, a global biodiversity hotspot. University of Western Australia Publishing, Crawley, pp 225–240

USDA (1998) Keys to soil taxonomy. United States Department of Agriculture, New York

Vasconcelos M (2011) O que são campos rupestres e campos de altitude nos topos de montanha do Leste do Brasil? Braz J Bot 34:241–246

Veldman JW, Buisson E, Durigan G, Fernandes GW, Le Stradic S, Mahy G, Negreiros D, Overbeck GE, Veldman RG, Zaloumis NP, Putz FE, Bond WJ (2015) Toward an old-growth concept for grasslands, savannas, and woodlands. Front Ecol Environ 13:154–162

Vergutz L, Manzoni S, Porporato A, Novais RF, Jackson RB (2012) Global resorption efficiencies and concentrations of carbon and nutrients in leaves of terrestrial plants. Ecol Monogr 82:205–220

Vickery B, Vickery M (1972) Fluoride metabolism in *Dichapetalum toxicarium*. Phytochem 11:1905–1909

Vincent RDC, Meguro M (2008) Influence of soil properties on the abundance of plant species in ferruginous rocky soils vegetation, southeastern Brazil. Rev Bras Bot 31:377–388

Vitousek PM, Cassman K, Cleveland C, Crews T, Field CB, Grimm NB, Howarth RW, Marino R, Martinelli L, Rastetter EB, Sprent JI (2002) Towards an ecological understanding of biological nitrogen fixation. Biogeochem 57–58:1–45

Walker TW, Syers JK (1976) The fate of phosphorus during pedogenesis. Geoderma 15:1–19

Walter H (1971) Ecology of tropical and subtropical vegetation. Oliver & Boyd, Edinburgh

West AG, Dawson TE, February EC, Midgley GF, Bond WJ, Aston TL (2012) Diverse functional responses to drought in a Mediterranean-type shrubland in South Africa. New Phytol 195:396–407

Williamson B (1973) Acid phosphatase and esterase activity in orchid mycorrhiza. Planta 112:149–158

Yao Q, Li X, Feng G, Christie P (2001) Mobilization of sparingly soluble inorganic phosphates by the external mycelium of an arbuscular mycorrhizal fungus. Plant Soil 230:279–285

Zemunik G, Turner BL, Lambers H, Laliberté E (2015) Diversity of plant nutrient-acquisition strategies increases during long-term ecosystem development. Nat Plants 1–4

Part III
Foreword 3 + Photo

The mosaic of soils of rupestrian grasslands has produced a rich flora, and an astonishing fauna has developed to feed on them, adapting to their behavior, availability, and idiosyncrasies. But the fauna of rupestrian grasslands are otherwise mostly unknown, with studies having been concentrated in just few areas.

This section of the book focuses mostly on the interactions among species of plants, animals, and microorganisms. The breathing of the rupestrian grassland is investigated for the very first time with plenty of information on the phenological patterns of plants. The timing of flowering along elevational gradients indicates not just how synchronized plants and pollinators are, but also highlights other unknown and rarely evaluated syndromes such as trends in flower colors and signaling to pollinators and fruit dispersers. As studies are spreading to different areas in this ecosystem, many new interactions and novelties have been described among insects and plants, and even among plant parasites and herbivores. Herbivory on environmentally stressed plants may augment the physiological stress of plants and cause a strong impact on host performance. Yet some organisms cannot be seen as only consumers, such as termites who play an important role in rupestrian grasslands by moving huge amounts of soil, thus increasing its quality for plants. The strong gradients in habitat quality along mountain chains represent powerful filters that have shaped plant and animal distributions.

Chapter 12
Phenology Patterns Across a Rupestrian Grassland Altitudinal Gradient

Nathália Miranda Walter Bretas Rocha, Daniel Wisbech Carstensen, G. Wilson Fernandes, Soizig Le Stradic, Elise Buisson and Leonor Patrícia Cerdeira Morellato

Abstract Phenology is the study of recurrent biological events in the life cycle of organisms. For plants, reproductive events such as flowering and fruiting are critical stages in their life cycles, which also greatly affect other organisms depending on these resources. Here, we present the first community level plant phenology study across an altitudinal gradient from cerrado through rupestrian grassland to higher altitudinal grasslands developed in the South Espinhaço Mountain range (Serra do Cipó, Brazil). We describe the seasonal patterns of flowering and fruiting phenology and discuss the dependency of reproductive events on the varying climate and

N.M.W.B. Rocha · D.W. Carstensen · L.P.C. Morellato (✉)
Departamento de Botânica, Universidade Estadual Paulista (UNESP),
Avenida 24A, 1515, 13506-900 Rio Claro, SP, Brazil
e-mail: pmorella@rc.unesp.br

N.M.W.B. Rocha
e-mail: nathymwbr@gmail.com

D.W. Carstensen
e-mail: daniel.carstensen@gmail.com

G.W. Fernandes
Ecologia Evolutiva & Biodiversidade/DBG, CP 486, ICB,
Universidade Federal de Minas Gerais, 31270-901 Belo Horizonte, MG, Brazil
e-mail: gw.fernandes@gmail.com

G.W. Fernandes
Department of Biology, Stanford University, Stanford, CA 94305, USA

S. Le Stradic
Gembloux Agro-Bio Tech Université de Liège, Biodiversité et Paysage,
Passage Des Déportés, 2, 5030 Gembloux, Belgium
e-mail: soizig.lestradic@gmail.com

E. Buisson
Université D'Avignon et Des Pays Du Vaucluse, IMBE, CNRS, IRD, Aix Marseille
Université, IUT D'Avignon, Agroparc BP 61207, 84911 Avignon Cedex 9, France
e-mail: elise.buisson@univ-avignon.fr

© Springer International Publishing Switzerland 2016

G.W. Fernandes (ed.), *Ecology and Conservation of Mountaintop Grasslands in Brazil*, DOI 10.1007/978-3-319-29808-5_12

habitat types across the gradient. We point out the importance of long-term monitoring and further detailed analyses of families and species contribution to unravel the structure of community phenology, the environmental drivers and sensitivity to climate changes.

Keywords Reproductive phenology · Seasonality patters · Altitudinal grasslands · Cerrado

Contents

12.1 Introduction

Phenology is a traditional research area studying the recurrent biological events in the life cycle of plants and animals and their relationship to climate, as well as other environmental, biotic and evolutionary factors (Staggemeier et al. 2010; Schwartz 2013). Phenological studies allow us to gain insight into the temporal organization of resources such as flowers, fruits and leaves essential for pollinators, frugivores and seed dispersers, herbivores and other animals relying on plants material for food (e.g. Morellato et al. 2000; Staggemeier et al. 2010, 2015). Moreover, leafing patterns define the dynamics of ecosystem processes such as water status, carbon balance and the primary productivity and resilience to disturbance (Morisette et al. 2009). Recently, phenology has gained relevance in the scenario of global warming research and is regarded as one of the simplest way to monitor and detect plant responses to climatic changes (Menzel et al. 2006; Rosenzweig et al. 2008). Phenology, thus, offers key information for conservation and management programs, the sustainable use of natural resources to ecosystem services provided by plants and animals and for tourism and education (Morellato et al. 2016). However, in spite of its importance and the increasing number of phenology papers during the last two decades in the Southern Hemisphere, and Brazil in particular, there is a lack of phenology research addressing open vegetations or life forms other than trees

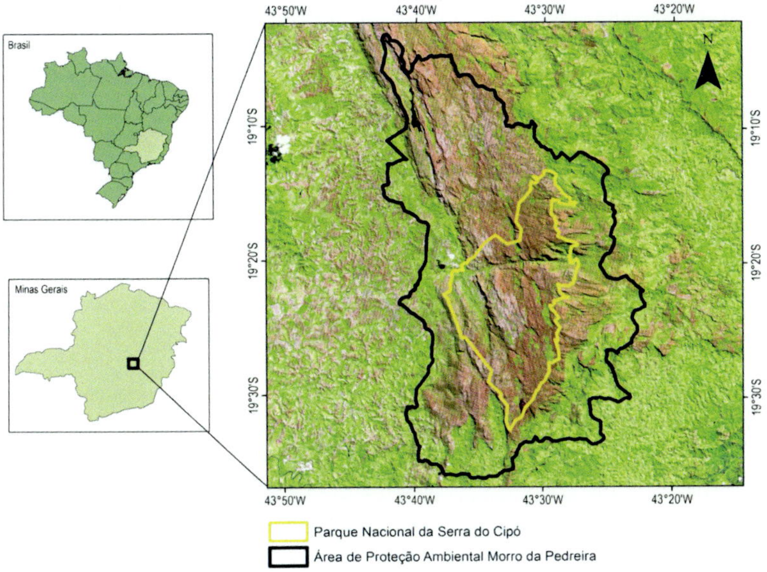

Fig. 12.1 Location of the study area in Brazil, Minas Gerais State, Southeastern Brazil, and the boundaries of National Park named Parque Nacional da Serra do Cipó (*yellow line*) and the Protection Area named Área de Proteção Ambiental Morro da Pedreira (APA Morro da Pedreira)

(Chambers et al. 2013; Morellato et al. 2013). Open, non-tree vegetation or grassland biomes have been largely overlooked and forgotten in the prospects for conservation and restoration (Veldman et al. 2015a, b).

Here, we bring to attention the seasonal patterns of the rupestrian grasslands, a spectacular and unique old-growth grassland (Veldman et al. 2015b) distributed mainly over the Espinhaço Moutain range (Fig. 12.1, Chap. 22) spanning the Southeast to Northeastern Brazil. The rupestrian grasslands comprise a high diversity of plant species from which an outstanding number are narrow endemic species (Silveira et al. 2016, Chap. 6). The ancient vegetation is shaped by the seasonally dry climate, frequent fires, soils, and herbivore pressures (Silveira et al. 2016). The deficiency of appropriate protection has resulted in severe loss of this invaluable ecosystem (Fernandes et al. 2014; Veldman et al. 2015a) and we still do not know the patterns and drivers of this vegetation phenology. Along the altitudinal gradient, changes in temperature, precipitation, wind speed, light incidence and soils, among other variables, define the modifications on species composition and likely, plant phenology (e.g. Giulietti et al. 1987; Callaway et al. 2002; Nunes et al. 2008; Venn et al. 2012). Most studies conducted so far in this ecosystem address the phenology of small groups of species and families (Madeira and Fernandes 1999; Fiaschi and Pirani 2005; Groppo and Pirani 2005; Kinoshita and

Simões 2005; Groppo et al. 2007; Neves 2009; Belo et al. 2013), but community wide inventories are still required (but see Le Stradic 2012).

The selective pressures imposed by seasonal climate and harsh local environmental conditions define the species' adaptation to local conditions and its occurrence on specific environments such as the rocky outcrops and likely shape their phenology (Giulietti and Pirani 1988; Batalha et al. 1997; Batalha and Mantovani 2000). The understanding of local environmental drivers of plant phenology may help to forecast plant responses and resilience to large scale climate changes (Wolkovich et al. 2014).

We present the first overview of the rupestrian grasslands vegetation phenology with a community approach, focused on the altitudinal gradients of Serra do Cipó, Minas Gerais State, in the South portion of the Espinhaço range. We aim to describe the reproductive phenology of the vegetation and to determine whether the reproductive phenology differs along the altitudinal gradient from Cerrado through rupestrian grasslands to altitudinal grasslands (summit areas mostly dominated by sandy grasslands with interspersed patches of Atlantic rain forests (see Chap. 7)), and whether the reproductive phenology is restricted to the warm and wet season or occurs throughout the year.

12.2 The Serra do Cipó

Serra do Cipó is located in the central part of Minas Gerais State, in the far South of Espinhaço mountain range (19° 10′–20′S and 43° 30′–40′W), encompassing three municipalities: Santana do Riacho, Jaboticatubas and Cardeal Mota, and includes also the National Park—Parque Nacional da Serra do Cipó and the Protection Area named Área de Proteção Ambiental Morro da Pedreira (APA Morro da Pedreira) (Fig. 12.1). The regional climate is classified as tropical altitudinal (Cwb) Köppen (1948), characterized by dry winters and rainy summers (Madeira and Fernandes 1999; Chap. 15). The climatic data were obtained from the National Institute of Meteorology (INMET) station at Conceição do Mato Dentro (MG), a city located approximately 60 km from the study area.

During the period of study (August 2012–December 2014) the temperature oscillated among years; the mean temperature was 21 °C with a mean minimum of 16 °C and a mean maximum of 28 °C. The rainy season spans from November to April, during the warmest months with the longest days, while the dry season occurred from May to October, during coldest months and short daylengths (Fig. 12.2).

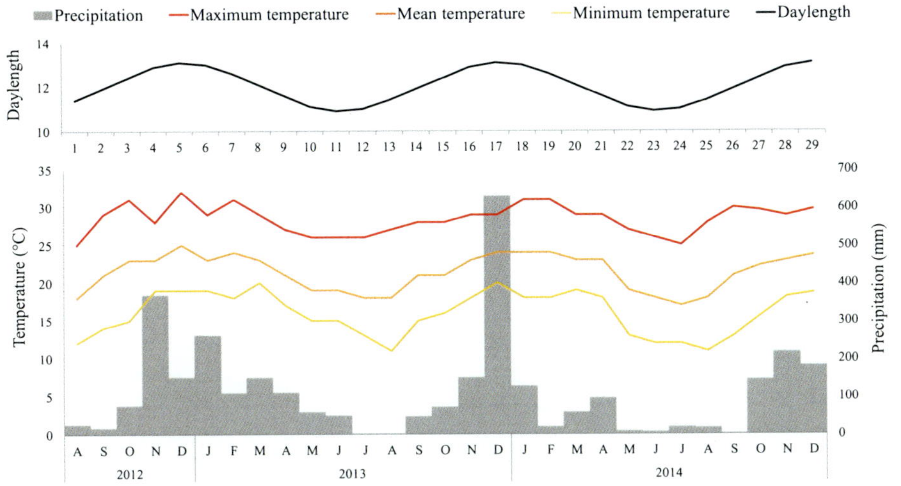

Fig. 12.2 Climate during the period of study at Conceição de Mato Dentro, MG. *Grey bars* Total monthly precipitation; maximum (*red line*), minimum (*yellow line*) and mean (*orange line*) monthly temperatures. *Source* Instituto Nacional de Meteorologia (INMET). The day length at 19° (*black line*) according to Pereira et al. (2002)

12.3 The Altitudinal Gradient of Rupestrian Grasslands

We sampled six sites within the protection area APA Morro da Pedreira, distributed over the altitudinal gradient (Fig. 12.3), named as follows (altitude within brackets): Cerrado, named Rio Cipó (RC) (821 m), Cerrado/Rupestrian grasslands, Usina (US) (977 m), Rupestrian grasslands: Cedro (CE) (1101 m), Pedra do Elefante (PE) (1255 m), Quadrante 16 (Q16) (1303 m) and Altitudinal grassland, Alto Palácio (AP) (1420 m).

(A) Cerrado, Rio Cipó (RC), 821 m: The lowest site, located in the Cipó river valley, at 821 m altitude (19° 21′29.0″S/43° 36′28.7″W), covered by cerrado, a Brazilian savanna growing on sandy soils, outcrops almost absent. The vegetation is dominated by medium to small trees and shrub species and a scattered herbaceous layer. The most common families are Fabaceae, Malpighyaceae, Asteraceae and Apocynaceae (Fig. 12.4a).

(B) Cerrado/Rupestrian grasslands, Usina (US), 977 m: This site is located at 977 m altitude (19° 16′10.2″S/43° 36′21.3″W) and is covered by a mix of cerrado vegetation and rupestrian grasslands growing on sandy soils and rocky outcrops. The transition from cerrado to rupestrian grasslands is marked by the small trees, shrubs and herbaceous layer well represented by the plant families Verbecaceae, Rubiaceae, Nyctaginaceae, Malpighyaceae and Fabaceae (Fig. 12.4b).

(C) Rupestrian grasslands, Cedro (CE), 1101 m: Located at 1101 m altitude (19° 13′55.48″S/3° 34′35.02″W), this site is dominated by rupestrian grasslands

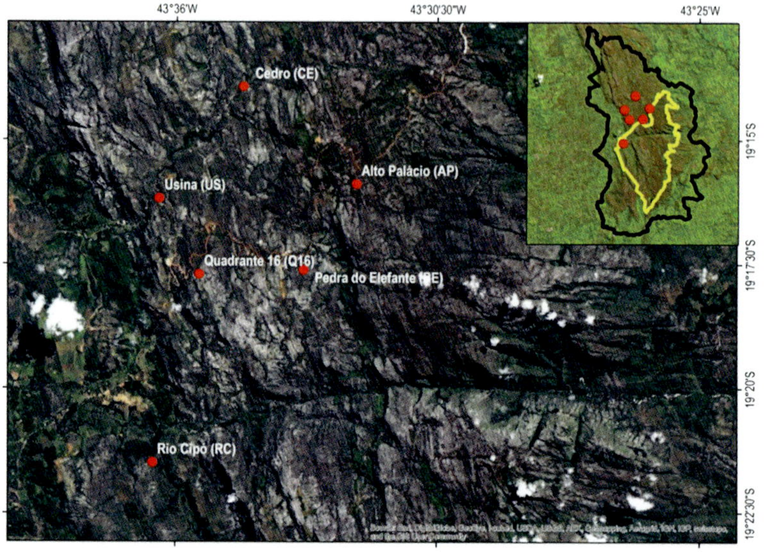

Fig. 12.3 Location of the six study sites at Serra do Cipó, Minas Gerais, Southeastern Brazil, along the altitudinal gradient. Local names were retained to facilitate the location of each site: Rio Cipó (821 m), Usina (977 m), Cedro (1101 m), Pedra do Elefante (1255 m), and Quadrante 16 (1303 m) and Alto Palácio (1420 m)

Fig. 12.4 Images from the six study sites at Serra do Cipó, Minas Gerais, Brazil: **a** Cerrado: Rio Cipó RC—821 m: overview of Rio Cipó valley and the cerrado in the lower front; **b** Cerrado/Rupestrian grasslands: Usina US—977 m, depicting the mountain outcrop and the mixed cerrado and grassland; **c–e** Rupestrian grasslands: **c** Cedro CE—1101 m showing the scattered woody component, **d** Pedra do Elefante PE—1255 m, where it is possible to see small trees and the outcrops and its typical vegetation; **e** Quadrante 16 Q16—1303 m showing an outcrop and the grassland (*left*) and another outcrop (*right*) in the background; and **f** Altitudinal grasslands: Alto Palacio AP—1420 m dominated by Poaceae and Eriocaulaceae. Photos N.M.W. B. Rocha

and the landscape is a mix of outcrops and sandy grasslands. The vegetation is characteristic of rupestrian grasslands composed by herbs and shrubs and occasional small treelets well represented by the plant families Velloziaceae, Eriocaulaceae, Primulaceae, Asteraceae and Melastomataceae (Fig. 12.4c).

(D) Rupestrian grasslands, Pedra do Elefante (PE), 1255 m: The second rupestrian grassland site is situated at 1255 m altitude (19° 13′55.48″S/3° 34′35.02″W), and includes sandy and rocky grasslands and the rocky outcrops. The typical vegetation of rupestrian grasslands is marked by herbs, Poaceae species and some small shrubs and trees, mostly species are from families Melastomataceae, Velloziaceae, Eriocaulaceae and Asteraceae (Fig. 12.4d).

(E) Rupestrian grasslands, Quadrante 16 (Q16), 1303 m: Located at 1303 m altitude (19° 17′42.0″S/43° 35′31.2″W), the rupestrian grasslands on Q16 grow on the top of dark clay soils rich in organic matter, dominated by sandy and rocky grasslands and the rocky outcrops are scattered across the landscape (Le Stradic 2012). The typical rupestrian grassland vegetation is covered by a rich herbaceous layer dominated by Poaeae and other herbs, shrubs mainly from Velloziaceae, Asteraceae and Eriocaulaceae and small treelets (Fig. 12.4e).

(F) Altitudinal grasslands, Alto Palácio (AP), 1420 m: The last study site is located at 1420 m altitude (19° 15′52.9″S/43° 32′12.1″W) and the grasslands grow on dark, wet, organic clay soils, sometimes covered by sand or small stones. The vegetation is typical of grasslands from higher altitudes, dominated by low diversity Poaceae and other herbaceous plants from the families Eriocalaceae, Asteraceae and Xyridaceae (Fig. 12.4f).

12.4 Plant Sampling and Phenological Observations

At each study site we set up 13 plots of 10 × 2 m along a linear transect, keeping 10 m distance between plots. All herbs, shrubs, and treelets larger than 1 cm diameter at soil level within the plots were sampled, tagged and observed for changes on phenology from August 2012 to December 2014. During the observations new individuals were included as they were identified in the plots and some were lost or died. Observations were carried out monthly for the presence and absence of flowering (flower buds and open flowers) and fruiting (unripe and ripe fruits). Two events of fire occurred during the study period: first on October 10th 2012, which reached site RC. The second, larger fire hit Serra do Cipó on October 2014 from 10th to 14th and burned three sites of rupestrian grasslands to the ground: CE, PE and Q16 as well as the altitudinal grassland site AP. The patterns presented here do not analyse the fire events, that will be discussed in further publications.

Vouchers of plant species from all study sites were collected near the plots during reproduction and are deposited in the Herbarium HRCB (Herbarium

Rioclarense—Universidade Estadual Paulista at Rio Claro) and Herbarium BHCB (Herbarium of Universidade Federal de Minas Gerais UFMG) and identified by comparison and specialists.

12.5 Changes in Species Richness Across the Altitudinal Gradient

We sampled a total of 2829 individuals belonging to 493 species and 54 families along the altitudinal gradient. We detected a reduction in the species number and an increase in the number of individuals per species from the 821 m cerrado towards the 1420 m altitudinal grasslands (Table 12.1). The 821 m cerrado at lower altitude presented the highest number of species (92) and lowest number of individuals (254) contrasting with the highest site, the 1420 m altitudinal grassland, where we surveyed the largest number of individuals (628) and the lowest species number (25). Additionally, at the 1420 m altitudinal grassland three species dominated the plots: *Actinocephalus polyanthus*, *Xyris nubigena* and *Xyris longiscapa*. The same pattern is found for a wide range of altitudinal gradients as described, for example, for a mountain gradient in east Nepal (Bhattarai and Vetaas 2003). In these environments, temperature is regarded as the main environmental constraint to species richness and we suggest that this would be the case for Serra do Cipó as well. Another explanation for the reduction in species richness as altitude increases, that also may apply to Serra do Cipó, is the requirement for plant species to be adapted to the more severe environmental conditions (Humboldt and Bonpland 1807; Lieberman et al. 1996; Borges 2011) and a harsher soil environment (Barrère 1992).

Table 12.1 Study sites, altitude, vegetation type and the number of individuals and species sampled across the altitudinal gradient of Serra do Cipó, MG, Southeastern Brazil

Site	Altitude (m)	Vegetation type	Number of individuals	Number of species
1. Rio Cipó (RC)	821	Cerrado	465[a]	53[a]
			254	92
2. Usina (US)	977	Cerrado/Rupestrian grasslands	376	89
3. Cedro (CE)	1101	Rupestrian grasslands	372	86
4. Pedra do Elefante (PE)	1255	Rupestrian grasslands	319	85
5. Quadrante 16 (Q16)	1303	Rupestrian grasslands	415	63
6. Alto Palácio (AP)	1420	Altitudinal grasslands	628	25

[a]Number of species and individuals sampled before fire on October 2012

12.6 Phenological Patterns Across the Altitudinal Gradient

We observed species flowering or fruiting at all sites throughout the year (Fig. 12.5). The only exception was the absence of fruiting in December or January at the highest site (1420 m) (Fig. 12.5f).

The 1420 m altitudinal grasslands was the most seasonal site, with well defined peaks of flowering in February–March (Fig. 12.5f), the warm and wet season, indicating that species of altitudinal grasslands are more constrained by the lower temperatures at higher altitudes, as observed in other mountain vegetation (Inouye 2008). Fruiting was also very seasonal and restricted mainly to the dry season, from June to September (Fig. 12.5f), and the fruit production concentrated in this period of the year likely favors the seed dispersal by wind presented by the majority of species at the 1420 m altitudinal grassland (unpublished data). The lowest site 821 m cerrado also showed flowering seasonality, but with an earlier peak in the transition from dry to wet season, while fruiting occurred continuously over the year (Fig. 12.5a). The patterns matched the reproductive phenology described for other areas of cerrado (Batalha et al. 1997; Batalha and Mantovani 2000). The transitional 877 m cerrado/rupestrian grassland presented a more continuous flowering and fruiting pattern, and seasonality was less evident (Fig. 12.5b),

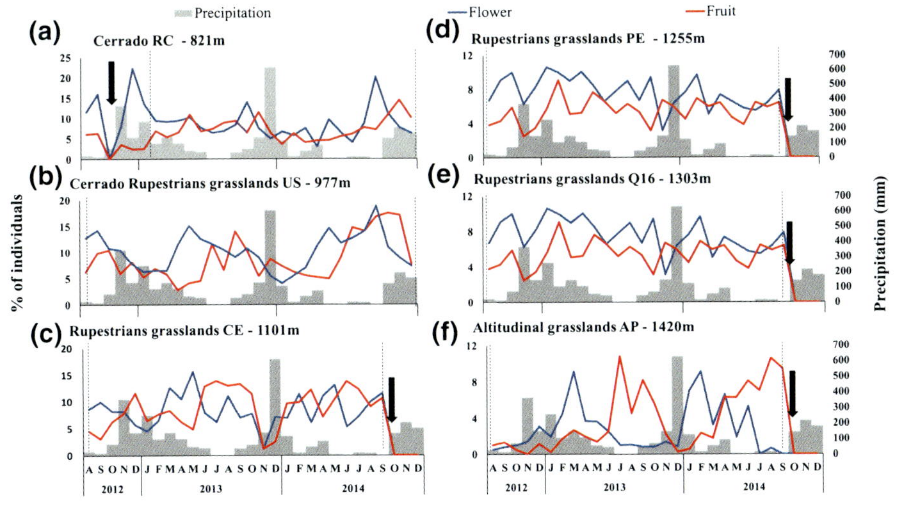

Fig. 12.5 Proportion of individuals flowering (*blue line*) and fruiting (*red line*) along the altitudinal gradient, from cerrado to rupestrian and altitudinal grasslands at Serra do Cipó, Southeastern Brazil. *Grey bars* Total monthly precipitation; *black arrow* fire event; *dotted vertical lines* core period for the description of phenological patterns at each site

although the flowering peak in the transition from dry to wet season in 2014 is similar to the one observed for the 821 m cerrado.

The dominant rupestrian grasslands above 1000 m altitude (CE, PE and Q16) presented individuals flowering and fruiting all year long (Fig. 12.5c–e). Our expectation of a seasonal reproduction constrained by the marked seasonality imposed by the cold, dry season was not confirmed and we argue that the long evolutionary history of species adapting to this old grassland (Veldman et al. 2015b; Silveira et al. 2016) may have allowed the evolution of a diversity of reproductive strategies that cannot be detected in the community level analyses conducted here. For instance, we may have differences in species phenology among sites that represent responses to specific local conditions. Inouye (2008) detected differences in the phenology of common species (*Delphinium barbeyi*, *Erigeron speciosus*, and *Helianthella quinquenervis*) among mountain sites over several years of observations that were related to specific local temperatures. In order to detect any potential recurring patterns in reproductive phenology of these communities we would likely need data collected over a much longer time period than the two years covered by the present study. However, as we did detect a seasonal pattern in the lowest (cerrado) and highest (altitudinal grassland) altitude sites this could indicate that such patterns are indeed either weak or absent from rupestrian grassland sites. Instead, plant phenology could be more strongly determined by other factors such as adaptation to pollinator or seed disperser availability (Lobo et al. 2003; Aizen 2003). Such variation over the year would likely differ between species and pollination/dispersal systems, creating a continuous pattern of flowering and fruiting at the community level.

The seeds of the majority of species in rupestrian grasslands are dispersed by wind or self-dispersed (Conceição et al. 2007; Dutra et al. 2009). A small proportion of species is dispersed by animals, mainly generalist frugivorous birds (Silveira et al. 2016; Chap. 13). Therefore, it would be advantageous for most species to set fruit during the dry, cold periods that favor seed dispersal by wind and the explosive opening of capsules of self-dispersed fruits (Frankie et al. 1974; Morellato and Leitão-Filho 1992; Oliveira and Moreira 1992; Morellato 1995). Our observations show an increase in the number of species fruiting in the dry season at 977 m cerrado/rupestrian grassland (US), the 1101 m rupestrian grassland (CE), and the 1420 m altitudinal grassland (AP) (Fig. 12.5b, c, f), but not over all grasslands sites. However, fruit phenology may be strongly linked to germination time, as already discussed for Melastomataceae from Serra do Cipó (Silveira et al. 2011). The germination onset would therefore determine the fruiting time of several species, as a necessary strategy to overcome the harsh environmental conditions (Silveira et al. 2015). A more complete understanding of biotic and abiotic factors affecting plant reproduction is likely necessary to untangle the determinants of plant phenology in rupestrian grasslands (Silveira et al. 2015, 2016).

12.7 Future Directions for Phenology Research in Rupestrian Grasslands

The ecology and evolution of rupestrian grasslands is poorly known (Silveira et al. 2016), and the data presented here are the first community level data on plant phenology from this type of vegetation and mountain gradient. The only other data available at the community level concerns the phenological patterns of the two main rupestrian grassland-types (i.e. sandy and stony grasslands as defined by Le Stradic et al. 2015) monitored for two consecutive years (Le Stradic 2012). While it might have been expected that the highly heterogeneous landscape of rupestrian grasslands would encompass different phenological patterns, these data mostly showed similar patterns in both communities (Fig. 12.6): flowering of all species in the community occurred all year long, with a slight increase between December and May, mostly in the wet season. Cyperaceae and Xyridaceae had the highest species contribution to overall phenology, while some dominant species belonging to Poaceae, among others, were not observed reproducing (Le Stradic 2012). However, the amplitude of phenophases, i.e. the net production of flowers and fruits per species, varied between grassland types according to each the species and its density as exemplified by the Eriocaulaceae *Paepalanthus geniculatus* (Fig. 12.7).

Long-term studies are now needed in order to gain insight into seasonal patterns and the response of plant communities to climate change. Detailed analyses that take into account plant functional groups and species may help to reveal the fine adaptation of the rupestrian grassland species' phenology to local environments, even in the absence of long-term data sets. Finally, the frequent and widespread fires as those observed during the present study, may also play a crucial role

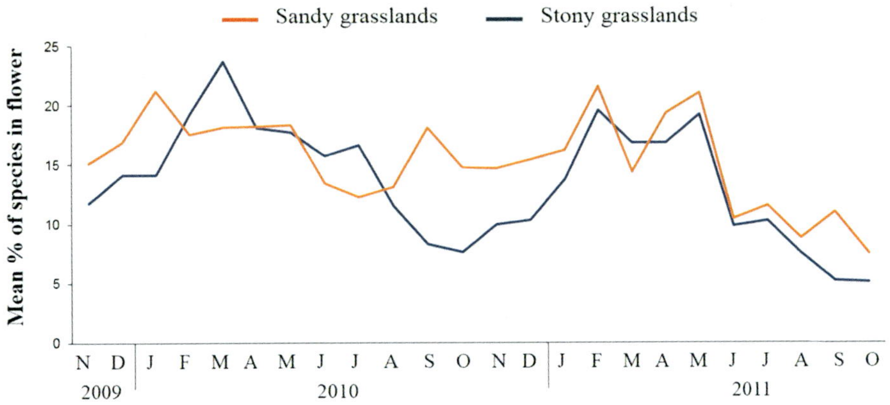

Fig. 12.6 Mean percent of species in flower in both sandy grasslands (*orange line*) and stony grasslands (*blue line*). Survey carried out on five sites for each grassland-type on ten permanent plots at Serra do Cipó, MG (Le Stradic 2012)

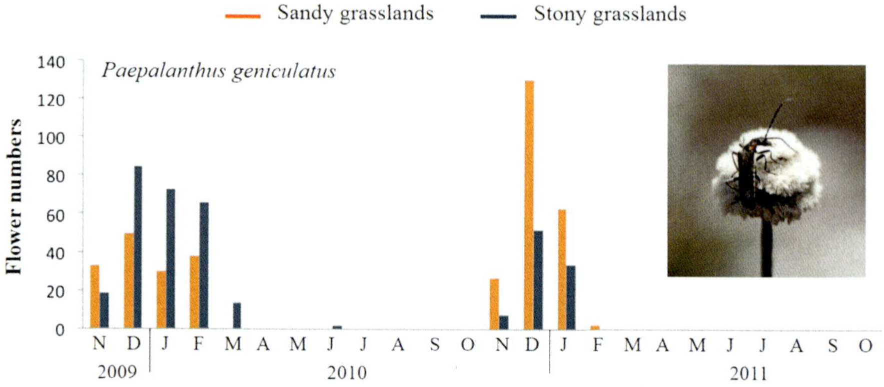

Fig. 12.7 Number of flowers of the Eriocaulaceae species *Paepalanthus geniculatus* in both sandy grasslands (*orange bars*) and stony grasslands (*blue bars*). Survey carried out on five sites for each grassland-type on ten permanent plots at Serra do Cipó, MG (Le Stradic 2012)

defining the phenologies across the rupestrian grasslands, as observed on other fire-prone vegetations (Alvarado et al. 2014), and need further and in depth investigations.

Acknowledgments We thank the many stakeholders that have allowed us to work in their lands and for logistical support: Serra Morena Inn, Cedro Cachoeira Textile Industry, Vellozia Reserve, Pedra do Elefante Inn, Eduardo C. Costa Jr, and Serra do Cipó National Park. We are grateful to Ana Cláudia Rodrigues, Valeria Campos, Natalia Costa Soares, Bruna Alberton, Irene Mendoza, Eduardo A. Athayde, Júlia Santos, Swanni Alvarado, Eli Garcia, Márcio Schafer Annia Susin, Thiago Horta, Leonardo F. Cancian, and Fernando Silveira for assistance in the field and to Ramiro Aguilar for valuable comments on the text. This research was supported by São Paulo Research Foundation FAPESP (FAPESP-VALE-FAPEMIG grant #2010/51307-0 and grant #2009/54208-6); Rocha, N.M.W.B. and Carstensen, D.W. received fellowships from FAPESP grants #2011/22635-2 and #2012/12429-9. Fernandes, GW thanks the grants and logistic supported provided by CNPq, Fapemig and Reserva Vellozia. Morellato, L.P.C. and Fernandes, GW received CNPq research productivity fellowship.

References

Aizen MA (2003) Influences of animal pollination and seed dispersal on winter flowering in a temperate Misteltoe. Ecology 84:2613–2627

Alvarado ST, Buisson E, Rabarison H, Rajeriarison C, Birkinshaw C, Lowry PP, Morellato LP (2014) Fire and the reproductive phonology of endangered Madagascar sclerophyllous tapia woodlands. South Afr J Bot 94:79–87

Barrère M (1992) Terra, patrimônio comum: a ciência a serviço do meio ambiente e do desenvolvimento. Nobel, São Paulo

Batalha MA, Mantovani W (2000) Reproductive phenological patterns of cerrado plant species at the Pe-de-Gigante reserve (Santa Rita do Passa Quatro, SP, Brazil): A comparison between the herbaceous and woody floras. Rev Bras Biol 60:129–145

Batalha MA, Aragaki S, Mantovani W (1997) Variações fenológicas das espécies do cerrado em Emas (Pirassununga, SP). Acta Bot Bras 11:61–78

Belo RM, Negreiros D, Fernandes GW, Silveira FAO, Ranieri BD, Morellato LPC (2013) Fenologia reprodutiva e vegetativa de arbustos endêmicos de campo rupestre na Serra do Cipó, Sudeste do Brasil. Rodriguésia 64:817–828

Bhattarai KR, Vetaas OL (2003) Variation in plant species richness of different life forms along a subtropical elevation gradient in the Himalayas, east Nepal. Glb Ecol Biogeogr 12:327–340

Borges DFdM (2011) Padrões de variação na riqueza de espécies em gradientes altitudinais: uma revisão multi-taxonômica. MSc Dissertation, Universidade de Brasília, Brasília

Callaway RM, Brooker RW, Choler P, Kikvidze Z, Lortiek CJ, Michalet R, Paolini L, Pugnaireq FI, Newingham B, Aschehoug ET, Armasq C, Kikodze D, Cook BJ (2002) Positive interactions among alpine plants increase with stress. Nature 417:844–848

Chambers LE, Altwegg R, Barbraud C, Barnard P, Beaumont LJ, Crawford RJM, Durant JM, Hughes L, Keatley MR, Low M, Morellato PC, Poloczanska ES, Ruoppolo V, Vanstreels RET, Woehler EJ, Wolfaardt AC (2013) Phenological changes in the southern hemisphere. Plos One 8:e75514. doi:10.71371/journal.pone.0075514

Conceição AA, Funch LS, Pirani JR (2007) Reproductive phenology, pollination and seed dispersal syndromes on sandstone outcrop vegetation in the "Chapada Diamantina", northeastern Brazil: population and community analyses. Rev Bras Bot 30:475–485

Dutra VF, Vieira MF, Garcia FCP, Lima HCL (2009) Fenologia reprodutiva, sindromes de polinização e dispersão em espécies de Leguminosae dos campos rupestres do Parque Estadual do Itacolomi, Minas Gerais, Brasil. Rodriguesia 60:371–387

Fernandes GW, Barbosa NP, Negreiros D, Paglia AP (2014) Challenges for the conservation of vanishing megadiverse rupestrian grasslands. Natureza Conservação 12(2):162–165

Fiaschi P, Jr Pirani (2005) Flora da Serra do Cipó, Minas Gerais: Araliaceae. Bol Bot Univ São Paulo 23:267–275

Frankie GW, Baker HG, Opler PA (1974) Comparative phenological studies of trees in tropical wet and dry forests in lowlands of Costa Rica. J Ecol 62:881–919

Giulietti AM, Jr Pirani (1988) Patterns of geographical distribution of some plant species from Espinhaço range, Minas Gerais and Bahia, Brazil. In: Vanzolini PE, Heyer WR (eds) Proceedings of a workshop on Neotropical distribution patterns. Academia Brasileira de Ciências, Rio de Janeiro, pp 39–69

Giulietti AM, Menezes NL, Pirani JR, Meguro M, Wanderley MGL (1987) Flora da Serra do Cipó, Minas Gerais: caracterização e lista de espécies. Bol Bot Univ São Paulo 9:1–151

Groppo M, Jr Pirani (2005) Flora da Serra do Cipó, Minas Gerais: Aquifoliaceae. Bol Bot Univ São Paulo 23:257–265

Groppo M, Amaral MM, Ceccantini GCT (2007) Flora da Serra do Cipó, Minas Gerais: Apodanthaceae (Rafflesiaceae s.l.), e notas sobre a anatomia de *Pilostyles*. Bol. Bol Bot Univ São Paulo 25:81–86

Humboldt AV, Bonpland A (1807) Essai sur la géographie des plantes. ccompagné d'um tableau physique des régions equinoxiales. Editorial Cultura, Mexico City

Inouye DW (2008) Effects of climate change on phenology, frost damage, and floral abundance of montane wildflowers. Ecology 89:353–362

Kinoshita LS, Simões AO (2005) Flora da Serra do Cipó, Minas Gerais: Apocynoideae s.str. (Rauvolfioideae e Apocynoideae). Bol Bot Univ São Paulo 23:235–256

Köppen W (1948) Climatologia. Fundo de Cultura Economica, Mexico

Le Stradic S (2012) Composition, phenology and restoration of campo rupestre mountain grasslands—Brazil. PhD thesis, Universidade Federal de Minas Gerais and Université d'Avignon et des Pays de Vaucluse, Belo Horizonte

Le Stradic S, Buisson E, Fernandes GW (2015) Vegetation composition and structure of some Neoptropical grasslands in Brazil. J Mt Sci 12:864–877

Lieberman D, Lieberman M, Peralta R, Hartshorn GS (1996) Tropical forest structure and composition on a large-scale altitudinal gradient in Costa Rica. J Ecol 84:137–152

Lobo JA, Quesada M, Stoner KE, Fuchs EJ, Herreiras-Diego Y, Rojas J, Saborio G (2003) Factor affecting phenological patterns of bombacaceous trees in seasonal forests in Costa Rica and Mexico. Am J Bot 90:1054–1063

Madeira JA, Fernandes GW (1999) Reproductive phenology of sympatric taxa of Chamaecrista (Leguminosae) in Serra do Cipo, Brazil. J Trop Ecol 15:463–479

Menzel A, Sparks TH, Estrella N, Koch E, Aasa A, Ahas R, Alm-Kuebler K, Bissolli P, Og Braslavska, Briede A, Chmielewski FM, Crepinsek Z, Curnel Y, Dahl A, Defila C, Donnelly A, Filella Y, Jatcza K, Mage F, Mestre A, Nordli O, Penuelas J, Pirinen P, Remisova V, Scheifinger H, Striz M, Susnik A, Van Vliet AJH, Wielgolaski F-E, Zach S, Zust A (2006) European phenological response to climate change matches the warming pattern. Glb Chan Biol 12:1969–1976

Morellato LPC (1995) As estações do ano na floresta. In: Morellato PC, Leitão-Filho HF (eds) Ecologia e preservação de uma floresta tropical urbana. Editora Unicamp, Campinas, pp 37–41

Morellato LPC, Leitão-Filho HF (1992) Padrões de frutificação e dispersão na Serra do Japi. In: Morellato LPC (ed) História natural da Serra do Japi: ecologia e preservação de uma área florestal no sudeste do Brasil. Editora da Unicamp - FAPESP, Campinas, pp 112–140

Morellato LPC, Talora DC, Takahasi A, Bencke CC, Romera EC, Zipparro VB (2000) Phenology of Atlantic rain forest trees: a comparative study. Biotropica 32:811–823

Morellato LPC, Gressler E, Camargo MGG (2013) A review of plant phenology in South and Central America. In: Schwartz MD (ed) Phenology: an integrative environmental science, 2nd edn. Springer, The Neederlands, pp 91–113

Morellato LPC, Alberton B, Alvarado ST, Borges BD, Buisson E, Camargo MGG, Cancian LF, Carstensen DW, Escobar D FE, Leite PTP, Mendoza I, Rocha NMWB, Silva TSF, Soares NC, Staggemeier VG, Streher AS, Vargas BC, Peres CA (2016) Linking plant phenology to conservation biology. Biol Cons 195:60–72

Morisette JT, Richardson AD, Knapp AK, Fisher JI, Graham EA, Abatzoglou J, Wilson BE, Breshears DD, Henebry GM, Hanes JM, Liang L (2009) Tracking the rhythm of the seasons in the face of global change: phenological research in the 21st century. Front Ecol Environ 7:253–260

Neves SPS (2009) Fenologia, biologia floral e polinização de espécies de Velloziaceae Endl. em área de campo rupestre na Chapada Diamantina, Bahia, Brasil. MSc Dissertation, Universidade Estadual de Feria de Santana, Feira de Santana

Nunes YRF, Landau EC, Veloso MDM (2008) Diversidade de Melastomataceae em diferentes altitudes de campos rupestres na Serra do Cipó, MG. Unim Cien 10:34–45

Oliveira PEAM, Moreira AG (1992) Anemocoria em espécies de cerrado e mata de galeria de Brasília, DF. Rev Bras Bot 15:163–174

Pereira AR, Angelocci LR, Sentelhas PC (2002) Agrometeorologia: fundamentos e aplicações práticas agropecuária. Editora Agopecuária, Guaíba, RS, 478 pp

Rosenzweig C, Karoly D, Vicarelli M, Neofotis P, Wu Q, Casassa G, Menzel A, Root TL, Estrella N, Seguin B, Tryjanowski P, Liu C, Rawlins S, Imeson A (2008) Attributing physical and biological impacts to anthropogenic climate change. Nature 453:353–357

Schwartz MD (2013) Phenology: an integrative environmental science, 2nd edn. Springer, The Neederlands

Silveira FAO, Ribeiro RC, Oliveira DMT, Fernandes GW, Lemos-Filho JP (2011) Evolution of physiological dormancy multiple times in Melastomataceae from Neotropical montane vegetation. Seed Sci Res 22:37–44

Silveira FO, Santos J, Franceschinelli EV, Morellato LPC, Fernandes GW (2015) Costs and benefits of reproducing under unfavorable conditions: an integrated view of ecological and physiological constraints in a cerrado shrub. Plant Ecol 216:963–974

Silveira F, Negreiros D, Barbosa N, Buisson E, Carmo F, Carstensen D, Conceição A, Cornelissen T, Echternacht L, Fernandes GW, Garcia Q, Guerra T, Jacobi C, Lemos-Filho JP, Le Stradic S, Morellato LPC, Neves F, Oliveira R, Schaefer C, Viana P, Lambers H (2016)

Ecology and evolution of the endangered campo rupestre: a neglected biodiversity conservation priority. Plant Soil. doi:10.1007/s11104-015-2637-8
Staggemeier VG, Diniz JAF, Morellato LPC (2010) The shared influence of phylogeny and ecology on the reproductive patterns of Myrteae (Myrtaceae). J Ecol 98:1409–1421
Staggemeier VG, Diniz-Filho JAF, Zipparro VB, Gressler E, de Castro ER, Mazine F, da Costa IR, Lucas E, Morellato LPC (2015) Clade-specific responses regulate phenological patterns in Neotropical Myrtaceae. Persp Plant Ecol Evol Syst 17:476–490
Veldman JW, Overbeck GE, Negreiros D, Mahy G, Le Stradic S, Fernandes GW, Durigan G, Buisson E, Putz FE, Bond WJ (2015a) Tyranny of trees in grassy biomes. Science 347:484–485
Veldman JW, Buisson E, Durigan G, Fernandes GW, Le Stradic S, Mahy G, Negreiros D, Overbeck GE, Veldman RG, Zaloumis NP, Putz FE, Bond WJ (2015b) Toward an old-growth concept for grasslands, savannas, and woodlands. Front Ecol Environ 13:154–162
Venn S, Pickering C, Green K (2012) Short-term variation in species richness across an altitudinal gradient of alpine summits. Biodiv Cons 21:3157–3186
Wolkovich EM, Cook BI, Davies TJ (2014) Progress towards an interdisciplinary science of plant phenology: building predictions across space, time and species diversity. New Phytol 201:1156–1162

Chapter 13
Mutualistic Interactions Among Free-Living Species in Rupestrian Grasslands

Tadeu J. Guerra, Daniel W. Carstensen, Leonor Patricia Cerdeira Morellato, Fernando A.O. Silveira and Fernanda V. Costa

Abstract Mutualisms such as animal pollination and seed dispersal, and protection of plants and insects by ants are ubiquitous in terrestrial ecosystems. Currently, mutualistic interactions among plants and animals are recognized for their paramount importance in biodiversity maintenance, especially in tropical ecosystems. In this chapter, we review the literature and present unpublished data on the ecology of mutualistic interactions among free-living species in Brazilian megadiverse montane Rupestrian Grasslands, the *Campo Rupestre*. We focus on interactions between plants and their pollinators and seed dispersers, and also interactions between ants, myrmecophilous plants and trophobiont insects. We provide basic information on mutualistic interactions, including data on natural history in addition to more advanced studies using network-based approaches. Only in the past decade mutualistic interactions have become intensively studied in *Campo Rupestre*, but information is still scattered and concentrated for a few localities. Pollination is an important process for reproduction of most plant species studied so far, with bees

T.J. Guerra (✉) · F.A.O. Silveira
Instituto de Ciências Biológicas, Departamento de Botânica, Universidade Federal de Minas Gerais, Avenida Antônio Carlos, 6627, Belo Horizonte, MG 31270-901, Brazil
e-mail: guerra.tj@gmail.com

F.A.O. Silveira
e-mail: faosilveira@icb.ufmg.br

D.W. Carstensen · L.P.C. Morellato
Departamento de Botânica, Universidade Estadual Paulista (UNESP), Avenida 24A, 1515, Rio Claro, SP 13506-900, Brazil
e-mail: daniel.carstensen@gmail.com

L.P.C. Morellato
e-mail: pmorella@rc.unesp.br

F.V. Costa
Instituto de Ciências Biológicas, Programa de Pós-Graduação em Ecologia, Conservação e Manejo da Vida Silvestre, Universidade Federal de Minas Gerais, Avenida Antônio Carlos, 6627, Belo Horizonte, MG 31270-901, Brazil
e-mail: fecostabio@gmail.com

© Springer International Publishing Switzerland 2016

G.W. Fernandes (ed.), *Ecology and Conservation of Mountaintop Grasslands in Brazil*, DOI 10.1007/978-3-319-29808-5_13

and hummingbirds acting as the main pollinators. Seemingly, animal pollination is more important than seed dispersal for gene flow across plant populations. Most vertebrate seed dispersal is carried out by non-specialized avian frugivorous, but also by short-distance seed dispersers such as ants and lizards. Moreover, ants are diversified and abundant group that seem to play a major role in these ecosystems, acting as secondary seed dispersers, as nectar thieves, and as bodyguards of plants and insects. We conclude this chapter by pointing out some gaps in our knowledge and proposing avenues for future research.

Contents

13.1 Introduction

Throughout the history of life, natural selection has repeatedly favored the evolution of traits that allow individuals to exploit the whole genomes of other species to survive and reproduce (Thompson 2010). In fact, coevolved interactions comprise the foundations of all species-rich ecosystems of the Earth and without such coevolved partnerships, highly diverse ecosystems would simply collapse (Thompson 2005). Organisms from all kingdoms are involved in mutually beneficial interactions among species, or mutualisms, which may take different forms in nature and vary from obligate to facultative (Bronstein 2009). The most well-known mutualistic interactions involving free-living species in terrestrial ecosystems are plant pollination and seed dispersal by animals and plant and insect protection, mostly by ants (Herrera and Pellmyr 2002). However, obligate and intimate relationships such as symbiotic mutualisms that include vertebrate and invertebrate gut symbionts, lichens, rhizobia and mycorrhizae are also important in ecosystems (Douglas 1994).

Mutualisms typically involve conflict of interests among interacting species and rise from mutual exploitation between selfish individuals (Axelrod and Hamilton 1981). Indeed, mutualistic systems are usually exploited by cheater species (Bronstein 2001). Even supposed mutualisms can be conditional, or context dependent, and they are usually contingent on the balance between costs and benefits for those individuals that interact, with the outcomes varying from positive to negative according to biotic and abiotic conditions (Bronstein 1994; Bronstein et al. 2003). Furthermore, other kinds of mutualisms may be observed when we consider the complexity of natural communities where species are also linked to each other by indirect interactions (van Ommeren and Whitham 2002; Romero and Vascocelos-Neto 2004; Verdú and Valiente-Banuet 2008).

In his classic book on orchid pollination, Charles Darwin established a naturalistic approach that led to the field of mutualism ecology (Darwin 1862). This mostly adaptive agenda were focused on understanding how traits of animals and plants favored survival and reproduction of interacting species. After the publication of a seminal paper by Janzen (1966), the mutualistic relationships among animals and plants became more intensively studied and the first landmark publications regarding mutualisms appeared in the 80s (Beattie 1985; Boucher 1985; Estrada and Fleming 1986). The knowledge accumulated over the past 50 years paved the current view that mutualistic interactions among plants and animals play a central role in the generation and maintenance of biodiversity (Bascompte and Jordano 2007). Moreover, mutualistic interactions such as pollination and seed dispersal are now recognized as important services providing processes that are essential for sustaining natural and agricultural ecosystems (Valiente-Banuet et al. 2015).

Understanding the identity of interacting species, what adaptations are involved in these interactions and the outcomes for the interacting partners is still of paramount importance in basic and applied ecology. Nevertheless, new ideas and approaches have emerged in the past years. For instance, network theory has played a major role in the understanding of mutualistic interactions among free-living species (Bascompte and Jordano 2014). The use of network metrics and structural properties such as nestedness, modularity, centrality, and link asymmetry allows the exploration, and comparison of patterns in structure and organization of ecological interactions within and between communities (Bascompte and Joradno 2007; Olesen et al. 2007; Gonzalez et al. 2010). Jordano et al. (2003) identified invariant properties of mutualistic networks that reveals some major ecological patterns of community organization. First, there is always a core of generalists in communities, those species that interact among themselves and with a set of more specialized species that consistently interact with that generalist core. Second, interactions are generally weak in terms of reciprocal dependence, but the few of them that are stronger tend to be quite asymmetric, which means that one partner is more dependent upon the interaction than the other (Bascompte et al. 2006). Finally, most networks show distinct modules or compartments, meaning that distinct subsets of species interact more strongly among themselves than with species form other modules (Olesen et al. 2007).

The Rupestrian Grasslands or *Campo Rupestre* comprise megadiverse montane fire-prone vegetation that occur on Brazilian ancient landscape mosaics (Fig. 13.1a), which include rocky outcrops of quartzite, sandstone or ironstone along with sandy, stony and waterlogged grasslands. *Campo Rupestre* ecosystems predominate in those areas above 900 m and up to 2000 m a.s.l. in disjoint areas, mostly within Espinhaço Mountain Range in Brazilian States of Minas Gerais and Bahia (Giulietti et al. 1997; Alves et al. 2014) and are recognized as an important center of biological diversity and endemism of flora (Giulietti et al. 1997; Rapini

Fig. 13.1 **a** A typical landscape encompassing *Campo Rupestre* ecosystems in Lapinha da Serra, southern Espinhaço Range. **b** The endemic bee-pollinated orchid *Constantia cipoensis*. The ornithocorich fruits of (**c**) *Miconia irwinii* and (**d**) *Struthanthus flexicaulis*. **e** The flowers of the hummingbird-pollinated mistletoe *Psittacanthus robustus*. **f** The aphid *Aphis spiraecola* attended by the ant *Cephalotes pusillus*. **g** The trophobiont stinkbug *Eurystethus microlobatus* attended by the ant *Camponotus rufipes*. **h** *Bionia coriaceum* with ant bodyguards associated to its extra-floral nectaries. (Photos by T.J. Guerra)

et al. 2008; Echternacht et al. 2011) and fauna (Alves et al. 2008; Leite et al. 2008; Chaves et al. 2015). Despite the relevance of *Campo Rupestre* for biodiversity conservation in Brazil (Fernandes et al. 2014; Silveira et al. in press), currently, there is no published compilation of studies concerning mutualistic interactions for these unique ecosystems.

13.2 Scope

In this chapter, we review the literature and present unpublished information on ecology of mutualistic interactions among free-living species in the *Campo Rupestre* (CR hereafter) ecosystems. We focus on interactions among plants and their pollinators and seed dispersers, and on interactions between ants, myrmecophilous plants and trophobiont insects, i.e. those that produce sugar- and amino acid-rich honeydew as rewards for their bodyguards. Our aim is to provide basic information on mutualistic interactions in CR, presenting data from those more naturalistic studies as well more advanced studies using network-based approaches. We conclude this chapter by pointing out some major gaps in the study of mutualisms in CR and propose avenues for future research.

13.3 Linking Plants to Their Pollinators

Biotic pollination, i.e., animals harvesting resources from flowers and in return providing pollination services, is ubiquitous to all terrestrial ecosystems. On average in tropical communities 94 % of the plant species are estimated to be pollinated by animals (Ollerton et al. 2011). There are few studies of animal and abiotic pollination mechanisms for the CR flora, but existing information are congruent with a predominance of biotic pollination. Jacobi and Carmo (2011) showed that insect pollinated species, mostly those pollinated by bees, comprise nearly 70 % of the plant species in the community, whereas bird pollinated plants represent almost 13 % and wind pollination 16 % in ironstone outcrop site in southern Espinhaço Range. In addition, Conceição et al. (2007) found that animal pollinated species also outweigh abiotic mechanisms of pollinationin Chapada Diamantina, northern Espinhaço Range. However, the authors observed a marked seasonal pattern, with entomophily occurring mostly from summer through autumn, ornithophily predominating during winter, whereas anemophily prevailed in the spring.

Existing work on pollination is sparse and restricted mostly to studies on the reproductive biology of a single or few species (e.g., Sazima 1977; Jacobi et al. 2000; Guerra et al. 2014). However, those studies offer valuable insight into the natural history of CR vegetation, including some extraordinary cases of mutualisms. For instance, the epiphytic orchid *Constantia cipoensis* (Fig. 13.1b), which grows only on two species of *Vellozia* (*V. piresisana* and *V. compacta*), is exclusively pollinated by the carpenter bee *Xylocopa artifex*, which also build its nests inside the branches of *V. piresiana* and *V. compacta* (Matias et al. 1996). This endemic and threatened orchid is restricted to growing on *Vellozia* trunks and several other poorly known orchids are found in connection with *Vellozia*, such as, *Cattleya brevipedunculata*, which grows on *V. gigantea*, which itself has a narrow geographical distribution (Lousada et al. 2011). Such highly specific requirements for growth and reproduction constrain gene flow and dispersal of individuals across populations and likely contribute to the strong heterogeneity in species distributions across CR landscapes.

Pollination systems of endemic species vary from highly specialized (e.g. few pollinator species within an animal group) to more generalized (e.g. many species from distinct taxonomic groups) and include both invertebrates and vertebrates as pollinators. For example, some endemic orchids are highly specific regarding their fly pollinator species and this specialization is consistent among populations (Borba and Semir 2001; Melo et al. 2011), with fly-pollination as the main mechanism favoring maintenance of high genetic variability in the populations studied by Borba et al. (2001). In contrast, the endemic bromeliad *Encholirium glaziovii* is pollinated by the nectarivorous bat *Lochophylla bokermani* in Serra do Cipó, but the authors also recorded visits by two hummingbird species during the day and noctuid and sphingid moths during the night (Sazima et al. 1989). Other CR species, *Paliavana sericiflora* (Gesneriaceae), represent an interesting case of transition between bat and hummingbird pollination. This species has features related to attraction of both groups, but in the field hummingbirds were observed as the main pollinators (San Martin-Gajardo and Sazima 2005). Some endemic species are pollinated exclusively by bees (Matias et al. 1996; Jacobi et al. 2000; Franco and Gimenes 2011), while other species have generalized systems involving bees and hummingbirds as pollinators (Sazima and Sazima 1990; Jacobi and del Sarto 2007; Jacobi and Antonini 2008; Ferreira and Viana 2010; Hipólito et al. 2013).

Interactions among hummingbirds and flowers are fairly common in CR and occur in plants from several families, including many endemic genera (Vasconcelos and Lombardi 2001; Machado et al. 2007; Santana and Machado 2010; Rodrigues and Rodrigues 2014). Nevertheless, the studies mentioned above raised two remarkable points related to interactions among plants and hummingbirds in CR. First, hummingbird species are more commonly reported as visitors of non-ornithophilous than ornithophilous species, e.g. plants with red, orange or yellow flowers that are tubular shaped and produce nectar as reward (Vasconcelos and Lombardi 2001; Machado et al. 2007; Santana and Machado 2010; Rodrigues and Rodrigues 2014). For example, the Espinhaço Range endemic hummingbird *Augastes scutatus* was commonly observed visiting species in the primarily insect pollinated family Asteraceae (Vasconcelos and Lombardi 2001; Rodrigues and Rodrigues 2014). Although only a subset of CR plants are considered ornithophilous, hummingbirds could play a major role in reproduction of some species, as thieves or pollinators. Second, nectar rich species in CR along the Espinhaço Region seem to concentrate some highly territorial hummingbird species that defend flower resources against other visitors (Sazima 1977; Sazima and Sazima 1990;

Vasconcelos and Lombardi 2001; Machado et al. 2007; Jacobi and Antonini 2008; Guerra et al. 2014). This common behavioral pattern of hummingbirds could result in restriction of gene flow to plants within some rock outcrop patches and, thus, reducing the variability in some hummingbird pollinated species. In fact, the bee pollinated *Vellozia epidendroides* has higher genetic variability than hummingbird pollinated *V. leptopetala* (Franceschinelli et al. 2006), suggesting that bees could be more effective in long distance dispersal of pollen than territorial hummingbirds.

The open, high altitude vegetation that characterizes the CR could be connected to an expectation of higher proportion of wind-pollinated plants (Culley et al. 2002). However, the spatial heterogeneity of the landscape, a mosaic of rocky outcrops embedded in a matrix of sandy or rocky soils, is another typical characteristic of CR. This causes patchy distribution patterns in most plant species, and has likely selected for animal pollination as an essential means to obtain gene flow between populations that are spatially constrained by environmental heterogeneity, reflected in substrate, hydrology, and local climate (see Giulietti et al. 1997; Silveira et al. in press). Even so, species confined to rocky outcrops seem to show lower genetic diversity than species associated with sandy soils (Franceschinelli et al. 2006; Lousada et al. 2011). The patchy distribution of populations translates into a high turnover of plants and pollinator species across space; this spatial variation being larger for plants than pollinators (Carstensen et al. 2014). Plant phenology patterns across the year are complex and likely affected by both regional and local climatic factors as well as occasional fires (Chap. 12). Pollinators, in turn, likely track this variation in flower resources in space and time, resulting in complex dynamics of local plant-pollinator interactions.

Plant-pollinator interactions within communities are analyzed and visualized using networks, in which plant and pollinator species (the network nodes) are connected (or linked) to each other if they are observed interacting (Jordano et al. 2003). While network-scale properties, such as nestedness and modularity, seem to be conserved, it has also been shown that detailed network structures, such as species and interaction composition (Dupont and Olesen 2009; Carstensen et al. 2014) and functional roles (Gonzalez et al. 2010), can change dramatically in both time and space (Trøjelsgaard et al. 2015). In Serra do Cipó rocky outcrops, communities seem to abide by the generalities of plant-pollinator networks found elsewhere (Fig. 13.2). The only existing community-scale dataset on plant-pollinator interactions from CR consists of seven sampled networks across rocky outcrop vegetation sites within CR landscape in Serra do Cipó (Carstensen et al. 2014). While network properties seem to be relatively constant across space (Carstensen et al. in prep) species and interaction turnover are large even across small distances (Carstensen et al. 2014). All sites combined showed a rich community with 101 flowering species visited by 199 pollinator species. In this dataset gathered in the beginning of the rainy season (from October to December), bees were dominant in terms of species richness and overall number of interactions, comprising 34 % of all species and 60 % of all interactions observed. Flies, butterflies, wasps, and beetles all had a strong presence in the data set, each group accounting for 13–18 % of the species observed. Hummingbirds and

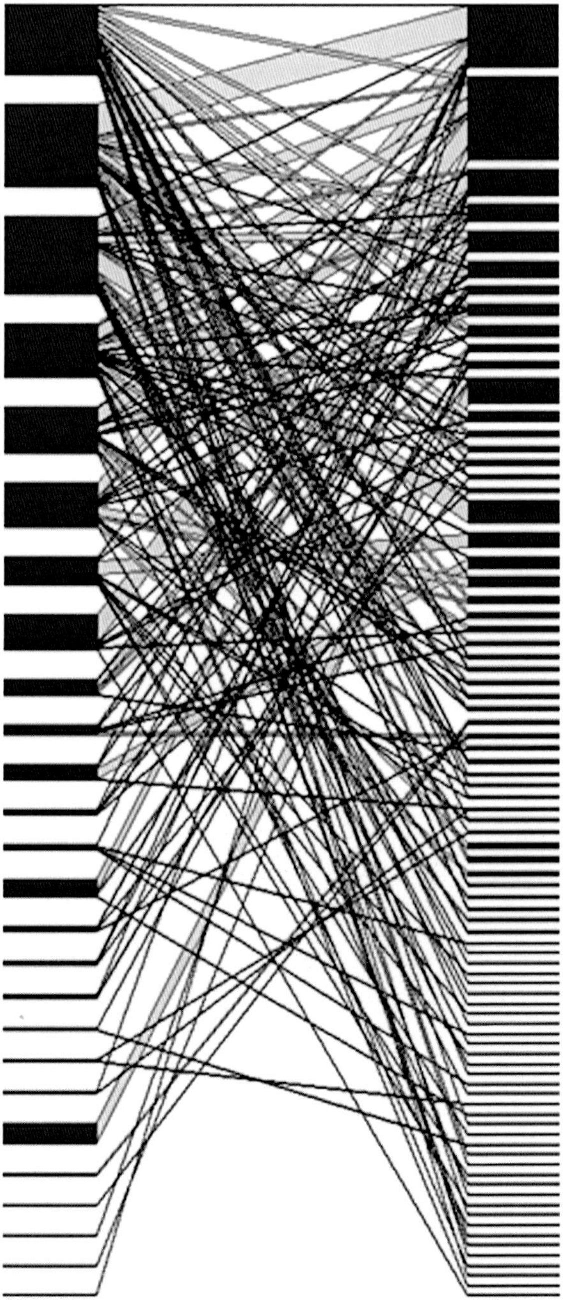

Fig. 13.2 The local plant-pollinator network sampled in the *Campo Rupestre* of Serra do Cipó, southeastern Brazil. *Black boxes* on the *left* represent plant species; *boxes* on the *right* represent pollinator species. The size of the boxes represents the total degree of the species in the network, that is, the number of interactions observed for a given species. Species are connected with lines if they were observed interacting, the thickness of the line representing the frequency of interaction. Notice the nested organization of interactions, where species with few interaction partners tend to interact with species with many interaction partners

ants were less diverse, only each comprising 6 % of the total species richness of the flower visitors. Only one plant species was found in all sites, *Cuphea ericoides* (Lythraceae), while others such as *Paepalanthus bromeliodes* (Eriocaulaceae) and *Aspilia jolyana* (Asteraceae) were widely common as well. On the other hand, several pollinators were found in all sites, e.g. *Apis mellifera*, *Bombus pauloensis* as well as six of other bee species, and hummingbirds such as *Chlorostilbon lucidus* and *Heliactin bilophus* were both observed in five out the seven sites.

In that unique dataset from Carstensen et al. (2014), flower visitors were regarded as pollinators if they touch the reproductive organs of the flowers, but not all flower visitors "follow the rules". Some can be classified as illegitimate visitors, that is, they visit the flower and harvest floral resources but do not return any immediate service in the form of pollination. This can be a result of either morphological mismatches or because they circumvent the floral system by puncturing the corolla and accessing nectar directly, without getting into contact with the reproductive organs of the plant. This puncturing, or an otherwise destructive behavior to access floral nectar, is termed nectar robbing (Inouye 1980) and can be observed in several types of flower visitors, but especially bees and hummingbirds. However, the role of flower visitors as pollinators and robbers or thieves in CR remains poorly known.

13.4 Frugivory and Seed Dispersal

To understand the ecology of seed dispersal in CR, first we need to understand the natural history and ecology of their fruits and frugivores. According to Silveira et al. (2016), the CR can be classified as a geologically old, climatically stable, infertile landscapes of the Earth (Hopper 2009). In those landscapes, reduced dispersal ability should be favored by natural selection, because recruitment is strongly controlled by soil conditions and the safe sites for establishment are usually near the parent-plants (Hopper 2009). As expected, most species from the CR have no conspicuous mechanism for seed dispersal. Unassisted dispersal is the prevalent dispersal mode among dominant (e.g. Poaceae, Cyperaceae, and Fabaceae) and herb families with many endemic species (e.g. Xyridaceae, Eriocaulaceae, and Velloziaceae). In fact, Conceição et al. (2007) observed that few species in CR site in Chapada Diamantina produce fleshy fruits, with those individuals representing 21 % of vegetation cover.

The production of fleshy fruits by CR plants seems to be restricted to a few Neotropical clades that are typically vertebrate-dispersed (e.g. Melastomataceae, Myrtaceae and Cactaceae), most of them which occur in woody vegetation, such as gallery forests and natural forest patches (locally known as *capões* de mata, see Chap. 7). However, plant species from at least 16 families that produce fleshy fruits were reported to occur in CR sites, mostly bird dispersed berries (Faustino and Machado 2006; Conceição et al. 2007; Silveira et al. 2012; Guerra and Pizo 2014).

In fact, even some endemic species (e.g. *Coccoloba cereifera*) typical from CR rocky outcrops produce fleshy fruits. Most of those fruits are small, and have dark-crimson or red color, but yellow, rose, brown, orange, and purple fruits are not uncommon (Faustino and Machado 2006; Guerra and Pizo 2014).

Birds and mammals are the predominant vertebrate seed dispersers worldwide (Fleming and Kress 2011). In CR, primates are virtually absent in the open vegetation areas, so seed dispersal by mammals is probably mostly carried out by bats, yet there is no study to support this idea. Although there are records of carnivorous species such as the maned wolf (*Chrysocyon brachyurus*) and the crab-eating fox (*Dusicyon thous*) serving as seed dispersers, the role of large mammals in seed dispersal is also unexplored. The scarce available evidence suggests that in CR birds are the main seed dispersers among vertebrates. However, the avifauna present in these ecosystems is characterized by the dominance of generalist species and the lack of specialized avian frugivores (Faustino and Machado 2006; Guerra and Pizo 2014). Seed dispersal by lizards is considered a predominantly oceanic island phenomenon (Olesen and Valido 2004). However, it has been recently observed in a CR site, where two cactus species are dispersed by two lizard species, but also by ants (Fonseca et al. 2012).

A useful starting point to unveil the complexity of seed dispersal in CR is the seminal paper by McKey (1975). He defined two opposing seed dispersal strategies that, rather than representing mutually exclusive categories, are best viewed as extremes of a continuum. The specialist plant syndrome comprises species that produce few, lipid rich fruits that contain a few (or a single) large seeds. The generalist syndrome, in turn, is observed in species with massive production of small-sized fruits with high sugar and low nutrient content, which often have several tiny seeds. The former are dispersed by effective, reliable and specialized frugivores, while the latter are dispersed by a wider range of non-specialized frugivorous species (Howe 1993). Studies focused on the Miconieae (Melastomataceae) species, plants which are the archetype of the generalist syndrome, have been the primary study model of seed dispersal in CR.

Studies on frugivory and seed dispersal in CR have focused on Miconieae not only because of their dominance and diversity, but also because of their key role in providing resources to several disperser groups (Silveira et al. 2013a). This combination offers the unique opportunity to compare seed dispersal effectiveness (sensu Schupp et al. 2010) among different taxonomic groups of frugivores. Miconieae is a species-rich Neotropical tribe of Melastomataceae that comprises nearly 2200 species (Goldenberg et al. 2008), which are commonly found in most Neotropical vegetation (Silveira et al. 2013a). Fruit morphology in Miconieae is extraordinarily diverse, with species showing marked differences in crop size, fruit size, color, water and sugar content, seed number, pulp/seed ratio, and contrasting displays. All those traits have been shown to directly influence removal rates (Blendinger et al. 2008; Christianini and Oliveira 2009; Camargo et al. 2013).

The primary seed consumers of those berries are a taxonomically diverse community of small-to-medium-sized birds (Silveira et al. 2013a) that include thrushes, tanagers, and *Elaenia*. The number of birds foraging on fruits of each

Miconia species ranges from two to ten (Guerra and Pizo 2014), which suggests a reduced number of seed dispersers for CR plants when compared to those of lowland tropical forests (Ellison et al. 1993). For instance, in tropical forest up to 36 bird species may feed on *Miconia* fruits (Galletti and Stotz 1996). Data on frugivory of two *Miconia* species in ironstone rocky outcrops in southern Espinhaço Range, indicate a reduced number of seed dispersers. The thrusher *Mimus saturninus* and the tanager *Schystoclamys ruficapillus* as the most effective dispersers of *M. ligustroides*, tough, most interestingly, the rufous-collared sparrow *Zonotrichia capensis*is an effective seed disperser of *M. pepericarpa* (Silveira, F.A.O. unpublished data). This result suggests that even typically granivorous birds may be important to the dispersal of small-sized seeds in CR vegetation, as suggested by Faustino and Machado (2006).

The probability that a dispersed seed will grow into a new reproductive individual depends on the site of seed deposition and on the quality of handling in the mouth and gut of the frugivore (Schupp et al. 2010). Therefore, we should disentangle the effects of seed cleaning (pulp removal), seed coat scarification, and fertilization effects (Samuels and Levey 2005). Seeds within berries of Miconieae cannot germinate because of the inhibition effect, i.e., the presence of germination inhibitors in the fleshy placentas and the dark pigmentation that prevents light from reaching the positively photoblastic seeds (Silveira et al. 2013b). It has been experimentally demonstrated that gut-passed seeds have similar germination percentages and germination time than hand-extracted seeds. Those results led to the conclusion that germination enhancement deriving from gut scarification does not play a role in recruitment of Miconieae. However, seed cleaning is an important service delivered by birds, and varies according plant and bird species involved in these mutualistic interactions (Silveira et al. 2012).

The large crop produced by Miconieae plants is usually not consumed by their primary seed dispersers. As a consequence, the fruits that fall to the ground become available for consumption by secondary seed dispersers (Christianini and Oliveira 2009). Myrmecochory (dispersal by ants) is a common phenomenon in sclerophyllous, fire-prone, nutrient-poor shrublands such as CR (Milewski and Bond 1982). Primary myrmecochory is rare in the Neotropics, with no reports for CR. However, a study on secondary seed dispersal by ants of *Miconia irwinii* (Fig. 13.1c), a primarily bird-dispersed commonly found in CR areas, suggests that ants in several genera could play an important role in seed fate (Lima et al. 2013). The ants transport fallen fruits to their nests, clean the seeds, and discard them outside the refuse piles, thereby rescuing many seeds that were not primarily dispersed by birds (Lima et al. 2013).

In CR, fallen fruits of Miconieae can also be exploited by marsupials, rodents (Lessa et al. 2013), and lizards. Marsupials could also provide important benefits to plants, since many of the small seeds can pass unharmed through their guts (Lessa et al. 2013). However, more work is necessary to evaluate the role of small mammals in Miconieae seed dispersal.

13.5 Mutualistic Interactions Between Birds and Mistletoes

Mistletoes comprise a polyphyletic group of parasitic plants that infect the aerial parts of their hosts, mostly trees and shrubs (Nickrent et al. 2010). Those plants are mostly hemiparasites, as they attach themselves to the host's shoots via haustorial roots, in order to obtain water and minerals, but are able to photosynthesize their own carbohydrates (Ehleringer et al. 1985). Mistletoes are important elements of natural communities, because they provide structural and nutritional resources within the canopy (Watson 2001), and their positive influence on diversity led to their recognition as keystone resources for animals (Watson and Herring 2013). Loranthaceae is the largest family among parasitic angiosperms with nearly 900 species distributed in approximately 70 genera, which occur in most terrestrial ecosystems of the world, including tropical and temperate forests, mangroves, savannas, semi-arid vegetation, and deserts (Calder and Bernhardt 1983). Most species produce nectar as a reward to pollinating birds (Kuijt 1969). The effectiveness of mistletoe pollinators, the quality of the floral reward, and the consequences of their interactions for plant reproduction are still poorly known (Aizen 2003). For example, plant breeding system, nectar availability and secretion dynamics are key traits to understand the behavior of pollinators, and may help us to determine costs and benefits of such interactions (Gill and Wolf 1975).

The mistletoe *Psittacanthus robustus* (Loranthaceae) has tubular and colorful flowers (Fig. 13.1e), which produce copious dilute nectar as the main reward for pollinators throughout the rainy season in the CR areas of Serra do Cipó (Guerra et al. 2014). Indeed, nectar of *P. robustus* represented an exceptional resource for birds. The local guild of nectarivorous birds that feed on *P. robustus* include eight hummingbirds and two passerines, which represents the largest guild reported for an ornithophilous plant species in the CR sites.

Mistletoe flowers secrete most of their nectar right after flower opening, with little sugar replenishment after experimental removal. The experiments with exclusion of flower visitors revealed that flowers quickly reabsorbed the nectar, which is evidence of the high costs of nectar secretion (Guerra et al. 2014). Surprisingly, flowers naturally exposed to pollinators produced less seeds than hand-, self-, and cross-pollinated flowers, which suggests some degree of pollination limitation. However, hummingbird-pollinated flowers still produce significantly more seeds than flowers excluded from pollinators. Because *P. robustus* cannot produce seeds spontaneously without pollinators, hummingbirds seem to be crucial for seed production. In effect, the low values of nectar recorded in open flowers are consistent the high hummingbird visitation rates reported. Thus, reproductive limitation in this mistletoe does not seem to be related to pollinator scarcity. Indeed, the high frequency of visitation by hummingbird pollinators seem to be costly, because they leave almost no nectar for reabsorption and the subsequent use of sugars to produce seeds in the plants. The pollination system involving the mistletoe *P. robustus* and hummingbirds in rocky outcrops illustrates that the costs and benefits of pollination mutualisms are influenced by the strength of the interaction among flowers and their pollinators (Guerra et al. 2014).

Mistletoes are known for their particular mutualistic associations with narrow bird guilds that disperse their seeds worldwide (Reid 1991). Birds obtain nutrients from mistletoe fruits, which in turn receive the service of directional seed dispersal onto the host twigs, which are safe sites for establishment (Roxburgh and Nicolson 2005). Seeds require pericarp removal by birds to germinate, as well as deposition on branches with specific diameters on live compatible host species (Sargent 1995). Thus, interaction with seed dispersers is obligate for mistletoes. Nevertheless, mutualism strength among mistletoes and their avian seed dispersers is still poorly known (Watson and Rawsthorne 2013). In this context, Guerra and Pizo (2014) evaluated simultaneously the effectiveness of the avian seed dispersers of the mistletoe *Struthanthus flexicaulis* (Loranthaceae) and the contribution of its fruits to their diets in a CR site at Serra do Cipó.

Fruits are yellow, small, lipid-rich pseudo-berries (Fig. 13.1d) produced asynchronously among individual plants and available throughout the year. Although four bird species were observed feeding on mistletoe fruits, the plain-crested elaenia *Elaenia cristata* (Tyrannidae) was the most effective disperser, responsible for more than 96 % of the seeds dispersed. This bird swallowed fruits whole, expelling and depositing undamaged seeds by regurgitation and bill wiping on perches. Seeds can be dispersed up to 100 m, but most were dispersed within 30 m from the parent-plants, because seed retention time in the disperser's gut is short ($\sim$6 min on average). Fifty-six percent of the dispersed seed land on safe sites: the thin live twigs of 38 potential host species. The avian seed dispersers were predominantly frugivorous, and feed on typically ornithocoric fruits of at least 12 species, but also on arthropods on the foliage. In fact, fruits represented 75 % of the plain-crested elaenia′s diet throughout the year, but the fruits of *S. flexicaulis* represented nearly 34 % of its diet. Those results, published by Guerra and Pizo (2014) corroborated predictions from network studies (Bascompte et al. 2006) highlighting the asymmetrical nature of seed dispersal mutualisms, with the mistletoe life cycle locally linked to one highly effective seed disperser that is weakly dependent on mistletoe fruits.

13.6 Linking Ant Bodyguards to Their Plant and Animal Partners

Ant-plant mutualisms have played a major role in shaping our broad understanding of mutualism (Bronstein et al. 2006). In the tropics, several ant species use plants as a foraging substrate to search for prey and food resources, such as extrafloral nectar and honeydew from trophobiont insects (Blüthgen et al. 2000; Rico-Gray and Oliveira 2007; Rosumek et al. 2009). Those associations between ants and plants can result in obligate or facultative mutualisms, whose mutual benefits will depend on the nature of the interacting species (Bronstein et al. 2006; Rico-Gray and

Oliveira 2007). Moreover, the outcomes of interactions between ants and plants bearing extrafloral nectaries (EFNs hereafter) or trophobiont insects are controversial, and most of them are facultative and dependent on biotic and abiotic conditions (Bronstein 1994).

Two main factors contribute to the high ant abundances found in tropical vegetation. First, the high availability of nesting sites, which are provided by galleries in trunks built by bark beetles and by unoccupied galls after the emergence of insects (Oliveira and Freitas 2004), and second, the high food availability provided by EFNs and insect trophobionts (Blüthgen et al. 2000, 2004). Studies carried out in the Cerrado (Brazilian savanna) showed that plants with EFNs (of at least 15 families) may represent up to 25 % of the species and individual trees in some vegetation (Oliveira and Leitão-Filho 1987; Oliveira and Freitas 2004) and around 30 % of the plants with insect trophobionts (Lopes 1995). Such conditions have made it possible to gather a large amount of information on multitrophic interactions involving ants in this biome (see Del-Claro et al. 2006; Rico-Gray and Oliveira 2007), although some vegetation within the Cerrado domain, such as CR, remain poorly studied. The first survey of extra-floral nectarines is the study from Morellato and Oliveira (1991) for iron-stone outcrops on North Brazil (Serra dos Carajás, Pará). They report 53 % of species (7 out of 13) bearing extrafloral nectaries. However the role of these plant resources on ant community organization still is in the beginning for CR.

The current knowledge on the ant-trophobiont mutualisms in CR is limited to a few studies (Fagundes et al. 2012; Guerra et al. 2011). The sap-feeding stinkbug *Eurystethus microlobatus* (Heteroptera: Pentatomidae) is exclusively associated with woody mistletoes in Serra do Cipó. These stinkbugs have cryptic coloration and maternal care, are gregarious, and produce allomone volatiles, but their most conspicuous defense strategy is ant protection (Guerra et al. 2011). In their study, stinkbugs produced honeydew and four ant species attended to their aggregations. However, *Camponotus rufipes* was the most frequent and aggressive ant species, and the only species that protected stinkbugs by night and day (Fig. 13.1 g). Other ants such *Cephalotes pusillus* and *Camponotus crassus* were less aggressive and mostly diurnal, whereas *Camponotus melanoticus* were was mostly nocturnal. The authors observed that aggregation size positively affect the number of ant attending stinkbugs, but per capita attendance were actually reduced in those very large aggregations. This remarkable interaction represents the only known case of trophobiosis between ants and stinkbugs, and it is the first report of such an association between heteropterans and ants in the American continent (Guerra et al. 2011).

Fagundes et al. (2012) found the trophobiont treehopper *Calloconophora pugionata* (Hemiptera: Membracidae) feeding on *Myrcia obovata* (Myrtaceae) in CR site. They observed that an increase in the number of trophobionts positively affected ant abundance, but negatively affected ant richness. They observed 10 ant species interacting with trophobionts, but *C. rufipes* and *C. crassus* were the most frequent and dominant, and had a negative influence on the occurrence of other species. In another study (Fagundes et al. 2013), the authors evaluated experimentally the role of ant protection against predators and showed that ants increased the survivorship and reproductive success of the trophobiont *C. pugionata*. Such case studies are highly important, as they describe a new and uncommon ant-hemipteran association and reveal a new kind of mutualism, although the benefits for ants are seldom evaluated (but see an example from the Cerrado, Byk and Del-Claro 2011).

Ants are assumed to be nectar robbers and have limited potential as pollinators (Hölldobler and Wilson 1990; Peakall et al. 1991). However, floral nectar is an essential food for many ant species (Davidson et al. 2003; Blüthgen et al. 2004), and studies on the relationship between ant activities and consumption of floral nectar in CR are scarce. Romero (2002) using termite baits to simulate herbivores on *Vochysia elliptica* (Vochysiaceae) found higher rates of removal by ants in leaves and inflorescences compared to plants without ant access, suggesting that even ant nectar thieves could protect those shrubs from herbivores. Flower-visiting ants may have positive or negative effects on the reproductive success of plants, depending on the context in which the interactions take place (Willmer et al. 2009).

Dáttilo et al. (2014) recorded 20 ant species associated with two plants with EFNs that occur in CR: *Chamaecrista mucronata* (Fabaceae) and *Stachytarpheta glabra* (Verbenaceae). The authors utilized a network approach to show how different ant species are linked to individuals of these two plant species and observed that the core of highly interacting species changed significantly between day and night. These results are important as they demonstrate daily partition of resources by ants, and highlight the need of nocturnal observations to understand the ecological dynamics of ant-plant interactions.

During ongoing studies on ant-plant interactions in CR sites in Serra do Cipó, Costa et al. (in prep.) recorded a multi-relation network formed by 5 types of interaction (subnetworks), 992 interaction events, 79 plant species, and 29 ant species. Seventy percent of those interaction events are neutral visits (ants foraging on the plant), whereas 15 % of interactions involve ants and EFNs, 7 % of interactions are between ants and nectar or pollen, 6 % of interactions are among ants and trophobiont insects, and 1 % of interactions is with fruits(frugivory or visit to extrafloral nectaries on fruits). This myriad of interaction types indicates that ants foraging on vegetation is a common phenomena and that ants could play a major role in interaction networks in CR.

A recent study carried out by Costa et al. (in prep.) in CR of Serra do Cipó indicates a total of 20 % of plants species with secretory structures considered extrafloral nectaries (16 out of 79-species), a result consistent with surveys from Cerrado, which indicate a total of 25 % of plants with have EFNs (Oliveira and Leitão-Filho 1987; Oliveira and Freitas 2004). Six plant families stand out due to the predominance of EFNs at CR: Fabaceae (5 species—e.g., *Bionia coriaceum* Fig. 13.1h), Malpighiaceae (4 species—e.g., *Peixotoa tomentosa* and *Banisteriopsis angustifolia*), Euphorbiaceae (3 species—e.g., *Sebastiania* sp), Asteraceae (2 species—*Symphyopapus reticulatus* and *Baccharis concinna*), Araceae (1 species—*Philodendron cipoense*), and Myrsinaceae (1 species—*Myrsine* sp.). The most common plant families utilized by ants, including all types of resources and neutral interactions were Asteraceae, Velloziaceae, Malpighiaceae, Fabaceae and Myrsinaceae.

At present, most studies focused in the organization of ant trophobiont interactions at the community level in tropical forests (Blüthgen et al. 2000), with no data available for CR sites. In areas of CR, 26 % of the plant species have established aggregations of hemipteran trophobionts (Costa et al. in prep), most of them members of the families Aphididae (e.g., *Aphis spiraecola* Fig. 13.1f) and Coccidae (e.g., *Parasaissetia nigra*). Considering together the ant fauna associated with EFNs and honeydew, a total of five subfamilies in which the most frequent ant species were *C. pusillus*, *C. crassus*, *C. rufipes* and two species of *Brachymyrmex* genus (unidentified). Together these five species were responsible for 73 % of the observed interactions during daily samplings, with *C. pusillus* and *C. crassus* being the most frequent species (29 and 13 % respectively) (Costa et al. in prep).

Structural organization of ant-EFNs interactions in CR (Fig. 13.3), suggest that network topology is not modular with low level of specialization. In the contrary,

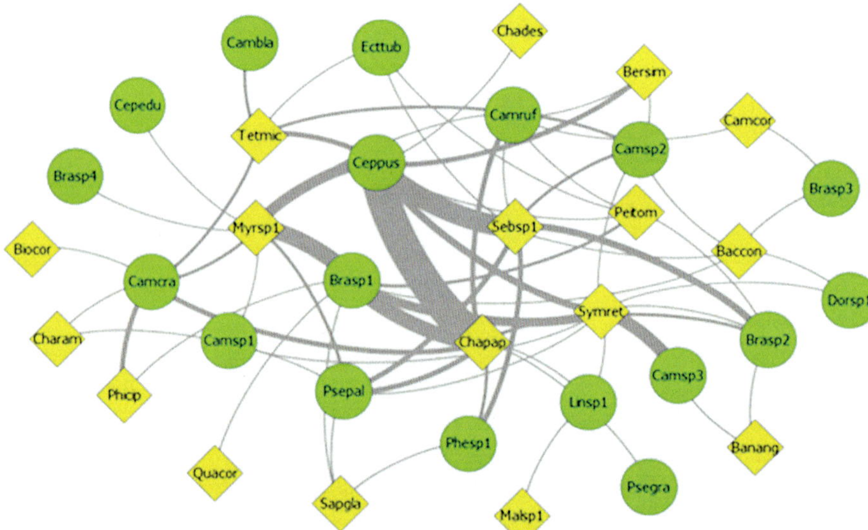

Fig. 13.3 The network formed between ants (*circles*) and plants with extrafloral nectaries (*diamonds*) in *Campo Rupestre*site atSerra do Cipó, southeastern Brazil. *Lines* represent interactions, and line thickness is proportional to interaction frequency. Plant species—Baccon: *Baccharis concinna*; Banaug: *Banisteriopsis angustifolia*; Bersim: *Bernardia similis*; Biocor: *Bionia coriacea*; Camcor: *Camptosema coriaceum*; Chades: *Chamaecrista desvauxii*; Chapap: *Chamaecrista papillata*; Charam: *Chamaecrista ramosa*; Malsp1: Malpiguiaceae sp1; Myrsp1: *Myrsine* sp1; Peitom: *Peixotoa tomentosa*; Phicip: *Philodendron cipoensis*; Quacor: *Qualea cordata*; Sapgla: *Sapium glandulatum*; Sebsp1: *Sebastiania* sp1; Symret: *Symphyopapus reticulatus*; Ant species—Brasp1: *Brachymyrmex* sp1; Brasp2: *Brachymyrmex* sp2; Brasp3: *Brachymyrmex* sp3; Brasp4: *Brachymyrmex* sp4; Camcra: *Camponotus crassus*; Camruf: *Camponotus rufipes*; Camsp1: *Camponotus(Tanaemyrmex)* sp1; Camsp2: *Camponotus* (*Myrmaphaenus*) sp2; Camsp3: *Camponotus*(Myrmaphaenus) sp3; Cepedu: *Cephalotes eduarduli*; Ceppus: *Cephalotes pusillus*; Dorsp1; *Dorymyrmex* sp1; Ecttub: *Ectatomma tuberculatum*; Linsp1; *Linepithema* sp1; Phesp1: *Pheidole* sp1; Psesp1: *Pseudomyrmex* sp1; Psesp2: *Pseudomyrmex* sp2

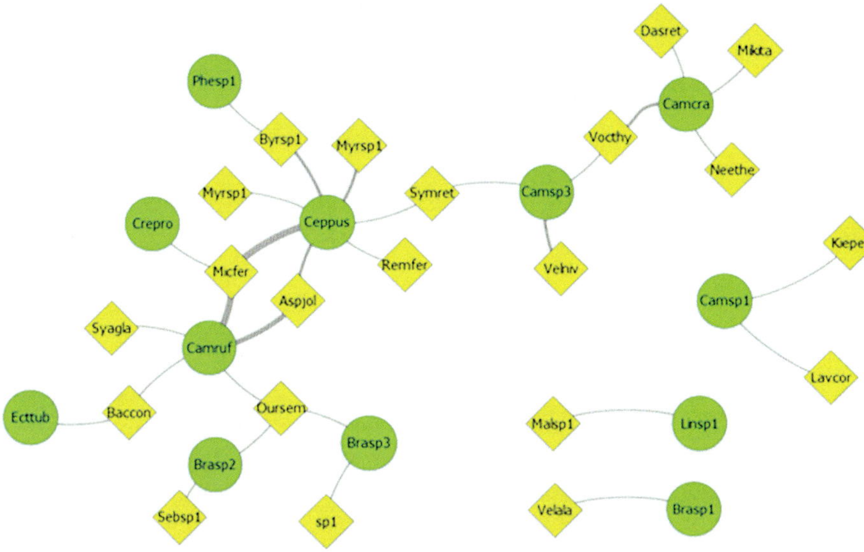

Fig. 13.4 The network formed between ants (*circles*) and plants with trophobiont insects (*diamonds*) in *Campo Rupestre* vegetation of Serra do Cipó, southeastern Brazil. *Lines* represent interactions, and line thickness is proportional to interaction frequency. Plant species—Aspjol: *Aspilia jolyana*; Baccon: *Baccharis concinna*; Byrsp1: *Byrsonima* sp1; Dasret: *Dasyphyllum reticulatum*; Kiepet: *Kielmeyera petiolaris*; Lavcor: *Lavoisiera cordata*; Malsp1: *Malpighiaceae* sp1; Micfer: *Miconia ferruginata*; Mikita: *Mikania itambana*; Mycsp1: *Myrcia* sp1; Myssp1: *Myrsine* sp1; Neethe: *Neeatheifera*; Oursem: *Ouratea semiserrata*; Remfer: *Remijia ferruginea*; Sebsp1: *Sebastiania* sp1; Sp1: unidentified; Syagla: *Syagrus glaucescens*; Symret: *Symphyopapus reticulatus*; Velala: *Vellozia alata*; Velniv: *Vellozia nivea*; Vocthy: *Vochysia thyrsoidea*; Ant species—Brasp1: *Brachymyrmex* sp1; Brasp2: *Brachymyrmex* sp2; Brasp3: *Brachymyrmex* sp3; Camcra: *Camponotus crassus*; Camruf: *Camponotus rufipes*; Camsp1: *Camponotus(Tanaemyrmex)* sp1; Camsp3: *Camponotus (Myrmaphaenus)* sp3; Ceppus: *Cephalotes pusillus*; Crepro: *Crematogaster*prox. *erecta*; Ecttub: *Ectatomma tuberculatum*; Linsp1: *Linepithema* sp1; Phesp1: *Pheidole* sp1

ant-trophobiosis subnetwork (Fig. 13.4) are modular and have higher specialization values. Through network structure it is possible to visualize that ant-trophobiont interactions, in opposition to ant-EFNs, forms distinct modules or groups of interactions, which suggest certain resource partitioning by ants. In fact, some field observations at Serra do Cipó, as well as previous studies performed in other ecosystems (Del-Claro and Oliveira 1993; Blüthgen et al. 2000), suggest that amino acid-rich honeydew is a more valuable resource than extrafloral nectar, being frequently monopolized by colonies of dominant ants.

Despite structural differences between EFN and trophobionts subnetworks, they share the five most central ant species (measured by number of interactions made by the species): *C. pusillus*, *C. rufipes*, *C. crassus*, and two unidentified species of *Brachymyrmex* genus. It seems that relatively few ant species are involved in those interactions with EFN-producing plants and trophobionts insects in CR sites, especially when we consider the huge richness recently proposed for these environments—around 288 species (Costa et al. 2015). In fact, the same ant species, *C. rufipes*, *C. crassus* and *C. pusillus* have consistently been associated to different liquid resources in different localities in CR and Cerrado sites, suggesting that they can be considered keystone species for ant-plant interactions, mainly mutualisms involving extrafloral nectar and honeydew. However, the role of these ants as bodyguards of plants and insect trophobionts, the benefits provided by these liquid sources for ant colony fitness as well the degree of interactions intimacy need further investigations in CR.

13.7 Concluding Remarks

The investigations regarding mutualisms among free-living species in CR started with pollination studies in the 70s, but only in the past decade have these interactions received more attention by ecologists. Despite an increase in the number of studies, the information remains scattered and concentrated for few localities, such Serra do Cipó and Chapada Diamantina mountains. The compilation presented in this chapter clearly indicates that the ecology of mutualisms in CR is still in its infancy and that there is lot of work ahead. However, some patterns have emerged and could serve as starting points for further investigations.

Despite the relatively crescent number of studies regarding interactions between flowers and theirs visitors in CR, the role of animals as effective pollinators or cheaters (e.g. thieves or robbers) remain poorly known. Bees and hummingbirds have been reported as the main flower visitors of CR plants, but pollination by other groups such, butterflies, moths, and bats need better evaluation in the future. Breeding biology of few species have been studied in detail so far, the available data indicate that pollination systems are diverse varying form highly specialized to more generalized. In the future it would be very important to determine the degree of specialization in pollination systems and the degree of plant dependence on their pollinators, especially for those endemic and threatened species.

Seed dispersal of CR plants is virtually unknown, as there is information available for too few species. For now we can only conclude that most vertebrate seed dispersal is carried out by birds, and to a lesser extent by ants and lizards. Because bird communities encompasses mostly non specialized frugivorous species, with few species relying only partially on fruits, we could expect a weak degree of interdependence among CR plants and avian frugivores, or at best, highly asymmetric degrees of interdependence among them. Nevertheless, the role of birds, lizards, mammals and ants as primary and secondary seed dispersers, or predators, remain poorly investigated. Additional studies determining structure of plant-frugivore networks in CR are deeply needed.

Ants seem to play a major hole in CR ecosystems, due their abundance and the diversity of interactions in which they are involved. Ants may act as seed dispersers, as nectar thieves, as bodyguards of myrmecophilous plants and trophobiont insects in CR. The studies encompassing a network approach applied to ant-plant and ant-insect interactions at the community level, associated with knowledge of natural history of these interactions, could bring insights on the ecology and evolution of protective mutualisms, particularly in megadiverse environments such as the CR. The next step is to extend the studies of ant-plant mutualisms to larger scales along the Espinhaço range and experimentally explore the possible outcomes of these interactions subjected to context-dependent variations.

The results of Carstensen et al. (2014) have provided important insight into the ecology of mutualisms in CR. The authors observed that even in small spatial scales there is a high site turnover in plant-pollinator interactions and this pattern could be a pervasive feature of all mutualistic interactions reported in this chapter, not only pollination. Many CR species have restricted distribution, while other species are more widespread. Besides, species distribution in CR are subject to high variation due the altitudinal and latitudinal gradients (Chap. 15). The matches and mismatches in partners` distributions could generate selection mosaics of interactions, thus CR could be a natural laboratory to explore predictions from the theory of coevolutionary mosaics proposed by Thompson (2005). To understand how coevolutionary processes shape the mutualistic interactions in CR is important to start long term studies of mutualistic interactions on broader spatial and temporal scales.

The study of community organization using mutualistic networks in CR is just at the beginning. Now is the time to start applying these ecological tools for biodiversity conservation. *Campo Rupestre* ecosystems are subject to intensification of fire regimes, mining, road construction and biological invasion (Fernandes et al. 2014). The way these negative impacts affect community organization and structure of mutualistic networks will require further investigations. Besides, mountain tops all around the world figure amongst the most threatened ecosystems by the global warming predicted for the next century (Colwell et al. 2008). Therefore, we suggest that standardized protocols for assessment and long term monitoring of mutualistic networks involving animals and plants in *Campo Rupestre* could be a useful tool for the evaluation of global warming effects on the biodiversity of tropical mountain tops.

Acknowledgments The authors would like to thank G.W. Fernandes for the invitation to write this chapter and for his inspiring passion about the Rupestrian Grasslands. Also Marco A. Mello and Beth Morrison for their suggestions and comments on the manuscript. T.J. Guerra would like to thank Marco A. Pizo and Wesley R. Silva for their guidance and constant support. T.J. Guerra is supported by postdoctoral fellowship from Coordenação de Aperfeiçoamento de Pessoal de Nível Superior (CAPES) and received a reseach grant from São Paulo Research Found (FAPESP) (Grant #07/59444-4). F.A.O. Silveira received a financial support from Conselho Nacional de Desenvolvimento Científico e Tecnológico (CNPq) (Grant #486742/2012-1) and Fundação de Amparo à Pesquisa do Estado de Minas Gerais (FAPEMIG) (Grant #APQ02231-12). F.V. Costa received a doctoral fellowship from CAPES and was supported by CNPq (Grant #478565/2012-7). L.P.C. Morellato and D.W. Carstensen were funded by FAPESP-VALE-FAPEMIG (Grant #2010/51307-0) and FAPESP (Grants #2011/22635-2, #2013/05920-0, BEPE, #2014/01594-4). L. P.C. Morellato and F.A.O. Silveira receive CNPq research productivity fellowship.

References

Aizen MA (2003) Influences of animal pollination and seed dispersal on winter flowering in a temperate mistletoe. Ecol 84:2613–2627

Alves CBM, Leal CG, Brito MFG, Santos ACA (2008) Biodiversidade e conservação de peixes do Complexo do Espinhaço. Megadiv 4:177–196

Alves RJV, Silva NG, Oliveira JA, Medeiros D (2014) Circumscribing campo rupestre: megadiverse Brazilian rocky montane savannas. Braz J Biol 74:355–362

Axelrod R, Hamilton WD (1981) The evolution of cooperation. Science 211:1390–1396

Bascompte J, Jordano P (2007) Plant–animal mutualistic networks: the architecture of biodiversity. Annu Rev Ecol Evol Syst 38:567–593

Bascompte J, Jordano P (2014) Mutualistic networks. Princeton University Press, Princeton

Bascompte J, Jordano P, Olesen JM (2006) Asymmetric coevolutionary networks facilitate biodiversity maintenance. Science 312:431–433

Beattie AJ (1985) The evolutionary ecology of ant–plant mutualisms. Cambridge University Press, Cambridge

Blendinger PG, Loiselle BA, Blake JG (2008) Crop size, plant aggregation, and microhabitat type affect fruit removal by birds from individual melastome plants in the upper Amazon. Oecol 158:273–283

Blüthgen N, VerhaaghM Goitía W, JafféK Morawetz W, Barthlott W (2000) How plants shape the ant community in the Amazonian rainforest canopy: the key role of extrafloral nectaries and homopteran honeydew. Oecol 125:229–240

Blüthgen N, Gottsberger G, Fiedler K (2004) Sugar and amino acid composition of ant-attended nectar and honeydew sources from an Australian rainforest. Aust Ecol 29:418–429

Borba EL, Semir J (2001) Pollinator specificity and convergence in fly-pollinated *Pleurothallis* (Orchidaceae) species: a multiple population approach. Ann Bot 88:75–88

Borba EL, Felix JM, Solferini VN, Semir J (2001) Fly-pollinated *Pleurothallis* (orchidaceae) species have high genetic variability: evidence from isozyme markers. Am J Bot 88:419–428

Boucher DH (1985) The biology of mutualisms. Croom Helm, London

Bronstein JL (1994) Conditional outcomes in mutualistc interactions. Trends Ecol Evol 9:214–217

Bronstein JL (2001) The exploitation of mutualisms. Ecol Lett 4:277–287

Bronstein JL (2009) The evolution of facilitation and mutualism. J Ecol 97:1160–1170

Bronstein JL, Wilson WG, Morris WF (2003) Ecological dynamics of mutualist/antagonist communities. Am Nat 162:24–39

Bronstein J, Alarcón R, Geber M (2006) The evolution of plant-insect mutualisms. New Phytol 172:412–428

Byk J, Del-Claro K (2011) Ant-plant interaction in the Neotropical savanna: Direct beneficial effects of extrafloral nectar on ant colony fitness. Popul Ecol 53:327–332

Calder M, Bernardt P (eds) (1983) The biology of mistletoes. Academic Press, Sidney

Camargo MG, Cazetta E, Schaefer HM, Morellato LPC (2013) Fruit color and contrast in seasonal habitats—a case study from a cerrado savanna. Oikos 122:1335–1342

Carstensen DW, Sabatino M, Trøjelsgaard K, Morellato LPC (2014) Beta diversity of plant-pollinator networks and the spatial turnover of pairwise interactions. PLoS ONE 9(11) e112903

Chaves AV, Freitas GHS, Vasconcelos MF, Santos FR (2015) Biogeographic patterns, origin and speciation of the endemic birds from eastern Brazilian mountaintops: a review. Syst Biodiv 13:1–16

Christianini AV, Oliveira PS (2009) The relevance of ants as seed rescuers of a primarily bird-dispersed tree in the Neotropical cerrado savanna. Oecol 160:735–745

Colwell RK, Brehm G, Cardelús CL, Gilman AC, Longino JT (2008) Global warming, elevational range shifts, and lowland biotic attrition in the wet tropics. Science 322:258–261

Conceição AA, Funch LS, Pirani JR (2007) Reproductive phenology, pollination and seed dispersal syndromes on sandstone vegetation in the "Chapada Diamantina", northeastern Brazil: population and community analyses. Rev Bras Bot 30:475–485

Costa FV, Mello R, Lana TC, Neves FDS (2015) Ant fauna in megadiverse mountains: a checklist for the rocky grasslands. Sociobiol 62:228–245

Culley TM, Weller SG, Sakai AK (2002) The evolution of wind pollination in angiosperms. Trends Ecol Evol 17:361–369

Darwin C (1862) On the various contrivances by which british and foreign orchids are fertilised by insects, and on the good effects of intercrossing. Murray, London

Dáttilo W, Fagundes R, Gurka CQ, Silva MSA, Vieira MCL, Izzo TJ, Díaz-Castelazo C, Del-Claro K, Rico-Gray V (2014) Individual-based ant-plant networks: diurnal-nocturnal structure and species-area relationship. PLoS ONE 9:e99838

Davidson DW, Cook SC, Snelling RR, Chua TH (2003) Explaining the abundance of ants in lowland tropical rainforest canopies. Science 300:969–972

Del-Claro K, Oliveira PS (1993) Ant-Homoptera interaction: do alternative sugar sources distract tending ants? Oikos 68:202–206

Del-Claro K, Byk J, Yugue G, Morato M (2006) Conservative benefits in an ant-hemipteran association in the Brazilian tropical savanna. Sociobiol 47:1–8

Douglas AE (1994) Symbiotic interactions. Oxford University Press, Oxford

Dupont YL, Olesen JM (2009) Ecological modules and roles of species in heathland plant–insect flower visitor networks. J Anim Ecol 78:346–353

Echternacht L, Trovó M, Oliveira CT, Pirani JR (2011) Areas of endemism in the Espinhaço range in Minas Gerais, Brazil. Flora 206:782–791

Ehleringer JR, Schulze ED, Ziegler H, Lange OL, Farquhar GD, Cowar IR (1985) Xilem-tapping mistletoes: water or nutrients parasites? Science 227:1479–1481

Elisson AM, Denslow JS, Loiselle BA, Brene D (1993) Seed and seedling ecology of neotropical Melastomataceae. Ecol 74:1733–1749

Estrada A, Fleming T (eds) (1986) Frugivores and seed dispersal. Junk Publishers, Dordrecht, Dr. W

Fagundes R, Del-Claro K, Ribeiro SP (2012) Effects of the trophobiont herbivore *Calloconophora pugionata* (Hemiptera) on ant fauna associated with *Myrcia obovata* (Myrtaceae) in a montane tropical forest. Psyche 2012:1–8

Fagundes R, Ribeiro SP, Del-Claro K (2013) Tending-Ants Increase Survivorship and Reproductive Success of Calloconophora pugionata Drietch (Hemiptera, Membracidae), a Trophobiont Herbivore of Myrcia obovata O. Berg (Myrtales, Myrtaceae). Sociobiol 60:11–19

Faustino TC, Machado CG (2006) Frugivoria por aves em uma área de campo rupestre na Chapada Diamantina, BA. Rev Bras Ornitol 14:137–143

Fernandes GW, Barbosa NPU, Negreiros D, Paglia AP (2014) Challenges for the conservation of vanishing megadiverse rupestrian grasslands. Nat Cons 12:162–165

Ferreira PA, Viana BF (2010) Pollination biology of *Paliavana tenuiflora* (Gesneriaceae: Sinningeae) in Northeastern Brazil. Acta Bot Bras 24:972–977

Fleming TH, Kress WJ (2011) A brief history of fruits and frugivores. Acta Oecol 37:521–530

Fonseca RBS, Funch LS, Borba EL (2012) Dispersão de sementes de *Melocactus glaucescens* e *M. paucispinus* (Cactaceae), no Município de Morro do Chapéu, Chapada Diamantina, BA. Acta Bot Bras 26:481–492

Franceschinelli EV, Jacobi CM, Drummond MG, Resende MFS (2006) The genetic diversity of two Brazilian *Vellozia* (Velloziaceae) with different patterns of spatial distribution and pollination biology. Ann Bot 97:585–592

Franco EL, Gimenes M (2011) Pollination of *Cambessedesia wurdockii* in Brazilian campo rupestre vegetation, with special reference to crepuscular bees. J Ins Sci 11:97

Galetti M, Stotz D (1996) *Miconia hypoleuca* (Melastomataceae) como espécie-chave para aves frugívoras no sudeste do Brasil. Rev Bras Biol 56:435–439

Gill FB, Wolf LL (1975) Foraging strategies and energetics of East African sunbirds at mistletoe flowers. Am Nat 109:491–510

Giulietti AM, Pirani JR, Harley RM (1997) Espinhaço range region, eastern Brazil. In: Davis SD, Heywood VH, Herrera-Macbryde O, Villa-Lobos J, Hamilton AC (eds) Centres of plant diversity: a guide and strategy for their conservation. IUCN Publication Unit, Cambridge, pp 397–404

Goldenberg R, Penneys DS, Almeda F, Judd WS, Michelangeli FA (2008) Phylogeny of *Miconia* (Melastomataceae): patterns of stamen diversification in a megadiverse neotropical genus. Int J Plant Sci 169:963–979

Gonzalez AMM, Dalsgaard B, Olesen JM (2010) Centrality measures and the importance of generalist species in pollination networks. Ecol Complex 7:36–43

Guerra TJ, Pizo MA (2014) Asymmetrical dependence between a Neotropical mistletoe and its avian seed disperser. Biotropica 46:285–293

Guerra TJ, Camarota F, Castro FS, Schwertner C, Grazia J (2011) Trophobiosis between ants and *Eurystethus microlobatus* Ruckes 1966 (Hemiptera: Heteroptera: Pentatomidae) a cryptic, gregarious and subsocial stinkbug. J Nat Hist 45:1101–1117

Guerra TJ, Galetto L, Silva WR (2014) Nectar secretion dynamic links pollinator behavior to consequences for plant reproductive success in the ornithophilous mistletoe *Psittacanthus robustus*. Plant Biol 16:956–966

Herrera C, PellmyrO (eds) (2002) Plant animal interactions: an evolutionary approach. Blackwell, Oxford

Hipólito J, Roque N, Galetto L, Viana BF, Kevan PG (2013) The pollination biology of *Pseudostifftia kingii* H. Rob. (Asteraceae), a rare endemic Brazilian species with uniflorous capitula. Braz J Bot 36:247–254

Hölldobler B, Wilson EO (1990) The ants. Belknaps/Havard University Press, Cambridge

Hopper SD (2009) OCBIL theory: towards an integrated understanding of the evolution, ecology and conservation of biodiversity on old, climatically buffered, infertile landscapes. Plant Soil 322:49–86

Howe HF (1993) Specialized and generalized dispersal systems: where does 'the paradigm' stand? Vegetatio 107(108):3–13

Inouye DW (1980) The terminology of floral larceny. Ecol 61:1251–1252

Jacobi CM, del Sarto MCL (2007) Pollination of two species of *Vellozia* (Velloziaceae) from high-altitude quartzitic grasslands, Brazil. Acta Bot Bras 21:325–333

Jacobi CM, Antonini Y (2008) Pollinators and defense of *Stachytarpheta glabra* (Verbenaceae) nectar resources by the hummingbird *Colibri serrirostris* (Trochilidae) on ironstone outcrops in south-east Brazil. J Trop Ecol 24:301–308

Jacobi CM, Carmo FF (2011) Life-forms, pollination and seed dispersal syndromes in plant communities on ironstone outcrops, SE Brazil. Acta Bot Bras 25:395–412

Jacobi CM, do Carmo RM, Oliveira RS (2000) The reproductive biology of two species of *Diplusodon* Pohl (Lythraceae) from Serra do Cipo, southeastern Brazil. Plant Biol 2:670–676

Janzen DH (1966) Co-evolution of mutualism between ants and acacias in Central America. Evolution 20:249–275

Jordano P, Bascompte J, Olesen JM (2003) Invariant properties in coevolutionary networks of plant-animal interactions. Ecol Lett 6:69–81

Kuijt J (1969) The biology of parasitic plants. University of California Press, Berkeley and Los Angeles

Leite FSF, Junca FA, Eterovick PC (2008) Status do conhecimento, endemismo e conservação de anfíbios anuros da Cadeia do Espinhaço, Brasil. Megadiv 4:182–200

Lessa LG, Geise L, Costa FN (2013) Effects of gut passage on the germination of seeds ingested by didelphid marsupials in a neotropical savanna. Acta Bot Bras 27:519–525

Lima MHC, Oliveira EG, Silveira FAO (2013) Interactions between ants and non-myrmecochorous fruits in *Miconia* (Melastomataceae) in a Neotropical Savanna. Biotropica 45:217–223

Lopes B (1995) Treehoppers (Homoptera, Membracidae) in southeastern Brazil: use of host plants. Rev Bras Zool 12:595–608

Lousada JM, Borba EL, Ribeiro KT, Ribeiro LC, Lovato MB (2011) Genetic structure and variability of the endemic and vulnerable *Vellozia gigantea* (Velloziaceae) associated with the landscape in the Espinhaco range, in southeastern Brazil: implications for conservation. Genetica 139:431–440

Machado CG, Coelho AG, Santana CS, Rodrigues M (2007) Beija-Flores e seus recursos florais em uma área de campo rupestre da Chapada Diamantina, Bahia. Rev Bras Ornitol 15:215–227

Matias LQ, Braga PIS, Freire AG (1996) Biologia reprodutiva de *Constantia cipoensis* Porto & Brade (Orchidaceae), endêmica da Serra do Cipó, Minas Gerais. Rev Bras Bot 19:119–125

Mckey D (1975) The ecology of coevolved seed dispersal systems. In: Gilbert LE, Raven PH (eds) Coevolution of animals and plants. University Texas Press, Austin, pp 159–191

Melo MC, Taucce PPG, Borba EL (2011) Reproductive biology and isolation mechanisms in rupicolous species of the *Acianthera prolifera* complex (Orchidaceae) occurring in southeastern Brazil. Plant Syst Evol 293:161–176

Milewski AV, Bond WJ (1982) Convergence of myrmecochory in mediterranean Australia and South Africa. In: Buckley RC (ed) Ant-plant interactions in Australia. Junk Press, The Hague, pp 89–98

Morellato LPC, Oliveira PS (1991) Distribution of extrafloral nectaries in different vegetation types of Amazonian Brazil. Flora 185:33–38

Nickrent DL, Malécot V, Vidal-Russell R, Der JP (2010) A revised classification of Santalales. Taxon 59:538–558

Olesen JM, Valido A (2004) Lizards as pollinators and seed dispersers: an island phenomenon. Trends Ecol Evol 18:177–181

Olesen JM, Bascompte J, Dupont YL, Jordano P (2007) The modularity of pollination networks. Proc Natl Acad Sci USA 104:19891–19896

Oliveira PS, Leitão-Filho H (1987) Extafloral nectaries: their taxonomic distribution and abundance on the woody flora of Cerrado vegetation in Southeast Brazil. Biotropica 19:140–148

Oliveira PS, Freitas AVL (2004) Ant–plant–herbivore interactions in the neotropical cerrado savanna. Naturwissenschaften 91:557–570

Ollerton J, Winfree R, Tarrant S (2011) How many flowering plants are pollinated by animals? Oikos 120:321–326

Peakall R, Handel SN, Beattie AJ (1991) The evidence for, and importance of, ant pollination. In: Huxley CR, Cutler DF (eds) Ant-plant interactions. Oxford University Press, Oxford, pp 421–429

Rapini A, Ribeiro PL, Lambert S, Pirani JR (2008) A flora dos campos rupestres da Cadeia do Espinhaço. Megadiv 4:15–23

Reid N (1991) Coevolution of mistletoes and frugivorous birds. Aust J Ecol 16:457–469

Rico-Gray V, Oliveira PS (2007) The ecology and evolution of ant-plant interactions. The University of Chicago Press, London

Rodrigues LC, Rodrigues M (2014) Flowers visited by hummingbirds in the open habitats of the southeastern Brazilian mountaintops: species composition and seasonality. Braz J Biol 74:659–676

Romero GQ (2002) Protection of *Vochysia elliptica* (Vochysiaceae) by a nectar-thieving ant. Braz J Biol 62:371–373

Romero GQ, Vasconcellos-Neto J (2004) Beneficial effects of flower-dwelling predators on their host plant. Ecol 85:446–457

Rosumek F, Silveira FAO, Neves FS, Barbosa NPU, Diniz L, Oki Y, Pezzini F, Fernandes GW, Cornelissen T (2009) Ants on plants: a meta-analysis of the role of ants as plant biotic defenses. Oecol 160:537–549

Roxburgh L, Nicolson SW (2005) Patterns of host use in two African mistletoes: the importance of mistletoe–host compatibility and avian disperser behavior. Func Ecol 19:865–873

Samuels IA, Levey DJ (2005) Effects of gut passage on seed germination: do experiments answer the questions they ask? Func Ecol 19:365–368

Sanmartin-Gajardo I, Sazima M (2005) Chiropterophily in Sinningieae (Gesneriaceae): *Sinningia brasiliensis* and *Paliavana prasinata* are bat-pollinated, but *P. sericiflora* is not. Not yet? Ann Bot-London 95:1097–1103

Santana CS, Machado CG (2010) Fenologia de floração e polinização de espécies ornitófilas de bromeliáceas em uma área de campo rupestre da Chapada Diamantina, BA, Brasil. Rev Bras Bot 33:469–477

Sargent S (1995) Seed fate in a tropical mistletoe: the importance of host twig size. Func Ecol 9:197–204

Sazima M (1977) Hummingbird pollination of *Barbacenia flava* (Velloziaceae) in the Serra do Cipó, Minas Gerais, Brazil. Flora 166:239–247

Sazima M, Sazima I (1990) Hummingbird pollination in two species of *Vellozia* (Liliiflorae: Velloziaceae) in southeastern Brazil. Bot Acta 103:83–86

Sazima I, Vogel S, Sazima M (1989) Bat pollination of *Encholirium glaziovii*, a terrestrial bromeliad. Plant Syst Evol 168:167–179

Schupp EW, Jordano P, Goméz JM (2010) Seed dispersal effectiveness revisited: a conceptual review. New Phytol 188:333–353

Silveira FAO, Mafia PO, Lemos-Filho JP, Fernandes GW (2012) Species-specific outcomes of avian gut passage on germination of Melastomataceae seeds. Plant Ecol Evol 145:350–355

Silveira FAO, Fernandes GW, Lemos-Filho JP (2013a) Seed and seedling ecophysiology of neotropical Melastomataceae: implications for conservation and restoration of savannas and rainforests. Ann Mo Bot Gard 99:82–99

Silveira FAO, Ribeiro RC, Soares S, Rocha D, Moura CO (2013b) Physiological dormancy and seed germination inhibitors in Melastomataceae. Plant Ecol Evol 146:290–294

Silveira FAO, Negreiros D, Barbosa NPU, Buisson E, Carmo FF, Carstensen DW, Conceição AA, Cornelissen TG, Echternacht L, Fernandes GW, Garcia QS, Guerra TJ, Jacobi CM, Lemos-Filho JP, Le Stradic S, Morellato LPC, Neves FS, Oliveira RS, Schaefer CE, Viana PL, Lambers H. (2016) Ecology and evolution of plant diversity in the endangered *Campo Rupestre*: a neglected conservation priority. Plant and soil (in press)

Thompson JN (2005) The geographic mosaic of coevolution. University of Chicago Press, Chicago

Thompson JN (2010) Four central points about coevolution. Evol EduOutreach 3:7–13

Trøjelsgaard K, Jordano P, Carstensen DW, Olesen JM (2015) Geographical variation in mutualistic networks: similarity, turnover and partner fidelity. P Roy Soc Lond B Bio 282.doi:10.1098/rspb.2014.2925

Valiente-Banuet A, Aizen MA, Alcántara JM, Arroyo J, Cocucci A, Galetti M, García MB, García D, Gómez JM, Jordano P, Medel R, Navarro L, Obeso JR, Oviedo R, Ramírez N, Rey PJ, Traveset A, Verdú M, Zamora R (2015) Beyond species loss: the extinction of ecological interactions in a changing world. Func Ecol 29:299–307

van Ommeren RJ, Whitham TG (2002) Changes in interactions between juniper and mistletoe mediated by shared avian frugivores: parasitism to potential mutualism. Oecol 130:281–288

Vasconcelos MF, Lombardi JA (2001) Hummingbirds and their flowers in the campos rupestres of southern Espinhaço range, Brazil. Melopsittacus 4:3–30

Verdú M, Valiente-Banuet A (2008) The nested assembly of plant facilitation networks prevents species extinctions. Am Nat 172:751–760

Watson DM (2001) Mistletoe—a keystone resource in forests and woodlands worldwide. Ann Rev Ecol Syst 32:219–249

Watson DM, Herring M (2013) Mistletoe as a keystone resource: an experimental test. P Roy Soc Lond B Bio 279:3853–3860

Watson DM, Rawsthorne J (2013) Mistletoe specialist frugivores: latterday 'Johnny Appleseeds' or self serving market gardeners? Oecol 172:925–932

Willmer PG, Nuttman CV, Raine NE, Stone GN, Pattrick JG, Henson K, Stillman P, Mcilroy L, Potts SG, Knudsen JT (2009) Floral volatiles controlling ant behaviour. Funct Ecol 23:888–900

Chapter 14
Antagonistic Interactions in the Rupestrian Grasslands: New Insights and Perspectives

Frederico S. Neves, Tatiana Cornelissen, Marcel S. Coelho, Milton Barbosa, Marco Antônio A. Carneiro, Sérvio P. Ribeiro and G. Wilson Fernandes

Abstract Antagonistic interactions are main ecological forces in terrestrial communities and include several examples involving plants and animals. The studies performed in the rupestrian grasslands have mostly focused on patterns of herbivory and on the effects of plant quality and natural enemies on insect herbivore abundance and distribution. Herbivory rates recorded for plants in this ecosystem are within the range proposed for open areas (ca. 15 %), although a wider range of studies are still needed. Galling insect richness is highly variable amongst sites

F.S. Neves (✉)
Laboratório de Ecologia de Insetos/DBG, ICB/Universidade Federal de Minas Gerais, Belo Horizonte, MG 30161 970, Brazil
e-mail: fredneves@icb.ufmg.br

T. Cornelissen
Departamento de Ciências Naturais, Universidade Federal de São João Del Rei, São João Del Rei, MG, Brazil
e-mail: tatiana@ufsj.edu.br

M.S. Coelho · G.W. Fernandes
Ecologia Evolutiva & Biodiversidade/DBG, ICB/Universidade Federal de Minas Gerais, Belo Horizonte, MG 30161 970, Brazil
e-mail: marcel.s.coelho@gmail.com

M. Barbosa
Department of Zoology, University of Oxford, South Parks Road, Oxford OX1 3PS, UK
e-mail: milton.barbosadasilva@zoo.ox.ac.uk

M.A.A. Carneiro · S.P. Ribeiro
Departamento de Biodiversidade, Evolução e Meio Ambiente, ICEB/Universidade Federal de Ouro Preto, Ouro Preto, MG 35400-000, Brazil
e-mail: mcarneirinho@gmail.com

M.A.A. Carneiro
Laboratório de Entomologia Ecológica, Universidade Federal de Ouro Preto, Ouro Preto, Brazil

© Springer International Publishing Switzerland 2016

G.W. Fernandes (ed.), *Ecology and Conservation of Mountaintop Grasslands in Brazil*, DOI 10.1007/978-3-319-29808-5_14

(between 18 and 241 species) and are mostly rich and abundant in xeric habitats compared to mesic habitats, corroborating the hypothesis of environmental harshness. Several new galling insect species have been discovered in this very diverse landscape, indicating its importance for insect radiation and evolution. Galling insect communities associated to host plants are influenced by host plant sex. Male plants generally host higher insect richness and herbivory levels than female plants. Gall insect richness sampled on 138 host plants in several rupestrian grassland sites indicates that larger plant families also host higher number of galling insects than smaller plant families or genera. Amongst the factors that influence host plant selection by herbivores in rupestrian grasslands aspects of plant quality such as leaf asymmetry, leaf sclerophylly and ant-interactions have been evaluated. However, mixed results for the effects of plant quality on insect abundance and performance have been found depending upon the insect-host system evaluated. Another antagonist interaction evaluated involves parasitism by two common species of mistletoes and results have shown that parasitism represents another important source of stress for host plants in the rupestrian grasslands. Finally, in this chapter we summarize several antagonistic, multitrophic and indirect interactions using plants in the genus *Baccharis* as a model, aiming to provide information and guidance for future studies.

Contents

S.P. Ribeiro
Laboratório de Ecologia Evolutiva de Insetos de Dossel e Sucessão Natural, Federal University of Ouro Preto, Ouro Preto, Brazil
e-mail: spribeiro@iceb.ufop.br

G.W. Fernandes
Department of Biology, Stanford University, Stanford, CA 94305, USA
e-mail: gw.fernandes@gmail.com

14.1 Antagonistic Interactions in Rupestrian Grasslands

Antagonistic interactions represent a type of interespecific interaction in which at least one interacting species is negatively impacted. The current ecological literature considers antagonistic interactions as a main structuring force in most tropical ecological communities (Garlaschelli et al. 2003; Dohn et al. 2013; Sutherland et al. 2013; Fayle et al. 2015). At least 30 studies involving antagonistic interactions have been reported so far in the rupestrian grasslands. However, there is a clear focus on a few plant-antagonist systems involving mainly, herbivory and insect herbivores on their host plants.

The first surveys date back to the end of the 80s (see Fernandes and Price 1988) and since then several studies have been conducted to estimate insect species richness and diversity, with particular interest in galling insects (see Fernandes et al. 1996; Carneiro et al. 2009a). Other antagonistic interactions investigated in the rupestrian grasslands include parasitism by mistletoes (Loranthaceae) and holoporasites (*Pilostyles*) on their host plants (Mourão et al. 2009; Bahia et al. 2015; Jacobi et al. 2015), and also predation by carnivorous plants [*Drosera* (Espírito-Santo and Werneck 1999)] and protocarnivorous plants [*Paepalanthus* (Nishi et al. 2013)]. Curiously, no study has yet addressed antagonistic interactions between vertebrates and plants in the rupestrian grasslands.

In this chapter, we aim to discuss what has been found in terms of insect-plant interactions and antagonistic patterns in the rupestrian grasslands as well as mechanisms of plant defence and selection by herbivores. We also present a synthesis on parasitic plants and their hosts, due to the frequency and importance of this type of association in this ecosystem. Finally, we portray the system *Baccharis*-insects as a model, due to the vast array of plant species in this genus and the rich insect community associated to them.

14.2 Herbivory in the Rupestrian Grasslands

One estimate of the ecological importance of plant-herbivore interactions within a community is the level of herbivory experienced by plants (Coley and Aide 1991). Global estimates of plant damage in dry areas of tropical vegetation are around 15 % of leaf area lost to herbivores (see Coley and Barone 1996), although Metcalffe et al. (2013) suggest rates of up to 30 % for open vegetation. Few studies regarding herbivory levels endured by plants in rupestrian grasslands have been conducted to date and there are no clear estimates of the percentage of herbivory when considering the plant community level. A study conducted in Cerrado areas surrounding rupestrian grasslands found herbivory rates to be ca. 10 % in *Solanum lycocarpum* (Solanaceae) (Cuevas-Reyes et al. 2013) while Ribeiro (2003) showed that herbivory rates in the Cerrado could reach 18 %.

Furthermore, even fewer studies have evaluated the likely impact of herbivores on a specific altitude or habitat. Rupestrian grasslands in the Espinhaço mountains are distributed throughout a large altitudinal gradient where the rupestrian sites are represented by a mosaic of habitats of different qualities to plants and herbivores. One exception is a study by Ribeiro et al. (1994a) on small black leaf beetles *Brachypnoea* spp. (Coleoptera, Chrysomelidae), which are normally associated to high altitudes in mountaintop vegetation. Previously, Fernandes and Ribeiro (1990) had shown the importance of Chrysomelidae herbivory on flowering, re-growth and damage compensation for a semelparous endemic species, the *Paepalanthus speciosus* (Eriocaulaceae). Species of the genus *Brachypnoea* were abundant at higher altitudes in Serra do Cipó (Ribeiro et al. 1994a), where most of *P. speciosus* populations occur. The plant is tall and has a large multi-capitulum inflorescence. In certain years, nearly half of the individuals can be damaged by the leaf beetles, which break the developing inflorescences, causing a delay in seed production due to re-growth of secondary smaller inflorescences.

Free-feeding insect herbivores are frequently found damaging rupestrian grassland plants (Fig. 14.1). Patterns of insect herbivore richness, abundance, and distribution have been evaluated by Ribeiro et al. (1998; see also Chap. 15). However, most studies done in the rupestrian grasslands were focused on gall-inducing insects, which alone represents more than half of all studies so far conducted (Fig. 14.1, Table 14.1).

Fig. 14.1 Insect herbivores in rupestrian grasslands from Serra do Cipó. Leaf chewing herbivores (**a**, **b**, **d**, **e**, **f**); sap sucking herbivore (**c**); galling herbivores (**g**, **h**, **i**). Photos **a–f** taken by L. Perillo and **g–i** by G.W. Fernandes

Table 14.1 Richness of gall-inducing insects in rupestrian vegetation sites (PE = Sate Park)

Author	Rupestrian field site	Gall Richness	Host plant richness
Maia and Fernandes (2004)	Serra de São José, MG	137	73
Maia (2014)	Several, including Brumadinho, Itamonte, Serra do Cipó, São João del-Rei and Tiradentes, MG	131	71
Urso-Guimarães et al. (2003)	Delfinópolis, MG	22	19
Coelho et al. (2013b)	Serra do Cabral, MG	47	89
Araújo et al. (2011)	Several, including rupestrian grasslands, GO	22	Not recorded for rupestrian grasslands separately
Fernandes et al. (1996)	Several cerrado and rupestrian grasslands	121	40
Carneiro et al. (2009a)	PE Rio Preto, MG	241	142
	RPPN Caraça		
	PE Biribiri		
	PE Itacolomi		
	PE Serra de Ouro Branco		
	PE Grão Mogol		
Coelho et al. (2013a)	PE Serra do Brigadeiro	102	50
	PN Caparaó		
	PE Ibitipoca		
	PN Itatiaia		

For all sites, Cecidomyiidae (Diptera) was the most common galling insect

These studies were primarily performed to test many hypotheses proposed to explain the high galling insect species richness worldwide. Fernandes and Price (1988, 1991), for example, raised the hypothesis that environmental harshness is the key factor determining galling species richness worldwide, with higher richness found at intermediate latitudes, warm habitats and sclerophyllous vegetation under hygrothermal and nutritional stress (but see Medianero et al. 2003; Julião et al. 2014). The adaptive radiation of this guild would be higher in this environment due to lowered mortality caused by pathogens, which are usually associated to moist vegetation. Although few studies have evaluated mortality factors associated to gall development in the ecosystem addressed here (e.g., Ribeiro-Mendes et al. 2002), the studies conducted have corroborated the hypothesis that gall richness and/or abundance are higher in xeric compared to mesic vegetation (e.g., Allain et al. 1998; Gonçalves-Alvim and Fernandes 2001; Ribeiro-Mendes et al. 2002; but see Fernandes et al. 1996; Araújo and Santos 2008; Carneiro et al. 2009a). The highly variable landscape found in the rupestrian grassland vegetation offers an unique area for the study of highly specialized insect herbivores, such as those represented by the galling insects. In there, their richness range from 18 to up to 241 species (Table 14.1).

Protection of the growing larva against natural enemies in xeric habitats as well as circumvention of plant defences have been suggested as potential mechanisms explaining the adaptive radiation of this insect guild in the warm and sclerophyllous rupestrian ecosystem. Nevertheless, the mechanism of higher mortality rates in mesic versus sclerophyllous vegetation seem to not be restricted to rupestrian grasslands, but common to any xeric-mesic pair, including upper canopy versus understory in tropical rainforests (Ribeiro and Basset 2007; Ribeiro et al. 2014). Studies on the mechanisms affecting patterns of insect distribution were only investigated in further details for galling insects in the rupestrian grasslands, but a broad view on general trends in the distribution of plants, animals and fungi in an altitudinal gradient in Serra do Cipó is presented in Chap. 15.

Gall inventories in Brazil have been extensively conducted over the past decades and several insects have been discovered as new to science (Coelho et al. 2013a, b; Fernandes and Santos 2014), indicating the importance of these studies to the knowledge of diversity patterns amongst vegetation types. However, evaluation of both bottom-up and top-down factors that explain variation between plants and within regions are still incipient and do not allow yet a clear recognition of their importance for insect-plant ecology and evolution. There is only one study that evaluated the impact of galls induced by the midge *Rhopalomyia* on growth and nutrient acquisition of the host plant *Diplusodon orbicularis* (Marini-Filho and Fernandes 2012), indicating a gap in our knowledge of the effects of galls as plant antagonists.

On the other hand, literature seems to have devoted much greater time in exploring the effects of host plants on associated insect herbivores at various ecological levels. Free-feeding insect herbivore richness decreases during the dry season and with increasing altitude in rupestrian grasslands indicating that plant sclerophylly may exert a strong negative influence on insect species richness (Ribeiro et al. 1998). A positive relationship has also been found between resource availability and herbivore abundance in rupestrian grasslands, as the number of flowers and height of the host plant influenced the abundance of herbivorous insects on *Vellozia nivea* (Velloziaceae) (Landau et al. 1998). Espírito-Santo et al. (2007), evaluating 17 species of *Baccharis* collected mainly in rupestrian grassland sites, suggested that interespecific differences in plant architecture might be responsible for differences in gall insect richness amongst species, promoting insect radiation among host plants of similar architecture through host shifts and/or sympatric speciation. Therefore, host plant characteristics such as morphology, biomass, height, chemistry and physiology might drive the patterns found for the variation in herbivore diversity and abundance in rupestrian grassland plants, influencing local and regional patterns of insect abundance. The next sections explore in details some of the most-known patterns of plant defences and plant-antagonistic insect interactions for this vegetation.

14.3 Plant Dioecy and Its Effects on Herbivory

Plant sex is suggested to be an important source of variation in patterns of growth, performance, and defence against herbivores in dioecious species (Boecklen et al. 1990) due to differential investment in vegetative growth and reproduction. Females, by investing more into reproduction than males, exhibit lower vegetative growth and more chemical and/or structural defences (Boecklen et al. 1990). Males, on the other hand, grow faster, invest less into chemical defences and exhibit better tissue nutritional quality (for a review see Cornelissen and Stiling 2005).

Sex-biased herbivory, often towards males, may be a result of differences in trade-offs between sinks of resource allocation in plants. Cornelissen and Stiling (2005) reviewed through meta-analysis the effects of sex on plant-herbivore interactions and found support for the hypothesis that increased growth in male plants coupled with reduced secondary chemistry compared to female plants leads to higher herbivory levels in the former. In that study, several guilds were evaluated, while for the rupestrian grasslands, mainly species within the genus *Baccharis* have been studied, and some conflicting results have been found (Araújo et al. 2003, 2006; Carneiro et al. 2005, 2006; Espírito-Santo et al. 2012). Carneiro et al. (2005), for example, examined the gall community associated to *Baccharis concinna* and found no support for the sex-biased hypothesis when the whole gall community was examined, but found support for the hypothesis when populations were examined separately, indicating the importance of searching for interactions between plant sex and local factors that influence insect richness and abundance. But Espírito-Santo et al. (2012) in a study conducted with dioecious shrub species of *Baccharis* found an absence of differential reproductive allocation and similar growth and shoot production between plant sexes.

In a study conducted by Araújo et al. (2003) in the dioecious shrub *Baccharis pseudomyriocephala* and its associated galling fauna, the authors found that this plant species can host up to 10 galling species, but they did not observe any effect of host plant sex on module size (an estimate of plant growth) and inflorescence number (an estimate of plant reproductive effort). Also, host sex did not show any influence on both the abundance and richness of galling insects, and gall survivorship was not influenced by module size. Results of this study showed that neither sex or vigor of this plant species influence the community structure of its galling herbivores. In another study with the same host, males of *B. pseudomyriocephala* exhibited significantly higher richness and abundance of galling and leaf mining insects than female plants (Cornelissen et al. 2013). The competition for establishment in those "optimal" male plants might be responsible for the patterns of insect assembly and organization in this dioecious species.

14.4 Plant Taxon Size and Herbivory

The galling guild is extremely diverse in the Neotropics. It is estimated to reach 21,000 species (Espírito-Santo and Fernandes 2007) although only a very small proportion has been described so far (Gagné 1994; Maia 2005). There is a large debate in the literature concerning the mechanisms that determine the diversity of galling insects (Fernandes and Price 1988, 1991; Price et al. 1998; Fernandes and Santos 2014). The most established pattern of distribution demonstrates a peak of richness at intermediate latitudes, in both up north and down south, associated to sclerophyllous vegetation such as the Mediterranean, Cerrado and Fynbos (Price et al. 1998). Amongst the plant formations with the greatest richness of galling species in the tropics, rupestrian grasslands stand out with the greatest species richness in the world so far documented (Lara and Fernandes 1996). As mentioned before, high galling species richness in this region could be explained by a lower mortality rate caused by natural enemies; but it is also believed to be associated to a greater availability of meristems or oviposition sites in those habitats (Fernandes and Price 1988, 1991; Espírito-Santo et al. 2007, 2012) or to a phenological synchronization of the plant communities (Mendonça 2001; Espírito-Santo et al. 2012).

The effects of taxon size—usually plant genera and families—on galling richness are hitherto understudied (Fernandes 1992; Mendonça 2007). According to the taxon size hypothesis ("The plant family size hypothesis", Fernandes 1992), larger plant taxa potentially offer more hosts and consequently bear a greater number of galling insects. Diverse groups of parasitic organisms are richer in species when larger host taxa are available (e.g. Poulin 2007). Gall inducing insects and other guilds of herbivores that obligatorily spend part of their life cycles on a single plant, however, may be considered a special type of parasitic organism (see Price 1980) and should respond to plant taxa size accordingly.

Data from Carneiro et al. (2009a) were used to address the question of whether there is a positive relationship between galling insect richness and plant taxon size. A total of 242 galling insects were found on 138 host plant species. The families that hosted the greatest number of galling species were Asteraceae (42 %), Melastomataceae (11 %) and Malpighiaceae (10 %). The genera which hosted the greatest number of galling species were *Baccharis* (16 %) and *Byrsonima* (7 %) (see Carneiro et al. 2009a) (Fig. 14.2).

Galling insect richness was positively correlated with plant taxon size as larger plant families hosted more galling species than smaller plant families ($r = 0.68$; $S = 8018.058$; $p < 0.001$; Fig. 14.3a). The same trend was found at the genus level ($r = 0.56$; $S = 190014.5$; $p < 0.001$; Fig. 14.3b).

Although some host taxa have shown an exceptionally higher than expected by chance number of insect galling species (e.g. Asteraceae, Melastomataceae, Malpighiaceae, the genera *Baccharis* and *Byrsonima*) there was a strong correlation between taxon size and galling insect richness, supporting the taxon size hypothesis. Considering the high species-specificity of galling insects to their host plants, an increase in plant taxon richness brings a potential increase in the number of host

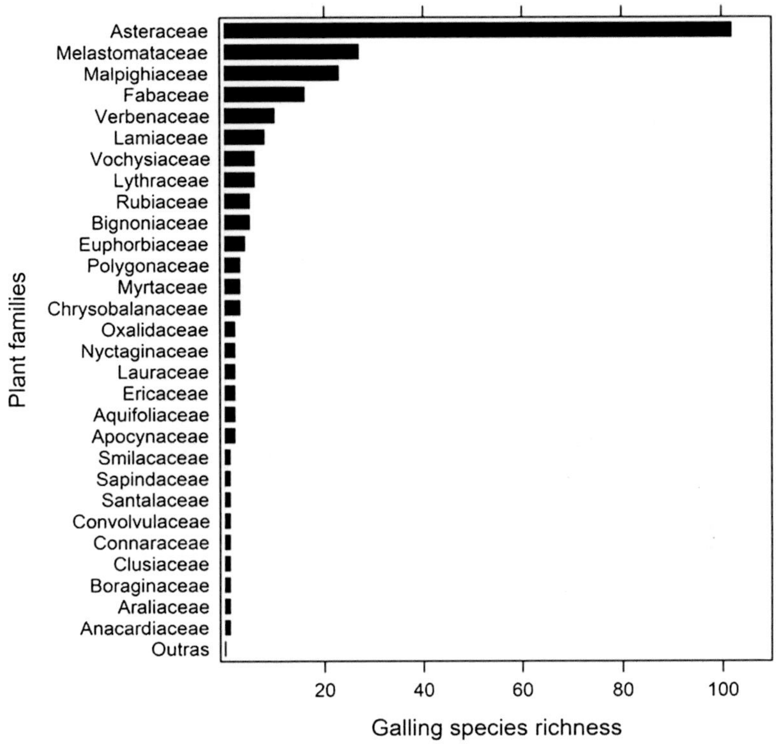

Fig. 14.2 Distribution of galling species richness per plant family in the rupestrian grasslands. Plant families in which no galling insect was recorded in the field were categorized together as "Others" (Erythroxylaceae, Gesneriaceae, Meliaceae, Velloziaceae, Annonaceae, Flacourtiaceae, Dilleniaceae, Eriocaulaceae, Malvaceae, Moraceae, Myrsinaceae, Ochinaceae, Proteaceae, Rutaceae, Campanulaceae, Caryocaraceae, Gentianaceae, Krameriaceae, Orquidaceae, Salicaceae, Scrophulariaceae, Solanaceae, Sterculiaceae, Winteraceae)

plants, or an increase in niches available for colonization by these insects (Fernandes 1992; Mendonça 2007; see also Strong et al. 1984).

This positive correlation between the increase in gall and host plant richness is a prediction of a more general hypothesis called taxonomic isolation (Southwood 1961; Lawton and Schoreder 1977; Strong et al. 1984; see also Fleck and Fonseca 2007). Host plants from related or closer lineages contain similar chemical compounds and are more easily subjugated by herbivorous insects (Ehrlich and Raven 1964). The occurrence of phylogenetically similar plant species, or rather, plants belonging to the same taxon in a sympatric condition would result in the speciation of galling insects via host replacement (Jermy 1984; Futuyma et al. 1995; Joy and Crespi 2007). However, some studies have shown that errors while choosing hosts may sometimes result in replacement by a plant species from another of a not so close lineage (Ohshima and Yoshizawa 2006). Species of galling insects on 'new host plants' could lead to a speciation process increasing therefore local richness

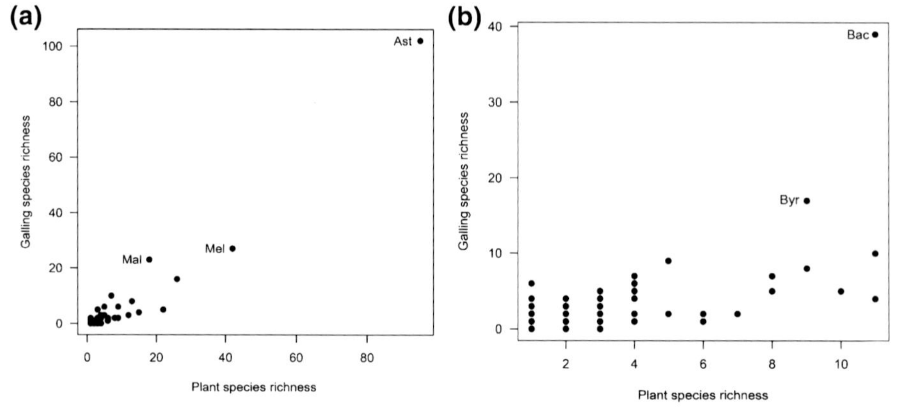

Fig. 14.3 Relationship between galling species richness and taxon size (=plant species richness). The galling species richness increases positively with the number of plant species in **a** plant families (Spearman: r = 0.68; S = 7909.624, $p < 0.001$); **b** plant genera (Spearman: r = 0.47; S = 237574.8, $p < 0.001$). *Ast* Asteraceae, *Mal* Malpighiaceae, *Mel* Melastomatacea, *Bac Baccharis Byr* Byrsonima

(Bush 1994; Feder 1998; Craig et al. 2001). This type of local speciation via host shift could be one of the main mechanisms driving the sympatric speciation of galling insects (Abrahamson et al. 2001). Nonetheless, speciation can also occur without host replacement via adaptive radiation in the same host plant (Cook et al. 2002; Després et al. 2002). In this case the adaptive radiation occurs via organ replacement on the same host plant with reproductive isolation involving phenological separation (Mendonça 2001; Ferdy et al. 2002; Joy and Crespi 2007). In both cases, there is a potential for positive correlation between galling insect richness and taxa (plant genus or family) size (Fernandes 1992), as observed for several sites in rupestrian grasslands.

14.5 Fluctuating Asymmetry and Herbivory

Fluctuating asymmetry (FA) represents small, random variation from symmetry in otherwise bilaterally symmetrical characters and it is a widely used measure of developmental instability in plants and animals (Diaz et al. 2004). FA occurs when otherwise bilateral traits show small variations in the size of the two sides of a character around a bilateral symmetric axis. Because the same genes control the development of both sides of bilaterally symmetric traits, the degree of FA represents developmental perturbations and reflects interference in the control of development processes when growing under given environmental conditions. Significant increases in FA have been found for several species of plants and animals in response to various stresses, including both environmental and genetic factors. FA has been widely used to compare developmental instability levels in populations subjected or not to a number of environmental stressors such as temperature, nutrition, radiation, chemicals, population density, parasites, light conditions, fire, predation risk, and habitat structure. Because it constitutes a quick, cheap and efficient measure, it has frequently been used for monitoring levels of ecological stress (see Beasley et al. 2013 for a review).

The relationship between FA and herbivory has been examined for several plant-herbivore systems involving a diverse array of insect guilds and results have supported a positive relationship between asymmetry and herbivory (Cornelissen and Stiling 2011). Such relationship might be explained by two conflicting hypotheses: (1) insects cause plant stress and increase asymmetry levels or (2) insects use asymmetry as a surrogate for tissue quality and feeding site selection. These two hypotheses have been both tested in the literature and there is support for both in different systems. Usually, support for the second hypothesis comes from studies where differences in nutritional quality are found between symmetric and asymmetric leaves (see Cornelissen and Stiling 2005, 2011).

Some tropical systems in rupestrian grasslands have been evaluated to test the relationship between FA and herbivory, with mixed results. No relationship between fluctuating asymmetry and insect herbivores were found for the endemic plant species *Coccoloba cereifera* (Polygonaceae) (Telhado et al. 2010), as well as for *Chamaecrista semaphora* in restored and natural environments (Costa et al. 2012). For four species of Melastomataceae, Telhado et al. (2015) found higher levels of asymmetry in damaged compared to undamaged leaves, but insect density was too low to generate any patterns between and within plants. While, the interactions of insect herbivores and asymmetry of leaves in the rupestrian plant species do not show any trends, more studies are called for as the generalized low quality of the plant material in this ecosystem may represent an interesting source of variation in leaf design and opportunity to further test the hypothesis.

14.6 Plant Defence and Selection by Herbivores

Hairston et al. (1960) pondered on "why the world is green" arguing that herbivores are not limited by resources but by natural enemies such as predators and parasites. Numerous criticisms have been made around this generalization and its simplistic assumptions (e.g. Murdoch 1966). Despite the fact that both bottom-up and top-down effects act synergistically there have attempts to formulate models contemplating one or the other as a main ecosystem driving force (Meserve et al. 2003). Generalizations are constrained by idiosyncratic patterns usually found in the literature and Hunter and Price (1992) suggested that each insect-plant system deserve a particular view of which factors are actually affecting herbivore survival, and suggest an even balance between host productivity and predation (see also Polis 1999).

A complex set of factors defines what is a safe place for insect herbivores to feed and develop on plant species. However, it is hard to detect or describe such factors for a tropical ecosystem. Besides plant gender, for species in the genus *Baccharis*, plant structural complexity might be a good predictor of herbivory by gall-formers (Espírito-Santo et al. 2007). The influence of structural complexity of *Baccharis pseudomyriocephala* on herbivore assemblage was investigated by Araújo et al. (2006). Plant architecture influenced positively species richness, abundance and survival of galls (Araújo et al. 2006; Espírito-Santo et al. 2007). The number of plant modules positively influenced the survival of gall-inducing insects, although architectural complexity did not significantly influence gall parasitism. Mortality of galling insects by parasitoids was low (13.26 %) suggesting that top-down control might be less important to the regulation of the community of gall-inducing insects than bottom-up control (see below).

In order to defend against herbivores, plants use physical, chemical and biological defences. The rupestrian grasslands occupy nutrient-poor soils, which constrain plant development and functional traits (Rizzini 1997; Negreiros et al. 2014). As a result, plant species are generally sclerophyllous, with dicot shrubs and herbs exhibiting long-lived leaves and several other traits related to herbivore deterrence (e.g., thick cuticle, trichomes, high lignin contents, phenolics, terpenoids, alkaloids or, most likely, dose-dependent compounds such as tannins (Fernandes and Price 1991; Oliveira et al. 2015; Silveira et al. 2016).

Toughness and sclerophylly may act in conjunction with high concentration of immobile defences to diminish the success of insect colonization in xeric habitats. Hence, sclerophylly has been assessed and suggested as a potential anti-herbivory strategy of rupestrian grassland plants (Ribeiro et al. 1998; Negreiros et al. 2014; Silveira et al. 2016). The harsh and severe leaf quality acts as a strong barrier refraining the feeding time of sap-sucking and leaf chewing insects (Ribeiro et al. 1994b). For tropical and subtropical forests, the distribution of surviving gall-formers was significantly influenced by sclerophylly, as predicted by Fernandes and Price (1991, 1992). Considering that forest canopies are sclerophyllous and xeric habitats similar to the open canopy of a savanna (sensu Ribeiro 2003; see also Bell et al. 1999; Medianero et al. 2003; Julião et al. 2014), gall survival has been found to be much higher in the canopy than in the mesic understory. In addition, Ribeiro and Basset (2007) and Ribeiro et al. (2014) found an inverse correlation between gall abundance and leaf area lost to herbivory by chewing herbivores, which is more intense in mesic than xeric habitats, increasing the mortality of endophagous insects in the understory due to accidental predation.

For chemical defenses, the seminal paper of Fraenkel (1959) has focused on the role of plant secondary compounds, as well as plant mechanical defenses, in shaping herbivory patterns found between and within plants. Evaluation of chemical defenses against herbivory for the rupestrian grassland plants has been rarely conducted, with the exception of quantification of phenolic compounds in a few species (see Moreira et al. 2002). Environmental factors, including low temperature and nutritional stress, which are typical of tropical highland environments, favour the evolution of several plant morphological traits (Smith and Young 1987; Safford 1999), with some potential impacts on leaf characteristics.In shrubs, such as species within the genus *Baccharis*, the leaf is often cupressoid, and leaves are clustered at the stem apex. *Baccharis* species occurring in rupestrian grassland are usually water and nutritionally stressed and the great number of clustered leaves should provide better protection against solar radiation and desiccation (see Herms and Mattson 1992).

Plants can ally to their natural enemies as predators or parasitoids as a defense strategy. However, top-down effects would be weakened in harsh environments, where parasitoids, pathogens and endophytic fungi would exhibit lower incidence (Fernandes and Price 1991, 1992; Ribeiro and Basset 2007). In addition to parasitoids, plants can attract ants (see Chap. 13), which may lead to the removal of herbivores (e.g. Del Claro 2004; Del Claro and Torezan-Silingardi 2009). Rupestrian grasslands are rich in examples of ants patrolling plants while seeking for resources with high levels of protein and carbohydrates as extra-floral nectaries and trophobiont insects (see review in Chap. 13, Fig. 14.4). Ant activity may explain variable herbivory rates, depending upon the predominance of bottom-up or top-down forces. In the tropics, confounding effects of such contradictory forces are even harder to investigate, as direct effects of interaction with microorganisms, vertebrate predation and other aspects have scarcely been investigated.

Another factor directly influencing the distribution of insect herbivore species is enemy-free space (sensu Jeffries and Lawton 1984). In many instances, the secondary effect of deterrent plant traits is to expose the insects to a greater chance of being attacked by natural enemies. Escaping predation and/or parasitism strongly influences galling species distribution. For endophagous insect species, avoiding leaves that are heavily chewed might be a way of escaping accidental predation, and this is a neglected effect of indirect competition amongst herbivore guilds. A simplified Azorean insular forest has proven to be a suitable woody system to study the effects of predators on insect herbivore distribution on tree crowns. In these forests, Ribeiro et al. (2005) found that insect herbivore distribution could hardly be related to any other negative effect but predation, as these insular tree species are usually poor in secondary chemical defences or other defensive barriers (Givnish 1998). However, the variety of natural enemies in an oceanic island is rather low compared with continental systems, as those studies revealed, for

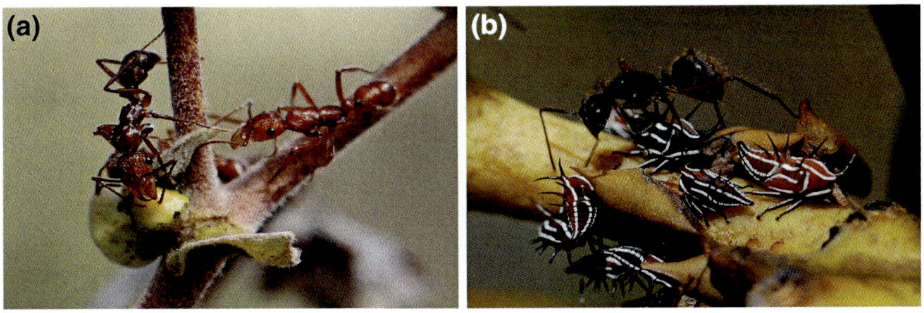

Fig. 14.4 Ants on plants in rupestrian grasslands from Serra do Cipó. *Ectatomma tuberculatum* foraging in extra-floral nectaries (**a**); *Camponotus rufipes* foraging and trophobiont insects (Membracidae) (**b**). Photo A taken by F. Costa and B by T. Guerra

example, the total absence of ants, lizards, and birds. Hence, a clear direct effect of other main predators such as spiders ought to be further examined (see Ribeiro and Borges 2010). Studies have shown that evolutionarily predictable and apparently safer tree species crowns influenced the distribution of a highly generalist herbivore community more than habitat size or plant taxonomy (Ribeiro et al. 2005, Ribeiro and Borges 2010).

Hence in the rupestrian grasslands patterns as those described for forests are unlikely to be detected, if they even exist. Actually, the widespread sclerophylly and very high pressure exerted by aggressive species of ants as predators must dominate the risk scenario for most insect herbivores, especially in shrubby rupestrian vegetation. However, so far no one has designed a proper experiment to explore causal effects of plant sclerophylly and other defense strategies against herbivores in rupestrian grasslands.

14.7 Mistletoe Interactions

Mistletoes are parasites occurring on some plant species (Kuijt 1969; Arruda et al. 2012). They can be holoparasites—when they do not synthesize chlorophyll and fully absorb all the nutrients they need from their host plants, or hemiparasites—when they actually photosynthesize, but attain from their host plants water and some inorganic compounds (Pennings and Callaway 2002). The majority of mistletoes are hemiparasites and use as hosts species within trees, shrubs, columnar cacti, and even other parasitic plants (Kuijt 1969; Calder and Bernhardt 1983). When installing on a host, mistletoes can alter host growth, physiology and reproduction of the individual being parasitized (Lüttge et al. 1998; Medel 2000; Aukema 2003; Mourão et al. 2009; Cameron et al. 2009; Bell and Adams 2011).

In rupestrian grasslands, species of mistletoes have been observed using different species of herbs, shrubs and trees as hosts. Two species of generalist mistletoes, *Psittacanthus robustus* and *Struthanthus flexicaulis,* have very wide distribution in the rupestrian grasslands in the southern Espinhaço mountains (e.g., Monteiro et al. 1992; Mourão et al. 2006; Guerra and Pizo 2014; Jacobi et al. 2015). These mistletoes use several species as hosts, and are most frequent on the most common and emerging plants in the system, but usually causing low mortality on their hosts (Jacobi et al. 2015; but see Bahia et al. 2015). *P. robustus* is a woody mistletoe and its main host species are in the Vochysiaceae and Melastomataceae families (Monteiro et al. 1992; Fig. 14.5), being very common on *Vochysia thyrsoidea* (Guerra 2010). *S. flexicaulis* has more flexible traits with a modular growth with shoots attaching to different points of the host crown and forming secondary epi-cortical root systems (Mourão et al. 2006; Fig. 14.5). It infects more the 60 species from different families (Mourão et al. 2006; Guerra and Pizo 2014). In ironstone outcrops, *Mimosa calodendron* (Fabaceae) is its most common host (Mourão et al. 2006). In the quartizitic rupestrian grasslands this species is becoming widespread and Bahia et al. (2015) have recorded strong top-down effects on the host and

Fig. 14.5 Generalist mistletoes (Loranthaceae) in rupestrian grasslands. *Struthanthus flexicaulis* in its most common host *Mimosa calodendron* (Fabaceae) (**a**); *Psittacanthus robustus* in *Vochysia thyrsoidea* (Vochysiaceae) (**b**). Photo A taken by F. Mourão and B by T. Guerra

perhaps on the entire community of associated organisms, including host mortality. More specialized plant mistletoes like *Tripodanthus acutifolius* (Jacobi et al. 2015) and *Phoradendron amplexicaule* (Cuevas-Reyes et al. 2011) in the rupestrian grasslands use less than 10 species as hosts (Jacobi et al. 2015).

The parasitism by mistletoes reduces host growth, photosynthesis and respiration rates, and also affect host architecture, water and nutrient balance (Tennakoon and Pate 1996; Meinzer et al. 2004; Bahia et al. 2015). Mistletoe parasitism represents an important factor of stress for host plants in the rupestrian grasslands as shown by Lüttge et al. (1998) (see also Schwartz et al. 2003; Fernandes et al. 1998). Crown deformation and host death by *S. flexicaulis* was reported by Bahia et al. (2015) (see also Pennings and Callaway 2002; Press and Phoenix 2005; Reblin et al. 2006). Parasitism by these plants may produce instability during the development of organs such as leaves due to the induced high stress (Press and Phoenix 2005), and was evaluated by Cuevas-Reyes et al. (2011) studying leaf fluctuating asymmetry of mistletoes' hosts as an indicative of stress. In rupestrian grasslands the presence of mistletoes on *Stachytarpheta glabra, Lychnophora pinastere* and *Mimosa calodendron* caused increased FA in parasitized compared to non-parasitized plants. On the other hand, these patterns were not maintained in different host systems also studied in the rupestrian grasslands. Leite (2015) found that FA in the host tree *Vochysia thyrsoidea* attacked by *P. robustus* did not change with the presence of mistletoes. However, this author noticed a reduction in 22 % in leaf size and 63 % increase in herbivory on parasitized *V. thyrsoidea* hosts.

The presence of mistletoes also influences other trophic levels. *Psytacanthus robustus* attracts birds (Guerra and Pizo 2014), herbivorous insects (Guerra et al. 2011) and ants (Freitas and Rossi 2015; Fig. 14.4). Jacobi et al. (2015) observed commensalism between the mistletoe *S. flexicaulis* and the ant *Camponotus crassus*, that cleans the seeds deposited by birds on the trunks of host (see also Guerra et al. 2011; Freitas and Rossi 2015; Chap. 13). In the presence of *P. robustus*, trophobionts attract some species of aggressive species of ants (e.g., *Camponotus rufipes* and *Cephalotes pusillus*) that attack potential mistletoe herbivore insects (Freitas and Rossi 2015; Leite 2015). However, the effects are apparently not extended to the host plant *V. thyrsoidea*, which are used only as bridges to access the trophobionts (Leite 2015). This association between trophobionts and ants in the mistletoes can further intensify the adverse effects of mistletoes on their hosts.

14.8 The Genus *Baccharis* as a Model for Multitrophic Interactions in the Rupestrian Grasslands

Baccharis (Asteraceae) is a genus comprising approximately 500 species distributed from the United States to Southern South America (Barroso 1976). The genus supports the richest galling insect fauna recorded so far in the Neotropics, with 121 galling species found on 40 host species (Fernandes et al. 1996, 2014; Carneiro et al. 2009a, b). The plant species *Baccharis dracunculifolia* D. C. (Asteraceae) is a perennial, evergreen and dioecious shrub, 2–3 m in height, which is widely distributed across the south-central portion of South America (Barroso 1976; Espírito-Santo et al. 2003), including the rupestrian grasslands. This species produces flowers twice a year, from March to June and again from November to December (Collevatti and Sperber 1997). *B. dracunculifolia* usually composes dense and well-defined clusters ranging from 18 to 12,000 m^2 (Collevatti and Sperber 1997). It hosts a speciose fauna of insect herbivores, including coleopterans, hemipterans and orthopterans (Fagundes et al. 2005; Fagundes and Fernandes 2011) and also the largest fauna of galling insects in the Neotropics, with 15 species (Fernandes et al. 1996).

Due to its diverse associated fauna and wide range of occurrence, usually in high frequency and across gradients (altitudinal, hygrothermal, and of habitat disturbance), *B. dracunculifolia* has been extensively used as a study model in ecological research. As a result, several interactions have been empirically demonstrated to date involving this plant species and associated insects. Many of them are multitrophic and indirect interactions. Here we summarize these interactions in a trophic interaction web (Fig. 14.6, Table 14.2) aiming to provide information and guidance for future studies based upon the *B. dracunculifolia* system.

Amongst the 15 species of gall inducing insects found on *B. dracunculifolia* (Fernandes et al. 1996), *Baccharopelma dracunculifoliae* (Sternorrhyncha: Psyllidae) is one of the commonest (Araújo et al. 1995; Fig. 14.6—Interaction 1). It induces a gall in the midrib of the leaf, which folds joining edges but not melting, forming an elliptical, green, glabrous, single-chambered gall (Fig. 14.7). This gall usually harbours one to four nymphs, although up to 21 have already been recorded (Lara and Fernandes 1994).

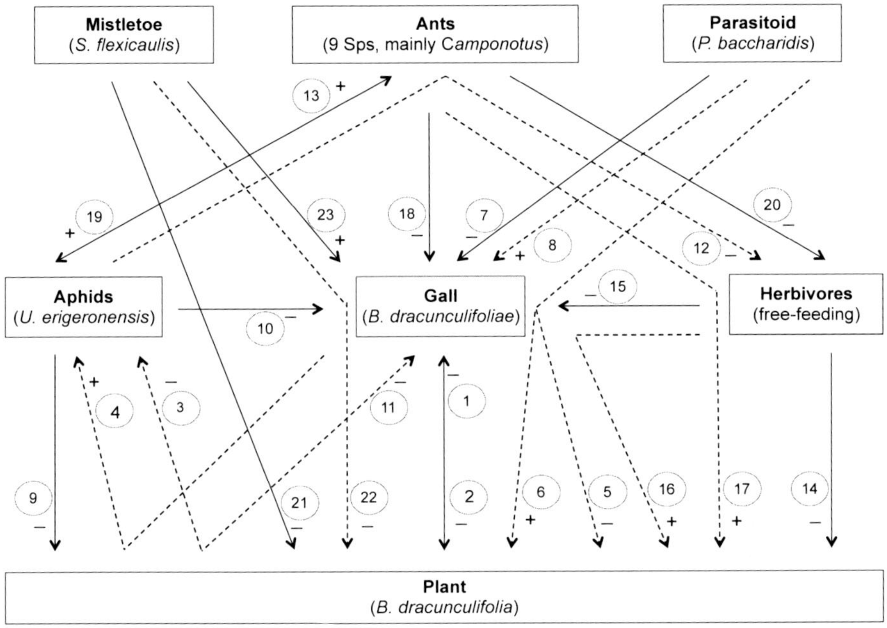

Fig. 14.6 Web of species and indirect interactions associated to the host plant *Baccharis dracunculifolia*. *Solid* and *dashed lines* show direct and indirect links, respectively. The numbers correspond to the description of each interaction found on

Galls of *B. dracunculifoliae* may be occupied by inquiline arthropods (Mani 1964) from the orders Diptera, Hemiptera (Aphididae), Neuroptera (Chrysopidae), Thysanoptera, and Acarina (Collevatti and Sperber 1997). These inquilines can use the chamber for protection and can consume silk produced by the galler and other plant products. They sometimes can also prey upon the galling larvae or on the gall tissue (Mani 1964; Espírito-Santo and Fernandes 2002). After dehiscence, the galls remain attached to the plant, sometimes for a few generations, and gradually become dry and woody (Espírito-Santo and Fernandes 1998).

According to Espírito-Santo and Fernandes (2002), the host plant *B. dracunculifolia* imposes strong bottom-up control over the galling population. Plant-mediated mortality in the galling insect *B. dracunculifoliae*, through gall dropping and plant resistance, represented 40.7 % of the initial cohort studied (Fig. 14.6—Interaction 2). However, top-down effects also play an important role in the performance of the galling psyllid. Espírito-Santo et al. (2004) report that, amongst the galls of *B. dracunculifoliae* that survived plant defences, parasitoid attack was the main cause of nymph mortality, with 45.2 % of the galls parasitized. Three parasitoid species are known to attack *B. dracunculifoliae*: *Lyrcus sp.* (Hymenoptera: Pteromlidae), *Brasema sp.* (Hymenoptera: Eupelmidae; Sperber and Collevatti 1996) and *Psyllaephagus baccharidis* (Hymenoptera: Encyrtidae; Tavares and Perioto 1993). *Psyllaephagus baccharidis* is the most common parasitoid of the

Table 14.2 Interactions empirically demonstrated involving the plant species *B. dracunculifolia* and associated species

Affecting	Affected				
	Baccharis dracunculifolia (Plant)	*Baccharopelma dracunculifoliae* (Gall)	*Uroleucon erigeronensis* (Sternorryncha: Aphididae; Aphid)	Free feeding herbivores	Ants—9 Sps, mainly *Camponotus* (Fagundes et al. 2005)
Baccharis dracunculifolia (Plant)		(1) ND—Bottom up effects (Espirito Santo and Fernandes 2002)			
Baccharopelma dracunculifoliae (Gall)	(2) ND—Nutrients sinking (Espirito Santo an Fernandes 2002)		(3) NI—Competition for plant photo assimilates (Fagundes et al. 2005); (4) PI—providing shelter/Inquilinism (Collevatti and Sperber 1997)		
Psyllaephagus baccharidis (Hymenoptera: Encyrtidae; Parasitoid)	(5) NI—Increase gall size (Espírito-Santo and Fernandes 2004); (6) PI—Gall parasitism (Espírito Santo et al. 2004)	(7) ND—Parasitism (Tavares and Perioto 1993); (8) PI—Health nymphs neighboring attacked ones are heavier (Espírito Santo et al. 2004)			
Uroleucon erigeronensis (Sternorryncha: Aphididae; Aphid)	(9) ND—Nutrients sinking affects shoot growth (Neves et al. 2011)	(10) ND—Inquilinism; kills nymphs (Espirito Santo and Fernandes 2002); (11) NI—Competition for plant photoassimilates (Fagundes et al. 2005)		(12) NI—Ants attracted reduce number of herbivores (Neves et al. 2011)	(13) PD—Trophobiosis (Fagundes et al. 2005; Neves et al. 2011)
Free feeding insect herbivores	(14) ND—Herbivory (Neves et al. 2011); (15) PI—Gall predation (Espirito Santo and Fernandes 2002)	(16) ND—Predation (Espirito Santo and Fernandes 2002)			

(continued)

Table 14.2 (continued)

Affecting	Affected				
	Baccharis dracunculifolia (Plant)	*Baccharopelma dracunculifoliae* (Gall)	*Uroleucon erigeronensis* (Sternorryncha: Aphididae; Aphid)	Free feeding herbivores	Ants—9 Sps, mainly *Camponotus* (Fagundes et al. 2005)
Ants (9 Sps, mainly *Camponotus*)	(17) PI—Herbivory reduction (Neves et al. 2011)	(18) ND—Less nymphs per gall as a result of interference with oviposition behaviour (Neves et al. 2011)	(19) PD—Protection against natural enemies (Fagundes et al. 2005)	(20) ND—predation or interference (Neves et al. 2011)	
Struthanthus flexicaulis (Mistletoe)	(21) ND—Parasitism (Bahia et al. 2015); (22) NI—Increase abundance of galls of *B. dracunculifoliae* (Bahia et al. 2015)	(23) PD—Facilitates gall incidence (Bahia et al. 2015)			

Numbers correspond to the links in the indirect interactions web (Fig. 14.3)
PD positive direct; *ND* negative direct; *PI* positive indirect; *NI* negative indirect

Fig. 14.7 *Baccharis dracunculifolia* hosting **a** the mistletoe *Struthanthus flexicaulis* and three stages of the gall induced by *Baccharopelma dracunculifoliae* (Sternorrhyncha: Psyllidae) in leaves of *B. dracunculifolia*; **b** intact gall hosting galling nymphs; **c** hatched gall; **d** gall hatched and dried. Photo A taken by T. Bahia, b-d by M. Barbosa

galling psyllid. In the survey carried out by Espírito-Santo et al. (2004), this parasitoid was responsible for 93 % of the parasitism recorded on galls of *B. dracunculifoliae* (Fig. 14.6—Interaction 7). *P. baccharidis* is a solitary koinobiont endoparasitoid which consumes the nymph, and pupates inside the host cuticle, causing so-called "nymph mummification" (Espírito-Santo et al. 2004). The attack occurs in the early stages of gall development increasing gall size and changing the colour and shape of the nymphs, which aids in the identification of mortality by parasitism (Espírito-Santo et al. 2004).

Apparently, the parasitoid *P. baccharidis* stimulates nymph feeding, increasing their size as well as the size of the gall induced by *B. dracunculifoliae* (Espírito-Santo et al. 2004). This can decrease plant performance as larger galls represent stronger nutrient sinks. Besides, the authors observed that healthy nymphs of *B. dracunculifoliae* growing in galls with parasitized nymphs were larger than healthy nymphs in unparasitized galls, regardless of nymph number. This might result in adults of the gall maker having increased survivorship and reproductive success (Espírito-Santo et al. 2004; Fig. 14.6—Interaction 8). Thus, the parasitoid *P. baccharidis* can have both a positive indirect effect on the host plant by controlling the population of the gall maker *B. dracunculifoliae* (Fig. 14.6—Interaction 6) as well as a negative indirect effect, by increasing gall nutrient sinking and galling insect fitness (Espírito-Santo et al. 2004; Fig. 14.6—Interaction 5).

Uroleucon erigeronensis (Sternorrhyncha: Aphididae) is the most abundant phloem sucking herbivore in *B. dracunculifolia* (Fagundes et al. 2005). This aphid feeds and reproduces on the apical meristems of the host plant (Fig. 14.6—Interaction 9) forming dense colonies that produce honeydew (sugary secretions). They attract at least nine species of ants, mainly from the genus *Camponotus*, to tend and protect them in a trophobiontic relationship (Fagundes et al. 2005; Neves et al. 2011; Fig. 14.6—Interactions 13 and 19). Fagundes et al. (2005) carried out an experiment in which ants, aphids or both were excluded from shoots of *B. dracunculifolia*. This study showed that ants had a direct negative effect on the performance of the galling insect *B. dracunculifoliae*. In the presence of ants, a smaller number of nymphs were found in each gall of *B. dracunculifoliae*, possibly as a result of ants interfering with the oviposition behaviour of the female gall maker (Fagundes et al. 2005; Fig. 14.6—Interaction 18). A recent survey by Monteiro (2014) of the insect herbivore community associated to *B. dracunculifolia* also found that the abundance of ants positively affected the abundance of trophobionts and negatively affected richness of chewing herbivores on the studied plants.

In addition, because gall size can be related to galling insect performance (Weis et al. 1988), aphids had a negative indirect effect on gall development: in their absence, galls were heavier. This suggests exploitative competition between *B. dracunculifoliae* galls and the sap-sucking aphid *U. erigeronensis* (Fagundes et al. 2005; Fig. 14.6—Interactions 3 and 11). Inquilinism by the aphid *U. erigeronensis* is also common and could be indirectly responsible for killing the nymphs of the gall maker *B. dracunculifoliae* (Espírito-Santo and Fernandes 2002; Fig. 14.6—Interaction 10).

In another exclusion experiment, Neves et al. (2011) observed that the presence of ants, aphids (*U. erigeronensis*) or both on *B. dracunculifolia* plants decreased the abundance of other free-feeding herbivores, and that the presence of the aphids decreased plant shoot growth (Fig. 14.6—Interaction 9). Ants tending aphids have a direct negative effect on herbivores by predating on or interfering with them to protect the aphids (Abe 1988; Fernandes et al. 1999; Fig. 14.6—Interactions 12 and 20). However the aphids by their own can also reduce abundance of sap-sucking and chewing insects due to exploitative competition or by altering the nutritional quality of the host plant (Fay et al. 1996; Larson and Whitham 1997). Therefore, the aphids can have a positive indirect effect not only on the host plant, by reducing herbivory, but also on the galling species *B. dracunculifolie*, as chewing insects can eat part of the gall wall and lead to death of the galling nymph (Espírito-Santo and Fernandes 2002; Fig. 14.6—Interaction 15).

Furthermore, the hemiparasitic plant *Struthanthus flexicaulis* often colonizes individuals of *B. dracunculifolia* (Fig. 14.7) and may have important effects on the *B. dracunculifolia* community of associated insects (Bahia et al. 2015). Bahia et al. (2015) report that although this mistletoe seems not to directly affect the host's physiology, it may reduce the number of leaves and cause the death of an occupied branch or even of the entire plant. In addition, *S. flexicaulis* may also negatively affect the host indirectly as branches occupied by the mistletoe exhibited higher abundance of galls induced by *B. dracunculifoliae* (Bahia et al. 2015).

14.9 Final Remarks

A fair amount of data has accumulated over the past years regarding antagonistic interactions in the rupestrian grassland ecosystem, but the information gathered so far and the patterns already described are still incipient compared to the richness of its fauna and flora and the myriad of possibilities of interactions among them. Also, studies are mainly concentrated in the southern portion of the Espinhaço range, where most of the works involving insect herbivores have been conducted. We have generated hypotheses to explain, for example, patterns of gall distribution and survivorship, and the importance of mortality factors, but most of the data comes from well-studied plant species such as those in the genus *Baccharis*. Although this vegetation is dominated by scherophyllous plant species and many long-standing hypotheses on plant-herbivore associations are related to plant tissue quality, the rupestrian grasslands offer many possibilities for further studies and hypotheses-testing at different spatial and temporal scales.Some patterns have also emerged on the importance of sclerophylly and association of ants as main defense mechanisms of rupestrian grassland plants; or the indirect effect of parasitism by mistletoes on other interactions. The same observation can certainly be extended to studies on network and food web interactions, the role of plant chemistry, taxonomical isolation and top-down *versus* bottom-up community organization. Despite its limitations, the knowledge acquired to date on rupestrian grasslands represents an important contribution to tropical ecology in general and more recently has served as basis for more mechanistic approaches. Furthermore, these possibilities are enormously augmented when one considers the length of the mountain and rupestrian environments provided by the Espinhaço range and isolated mountaintops.

Acknowledgments The authors would like to thank CNPq and FAPEMIG and PELD/CRSC, ComCerrado for several grants that have funded studies over the past years, especially at Serra do Cipó and to the Reserva Vellozia for the logistical support.

References

Abe Y (1988) Trophobiosis between the gall wasp *Andricus symbioticus*, and the gall-attending ant, *Lasius niger*. Appl Entomol Zool 23:41–44

Abrahamson WG, Eubanks MD, Blair CP, Whippie AV (2001) Galls flies, inquilines, and goldenrods: a model for host-race formation and sympatric speciation. Am Zool 41:928–938

Allain RL, Lara ACF, Fernandes GW, Marques EA (1998) Leaf tannin concentration, toughness, and sclerophylly in some typical woody species of the savannas of southeastern Brazil. Bios 6:5–10

Araújo WS, Santos BB (2008) Efeitos do habitat e da sazonalidade na distribuição de insetos galhadores na Serra dos Pirineus, Goiás, Brasil. Rev Biol Neotr 5:33–39

Araújo AM, Fernandes GW, Bede LC (1995) Influência do sexo e fenologia de *Baccharis dracunculifolia* DC. (Asteraceae) sobre insetos herbívoros. Rev Bras Entomol 39:347–353

Araújo APA, Carneiro MAA, Fernandes GW (2003) Efeitos do sexo, do vigor e do tamanho da planta hospedeira sobre a distribuição de insetos indutores de galhas em *Baccharis pseudomyriocephala* Teodoro (Asteraceae). Rev Bras Entomol 47:483–490

Araújo APA, Paula JD, Carneiro MAA, Schoereder JH (2006) Effects of host plant architecture on colonization by galling insects. Austral Ecol 31:343–348

Araújo WS, Santos BB, Gomes-Klein VL (2011) Insect galls from Serra dos Pirineus, GO, Brazil. Biota Neotrop 11:357–365

Arruda R, Fadini RF, Carvalho LN, Del-Claro K, Mourão FA, Jacobi CM, Teodoro GS, van den Berg E, Caires CS, Dettke GA (2012) Ecology of neotropical mistletoes: an important canopy-dwelling component of Brazilian forests and savannas. Acta Bot Bras 26:264–274

Aukema JE (2003) Vectors, viscin, and Viscaceae: mistletoes as parasites, mutualists, and resources. Front Ecol Environ 1:212–219
Bahia TO, Zúñiga IG, Souza ML, Coutinho ES, Quesada M, Fernandes GW (2015) Hemiparasitism effect on *Baccharis dracunculifolia* DC. and consequences to its major galling herbivore. Acta Bot Bras 29:339–345
Barroso GM (1976) Compositae – subtribo Baccharidinae Hoffman: estudo das espécies ocorrentes no Brasil. Rodriguesia 40:3–273
Beasley DAE, Bonisoli-Alquati A, Mousseau TA (2013) The use of fluctuating asymmetry as a measure of environmentally induced developmental instability: a meta-analysis. Ecol Indic 30:218–226
Bell TL, Adams MA (2011) Attack on all fronts: functional relationships between aerial and root parasitic plants and their woody hosts and consequences for ecosystems. Tree Physiol 31:3–15
Bell AD, Bell A, Dines TD (1999) Branch construction and bud defence status at the canopy surface of a West African rainforest. Biol J Linn Soc 66:481–499
Boecklen WJ, Price PW, Mopper S (1990) Sex and drugs and herbivores: sex-biased herbivory in arroyo willow (*Salix lasiolepis*). Ecology 71:581–588
Bush GL (1994) Sympatric speciation in animals: new wine in old bottles. Tree 9:285–288
Calder M, Bernhardt P (1983) The biology of mistletoes. Academic Press, New York
Cameron DD, White A, Antonovics J (2009) Parasite-grass-forb interactions and rock-paper-scissor dynamics: predicting the effects of the parasitic plant *Rhinanthus minor* on host plant communities. J Ecol 97:1311–1319
Carneiro MAA, Fernandes GW, Souza OFF (2005) Convergência na variação local e regional da riqueza de espécies de insetos galhadores. Neotrop Entomol 34:547–553
Carneiro MAA, Souza WVM, Fernandes GW, DeSouza OFF (2006) Sex-mediated herbivory by galling insects on *Baccharis concinna* (Asteraceae). Rev Bras Entomol 50:394–398
Carneiro MAA, Borges RAX, Araújo APA, Fernandes GW (2009a) Insetos indutores de galhas da porção sul da Cadeia do Espinhaço, MG. Rev Bras Entomol 53:570–592
Carneiro MAA, Branco CSA, Braga CED, Almada ED, Costa MBM, Fernandes GW, Maia VC (2009b) Are morphologically similar galls really induced by the same gall midges species? Rev Bras Entomol 53:365–378
Coelho MS, Mourão Carneiro MAA, Branco CSA, Borges RAX, Fernandes GW (2013a) Gall-inducing insects from campos de altitude, Brazil. Biota Neotrop 13:139–151
Coelho MS, Carneiro MAA, Branco CSA, Borg es RAX, Fernandes GW (2013b) Gall-inducing insects from Serra do Cabral. Biota Neotrop 13:102–109
Coley PD, Aide TM (1991) Comparison of herbivory and plant defenses in temperate and tropical broad-leaved forests. In: Price PW, Lewinsohn TM, Fernandes GW, Benson WW (eds) Comparison of herbivory and plant defenses in temperate and tropical broad-leaved forests. Wiley, New york, pp 25–40
Coley PD, Barone JA (1996) Herbivory and plant defences in tropical forests. Ann Rev Ecol Syst 27:305–335
Collevatti RG, Sperber CF (1997) The gall maker *Neopelma baccharidis* Burk. (Homoptera: Psyllidae) on *Baccharis dracunculifolia* (Asteraceae): individual, local and regional patterns. An Soc Entomol Bras 26:45–53
Cook JM, Rokas A, Pagel M, Stone GN (2002) Evolutionary shifts between host oak sections and host plant organs in *Andricus* gall wasps. Evol 56:1821–1830
Cornelissen T, Stiling P (2005) Perfect is best: low leaf fluctuating asymmetry reduces herbivory by leaf miners. Oecol 142:46–56
Cornelissen T, Stiling P (2011) Similar responses of insect herbivores to leaf fluctuating asymmetry. Arthrop Plant Inter 5:59–69
Cornelissen T, Guimarães CD, Viana JPR, Silva B (2013) Interespecific competition influences the organization of diverse sessile insect community. Acta Oecol 52:15–18
Costa FV, Azevedo IFP, Braga LL, Perillo LN, Neves FS, Leite LO, Silva BL, Ribeiro LC, Fernandes GW, Cuevas-Reyes P (2012) Fluctuating asymmetry and herbivory in two

ortogenetical stages of *Chamaecrista semaphore* in restored and natural environments. J Plant Inter 8:179–186
Craig TP, Horner JD, Itami JK (2001) Genetics, experience, and host–plant preference in *Eurosta solidaginis*: implications for host shifts and speciation. Evol 55:773–782
Cuevas-Reyes P, Fernandes GW, González-Rodríguez A, Pimenta M (2011) Effects of generalist and specialist parasitic plants (Loranthaceae) on the fluctuating asymmetry patterns of ruprestrian host plants. Basic Appl Ecol 12:449–455
Cuevas-Reyes P, Gilberti L, González-Rodríguez A, Fernandes GW (2013) Patterns of herbivory and fluctuating asymmetry in *Solanum lycocarpum* St. Hill (Solanaceae) along an urban gradient in Brazil. Ecol Ind 24:557–561
Del-Claro K (2004) Multitrophic relationships, conditional mutualisms, and the study of interaction biodiversity in tropical savannahs. Neotrop Entomol 33:665–672
Del-Claro K, Torezan-Silingardi HM (2009) Insect-plant interactions: new pathways to a better comprehension of ecological communities in Neotropical savannas. Neotrop Entomol 38: 159–164
Després L, Pettex EV, Plaisance V, Pompanon F (2002) Speciation in the globeflower fly *Chiastocheta* spp. (Diptera: Anthomyiidae) in relation to host plant species, biogeography, and morphology. Mol Phylogenet Evol 22:258–268
Díaz M, Pulido FJ, Moller AP (2004) Herbivore effects on developmental instability and fecundity of holm oaks. Oecol 139:224–234
Dohn J, Dembélé F, Karembé M, Moustakas A, Amévor KA, Hanan NP (2013) Tree effects on grass growth in savannas: competition, facilitation and the stress- gradient hypothesis. J Ecol 101:202–209
Ehrlich PR, Raven PH (1964) Butterflies and plants: a study in coevolution. Evol 18:586–608
Espírito-Santo MM, Fernandes GW (2002) Host plant effects on development and survivorship of the *Neopelma baccharidis* (Homoptera: Psyllidae). Aust Ecol 27:249–257
Espírito-Santo MM, Fernandes GW (1998) Abundance of Baccharopelma baccharidis (Homoptera: Psyllidae) galls on the dioecious shrub Baccharis dracunculifolia (Asteraceae). Environ Entomol 27:870–876
Espírito-Santo MM, Fernandes GW (2007) How many species of gall-inducing insects are there on earth, and where are there? Ann Entomol Soc Am 100:95–99
Espírito-Santo MM, Werneck MS (1999) Efeitos da umidade do solo e da cobertura vegetal na distribuição e abundância de *Drosera montana*. Acta Bot Bras 13(2):99–305
Espírito-Santo MM, Madeira BG, Neves FS et al (2003) Sexual differences in reproductive phenology and their consequences to the demography of *Baccharis dracunculifolia* (Asteraceae), a dioecious tropical shrub. Ann Bot 91:13–19
Espírito-Santo MM, Faria ML, Fernandes GW (2004) Parasitoid attack and its consequence of the galling psyllid *Baccharopelma dracunculifoliae*. Bas Appl Ecol 5:475–484
Espírito-Santo MM, Neves FS, Andrade-Neto FR, Fernandes GW (2007) Plant architecture and meristem dynamics as the mechanism determining the diversity of gall-inducing insects. Oecol 153:353–364
Espírito-Santo MM, Neves FS, Fernandes GW, Silva JO (2012) Plant phenology and absence of sex-biased gall attack on three species of *Baccharis*. PLoSONE 7:e46896. doi:10.1371/journal.pone.0046896
Fagundes M, Fernandes GW (2011) Insect herbivores associated *with Baccharis dracunculifolia* (Asteraceae): responses of gall-forming and free-feeding insects to latitudinal variation. Rev Biol Trop 59:1419–1432
Fagundes M, Neves FS, Fernandes GW (2005) Direct and indirect interactions involving ants, insect herbivores, parasitoids and the host plant *Baccharis dracunculifolia* (Asteraceae). Ecol Entomol 30:28–35
Fay PA, Hartnett DC, Knapp AK (1996) Plant tolerance of gall-insect attack and gall-insect performance. Ecol 77:521–534

Fayle TM, Eggleton P, Manica A, Yusah KM, Foster WA (2015) Experimentally testing and assessing the predictive power of species assembly rules for tropical canopy ants. Ecol Lett 18:254–262
Feder JL (1998) The apple maggot fly, *Rhagoletis pomonella*: flies in the face of conventional wisdom about speciation. In: Howard DJ, Berlocher SH (eds) Endless forms: species and speciation. Oxford University Press, New York, pp 130–144
Ferdy JB, Després L, Godelle B (2002) Evolution of mutualism between globeflowers and their pollinating flies. J Theor Biol 217:219–234
Fernandes GW (1992) A gradient analysis of plant forms from Northern Arizona. J Az-Nv Acad Sci 24–25:21–30
Fernandes GW, de Mattos EA, Franco AC, Lüttqe U, Ziegler H (1998) Influence of the parasite Pilostyles ingae (Rafflesiaceae) on some Physiological Parameters of the Host Plant, *Mimosa naguisei* (Mimosaceae). Acta Bot 111:54–58
Fernandes GW, Price PW (1988) Biogeographical gradients in galling species richness: tests of hypotheses. Oecol 76:161–167
Fernandes GW, Price PW (1991) Comparison of tropical and temperate galling species richness: the roles of environmental harshness and plant nutrient status. In: Price PW, Lewinsohn TM, Fernandes GW, Benson WW (eds) Plant–animal interactions: evolutionary ecology in tropical and temperate regions. Wiley, New York, pp 91–115
Fernandes GW, Price PW (1992) The adaptive significance of insect gall distribution: survivorship of species in xeric and mesic habitats. Oecologia 90:14–20
Fernandes GW, Ribeiro SP (1990) Plant response to herbivory: two examples from the Neotropics. Ecotrópicos 3:77–86
Fernandes GW, Santos JC (2014) Neotropical insect galls. Springer, New York
Fernandes GW, Carneiro MAA, Lara ACF, Allain LR, Andrade GI, Julião GR, Reis TR, Silva IM (1996) Galling insects on Neotropical species of *Baccharis* (Asteraceae). Trop Zool 9:315–332
Fernandes GW, Fagundes M, Woodman RL, Price PW (1999) Ant effects on three-trophic level interactions: plant, galls, and parasitoids. Ecol Entomol 34:411–415
Fernandes GW, Silva JO, Espírito-Santo MM, Fagundes M, Oki Y, Carneiro MAA (2014) *Baccharis*: a neotropical model system to study insect plant interactions. In: Fernandes GW, Santos JC (eds) Neotropical insect galls. Springer, Dordrecht, pp 193–219
Fleck T, Fonseca CR (2007) Hipóteses sobre a riqueza de insetos galhadores: uma revisão considerando os níveis intra-específico, interespecífico e de comunidade. Neotrop Biol Cons 2:36–45
Fraenkel GS (1959) The raison d'Être of secondary plant substances. Science 129:1466–1470
Freitas JD, Rossi MN (2015) Interaction between trophobiont insects and ants: the effect of mutualism on the associated arthropod community. J Ins Cons (in press)
Futuyma DJ, Keese MC, Funk DJ (1995) Genetic constraints on macroevolution: the evolution of host affiliation in the leaf beetle genus *Ophraella*. Evol 49:797–809
Gagné RJ (1994) The gall midges of the neotropical region. Cornell University Press, Ithaca
Garlaschelli D, Caldarelli G, Pietronero L (2003) Universal scaling relations in food webs. Nature 423:165–168
Givnish TJ (1998) Adaptative plant evolution on islands: classical patterns. In: Grant PR (ed) Evolution on islands. Oxford University Press, pp 281–304
Gonçalves-Alvim SJ, Fernandes GW (2001) Biodiversity of galling insects: historical, community and habitat effects in four neotropical savannas. Biodiv Cons 10:79–98
Guerra TJ (2010) História natural da erva-de-passarinho *Psittacanthus robustus* (Loranthaceae) em uma área de campo rupestre do sudeste brasileiro: interações com hospedeiras, dispersores, polinizadores e insetos herbívoros. Universidade Estadual de Campinas, Thesis
Guerra TJ, Pizo MA (2014) Asymmetrical dependence between a Neotropical mistletoe and its avian seed disperser. Biotropica 46:285–293

Guerra T, Camarota F, Castro FS, Schwertner CF, Grazia J (2011) Trophobiosis between ants and *Eurystethus microlobatus* Ruckes 1966 (Hemiptera: Heteroptera: Pentatomidae) a cryptic, gregarious and subsocial stinkbug. J Nat Hist 45:1101–1117

Hairston NG, Smith FE, Slobodkin LB (1960) Community structure, population control, and competition. Am Soc Nat 94:421–425

Herms DA, Mattson WJ (1992) The dilemma of plants: to grow or defend. Quart Rev Biol 67: 283–335

Hunter MD, Price PW (1992) Playing chutes and ladders: heterogeneity and the relative roles of bottom-up and top-down forces in natural communities. Ecol 73:724–732

Jacobi CM, Mourão FA, Antonini Y, Fagundes R, Kumagai AF (2015) Interações ecológicas em ambientes de canga do quadrilátero ferrífero, Minas Gerais. In: Carmo FF, Kamino LHY (ed) Geossistemas ferruginosos do Brasil: áreas prioritárias para conservação da diversidade geológica e biológica, patrimônio cultural e serviços ambientais, Editora 3i, Belo Horizonte, p 411–428

Jeffries MJ, Lawton JH (1984) Enemy free space and the structure of ecological communties. Biol J Linn Soc 23:269–286

Jermy T (1984) Evolution of insect-host plant relationships. Am Nat 124:609–630

Joy JB, Crespi BJ (2007) Adaptative radiation of gall-inducing insects within a single host-plant species. Evol 61:784–795

Julião GR, Venticinque EM, Fernandes GW, Price PW (2014) Unexpected high diversity of galling insects in the amazonian upper canopy: the savanna out there. PLoS ONE 9:e114986. doi:10.1371/journal.pone.0114986

Kuijt J (1969) The biology of parasitic flowering plants. Univ Calif Press, Berkeley

Landau EC, Gonçalves-Alvim SJ, Fagundes M, Fernandes GW (1998) Riqueza e abundância de herbívoros em flores de *Vellozia nivea* (Velloziaceae). Acta Bot Bras 12:403–409

Lara ACF, Fernandes GW (1994) Distribuição de galhas de *Neopelma baccharidis* (Homoptera: Psyllidae) em *Baccharis dracunculifolia* (Asteraceae). Rev Bras Biol 54:661–668

Lara ACF, Fernandes GW (1996) The highest diversity of galling insects: Serra do Cipó, Brazil. Biodiv Lett 3:111–114

Larson KC, Whitham TG (1997) Competition between gall aphids and natural plant sinks: plant architecture affects resistance to galling. Oecol 109:575–582

Lawton JH, Schroder D (1977) Effects of plant type, size of geographical range and taxonomic isolation on number of insects associated with British plants. Nature 265:137–140

Leite AC (2015) Impactos da erva-de-passarinho *Psitacanthus robustus* (Lorantaceae) na cadeia trófica em campos rupestres. Dissertation, Universidade Federal de Minas Gerais

Lüttge U, Haridasan M, Fernandes GW, Mattos EA, Trimborn P, Franco AC, Caldas LS, Zielgler H (1998) Photosynthesis of mistletoes in relation to their host at various sites of tropical Brazil. Trees 12:167–174

Maia VC (2005) Catálogo dos Cecidomyiidae (Diptera) do estado do Rio de Janeiro. Biota Neotrop 5:1–26

Maia VC (2014) Insect galls of Itamonte (Minas Gerais, Brazil): characterization and occurrence. Biota Neotrop 14:1–17

Maia VC, Fernandes GW (2004) Insect galls from Serra de São José (Tiradentes, MG, Brazil). Rev Bras Entomol 6:423–445

Mani M (1964) The ecology of plant galls. W. Junk, Netherlands, The Hague

Marini-Filho OJ, Fernandes GW (2012) Stem galls drain nutrients and decrease shoot performance in *Diplusodon orbicularis* (Lithraceae). Arthrop Plant Inter 6:121–128

Medel R (2000) Assessment of parasite-mediated selection in a host-parasite system in plants. Ecol 81:1554–1564

Medianero E, Valderrama A, Barrios H (2003) Diversidad de insectos minadores de hojas y formadores de agallas en el dosel y sotobosque del bosque tropical. Acta Zool Mex 89:153–168

Meinzer FC, Woodruff DR, Shaw DC (2004) Integrated responses of hydraulic architecture, water and carbon relations of western hemlock to dwarf mistletoe infection. Plant Cell Environ 27:937–946

Mendonça MS (2001) Galling insect diversity 'patterns: the resource synchronization hypothesis. Oikos 95:171–176

Mendonça MS (2007) Plant diversity and galling arthropod diversity searching for taxonomic patterns in an animal-plant interaction in the neotropics. Bol Soc Argent Bot 42:347–357

Meserve PL, Kelt DA, Milstead WB, Gutiérrez JR (2003) Thirteen years of shifting top-down and bottom-up control. Bioscience 53:633–646

Metcalffe DB, Asner GP, Martin RE, Espejo JES, Huasco WH, Amézquita FFF, Carranza-Jimenez L, Cabrera DFG, Baca LD, Sinca F, Quispe LPH, Taype IA, Mora LE, Dávila AR, Solórzano MM, Vilaca BLP, Román JML, Bustios PCG, Revilla NS, Tupayachi R, Girardin CAJ, Doughty CE, Malhi Y (2013) Herbivory makes major contributions to ecosystem carbon and nutrient cycling in tropical forests. Ecol Lett 17:324–332

Monteiro GF (2014) Padrão altitudinal de distribuição de insetos herbívoros associados ao arbusto *Baccharis dracunculifolia* (Asteraceae). Dissertation, Universidade Estadual de Montes Claros

Monteiro RF, Martins RP, Yakamoto K (1992) Host specificity and seed dispersal of *Psittacanthus robustus* (Loranthaceae) in south-weast Brazil. J Trop Ecol 8:307–314

Moreira DL, Engelhardt RL, Reis AS, Sanches EM, Leitão SG, Leitão GG (2002) Substâncias fenólicas com atividade antioxidante de *Pseudopiptadenia contorta* (Leguminosae-Mimosoideae). Rev Bras Farmacogn 12:124–125

Mourão FA, Carmo FF, Ratton P, Jacobi CM (2006) Hospedeiras da hemiparasita *Struthanthus flexicaulis* (Mart.) Mart. (Loranthaceae) em campos rupestres ferruginosos do Quadrilátero Ferrífero. Minas Gerais. Lundiana 7:103–109

Mourão FA, Jacobi CM, Figueira JEC, Batista EKL (2009) Effects of the parasitism of *Struthanthus flexicaulis* (Loranthaceae) on the fitness of *Mimosa calodendron* (Fabaceae), and endemic shrub from rupestrian fields over ironstone outcrops, Minas Gerais state, Brazil. Acta Bot Bras 23:820–825

Murdoch WW (1966) Community structure, population control, and competition—a critique. Am Nat 100:219–226

Negreiros D, Le Stradic S, Fernandes GW, Rennó HC (2014) CSR analysis of plant functional types in highly diverse tropical grasslands of harsh environments. Plant Ecol 215:379–388

Neves FS, Fagundes M, Sperber CF, Fernandes GW (2011) Tri-trophic level interactions affect host plant development and abundance of insect herbivores. Arthrop Plant Inter 5:351–357

Nishi AH, Vasconcellos-Neto J, Romero GQ (2013) The role of multiple partners in a digestive mutualism with a protocarnivorous plant. Ann Bot 111:143–150

Ohshima I, Yoshizawa K (2006) Multiple host shifts between distantly related plants, Juglandaceae and Ericaceae, in the leaf-mining moth *Acrocercops leucophaea* complex (Lepidoptera: Gracillariidae). Mol Phylogenet Evol 38:231–240

Oliveira RS, Galvão HC, De Campos MCR, Eller CB, Pearse SJ, Lambers H (2015) Mineral nutrition of campos rupestres plant species on contrasting nutrient-impoverished soil types. New Phytol 205:1183–1194

Pennings S, Callaway R (2002) Parasitic plants: parallels and contrasts with herbivores. Oecol 131:479–489

Polis GA (1999) Why are parts of the world green? Multiple factors control productivity and the distribution of biomass. Oikos 86:3–15

Poulin R (2007) Are there general laws in parasite ecology? Parasitol 134:763–776

Press MC, Phoenix GK (2005) Impacts of parasitic plants on natural communities. New Phytol 166:737–751

Price PW (1980) Evolutionary biology of parasites. Princeton University Press, New Jersey

Price PW, Fernandes GW, Lara ACF, Brawn J, Gerling D, BarriosH Wright M, Ribeiro SP, Rothcliff N (1998) Global patterns in local number of insect galling species. J Biogeogr 25:581–592

Reblin JS, Logan BA, Tissue DT (2006) Impact of eastern dwarf mistletoe (*Arceuthobium pusillum*) infection on the needles of red spruce (*Picea rubens*) and white spruce (*Picea glauca*): oxygen exchange, morphology and composition. Tree Physiol 26:1325–1332

Ribeiro SP (2003) Insect herbivores in the canopies of savannas and rainforests. In: Basset Y, Novotny V, Miller S, Kitching R (eds) Arthropods of tropical forests spatio-temporal dynamics and resource use in the canopy. Cambridge University Press, Cambridge, pp 348–359

Ribeiro SP, Basset Y (2007) Gall-forming and free-feeding herbivory along vertical gradients in a lowland tropical rainforest: the importance of leaf sclerophylly. Ecography 30:663–672

Ribeiro SP, Borges PP (2010) Canopy habitat area effect on the arthropod species densities in the Azores: pondering the contribution of tourist species and other life histories. In: Serrano ARM, Borges PAV, Boieiro M, Oromí P (eds) Terrestrial arthropods of Macaronesia—biodiversity, ecology and evolution. Sociedade Portuguesa de Entomologia, Lisboa, pp 81–106

Ribeiro SP, Carneiro MAA, Fernandes GW (1994a) Distribution of *Brachypnoea* (Chrysomelidae, Coleoptera) in an altitudinal gradient in a Brazilian savanna vegetation. Phytophaga 6:1–6

Ribeiro SP, Pimenta HR, Fernandes GW (1994b) Herbivory by chewing and sucking insects on *Tabebuia ochracea*. Biotropica 26:302–307

Ribeiro SP, Carneiro MAA, Fernandes GW (1998) Richness and distribution of free-feeding insect herbivores in Serra do Cipó, on old quartzite Brazilian mountain. J Insect Conserv 2:1–12

Ribeiro SP, Borges PAV, Gaspar C, Melo C, Serrano ARM, Amaral J, Aguiar C, André G, Quartau JA (2005) Canopy insect herbivores in the Azorean Laurisilva forests: key host plant species in a highly generalist insect community. Ecography 28:315–330

Ribeiro SP, Basset Y, Kitching R (2014) Density of insect galls in the forest understorey and canopy: Neotropical, Gondwana or global patterns? In: Fernandes GW, Santos JC (eds) Neotropical insect galls. Springer, New York, pp 129–141

Ribeiro-Mendes H, Marques ESA, Silva IM, Fernandes GW (2002) Influence of host-plant sex and habitat on survivorship of insect galls within the geographical range of the host-plant. Trop Zool 15:5–15

Rizzini CT (1997) Tratado de fitogeografia do Brasil: aspectos ecológicos, sociológicos e florísticos. Âmbito Cultural Edições LTDA, Rio de Janeiro

Safford HD (1999) Brazilian páramos I. An introduction to the physical environment and vegetation of the campos de altitude. J Biogeogr 26:693–712

Schwartz G, Hanazaki N, Silva MB, Izzo TJ, Bejar MEP, Mesquita MR, Fernandes GW (2003) Evidence for a stress hypothesis: Hemiparasitism effect on the colonization of *Achornea castaeneaefolia* A. Juss. (Euphorbiaceae) by galling insects. Acta Amaz 33:275–280

Silveira FAO, Negreiros D, Barbosa NPU, Buisson E, Carmo FF, Carstensen DW, Conceição AA, Cornelissen TG, Echternacht L, Fernandes GW, Garcia, QS, Guerra, TJ, Jacobi CM, Lemos-Filho JP, LeStradic S, Morellato LPC, Neves FS, Oliveira RS, Schaefer CE, Viana PL, Lambers H (2016) Ecology and evolution of plant diversity in the endangered campo rupestre: a neglected conservation priority. Plant soil (in press). doi 10.1007/s11104-015-2637-8

Smith AP, Young TP (1987) Tropical alpine plant ecology. Annu Rev Ecol Syst 18:137–158

Southwood TRE (1961) The number of species of insect associated with various trees. J Anim Ecol 30:1–8

Sperber CF, Collevatti RG (1996) The gall maker *Baccharopelma dracunculifoliae* Burck. (Homoptera: Psyllidae) on *Baccharis dracunculifolia* (Asteraceae): success and parasitoidism density dependence. An Soc Entomol Bras 25:59–63

Strong DR, Lawton JH, Southwoon R (1984) Insects on plants. Blackwell Scientific Publications, Oxford, Community patterns and mechanisms

Sutherland WJ, Freckleton RP, Godfray HCJ, Beissinger SR, Benton T, Cameron DD, Carmel Y, Coomes DA, Coulson T, Emmerson MC, Hails RS, Hays GC, Hodgson DJ, Hutchings MJ, Johnson D, Jones JPG, Keeling MJ, Kokko H, Kunin WE, Lambin X, Lewis OT, Malhi Y, Mieszkowska N, Milner-Gulland EJ, Norris K, Phillimore AB, Purves DW, Reid JM,

Reuman DC, Thompson K, Travis JMJ, Turnbull LA, Wardle DA, Wiegand T (2013) Identification of 100 fundamental ecological questions. J Ecol 101:58–67
Tavares MT, Perioto NW (1993) *Psyllaephagus baccharidis*, sp. n., (Hymenoptera, Encyrtidae), a nymphal parasitoid of *Neopelma baccharidis* Burckhardt (Homoptera, Psyllidae). Rev Bras Entomol 37:317–320
Telhado C, Esteves D, Cornelissen T, Fernandes GW, Carneiro MA (2010) Insect herbivores of Coccoloba cereifera do not select asymmetric plants. Environ Entomol 39:849–855
Telhado C, Silveira FAO, Fernandes GW, Cornelissen T (2015) Fluctuating asymmetry in leaves and flowers of sympatric species in a harsh environment. Plant Species Biol (in review)
Tennakoon KU, Pate JS (1996) Heterotrophic gain of carbon from hosts by the xylem tapping root hemiparasite *Olax phyllanthi* (Olacaceae). Oecol 105:369–376
Urso-Guimarães MV, Scareli-Santos C, Bonifácio-Silva AC (2003) Occurrence and characterization of entomogen galls in plants from natural vegetation areas in Delfinópolis, MG, Brazil. Braz J Biol 63:705–715
Weis AE, Walton R, Grego CL (1988) Reactive plant tissue sites and the population biology of gall makers. Annu Rev Entom 33:467–486

Chapter 15
Cerrado to Rupestrian Grasslands: Patterns of Species Distribution and the Forces Shaping Them Along an Altitudinal Gradient

G. Wilson Fernandes, Hernani A. Almeida, Cássio A. Nunes, João Henrique A. Xavier, Neil S. Cobb, Marco Antônio A. Carneiro, Tatiana Cornelissen, Frederico S. Neves, Sérvio P. Ribeiro, Yule Roberta F. Nunes, Ana Carolina V. Pires and Marina V. Beirão

Abstract Due to clear variations in a relatively small area, mountains represent natural laboratories for evolutionary and ecological studies. In a large degree, these variations are driven by changes in climate and soil that occur along altitudinal gradients and influence the ecology, evolution and geography of species. In spite of

G.W. Fernandes (✉) · H.A. Almeida · C.A. Nunes · J.H.A. Xavier · A.C.V. Pires · M.V. Beirão
Ecologia Evolutiva & Biodiversidade/DBG, CP 486, ICB/Universidade Federal de Minas Gerais, Belo Horizonte, MG 31270-901, Brazil
e-mail: gw.fernandes@gmail.com

G.W. Fernandes
Department of Biology, Stanford University, Stanford, CA 94305, USA

N.S. Cobb
Merriam-Powell Center for Environmental Research Peterson Hall, Northern Arizona University, Bldg 22, Rm 330, Box 6077, Flagstaff, AZ 86011, USA
e-mail: Neil.Cobb@nau.edu

M.A.A. Carneiro
Entomologia Ecológica/DEBIO, ICEB, Universidade Federal de Ouro Preto, Ouro Preto, MG 35400-000, Brazil
e-mail: mcarneirinho@gmail.com

T. Cornelissen
Departamento de Ciências Naturais, Campus Dom Bosco, Universidade Federal de São João del-Rei, São João del-Rei, MG 36301-160, Brazil
e-mail: tatiana@ufsj.edu.br

F.S. Neves
Departamento de Biologia Geral, ICB/Universidade Federal de Minas Gerais, Belo Horizonte, MG 31270-901, Brazil
e-mail: fred.neves@gmail.com

© Springer International Publishing Switzerland 2016

G.W. Fernandes (ed.), *Ecology and Conservation of Mountaintop Grasslands in Brazil*, DOI 10.1007/978-3-319-29808-5_15

being old and eroded, the southern Brazilian mountains provide enough variation and heterogeneity to influence species distribution and diversity. The best-known Brazilian mountain range is the Espinhaço (the Backbone mountains). The Espinhaço is a large natural watershed divider of major ecological importance in eastern Brazil. The altitudinal gradient in the Espinhaço Mountains is low when compared to other tall mountains in the world as it only varies from ca. 650 to 2.072 m a.s.l. at the Sun Peak (Pico do Sol). This chapter synthesizes the results of 10 studies that have collectively examined the soil, climate to better understand patterns and processes associated with biodiversity of key groups of organism, including of plants, termites, dung beetles, ants, butterflies, arbuscular mycorrhizal fungi, insect herbivores and birds in an altitudinal gradient (from 800 to 1400 m a.s.l.) at Espinhaço mountain range. In this mountain range, the soils are primarily poor and the vegetation is dominated by rupestrian grasslands, and both are known to vary along the altitudinal gradient hence providing opportunities for variation in the associated organisms.

Contents

S.P. Ribeiro
EcologiaEvolutiva de Insetos de Dossel e Sucessão Natural/DEBIO, ICEB, Universidade Federal de Ouro Preto, Ouro Preto, MG 35400-000, Brazil
e-mail: spribeiro@iceb.ufop.br

Y.R.F. Nunes
Ecologia Vegetal/DBG, CP 126, CCBS/Universidade Estadual de Montes Claros, Montes Claros, MG 39401-089, Brazil
e-mail: yule.nunes@unimontes.br

15.1 Species Diversity Along Altitudinal Gradients

Due to clear variations in a relatively small area, mountains represent natural laboratories for evolutionary and ecological studies. The most common pattern of species distribution in mountains is the decline of richness as altitude rises, and the same is true for the Espinhaco Mountains described here. As a consequence, lowlands are relatively richer in number of species compared with highlands (e.g., Terborgh 1977; Wolda 1987; Fernandes and Price 1988, 1991). This pattern is likely due to abiotic factors and processes associated with the increasing altitude and the decreasing habitat area. However, not all taxa respond to altitude in the same way (e.g., McCoy 1990), and species richness peaks can be found at high, medium, and even low altitudes, depending upon the community and the system of study (Rahbek 2005).

To a large degree, these variations are driven by changes in climate and soil that occur along altitudinal gradients and influence the ecology, evolution and geography of species (e.g., Fernandes and Price 1988; Lomolino 2001; Korner 2007; Joly et al. 2012). In spite of being old and eroded, the southern Brazilian mountains provide enough variation and heterogeneity to influence species distribution, diversity and community composition. The best-known Brazilian mountain range is the Espinhaço (the Backbone mountains) (see Chaps. 1 and 2). The Espinhaço is a large natural watershed divider of major ecological importance in eastern Brazil (Chaps. 1 and 5).

The altitudinal gradient in the Espinhaço Mountains varies from ca. 650 to 2072 m a.s.l. at the Sun Peak (Pico do Sol) (Vasconcelos et al. 2007), which is a relatively short range in altitude and very low for the highest point compared to other mountains in the world. The soils of the Espinhaco Mountains are shallow and low in nutrients and the vegetation is dominated by rupestrian grasslands. Both soils and vegetation vary along the altitudinal gradient hence providing opportunities for variation in the associated organisms (see Fig. 15.1). Between 900 and 1100 m, there is an ecotone, where plant species of the cerrado (lowest vegetation type) and rupestrian grassland are intermixed. In xeric habitats between 1300 and 1400 m vegetation structure becomes much simpler with few sparsely distributed trees, whereas over 1400 m the vegetation is dominated by sandy and stone grassland habitats in spite of the presence of natural patches of Atlantic rain forests which due to higher structural complexity could positively influence patterns of species richness (Chap. 7).

15.2 Mountains, Altitudinal Gradients and Species Geography

Mountains occur within all major world biomes, occupying 25 % of the area of the planet, and are also taken by 10 % of the world's human population. Fifty percent of the world population makes direct use of their resources. In addition, mountains

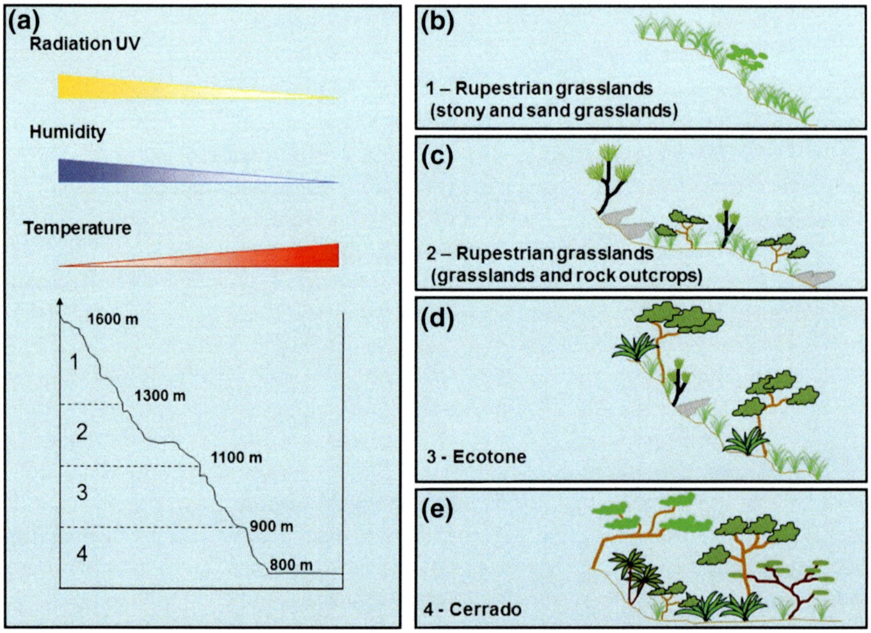

Fig. 15.1 Illustration showing the variation in the altitudinal gradient in Serra do Cipó-MG, and the most representative vegetation found at each elevation interval. There is however, great variation within a single elevation and habitats may change over a scale of a few centimeters. Overall, humidity and radiation decrease at higher elevations whereas temperature increases to the lowland direction (**a**). The vegetation changes along the elevation with more open, sandy and stony grasslands dominating at higher elevations (above 1300 m a.s.l.). Grasslands and rock outcrops are mostly common at 1100 and 1300 m a.s.l., and an ecotone between the rupestrian grassland and cerrado is found between 900 and 1100 m a.s.l. Cerrado vegetation prevails at 900 and lower elevations

house springs of the major rivers of the world and are also where biodiversity and endemism reach highest levels (Lomolino 2001, see also Chap. 5). They represent singular environments to detect and record the impacts of global changes, but also exhibit sharp gradients in many biotic and abiotic factors that provide variation for ecologists to search for footprints of climate change (e.g. Beniston et al. 1997; Pounds et al. 1999, 2006; Hughes 2000). Studies on mountain ecosystems are urgently needed due to their fragility, their appropriateness to record climate change, and because a large proportion of ecosystem services and populations depend upon the health and integrity of the mountain ecosystems (Schroter et al. 2005; Thuiller et al. 2005; Loarie et al. 2009; Araújo et al. 2011).

Mountain systems are natural laboratories, acting as barometers of climate change effects on species and populations, where time is replaced by space (i.e. elevation gradients) to understand how species may respond to climate change. Our long-term research program has included scores of studies on several plant and animal groups along an altitudinal gradient in Serra do Cipó (Brazil) for the past 25 years. Serra do Cipó is perhaps the most well known region of the Espinhaço range and is located at its southern portion. Studies have been concentrated on an altitudinal gradient that varies from cerrado (savanna) (800 m a.s.l.) to mountaintop vegetation of grasslands and natural patches of Atlantic rain forests (1.400–1500 m a.s.l.) in the western slope of the Espinhaço range. Altitudes above this range were not studied due to long distances, access and logistical support. The western slope of the mountain range is drier and dominated by rupestrian grasslands, whereas the eastern slope is wetter and dominated by Atlantic rain forest.

15.3 Sample Design

An altitudinal range in Serra do Cipó in the state of Minas Gerais (MG) was evaluated to address the influence of the altitudinal gradient at different groups of organisms. In this gradient, we concentrated our efforts to find out the influence of altitude and correlated variables on the distribution of major taxa of organisms, including plants, birds, ants, galling, mining and free-feeding insect herbivores, termites, dung beetles, and mycorrhizae. Along this altitudinal gradient, seven locations at different elevations were selected at every 100 m intervals (800–1400 m a.s.l.). The study sites were, at least, 2.5 km apart as longer distances would render the work much more difficult due to road distances. At each location, we settled one complete climatic station (Onset climatic stations). Climatic data were collected from 2011 to 2014. We recorded air temperature, photosynthetically active radiation, atmospheric pressure, soil humidity and soil temperature at 5 and 20 cm depth, wind speed and direction, air and foliar humidity, solar radiation, and accumulated annual precipitation. At each site, three 200 m long parallel transects were installed. They were positioned 200 m apart from each other, in a north-south orientation. At every 50 m along each transect, a sample point was marked and at this point the different groups of organisms were sampled. Some taxa, such as plants, had a different sampling design, although all sampling occurred in the same region. Figure 15.2 shows the sample sites along the altitudinal gradient.

Fig. 15.2 Distribution of the long-term ecological research sites (PELD-CNPQ Site 17) where studies have been carried out in Serra do Cipó, Brazil. (Map by Felipe Carvalho)

15.4 Variation of Abiotic Factors Along the Altitudinal Gradient

15.4.1 Climate

We here provide a brief overview of key climatic factors taken from the climatic stations that promote the great biodiversity of the Espinhaço Mountains; these include (1) average annual air temperatures, (2) PAR (Photosynthetically active radiation), (3) atmospheric pressure, (4) soil humidity, (5) soil temperature at 5 and 20 cm depth, (6) wind speed, (7) air and foliar humidity, (8) solar radiation, and (9) accumulated annual precipitation.

As altitude increases, average annual temperature decreases predictably ($p < 0.001$; Fig. 15.3a). The rising of 700 m in altitude caused a reduction of ~5 °C in the average temperature. On the other hand, the relationship between average annual precipitation and altitude was not statistically significant ($p = 0.5865$; Fig. 15.3b). Higher average annual precipitation was found at intermediate altitudes, where rainfall reached 862 mm at 1100 m a.s.l. and 868 mm at 1200 m a.s.l. The higher precipitation at this elevation could be influenced by the saturated air that cannot cross the mountain and condensates, generates fog, and ultimately results in higher precipitation (see Milanesi and Galvani 2011). On the other hand, Glarizin (1997) showed that precipitation distribution along altitudinal gradients may assume several shapes depending on the season and locality. Even though we did not detect a significant relationship between precipitation and altitude, air humidity increased with increasing altitude ($p = 0.001$; Fig. 15.3c), with averages ranging between 69 % at 800 and 88 % at 1400 m a.s.l. Regardless of precipitation, at the altitude of 1400 m, the average humidity was higher. This can be explained by the local lower temperatures (μ = 17 °C) and by the low atmospheric pressure (μ = 862 mbar). On the other hand, a deeper understanding of these interactions is needed.

Radiation and PAR exhibited the same pattern of no significant variation with altitude ($p = 0.1894$, and $p = 0.069$, respectively) (Fig. 15.3d). Both were high at all altitudes, except at 1400 m, where the measurements dropped considerably (solar radiation = 164.70 W/m^2 and PAR = 314.59 u/E). This trend can be explained by increased clouds at high altitudes, generated by the humidity accumulated by the air in this region (see also Chap. 7).

We found a significant relationship between soil temperature and altitude. At the depths of 5 and 20 cm, soil temperature increased with decreasing altitude (5 cm: $p = 0.014$; 20 cm: $p = 0.013$; Fig. 15.3e). On the other hand, the average soil temperatures at both depths were alike. At 1400 m, the average soil temperature at 5 cm deep was 18.89 °C, while at 20 cm deep temperature was 19.19 °C, despite the average humidity of the soil of 0.11 % 5 cm deep and of 0.79 % 20 cm deep at this altitude.

The variation in altitude did not affect soil moisture at both depths (at 5 cm $P = 0.5538$ and at 20 cm $p = 0.1339$) (Fig. 15.3f). At the altitude of 1200 and 1400 m, soil at the depth of 20 cm exhibited higher moisture (0.79 %) than soil at

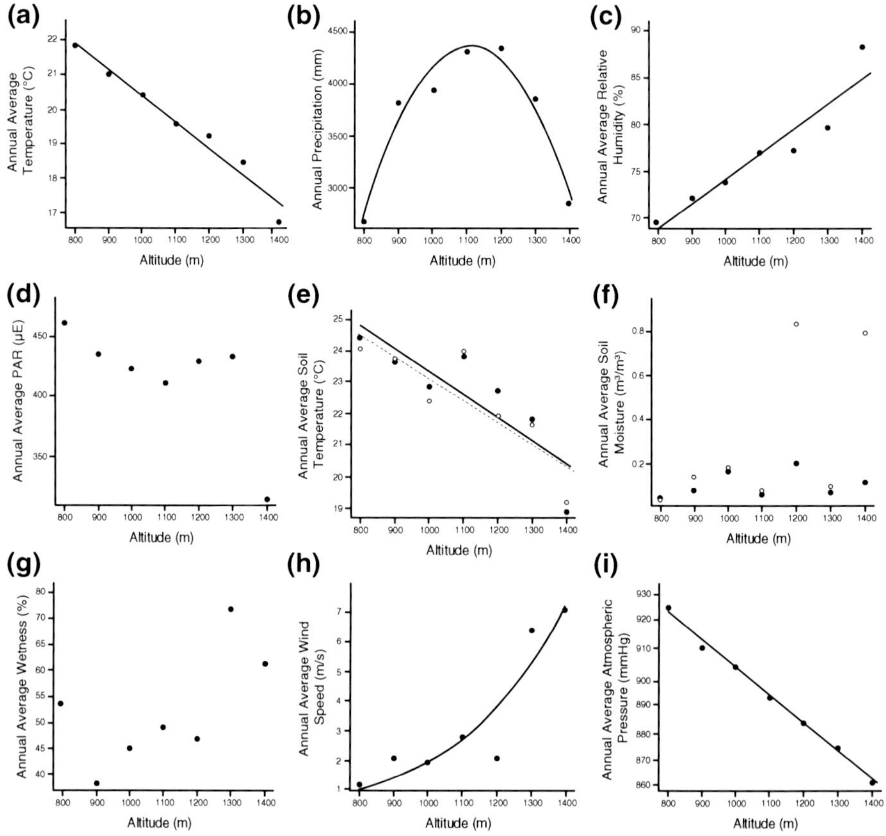

Fig. 15.3 Annual average of climatic variables from Serra do Cipó, Brazil collected from the Onset climatic stations. **a** Air temperature. **b** Precipitation. **c** Air relative humidity. **d** Photosynthetically Active Radiation (PAR). **e** Soil temperature. **f** Soil moisture; (*filled circles* 5 cm depth and *open circles* 20 cm depth). **g** Humidity. **h** Wind speed. **i** Atmospheric pressure

5 cm deep (0.11 % of moisture). This may be explained by soil particle size, which does not provide the water accumulation on its surface, always leaving it to a greater depth. The air moisture (Fig. 15.3c), was higher at upper regions, maintaining the soil wetter. Leaf humidity (dew point) did not vary with altitude (p = 0.1306; Fig. 15.3g). Average wind speed increased with increasing altitude (p = 0.012; Fig. 15.3h), reaching 6.38 and 7.05 m/s at 1300 and 1400 m a.s.l., respectively. The average atmospheric pressure was inversely related to altitude ($p < 0.001$; Fig. 15.3i), ranging from 925 mbar at 800 m to 862 mbar at 1400 m. The average atmospheric pressure fell around 10.45 mbar at every 100 m increase. With this pressure reduction, there is an accumulation of humidity at higher altitudes. Thus, there is a higher chance for cloud accumulation and a consequent fall at solar radiation and PAR at such altitudes. The biological importance of it has been recently discussed by Barbosa et al. (2015) (see also Chap. 7).

Most of the climatic patterns recorded for the altitudinal gradient in this rupestrian grassland mountain are similar to patterns described in the literature for other mountains. Precipitation, for example, showed a decline above 1000 m, as described in many other studies (e.g., Veneklaas and Van Ek 1990; Leuschner et al. 2007; Rapp and Silman 2012). Temperature and PAR also declined with altitude, as recorded in other studies (e.g., Rapp and Silman 2012; Gerold et al. 2008). On the other hand, we need a much longer data set to better understand climatic variation within this mountain range. Climate is an important component of the environment that conditions many natural processes and influences species development and geography (e.g., Fernandes and Price 1988, 1991; Ayoade 2010; Ricklefs 2013). In tropical and subtropical regions small altitudinal differences might greatly influence climatic variables (Fritzsons et al. 2008).

The climograph Fig. 15.4 provides an overview of monthly rainfall and temperature in Serra do Cipó, indicating that most precipitation and highest temperatures occur from November to February. The months of April and May represent the "post-rain" period, when average temperatures begin to fall in the region. The months of April and May represent the "post-rain" period, when average temperatures begin to fall in the region. The months with lowest average rainfall and temperatures are those from June through August, while September and October represent a "post-dry" period, during which the rains start to increase and temperatures begin to rise again. Madeira and Fernandes (1999) also reached similar conclusion after analyzing 20 years of data form a meteorological station located at least 50 km away.

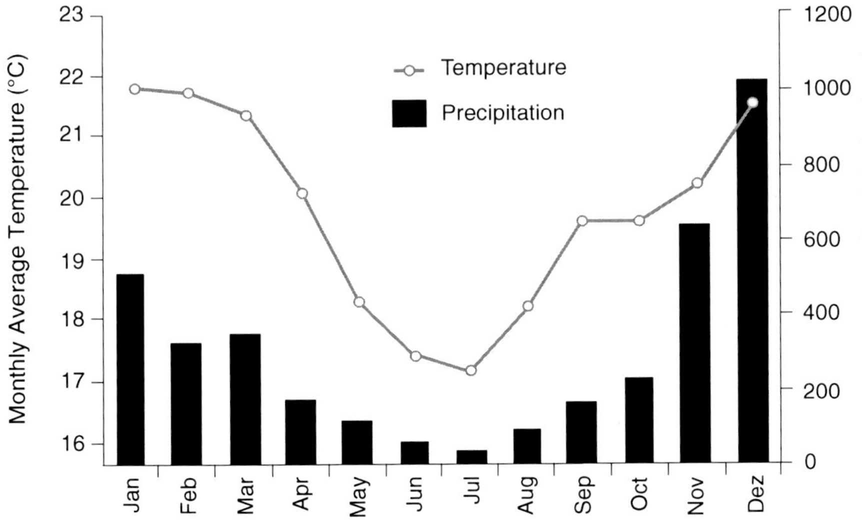

Fig. 15.4 Climograph from Serra do Cipó, Brazil, showing the monthly average temperature, and precipitation, collected from the Onset climatic stations

15.4.2 Soils

We performed several soil studies along this altitudinal gradient several times over the past decades. Generally, soils at Serra do Cipó are podzolic, with quartzitic rocks of the Middle Pre-Cambrian, generally shallow, of coarse texture, with varying amounts of organic matter and with high concentration of aluminum (see also Chap. 3 for a review). A marked feature of the rupestrian grassland soils is their nutrient deficiency and the low organic carbon content. As such, they are classified as Dystrophic Entisol or Psamitic Entisol (high sand content). They have a moderate A horizon overlying a coarse textured mineral layer of up to 50 cm, which rests on rock (quartzite) or deep saprolite (schists/pellitic rocks) (see algo Chap. 3).

The relationship between soil characteristics and altitude was addressed by Coutinho et al. (2015) in their study of the diversity of mycorrhizae (see also Carvalho et al. 2012, 2014). Soil samples were taken at each sample site as described previously in our sampling protocol. Thirteen square plots with 10-meter sides were defined along each transect for a total of 91 plots. Five soil samples were collected from the surface layer (0–0.2 m in depth) in each plot at each altitude for soil analysis, totaling 91 samples.

Generally, soils in the altitudinal gradient were acidic (4.7–5.2 pH/H_2O) and poor in all of the analyzed macronutrients (Fig. 15.5a). The average amount of organic matter was low (3.51–6.90 dag/kg) (Fig. 15.5b). The mean magnesium content was 0.11 cmol/dm^3 (Fig. 15.5c), while Calcium level was highest at the highest altitude of 1400 m (0.90 cmol/dm^3) and lower at 900, 1000, and 1300 m (0.20 cmol/dm^3) (Fig. 15.5c). The concentration of magnesium, calcium, phosphorous, potassium and organic matter did not exhibit a pattern related to altitude (Fig. 15.5d–f). With regard to texture, soils were predominantly sandy along the gradient (Fig. 15.5g). The trends we describe here corroborate the findings of Dossin et al. (1990) that described the soils of the entire Espinhaço mountain range as nutrient deficient.

15.5 Understanding Distribution of Species Along the Altitudinal Gradient

Although encompassing one of the most species-rich flora and fauna of the world, very few studies involving altitudinal gradients have been carried out in Brazil until very recently (but see Holt 1928; Scott and Brooke 1985; Fernandes and Price 1988, 1991). This is especially true for the semi-arid cerrado vegetation, that although represents the second most extensive vegetation domain in South America (Eiten 1972), has not yet been studied to the degree it deserves. Bellow we summarize the trends in the distribution of plants, animals and fungi along the rupestrian grassland altitudinal gradient at Serra do Cipó. Our studies occurred in two different phases: the first one took place in the early 1990s whereas the second

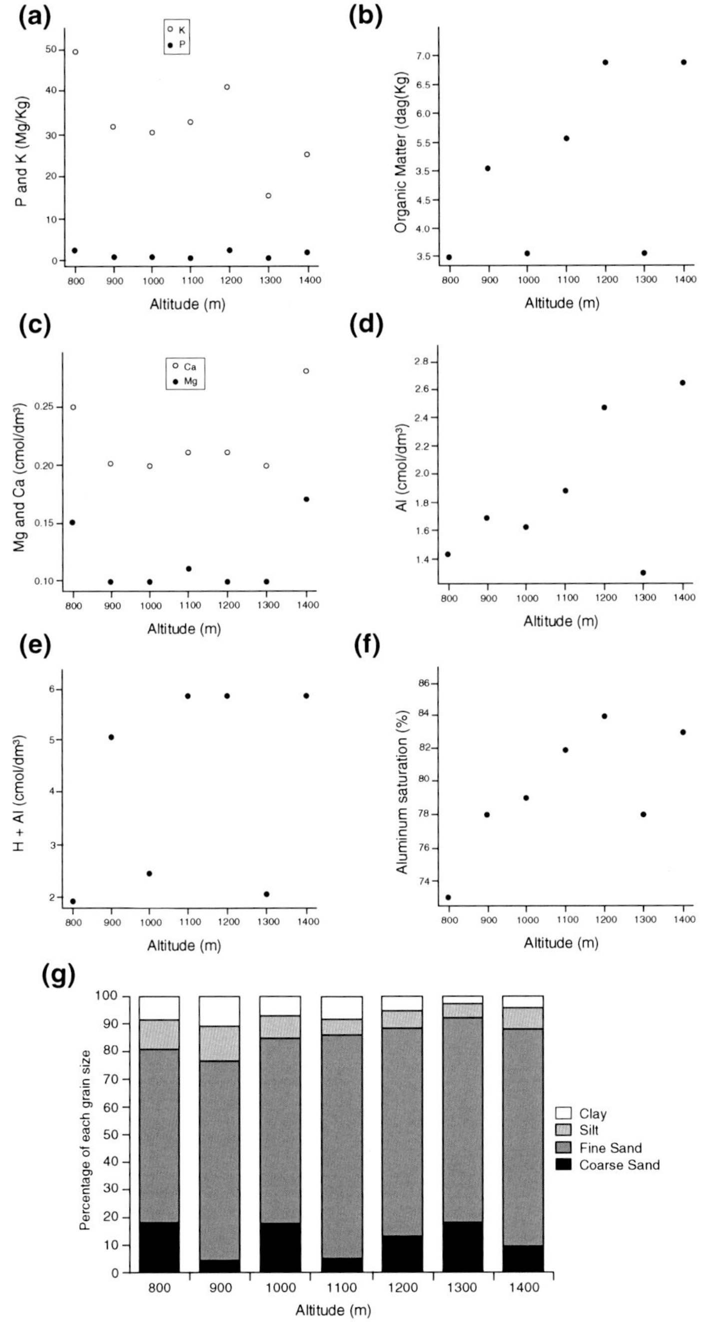

Fig. 15.5 Soil variables along the altitudinal gradient at Serra do Cipó, Brazil. **a** Aluminum. **b** Hydrogen + Aluminum. **c** Aluminum saturation. **d** *Mg* Magnesium, *Ca* Calcium. **e** Organic matter. **f** *P* Phosphorus, *K* Potassium. **g** Soil Granulometry

phase began in 2012 and is ongoing with many data being presented for the first time in this chapter.

15.5.1 Vegetation

For the first time, we present data on plant species richness, encompassing woody and herbaceous species along the altitudinal gradient in Serra do Cipó. Sampling was performed along the same altitudinal gradient described above, in 13 plots of various sizes (100 m^2 for woody established species here called 'adults') and 1 m^2 for woody regenerating and herbaceous species (here called 'juveniles'). For further details see Mota et al. (2016). All woody individuals with diameter at ground height (DGH) larger than 1.0 cm in 100 m^2 plots and the regenerating woody plants with DGH equal or less than 1.0 cm in 1 m^2 plots were inventoried.

Plant species richness generally declined with increasing altitude ($p < 0.05$) (Fig. 15.6). For woody species, the highest richness was found at the altitude of 900 m (119 species) whereas for herbaceous species the highest richness was found at 1000 m (153 species). Altitude accounted for 74 % of the variation in woody species ($r^2 = 0.74$) and for 66 % of the variation in herbaceous species ($r^2 = 0.66$).

One of the most well documented biogeographic patterns on Earth is the increase in flora diversity with decreasing altitude (e.g., Whittaker 1956; Hamilton 1975; Gentry 1988; Aiba and Kitayama 1999; Hussain and Malik 2012). However, several studies have shown non-unimodal patterns for plant richness along altitudinal gradients (e.g., Fernandes 1992; Liberman et al. 1996; Wang et al. 2002; Joly et al. 2012) while others report diversity peaks at intermediate elevations (e.g., Lomolino 2001; Sang 2009).

Multiple environmental factors vary along altitudinal gradients, but an altitudinal gradient basically corresponds to a temperature gradient (Kitayama 1992), where

Fig. 15.6 Regression of altitude and woody plants (*dotted lines, open circles*) ($r^2 = 0.66$) and herbaceous plants (*solid lines, black circles*) ($r^2 = 0.74$)

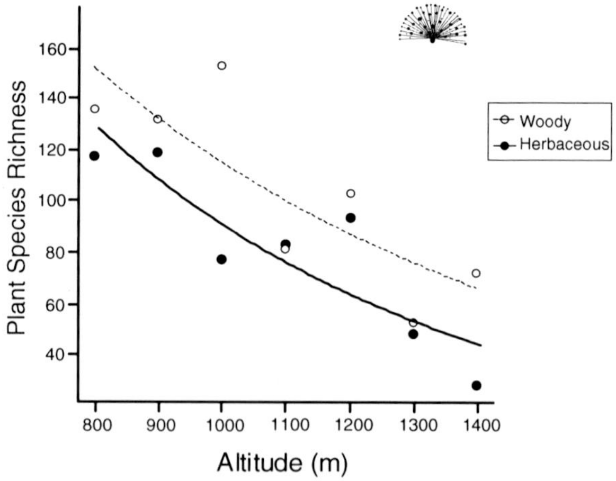

the reduction of productivity is associated with the low temperatures and shorter growing seasons at higher altitudes (Begon et al. 2006). Within the same altitudinal range, co-acting factors such as precipitation, topography, slope and soil types also regulate plant species distribution (Holland and Steyn 1975). The effects of area and isolation are also likely to contribute to observed decreases in species richness with altitude (Brown and Lomolino 1998; Korner 2007). The study of Coutinho et al. (2015) argued that the highly adverse soil conditions that prevail in this montane environment of the Espinhaco should represent a strong environmental filter for the flora. The extremely low fertility status of these soils, combined with high acidic and levels of aluminum that collectively mediate evolutionary strategies by plant species.

15.5.2 Distribution of Species from a Climatic-Vegetation Axis

Our ongoing studies examining patterns and processes driving biodiversity of key taxa including termites, dung beetles, and mycorrhizae used Principal Components Analysis (PCA) to summarize climatic and vegetation variables (Climatic-Vegetation Axis) that are correlated and respond to elevation. Ten variables were selected to build the climatic scenario of each altitude: average air temperature, air humidity, soil humidity and PAR; variation of air temperature and air humidity; maximum of air temperature; minimum of air temperature and air humidity; and sum of precipitation. We also selected four vegetation variables: plant abundance, richness, height and basal area. Figure 15.7a shows the correlations of each variable with the axis and Fig. 15.7b shows the regression of the axis with the altitude. The combination of factors expressed in this variable was not relevant for some taxa and only variables that most explained the variation are presented Fig. 15.7.

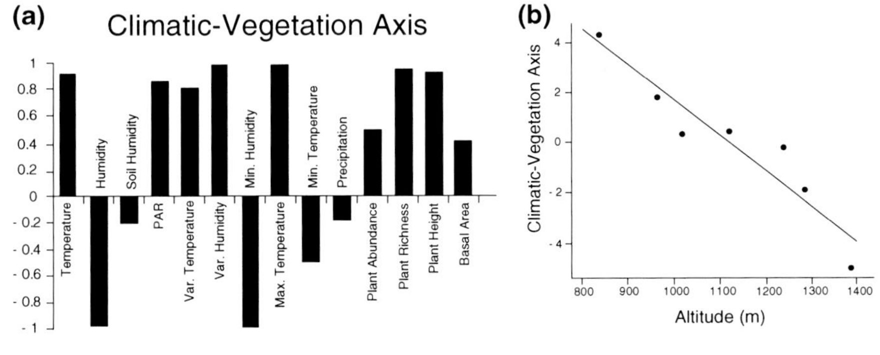

Fig. 15.7 **a** Correlations of all variables with the values of climatic-vegetation axis. **b** Regression of climatic-vegetation axis values with altitude [$p = 0.001$, $R^2 = 0.89$, (Axis value = (15.82 − 0.143 X Altitude)]

15.5.3 Termites

Termites (Blattodea: Termitoidea) are an important group of soil insects in tropical ecosystems and can be found during all seasons of the year. However, there are few studies considering their distribution along altitudinal gradients (but see Donovan et al. 2002; Palin et al. 2011). Sampling for this target group was performed according to Jones and Eggleton (2000; for details see Quintino 2014).

Overall, 50 species of termites were recorded along the altitudinal gradient in Serra do Cipó. Termite richness decreased with altitude (Fig. 15.8a) and altitude accounted for 80 % of the variation in termite richness (R^2 = 0.80). The climatic-vegetation PCA axis accounted for 84 % of the termite richness variation (Fig. 15.7b; R^2 = 0.84). Termite richness declined with decreasing in temperature, but also with a decrease in plant richness, abundance and height. Also, termite richness was higher where there were higher proportions of wood in the litter. Termites, as almost all insects, are less active and avoid cold conditions (e.g., Speight et al. 2008; Davies et al. 2015). They are also highly dependent upon vegetation to survive, sometimes not even being considered true herbivores. Vegetation structure and litter composition influence the habitat exploitation by termites and nest construction, influencing richness and community composition (see Miura and Matsumoto 1997; Alves et al. 2011; Bezerra-Gusmão et al. 2013).

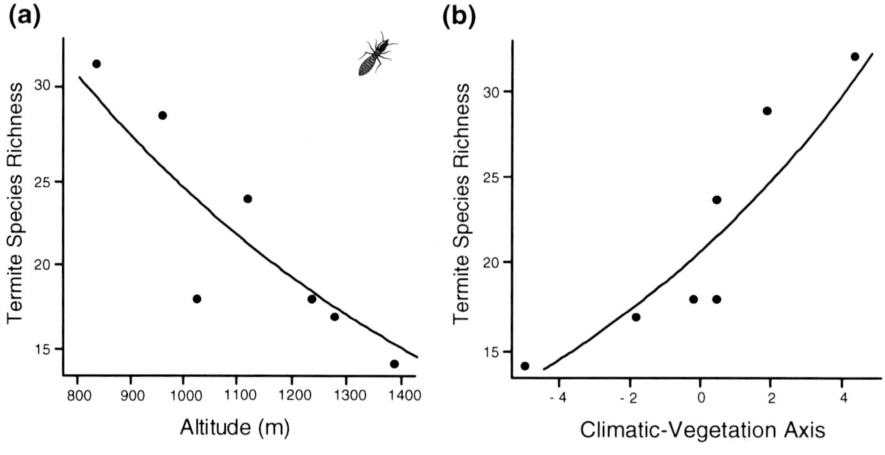

Fig. 15.8 **a** Regression of termite species richness against altitude (R^2 = 0.80). **b** Regression of termites species richness against climatic-vegetation summarized variables (R^2 = 0.84)

15.5.4 Dung Beetles

Most species of dung beetles (Coleoptera: Scarabaeidae: Scarabaeinae) feed on the microorganism-rich liquid component of mammalian dung and uses the fibrous material of dung to brood their larvae (Halffter and Edmonds 1982). Some species feed on rotting fruits, fungus and also carrion. Because of their extreme olfactory sensibility, moderate flight and muscular power and dung burrowing behavior, they play a major role in several ecological functions such as nutrient recycling, soil turnover, enhancement of plant growth, and secondary seed dispersal (Nichols et al. 2008). Although there are several studies showing a consistent decrease in the number of species with increasing altitude (Escobar et al. 2005; Herzog et al. 2013; Lobo and Halffter 2000), little is known about dung beetles in the tropical mountains of southeastern and northeastern of Brazil.

During four dry and wet seasons dung beetles were sampled with baited pitfall traps. After 48 h of exposure beetles were collected, preserved and transported to the laboratory where all individuals were counted and identified to the lowest taxonomical level (for futher details see Nunes 2015). Dung beetle richness decreased with altitude (Fig. 15.9a) and altitude accounted for 82 % of the variation in dung beetle richness ($r^2 = 0.82$). The climatic-vegetation PCA axis accounted for 92 % of variation in richness (Fig. 15.9b; $R^2 = 0.92$). Dung beetle richness declined with a decrease in temperature and also with a decrease in plant richness, abundance and height. This trend was somewhat expected, as Scarabaeinae is a group comprised of warm-adapted species (Lobo and Halffter 2000). Their diversity is higher in tropical forests and savannas with warm summer rainfall climates (Davis and Scholtz 2001). As dung beetles feed mostly on dung of mammals, and mammal richness also diminishes with elevation (McCain 2005, 2007), there is a reduction in food availability with

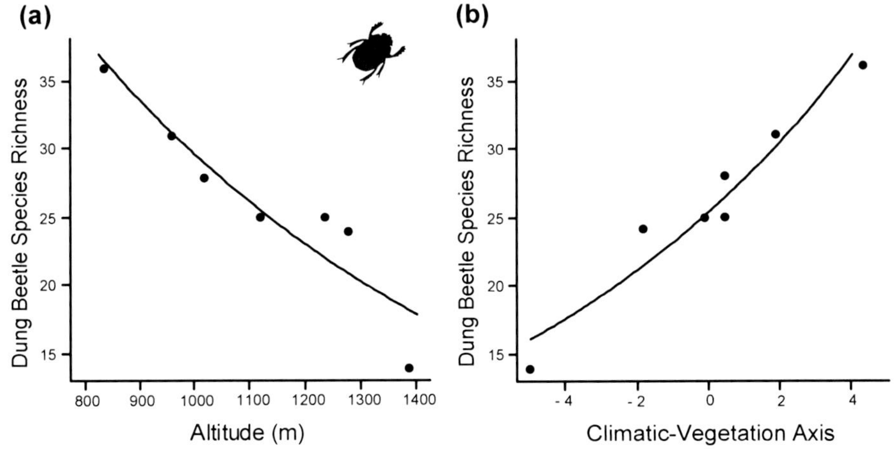

Fig. 15.9 **a** Regression of dung beetle richness against altitude (800–1400 m a.s.l.) ($R^2 = 0.82$). **b** Regression of dung beetles species richness against climatic-vegetation summarized variables ($R^2 = 0.92$)

increasing altitude. However, we have no studies on mammal distribution in this altitudinal gradient. Vegetation structure also affects dung beetles directly, by acting as regulators of microclimatic condition, or indirectly affecting vertebrate fauna and consequently food availability (Nichols et al. 2009; Louzada et al. 2010).

15.5.5 Ants

The distribution of ants along altitudinal gradients exhibit at least two general trends: a higher species richness at intermediate altitudes (e.g., Fisher 1998; Samson et al. 1997; Munyai and Foord 2012; Bharti et al. 2013; Bishop et al. 2014) or a decline in the number of ant species with increasing altitude (Bruhl et al. 1999; Longino and Colwell 2011; Bharti et al. 2013). Costa et al. (2015) recorded 288 ant species of 53 genera and eight subfamilies in the rupestrian grassland ecosystem. In the rupestrian grasslands of Serra do Cipó, Araújo and Fernandes (2003) recorded 42 morphospecies of ants using baits. Their study showed that ground dwelling ant richness was negatively correlated with altitude, both in mesic and xeric habitats. The authors argued that the mechanism producing these patterns were increased habitat structure at lower elevation and at mesic environments, where more woody vegetation prevailed. It is important to point, however, that bait sampling has been shown to be selective and hence not capable of sampling the entire ant fauna (Delabie et al. 2000). For this reason, we resampled the entire altitudinal gradient using a more robust protocol with different complementary techniques and sampling periods in the before mentioned sampling sites. Samples were performed with the use of pitfall traps, sweeping, and Winkler extractor, in four distinct periods: during two rainy seasons and during two dry seasons.

Ant species diversity along the altitudinal gradient was exceptionally high; overall, we recorded 170 ant species belonging to eight subfamilies. The most representative subfamily was Myrmicinae, with the genus *Pheidole* being the most species rich. The number of ant species recorded in the rupestrian grasslands is very high compared to other studies on ground dwelling ants of the Cerrado vegetation (Brandão et al. 2011; Frizzo et al. 2011), and reinforces the need of complementary methods to sample ant species in mountain environments.

Ant species composition changed along the altitudinal gradient and a negative correlation was found between ant species richness and altitude (Fig. 15.10a), with altitude accounting for 52 % of the variation in ant species richness ($R^2 = 0.52$). The decline in temperature over altitudinal gradients has been shown to be one of the most important factors of species richness decline with increasing altitude (e.g., Fernandes and Price 1988, 1991; Malsch et al. 2008; Longino and Colwell 2011; Bharti et al. 2013; Bishop et al. 2014). Although natural history studies are needed, these findings indicate that the limited distribution of these species must be linked to thermal tolerance. Longino and Colwell (2011) argued that temperature is the best predictor of ant species richness and also worker density. However, we found that the main driver for the ant community structure in our altitudinal gradient was

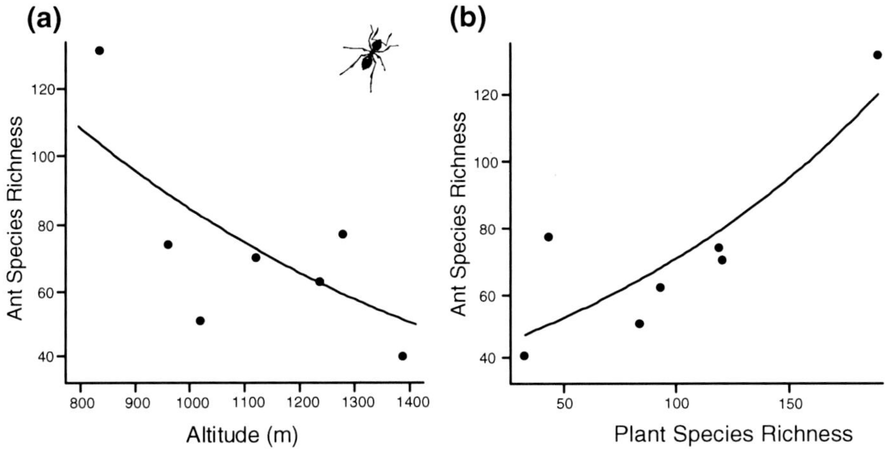

Fig. 15.10 **a** Regression of ant richness against altitude (800–1400 m a.s.l.) (R^2 = 0.52). **b** Regression of ant species richness against plant species richness (R^2 = 0.71)

vegetation structure. In spite of the decrease of the average temperature along the altitudinal gradient in the rupestrian grasslands, temperature did not correlate with ant species richness ($p > 0.05$). This new and more complete ant sampling corroborates the earlier study of Araújo and Fernandes (2003) and indicates that plant species richness is the most important factor correlating with ant species richness along the altitudinal gradient in the rupestrian grasslands (R^2 = 0.71, Fig. 15.10b). The richness of plants and habitat heterogeneity represent important mechanisms that influence the community structure of ants (Ribas et al. 2003; Pacheco and Vasconcelos 2012). Plants provide ants with food resources, nesting sites, shelter, and space to interact with other organisms (e.g. throphobionts and preys).

15.5.6 Butterflies

Different patterns of distribution of Lepidoptera (butterflies and moths) along altitudinal gradients have been found: a negative relationship with elevation (Austria: Schnepf 2010; Brazil: Carneiro et al 2014a); a positive relationship (Equador: Pyrcz et al. 2009); a mid-elevation relationship (Spain: Sanchez-Rodriguez and Baz 1995; Wilson et al. 2007; Israel: Levanoni et al. 2011); a low plateau, where richness decreases after certain low altitudes (Mexico: Molina-Martinez et al. 2013); a low richness at mid-elevation (Indonesia: Tati-Subahar et al. 2007); and the absence of pattern (Andes: Arroyo et al. 1982). Differences in life history traits amongst Lepidoptera might account for the differences found, as observed in the Comores island, where Lewis et al. (1998) found distinct patterns for endemic and non-endemic butterflies. Endemic species were more diverse at intermediate altitudes whereas the non-endemic group was more diverse at the base of the mountains.

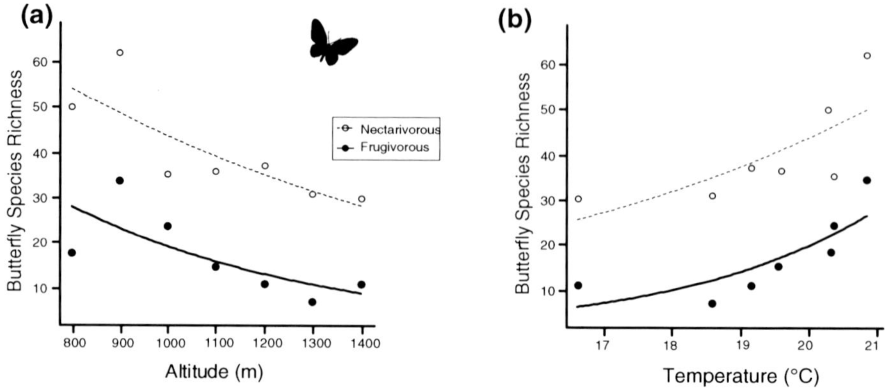

Fig. 15.11 Regression of nectarivorous (*dotted lines*, *open circles*) (R^2 = 0.68) and frugivorous (*solid lines*, *black circles*) (R^2 = 0.58) butterfly species richness against altitude (800–1400 m a.s.l)

We studied nectar-feeding butterflies (Papilionidae, Pieridae, Lycaenidae, Hesperiidae and part of the Nymphalidae family) and fruit-feeding butterflies, those that feed upon rotting fruits, fermenting sap and animal waste (part of the Nymphalidae family: Satyrinae, Biblidinae, Charaxinae and part of the Nymphalinae subfamilies) (see DeVries 1987) in the altitudinal gradient at Serra do Cipó. The richness of the two butterfly guilds was lower at higher altitudes (p = 0.003; Fig. 15.11). Temperature strongly regulates and acts directly on lepidoptera physiology, metabolism, life cycle, behavior, development and survivor (see Bale et al. 2002; Sanders et al. 2007; Speight et al. 2008) and ultimately influences their distribution (Molina-Martinez et al. 2013). The richness (p = 0.03) of butterfly communities correlated positively with temperature along the altitudinal gradient studied. Butterflies are extremely sensitive to changes in the composition of the vegetation and they are involved into many ecological processes in the ecosystems (see Arroyo et al. 1982; Sawchik et al. 2003). The lepidopteran community can be affected by temperature, humidity, rain, and wind speed (e.g., Stefanescu et al. 2004; Bhardwaj et al. 2012). Besides, seasonality and spatial and temporal variation of the food resources are crucial factors determining their diversity, composition, and dispersion (e.g., Abrahamczik et al. 2011; Baguette et al. 2011).

15.5.7 Arbuscular Mycorrhizal Fungi

Mycorrhizal diversity and abundance are strongly influenced by soil properties (Bryant et al. 2008; Fierer et al. 2011). The high degree of endemism and biodiversity in the Espinhaço Mountains have often been attributed to the mosaic vegetation formation, rugged landscape, microclimatic variation and properties of several soil types. Carvalho et al. (2012) suggested that a positive feedback might exist between Arbuscular Mycorrhizal Fungi (AMF) diversity and the high endemism and diversity of plant species in the nutrient-poor soils of these mountains.

The composition, richness, density, and relative frequency of species of AMF communities along the altitudinal gradient were studied by Coutinho et al. (2015). Thirteen plots of 100 m^2 (10 × 10 m) were defined along each transect, for a total of 91 plots (0.91 ha). AMF was studied from a mixed sample of soils in the studied areas, as described previously. Fifty-one species of AMF were recorded, with 14 species being possibly new to science and nine species representing new records for Brazil. This study, together with that of Carvalho et al. (2012) (see also Chap. 8) indicates that rupestrian grasslands constitute a source of AMF's diversity and probably the most diverse region for AMF in the world. The species recorded belong to 15 genera and 11 families, with a large proportion of the species belonging to the genera *Acaulospora* (14 species) and *Glomus* (14 species). It seems that these genera could have experienced an intensive speciation in the rupestrian grasslands, although detailed studies are needed to confirm such hypothesis.

The correlation of altitude with richness of AMF spores did not show a linear relationship but instead a parabolic one (R^2 = 0.89; Fig. 15.12). Coutinho et al. (2015) also sowed that the total density of the spores varied from 5095 at 800 m to 13,510, and that the highest densities and greatest numbers of spores were observed at the intermediate altitudes of 1100 and 1200 m. Their study corroborated the hypothesis that plant species benefit from the symbiosis with AMF in low fertility environments (e.g., Smith and Read 1997; Berbara et al. 2006; Lisboa et al. 2014). In summary, the distribution of AMF along the altitudinal gradient was higher at intermediate altitudes, characterized by numerous habitat types and the ecotone cerrado (savanna) *sensu structo*/rupestrian grassland physiognomies.

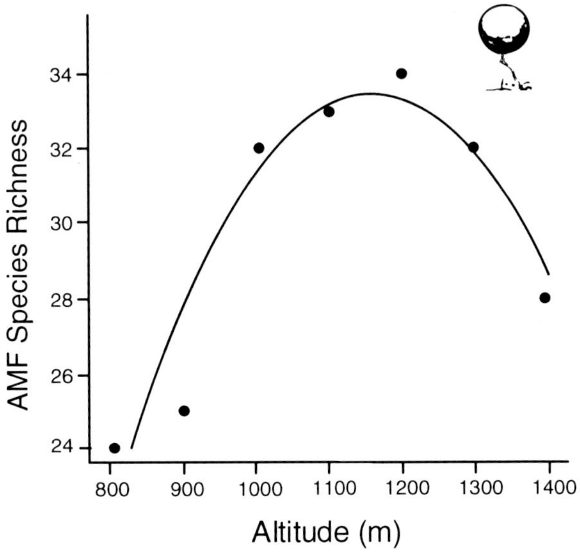

Fig. 15.12 Regression of arbuscular mycorrhizal fungi species richness against altitude (800–1400 m a.s.l.) (R^2 = 0.89)

15.5.8 *Insect Herbivores*

Free feeding, galling, and mining insect herbivores were sampled along an altitudinal gradient at three sites widely distant from each other (>1 km) that were established every 100 m of altitude from 800 to 1500 m. Each sample site included one xeric habitat, comprising small trees and shrubs in the cerrado (800–1100) or on grasses and herbs in grasslands, up to 1100 m. We also sampled the crowns of gallery forests in nearby mesic sites. Hence, the design compared upper canopy structures from both xeric cerrado and wet gallery forest (Ribeiro 2003).

15.5.8.1 Free Feeding Herbivores

A total of 4351 specimens of free-feeding insect herbivores, belonging to 38 families, were sampled for the dry and wet seasons, habitats and altitudes. In the wet season, insect species richness decreased with increasing altitude in the xeric habitats ($R^2 = 0.33$; Fig. 15.13), but no trend was found in mesic habitats. In the dry season, altitude did not affect insect richness in xeric habitats, a result strongly related to an overall decreasing in the average number of individuals in the dry season (Fig. 15.13). Then, as expected, in the mesic habitats such seasonal overall decreasing was not observed, but the distribution pattern was contrary of expectation: species richness increased with increasing altitude in the mesic habitats in the dry season ($R^2 = 0.39$, $p < 0.01$, Fig. 15.13).

The whole altitudinal pattern was, therefore, non-consistent along the seasons or between habitats, and reflected specific effects of dramatic fauna reduction in certain climatic conditions. As in the case of leaf mining (see below), which were more diverse and abundant in habitats where predation pressure must be more intense but climate and leaf hardness milder, free-feeding herbivores also declined when

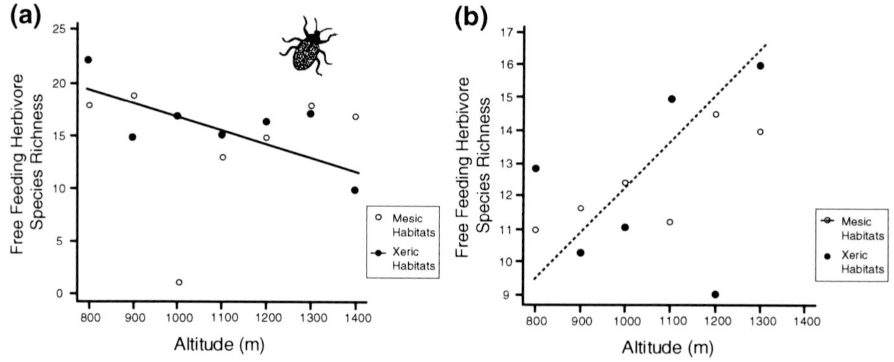

Fig. 15.13 Regression of free feeding herbivore species richness against altitude (800–1500 m) in xeric (*solid lines, closed circles*) and mesic (*dotted lines, open circles*) habitats in the **a** wet and **b** dry season at Serra do Cipó, Brazil

conditions got more extreme, both in time or space. For instance, the altitudinal effect on xeric habitats was strongly influenced by a severe decline in insects sampled in the two higher altitudes, but also affected by the presence of fragmented habitats such as woody rupestrian grasslands that occur entangled with more grassy continuous grasslands. A new model omitting values from rupestrian fields and leaving only large, continuous woody grassland sites, explained 59 % of variance in comparison with 33 % of the previous model.

15.5.8.2 Galling Herbivores

Galling insects and their host plants are by far the most studied group in rupestrian grasslands. At a local scale, the majority of the studies found a negative relation between galling species richness and altitude in different biogeographic regions (Fernandes and Price 1988; Fernandes and Lara 1993; Fernandes et al. 1994; Price et al. 1998; Waring and Price 1990). Studies developed in the altitudinal gradient in the rupestrian grasslands of Serra do Cipó (Price et al. 1998; Price 1991) indicated a strong and negative interaction of galling insect species richness and altitude. Lara et al. (2002) collected galling insects in the same sites as free-feeding insects were studied (above) (44 sample sites, and 38,270 plants surveyed).

Galling insect richness was higher and galls were more abundant in xeric habitats compared to mesic habitats. The total galling insect species richness decreased with altitude (Fig. 15.14) and altitude explained 94 % of the variation in galling species richness (R^2 = 0.94).

Other factors like host plant species richness were considered less relevant to explain the variation in galling insect species richness (Fernandes and Price 1988;

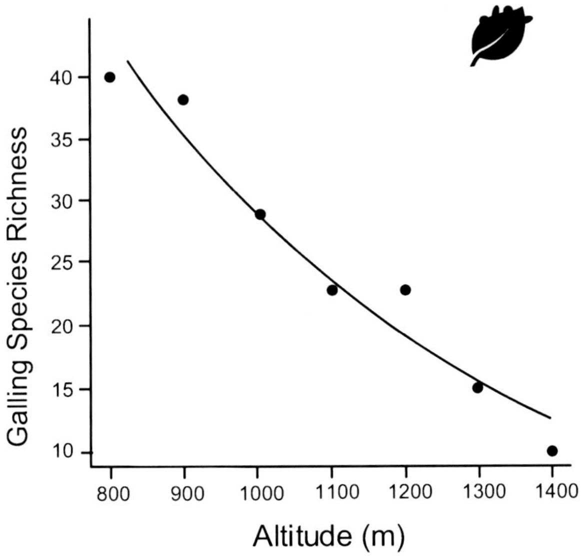

Fig. 15.14 Regression of galling species richness against altitude gradient (800–1400 m a.s.l.) (R^2 = 0.94)

Lara et al. 2002). Some studies have shown, however, that host plants in different architectonic groups (e.g., shrubs) can strongly influence the variation in the species richness of galling insects (e.g., Wright and Samways 1996, 1998; Blanche and Ludwig 2001; Gonçalves-Alvim and Fernandes 2001; Carneiro et al. 2014b). The total species richness of host plants was rejected as a determinant factor of the species richness of galling insects because most of the galling insects species were mostly concentrated on some plant species, commonly called 'superhosts', representing some plant taxa that exhibit a high species richness of galling insects in different biogeographic regions. Examples include the Cecidomyiidae on *Baccharis* at the Neotropical region (Fernandes et al. 1996, 2014) and on *Solidago* in the Neartic region (Gagné 1994); Cynipidae on *Quercus* species on the Neartic region (Stone et al. 2002) and on *Acacia species* at Etiopic region, Chalcidoidea and Coccoidea on Myrtaceae (specially *Eucalyptus*) in Australian region (Blanche 1994). In this situation, the taxonomic composition of the vegetation has a strong effect in the species richness of galling insects.

15.5.8.3 Mining Herbivores

Leaf miners exhibit several spatial distribution patterns (e.g., Needham et al. 1928; Hering 1951; Faeth et al. 1981; Cumbrera 1998; Tantowijoyo and Hoffman 2010). The adults are generally host specific and females are very selective on their oviposition sites (Whitfield et al. 1985). Furthermore, leaf miners have an intermediate form of host use between gallers and free-feeders, i.e., their larvae are internal tissue feeders, but do not develop galls. Therefore, they should present strong habitat preference, mainly considering enemy free space and host plant quality.

Leaf miners were studied exactly at the same sites and same plant species as those described for galling insects. Mining insect species richness was negatively correlated with altitude in xeric habitats (Fig. 15.15). Altitude accounted for by

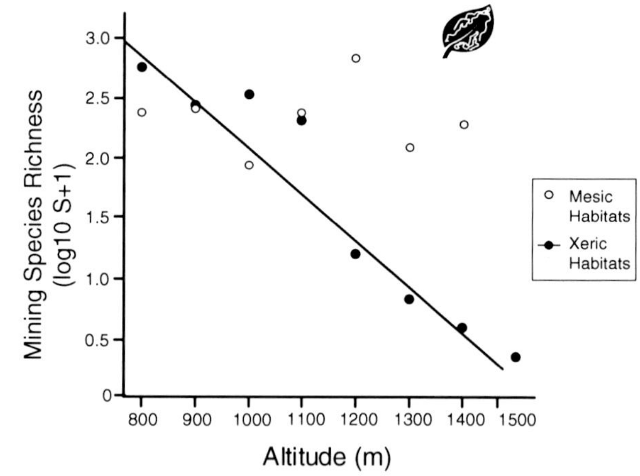

Fig. 15.15 Regression of mining species richness against an altitudinal gradient (800–1400 m a.s.l.) in xeric (*solid lines, closed circles*) ($R^2 = 0.76$) and mesic habitats (*opened circles*) at Serra do Cipó, Brazil. The "y" axis values are on a logarithmic scale

76 % of the variation in mine richness on all plants (herbs, shrubs and trees) in xeric habitats (R^2 = 0.76) (see Fernandes et al. 2004). In mesic habitats, on the other hand, mining insect richness did not show any statistically significant relationship with elevation (Fig. 15.15). Plant species richness positively influenced mining species richness in xeric habitats. In these habitats, plant species richness influenced 76 % of the variation in the number of mining species on all plants (R^2 = 0.76). In mesic habitats plant species richness did not influence on the richness of insect mining species.

Leaf mining species richness decreased with increasing elevation in mesic habitats, while leaf miner richness was relatively constant and high in xeric habitats along the altitudinal gradient. The mechanisms driving the patterns here found are still unknown for leaf mining insects. However, an important issue is plant species richness. The association between mining insects and host plants is very specific (Hering 1951), and can drive the richness pattern found. The negative correlation between mining richness and altitude can be explained by the positive correlation between plant species richness and altitude at lower elevations. Within xeric habitats, sites at lower altitudes presented more plant species than habitats at higher altitudes. Another important factor that may account for the lower diversity of miners in xeric habitats is the high degree of sclerophylly of cerrado host plants, which would reduce larval establishment and success. Furthermore, sclerophylly is said to be strongly correlated to high levels of chemical defences, particularly tannins, in the region (Ribeiro and Fernades 2000).

15.5.9 Birds

The influence of habitat structure on patterns of species richness and composition are particularly well documented for birds (e.g., MacArthur et al. 1966; Terborgh 1971; Stirnemann et al. 2015). Melo-Júnior et al. (2001) for the first time described the distribution of bird species along the altitudinal gradient in the rupestrian grasslands. Their study was also performed along an altitudinal gradient between 800 and 1400 m a.s.l. but at different localities from the ones here described. For this study, at each altitudinal interval (every 100 m in elevation), three sites were selected, each containing a paired xeric and mesic habitats, except at 1400 m, in which only two sites were found, totaling 20 sites, and totaling 110 h of observation.

Bird species richness correlated negatively with elevation both in xeric and mesic habitats. In xeric habitats 45.0 % of the variation in the number of bird species was explained by elevation (R^2 = 0.45, $p < 0.001$), whereas in mesic habitats elevation explained only 30.0 % of the variation in bird species number ($R^2 = 0.30$, $p < 0.01$) (see Melo-Júnior et al. 2001 for details). Bird species richness in the rupestrian grasslands in Serra do Cipó did not differ statistically between habitat types ($p > 0.05$). Therefore, Fig. 15.16 shows the combined distribution of birds in the two habitat types. At lower elevations (800–900 m), 15 species were

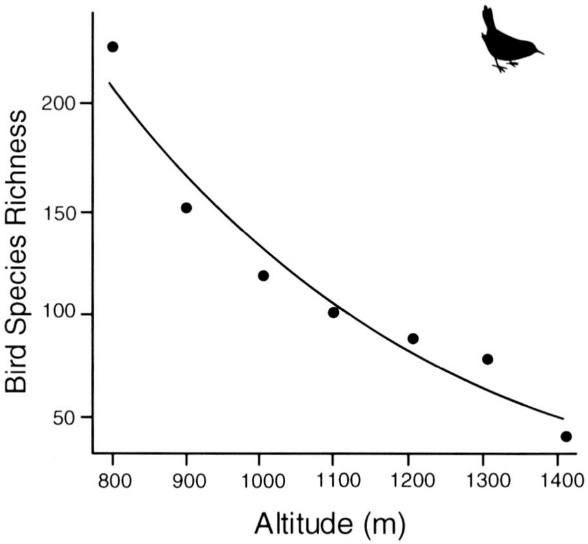

Fig. 15.16 Regression of bird species richness against altitude (800–1400 m a.s.l.) (R^2 = 0.95)

found exclusively in xeric habitats, and 28 species were found exclusively in mesic habitats. Thirteen species were exclusively found at higher elevation (1200–1400 m) in xeric habitats, while in mesic habitats these were represented by nine species.

The study by Melo-Júnior et al. (2001) corroborates the overall pattern found worldwide where bird species richness decreases at higher elevations (e.g., Terborgh 1977: Cordillera Vilcabamba—Peru; Loiselle and Blake 1991: Costa Rican rain forest; Navarro 1992: Sierra Madre del Sur—Mexico; Ambarli and Bilgin 2014: Anatolian steppes—Turkey). Their results also indicated that bird species distribution pattern in Serra do Cipó is associated with the vegetation types encountered along the gradient. The increased number of species found at 1100 m in xeric habitats might has been influenced by the mosaic of vegetation types and habitats there found (see also Carvalho et al. 2012). Between 1000–1100 m, there is an ecotone with plant species of the Cerrado and rupestrian grassland intermixed in the area. In xeric habitats above 1300 m, vegetation structure becomes much simpler with few sparsely distributed trees, whereas over 1400 m the vegetation is dominated by grasslands, in spite of the presence of natural islands of Atlantic rain forests which due to higher structural complexity could positively influence their richness patterns (Chap. 7).

No significant differences in bird species richness between mesic and xeric habitats were reported and the authors argued that the gallery forests of the region might not be large enough to support a more diversified bird fauna. Actually, gallery forests at Serra do Cipó are rather narrow, ranging from 10 to 30 m. Therefore, the width of gallery forests may represent a relevant factor that explains the similarity in bird species richness encountered in the various habitats in the rupestrian grasslands.

Since there is no significant difference in the bird species richness between mesic and xeric habitats, a new analysis was performed with the total bird species richness against altitude. A negative correlation was found (Fig. 15.16), with altitude accounting for 95 % of the variation in bird species richness ($R^2 = 0.95$).

15.6 Synthesis

One of the most striking results from our studies is that the degree of ecological changes occurring along the altitudinal gradient of the Espinhaço Mountains are as pronounced as those reported for other mountain gradients, and yet the Espinhaço gradient studied by us only spans 600 m as opposed to several thousand meters in other systems. Additionally, the highest point of our gradient is very low compared to other mountains, it does not represent extreme low-temperature environments found in most mountain systems and it is still relatively warm compared to other mountaintops. The third characteristic that seems to be unique for the Espinhaço gradient is the nonlinear pattern of precipitation, where the most precipitation occurs in the points of mid-elevation (but see Glarizin 1997). This mid-elevation peak in precipitation does not generally correlate with ecological patterns we have observed to date, all of which show a decrease/increase from low to high altitudes Fig. 15.17. This suggests that precipitation is not as limiting as other factors in mediating species richness patterns. Arbuscular mycorrhizal fungi diversity was the only group of organisms that appeared to track precipitation, with highest species diversity at mid-elevation. While precipitation exhibited a dome-shaped distribution, moisten laid clouds may likely be responsible for the higher average humidity at higher elevations. Regardless, it is one of the few situations we know where humidity does not correlate with precipitation and warrants further studies to better understand the response of dominant plant species and their associated communities.

In the short altitudinal range here described we observed a decrease in ca. 6 °C in 600 m elevation. This is within what is expected in the literature (Fritzsons et al. 2008). Temperature can vary for a number of different reasons in a landscape. Shade, sun, and inversions provoked by colder air sinking into valleys can all influence the temperature. Of great importance in the Espinhaço Mountains are the presence of cool air pooling. Barbosa et al. (2015) have already suggested the importance these resident clouds could have on vegetation structure and composition (see also Chap. 7). Under a cloudy or rainy environment, temperature decreases by about 6 °C per 1000 m (Dury 1972; Mendonça and Danni-Oliveira 2007).

The response of the community of many tropical organisms to variation in temperature, humidity, wind or a combination of them is mostly unknown. For instance, lepidopterans and galling insects are strongly influenced by temperature (Fernandes and Price 1988; Stefanescu et al. 2004; Bhardwaj et al. 2012). Therefore, not just detailed knowledge on the climatic variation but also on the synergistic effects of several variables that constitute climate for particular groups are necessary

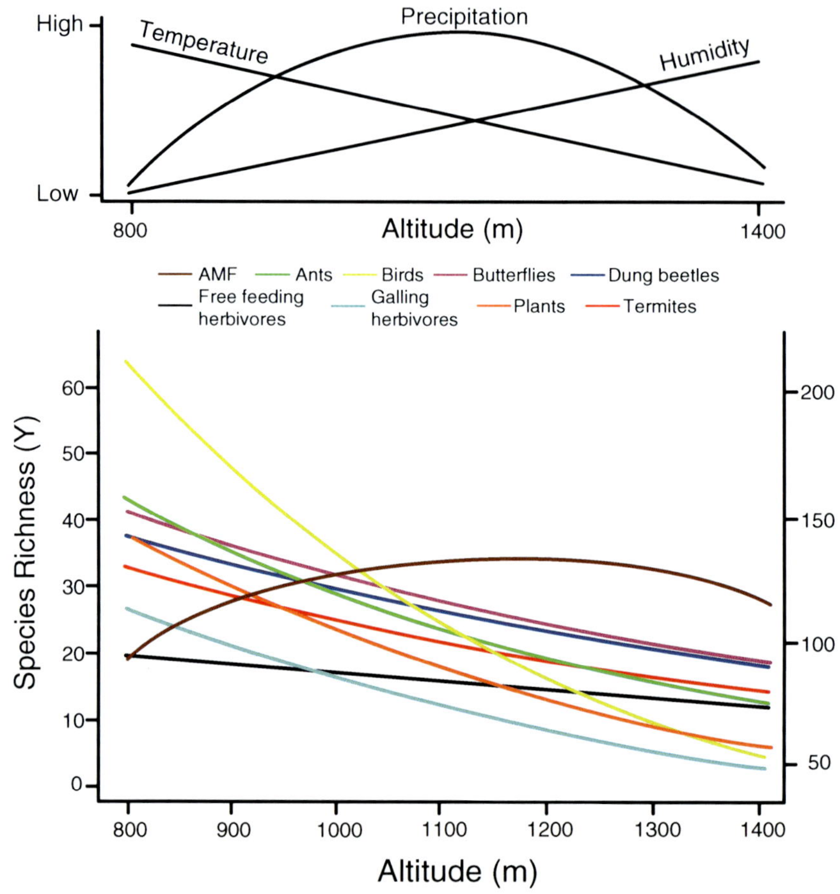

Fig. 15.17 Overall plot of species richness along the altitudinal gradient in the rupestrian grassland at Serra do Cipó, Brazil. As the richness of each taxon is different the figure depicts two different scales (y and z) for richness. Axis y depicts the richness (0–60 species) for mycohrrizae, butterflies, dung beetles, free-feeding herbivores, and termites. Axis z depicts richness (50–200 species) for ants, birds, galling herbivores, and plants

to further understand climatic effects in species richness, probably through effects on their metabolism and physiology. This is not a trivial question as these are the climatic conditions predicted to change due to climate changes (IPCC 2014).

Another aspect that is of great interest is the extended effects of climate on vegetation attributes as well as on animal communities along elevation. For instance, what are the effects of immobile defenses on the success of herbivore colonization at different habitats along the altitudinal gradient? Many herbivores (e.g. mining and galling insect communities) may be strongly influenced by the quality of their hosts and climatic factors such as high temperature and solar radiation may positively or negatively influence them (e.g., Fernandes and Price 1991; Connor and Taverner

1997; Cumbrera 1998; Julião et al. 2014). In the case of the Cerrado vegetation, the harsh and severe leaf quality act as strong barriers to colonization by many herbivores (e.g., Ribeiro et al. 1994; Barbosa and Fernandes 2014).

The compilation of the general information provided by studies performed over the past 25 years along the same altitudinal gradient is of major importance, and it stands as one of the most comprehensive studies examining the dynamics of biodiversity along an altitudinal mountain gradient. We are expanding this work to study other key taxa (e.g., bees, grasshoppers, amphibians, reptiles, and wasps), establish reciprocal transplant experiments, and extend the scale of our research to other altitudinal gradients in South America and collaborative research on mountain systems on other continents. A key goal is to develop synthetic hypotheses to explain general patterns of global biodiversity for all mountainous regions. Finally, although we have not mentioned the socioeconomic impacts of biodiversity on local communities that depend on the ecosystem services of mountain habitats, we continue to solicit involvement of stakehoders and policymakers in conserving biodiversity and developing management policies that include sustainability of biodiversity.

Acknowledgements We thank the many stakeholders that have actively participated in this study by providing the land for the research, logistics, and even engaged in seminars and workshops: Serra Morena Inn, Cedro Cachoeira Textile Industry, Vellozia Reserve, Pedra do Elefante Inn, Eduardo C. Costa Jr, Planta Ltda, and Serra do Cipó National Park for logistical support. We are thankful to CNPq/PELD, ComCerrado/MCT and FAPEMIG for several grants that have supported this study, to José Eugênio C. Figueira for statistical advice, to Campestris for support in the editing and formatting some of the illustrations, and to OnsetBrasil for continued field support with the climatic stations.

References

Abrahamczyk S, Kluge J, Gareca Y, Reichle S, Kessler M (2011) The influence of climatic seasonality on the diversity of different tropical pollinator groups. PLoS ONE 6:1–9

Aiba S, Kitayama K (1999) Structure, composition and species diversity in an altitude-substrate matrix of rain forest tree communities on Mount Kinabalu, Borneo. Plant Ecol 140:139–157

Alves WDF, Mota AS, Lima RAA, Bellezoni R, Vasconcellos A (2011) Termites as bioindicators of habitat quality in the caatinga, Brazil: is there agreement between structural habitat variables and the sampled assemblages? Neotrop Entomol 40:39–46

Ambarli D, Bilgin CC (2014) Effects of landscape, land use and vegetation on bird community composition and diversity in Inner Anatolian steppes. Agr Ecosyst Environ 182:37–46

Araújo LM, Fernandes GW (2003) Altitudinal patterns in a tropical ant assemblage and variation in species richness between habitats. Lundiana 4:103–109

Araújo MB, Alagador D, Cabeza M, Nogués-Bravo D, Thuiller W (2011) Climate change threatens European conservation areas. Ecol Lett 14:484–492

Arroyo MTK, Primack R, Armesto J (1982) Community studies in pollination ecology in the high temperate Andes of central Chile. I. pollination mechanisms and altitudinal variation. Am J Bot 69:82–97

Ayoade JO (2010) Introdução a Climatologia para os Trópicos. Bertrand, Rio de Janeiro

Baguette M, Clobert J, Schtickzelle N (2011) Metapopulation dynamics of the bog fritillary butterfly: experimental changes in habitat quality induced negative density-dependent dispersal. Ecography 34:170–176
Bale J, Masters G, Hodkinson ID, Awmack C, Bezemer TM, Brown VK, Butterfield J, Buse A, Coulson JC, Farrar J, Good JE, Harrington R, Hartley S, Jones TH, Lindroth RL, Press MC, Symrnioudis I, Watt A, Whittaker JB (2002) Herbivory in global climate change research: direct effects of rising temperature on insect herbivores. Glob Chan Biol 8:1–16
Barbosa M, Fernandes GW (2014) Bottom-up effects on gall distribution. In: Fernandes GW, Santos JC (eds) Neotropical insect galls. Springer, Dordrecht, pp 99–113
Barbosa NPU, Fernandes GW, Sanchez-Azofeifa A (2015) A relict species restricted to a quartzitic mountain in tropical America: an exemplo of microrefugium? Acta Bot Bras 29:299–309
Begon M, Townsend CR, Harper JL (2006) Ecology: from individuals to ecosystems. Blackwell, Hong Kong
Beniston M, Diaz HF, Bradley RS (1997) Climate change at high elevation sites: an overview. Clim Chan 36:233–251
Berbara RLL, De Sousa FA, Fonseca HMA (2006) Fungos micorrízicos arbusculares: muito além da nutrição. In: Fernandes MS (ed) Nutrição Mineral de Plantas. SBCS, Viçosa, pp 53–88
Bezerra-Gusmão MA, Marinho RA, Kogiso KA, Bandeira AG, Barbosa MRV (2013) Nest dynamics of *Constrictotermes cyphergaster* (Termitidae, Nasutitermitinae) and its association with the supporting vegetation in a semiarid area, northeast, Brazil. J Arid Environ 91:1–6
Bhardwaj M, Uniyal VP, Sanyal AK, Singh AP (2012) Butterfly communities along an elevational gradient in the Tons Valley, Western Himalayas: implications of rapid assessment for insect conservation. J Asia Pac Entomol 15:207–217
Bharti H, Sharma YP, Bharti M, Pfeiffer M (2013) Ant species richness, endemicity and functional groups, along an elevational gradient in the Himalayas. Asian Myrmecol 5:79–101
Bishop TR, Robertson MP, Rensburg BJ, Parr CL (2014) Elevation—diversity patterns through space and time: ant communities of the Maloti-Drakensberg mountains of southern Africa. J Biogeogr 41:1–13
Blanche KR (1994) Insect induced galls on Australian vegetation. In: Price PW, Mattson WJ, Baranchikov Y (eds) The ecology and evolution of gall-forming insects. Forest Service US Department of Agriculture, St. Paul, pp 49–55
Blanche KR, Ludwig JA (2001) Species richness of gall inducing insects and host plants along an altitudinal gradient in Big Bend National Park, Texas. Am Midl Nat 145:219–232
Brandão CRF, Silva RR, Feitosa RM (2011) Cerrado ground–dwelling ants (Hymenoptera: Formicidae) as indicators of edge effects. Zool 28:379–387
Brown JH, Lomolino MV (1998) Biogeography. Sinauer, Sunderland
Bruhl CA, Mohamed M, Linsenmair KE (1999) Altitudinal distribution of leaf litter ants along a transect in primary forests on Mount Kinabalu, Sabah, Malaysia. J Trop Ecol 15:265–277
Bryant JA, Lamanna C, Morlon H, Kerkhoff AJ, Enquist BJ, Green JL (2008) Microbes on mountainsides: contrasting elevational patterns of bacterial and plant diversity. Proc Natl Acad Sci USA 105:11505–11511
Carneiro E, Mielke OHH, Casagrande MM, Fiedler K (2014a) Skipper richness (Hesperiidae) along elevational gradients in Brazilian Atlantic Forest. Neotrop Entomol 43:27–38
Carneiro MAA, Coelho MS, Fernandes GW (2014b) Galls in Brazilian mountains: new reports and perspectives. In: Fernandes GW, Santos JC (eds) Neotropical insect galls. Springer, Dordrecht, pp 273–293
Carvalho F, Souza FA, Carrenho R, Moreira FMS, Jesus EC, Fernandes GW (2012) The mosaic of habitats in the high–altitude Brazilian rupestrian fields is a hotspot for arbuscular mycorrhizal fungi. Appl Soil Ecol 52:9–19
Carvalho F, Godoy EL, Lisboa FJG, Moreira FMS, Souza FA, Berbara RLL, Fernandes GW (2014) Relationship between physical and chemical soil attributes and plant species diversity in tropical mountain ecosystems from Brazil. J Mt Sci 11:875–883
Connor EF, Taverner MP (1997) The evolution and adaptive significance of the leaf–mining habit. Oikos 79:6–25

Costa FV, Mello R, Lana CT, Neves FS (2015) Ant fauna in megadiverse mountains: a checklist for the rocky grasslands. Sociobiol 62:228–245

Coutinho ES, Fernandes GW, Berbara RLL, Valério HM, Goto BT (2015) Variation of arbuscular mycorrhizal fungal communities along an altitudinal gradient in rupestrian grasslands in Brazil. Mycorrhiza 25:627–638

Cumbrera AV (1998) Diversidad de insectos minadores en un bosque tropical. Dissertation, Universidad de Panama

Davies AB, Eggleton P, van Rensburg BJ, Parr CL (2015) Seasonal activity patterns of African savanna termites vary across a rainfall gradient. Ins Soc 62:157–165

Davis ALV, Scholtz CH (2001) Historical vs. ecological factors influencing global patterns of scarabaeine dung beetle diversity. Div Dist 7:161–174

Delabie JH, Fisher BL, Majer Wright IW (2000) Sampling effort and choice of methods. In: Agosti D, Majer JD, Alonso LE, Schultz TR (eds) Ants: standard methods for measuring and monitoring biodiversity. Smithsonian institution press, Washington and London, pp 145–154

DeVries PJ (1987) The butterflies of Costa Rica and their natural history. Vol I: Papilionidae, Pieridae, Nymphalidae. Princeton University Press, New Jersey, p 456

Donovan SE, Eggleton P, Martin A (2002) Species composition of termites of the Nyika plateau forests, northern Malawi, over an altitudinal gradient. Afr J Ecol 40:379–385

Dossin IA, Dossin TM, Chaves MLSC (1990) Compartimentação estratigráfica do supergrupo Espinhaço em Minas Gerais - os grupos Diamantina e Conselheiro Mata. Rev Bras Geoc 20:178–186

Dury GH (1972) High temperature extremes in Austrália. Ann Assoc Am Geo 62:388–400

Eiten G (1972) The cerrado vegetation of Brazil. Bot Rev 38:201–341

Escobar F, Lobo JM, Halffter G (2005) Altitudinal variation of dung beetle (Scarabaeidae: Scarabaeinae) assemblages in the Colombian Andes. Glob Ecol Biogeogr 14:327–337

Faeth SH, Mopper S, Simberloff D (1981) Abundances and diversity of leaf–mining insects on three oak host species: effects of host plant phenology and nitrogen content of leaves. Oikos 37:238–251

Fernandes GW (1992) A gradient analysis of plant forms from northern Arizona. J Ariz-Nev Acad Sci 25:21–30

Fernandes GW, Lara ACF (1993) Diversity of Indonesian gall–forming herbivores along altitudinal gradients. Biodiv Lett 1:186–192

Fernandes GW, Price PW (1988) Biogeographical gradients in galling species richness: tests of hypotheses. Oecologia 76:161–167

Fernandes GW, Price PW (1991) Comparison of tropical and temperate galling species richness: the roles of environmental harshness and plant nutrient status. In: Price PW, Lewinsohn TM, Fernandes GW, Benson WW (eds) Plant–animal interactions: evolutionary ecology in tropical and temperate regions. Wiley, New York, pp 91–115

Fernandes GW, Lara ACF, Price PW (1994) The geography of galling insects and the mechanisms resulting in patterns. In: Price PW, Mattson WJ, Baranchikov Y (eds) Gall–forming insects: ecology, physiology and evolution. Forest Service US Department of Agriculture, St. Paul, pp 42–48

Fernandes GW, Carneiro MAA, Lara ACF, Allain LR, Andrade GI, Julião GR, Reis TR, Silva IM (1996) Galling insects on neotropical species of *Baccharis* (Asteraceae). Trop Zool 9:315–332

Fernandes GW, Castro FMC, Faira ML, Marques ESA, Greco MKB (2004) Effects of hygrothermal stress, plant richness, and architecture on mining insect diversity. Biotropica 36:240–247

Fernandes GW, Silva JO, Espírito-Santo MM, Fagundes M, Oki Y, Carneiro MAA (2014) *Baccharis*: A neotropical model system to study insect plant interactions. In: Fernandes GW, Santos JC (eds) Neotropical insect galls. Springer, Dordrecht, pp 193–219

Fierer N, McCain CM, Meir P, Zimmermann M, Rapp JM, Silman MR, Knight R (2011) Microbes do not follow the elevational diversity patterns of plants and animals. Ecology 92:797–804

Fisher BL (1998) Ant diversity patterns along an elevational gradient in the Reserve Special d'Anjanaharibesud and on the western Masoala Peninsula, Madagascar. Field Zool 90:39–67

Fritzsons E, Mantovani LE, Aguiar AV (2008) Relação entre altitude e temperatura: uma contribuição ao zoneamento climático no estado do Paraná. Rev Est Amb 10:49–64
Frizzo TLM, Campos RI, Vasconcelos HL (2011) Contrasting effects of fire on arboreal and ground–dwelling ant communities of a neotropical Savanna. Biotropica 44:1–8
Gagné RJ (1994) The gall midges of the neotropical region. Cornell University Press, Ithaca
Gentry AH (1988) Changes in plant community diversity and floristic composition on environmental and geographical gradientes. Ann Mo Bot Gard 75:1–34
Gerold G, Schawe M, Bach K (2008) Hydrometeorologic, pedologic and vegetation patterns along an elevational transect in the montane forest of the Bolivian Yungas. Erde 139:141–168
Glarizin GE (1997) Precipitation distribution with altitude. Theor Appl Climatol 58:141–145
Gonçalves-Alvim SJ, Fernandes GW (2001) Biodiversity of galling insects: historical, community and habitat effects in four neotropical savannas. Biodiv Cons 10:79–98
Halffter G, Edmonds WD (1982) The nesting behaviour of dung beetles (Scarabaeinae): an ecological and evolutive approach. Instituto de Ecología, México
Hamilton AC (1975) A quantitative analysis of altitudinal zonation in Uganda forests. Vegetatio 30:99–106
Hering EM (1951) Biology of the leaf miners. Junk, The Hague
Herzog SK, Hamel-Leigue AC, Larsen TH, Mann DJ, Soria-Auza RW, Gill BD, Edmonds WD, Spector S (2013) Elevational distribution and conservation biogeography of Phanaeine dung beetles (Coleoptera: Scarabaeinae) in Bolivia. PLoS ONE. doi:10.1371/journal.pone.0064963
Holland PG, Steyn DG (1975) Vegetational responses to latitudinal in slope angle and aspect. J Biogeog 179–183
Holt EG (1928) An ornithological survey of the Serra do Itatiaya, Brazil. Bull Amer Mus Nat Hist 57:251–326
Hughes L (2000) Biological consequences of global warming: is the signal already apparent? Tren Ecol Evol 15:56–61
Hussain MZ, Malik NZ (2012) High altitude forest composition diversity and its component in a part of Ganga Chotti and Bedori Hills District Bagh. Azad Jammu and Kashmir Pakistan. AGD Land Environ 6:31–40
IPCC (2014) Climate change 2014: synthesis report. Core writing team. In: Pachauri RK, Meyer LA (eds) Contribution of working groups I, II and III to the Fifth Assessment report of the Intergovernmental panel on climate change. IPCC, Switzerland
Joly CA, Assis MA, Bernacci LC, Tamashiro JY, Campos MCR, Gomes JAMA, Lacerda MS, Santos FAM, Pedroni F, Pereira LS, Padgurshi MCG, Prata EMB, Ramos E, Torres RB, Rochelle A, Martins FR, Alves LF, Vieira SA, Martinelli LA, Camargo PB, Aidar MPM, Eisenlohr PV, Simões E, Villani JP, Belinello R (2012) Florística e fitossociologia em parcelas permanentes da Mata Atlântica do sudeste do Brasil ao longo de um gradiente altitudinal. Biota Neotrop 12:125–145
Jones DT, Eggleton P (2000) Sampling termite assemblages in tropical forests: testing a rapid biodiversity assessment protocol. J Appl Ecol 37:191–203
Julião GR, Venticique EM, Fernandes GW, Price PW (2014) Unexpected hight diversity of galling insects in the Amazonian upper canopy: the savanna out there. PLoS ONE 9(12):e114986
Kitayama K (1992) An altitudinal transect study of the vegetation on Mount Kinabalu, Borneo. Vegetatio 102:149–171
Korner C (2007) The use of "altitude" in ecological research. Trend Ecol Evol 22:569–574
Lara ACF, Fernandes GW, Gonçalves-Alvim SJ (2002) Tests of hypotheses on patterns of gall distribution along an altitudinal gradient. Trop Zool 15:219–232
Leuschner C, Moser G, Bertsch C, Roderstein M, Hertel D (2007) Large altitudinal increase in tree root/shoot ratio in tropical mountain forests of Ecuador. Bas Appl Ecol 8:219–230
Levanoni O, Levin N, Pe'er G, Turbé A, Kark S (2011) Can we predict butterfly diversity along an elevation gradient from space? Ecography 34:372–383
Lewis OT, Wilson RJ, Harper MC (1998) Endemic butterflies of Grande Comore: habitat preferences and conservation priorities. Biol Cons 85:113–121

Lieberman D, Lieberman M, Peralta R, Hartshorn GS (1996) Tropical forest structure and composition on a large-scale altitudinal gradient in Costa Rica. J Ecol 84:137–152

Lisboa FJG, Chaer G, Fernandes MF, Berbara RLL, Madari BE (2014) The match between microbial community structure and soil properties is modulated by land use types and sample origin within an integrated agroecosystem. Soil Biol Biochem 78:97–118

Loarie SR, Duffy PB, Hamilton H, Asner GP, Field CB, Ackerly DD (2009) The velocity of climate change. Nature 462:1052–1057

Lobo JM, Halffter G (2000) Biogeographical and ecological factors affecting the altitudinal variation of mountainous communities of coprophagous beetles (Coleoptera: Scarabaeoidea): a comparative study. Ann Entomol Soc Am 93:115–126

Loiselle BA, Blake JG (1991) Temporal variation in birds and fruits along an elevational gradient in Costa Rica. Ecology 72:180–183

Lomolino MV (2001) Elevation gradients of species–density: historical and prospective views. Glob Ecol Biogeogr 10:3–13

Longino JT, Colwell RK (2011) Density compensation, species composition, and richness of ants on a neotropical elevational gradient. Ecosphere 2:1–20

Louzada J, Lima AP, Matavelli R, Zambaldi L, Barlow J (2010) Community structure of dung beetles in Amazonian savannas: role of fire disturbance, vegetation and landscape structure. Land Ecol 25:631–641

MacArthur RH, Recher H, Cody M (1966) On the relation between habitat selection and species diversity. Am Nat 100:319–332

Madeira AJ, Fernandes WG (1999) Reprodutive phenology of sympatric taxa of *Chamaecrista* (Leguminosae) in Serra do Cipó, Brasil. J Trop Ecol 15:463–479

Malsch AKF, Fiala B, Maschwitz U, Mohamed M, Nais J, Linsenmair KE (2008) An analysis of declining ant species richness with increasing elevation at Mount Kinabalu, Sabah, Borneo. Asian Myrmecol 2:33–49

McCain CM (2005) Elevational gradients in diversity of small mammals. Ecology 86:366–372

McCain CM (2007) Area and mammalian elevational diversity. Ecology 88:76–86

McCoy ED (1990) The distribution of insects along elevational gradients. Oikos 58:313–322

Melo-Júnior TA, Vasconcelos MF, Fernandes GW, Marini M (2001) Bird species distribution and conservation in Serra do Cipó, Minas Gerais, Brazil. Bird Cons Int 11:189–204

Mendonça F, Danni-Oliveira IM (2007) Climatologia – Noções básicas e climas do Brasil. Oficina e Textos, São Paulo

Milanesi MA, Galvani E (2011) Efeito orográfico na Ilha de São Sebastião (Ilhabela – SP). Rev Bras Climatol 9:68–79

Miura T, Matsumoto T (1997) Diet and nest material of the processional termite Hospitalitermes, and cohabitation of *Termes* (Isoptera, Termitidae) on Borneo Island. Ins Soc 44:267–275

Molina-Martinez A, León-Cortés JL, Regan HM (2013) Climatic and geometric constraints as driving factors of butterfly species richness along a neotropical elevational gradient. J Ins Cons 17:1169–1180

Mota NM, Rezende VL, Mota GS, Fernandes GW, Nunes YRF (2016) Driving forces along an altitudinal gradient in the regeneration component of the rupestrian grassland complex at Serra do Cipó, Brazil. Rev Braz Bot (in review)

Munyai TC, Foord SH (2012) Ants on a mountain: spatial, environmental and habitat associations along an altitudinal transect in a centre of endemism. J Ins Cons 16:677–695

Navarro AGS (1992) Altitudinal distribution of birds in the Sierra Madre del Sur, Guerrero, Mexico. Condor 94:29–39

Needham JG, Frost SW, Tothill BH (1928) Leaf-mining insects. Williams and Williams, Baltimore

Nichols E, Spector S, Louzada J, Larsen T, Amezquita S, Favila ME (2008) Ecological functions and ecosystem services provided by Scarabaeinae dung beetles. Biol Cons 141:1461–1474

Nichols E, Gardner TA, Peres S, Spector S (2009) Co–declines in large mammals and dung beetles: an impending ecological cascade. Oikos 118:481–487

Nunes CA (2015) Besouros rola-bostas (Coleoptera: Scarabaeidae: Scarabaeinae) e suas funções ecológicas ao longo de um gradiente altitudinal tropical. MSc Dissertation, Universidade Federal de Minas Gerais

Pacheco R, Vasconcelos HL (2012) Habitat diversity enhances ant diversity in a naturally heterogeneous Brazilian landscape. Biodiv Cons 21:797–809

Palin OF, Eggleton P, Malhi Y, Girardin CAJ, Rozas-Dávila A, Parr CL (2011) Termite diversity along an Amazon-Andes elevation gradient, Peru. Biotropica 43:100–107

Pounds JA, Fogden MPL, Campbell JH (1999) Biological response to climate change on a tropical mountain. Nature 398:611–615

Pounds JA, Bustamante MR, Coloma LA, Consuegra JA, Fogden MPL, Foster PN, La Marca E, Masters KL, Merino-Viteri A, Puschendorf R, Ron SR, Sánchez-Azofeifa GA, Still CJ, Young BE (2006) Widespread amphibian extinctions from epidemic disease driven by global warming. Nature 439:161–167

Price PW (1991) Patterns in comunities along latitudinal gradients. In: Price PW, Lewinsohn TM, Fernandes GW, Benson WW (eds) Plant–animal interactions: evolutionary ecology in tropical and temperate regions. Wiley, New York, pp 51–69

Price PW, Fernandes GW, Lara ACF, Brawn J, Barrios H, Wright MG, Ribeiro SP, Rothcliff N (1998) Global patterns in local number of insect galling species. J Biogeogr 25:581–592

Pyrcz TW, Wojtusiak J, Garlacz R (2009) Diversity and distribution patterns of Pronophilina butterflies (Lepidoptera: Nymphalidae: Satyrinae) along an altitudinal transect in North-Western Ecuador. Neotrop Entomol 38:716–726

Quintino AV (2014) Riqueza de espécies de térmitas e sua relação com a distribuição de recursos ao longo de um gradiente altimontano adverso nos trópicos. Dissertation, Universidade Estadual de Montes Claros

Rahbek C (2005) The role of spatial scale and the perception of large-scale species-richness patterns. Ecol Lett 8:224–239

Rapp JM, Silman MR (2012) Diurnal, seasonal, and altitudinal trends in microclimate across a tropical montane cloud forest. Clim Res 55:17–32

Ribas CR, Schoereder JH, Pic M, Soares SM (2003) Tree heterogeneity, resource availability, and larger scale processes regulating arboreal ant species richness. Aust Ecol 28:305–314

Ribeiro SP (2003) Insect herbivores in the canopies of savannas and rainforests. In: Basset Y, Novotny V, Miller S, Kitching R (eds) Arthropods of tropical forests: spatio-temporal dynamics and resource use in the canopy. Cambridge press, New York, pp 348–359

Ribeiro KT, Fernandes GW (2000) Patterns of abundance of a narrow endemic species in a tropical and infertile montane habitat. Plant Ecol 147:205–218

Ribeiro SP, Pimenta HR, Fernandes GW (1994) Herbivory by chewing and sucking insects on *Tabebuia ochracea*. Biotropica 26:302–307

Ricklefs RE (2013) A economia da natureza. Guanabara Koogan, Rio de Janeiro

Samson DA, Rickart EA, Gonzales PC (1997) Ant diversity and abundance along an elevational gradient in the Philippines. Biotropica 29:349–363

Sanchez-Rodriguez JF, Baz A (1995) The effects of elevation on the butterfly communities of a Mediterranean mountain, Sierra de Javalambre, central Spain. J Lepid Soc 49:192–207

Sanders NJ, Lessard JP, Fitzpatrick MC, Dunn RR (2007) Temperature, but not productivity or geometry, predicts elevational diversity gradients in ants across spatial grains. Glob Ecol Biogeogr 16:640–649

Sang W (2009) Plant diversity patterns and their relationships with soil and climatic factors along an altitudinal gradient in the middle Tianshan Mountain area, Xinjiang, China. Ecol Res 24:303–314

Sawchik J, Dufrêne M, Lebrun P (2003) Estimation of habitat quality based on plant community, and effects of isolation in a network of butterfly habitat patches. Acta Oecol 24:25–33

Schnepf J (2010) Diversity of diurnal Lepidoptera across an elevational gradient above the timberline: a case study from the Austrian Alps. Dissertation, University Wein

Schroter D, Cramer W, Leemans R, Prentice IC, Araujo MB, Arnell NW, Bondeau A, Bugmann H, Carter TR, Gracia CA, de la Vega-Leinert AC, Erhard M, Ewert F, Glendining M, House JI, Kankaanpaa S, Klein RJT, Lavorel S, Lindner M, Metzger MJ, Meyer J, Mitchell TD, Reginster I, Rounsevell M, Sabate S, Sitch S, Smith B, Smith J, Smith P, Sykes MT, Thonicke K, Thuiller W, Tuck G, Zaehle S, Zierl B (2005) Ecosystem service supply and vulnerability to global change in Europe. Science 310:1333–1337

Scott DA, Brooke M (1985) The endangered avifauna of southeastern Brazil: a report on the BOU/WWF expeditions of 1980/81 and 1981/82. ICBP Tech Pub 4:115–139

Smith SE, Read JD (1997) Mycorrhizal symbiosis, 2nd edn. Academic press, New York

Speight MR, Hunter MD, Watt AD (2008) Ecology of insects: concepts and applications, 2nd edn. Blackwell science, Oxford

Stefanescu C, Herrando S, Páramo F (2004) Butterfly species richness in the north-west Mediterranean basin: the role of natural and human-induced factors. J Biogeogr 31:905–915

Stirnemann IA, Ikin K, Gibbons P, Blanchard W, Lindenmayer DB (2015) Measuring habitat heterogeneity reveals new insights into bird community composition. Oecologia 177:733–746

Stone GN, Schönrogge K, Atkinson RJ, Bellido D, Pujade-Villar J (2002) The population biology of oak gall wasps (Hymenoptera: Cynipidae). Annu Rev Entomol 47:633–668

Tantowijoyo W, Hoffman AA (2010) Identifying factors determining the altitudinal distribution of the invasive pest leafminers *Liriomyza huidobrensis* and *Liriomyza sativae*. Entomol Exp Appl 135:141–153

Tati-Subahar SS, Amasya AF, Choesin DN (2007) Butterfly (Lepidoptera: Hopalocera) distribution along an altitudinal gradient on mount Tangkuban Parahu, west Java, Indonesia. Raffles B Zool 55:175–178

Terborgh J (1971) Distribution on environmental gradients: theory and a preliminary interpretation of distributional patterns in the avifauna of Cordillera Vilcabamba. Ecology 52:23–40

Terborgh J (1977) Bird species diversity on an Andean elevational gradient. Ecol 58:1007–1019

Thuiller W, Layorel S, Araújo MB, Sykes MT, Prentice IC (2005) Climate change threats to plant diversity in Europe. Proc Natl Acad Sci USA 102:8245–8250

Vasconcelos MF, Lopes LE, Hoffmann D (2007) Dieta e comportamento deforrageamento de *Oreophylax moreirae* (Aves: Furnariidae) na Serra do Caraça, Minas Gerais, Brasil. Rev Bras Ornitol 15:439–442

Veneklaas EJ, Van Ek R (1990) Rainfall interception in two tropical montane rain forests, Colombia. Hydrol Proc 4:311–326

Wang G, Zhou G, Yang L, Li Z (2002) Distribution species diversity and life-form spectra of plant communities along an altitudinal gradient in the northern slopes of Qilianshan Mountains, Gansu, China. Plant Ecol 165:169–181

Waring GL, Price PW (1990) Plant water stress and gall formation (Cecidomyiidae: *Asphondylia spp.*) on creosote bush. Ecol Entomol 15:87–95

Whitfield GH, Carruthers RI, Lampert EP, Haynes DL (1985) Spatial and temporal distribution of plant damage caused by the onion maggot (Diptera: Anthomyiidae). Environ Entomol 14:262–266

Whittaker RH (1956) Vegetation of the great Smoky Mountains. Ecol Monogr 26:1–80

Wilson RJ, Gutiérrez D, Gutiérrez J, Monserrat VJ (2007) An elevational shift in butterfly species richness and composition accompanying recent climate change. Glob Chan Biol 13:1873–1887

Wolda H (1987) Altitude, habitat and tropical insect diversity. Biol J Linn Soc 30:313–323

Wright MG, Sanways MJ (1996) Gall–insect species richness in African Fynbos and Karoo vegetation: the importance of plant species richness. Biodiv Lett 3:151–155

Wright MG, Sanways MJ (1998) Insect species richness in a diverse flora: gall–insects in the Cape Floristic Region, South Africa. Oecologia 115:427–433

Chapter 16
Termite Role in Soil Nutrient Cycling in Ironstone Rupestrian Grasslands (*Canga*) in Carajás, Brazilian Amazonia

Carlos Ernesto G.R. Schaefer, Alessandra Marins, Guilherme Resende Corrêa, Og deSouza and Jaquelina Alves Nunes

Abstract Termites are key organisms in tropical soil processes, accounting for a marked role in pedobioturbation and nutrient cycling. We report on the mound-building termite effects across a representative soil/vegetation gradient in Ironstone Rupestrian Grassland (*canga*) in Carajás, Brazilian Amazonia. This complex is formed by three main different environments: Grassland, Low forest and Tall forest. The chemical composition of termite mounds was compared with adjacent soil in all environments; also the termite species richness and abundance were estimated. Termite richness varied across the Ironstone Rupestrian Grassland complex: 24 genera in Tall forest >18 genera in Low forest >9 genera in Grassland. Termite activity significantly increased the concentration of organic matter, available P and exchangeable Ca, K, N, Mg in the soil. Mound-building termites comprised 70 % of termite species in Grassland decreasing to 40 % in deeper forested soils. The chemical improvements resulting from mound-building termite activities were relatively more relevant to Grassland compared with Forests. Mound-building termites are key organisms in structuring the Ironstone Rupestrian Grassland complex in Amazonian ironstone.

C.E.G.R. Schaefer (✉) · A. Marins
Departamento de Solos, Universidade Federal de Viçosa, Av. P.H. Rolfs, S/N, Campus, Viçosa, MG 36570-000, Brazil
e-mail: carlos.schaefer@ufv.br

A. Marins
e-mail: alemarins@gmail.com

G.R. Corrêa
Instituto de Geografia da Universidade Federal de Uberlandia, 38400-902 Uberlandia, Brazil
e-mail: guilhermeudi@yahoo.com.br

O. deSouza
Departamento de Entomologia, Universidade Federal de Viçosa, Viçosa, MG, Brazil
e-mail: og.souza@ufv.br

J.A. Nunes
Unidade Carangola, Praça dos Estudantes, Universidade do Estado de Minas Gerais (UEMG), 23-Santa Emília, Carangola, MG CEP36800-000, Brazil
e-mail: jaquelinabiologa@yahoo.com.br

© Springer International Publishing Switzerland 2016

G.W. Fernandes (ed.), *Ecology and Conservation of Mountaintop Grasslands in Brazil*, DOI 10.1007/978-3-319-29808-5_16

Contents

16.1 Termites and Soil

Termites, ants and earthworms are amongst the most important organisms responsible for soil formation (Six et al. 2004; Bronick and Lal 2005). In seasonal tropical climates, termites are widespread, since earthworms and other organism have their activities diminished in such conditions (Paoletti 1999). Termites are able to digest cellulose, and due to the efficiency of their digestive system (Watanabe and Tokuda 2010) and high biomass, they recycle a great part of plant material into soil organic carbon. Through underground galleries construction and cellulose feeding, termites affect soil structure, both physically and chemically (Lee and Wood 1971).

Termite mounds concentrate more nutrients than the surrounding soil (Brossard et al. 2007), with a direct effect on plants growth and production (Joseph et al. 2011), and an indirect effect on fauna richness and abundance in savanna ecosystems (Mobaek et al. 2005; Pringle et al. 2010). In man-made environments, such as Australian dryland agriculture, termites are responsible for increasing growth and production (Evans et al. 2011). The importance of termites for Brazilian soils has been described in Schaefer (2001) showing that termites are involved in the genesis of highly stable microstructure, typical of Brazilians oxisols (latosols). There is also empirical evidence that the soil of a termite mound presents nutrient accumulation (Sarcinelli et al. 2009; Rückamp et al. 2012) and physical modification (Sarcinelli et al. 2009) on deep soils from Brazil. In rupestrian grassland (*Campo rupestre*) formations, termite mounds play an important role (Schaefer et al. 2015). However, little information is available on how mound building termites affect soil structure in the Ironstone Rupestrian Grassland (IRG) in the Amazonian region of Carajás, the largest iron mining area in Brazil.

Although termites are long known as ecosystem engineers (Jouquet et al. 2011 and references therein) quantitative measurements of their direct effect directly on Brazilian soils are recent (Sarcinelli et al. 2009, 2013). On tropical Latosols, there is strong empirical evidence that termite mound presents nutrient accumulation (of organic carbon, nitrogen, phosphorous calcium and magnesium) and physical modification (by galleries constructions). This accumulation eventually returns to soil due to mound erosion (Sarcinelli et al. 2009). According to these authors' estimative, they can bring up to the surface 13 t^{-1} ha^{-1} year of deep soil layers. The nutrient accumulation in termite mounds compared with adjacent soil depends on soil nutrient status and clay content (Sarcinelli et al. 2013). In oligotrophic soils, such as IRG, termites are responsible for concentrating very limiting amounts of nutrients in their mounds, which make their presence even more crucial to ecosystem function.

We investigated the mound-building effects on nutrient accumulation and cycling in an extreme oligotrophic environment: the Ironstone Rupestrian Grassland (IRG) from Carajás, Brazilian Amazonia. The study was conducted on the Ironstone (*canga*) plateaux in the Carajás National Forest, southwestern Pará State. The region is part of the Amazonian morphoclimatic domain (sensu Ab'Saber 2003). The regional climate is tropical, hot and humid ("Aw" type Köppen system) (Ab'Saber 1986). Four distinct vegetation physiognomies were considered as making up the gradient of Ironstone Rupestrian Grassland complex on the iron-rich *canga*. The IRG ranges from herbaceous open grassy environments to forested environments. The grassy environments were categorized as Grasslands with *Vellozia* and Grassland (without *Vellozia*—Hydrophilous); and the forest environments as Low forest (Shrub and Low forest) and Tall forest. In the forests, low and tall, the ironstone (*canga*) is fragmented and weathered to greater depth (Nunes et al. 2015).

16.2 Sampling Termites

Termite species diversity was sampled in six areas throughout IRG complex; in each area, we selected three collection sites across the soil vegetation gradient with different nutrient status: Grasslands (high oligotrophism), Low forest (medium oligotrophism) and Tall forest (low oligotrophism). In each collection site, tree quadrants measuring 2 m × 5 m were sampled by two people for 20 min. Termite individuals were manually sampled in mounds, logs, trees (up to 2 m), litter and soil, and placed in 80 % alcohol for further identification. Individuals were identified according to taxonomic keys.

16.3 Trends in Species Richness and Distribution

The overall termite richness comprised 27 genus, a lower richness than that found in quartzitic rupestrian grassland—*Cerrado altitudinal* gradient in Serra do Cipó (Chap. 15). This lower richness should be expected due to the isolation of the Carajás mountain complex. Termite richness differed between environments: Grasslands had nine genera, Low forest, 18, and Tall forest 24 genera (Table 16.1). However, the number of samples collected was comparable: (106 in Grasslands, 121 in Low forest

Table 16.1 Abundance of termites (number of records) collected at Pará State, North of Brazil, sorted by type of environment and locations (north N, west W, tarzan T and south S)

Genus	Environment														
	Grasslands					Low forest					Tall forest				
	Location					Location					Location				
	N	W	T	S	All	N	W	T	S	All	N	W	T	S	All
Angularitermes													2		2
Anoplotermes	2			3	5	3		1	8	12	4		2	1	7
Armitermes	1		8	3	12				9	9	1				1
Atlantitermes						2		3	4	9			1	1	2
Cavitermes						1	1		4	6				3	3
Coptotermes											1				1
Cornicapritermes									1	1		1			1
Cornitermes				1	1										
Curvitermes				3	3				1	1		1			1
Cylindrotermes						3				3	1				1
Cyranotermes							1			1					
Cyrilliotermes											2				2
Dihoplotermes											1				1
Silvestritermes	1			1	2	5	1	1	13	20	5	2	6	11	24
Genuotermes												1			1
Grigiotermes				5	5			1	1	2					
Heterotermes						2				2	1		1	2	4
Labiotermes								3	3	6				1	1
Microcerotermes						2	1	1	3	7				2	2
Nasutitermes	17	1	2	53	70	3	3	5	23	34	5	2	6	19	32
Neocapritermes											1				1
Planicapritermes													1	1	2
Rhinotermes						1				1		1			1
Spinitermes		1			1				2	2		1		2	3
Subulitermes														1	1
Syntermes									1	1	1		1		2
Termes			1	3	4	2		1	1	4	2		1	4	7
Total samples	21	2	11	72	106	24	7	16	64	111	25	9	21	48	103
Total richness	4	2	3	8	9	10	5	8	14	18	12	7	9	12	24

Note that in the south (S) there was a bigger sampling effort than other locations, hence the higher termite richness and abundance

and 103 in Tall forest). This implies that, despite fewer species are observed in shallow soils under Grasslands, the abundance was similar. Since only a few species are able to colonize the shallow soils of Grasslands, they are extremely well-adapted, so that they show high numerical dominance and abundance (the dominant species: *Nasutitermes* cf. *ephratae* account for 70 % of species abundance). In Forests, conversely, a greater number of species are able to colonize, so that the dominant species *N.* cf. *ephratae* account for only 30 % of the total samples.

16.4 Termite Community Variation Among Habitat

In Carajás complex, the mound-building termite *N*.cf. *ephratae* usually builds dark OM-rich mounds directly on top of the indurated ironstone, feeding mainly on twigs, branches, and other dead plant matter. Although this species is very dominant in IRG, it is common to find other termite species living inside their mounds (with the builder colony in the same mound) as inquilines (such as *Termes* and *Silvestritermes*). These inquilines have a facultative association with Nasutitermes and feed from humus (Rocha et al. 2012), either found in *Nasutitermes*' mound or in the adjacent soil, especially on forest soils. The most common genera in Grasslands are *Anoplotermes*, *Grigiotermes*, *Silvestritermes*, and *Nasutitermes*. The species of the first three genera are soil feeders and may live in galleries dispersed throughout the soil, or as inquilines inside other termite mounds; *Nasutitermes* cf. *efratae* are wood feeders and live in its own mound.

The mound-building activity in Grasslands has the potential to facilitate not only the presence of other termite species, such as inquilines but also, other soil fauna and plant species. Termites are considered ecosystem engineer (Jones et al. 1994; Pringle et al. 2010) that enhance other species establishment through habitat modification. We postulate that termite effect on IRG ameliorate its harsh chemical condition and allow plants and soil fauna colonization due to their effects on soil chemistry and structure, as observed by the concentration of *Vellozia* sp. on the vicinity of termite mounds, active or not.

16.5 Soil Characterization

16.5.1 Characterization of Termite Mounds and Adjacent Soil

Samples were collected in all different environments (Hydrophilous Grasslands, Shrub, Grasslands with *Vellozia*, Low forest, Tall forest, and Burnt grasslands) in different locations across the Carajás Plateaux, on Pará state, Northern Brazil. Eventually, four vegetation categories based on the above physiognomies were grouped, to allow a more focused discussion. The four categories were: Grasslands with *Vellozia*, Grasslands (Hydrophilous Grasslands), Low forest (Shrub and Low forest) and Tall forest. A similarity analysis between termite mounds and adjacent soils was performed. In total, 45 areas with three replicates for termite mounds and three replicates for adjacent soils were sampled (135 samples each).

All samples were air-dried and passed through a 2 mm sieve and then submitted to chemical and physical analyses. Chemical analyses comprised pH in water, total N (Kjeldahl method), organic C (Walkley–Black method), exchangeable Ca, Mg and Al^{3+} (extractantKCl 1 mol/L), available K, P, Fe, Zn, Mn and Cu (extractantMehlich–1) and H + Al (extractantCa acetate 0,5 mol/L) according to

Embrapa-Centro Nacional de Pesquisa de Solos (1997). The concentration of phosphorus in the equilibrium solution (P-equil) was determined after shaking 5 cm^3 of soil sample with 50 mL of a solution 10 mmol/L $CaCl_2$, containing 60 mg/L of P, during 1 h (Alvarez et al. 2000).

16.5.2 Similarity Between Soil Variables

The similarity analysis of the mounds variables (Fig. 16.1), showed that the cation exchange capacity (CEC), exchangeable aluminum and H + Al group together, indicating high nutrient contents in termite mounds (that presents high levels of acidity). Phosphorous grouped with Cu, suggesting a relationship between Cu and organic P cycling, once Cu is associated with soil organic matter. The bases sum and bases saturation grouped with Ca, Mg, K and Na, amount as expected. The fact that pH also grouped indicate that greater nutrients contents may account for increasing pH levels since CEC is very low. Also, organic matter (OM) and remaining phosphorus were grouped showing the effects of OM in decreasing phosphorus adsorption, since soils are Fe-rich and have high P retention capacity. These results are consistent with analyses of adjacent soil.

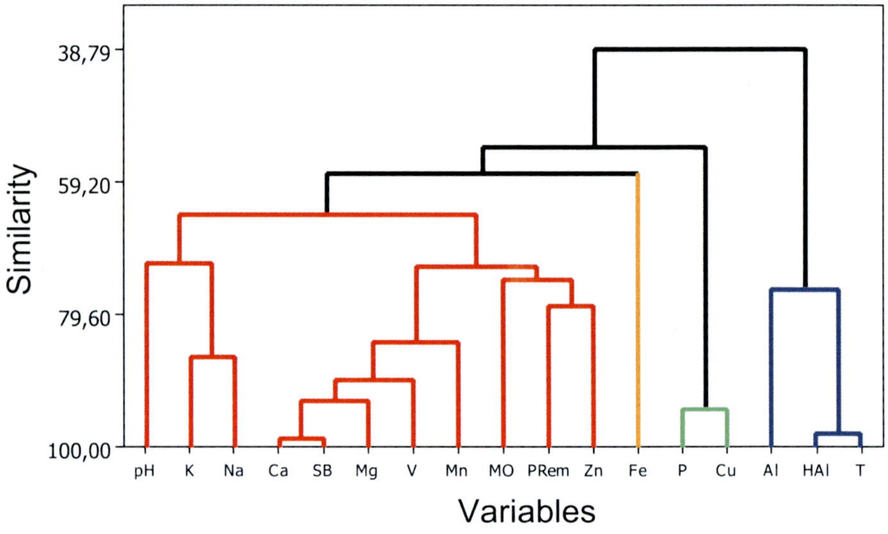

Fig. 16.1 Mean similarity dendrogram for termite mound samples on Pará state, North of Brazil. Grouping index Pearson's distance, unweighted pair group method with arithmetic mean (UPGMA). Different colors show groups with 60 % of similarity. This analysis shows that most of the variables are grouped according to the functional group. *Blue* Cation exchange capacity (CEC), Al^+ and H^+ Al; *Green* Phosphorous and Cu; *Yellow* Iron; *Red* All nutrient content grouped with pH and organic matter. This grouping shows how soil variables are correlated to one another in termite mounds

16.5.3 Similarity Between Environments

The similarity analysis of termite mounds (Fig. 16.2), showed that most of the samples are grouped according to the type of environment. The nutrients in ashes of termite mounds differ from all others samples, due to its high concentration of phosphorus and calcium. In another group (blue), samples from Grassland formations with or without *Vellozia* were kept together, showing that the low variability in nutrient concentration from different or distant places. In a third group, most samples of Tall forest were separated with Low forest, showing a clear trend in nutrient concentration for forest-dwelling termites. In the red group, we had samples from all other formations. Both extreme physiognomies (Grasslands and Tall forest) formed distinct groups regardless of their location.

The similarity results are in agreement with analyses of each termite mounds nutrient across the environments. As a general trend, termite mounds from Grasslands with and without *Vellozia* were statistically similar (10 out of 16 measurements: Figs. 16.3 and 16.4). Also, only four of these measurements were similar between Grasslands and Low forest; also, four measurements were statistically the same for Low and Tall forest (Figs. 16.3 and 16.4). Although termites have the ability to select the soil particles they use to build their mounds, there is a big contribution of the environment—and hence the adjacent soil used as matrix—in

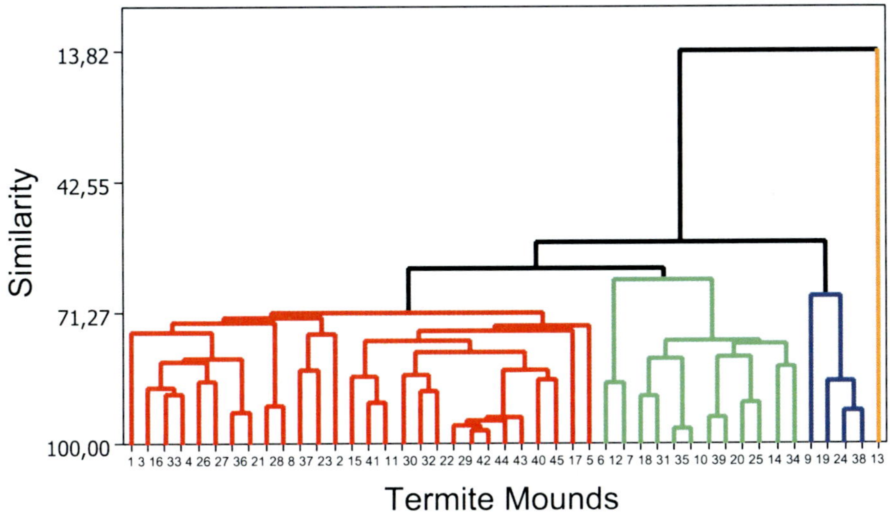

Fig. 16.2 Mean similarity dendrogram for soil sample variables of termite mounds on Pará state, North of Brazil. Grouping index Pearson's distance, unweighted pair group method with arithmetic mean (UPGMA). Different colors show groups with 60 % of similarity. This analysis shows that most of the samples are grouped according to the type of environment. *Yellow* termite ashes from burned grasslands; *Blue* grasslands with or without *Vellozia*; *Green* Low and Tall forests; *Red* all grasslands. This grouping shows that termite mounds are similar regarding nutrients content within each environment type

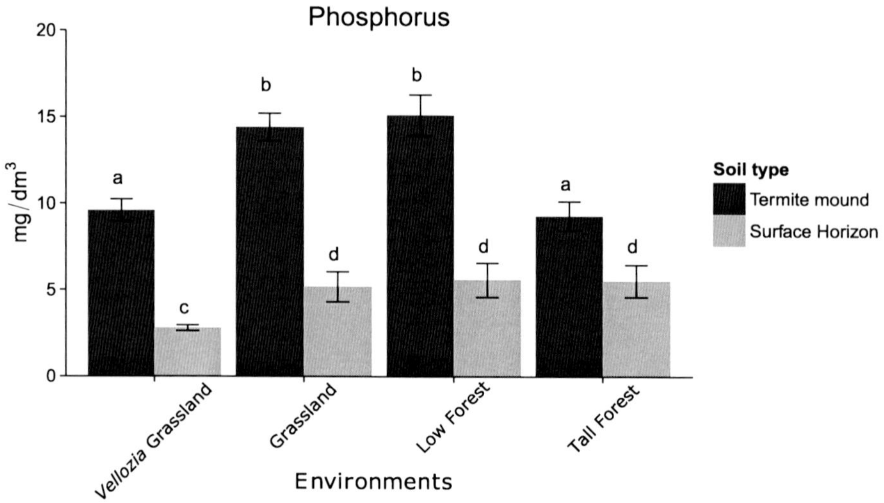

Fig. 16.3 Average values levels on termite mounds and soil samples on Pará state, North of Brazil. Error bars shows standard errors. Soil type differs statistically, and environment type also differs, same letters are statistic similar at $p > 0.05$ on post hoc Anova test. Values refer to: pH H_2O; pH KCl; Sum of Bases; H + Al; Reminiscent Phosphorous; Zi

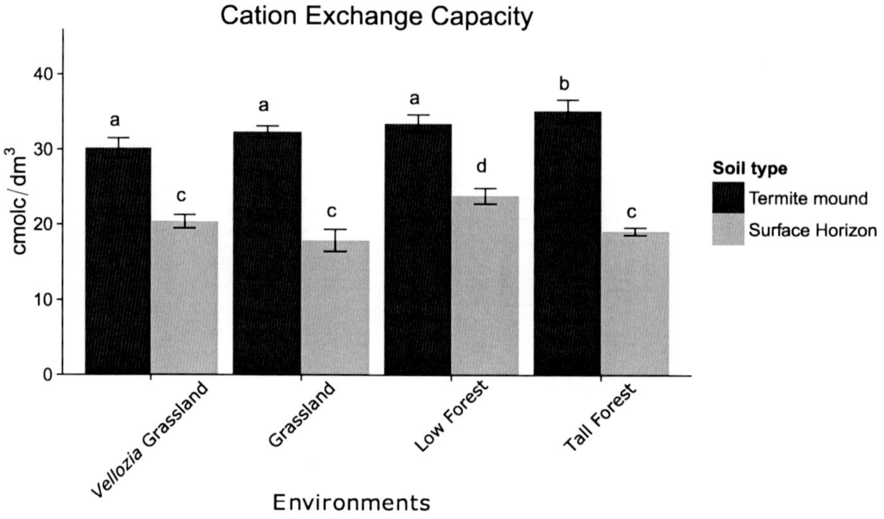

Fig. 16.4 Average values levels on termite mounds and soil samples on Pará state, North of Brazil. Error bars shows standard errors. Soil type differs statistically, and environment type also differs, same letters are statistic similar at $p > 0.05$ on post hoc Anova test. Values refer to: Fe; Mn; Cu; Mg; K; Ca; Al

determining nutrients composition in the mounds. Another concomitant explanation is that the blend of mound-building species in each environment is different. In Grassland with and without *Vellozia* we had *Nasutitermes* cf. *ephratae* as the only mound builder (except for a *Cornitermes* mound found once), whereas in forests we had also *Microcerotermes*, *Labiotermes* and *Syntermes*, occurring in average abundance (although *Nasutitermes* had still a greater frequency). It is important to note that each termite species have its own specific requirements for its establishment so that different soil composition may result in different termite fauna, with different nutrient accumulation.

16.5.4 Termite Mounds and Adjacent Soils

We performed chemical analyses on termite mounds to compare their chemical status among environments, location and in relation to adjacent soil. Such analyses revealed important ecological aspects of the termite—soil interaction.

In Grasslands with *Vellozia*, termite mound concentrates more phosphorous than adjacent soils—from two to eight times higher in mounds than in surface horizons. Calcium values in mounds are consistently higher than in soils, illustrating the role termite has in cycling nutrients. Exchangeable [K, Mg, and K] were also greater in mounds (Fig. 16.4). In Grasslands, the P concentration in mounds is seven to ten times higher than adjacent soil, as well as in the Low forests. Given that Ca amounts in this soil are negligible, Ca accumulation in mounds is extremely

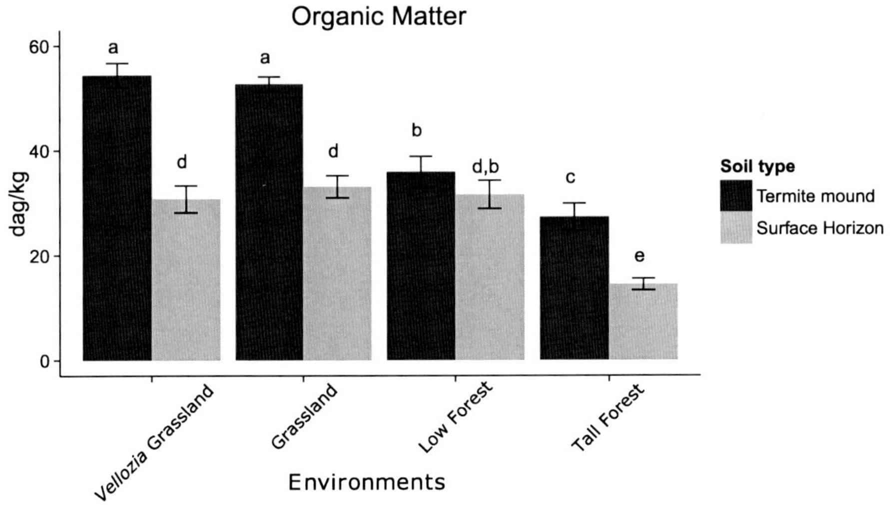

Fig. 16.5 Average Phosphorus levels on termite mounds and soil samples on Pará state, North of Brazil. Error bars shows standard errors. Environment type (Gaussian GLM, n = 251, $\chi^2 = 851.3$, $p < 0.001$) and soil type (Gaussian GLM, n = 250, $\chi^2 = 3479.3$, $p < 0.001$) differ statistically, same letters are statistic similar at $p > 0.05$ on post hoc Anova test

relevant, since they may account for a significant source of this nutrient. In Forests, P values in mounds are equivalent to mounds in *Vellozia*, but the difference between mounds and adjacent soil samples decreased, since forested soils present greater P amounts than Grasslands with *Vellozia* soils (Fig. 16.5). Potassium and Mg values are higher in mounds, also exchangeable Al and acidity. The other macronutrients analyzed (K and Mg) were also higher in mounds, although the acidity and exchangeable Al also increased in mounds. The percentage of Clay is similar between the soils (17–34 %), and there is coarse sand dominance in all mounds. The organic matter amounts in the mounds were almost always higher than in adjacent soils except in Low forests that presented similar amounts of organic matter in soil and termite mounds (Fig. 16.6), showing that termite feeding activities is able to increase the uptake of humic substances with higher cation exchange capacity than the local soils.

The mounds' material and adjacent soil comparison revealed the important ecological role that mound-building termite play in Ironstone Rupestrian Grassland, especially regarding phosphorous concentration. It is noteworthy to observe that the

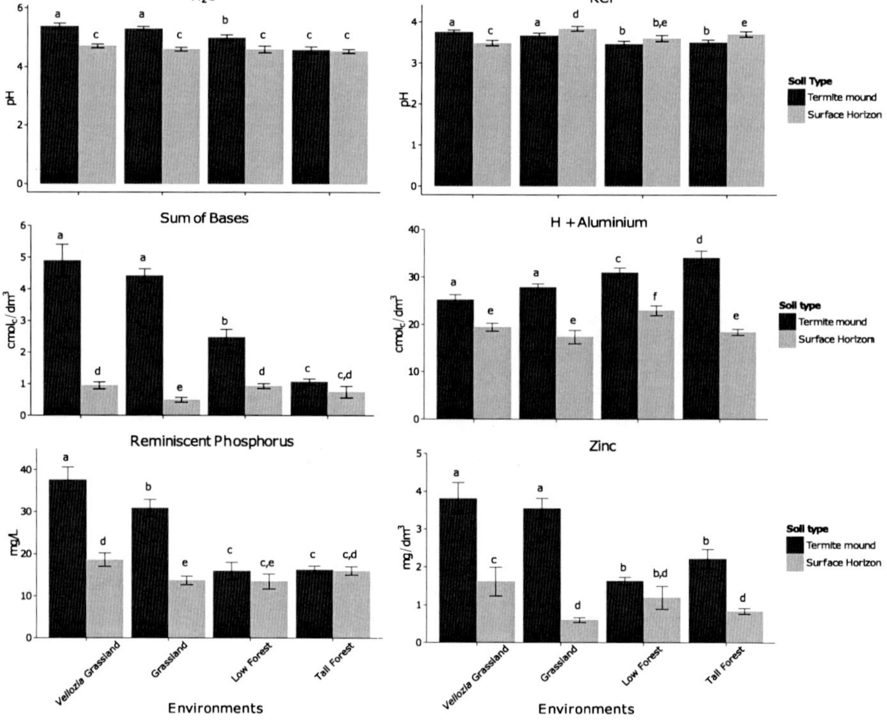

Fig. 16.6 Average Organic Matter (OM) levels on termite mounds and soil samples on Pará state, North of Brazil. Error bars shows standard errors. Soil type differ statistically (Gaussian GLM, n = 200, χ^2 = 16,283, p < 0.01), and environment type also differ (Gaussian GLM, n = 197, χ^2 = 21,356, p < 0.01), same letters are statistic similar at p > 0.05 on post hoc Anova test

P content in termite mounds in *Vellozia* Grasslands and Tall forest are statistically similar (c.a. 10 mg/dm^3), whereas the P content in adjacent soils are very different for each environment: 2.79 ± 0.83 mg/dm^3 for *Vellozia* Grasslands, and 5.51 ± 4.91 mg/dm^3 in Tall forest (Fig. 16.5). The Cation Exchange Capacity in mounds is always higher than adjacent soils (Fig. 16.7). For organic matter, the greatest enrichment was observed in *Vellozia* Grasslands: from 30.7 dag/kg in adjacent soil to 54.3 dag/kg in mounds, however, this enrichment was not observed in the deeper soils under low and tall forests (Fig. 16.6). Hence, termite importance seems greater in Grassland with and without *Vellozia*, whereas their relevance as

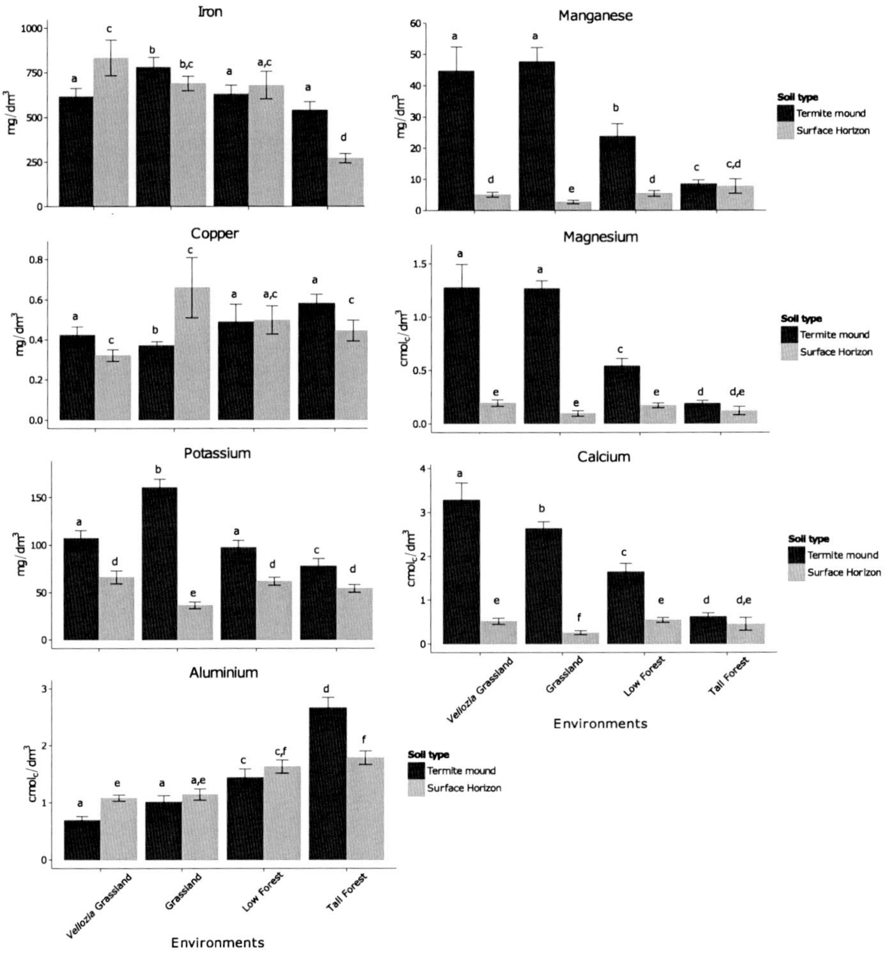

Fig. 16.7 Average Cation Exchange Capacity (CEC) levels on termite mounds and soil samples on Pará state, North of Brazil. Error bars shows standard errors. Soil type differ statistically (Gaussian GLM, n = 200, χ^2 = 10,694, $p < 0.01$), and environment type also differ (Gaussian GLM, n = 197, χ^2 = 681, p = 0.002), same letters are statistic similar at $p > 0.05$ on post hoc Anova test

soil modifier relatively decreased in deeper forest soils. Therefore, termite habitat modifications are more important in shallow soil environments under open vegetation and lower clay content.

16.6 Conclusions

Mound-building termites are key players in nutrient cycling in IRG (Ironstone Rupestrian Grassland) with a significant concentration of P, Ca, K, Na, Mg and OM. Termite mound-building activity is responsible for greater soil modification than non-building activities. Mound-building species comprised 70 % of termite fauna in shallow soils, and only 40 % in deeper soils, highlighting the termite importance in such environments.

There is a gradient in termite richness between the most oligotrophic environment (Grassland) and the least one (Tall forest): nine species in Grasslands; 18 in Low forest; and 24 in Tall forest, even though the abundance is almost the same. Hence, only a few well-adapted species is able to colonize the shallow soils under Grasslands.

The soil chemical modification performed by termite mound builders is more relevant in shallow soils under Ironstone Rupestrian Grasslands than in deeper soils under Forests. Such well adapted species performed a crucial role in structuring the Ironstone Rupestrian Grassland complex.

Acknowledgments To Vale S.A. (CVRD) for the logistic support. Research project funding by The State Funding Agency of Minas Gerais (FAPEMIG) (Project: Geodiversidade dos Ambientes de Canga na Região de Carajás, PA, Brasil). AM post-doctoral grant is held by National Council for Scientific and Technological Development (CNPq 150255/2015-2). ODS holds a CNPq Fellowship (305736/2013-2).

References

Ab'Saber AN (1986) Geomorfologia da região Corridor Carajás-São Luiz. In: Carajás. Desafio Político, Ecologia e Desenvolvimento. CNPq, São Paulo, pp 88–123

Ab'Saber AN (2003) Os domínios de natureza no Brasil: potencialidades paisagísticas. Ateliê Editorial, São Paulo

Alvarez VVH, Novais RF, Dias LE, Oliveira JA (2000) Determinação e uso do fósforo remanescente. Bol Inf Soc Bras 25:27–32

Bronick CJ, Lal R (2005) Soil structure and management: a review. Geoderma 124:3–22

Brossard M, Lopez-Hernandez D, Lepage M, Leprun JC (2007) Nutrient storage in soils and nests of mound-building *Trinervitermes* termites in Central Burkina Faso: consequences for soil fertility. Biol Fertil Soils 43:437–447

EMBRAPA (1997) Manual de Métodos de Análise de Solos. CNPS-Rio de Janeiro 212p

Evans TA, Dawes TZ, Ward PR, Lo N (2011) Ants and termites increase crop yield in a dry climate. Nat Commun 2:262

Jones CG, Lawton JH, Shachak M (1994) Organisms as ecosystem engineers. Oikos 69:373–386

Joseph GS, Cumming GS, Cumming DHM, Mahlangu Z, Altwegg R, Seymour CL (2011) Large termitaria act as refugia for tall trees, deadwood and cavity-using birds in a miombo woodland. Landsc Ecol 26:439–448

Jouquet P, Traoré S, Choosai C, Hartmann C, Bignell D (2011) Influence of termites on ecosystem functioning: ecosystem services provided by termites. Eur J Soil Biol 47:215–222

Lee KE, Wood TG (1971) Termites and soils. Academic Press of London, London

Mobaek R, Narmo AK, Moe SR (2005) Termitaria are focal feeding sites for large ungulates in Lake Mburo National Park, Uganda. J Zool 267:97–102

Nunes JA, Schaefer CEGR, Ferreira Junior WG, Neri AV, Correa GR, Enright NJ (2015) Soil-vegetation relationships on a banded ironstone 'island', Carajás Plateau, Brazilian Eastern Amazonia. An Acad Bras Ciênc 87:2097–2110 (Impress)

Paoletti MG (1999) The role of earthworms for assessment of sustainability and as bioindicators. Agric Ecosyst Environ 74:137–155

Pringle RM, Doak DF, Brody AK, Jocqué R, Palmer TM (2010) Spatial pattern enhances ecosystem functioning in an African savanna. PLoS Biol 8:e1000377

Rocha MM, Cancello EM, Carrijo TF (2012) Neotropical termites: revision of *Armitermes*Wasmann (Isoptera, Termitidae, Syntermitinae) and phylogeny of the Syntermitinae. Syst Entomol 37:793–827

Rückamp D, Martius C, Bornemann L, Kurzatkowski D, Naval LP, Amelung W (2012) Soil genesis and heterogeneity of phosphorus forms and carbon below mounds inhabited by primary and secondary termites. Geoderma 170:239–250

Sarcinelli TS, Schaefer CEGR, FernandesFilho EI, Mafia RG, Neri AV (2013) Soil modification by termites in a sandy-soil vegetation in the Brazilian Atlantic rain forest. J Trop Ecol 29:439–448

Sarcinelli TS, Schaefer CEGR, Lynch LS, Arato HD, Viana JHM, Filho MRdA, Gonçalves TT (2009) Chemical, physical and micromorphological properties of termite mounds and adjacent soils along a toposequence in Zona da Mata, Minas Gerais State, Brazil. Catena 76:107–113

Schaefer C (2001) Brazilian latosols and their B horizon microstructure as long-term biotic constructs. Aust J Soil Res 39:909–926

Schaefer CE, Cândido HG, Corrêa GR, Pereira A, Nunes JA, Souza O, Marins A, Fernandes-Filho E, Ker JC (2015) Solos desenvolvidos sobre canga ferruginosa no Brasil: uma revisão crítica e papel ecológico de termiteiros. In: Carmo FF, Kamino LHY (eds) Geossistemas Ferruginosos do Brasil: áreas prioritárias para conservação da diversidade geológica e biológica, patrimônio cultural e serviços ambientais. 3i, Belo Horizonte—BH, pp 77–102

Six J, Bossuyt H, Degryze S, Denef K (2004) A history of research on the link between (micro) aggregates, soil biota, and soil organic matter dynamics. Soil Tillage Res 79:7–31

Watanabe H, Tokuda G (2010) Cellulolytic systems in insects. Annu Rev Entomol 55:609–632

Part IV
Foreword 4 + Photo

While ecologists continue to develop their science by looking at, among many other themes, patterns of species diversity and the mechanisms that produce them, and searching for instances of natural selection shaping species behavior and distribution, the need for integration of the sciences to save the world and its beings for future generations calls for a multidisciplinary approach. We need to actively provide society with the most recent findings in ecology and natural history so that wise decisions on conservation and management of the rupestrian grasslands are made and that actions are taken at the needed pace given the ever-growing pressures. For instance, the growing threats posed by afforestation with exotic plantations of eucalyptus in the Espinhaço need detailed study and rapid and efficient communication of those results to stakeholders and policy makers. While we are aware of the negative impacts that human-induced fire has on global climate change and the direct impact on vegetation and animal life, we must understand its general influence on the functioning of the rupestrian grassland and how to mediate other threats such as promoting biological invasion. Restoration of rupestrian grasslands cannot continue to be done with exotic species that can escape from revegetated areas and disseminate into protected and remnant areas. Restoration must be scientifically driven and not simply a forest plantation. Therefore, there is much to solve in this regard if it is to be applied in the low resilient rupestrian grasslands. Humans are an important component of the rupestrian grassland, as they have exploited such territory for at least 12,000 years, yet the need for detailed understanding of the relationship between humans and nature in this harsh environment is crucial. While these landscapes and terrains have remained stable for at least the last 20,000 years (i.e., since the last glaciation maximum), the future may not be as friendly. Any prediction obtained from current climatic models indicates that the rupestrian grassland ecosystem will change dramatically in the forthcoming decades, with serious consequences to humans, wildlife, and ecosystem services. Most of the threats imposed on the rupestrian grasslands are caused by incipient and fragmentary knowledge and misleading conservation strategies, in addition to direct human impacts.

Chapter 17
Afforestation in the Rupestrian Grasslands: The Augmenting Pressure of *Eucalyptus*

Rodrigo Pinheiro Ribas, Rogério Machado Caetano, Bernardo Machado Gontijo and João Henrique de Azevedo Xavier

Abstract For at least 30 years, the expansion of eucalyptus plantations in municipalities like Grão-Mogol and Itamarandiba, in Minas Gerais State, has been provoking considerable impacts on ecosystems and human communities in the rocky fields of the Espinhaço Mountain Range. From an environmental point of view, eucalyptus monocultures fragment the areas hitherto preserved of the Cerrado (savannah) biome, leading to loss of connectivity and hampering reproduction of plants and animals. The removal of native vegetation reduces the reproductive capacity of fauna, which in turn, interferes with the resilience capacity of savannah vegetation areas still preserved. In socio-economic terms, the afforestation contributes to the concentration of income and boosts the rural exodus and uncontrolled urbanization process. While afforestation projects can be seen, individually, as economically viable activities, systemically they can result in significant externalities, especially when conducted on large scale, with environmental and social losses that sooner or later will lead to economic unsustainability. This article highlights how the expansion of eucalyptus plantations can be harmful when it produces a homogeneous landscape. The authors hope that this can be an issue for the discussion about land management in its broadest sense.

R.P. Ribas (✉) · R.M. Caetano · B.M. Gontijo
Departamento de Geografia, Universidade Federal de Minas Gerais,
Instituto de Geociências, Abel Araújo Street, 11, Apartment 302,
Santa Lúcia Neighborhood. 30.350-050, Belo Horizonte, Minas Gerais, Brazil
e-mail: ribasgeo@gmail.com

J.H. de Azevedo Xavier
Instituto de Ciências Biológicas, Laboratório de Ecologia Evolutiva E Biodiversidade,
Universidade Federal de Minas Gerais, Belo Horizonte, Brazil

© Springer International Publishing Switzerland 2016

G.W. Fernandes (ed.), *Ecology and Conservation of Mountaintop Grasslands in Brazil*, DOI 10.1007/978-3-319-29808-5_17

Contents

17.1 Introduction

The great mountain range called Espinhaço lies in Brazil from central Minas Gerais to north Bahia states, covering 1200 km. The relief is marked by steep scarps, constituting a great watershed between São Francisco river and Brazilian East Coast hydrographic basins. Researchers divide Espinhaço Mountain Range in two areas: Southern Espinhaço, around 300 km long in north-south orientation in Minas Gerais, and Northern Espinhaço, also called Chapada Diamantina, at Bahia state. These portions are separated by a system of sinistral strike-slip faults in northeast direction, composing the Araçuaí band. Southern Espinhaço extends from Quadrilátero Ferrífero in the south till a region near Olhos D'água county in the north (Fig. 17.1). It is home to great geobiodiversity, supporting a very reach biodiversity, covering two great Brazilian biomes (Cerrado and Atlantic Forest), which are divided by the mountain range itself. Beyond a biome divider, for many researchers Serra do Espinhaço represents a distinct biome. According to Gontijo (2008), even been conceived as a transition band between biomes, it is necessary to consider the Espinhaço Mountain Range a biome itself, the rupestrian grasslands, with characteristics that go way beyond a mere floristic arrangement.

It is a fact that people have been transforming the landscapes according to the relationship between human communities and the environment. The tendency of fragmentation in natural habitats prevails in most of the natural landscapes around the world. In Brazil, the world's most biodiverse country, the agricultural expansion, the exploitation of natural resources and the cities growth are driving a great landscape transformation. The north central region of Minas Gerais state has been undergoing striking transformations. One of the most dramatic is the implementation of great eucalyptus monocultures. In Southern Espinhaço and its surroundings, the landscape is been invaded by eucalyptus monocultures, losing its connectivity, which implies in biodiversity reduction. In this critic context, the eucalyptus monocultures expansion is discussed in this chapter.

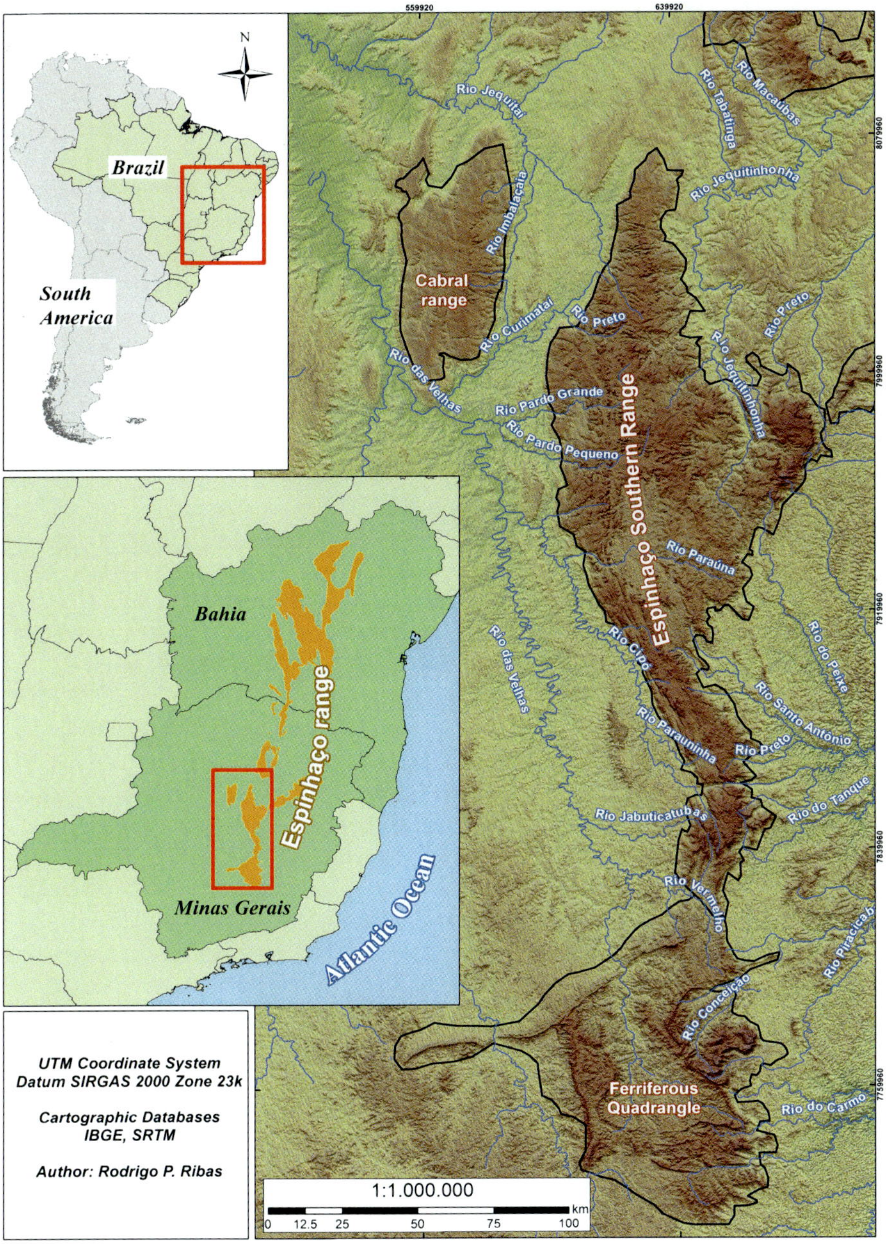

Fig. 17.1 The espinhaço southern range in Minas Gerais state

17.2 Losing Connectivity and Diversity

The consensus among researchers that the habitat fragmentation process is severely threatening the world's biodiversity is notorious. (Bierregard et al. 1992; Fahrig 2003; Tabarelli et al. 2004). The connectivity expresses, in many aspects, the opposite process (Noss 1992). Around the 80s, discussions about the concept of connectivity started (Merriam 1984; Fahrig and Merriam 1985; Forman and Godron 1986). Since then, the term is been passing through conceptual evolutions, especially when it comes to the subdivision in structural and functional connectivity.

In general, evaluating the landscape's connectivity consists in identifying and characterizing the aspects that favor or complicate the connection between the fragments. According to Forman and Baudry (1984), the landscapes' connectivity emerges from the spatial arrangement of habitat fragments, demonstrating the landscape's capacity of ameliorating biological flows and the intensity of organisms' movement between habitats. Therefore, it represents in the landscapes, the connection of ecological processes in multiple spatial scales. The connectivity is also defined as the organisms' capacity of moving between habitat patches and the functional relations in between (With et al. 1997). According to Taylor (1993), connectivity means one organism's ability to move along separated habitats. Therefore, we can think of connectivity as the existence of a localized spatial path between two habitat patches that permits the passage and the functional relations of biodiversity.

When it comes to the dissociation between structural and functional connectivity, there are authors that do not consider this division pertinent, understanding that the connectivity covers all the connection possibilities between habitats. For them, the difference is only in the quantification, directed to the functional relations between organisms or to the landscape's elements (Taylor 1993). However, there is a current tendency to dissociate it, and many authors are focusing at functional connectivity (Jórdan et al. 2007; Saura and Torné 2009). These studies investigate the habitats of a matrix together with the landscape's connective elements, in a target species' point of view. Therefore, the studies analyze the behavior that determines species and their habitat peculiarities, providing the understanding of the dynamics of this species' movement in heterogeneous habitats. According to Lang and Blaschke (2009), is spite of their fundamental role, these studies are hardly applied in bigger scales, since they usually study a focal species. Since planning interventions in great areas implies diverse landscape elements, this restriction is not desirable. Searching for alternatives, studies with coexistent population groups were developed. As one well structured example, there's the metapopulations theory (Hanski and Gaggiotti 2004) pointing that, if an specific isolated reminiscent patch is not able to maintain a species, a group of this patches, if well connected, may be viable.

Researches focused on structural connectivity are directed by the physical attributes of the landscape (size, shape, etc.), without a specific study of the dispersion ability of the organisms (Hargrove et al. 2004; Goez et al. 2009). These studies cover the metric applications to analyze the landscape's structural elements.

Fahrig and Merriam (1994) affirmed that the understanding of the spatial structure of the landscapes has central importance on the comprehension of the fragmentation effects on the populations in various habitats and the spatial structure of the landscape consists in a spatial relation between the habitat parts and the matrix where they are inserted. According to Rouget et al. (2006), the connectivity of fragmented landscapes occurs in an asymmetric pattern. They observed that the preferential direction of the movement is related to the combination of different factors, like the movement difficulty originated by the relief and the proximity of the priority natural resource. This concept was criticized because the evaluation of the structural elements of the landscape would be insufficient to demonstrate the real connectivity in biological terms.

Calabrese and Fagan (2004) present an interesting view of the connectivity concept when distinguishing three different approaches: (1) structural connectivity (focusing on the landscape's attributes, not on the dispersion of organisms); (2) potential connectivity (characterized as a combination of landscape's attributes and superficial information about species' dispersal); and, at last, (3) the real connectivity (based on the observation of individual dispersal and biotic dynamics, with estimates of the interaction between the landscape and the populations).

According to Viana (1995), the major impact of the fragmentation process and the consequent reduction at the spatial connectivity is the loss of regional biodiversity. With more homogeneous abiotic conditions, fauna and flora develop themselves in a particular inter-related way. The fragmentation process, by removing the natural vegetation and occupying the land with a different system, triggers a series of biotic and abiotic alterations. When we think at this system acting in an integrated way, we can predict the instability chain reaction that will be created due to the alteration of the energy flow in the area The perception of this process' consequences is a fundamental subject because, the more fragmented are the landscapes, bigger will be the biodiversity conservation effort. Since the fragmentation process is growing at some regions, conservation and environmental management strategies cannot ignore its consequences.

Tabarelli and Gascon (2005) present empirical support with six guidelines for the environmental management of fragmented areas, expecting that they are capable of mitigate or eliminate the negative impacts caused by the anthropic transformation of the landscapes. (1) Incorporating protection actions as part of the project development process, (2) protecting extensive areas and avoiding the fragmentation of the remaining continuous areas of native vegetation, (3) managing the borders of the fragments since the creation moment, (4) protecting the riparian forests to connect isolated forest fragments, (5) controlling the fire use and the introduction of exotic plants and limiting the use of biocides in the landscape and (6) promoting the reforestation and extension of the forest cover in critic areas of the landscape.

After the reduction of the natural habitats, the main reason for the biodiversity reduction is the introduction of invasive species (McNeely et al. 2001). After introduced, invasive species can stabilize and disperse prapagules, with the possibility of triggering a biological contamination process that can change local ecological peculiarities, modifications that can affect the nutrient cycling, the vegetation productivity, the trophic chains and the density, distribution and functions of the ecosystems' species, also modifying the fire and hydrological regime (Ziller 2001). Although not all the exotic species become invasive in a habitat, extreme precaution, with rigorous control to avoid negative impacts is needed.

Large areas in Brazil and around the world are being afforested and fragmented for wood production, with the introduction of eucalyptus monocultures. The great dispersal force of eucalyptus in the neotropics is mainly economical, due to the combination of economic stimulus and the common opinion that eucalyptus monocultures are "sustainable" alternative for other land uses, and often called "planted forests", as if they could substitute the native Atlantic Forest and Cerrado. In this chapter, we explore the dissemination of these monocultures and their impacts on native vegetation (Cerrado, Atlantic Forests, and Rupestrian Grasslands) at Minas Gerais, Brazil.

The extensive eucalyptus and pine plantations, as any monoculture, are often called green deserts by naturalists, duo to their homogeneity, the depleted soil and the severely reduction in the biodiversity. Furthermore, the expected social benefits are not achieved due to the low employability of these mechanized enterprises that deplete the natural resources and degrade the culture for monetary reasons. According to Shiva (2003), there's a tendency to think and plan initiatives in an uniform, homogenized way making diversity disappear from the perception and consequently from the world, the so called "monocultures of the mind".

The eucalyptus, native from Australia, invaded many regions in the world intending to supply the demand of materials for construction, furniture, coal, cellulose, among others. The big companies' speech always highlights that this activity reduces the pressure on the natural ecosystems, allows a better use of depleted lands, contributes with the reduction of global warming and generates jobs for local population. The "planted forests" are presented as a great business opportunity, which can promote social and environmental sustainability. However, this advertisement is fallacious, with disastrous experiences around the world, especially in the southern hemisphere (Carrere and Lohmann 1996).

An eucalyptus monoculture is nothing alike the megadiverse native forests from Brazil, since it is a geometric plantation of one single species. A native ecosystem has innumerous plant and animal species interacting in a biogeochemical cycle. There are also the traditional human populations that customarily extract forest products. An eucalyptus monoculture does not establish any of these integrating relations that characterize a native forest, and can vanish in a matter of hours at harvest time. The fast-growing artificial forests, as they generate monotone landscapes, replacing grasslands, forests, savannas, prairies and mountainous ecosystems, could never be environmentally viable. The geographic space evolves from a group of relations (Dollfus 1970). Every landscape has its own biogeochemical intrinsic peculiarities, sustaining an adapted biodiversity.

Eucalyptus species are not very demanding when it comes to soil. Open fields, like grasslands, are often used to plant eucalyptus monocultures, since eucalyptus are well adapted to the pedology of these areas. Usually poor in soil nutrients, these areas are culturally seen as non-forested, with low biodiversity. This situation is aggravated because these areas are not recognized formally as important preservation regions by the United Nations (UN) Framework Convention on Climate Change, the program for Reducing Emissions from Deforestation and Forest Degradation (REDD+) or the UN Food and Agriculture Organization.

The economic, ecological, agricultural and agroclimatic zoning options, among other ways of planning landscapes, aim to generate knowledge for the introduction of species in an environment. The scope of these studies is related to the analyses of the landscape's peculiarities, identifying and mapping locations with higher productive potential and, consequently, more viability to introduce specific plantations. For Minas Gerais state, the works of Golfari (1975) and Scolforo et al. (2008) are good references. Beyond these initiatives, other zoning proposals can be found in various technical works and researches.

Recently, a new work was published as a contribution to the Global Partnership on Forest Landscape Restoration: The Atlas of Forest Landscape Restoration Opportunity (WRI 2014a), produced by the World Resources Institute (WRI) in collaboration with the University of Maryland and International Union for Conservation of Nature (IUCN). The maps in this atlas were produced with 1 km resolution, considered adequate since it proposes to embrace a huge part of the globe. Data from MODIS sensor were utilized to map the different kinds of vegetation and their density. The result was divided in three types: (1) "closed forest (canopy density greater than 45 %); (2) open forest (canopy density between 25 and 45 %) and (3) woodlands (sparse, savanna type forest with canopy density between 10 and 25 %)". This map was combined with climatic, pedological and topographic data and with the ecoregions mapping, allowing the creation of a forest condition map. The second step consisted in mapping areas suffering great anthropic pressure, which were directed to the creation of the "world map of restoration opportunity areas" (WRI 2014b).

Probably due to the generalist scale in which the mapping was produced, inconsistencies were observed at the mapping of the "grassland areas". Among the 23 million km^2 mapped areas considered highly suitable for tree planting, a great area correspond to the world's ancient grassy biomes (Veldman et al. 2015). The Rupestrian Grasslands, at southern Espinhaço Mountain were mapped as fragmented/managed areas, with woodlands that are fragmented by roads and/or managed for timber production and formerly forested areas, considered an opportunity for returning forests and trees to the landscape. The restoration proposal for Rupestrian Grasslands was a *Wide-scale restoration*. This type of restoration is more likely in deforested or degraded landscapes with low population density (<10 people/km2) that are also areas where closed forests formerly dominated the landscape (WRI 2014b). Another inconsistency observed at WRI's atlas was the misclassification of great eucalyptus monoculture areas at Southern Espinhaço as natural forests. This mapping and its inconsistencies, when classifying the Rupestrian Grasslands and the native forests, represent a critic situation for the region, especially when this atlas is aligned with worldwide actions like the Bonn challenge, that projects the restoration of 150 million hectares in the whole world till 2020.

17.3 Scale Approach: Cartography of the Eucalyptus Versus Native Vegetation in Minas Gerais

In a scale approach effort, a series of analysis is presented based on the Mapping of the Native Flora and Forest Inventory of Minas Gerais, of 2006 and 2008, regarding the period of 2004–2005 and 2006–2007, respectively (Table 17.1). Those editions suffered changes in their presentation, regionalization, photointerpretation and dendometric calculation, and in the 2008 edition, there was a rectification of the data regarding the deforestation and afforestation occurred in 2005[1]. The differences in methodology between one and the other make the harmonisation of data difficult and compromise the analysis coherence of the addressed four years. Regardless, it is a valuable material for future monitoring.

According to the Atlas, there was a loss of native vegetation in Minas Gerais of 94.856 ha for each year in the 2004–2005 period, and f 54.878 ha in each year of the following 2006–2007 period. The annual average was of 74.867 ha, summing up to 299.466 ha in all four years considered (Fig. 17.2a).

The following map was made based on the data from the 2008 edition of the Mapping made by Scolforo et al. (2008), showing a substitution of the class intervals, excluding the class over 4000 ha and including a class from 200 to 750 ha (Fig. 17.2b).

It follows from the contrast between the two maps that there has been a reduction in the deforestation in absolute terms; however, a spatial pattern is maintained, in which the deforestation is concentrated in the municipalities of João Pinheiro, Grão Mogol, Paracatu, Itamarandiba and Pompéu, in darker colours in each map (Fig. 17.2c).

It is noted that the 47 municipalities represented in this map concentrate almost 50 % of the deforestation occurred in the 2004–2007 period. When overlapping a schematic road map it is observed that the deforestation is concentrated along the main axes of the federal road's network in Northern Minas Gerais, especially in a NE-SW arch that borders the BR-365 and BR-251 roads.

Grão Mogol and Itamarandiba are within the Espinhaço region, both in the high Jequitinhonha area. In these municipalities, large areas of eucalyptus are been planted, a result of public policies intending to maintain a metallurgy complex implanted in the area. Several incentive mechanisms were created to enable results

Table 17.1 Recent evolution of the vegetation stocks in Minas Gerais—in hectares

Biome	Native forest remnants			Losses by period	
	2003	2005	2007	2004–2005	2006–2007
SAVANNAh/cerrado	13.570.311	13.431.119	13.347.174	139.192	83.945
Forest	5.730.345	5.681.257	5.656.227	49.088	25.030
Caatinga	654.041	652.609	651.829	1.432	780
Total	19.954.696	19.764.985	19.655.230	189.712	109.755

Source IEF/Ufla (2006 e 2008)

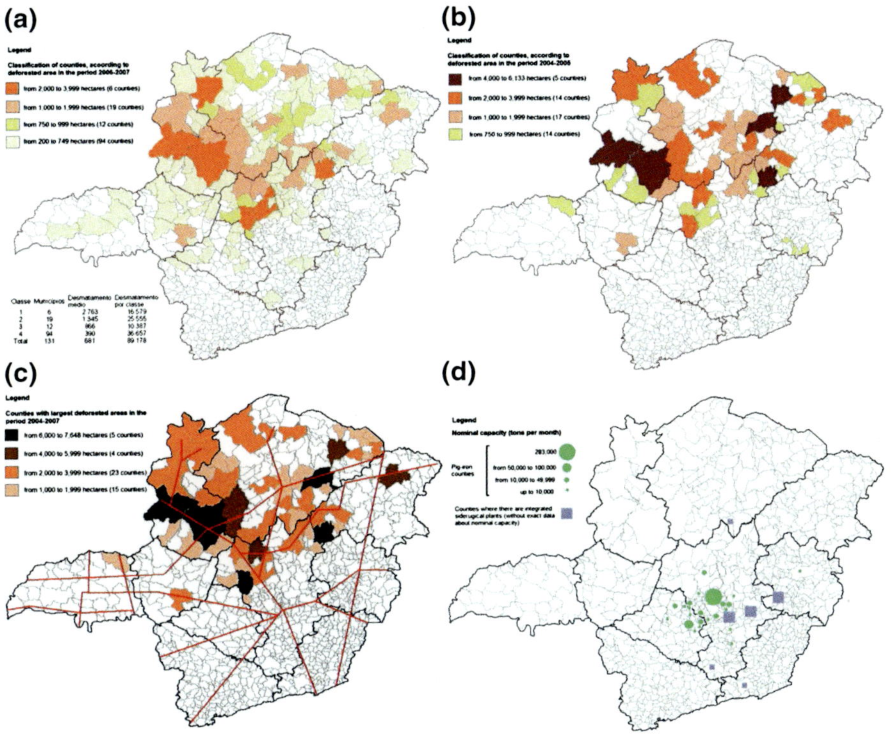

Fig. 17.2 **a** Municipalities with deforestation over 750 ha in the 2004–2005 period. *Source* Scolforo and Carvalho (2006). **b** Municipalities with deforestation over 200 ha in the 2006–2007 period. *Source* Scolforo et al. (2008). **c** Municipalities with deforestation over 1.000 ha in the 2004–2007 period. *Source* Scolforo et al. (2008). **d** Integrated pig-iron centres and metallurgies in Minas Gerais. *Source* Sindifer/MG (2009)

from this initiative: wastelands in suitable areas were occupied in the North and Northeast areas of the state, taxes were forfeit, logistics infrastructure for support and distribution of the production was created and lending contracts were renewed. These mechanisms are, at last, to avoid the "green blackout" at all costs, as warned by the charcoal industry facing the imminent downfall of the forester model implemented since the 1970.

The fact that the most intense deforestation is not in the extreme municipalities confirms the tendency of getting the charcoal in locations closer to the pig-iron centres (Fig. 17.2d).

A correlation between the deforestation and the expansion of the agricultural frontier is made, aided by the Universidade Federal de Lavras (UFLA) mapping (Scolforo et al. 2008). Currently in Minas Gerais, due to the low occurrence of species that are more valuable and the pressure from the pig iron and steel industries, the native forests are turned into charcoal and afterwards the soil is prepared for the expansion of agricultural activity. It is not known for sure which sector has a higher influence on the deforestation, if simply the extraction for charcoal production or the search for new and more fertile, less degraded and less eroded areas for the development of the agricultural business. However, it can be verified that the existing anthropized area is enough for production, making investments necessary to recover degraded areas and for the development of high end technology to increase productivity (Scolforo et al. 2008)

According to UFLA's atlas Scolforo et al. (2008), 7.848 deforestation polygons were counted in Minas Gerais, summing up a total of 109 thousand hectares in the 2006–2007 period. From this total, the 650 largest polygons were selected to be checked in the field, with visits or fly-overs. These polygons total a third of the deforested area in the period (33.715 ha), concentrated in regions that "present large areas covered by native vegetation" (Scolforo et al. 2008: 331). When overlapping the map with the 7.848 polygons and the map that highlights the largest 650 polygons, it gives a notion of the zones with the largest magnitudes of deforestation, (Fig. 17.3a).

The blue spots area referent to the 33.715 ha visited or flown-over by IEF, converted into grassland (48.4 %); afforestation (21.2 %); bare ground (12.4 %); agriculture (11.4 %); regeneration (6.3 %); and others (0.3 %). The categories of "bare ground" and "regeneration" could be abandoned areas, futurely convertible. The tie between the charcoal industry and the expansion of the agricultural frontier is visible in the landscape as the charcoal furnaces were identified amidst the deforested areas.

From the data mentioned above, 80 % of the deforested areas were taken in the advancement of the agricultural frontier in the 2006–2007 period. From the 650 polygon sample, it can be statistically extrapolated the number of deforested hectares for the immediate implementation of afforestation. If 21.2 % of the deforested areas were substituted by eucalyptus plantations, it would correspond to approximately 23.000 ha.

These 23.000 ha represent a 25 % fraction of the implemented afforestation in the period, leading to the conclusion that the largest part of the eucalyptus

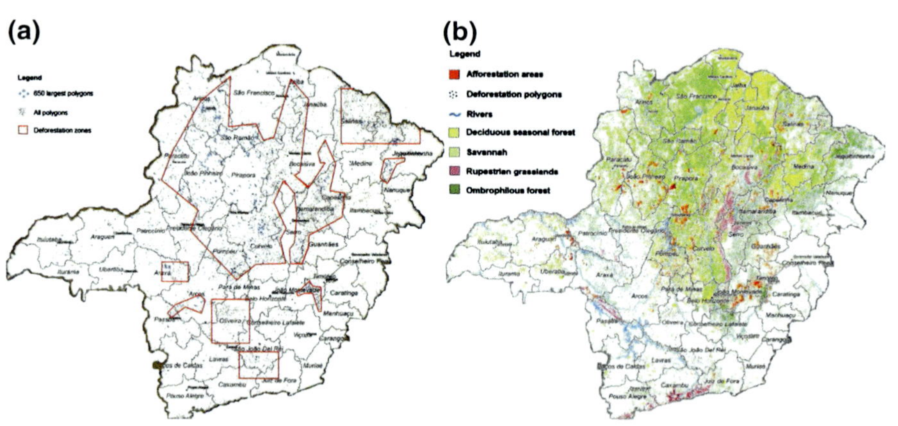

Fig. 17.3 **a** Overlap of the deforested areas and the polygons checked by IEF. **b** Overlapping of the deforestations, biomes and afforestation

plantations has advanced over grasslands, bare ground areas or regenerating cerrado that was deforested previously. In other words, only 25 % of the eucalyptus plantations advances over native vegetation, whereas the other 75 % over concurrent uses of the territory (Table 17.2).

The São Francisco river basin has a moderately paced increasing afforestation, also occurring in the Rio Doce and Paranaíba basins. In absolute terms, the frontier advances more in the higher Vale do Jequitinhonha area, part of the Espinhaço landscape. In relative terms, the Vale do Mucuri area experiences great afforestation expansion, while in Pardo basin it is literally stagnated (Table 17.2).

As previously said, although the implementation of eucalyptus plantations was almost equal to the deforestation, it does not mean that the forest plantations advanced directly over deforested areas in the period. However, future expansions of the eucalyptus plantations can take advantage of the presently deforested areas, adjacent to the afforestation. This is something already observed at the tops of the Serra do Cabral mountain, inside the Biosphere Reserve of the Espinhaço mountain range. In this sense, the overlapping of the eucalyptus plantation map and the biome and deforestation polygons map allows to infer future areas for the eucalyptus expansion in the adjacent areas to the afforestation (Fig. 17.3b).

The overlapping map analysis shows deforestation areas that are apparently autonomous and others that are very close to the eucalyptus plantation areas, the called deforestation corridors, associated with forest plantations. In Espinhaço mountain range and Jequitinhonha's case, cerrado deforestation zones are probably associated with the metallurgy centres in the state (including the São João del Rei

Table 17.2 Expansion of eucalyptus plantations and deforestation in the main river basins of Minas Gerais, between 2005 and 2007—in hectares

River basin	Forest plantation stocks		Expansion of plantations (C)	Increase percent (C/A)	Deforestation between 2005 and 2007
	2005 (A)	2007 (B)			
S. Francisco	579.750	604.380	24.630	4	−64.884
Doce	252.668	259.537	6.869	3	−5.869
Jequitinhonha	220.054	259.015	38.961	18	−17.040
Paranaíba	96.722	100.455	3.733	4	−7.800
Grande	50.275	60.515	10.240	20	−5.175
Pardo	46.407	46.410	3	0	−6.883
Paraíba do Sul	12.383	12.707	324	3	−269
Mucuri	2.715	10.659	7.944	293	−1.407
Piracicaba	8.078	7.807	−271	−3	−90
Others	122	122	0	–	–
Totals	1.269.174	1.361.607	92.433	–	−109.417

The regionalization proposed by Scolforo et al. (2008) differs from the one made in Scolforo and Carvalho (2006), making it difficult to understand the monitoring
Source Scolforo et al. (2008)

region). The regions that have a higher occurrence of deforestation (North, Northeast and Jequitinhonha) are the ones with largest native vegetation coverage.

Regarding the distribution of the afforestation areas according to the altitude, it is important to highlight that they generally occur in more elevated areas, interfluves, being observed in the Conservation Units Mosaic of the Espinhaço mountain range region, at Serra do Cabral Mountain, in higher Jequitinhonha. There are a few plantations close to the river channels, except for the afforestation that reaches the Rio Doce riverbanks, the Três Marias Reservoir and the Rio Paracatu river (Fig. 17.4a).

The main frontiers for the eucalyptus expansion are the Northern Minas Gerais and the Mucuri and Jequitinhonha valleys, and in a lesser degree, the Rio Doce river, municipalities at the northeast from Sete Lagoas and also the vicinities of São João del Rei. The municipalities with higher increases were Itamarandiba (12.000 ha planted), Rio Pardo de Minas (7.700 ha), Turmalina, Grão Mogol and Nanuque, (6.200 ha average). Itamarandiba and Turmalina are within the referred Protected Areas' Mosaic, being Grão Mogol right up North, but inserted in the Espinhaço region, as seen before.

If the cumulative maps of the eucalyptus and deforestation maps are overlapped, the simultaneous occurrence of both phenomena is observed in several municipalities. It cannot be confirmed that the eucalyptus plantations are undeniably the direct cause of the deforestation, because the planting could be occurring over grassland areas, for example. However, it can be said that the changes in the Minas Gerais territory in the order of 100000 ha occurred when added the deforestation and eucalyptus plantation areas (Fig. 17.4b).

The eucalyptus plantation areas expansion can be divided into two moments. The second moment shows a more pronounced expansion, with higher concentration at the south of the parallel that passes through Montes Claros, in the central region of the Espinhaço mountain range in Minas Gerais, including the municipalities that now are part of the Conservation Unit's Mosaic of the Espinhaço at Serra do Cabral Mountain, in Higher Jequitinhonha. One of the reasons for this concentration is that the Northern portion has been receiving incentives from Banco do Nordeste[1] for afforestation. The occurrence of less significant advances (Fig. 17.4c), stable areas and eucalyptus losses (Fig. 17.4d) can be seen mapped separately.

The expansion of the eucalyptus plantations in the 111 municipalities sums 14.880 ha, a little over what is verified in the Itamarandima municipality only. These occurrences form a belt around the Metallurgy area. The more expensive prices in the Central region explain the tendency of increase in the further distances from the steelyards, but also by a tendency initiated at the time of the forestry promotion initiatives, according to Gontijo (2001). The author reminds that the return of those projects was at the time of the planting, by tax exemptions. These could be located in regions where the land was less valuable and not necessarily

[1]According to the institution, the bank has given, in the 2006–2008 period, approximately R $100 million for the planting of 44 thousand hectares in Northern Minas Gerais. Of this total, 29 % were commissioned in the years 2006 and 2007. Thus, the financing concessions could be detected in the IEF analysis, even if some plantations take up to five years to be done.

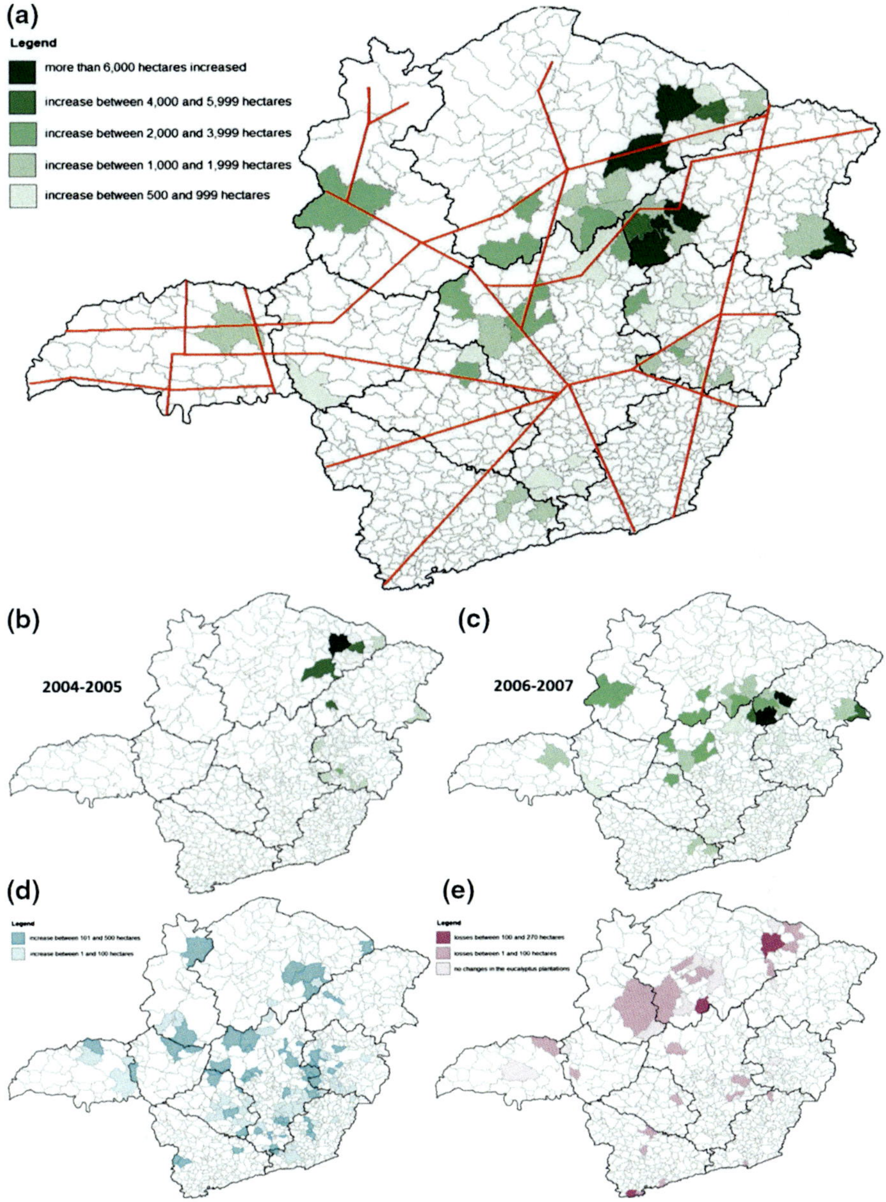

Fig. 17.4 **a** Afforestation advances in Minas Gerais (Municipalities with absolute increases over 500 ha in the 2004–2007 period). **b** Afforestation advances in MG (Absolute increase in a two year period) 2004–2005. **c** Afforestation advances in MG (Absolute increase in a two year period) 2006–2007. **d** Moderate advances of afforestation in Minas Gerais (2006–2007). **e** Stability and losses in the afforestation stocks in Minas Gerais (2006–2007)

close to the consumer areas. The Vale do Jequitinhonha region, due to the low cost of the land and the existence of wastelands favourable to planting, was one of the preferable areas in these projects, mainly at the top of the plateaus where the native vegetation coverage varied between cerradão and arboreal caatinga (Gontijo 2001).

The selection of the planting areas was not necessarily the "industry's choice", but the choice of individual producers. In other words, the afforestation businesses were not evaluated concerning the production chain, but autonomously. In this sense, the "economical rationality" of the time leaded to the removal of the afforestation in relation to the consumer centres, revealing a misconception, which in the last decades, is being mitigated by productivity gains in the forestry handling and the charcoal processing.

Concerning the deforestation in the 2004–2007 period, it is interesting to notice that it does not occur in the Central region. That could be explained by at least three reasons: (1) fewer cerrado stocks available; (2) more proximity to the inspection and (3) larger probability for complaints (Fig. 17.2c).

The net loss of eucalyptus plantation areas was practically none. The decrease of eucalyptus plantations in the 25 municipalities in classes 1 and 2 was of 967 ha, a number much lower to the one accounted for in the 2006 edition of the Atlas. This fact allowed the observation of a more severe rear, with 6000 ha in losses in some municipalities. Of those, 1263 ha were abandoned only in the Bom Despacho municipality (Fig. 17.4d).

The cerrado deforestation is an important aspect in the financial constitution of a charcoal forest business. Considering that the first crops were initiated in 1974, the first reforms happened in 1995, in case the cultures obeyed three seven-year cycles. In 2001, Gontijo wondered "what to do with the eucalyptus plantations" of the Vale do Jequitinhonha region, questioning the economic viability of keeping large forested areas so far from consumer centres. If these plantations were cut down questions regarding the environmental consequences of this cutting are brought up, especially those referring to soil erosion, air moisture and to the definite ownership of the land itself. (Gontijo 2001). It occurs that the crops are cut down or managed gradually. Each year's crop constitutes, on average, a seventh of the areas available in the state. There are still doubts regarding the occurrence of reform, planting or discontinuation. Data given by companies and representative entities of the eucalyptus plantation segment show an inconsistency with the numbers of the Forest Inventory presented by IEF/UFLA (Table 17.3).

Table 17.3 Afforestation made yearly in Minas Gerais —in hectares

	2004	2005	2006	2007*
Charcoal	96.580	121.414	119.872	–
Cellulose	29.370	20.913	22.900	–
Others	14.600	15.500	16.900	–
Total	140.500	157.900	159.600	169.000

Source UFLA and IEF, with data from companies and representative entities of the eucalyptus plantation segment

*Estimative related to the amounts of previous years

If the eucalyptus plantations in Minas Gerais during the 2006–2007 period yielded around 330 thousand hectares, why does the IEF survey accounted only 92.433 ha? To answer this question, the following hypothesis were raised: (1) overestimation of the forestry segment in relation to the planting done by companies and small and medium size producers; (2) the discontinuation of eucalyptus plantations unaccounted by IEF; (3) reformed areas being interpreted by IEF as deforestation; (4) accounting of reforms as new plantations by the forestry segment; (5) a combination of all previous alternatives.

A haze persists making difficult a middle ground between what the satellites and the companies say; a partial eclipse, a relative "blackout", endorsed by offices such as the Brazilian Agricultural Research Company (Empresa Brasileira de Pesquisa Agropecuária—Embrapa). According to a pamphlet produced by the institution, the wood productivity in Brazil from planted forests surpasses countries such as Canada and the United States of America, which gives it distinction and competitiveness in the global cellulose Market and credits it to consolidate its calling also in the other fields of forestry goods production and environmental services. The great competitiveness of the country in the forest agricultural business is guaranteed mainly by its climate and soil characteristics and by the technological advances in the forestry area.

In accordance with ABRAF (Anuário Estatístico 2009) the planted area of industrial forests (eucalyptus and pinus) is increasing, although, not at the rates necessary to supply the demand of forestry products (EMBRAPA 2009).

Beforehand, is of utmost importance to try to understand what those demands are: hypothetical, real, present, and future? If the demands are current, why is the planted area growing? In any case, what is expressed in this passage is the advocacy of the Brazilian option for the forestry biomass, with higher productivity and lower cost (in the case of eucalyptus) to supply a (probably external) "demand" and "consolidate" this country's calling.

In Embrapa's text is the information that the National Forest Program established a goal of planting 600 thousand hectares/year between 2004 and 2007, all over the country. However, an undeniable fact is the difference among what was planned and what was done. However, it goes from the numbers to the contradicting explanations and the reiteration of the "blackout" theory: the annual increases were below the established goals of the National Forests Plan (Plano Nacional de Florestas—PNF), of 600 thousand mil hectares in the 2004–2007 period, generating an annual deficit in demand of forested areas. In this scenario, the eventual resumption of the forest biomass market will generate a deficit in the wood supply in the next decade, due to the long maturing period of the afforestation projects and also a lower planting rate compared to use rate (EMBRAPA 2009) (Table 17.4).

If the planting rate is lower than the use rate, why do the planted areas have increased? The publication goes on concluding that the deficit, sustained in current levels is a strong indicator of a possible "forest blackout" in the next decade. Making it necessary to implement public policies to foster forest planting, as well as increasing investments in RD&I to increase efficiency in raw product use (EMBRAPA 2009). The "blackout" occurs since the beginning of charcoal fuelled metallurgy and what is being done with the expansion of the eucalyptus plantations

Table 17.4 Total planted area evolution in the 2004–2008 period and calculated deficit for an estimate 600 thousand new

	2004	2005	2006	2007	2008
Planted area (ha)	4.963.511	5.294.204	5.632.080	5.836.610	6.126.384
Increase (ha)		330.693	337.876	204.530	289.774
Annual Deficit (ha)		269.307	262.124	395.470	310.226
Accumulated Deficit (ha)		269.307	531.431	926.901	1.237.127

Source Embrapa, with data from the Statistical Yearbook of ABRAF 2009

is to try and diminish it. An important aid in this sense was the crisis initiated in 2008 (and the 2014/2015 crisis), that reduced the pressures over the native vegetation and gave more time for the growth of forest plantations.

According to numbers given by the Brazilian Trees Institute (Instituto Brasileiro de Árvores—IBA), published in the "Estado de Minas" paper of June, 22nd 2015, the planted forest area of Minas Gerais in 2013 was of 1.4 million hectares, a little over the 2007 numbers, almost a decade later. Such data corroborates the increase, in a slower pace than what is desirable, of the planted forest areas.

The approval of the state Law 18.365, in 2009 (Minas Gerais state Law nº 18.365/2009), known as the Forestry Law, determined as a goal to be reached the reduction of charcoal from native vegetation to a maximum of 5 % of the total consumption by year beginning in 2019. It may cause a decrease in the fragmentation of natural forest habitats, however creates also an even higher expectation of growth in this sector. According to data from IEF (2014), in 2008 the consumption of charcoal from native vegetation in Minas Gerais was of 8252160.97 cubic meters. In 2011 this consumption decreased to 3.160.981.10 cubic meters. In 2009 the consumption was of 6.278.903.29 and in 2010 of 4.325.823.95 cubic meters of charcoal from native vegetation. In 2011, from the total of charcoal produced and consumed, only 4.4 % was from native vegetation, demonstrating a continuous reduction of the fraction produced in the state. In this race towards proving the consumption at 95 % rate of raw material of planted forest origin, the intention to preserve the native forests is highly praised, but there is also a fear of an effective monitoring. Will enough areas remain for the native forests recuperation? Alternatively, will any tract of land be identified as potential to introduce eucalyptus monoculture?

17.4 Persisting Dilemmas—Multiple Determinations, Interests and Visions

Having the numbers related to the eucalyptus productivity and native vegetation, it is possible to extrapolate, in a local scale, in regard to the pressure suffered by the Espinhaço mountain range and that can go on in long term, even if the rhythm of the eucalyptus plantations expansion in the state is relatively modest.

The pig iron production in Brazil is in the order of 33 million tons, being two thirds of it produced with the use of mineral coal (coke) and the other third from charcoal from eucalyptus and native vegetation. The main consumer of the eucalyptus production in Minas Gerais (approximately 60 %) is this industry segment.

Currently, all the eucalyptus area in the country does not pass the 6 million hectares, being 1.5 million hectares designated to the metallurgy industry. Only in Minas Gerais, the destined area for the metallurgy industry is approximately 1-million hectares of planted forests. To let the metallurgy industry's energy matrix be sustained only by planted forests, it would be necessary to mobilize around 7.5 million hectares of land exclusively to plant eucalyptus.

The substitution of coke by charcoal in the national metallurgy matrix is an almost unimaginable fact at least until the middle of the present century, for logistic and economic reasons, as well as the sluggishness in the increment of planted forests in Brazil. It is estimated that the production of eucalyptus plantations in Minas Gerais should increase in approximately 3 million hectares to meet the scenario of total conversion of the metallurgy matrix.

The territorial impact of a possible metallurgy's energetic reconversion for the state can be visualized in at least three ways: as a square with sides of 213 km (Fig. 17.5a); as a fractioning and dispersion of this square towards the main producing regions (Fig. 17.5b); or as an unified spot of the main producing regions (Fig. 17.5c). In this scenario, where the demand for eucalyptus by other sectors (cellulose, for example) in Minas Gerais (400.000 ha) is not being considered, a seventh of each group of eucalyptus plantations should be harvested annually (6.480 km^2 or 648.000 ha), just to meet the demands of the metallurgical sector.

It is important to bear in mind that the spots in these maps mean the clustering of the plantations, because if the possibility of spatial dispersion due to cities, conservation units, bad soils and even agroforestry systems were considered, there would be a tendency to perceive the eucalyptus much more disseminated that it actually would be.

Still regarding the spots that appear in the figure, they signal to a even larger pressure towards the Espinhaço range, especially in its central-north portion, from

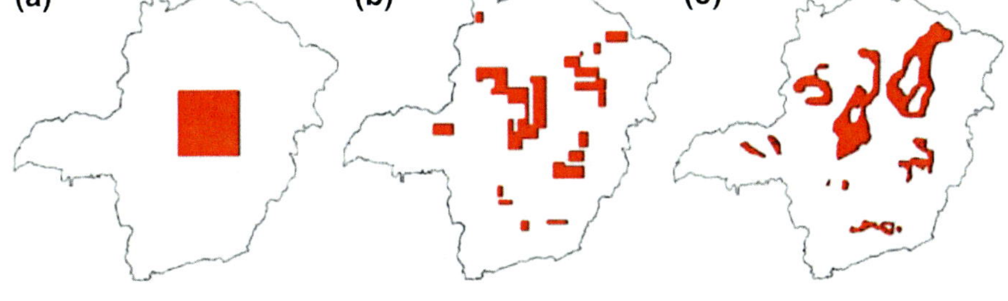

Fig. 17.5 abc Areas to be hypothetically occupied by planted forests in Minas Gerais if only eucalyptus were designed to meet the current national demand for charcoal in the metallurgical sector

the São Francisco river basin (where the expansion of eucalyptus in the Serra do Cabral mountain is very typical), as well as from the Upper Jequitinhonha region (including tributaries like the Araçuaí river). The vulnerable position of the eastern and western borders (upper Jequitinhonha and Serra do Cabral, respectively) of the Conservation Units Mosaic of Espinhaço: Serra do Cabral/Alto Jequitinhonha is determined, respectively anchored by the State Parks of the Serra Negra Mountain and Serra do Cabral Mountain. This fact reinforces the importance of initiatives such as the creation and effective integrated management of the Mosaic. In another way, both conservation units, located inside an ever growing eucalyptus plantation's matrix, would see themselves hopelessly isolated and without any perspective of connection to the nearby protected areas.

Considering a more detailed scale when zooming into the Espinhaço mountain range region, new analysis possibilities appear. In an analysis of the time dynamics of the landscape in a protected areas' mosaic in the Southern Espinhaço between 1984 and 2013 (Ribas and Gontijo 2014a) was verified a fragmentation of the natural habitats and a large advance in the eucalyptus plantations in the region. The plantation areas doubled in the analysed period, with an approximate increase of 30 thousand planted hectares.

Bearing in mind that the indiscriminate growth of the eucalyptus plantations represents a risk for the native vegetation, as well as the potential to generate land use conflicts, the authors produced a study (Ribas and Gontijo 2014b) with the purpose of sizing the eucalyptus' growth rate in the conservation units' mosaic and creating a prediction model for the region's scenario. The model indicated a tendency and locations with higher probability for the expansion of eucalyptus, demonstrating a situation capable of generating several environmental and territorial conflicts, as well as the connectivity loos in the landscape.

17.5 Final Considerations

The eucalyptus plantations in northern Minas Gerais State are settled in areas where residents are underserved by public services and have no social organization to oppose or, at best, to take part in these enterprises, precisely at a place where the material or technical development would be somewhat welcome. The spread of monocultures is promoting landscape homogenization and contributing to accelerate the already disorganized urbanization process in small and medium-size cities in ESR region. (Baggio 2002; Bethônico 2002; Calixto 2006; Gama 2006; Melo 2008). Focusing on Upper Jequitinhonha River areas superposed to the ESR, Calixto (2006) questions "the effectiveness of big reforestation projects (prompted by tax incentives) as effective drivers of development". The author stresses "the productive and occupational importance of family farming—an activity often considered only as subsistence—for the Jequitinhonha region, and the need to address this sector as a potential generator of wealth, not just as a fragile industry that needs financial resources not to succumb" (Calixto 2006). In a zero-sum game, the price paid for the transformation of idle areas in "productive land" is the formation of an homogeneous landscape, bordered on its possibilities and closed in his senses; unlike the interior, which "lacks closures", in Guimarães Rosa's words. Certainly, this "closing of senses" is not an inescapable option, for new possibilities are always emerging, as a result of the synthesis between "productive territories" and "unproductive places."

References

Baggio Filho H (2002) Alterações na paisagem natural e agrícola do município de Buritizeiro – MG: implicações do plantio generalizado de eucaliptos e pinus no meio ambiente físico, biológico e sócio-econômico. 139p. Dissertation, Universidade Federal de Minas Gerais

Bethonico MB de M (2002) Produção de carvão vegetal no município de Montezuma: impactos sócio-ambientais. 204p. Dissertation Universidade Federal de Minas Gerais

Bierregard RO, Lovejoy TE, Kapos V, Santos AA, Hutchings W (1992) The biological dynamics of tropical rain forest fragments. Bioscience 42:859–866

Calabrese JM, Fagan WF (2004) A comparison-shopper's guide to connectivity metrics. Front Ecol Environ 2:529–536

Calixto JS (2006) Reflorestamento, terra e trabalho: análise da ocupação fundiária e da força de trabalho no Alto Jequitinhonha, MG. 135p. Dissertation, Universidade Federal de Lavras

Carrere R, Lohmann L (1996) Pulping the south: industrial tree plantations in the world paper economy. Zed Books Ltd, London and New Jersey

Dollfus O (1970) L'espace géographique. Presses Universitaire de France, Paris

Embrapa Agroenergia (2009) Florestas energéticas: inovação para sustentabilidade. Colombo, Embrapa Florestas, Brasília

Fahrig L (2003) Effects of habitat fragmentation on biodiversity. Annu Rev Ecol Evol S 34: 487–515

Fahrig L, Merriam G (1985) Habitat patch connectivity and population survival. Ecol 66: 1762–1768

Fahrig L, Merriam G (1994) Conservation of fragmented populations. Cons Biol 8:50–59

Forman RT, Baudry J (1984) Hedgerows and hedgerow networks in landscape ecology. Environ Manage 8:499–510

Forman RT, Godron M (1986) Landscape ecology. Wiley, New York

Gama M das GCC (2006) Água, vereda, veredeiro: um estudo sobre as agriculturas camponesa e comercial nas cabeceiras do Rio Formoso, em Buritizeiro/MG. 118p. Dissertetation, Universidade Federal de Uberlândia

Goez SJ, Jantz P, Jantz CA (2009) Connectivity of core habitat in the northeastern United States: parks and protected areas in a landscape context. Remote Sens Environ 113:1421–1429

Golfari L (1975) Zoneamento ecológico do estado de Minas Gerais para reflorestamento, 65p. Centro de Pesquisa Florestal da Região do Cerrado, Belo Horizonte

Gontijo BM (2001) Implicações do plantio generalizado de *eucaliptus* no empobrecimento social e da biodiversidade do Alto/Médio Jequitinhonha-MG. Bol Paul Geogr 77:57–77

Gontijo BM (2008) Uma geografia para a Cadeia do Espinhaço. Megadiv 4:1–2

Hanski I, Gaggiotti OE (2004) Metapopulation biology: past, present and future. In: Hanski I, Gaggiotti OE (eds) Ecology, genetics and evolution of metapopulations. Elsevier academic press, San Diego

Hargrove WW, Hoffman EM, Efroymson RA (2004) A practical map-analysis tool for detecting potential dispersal corridors. Land Ecol. 20:361–373

Instituto Estadual de Florestas. IEF (2014) http://www.ief.mg.gov.br/noticias/1/1354-consumo-de-carvao-vegetal-de-origem-nativa-cai-61-em-minas. Accessed 25 Mar 2014

Jórdan F, Magura T, Tóthmérész B, Vasa V, Ködöböcz V (2007) Carabids (coleoptera: carabidae) in a forest patchwork: a connectivity analysis of the bereg plain landscape graph. Land Ecol 22:1527–1539

Lang S, Blaschke T (2009) Análise da paisagem com SIG. Ed. Oficina de Textos, São Paulo

McNeely JA, Mooney HA, Neville LE, Schei P, Waage JK (eds) (2001) A global strategy on invasive alien species. IUCN Gland, Switzerland, and Cambridge, UK

Melo DR de (2008) Evolução das veredas sob impactos ambientais nos geossistemas planaltos de Buritizeiro/MG. Thesis, Universidade Federal de Minas Gerais, Brazil

Merriam G (1984) Connectivity: a fundamental ecological characteristic of landscape pattern. In: Proceedings of the first international seminar on methodology in landscape ecological research and planning vol 1, pp 5–15. Roskilde BP, Roskilde AP (eds) Universitesforlag Georuc, Roskilde, Denmark

Ribas RP, Gontijo BM (2014a) Análise multitemporal da evolução estrutural da paisagem por meio de técnicas de sensoriamento remoto e métricas de paisagem. Rev Asoc Arg Ecol Pais 5:38–44

Ribas RP, Gontijo BM (2014b) Detecção e previsão da expansão de eucalipto em mosaico de áreas protegidas utilizando técnicas de sensoriamento remoto e regressão logística. In: XVI Simpósio Internacional Sociedad Latinoamericana en Percepción Remota y Sistemas de Información Espacial, SELPER, Medellin. Memórias SELPER, Colombia, 2014

Rouget M, Cowling RM, Lombard AT, Knight AT, Kerley GIH (2006) Designing large-scale conservation corridors for pattern and process. Cons Biol 20:549–561

Saura S, Torné J (2009) Conefor sensinode 2.2: a software package for quantifying the importance of habitat patches for landscape connectivity. Environ Modell Softw 24:135–139

Scolforo JRS, Carvalho LMT (eds) (2006) Mapeamento e inventário da flora nativa e dos reflorestamentos de Minas Gerais. Editora Universidade Federal de Lavras, Lavras

Scolforo JRS, Carvalho LMT, Oliveira AD (2008) Zoneamento ecológico-econômico de Minas Gerais—componentes geofísico e biótico. Editora Universidade Federal de Lavras, Lavras

Sevá Filho O (2003) Renovação e sustentação da produção energética. In: Cavalcanti C (ed) Desenvolvimento e natureza: estudos para uma sociedade sustentável, 1st edn. Cortez, São Paulo p, pp 345–365

Shiva V (1993) Monocultures of the mind: biodiversity, biotechnology and agriculture. Zed Press, New Delhi

Sindifer/IBS. Anuário estatístico (2009) wttp://www.sindifer.org.br. Accessed 26 May 2009

Tabarelli M, Silva JMC, Gascon C (2004) Forest fragmentation, synergisms and the impoverishment of neotropical forests. Biodiv Cons 13:1419–1425

Tabarelli M, Gascon C (2005) Lições da pesquisa sobre fragmentação aperfeiçoando políticas e diretrizes de manejo para a conservação da biodiversidade. Megadiv 1:181–188

Taylor PD (1993) Connectivity is a vital element of landscape structure. Oikos 68:571–572

Veldman JW, Overbeck GE, Negreiros D, Mahy G, Le Stradic S, Fernandes GW, Durigan G, Buisson E, Putz FE, Bond WJ (2015) Tyranny of trees in grassy biomes. Science 347:484–485

Viana VM (1995) Conservação da biodiversidade de fragmentos florestais em paisagens tropicais intensamente cultivadas. Abordagens interdisciplinares para a conservação da biodiversidade biológica e dinâmica do uso da terra, 1st edn. Universidade Federal de Minas Gerais, Belo Horizonte, pp 135–154

With KA, Gardner RH, Turner MG (1997) Landscape connectivity and population distribution in hetereogeneous environments. Oikos 78:151–169

WRI (2014a) Atlas of forest and landscape restoration opportunities. http://www.wri.org/applications/maps/flr-atlas/. Accessed 15 Jun 2015

WRI (2014b) Methodology of forest and landscape restoration opportunities. http://www.wri.org/methodology-about-atlas-forest-landscape-restoration-opportunity. Accessed 15 Jun 2015

Ziller SR (2001) Os processos de degradação ambiental originados por plantas invasoras. Cien Hoje 178:77–79

Chapter 18
Fire in Rupestrian Grasslands: Plant Response and Management

José Eugênio Côrtes Figueira, Katia Torres Ribeiro, Marilene Cardoso Ribeiro, Claudia Maria Jacobi, Helena França, Ana Carolina de Oliveira Neves, Abel Augusto Conceição, Fabiana Alves Mourão, Jumara Marques Souza and Carlos Abraham de Knegt Miranda

> Like a giant version of an amoeboid creature, fire emits pseudopodia on vegetation and engulfs it, drawing its energy to incorporate into its igneous body, excreting ash, smoke, carbon dioxide, water vapor and other gases, leaving behind, in the form of charred wood, only the parts of its "food" that eventually could not digest. In this feeding process, fire takes on the characteristics of the vegetation that sustains it, the physical environment that receives it, the winds that propel it and the topography that leads it (Miranda 2002).

Abstract Plant communities and species composing rupestrian grassland complexes within Brazilian savannas show varied responses to fire, and display similarities with other fire-prone ecosystems. Shallow, nutrient-poor soils subjected to severe water stress favor grasslands which carry fire in a landscape punctuated or crossed by rocky outcrops, riverine forests and other forest patches that act as

J.E.C. Figueira (✉) · C.M. Jacobi · A.C. de Oliveira Neves · F.A. Mourão
Instituto de Ciências Biológicas, Depto. Biologia Geral, Universidade Federal de Minas Gerais (UFMG), Av. Presidente Antônio Carlos, 6627, Pampulha, 30161970 Belo Horizonte, MG, Brazil
e-mail: cortes@icb.ufmg.br

C.M. Jacobi
e-mail: jacobi@icb.ufmg.br

A.C. de Oliveira Neves
e-mail: ananeves@gmail.com

F.A. Mourão
e-mail: fabimourao@gmail.com

K.T. Ribeiro
Instituto Chico Mendes de Conservação da Biodiversidade (ICMBio), Brasília, Brazil
e-mail: katia.ribeiro@icmbio.gov.br

M.C. Ribeiro
Instituto Biotrópicos, Diamantina, Brazil
e-mail: mcardosoribeiro@gmail.com; marilene@biotropicos.org.br

© Springer International Publishing Switzerland 2016

G.W. Fernandes (ed.), *Ecology and Conservation of Mountaintop Grasslands in Brazil*, DOI 10.1007/978-3-319-29808-5_18

barriers to fire. The long-term fire regime in these physiognomies is largely unknown, but, in post-European colonization times, the use of fire to livestock (cattle) pastures management and arson, most commonly in the dry season, have dominated the fire dynamics of these ecosystems. Several traits of rupestrian grassland plant species allow them to survive some fire regimes and/or take advantage of the post-burn environment, suggesting an ancient role of fire in their evolution and in defining agricultural practices. Fire management must consider the intrinsic heterogeneity and socioeconomic complexities of rupestrian grasslands landscapes. Evaluation and adaptation of strategies, ranging from protecting target areas from fire to applying prescribed patchy burning, and guaranteeing dialogue regarding people's needs, practices and knowledge, like those related to cattle raising and everlasting flowers harvesting, must be assured for the accomplishment of biodiversity conservation and sustainability goals. Decision-makers are encouraged to work in partnership with ecologists, policy-makers, and local communities, in an adaptive management approach.

Contents

H. França
Centro de Engenharias, Modelagem e Ciências Sociais Aplicadas,
Universidade Federal do ABC—UFABC, Santo André, Brazil
e-mail: helena.franca@ufabc.edu.br

A.A. Conceição
Depto Ciências Biológicas, Universidade Estadual de Feira de Santana (UEFS),
Feira de Santana, Brazil
e-mail: abel18@gmail.com

J.M. Souza
Programa de Pós-Graduação em Botânica, Universidade Estadual
de Feira de Santana (UEFS), Feira de Santana, Brazil
e-mail: jumarasouza@yahoo.com.br

C.A. de Knegt Miranda
Petróleo Brasileiro S.A.—PETROBRAS, Rio de Janeiro, Brazil
e-mail: cakmiranda@gmail.com

18.1 The Rupestrian Grassland Environment

The rupestrian grasslands of Brazil are a mosaic of highly diverse, fire-prone vegetation on poor-soils, encompassing grasslands to shrublands (Giulietti et al. 1997; Conceição and Pirani 2005; Neves and Conceição 2010; Chaps. 1, 6; Silveira et al. 2016). They display similarities with other fire-prone ecosystems, such as kwongan in Australia (Pignatti et al. 1993), and fynbos in South Africa (Manning 2007), as well as other features that support the OCBIL theory—Old, Climatically Buffered, Infertile Landscapes—(Hopper 2009) (Chap. 6; Silveira et al. 2016).

Shrubs and small trees are rare in rupestrian grasslands, in spite of regional annual rainfall levels (mean around 1500 mm), and a dry period length (around six months) that are able to support woody vegetation. The rupestrian grasslands are primarily determined by bottom-up edaphic control imposed by shallow, dry, hard, coarse-textured dystrophic soils derived from the nutrient-poor quartzitic and ironstone bedrock. Furthermore, these soils are rich in aluminum and subjected to intense sunlight, adding to the leaching and carbon loss due to fire (Giulietti et al. 1987; Rizzini 1992; Benites et al. 2007; Negreiros et al. 2014; Veldman et al. 2015). On these soils, woody vegetation is sclerophyllous and size-limited, as strong winds and gravity increases uprooting, and thus favoring nanophanerophytes and even underground shrubs and trees (Kolbek and Alves 2008). Cerrado vegetation occurs in more fertile soils associated with intrusions of metabasic rocks and metabasalt (Almeida-Abreu 1995), which also sustain forest patches (with floristic affinity to riverine forests) in drainage areas (ICMBio 2009). On the other hand, rocky outcrops often interrupt the grassy matrix allowing for root anchorage and favoring herbs, shrubs, small trees and even "giant" Velloziaceae rarely or never found in the surroundings, as well as many plant species sensitive to fire, such as orchids and ferns associated with the Atlantic Forest domain (Lousada et al. 2011). Partially supporting Bond and Keeley (2005) and Bond et al. (2005), the synergy between fast re-growth after fire and flammability allows grasses to outcompete the rare shrub and tree seedlings trying to establish themselves in open grasslands, reduces density within savanna patches and helps to maintain a well-defined boundary around rock outcrops, riverine forests and forest patches.

The rupestrian grasslands of Brazil are highly threatened by frequent man-made fires (Fig. 18.1) whose ecological consequences and environmental impacts have been poorly studied, thus motivating this revision. The aim of this chapter is to explore the historic role of fire in creating and maintaining Brazilian rupestrian grasslands and their associated biodiversity, and the challenges of identifying and applying appropriate fire management in the future, so as to assure continued ecological protection as well as sustainable land uses. We address the following questions: (1) What is the current fire regime and does it lie outside the range of historic fire regimes? (2) How are humans contributing to the current fire regime and why? (3) How do plant species and communities respond to fire? (4) Can we identify fire-adaptive plant traits? (5) What sort of fire regimes should be applied

Fig. 18.1 **a** A lightning fire in a rupestrian grassland in Serra do Cipó National Park (SCiNP) in January 2011. Notice the rocky outcrop and forest patch acting as fire barriers; **b** large rosettes of *Actinocephalus polyanthus* emerge above a dense grass layer (SCiNP); **c** mass flowering of *A. polyanthus* around a ravine (SCiNP); **d** topkill of small trees on an ironstone outcrop in the Serra do Rola Moça State Park (SRMSP); **e** vigorous sprouts at the base of burned stems of a small shrub (SRMSP); **f** a boom of flowering herbs in contrast with the carbonized *Vellozia sp.* trunks (SRMSP); **g** *Melinis minutiflora* covering a valley bordering ironstone outcrops (SRMSP); **h** mass flowering of Velloziaceae (*Vellozia* aff. *sincorana*) in the Chapada Diamantina National Park (CDNP); **i** a snake and a scorpion killed by fire (CDNP). Credits: Edward Elias Junior (**a**), José E. C. Figueira (**b**–**g**), Abel A. Conceição (**h**-**i**)

(or allowed) so as to maintain the ecosystem characteristics and biodiversity? (6) What land management strategies might be applied (or at least tested) to ensure sustained conservation of the natural values of this ecosystem?

18.2 The Burning Regimes—Past and Present

The evolutionary history of rupestrian grassland flora is strongly intermingled, generally, with that of Cerrado vegetation, and fire is a very important characteristic and evolutionary factor of both. Simon et al. (2009), through analyses of time-calibrated phylogenies, argued that the diversification of many lineages characteristic of the Cerrado flora began about 4 million years ago, coinciding with the rise to dominance of flammable C4 grasses and the expansion of the savanna biome worldwide. Pollen data and charcoal/carbon analyses indicate that natural fires were common before human occupation in the region (Fidelis and Pivello 2011). Intensification of the fire regime during the Holocene seems to be related to climatic changes but also to the arrival and population expansion of humans (Behling 1995; Pausas and Keeley 2009). Records of heavy usage of fire by human communities in the Cerrado region are as old as 4000–5000 years before present (Prous 1992). Pre-European groups of humans set fire to vegetation as a hunting strategy, for "cleaning up", or for attracting animals, among many other reasons - probably very similar to the practices of many indigenous groups today (Mistry et al. 2005). Fire still plays an important role in the management of land by indigenous people and farmers in many regions of the world and also in Brazil. The origin of the practice of burning rupestrian grassland is largely unknown, but the intensification of the use of fire is clearly related to European colonization of many Brazilian areas (e.g., Behling 1995, 1998). The frequent and widespread use of fire to clear land and to manage native pastures has already made a strong deleterious effect on Cerrado biodiversity, despite being a fire-prone ecosystem (Klink and Machado 2005).

The present use of fire by indigenous people in Cerrado has been described (e.g. Mistry et al. 2005), as well as traditional use of fire by non-indigenous communities in rupestrian grassland sites (e.g. Bedê 2006). Some aspects are being recovered by palynological studies (Behling 1998). There are reports about wiser use of fire in land management in the past, before prohibitions imposed by environmental policies. Fire regimes would have been worsened, with severe wildfires in the late dry season, due to clandestinity in the use of fire and the impossibility of excluding fire in these ecosystems. Renewal of natural pastures for livestock, management of flowering of everlasting plants, hunting and accidents have been the main human causes of fire in rupestrian grasslands (Mesquita et al. 2011; Ribeiro and Figueira 2011). These fires also frequently escape neighboring farms and invade protected areas, which have become especially threatened in recent decades by fuel accumulation due to zero-fire policy. Some rupestrian grasslands are also affected by a high density of lightning strike per year (França et al. 2007).

There are few rupestrian grassland sites for which there are good descriptions of recent fire regimes. Here we describe the fire regime and landscape/socioeconomic context for two sites located in Minas Gerais, southwestern Brazil: Serra da Canastra National Park (SCaNP), with 85,390 ha of regularized area (46°40′W, 20°12′S) and Serra do Cipó National Park (SCiNP), with 31,700 ha (43°32′W, 19°22′S). Both protected areas have at least 90 % of their lands at elevations that exceed 900 m (Fig. 18.2), where the prevailing climate is temperate highland tropical with dry winters (Cwb), according to Köppen's classification. In both areas, mean monthly temperatures vary between 18 and 22 °C and the mean annual precipitation is approximately 1500 mm, a combination that allows for rapid biomass production. The rains, however, are concentrated (85 %) during the period of October to March, along with the associated lightning, while the months from April to September are severely dry (ICMBio 2009; IBAMA 2005).

In the SCaNP, the open physiognomies that dominate the landscape are located on top of a rocky quartzite block, locally known as "*Chapadão da Canastra*", that

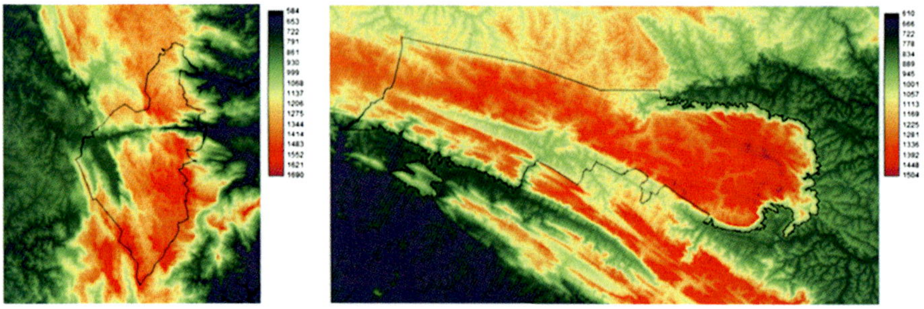

Fig. 18.2 Digital elevation models and regularized boundaries of Serra do Cipó National Park (**left**) and Serra da Canastra National Park (**right**). (França, unpublished data)

extends approximately 65 km in its longest axis (NW-SE direction), and to a maximum 20 km in width, where the planed surfaces, with slopes of less than 10°, correspond to 65 % of the area. Rupestrian grasslands do not prevail in the regional landscape, but occur in the park as a whole at elevations above 900 m, and occupy between 15 and 20 % of the park (Couto Júnior et al. 2010). The rupestrian grasslands, together with other open physiognomies, locally known as *campo limpo* and *campo sujo* (Coutinho 1990; Rizzini 1992), harbor an extensive layer of fine fuel—biomass with small thickness—throughout almost the entire park. In *campos limpos* the fine fuel accumulates rapidly after fires, reaching 3 ton/ha in the first year and >6.5 ton/ha in the fourth year without indication of stabilization, when the dead:live biomass ratio increases from 58 to 85 % (França, unpublished data), increasing the probability of fire spreading to the rupestrian grassland patches.

The SCiNP is located in the southern portion of the Serra do Espinhaço, and is largely situated on a massive quartzite formation with complex relief and steep slopes, cliffs and rock outcrops. The predominant vegetation is rupestrian grassland which, in this case, compose a continuous area of 26,500 ha in the regions of the Park above 900 m (Vasconcelos 2011), of which 60 % have slopes greater than 10°. In open inflammable landscapes, Poaceae (35–41 %) and Cyperaceae (29–54 %) are the main components of fine fuel biomass in sandy soils, while Poaceae (36 %), Velloziaceae (30 %) and Cyperaceae (22 %) predominate on the sandy-rocky hillsides (Vitta 2002). A tentative chronosequence of fine fuel (biomass) accumulation in areas of the park burned at different times suggests that, only two years after a fire, biomass can be as high as 5.5 ton/ha, and up to almost 10 ton/ha in some patches (the broad range reflecting environmental landscape heterogeneity within and among sampled areas), and the dead:live biomass ratio, indicating rapid desiccation, equaling or exceeding 50 % (Miranda 2002). Biomass accumulation decelerates, tending to level off at around 10–15 ton/ha in areas free of fire for decades, where the dead:live biomass ratio varied from 3 to 6.

In the dry season, mainly during El Niño events (e.g. Mesquita et al. 2011), a combination of an almost continuous layer of fine fuel (the grassy matrix with high levels of dry biomass), high temperatures and winds favor the spread of fire in the open landscapes of rupestrian grasslands, subjecting them, frequently, to very intense, out of control fires over large areas. In both parks current burnings are mainly anthropogenic and are superficial, consuming primarily fine fuels (<6 mm of thickness) of the herbaceous layer.

The burning regimes in SCaNP and SCiNP were determined by examination of fire scar maps and by visual interpretation of Landsat images (TM-5, ETM-7, OLI-8), ResourceSat (LISS-3) and CBERS (CCD2, CCD-2B) from 1984 to 2014, totalizing 177 images for SCaNP (França unpublished data) and 65 images for SCiNP (França and Ribeiro 2008). For SCaNP the month of burnings was determined, whenever possible, from reports and consultation with park management, by the eventual presence of active fire in images, by the presence of an active fire focus detected by the burning monitoring system of the National Institute for Space Research (INPE 2015), or estimated by the nature of burn scars in images. Determining the month of fires using this set of procedures was not applied to the SCiNP because there were no reports for nearly all of the study period, and the duration of the burn scars was too short to study using images. Maps were generated and analyzed using Geographic Information Systems (GIS) together with topographic data from the Topodata database (Valeriano and Rossetti 2012). Data on hydrography was obtained from topographic maps at a scale of 1:50,000 and 1:100,000 (IBGE 2015). A total of 31 annual burning maps were produced for each study area and burn areas were measured and calculated for each. Using this data set, the burning regime in both areas was described by characterization of fuel type, extent, length of occurrence and the burn frequency (Whelan 1995). The Weibull probability distribution function (Grissino-Mayer 1999) was the parametric model used to characterize the frequency of burnings, which was understood to be the period of time between two burning events in the same place.

In the SCaNP, the greatest number of burnings occur in August, however, it is in September, during the days preceding the first rains, that fires tend to burn larger areas. In the SCiNP, burnings are more numerous, although of smaller size when compared to those of the SCaNP. Small burning events are also significant in the SCaNP, but it is the large fires that characterize the fire regime (Fig. 18.3). This difference in size of burnings is a function of both landscape characteristics and the occupation history of these areas, plus conflicts in the relationships between dwellers and the administration of the parks. In the SCaNP, the fine fuel is arranged in extensive continuous areas, and the vegetation associated with more rocky areas are subjected to the same burning regime as the *campo limpo* and the *campo sujo* which occupy most of the park. In the area of SCiNP, the widespread rock scarps and outcrops create discontinuities in the fuel layer, which, in general, prevent fire spread. In addition, rivers associated with wet riparian forests serve as barriers to the spread of fire. The density of the drainage network of 2.4 km of rivers/km^2 in the SCiNP is 35 % greater than that in the SCaNP (1.8 km rivers/km^2), thus contributing to the separating of areas of fuel and the reduction in the number of

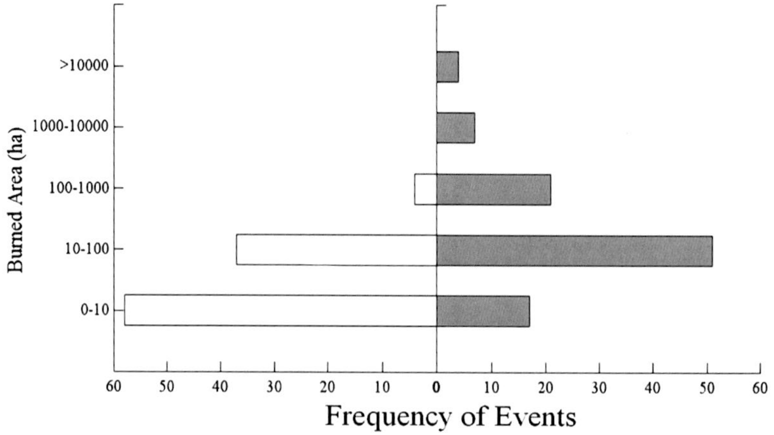

Fig. 18.3 Frequencies of burning events of different sizes in the Serra do Cipó National Park (left) and Serra da Canastra National Park (right), from 1984 to 2014 (França, unpublished data)

wildfires of comparable dimensions to those of the SCaNP. These natural fire barriers, as well as recently burned patches, become less effective as the dry season advances (see also Mesquita et al. 2011 for the case of Chapada Diamantina National Park).

The frequency of fire is quite different between the SCaNP and the SCiNP. In the SCaNP, the most common interval between two fire events is two years, reflecting the occurrence of large fires in shorter time intervals. In SCiNP, the peak frequency is seven years, indicating that the area is less susceptible to fire, considering ignition sources and spread conditions, not only fuel accumulation (Fig. 18.4). In addition, during the study period, 15 % of rupestrian grasslands in SCiNP did not burn and 21 % burned only once, whereas in SCaNP these values were 1 and 3 %, respectively.

The traditional agriculture practices of the population that lives in the region around the SCiNP, as well as the conflicts that developed between the State and these populations during and after the implementation of the park, are strongly related to the fire regimes of the last few decades. Until 2004, small ranchers/farmers that lived in the interior of the park used fire during the dry period for pasture management and opening of areas for plantation (ICMBio 2009). Such practices resulted in small and numerous burned areas with traditional techniques keeping fires under control and within areas pre-established by the residents. Very few dwellers still remain living inside the SCiNP, while in the SCaNP the last residents left the park in 1980, when the area was considered regularized and all cattle removed definitively (IBAMA 2005). Paradoxically, large wildfires covering 40–70 % of the area are common in the SCaNP, while in the SCiNP even the largest fire events never exceed 30 % of the rupestrian grassland area. In both parks, the larger wildfires normally originate externally and spread accidentally, or intentionally, to the park interior, but intentional fires beginning within the parks also occur. In both parks, fires have been employed as a means of protest against State

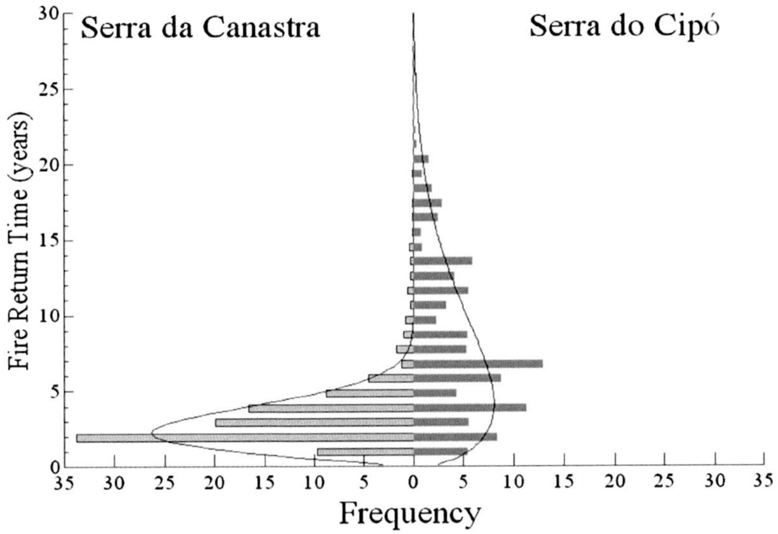

Fig. 18.4 Frequencies of fire return time in the Serra da Canastra National Park and Serra do Cipó National Park, from 1984 to 2014. The curves are Weibull distribution functions fitted to the data (França, unpublished data)

actions of expropriation and expulsion of the traditional dwellers or the impediment of their subsistence activities (Ferreira 2013), but the severity of conflict is different between them.

In SCiNP, cattle removal increased the vigilance and control of fires by fire fighters, thus contributing to a decline in the frequency of fires after 2003 (ICMBio 2009; Ribeiro and Figueira 2011). However, these procedures have led to an increase in the amount of fine fuel, predisposing the rupestrian grasslands to larger fires that are more difficult to control. França and Ribeiro (2008) noticed a tendency for large scale fires to occur every 6–8 years, which would correspond, based on the chronosequence, to a mean fine fuel biomass of above 9 ton/ha and a ratio of dead to live biomass of around 3.5.

Surveys of the fire reports for the SCaNP for the period of 2004–2011 found that lightning caused 35 fires in the regulated area of the park (whereas the number of anthropogenic fires was 62). The largest number of natural burnings occurred in October, a month that combines the most propitious conditions for natural fire because the vegetation is still dry when the rainy season begins, which also brings the increase in frequency of lightning. The size of areas burnt by lightning may equal the areas burnt by some anthropogenic fires, and apparently a single lightning strike during the dry season generated a large-scale fire ($\sim$13,400 ha), but still 5.5 times smaller than out-of-control anthropogenic fires (74,151 ha, Fig. 18.5). However, natural burnings are smaller and are being influenced by anthropogenic burnings in the Serra da Canastra, since anthropogenic fires consume the fuel that would then not be available for lightning burnings during the rainy season.

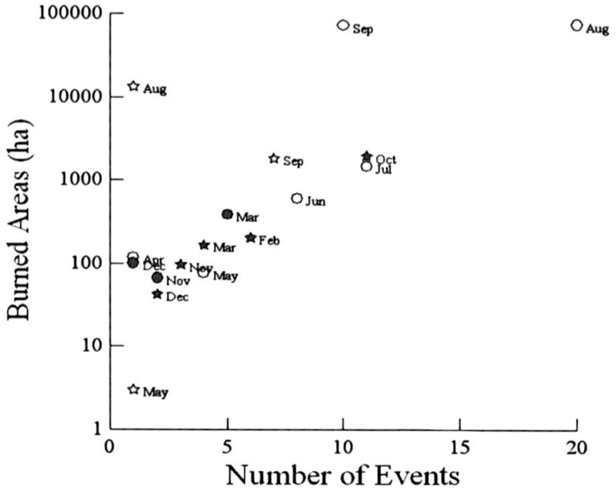

Fig. 18.5 Contrast between anthropogenic (circles) and natural (lightning fires; stars) fire regimes of Serra da Canastra National Park. The values are total burned area and total number of burn events from 2004 to 2011. Closed symbols indicate wet months (França, unpublished data)

In the SCiNP, it seems that natural burnings are less frequent, limited to one or two events per year; however, they usually do not exceed 1 ha (they have been noticed in isolated areas that have escaped fire for years, suggesting a slow return to a natural fire regime—Ribeiro and Figueira 2011). This difference in natural fire frequency between the two parks may be due to both the less frequent lightning in the SCiNP (5.5 lightning/km^2 year in comparison to the SCaNP with 10 lightning/km^2 year—ELAT 2015), and a landscape configuration in SCiNP that does not favor the development of large burnings. In both parks, the frequency of fires initiated by lightning are certainly underestimated, since most are extinguished by the rain that follows the lightning, and thus not reaching sufficient dimensions for local detection or detection by satellite images. It should be noted that neither of the parks employ an effective monitoring system that employs observation, smoke sensors or video cameras.

A characteristic practice in rupestrian grasslands is the use of fire for stimulating the blossoming of flowers of economic importance, mainly the many species of the so-called *sempre vivas* (everlasting) flowers (e.g. Neves et al. 2011). The flower heads and floral scapes are harvested from native grasslands by low-income workers from rural areas. After being dehydrated, to maintain their shape and color, these flowers are sold in national and international markets to be used as decorative objects (Instituto Terra Brasilis 1999; Parra et al. 2010). Amongst the harvesters there is the perception that fire encourages flowering of some of these commercial species (Bedê 2006). For example, 85 % of harvesters in the city of Diamantina (Minas Gerais State) said they use fire as a management tool (Instituto Terra Brasilis 1999), and harvesters of *S. elegans* in Jalapão (Tocantins State) said they use prescribed burning in alternate years to promote floral scape production (Schmidt et al. 2007). Fire is, and has been over the last few decades, employed for pasture management everywhere within rupestrian grasslands, but usually only outside of the parks.

The progressive insularization of these protected areas within a matrix of farms, cities, roads, and mines, combined with unmanaged tourism, urban expansion and the transit of humans, predispose the frontiers of these parks to growing disturbances, including fire from a variety of sources (substitution of native pastures by exotic grasses requiring less use of fire is in course, but constitutes another threat). The lack of small-scale fire management by farmers, related to the many restrictions on the use of fire and thus the employment of clandestine practices, combined with an increase in fine biomass and ignition sources, calls for new management approaches, since sensitive habitats are being destroyed and severe landscape homogenization is occurring as a result of the current fire regime.

18.3 Plant Response to Fire

In relation to vegetation in rupestrian grasslands, it is useful to consider that these heterogeneous landscapes harbor a mosaic of physiognomies that are very close to each other, with some being sensitive to fire, such as forest patches, and others being highly resistant and even dependent on fire, in a dynamic relationship with other vegetation types, and many gradations between these extremes. Within these physiognomies plants show a large array of responses to fire. Generally, plants may exhibit morphological (specialized structures such as lignotubers), physiological (induced sprouting, flowering, fruit dehiscence, dispersion or breaking dormancy of seeds) and demographic (recruitment or adult mortality) responses to fire.

Here we explore deeper the case of the family Eriocaulaceae, which represents an excellent group for the study of the effect of fire on vegetation in this context because it is one of the most speciose and characteristic families of the rupestrian grasslands of the Espinhaço Mountain Range, comprising 380 species (Giulietti et al. 1987, 2005), with a wealth of life forms and reproductive strategies. Many Eriocaulaceae species are threatened and are harvesting targets for everlasting flowers. The basic form of these plants is a rosette ranging from a few millimeters to about 0.8 m in diameter and almost one meter high. Life forms may include prostrate to caulescent rosettes, with solitary genets or genets composed of a few to many rosettes (ramets). Propagation forms include the production of sexual seeds, agamospermy, budding via rhizomes and pseudovivipary, which is the production of vegetative buds in inflorescences, however, many species combine more than one propagation mode (Coelho et al. 2008; Oriani et al. 2008). Eriocaulaceae exhibits both semelparous (the meristem that produces the inflorescence is central) and iteroparous (the meristem that produces the inflorescence is radial) life forms.

Neves et al. (2011) reviewed the effect of fire on natural populations of seven species of Eriocaulaceae and found distinct responses. Fire caused an increase in the number of reproductive individuals in all the species studied (*Actinocephalus polyanthus*, *Comanthera elegantula*, *Syngonanthus nitens*), an increase in inflorescence number per individual in half of the studied species (*C. elegantula, Leiothrix crassifolia*) and an increase in the number of seeds per flowerhead in a single studied species (*S. nitens*). Fire stimulated recruitment of seedlings in three out of four species (*A. polyanthus*, *C. elegantula*, *L. arrecta*) and by rhizome sprouts in *S. nitens* by means of eliminating the competing surrounding vegetation. Such apparent increase in reproductive effort in response to fire can negatively impact the production of inflorescences in subsequent years (e.g., *C. elegantula* and *S. nitens*), as well as the growth and survival of parental plants (e.g., *C. elegantula*) (Table 18.1).

In four species of rosette plants in Chapada Diamantina (Bahia State), mortality rate, the timing of flowering and the proportion of flowering plants triggered by fire may be associated with microhabitat heterogeneity (Table 18.2). Thus, fire susceptibility and severity may be key factors in selecting for different life history and demographic traits, at least in these life forms.

Therefore, mass flowering is a widespread response of rupestrian grassland plants to fire (Giulietti et al. 1987), that could be interpreted as an evolutionary response to the amelioration of microclimate, temporary increase in soil nutrient supply and light, removal of competitors and attraction of pollinators (see Whelan 1995 for a short review), since other disturbance types appear to be rare in rupestrian grasslands. However, Coutinho (1976) showed that cutting Cerrado herbaceous plants near level soil, exposing them to drought, or killing their epigeous parts also resulted in a high flowering percentage.

Native grasses resprout a few days after high-severe fires, while small hemicryptophyte rosette plants, like *Leiothrix* and *Syngonanthus*, may survive even moderated fires on sparse fuel beds generally covering hilltops and sandy-gravel soils. Bare soil patches within the grass layer also allow some escape from fire. These bare soil patches seem to represent micro-scale alternative states (Beisner et al. 2003) of the rupestrian grassland landscape. They have been bare for, at least, decades, suggesting strong hysteresis, and thus precluding the expansion of the grassy layer and/or the establishment of other monocot or dicot seedlings (Le Stradic et al. 2013), due to intense sunlight and strong hydric stress compared to the humid and shady grassy soils that border them. The small species of *Leiothrix* with a pseudoviviparous reproductive mode, such as *L. spiralis*, can circumvent this barrier to microenvironmental establishment since their clonal seedlings derived from their flower heads are connected to mother rosettes growing within the grass layer or even on bare soils through long lasting (Coelho et al. 2006; Coelho et al. 2008) (from 40 to 280 days) flexible stalks (up to 60–90 cm). However, part of these clonal pseudoviviparous propagules are suspended above the grass layer before rooting and when partially rooted, and so suffer heavy losses during fire due to direct burning or stalk breakage. The impact of the loss of these propagules, which carry large elasticities, on population growth rate (lambda) is stronger during the dry season when growth rate is low (Figueiredo 2014), and man-induced fires become common. On the other hand, old, rooted, and isolated rosettes or clones containing up to six rosettes can escape or survive fire, even when covered by a moderate grass layer since they grow at soil level rooted among (or away from) the base of monocot tussocks. Even in gravel soil, where grasses and other herbaceous plants are practically absent, the large density of stalks of *L. flagellaris* (Figueira

Table 18.1 Response of rupestrian grassland plants to fire (modified from Neves et al. 2011)

Family	Species	Study area	Reproductive season after burn					Years following burn		References
			No of reprod. Ind.	No of reprod. structures	Size of reprod. Structures	Recruit.	Mortality	Growth	Reprod. investment	
Amaranthaceae	*Xerosiphon aphyllus* (Pohl ex Moq.) Pedersen	QRG—CDNP	↑	–	–	–	–	–	–	A.A. Conceição, unpublished
Amaryllidaceae	*Habranthus irwinianus* Ravenna	IRG—SMNM	↑	↑		↑		↑	↑	Carmo et al. unpublished
Apocynaceae	*Mandevilla tenuifolia* (J.C. Mikan) Woodson	QRG—CDNP	↑	–	–	–	–	–	–	A.A. Conceição, unpublished
Asteraceae	*Ageratum fastigiatum* (Gardn.) R.M. King & H. Rob.	IRG—SMNM, SRMSP	↑			↑			↑	Mourão et al. unpublished; Carmo et al. unpublished
Asteraceae	*Lychnophora pinaster* Mart.	IRG—SRMSP	↓	–	–	↓	↑	–	–	Mourão et al. unpublished
Asteraceae	*Symphyopapus brasiliensis* (Gardner) R. M. King & H. Rob.	IRG—SRMSP	↑	–	–	↑	↓	–	–	Mourão et al. unpublished
Bromeliaceae	*Dyckia dissitiflora* Schult.f.	QRG—CDNP	↑	–	–	–	–	–	–	Souza (2011)

(continued)

Table 18.1 (continued)

Family	Species	Study area	Reproductive season after burn					Years following burn		References
			No of reprod. Ind.	No of reprod. structures	Size of reprod. Structures	Recruit.	Mortality	Growth	Reprod. investment	
Bromeliaceae	*Cottendorfia florida* Schult.f.	QRG—CDNP	↑	–	–	–	–	–	–	Souza (2011)
Bromeliaceae	*Orthophytum burle-marxii* L. B.Sm & Read	QRG—CDNP	Not change	–	–	–	–	–	–	Souza (2011)
Cleomaceae	*Dactylaena microphylla* Eichler	QRG—MPI	↑	–	–	–	–	–	–	A.A. Conceição unpublished
Cyperaceae	*Bulbostylis paradoxa* (Spreng.) Lindm.	QRG—CDNP	↑	–	–	–	–	–	–	A.A. Conceição, unpublished
Eriocaulaceae	*Actinocephaulus polyanthus* (Bong.) Sano	QRG—SCiNP, MPEPA	↑	–	–	↑	↑	–	–	Figueira (1998)
Eriocaulaceae	*Comanthera elegantula* (Ruhland) L.R. Parra & Giul.	QRG—RPSP	–	↑	Average scape weight do not change	↑	↑	↓	↓	Bedê (2006)
Eriocaulaceae	*Leiothrix crassifolia* (Bong.) Ruhl.	QRG—MPEPA	–	↑	Capitulum diameter do not change, Flower scape length ↓	–	–	–	–	Neves et al. unpublished
Eriocaulaceae			–			–	–	–	–	

(continued)

Table 18.1 (continued)

Family	Species	Study area	Reproductive season after burn					Years following burn		References
			No of reprod. Ind.	No of reprod. structures	Size of reprod. Structures	Recruit.	Mortality	Growth	Reprod. investment	
	Leiothrix curvifolia (Bong.) Ruhland	QRG—MPEPA		Not change	Capitulum diameter do not change, flower scape length ↑					Neves et al. unpublished
Eriocaulaceae	*Leiothrix spiralis* (Körn.) Bong.	QRG—MPEPA	–	Not change	Capitulum diameter do not change, flower scape length do not change	–	–	–	–	Neves et al. unpublished
Eriocaulaceae	*Leiothrix arrecta* Ruhland	QRG—MPEPA	↑	–	–	↑	–	–	–	Neves (2012)
Eriocaulaceae	*Actinocephalus ramosus* (Wikstr.) Sano	QRG—CDNP	↑	–	–	–	–	–	–	Souza (2011)
Fabaceae	*Microstachys daphnoides* (Mart.) Müll. Arg.	IRG—SRMSP	↑	–	–	↑	↓	–	–	Mourão et al. unpublished
Lamiaceae	*Eriope macrostachya* Mart. ex Benth.	IRG—SRMSP	↓	–	–	↓	↑	–	–	Mourão et al. unpublished
Lamiaceae		QRG—CDNP	↑							

(continued)

Table 18.1 (continued)

Family	Species	Study area	Reproductive season after burn					Years following burn		References
			No of reprod. Ind.	No of reprod. structures	Size of reprod. Structures	Recruit.	Mortality	Growth	Reprod. investment	
	Eriope obovata Epling									A.A. Conceição, unpublished
Iridaceae	*Pseudiris speciosa* Chukr & A.Gill	QRG—CDNP	↑	–	–	–	–	–	–	A.A. Conceição, unpublished
Leguminosae	*Aeschynomene carvalhoi* G.P. Lewis	QRG—CDNP	↑	–	–	–	–	–	–	A.A. Conceição, unpublished +J38
Ochnaceae	*Sauvagesia paniculata* D. Cardoso & A.A. Conc.	QRG—CDNP	↑	–	–	–	–	–	–	A.A. Conceição, unpublished
Velloziaceae	*Vellozia gigantea* Menezes e Mello-Silva	QRG—MPEPA, SCiNP	↑	↑	–	–	↑	–	–	Ribeiro (2007)
Velloziaceae	*Vellozia* aff. *sincorana* L.B. Sm. & Ayensu	QRG—CDNP	↑	↑	–	↑	–	–	–	Conceição and Orr (2012), Conceição et al. (2013)

QRG Quartzitic rupestrian grasslands, *IRG* Ironstone rupestrian grasslands, Morro da Pedreira Environmental Protection Area (MPEPA) Serra do Cipó National Park (SCiNP), Rio Preto State Park (RPSP), Chapada Diamantina National Park (CDNP), Serra da Moeda Natural Monument (SMNM), Serra do Rola Moça State Park (SRMSP), Morro do Pai Inácio (Chapada Diamantina) (MPI)

Table 18.2 Flowering (F) and mortality (M) in four rosette plant species monitored during two years in Chapada Diamantina National Park, Bahia, Brazil

Micro-habitat	Fi	Rosette plant species	Unburned			Burned		
			F (%)	D.F.	M	F. (%)	D.F.	M
Rock outcrop (Rock)	L	*Othophytum burle-marxii*	18	180	0 (n = 18)	3	180	0 (n = 30)
Grassland (Grass.)	H	*Actinocepha-lus ramosus*	20	90	25 (n = 20)	100	90	100 (n = 15)
Mixture Rock/Grass.	I	*Dyckia dissitiflora*	0		0 (n = 11)	100	30	11 (n = 18)
Mixture Rock/Grass.	I	*Cottendorfia florida*	20	90	25 (n = 20)	100	90	100 (n = 15)

Fi: relative fire severity (L = low, H = high, I = intermediate). D.F.: days to flowering (Souza 2011)

and Del Sarto 2007) conduce fire flames, allowing fire to spread over large areas (Figueira, personal observation). Canopy forming species occur in habitats with the densest herbaceous cover of species of *Leiothrix* (Coelho et al. 2008—but see Figueira and Del Sarto 2007), which are extremely affected by burning. In fire events, old and dense populations of *Leiothrix arrecta*, one such species, lose all their pseudoviviparous rosettes permanently suspended above the grass layer by their long lasting and fragile stalks. Since they lack underground meristems, adults are completely eliminated by fire, relying on the seed bank for their replacement. This explains the high seed-based reproductive effort of canopy forming species, as observed by Neves et al. (2014).

Massive rosettes (up to 80 cm in diameter) characterize some larger species of Eriocaulaceae, as is the case of the hemicryptophyte *Paepalanthus bromelioides*. Several layers of green leaves protect its central meristem from intense heat. Even their small rosettes survive after moderate to high severe fires. The chamaephyte species of this family, such as *P. vellozioides*, *Actinocephalus robustus* and *A. polyanthus,* are caulescent rosette plants. Their rosettes have indeterminate growth and also grow vertically, maintaining the dry leaves attached around their elongating vascular systems, thus forming small stems. The stems of *A. polyanthus* reach 90 cm in height, with a mean estimated age of more than 30 years. The dry leaves of these chamaephytes burn, leaving a residual layer of leaf sheaths compact enough to isolate their vascular system from high temperatures (Figueira 1998). This also occurs among caulescent Velloziaceae, where a much more compact layer of leaf sheaths has its origin in the breakage of dry leaves at their base during vertical growth (Giulietti et al. 1987).

Fire-induced mass flowerings disrupt the population size structures of *A. polyanthus*, causing population crashes. This is a demographic consequence of flowering probabilities stimulated by fire which increases rapidly with plant size, compared to natural flowering which are generally very low (a few rosettes among several thousand during a year). As every flowering plant dies, the probability of persistence, due to the permanence in vegetative state is inversely related to the flowering probability (Fig. 18.6). Populations of *A. polyanthus* that have been

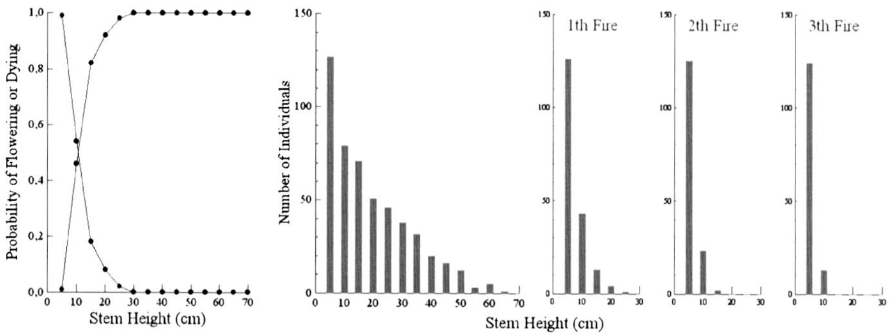

Fig. 18.6 Fire induced flowering probabilities (*upper curve*) and reproduction induced risk of dying (*lower curve*) of *Actinocephalus polyanthus* as a function of stem height. Changes in population structure after a first real fire and two sequential simulated fires. Notice the population crash after the first fire and small changes onwards (after Figueira 1998)

protected from fire for a long period of time acquire a typical "gradual reversed J" size distribution, where the largest individuals attain up to 90 cm in height. After a fire, these populations crash since they contain a large proportion of individuals with high flowering probabilities (losses may be as high as 63 %). The resulting size structure lacks all the previous reproductive individuals and acquires an "abrupt reverse J" size distribution. The predominance of small individuals in these post-reproductive populations results in minor changes after subsequent fires, were estimated losses can be as low as 12 % (Figueira 1998). Thus, population structures of *A. polyanthus* are bioindicators of fire in the rupestrian grasslands. In fact, an age-size relationship obtained for this plant was used to estimate how long some areas have escaped fires, fine biomass load and dead:live biomass ratio (Miranda 2002). The peculiar demography and fire resistant rosettes of *A. polyanthus* allow the maintenance of a "core of individuals" essential for population persistence and recovery, making this species resilient in the fire-prone environment of rupestrian grasslands. This is observed in patches of this plant outside the borders of the SCiNP that burn almost every two years.

However, post-fire stressors can kill non-reproductive plants. After being burned, some populations of *A. polyanthus* are attacked by caterpillars of the genus *Paracles*, that feed on inflorescence stems and rosette leaves, causing heavy reproductive losses without killing the plants. In one case, a rare and voracious attack of this caterpillar killed 33 % of non-reproductive rosettes after a fire. Their leaves were heavily eaten, and most of the dead rosettes were growing on bare soil patches (as described above for *Leiothrix*), suggesting a combined effect of herbivory and dry soil stress (Figueira 1998).

The interval between fire events is extremely important for individual plants to recover physiologically and demographically. Frequent fires can cause populations to decline by the exhaustion of seed banks, mortality (mainly of seedlings), induced reproduction of young/small individuals (in which the production of flower heads is usually small) and early death in semelparous species (but see *A. polyanthus* above). Furthermore, the exclusion of fire for long periods can result in aging/declining populations due to limited recruitment and increased mortality caused by increased competition with surrounding grasses(although *Leiothrix vivipara* and *L. arrecta* seem to take advantage of the presence of a grass layer by multiplying their pseudoviviparous photosynthesizing rosettes and seed output (Coelho et al. 2008; Neves et al. 2014).

Studies focusing on post-fire recolonization strategies of outcrop plant communities are still needed to determine key aspects of their dynamics, such as their level of resilience. The issue is important because fire effects on rupestrian grasslands depend on the habitat impacted. Strong differences have been reported between rock outcrop and grassland communities. Rock outcrops are important landscape features of rupestrian grasslands and host unique species-rich plant communities. Because they are naturally isolated and endure several abiotic stressors, these saxicolous communities are deemed very sensitive to factors that may promote further isolation or have a direct impact, such as fire (Clarke 2002). It is difficult for fire to spread through rock outcrops, but when it does it can kill several sensitive endemic species (Neves and Conceição 2010).

Fire also acts on seed bank dynamics and directly affects plant survival, growth, resprouting and reproduction. Plants that colonize a habitat after a fire adopt strategies according to their life history. They can be obligate seeders, resprouters, or rely on both strategies, i.e. are facultative resprouters (Keeley and Zedler 1978; Noble and Slatyer 1980; Ojeda et al. 2005; Buhk et al. 2006). Initially, there is little, if any, competition for resources (space, light, water), favoring species that are resistant to more severe environmental conditions. Competition turns progressively significant with time, selecting for species that are strong space-competitors (Ghermandi et al. 2004).

Changes in fire regimes may alter the structure of communities that occupy burnt areas by favoring some strategies over others. In Mediterranean climates, mild conditions and low fire frequencies tend to enhance recruitment by obligate seeders, while more severe conditions and high fire frequencies favor resprouters (Ojeda et al. 2005). Resprouting seems to be the most common post-fire colonization strategy among savanna plants (Bond and Midgley 2001), and is also widespread among Cerrado shrubs and trees, which also rely on bark protection (Hoffmann et al. 2009; Simon et al. 2009). However, this ability may be hampered by the fire regime, particularly fire frequency. Annual fires, for example, reduce the height and diameter of resprouters and increase their mortality rate (Medeiros and Miranda 2008).

Recent studies of ironstone outcrop (*canga*) plant communities in the Iron Quadrangle Region in southeast Brazil, suggest that some aspects of the fire regime (frequency and season) may promote important changes in parameters such as succession speed and species composition and dominance, thus affecting resources for pollinators and frugivores. Canga vegetation is established directly on the ferrugineous laterite or on debris and protosoils which are characteristically nutrient-poor and with low water retention (Klein 2000; Rosière and Chemale 2000; Vilela et al. 2004), a situation of stress aggravated by high UV radiation, constant winds and heavy metals (Benites et al. 2007). Because of the restrictions to growth caused by the bedrock, notably root development, vegetation is mostly herbaceous and/or shrubby in these areas, and with fewer arboreal species (restricted to patches where the ironstone outcrop is densely fractured) and less graminoid biomass than in the surrounding grasslands. Among the eudicots, Asteraceae, Fabaceae and Myrtaceae dominate, while Poaceae, Cyperaceae, and Orchidaceae are the prevalent monocot families (Viana and Lombardi 2007; Jacobi et al. 2007).

Vegetation structure and recolonization strategies were studied after anthropic fire events on two ironstone outcrops in the Iron Quadrangle that had been previously surveyed: Serra da Moeda and Serra do Rola Moça. Early-succession monitoring started some weeks after the event in each case, and continued for two years. Anthropic-related fires in the region usually occur in the dry season. Therefore, with the absence of the rain necessary for germination, the early stages of post-fire succession are usually characterized by fast resprouting and almost no seeding. Seedlings appear only after the onset of the rainy season. The recolonization strategies in the initial stages differ among sites, and this may be attributed to the season in which the fire took place. In spite of this different initial succession, resprouting in both cases decreased with time and after five months new resprouters were hardly seen, and the great majority of the new species and individuals were seeders.

In the Rola Moça pre-fire community 36 species from 22 families were identified, of which the shrubs *Mimosa calodendron* (Fabaceae) and *Lychnophora pinaster* (Asteraceae) dominated plant cover. Although several of these species had clumped distributions, many of these shrubs sustained populations of the mistletoe *Struthanthus flexicalis* (Loranthaceae), which connected their crowns and aided the spread of fire. All shrubs and their associated hemiparasites died, and their recovery was dependent on the rare event of seed germination within ironstone crevices. Regrowth of aerial biomass only became apparent after five months, coinciding with the wet season, and almost 70 % of these individuals recruited from seeds. The first species to recolonize were *Microstachys daphnoides, Symphyopappus brasiliensis*, and several species of Poaceae and Cyperaceae. The pre-fire dominant *M. calodendron* and *L. pinaster* were initially scarce and did not fully recover in 36 months (Fig. 18.7).

Only 18 species (13 families) reestablished themselves six months later and 19 species after 12 months. The obligate seeder *Ageratum fastigiatum* (Asteraceae) dominated during the first year. It was gradually substituted by resprouters and slower-growing eudicots, which became apparent after the second year. Of the 36 species in the pre-fire community, only 19 (13 families) had recolonized the burnt area after two years, and 11 were never seen. The reestablishment of four obligate-seeder dicots (*Chromolaena multiflosculosa, Ditassa mucronata, Microlicia* sp. and *Portulaca hirsutissima*) was short lasting. Notably, only two new species appeared, and both with the potential to increase the risk of fire if in high density: the mistletoe *Tripodanthus acutifolia* (Loranthaceae) and the invasive molasses grass *Melinis minutiflora* (Poaceae). *Tripodanthus acutifolia* xylem-taps on woody shrubs and treelets, rendering them more vulnerable. The molasses grass grows faster than native grasses, and its blades contain a viscid oil which increases

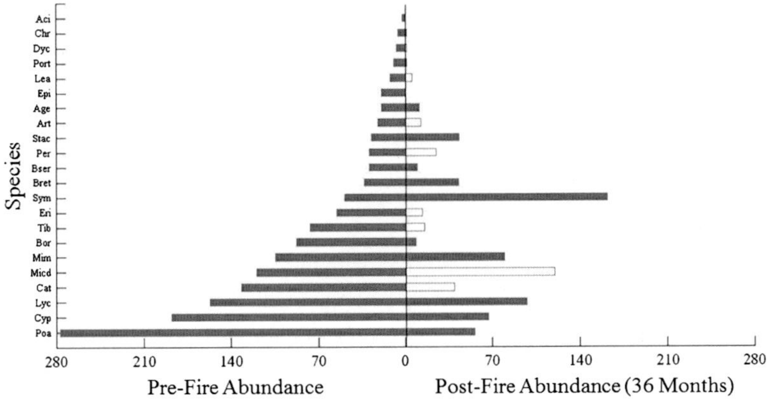

Fig. 18.7 Plant species abundance of an ironstone outcrop of Serra do Rola Moça State Park before and 36 months after fire. *Open bars* indicate resprouting species. Aci = *Acianthera teres*; Age = *Ageratum fastigiatum*; Art = *Arthrocereus glaziovii*; Bor = *Borreria cf. capitata*; Bret = *Baccharis reticularia*; Bser = *Baccharis serrulata*; Cat = *Cattleya caulescens;* Chr = *Chromolaena lychnophorioides*; Cy = Cyperaceae; *Dyckia cf. simulans*; Epi = *Epidendrum secundum;* Eri = *Eriope macrostachya*; Lea = *Leandra australis*; Lyc = *Lychnophora pinaster*; MicD = *Microstachys daphnoides*; Min = *Mimosa calodendron*; Per = *Periandra mediterranea*; Poa = Poaceae; Port = *Portulaca hirsutissima*; Stac = *Stachytarpheta glabra*; Sym = *Symphyoppapus brasiliensis*; Tib = *Tibouchina heteromalla*

the frequency and intensity of fires (Brooks et al. 2004). A fire at the end of the dry season in an area of Cerrado that had been invaded by *M. minutiflora* had peak temperatures of 800 to more than 1000 °C, flames over 6 m high, and a resident time that lasted over 3 min at average temperatures of 300 °C (see Mistry and Berardi 2005).

Four microhabitats in Moeda were investigated following a fire and ranked according to their degree of organic matter accumulation and water retention: (1) CD—cracks or depressions in the bedrock, (2) LY—*Lychnophora pinaster* (Asteraceae), (3) LG—*Lagenocarpus rigidus* (Cyperaceae) and 4) MM—monocot mats (particularly *Vellozia* spp.). With the exception of the first habitat type, accumulated organic matter was shed mostly by the dominant species.

Two months after the fire, practically all regrowth in Moeda came from resprouting. The dominant species were *Bulbostylis fimbriata* in CD, *Axonopus siccus* and *L. rigidus* in LG, *A. siccus* in LY, and *Habranthus irwinianus* in MM. After two years, 13 of the original species were still absent, among which were three Poaceae, three Asteraceae and two medium-sized Velloziaceae. In contrast to the early succession at Rola Moça, 27 new species appeared after the fire at Moeda. Of these, 10 were obligate-seeder species of Asteraceae, a group with long-distance seed dispersal capability. Among the grassy species, five new Cyperaceae were recruited.

The different outcomes of the two cases discussed here indicate that with each new fire event new succession patterns may be generated. Although the plant cover in both sites was reestablished in a few months, changes in species composition suggest that pre-fire community structure, particularly in Moeda, may take much longer to recover, or may even be part of a multiple-state fire dynamic. In general, species of groups such as Asteraceae and eudicots need time between fires to reproduce and replace the seed bank or else they jeopardize becoming locally extinct (Buhk et al. 2006).

If fire frequency increases, resprouters and fast-growing species, especially grasses and sedges, will be favored and, in turn, their biomass would enhance the risk of additional fires. The ironstone outcrops of the Rola Moça site are bordered by Cerrado areas invaded by molasses grass, where fine fuel biomass can be more than three times the 5.6 ton/ha estimated in non-invaded areas. The presence of this species increases fire frequency and destructiveness (Rossi et al. 2014), thereby increasing the risk of spread to the sensitive ironstone outcrops.

Fire tends to be more harmful to animals than to plants (e.g., Whelan 1995; Lyon et al. 2000; Koproski et al. 2006; Frizzo et al. 2011). The z value of the species-area relationship for medium to large sized mammals in several parks of the Iron Quadrangle Region, including Rola Moça, was within the range of very isolated continental remnants. In addition, their mammalian assemblages exhibited a nested occurrence pattern related to body sizes, suggesting non-random extinction caused by habitat loss (Morcatty et al. 2013). Fire is an additional non-computed threat to these animals of the parks. However, the bulk of medium to large sized mammals avoid the open landscapes of rupestrian grasslands, being concentrated in riverine forests, savanna vegetation, and forest patches (Oliveira 2004). As a consequence, the local effects of fire on these animals are poorly known (e.g. Frizzo et al. 2011). On the other hand, small groups of birds of prey, such as the southern crested caracara (*Polyborus plancus*), hunt and feed on hurt or dying vertebrates and invertebrates along fire lines, and are likely attracted by the smoke that is visible from distant places. In addition, the massive flowerings of resprouts immediately after fires (within 40 days), such as with *Vellozia* aff. *sincorana* L.B.Sm. & Ayensu (an endemic of Chapada Diamantina), provide an important source of nectar for insects (Conceição and Orr 2012; Conceição et al. 2013).

18.4 Fire and Management

Catastrophic wildfires are strongly related to the fronts of agriculture colonization, both in fire-prone grasslands and savannas and in sensitive biomes (like tropical and subtropical forests). A worldwide perception and conviction that fire could, and should, be suppressed has led to the development of strong policies to combat fire (Myers 2006; see Berlinck and Ribeiro 2012 for an analysis in Brazil). This approach stems from the traditional strong command-and-control paradigm that is slowly being replaced by a more comprehensive "learning by doing" approach on environmental management and by the acceptance of uncertainty and complexity in environmental management (see Keith et al. 2011; Penman et al. 2011 for challenges with implementing adaptive management). Fire-banning approaches have resulted in a large accumulation of "fuel" and consequent severe wildfires throughout the world (see Van Wilgen et al. 1998; Myers 2006). The prohibition has also resulted in clandestine use of fire by local communities (Mistry and Bizerril 2011). A paradigm shift is underway (Howe 1994; Bowman 2003; Mabunda et al. 2003; Whelan 2004; Myers 2006; França et al. 2007; Berlinck and Ribeiro 2012), based on a better understanding of the processes of evolution and the human influence on them (e.g. Trauernicht et al. 2015), which is reinforced by an appreciation of the deep complexity of social and conservation relationships.

Nevertheless, the start of a fire management program using fire as a conservation tool is not simple, since it deals with a large array of perceptions, expectations and knowledge and also with the inertia of institutions. Furthermore, there is the so common argument that we still do not understand well enough the ecological system and its constituent species specificities in order to apply an often polemic approach. The key to action relies on assuming the inescapable uncertainty related to fire management in fire-prone ecosystems and establishing common initial goals that deal with the main threats already described—this is the basis for fire management based on adaptive management principles (e.g. Bond and Archibald 2003). Some rationales are useful in this context, allowing decision despite the lack of detailed information. For example, the concept of vegetation sensitivity to fire, as proposed by Myers (2006). According to this author, there are physiognomies that are resistant to, or that even profit from, fire, such as most grasslands (although constituent species have a large spectrum of responses). There are physiognomies that oscillate in extension according to fire regime, and others that are very sensitive to it, such as riverine forests. Some physiognomies are very sensitive to fire but are well protected from it, like those on isolated rock outcrops. The use of fire to protect sensitive vegetation is a goal easily agreed to by many stakeholders. Avoidance of known deleterious fire regimes is also a common starting point. The season and extension of fire can be modulated using prescribed burns or openly dealing with the needs of rural communities, as an approach to cope with late dry season fires. The idea is that these first simple goals, related to landscape patterns, allow for some agreement and the beginning of fire management, before considering the myriad of questions posed when considering species dynamics and other aspects.

Nevertheless, the use of fire as an ecological tool may also be directed to more specific goals, for example, the creation of new open sites for the establishment of target species; providing habitat for target animal species; stimulating vegetative spread as well as mass blossoming and/or seed dispersal/recruitment of fire-adapted species; diminishing population/cover/biomass of fire-sensitive species; and the management of invasive exotic species (Whelan 1995).

Goals not based on the characteristics and requirements of individual species include the prevention of fuel accumulation, thus minimizing high intensity and out of control wildfires ('fuel reduction burning'—Penman et al. 2011); the prevention of the burning of sensitive vegetation/habitats; and, increasingly in evidence, the promotion of a mosaic of vegetation types and fire regimes since landscape and dynamic heterogeneity are themselves surrogates for ecological and evolutionary diversity-i.e. strategies generally based on avoiding homogenization (Whelan 1995; Bradstock et al. 2002; Beale et al. 2013). Bolder management proposals are being implemented nowadays in Kruger National Park, South Africa. For example, after the decision to reduce vegetation encroachment and forest spread, intense fires are being applied—in the dry season, in the middle of the day, with low air humidity—in order to favor open physiognomies. This kind of decision is only possible when the discussion among stakeholders about management practices is mature.

In the Cerrado biome as a whole, dry biomass accumulation and the avoidance of fire in humid/early dry seasons has resulted in wildfires that eventually affect all kinds of phytophysiognomies (França et al. 2007), including rock outcrops (Ribeiro and Figueira 2011) and riverine forests (Schmidt et al. 2011), which are usually protected from fire. Wildfires can also result in the burning of more than 50 % of a protected area in a single event, killing a large number of animals, homogenizing the vegetation and affecting sensitive physiognomies (Silveira et al. 1999; França et al. 2007; Santos et al. 2014). Although extensive fires seem to have been common due to lightning in the past (Myers 2006), the insularization of many protected areas adds an additional threat factor since now the whole available habitat or the entire local population of a species may be affected at once, with no more refuges or recruitment sites. Many protected areas in the Cerrado that have experienced a combination of cattle removal (i.e. local herbivory rate shift) and fire suppression policies ended up going through cyclic catastrophic events as described above. The absence of planned and negotiated strategies regarding both the use and the avoidance of fire can lead to environmental impoverishment or loss of species-besides the accidents, people at the surroundings of protected areas will eventually use fire for their own personal needs (e.g. towards a massive bloom of everlasting flowers or a massive resprouting of grasses for cattle grazing) not including any further conservation aim. Such a scenario has been noticed by Daniel Rios M. Borges (personal communication) at Sempre-Vivas National Park regarding the intentional annual fires locally called 'fogo das flores (flower fire)' and 'fogo do gado (cattle fire)' set by locals, calling for a better dialogue about goals and concerns.

Prescribed burning is still very controversial in Brazil, although rapid evolution in the debate is being observed. It is increasingly clear for many stakeholders in Brazil that fire should be managed as a necessary conservation tool, and not simply suppressed. This view is reflected in the text of the most recent Forest Code (Federal Law 12.727, 2012): "The use of fire in native vegetation is forbidden, except: in places or regions whose specificities justify the use of fire in rural practices (…), use of prescribed burn for conservation management of native vegetation whose ecological characteristics are evolutionarily associated with fire occurrence; and research projects."

Despite the many losses for environmental conservation in the negotiation of this new Forest Code, this specific issue was a victory for those who defend the use of fire for management purposes, always reinforcing that the methods go well beyond the use of just prescribed burns, as will be discussed below.

In parallel, there is a legitimate concern regarding greenhouse gas emissions from burning and deforestation of Brazilian biomes. This has resulted in the Action plan for prevention and control of deforestation and burnings in the Cerrado biome (PPCerrado—MMA 2011), with many goals related to reducing deforestation and fire. In consonance with this plan, a comprehensive Brazilian-German partnership, the Cerrado Jalapão project, began in 2012 with a clear goal of reducing burnings and emissions in the central region of the Cerrado biome. This project involves the Ministry for the Environment and directly linked national (ICMBio, Ibama) and state agencies, the Brazilian Institute for Spatial Studies (INPE) and universities, mainly University of Brasilia (UnB). Since the beginning, evolutionary and ecological factors have been valued alongside the culture, expectations, needs and rights of local populations (Lúcio et al. 2014). Although not initially dealing with rupestrian grasslands, deep discussions and exchanges promoted by the project have shed light on some of the problems and their potential solutions shared by sites with fire-prone and fire-adapted native vegetations.

It is worth noting that a successful strategy for fire management in fire-prone ecosystems with the aim of biodiversity conservation ('ecological burning' according to Penman et al. 2011) must rely on the involvement of people directly related to the subject, as well as many other stakeholders—and this should be taken as a cornerstone to achieve that goal (Myers 2006; Mistry and Bizerril 2011; Beale et al. 2013; Lúcio et al. 2014). This is not only because there is always a benefit in involving people, but also because an objective solution based only on science should not be expected (Keith et al. 2011; Cundill et al. 2012), since the issue lies in the realm of both biological and social complexity (Bold and Archibald 2003, Fariñas and Silva 2009).

In other countries—like Australia (Cawson and Muir 2008), New Zealand (Allen et al. 1996), Ethiopia (Johansson et al. 2013), and South Africa (Parr et al. 2009, Van Wilgen et al. 2011)—major initiatives are testing and implementing adaptive strategies concerning distinct fire regimes based on local specificities, well-established goals (including progressively higher complexity but beginning simply) and, mainly, following the adaptive management approach (Whelan 2004; Van Wilgen et al. 2011).

Some features of rupestrian grasslands, such as rapid fine biomass (fuel) accumulation, high wind speed and remoteness, and high species diversity and endemism make fire management in this ecosystem a tricky issue. To start with, fire can easily become out of control in these areas and, as addressed in *Plant Response to Fire* (above in this chapter), the sensitivity of rupestrian grassland species to fire can also vary remarkably within the same area. Further complicating matters are: (i) the spread of alien grasses such as *Melinis minutiflora* and species of *Brachiaria* and *Urochloa* over native grass species facilitated by fire (Alves and Silva 2011) and (ii) what humans value this ecosystem to provide as a parameter for successful burning plans (e.g. biology conservation perspective vs. economic perspective). Fire management proposals for rupestrian grasslands have many similarities with those posed for management of Fynbos in South Africa, as described by Van Wilgen (2013).

Despite their remarkable diversity and relative low value for agriculture, rupestrian grasslands are still underrepresented in protected areas (PAs) comprising only 2.6 % of 18.2 million km^2 protected, mostly settled in parks (Silva et al. 2008). As acknowledged above, many uses of rupestrian grasslands by local populations, including its biodiversity, involve burning. Even within the more restrictive Brazilian parks there is still the pressure of everlasting flower harvesting and cattle raising, at different degrees of legal provision. Thus, a fire regime strongly associated with human economic goals still prevails in rupestrian grasslands, and in most cases it is implemented in a clandestine way. Nevertheless, many other areas no longer have dwellers or people using the resources directly, and fire management decisions are more clearly related to biodiversity conservation issues, which is by no means simpler since establishing goals for biodiversity conservation of highly diverse systems may be more challenging than dealing with rural demands.

18.5 Suggested Directions for Fire Management in Rupestrian Grasslands

Every distinct area of rupestrian grassland seems to respond differently to fire—contrasting habitats coexist within small distances from each other, and fire regimes and behavior are very different between sites. Thus, local knowledge must be considered along with an associated research program in order to accelerate learning and enlighten management decisions. The ideal fire regime is far from being proposed for any site, and it seems to be an unreachable expectation due to the intrinsically complex nature of socio-ecological systems (Phillips 1930; Mabunda et al. 2003; Mistry and Berardi 2006). Moreover, management is strongly related to human aims, and these aims are quite variable among and within groups (Biggs et al. 2012). Even among biologists there is divergence in opinion on what fire management approach should be implemented. Opting for trying to ban any fire or rather using specific patchy prescribed fire regimes are decisions related to a dynamic process in which time, biotic, abiotic, cultural, social, economic and historical factors are interwoven. In spite of all the differences in opinion and conflicts, there is evidence that: (i) fire will occur, due to fine fuel accumulation and many ignition sources, and that (ii) the absence of wise management is leading to biological impoverishment.

A reasonable approach for the use of prescribed burns in some rupestrian grasslands sites would be the protection of sensitive vegetation. In the past, protection of forest patches with black firebreaks was commonly adopted by farmers, in order to protect timber, watersprings and to avoid recruitment of plants poisonous to cattle at the forest borders. In SCiNP the use of fire to protect the "giant" Velloziaceae are beginning to be applied (Christian Berlinck, personal communication). This goal can easily be established with many stakeholders.

The use of prescribed burnings in the early dry season in order to avoid severe wildfires is also widely understood to be a useful first approach that can be subsequently modulated according to its consequent outcomes—this is the main management goal in two of the most widely burned Brazilian Parks since 2014—Chapada das Mesas and Serra Geral do Tocantins (Cerrado, but not with rupestrian grassland formations). As discussed above, landscape characteristics affect strongly the spread of fire, and consequently the diversity of fire regimes and degree of vegetation homogenization. In Chapada Diamantina National Park, natural barriers to fire propagations have been mapped and Gonçalves et al. (2011) have demonstrated that their effectiveness declines as the dry season progresses. Fire sectors were defined and they provide the basis for many fire management proposals—there are sectors with rare cases of fire ignition, sectors in which fires are concentrated, and sectors where management efforts are concentrated and include deals with dwellers.

Some managers do not combat fire related to lightning, but usually this practice is not assumed and registered and consequently it is not studied and evaluated in relation to its effectiveness. Goals are usually not clear either in such cases. Another management strategy could be "mimicking" natural fire regimes, such as the fires triggered by lightning at Serra da Canastra National Park (SCaNP). These spontaneously triggered fires occur mainly in the early rainy season, burning small areas that are readily extinguished by subsequent rainfall (see also Ramos-Neto and Pivello 2000). The reason for choosing this distinct management strategy (with its main objective of boostering biodiversityand ecological processes) is based on the assumption that 'unnatural' fire regimes performed in a context different from that of natural fires (i.e. ecological burning conducted in the early dry season instead of mimicking the lightning season) would be deleterious because elements of the biota have evolved under a different set of fire conditions (Whelan 1995).

A bolder but reasonable strategy, considering the highly heterogeneous constitution of rupestrian grasslands, would be the use of patchy fires conducted in the early and middle dry season (taking into account the effectiveness of natural firebreaks during the season), at an interval of 3–4 years. The intentional burning of patches that have been protected from fire for a long period of time on fire-prone vegetation can also improve diversity. The suggestion of a 3–4 year interval between fires in more frequently burned patches is based on accumulation rates of fine biomass and probability of burning. Studies by Oliveiras et al. (2012) and Brito (2011), for example, suggested that an interval of at least 3 to 4 years between each burn is necessary for the reestablishment of some Cerrado and rupestrian grassland plant communities to a pre-burn condition.

Fire should not be prescribed by rigid protocols (e.g. fixed spaces delimited by permanent firebreaks, for example). Instead, free running burns performed before the critical months, taking into consideration places where it will naturally stop, should be adopted, which would also facilitate field work. This strategy is usually adopted in order to further promote diversity, even among populations of the same species.

With further knowledge (and/or local demands), specific practices can be established for target species. For example, the requirements of threatened species, like some Eriocaulaceae and Velloziaceae described above, can drive some management decisions in order to promote habitat and population recovery. In these cases, it is important to monitor the effects on other community components (e.g. Schmidt et al. 2011).

From the socioeconomic perspective, considering the sustainable use of rupestrian grassland resources by human beings, Bedê (2006) studied the effects of extractive management on the life history of *C. elegantula* (an everlasting flower species) in an experiment with eight possible combinations of the treatments 'fire', 'without fire', 'harvesting' and 'without harvesting' at a two year intervals. He found that the combinations of treatments that included harvesting after fire, such as currently used by flower harvesters, increased reproduction and allowed the long-term maintenance of plant populations. The harvesting of inflorescences before seed production seems to prevent plants from allocating an important part of their biomass, energy and nutrients to seed production, and thus reducing subsequent mortality of parental plants.

Little is currently known concerning the supposed optimal fire regime for conservation of animal populations and communities (Silveira et al. 1999).

Being aware of all the uncertainties acknowledged herein, but also considering the necessity of imminent attitudes decision-makers must take in order to assure a sustainable future for the rupestrian grasslands in face of the current fire regime scenario in Brazil, actions guided by adaptive management are desired. Managers should be allowed to focus on their local context and work together with the different stakeholders—policy-makers, scientists and local groups—to design the most suitable strategies/actions for their local situation. This said, managers must also follow their institutions general guidelines so as to not place themselves in an unsecure and excessively exposed position, and to assure continuity in management in the long term, another challenge in Brazil.

Here we list some additional considerations that managers should consider in order to perform proper adaptive management using fire in rupestrian grasslands: (i) well-defined and measurable aims:—for example—is the aim to avoid severe wildfires? Is it to maintain or improve vegetation heterogeneity within a specific rupestrian grassland? Is it to increase the population size of an endangered species? Is it to ensure the livelihood for the harvesters of everlasting flowers, yet prevent the dismantling of both the plant community structure and their associated species population structure?; (ii) the presence of specific target species, such as threatened species; (iii) probable sources of ignition; (iv) the dynamics of dry matter (fuel) accumulation; (v) the organization of elements in the landscape, especially sensitive and tolerant phytophysiognomies and natural barriers, creating a mosaic by sectioning the landscape; (vi) the changes that are expected to occur in seasonality as well as in the composition/distribution/behavior of the fauna and flora due to global warming—for example, forest advance into species rich grasslands or the opposite; (vii) the local interaction between distinct human groups and the land (as well as divergent interests); and (viii) the spread of alien plant species, especially African grasses.

Other adaptive management steps should be carefully considered: (xix) short and long-term monitoring of the results of the strategies chosen; (x) possible partnerships and protocols to assure the processes of monitoring and evaluation of the efficiency of the applied strategy; (xi) local involvement in the planning, burning (or not burning), evaluation, and decision-making processes, including private landowners; (xii) and discussion of results and continuous management based on what has been learned from an implemented strategy.

In summary, the restitution of an original fire regime is a common expectation, although it is difficult to assess and difficult to accomplish, as discussed above. Nevertheless, a temporal and spatial mosaic of burned areas, protecting more sensitive formations, seems to be an acceptable initial fire regime for a healthy balance between plant demography and diversity in rupestrian grasslands. Furthermore, such a regime is highly achievable in comparison to the elevated annual costs of fighting wildfires if wise use of prescribed burns and other management tools are implemented together with local knowledge, science and technology.

Acknowledgements We gratefully acknowledge Dr. Robert Whelan for inspiration, constructive comments and very useful suggestions on this chapter. We thank CAPES, CNPq, FAPEMIG, FAPESP, FAPESB, and FUNDAÇÃO O BOTICÁRIO for several types of financial support to the authors, that include scholarships and grants, and Erik Wild for revising the English version of this chapter.

References

Allen RB, Basher LR, Comrie J (1996) The use of fire for conservation management in New Zealand. N.Z. Dept of Conservation, Wellington

Almeida-Abreu PA (1995) O Supergrupo Espinhaço: o rifte, a bacia e o orógeno. Geonomos 3:1–18

Alves RJV, Silva NG (2011) O fogo é sempre um vilão nos campos rupestres? Biodiv Bras 2:120–127

Beale CM, van Rensberg S, Bond W, Coughenour M, Fynn R, Gaylard A, Grant R, Harris B, Jones T, Mduma S, Owen-Smith N, Sinclair ARE (2013) Ten lessons for the conservation of African savanna ecosystems. Biol Conserv 167:224–232

Bedê LC (2006) Alternativas para o uso sustentado de sempre-vivas: efeitos do manejo extrativista sobre *Syngonanthus elegantulus* Ruhland (Eriocaulaceae). Dissertation, Universidade Federal de Minas Gerais, Brasil

Behling H (1995) A high resolution Holocene pollen record from Lago do Pires, SE Brazil: vegetation, climate and fire history. J Paleolimnol 14:253–268

Behling H (1998) Late Quaternary vegetational and climatic changes in Brazil. Rev Palaeobot Palyno 99:143–156

Beisner BE, Haydon DT, Cuddington K (2003) Alternative stable states in ecology. Front Ecol Environ 1:376–382

Benites VM, Shaefer CEGR, Simas FNB, Santos HG (2007) Soils associated with rock outcrops in the Brazilian mountain ranges Mantiqueira and Espinhaço. Revista Brasil Bot 4:569–577

Berlinck CN, Ribeiro KT (2012) Pesquisa e manejo do fogo em áreas protegidas federais. In: Lima GS, Bontempo G, Almeida M, Gonçalves W (org.) Gestão, Pesquisa e Conservação em Áreas Protegidas. Viçosa, pp 119–136

Biggs R, Schlüter M, Biggs D, Bohensky EL, Silver SB, Cundill G, Dakos V, Daw TM, Evans LS, Kotschy K, Leitch AM, Meek C, Quinlan A, Raudsepp-Hearne C, Robards MD, Schoon ML, Schultz L, West PC (2012) Toward principles for enhancing the resilience of ecosystem services. Annu Rev Environ Resour 37:421–448

Bond WJ, Archibald S (2003) Confronting complexity: fire policy choices in South African savanna parks. Int J Wildland Fire 12:381–389

Bond WJ, Keeley JE (2005) Fire as a global 'herbivore': the ecology and evolution of flammable ecosystems. Trends Ecol Evol 20:387–394

Bond WJ, Midgley JJ (2001) Ecology of sprouting in woody plants: the persistence niche. Trends Ecol Evol 16:45–51

Bond WJ, Woodward FI, Midgley GF (2005) The global distribution of ecosystems in a world without fire. New Phytol 165:525–538

Bowman D (2003) Bushfires: a Darwinian perspective. In: Cary G, Lindemayer D, Dovers S (eds) Australia burning: fire ecology, management and policy issues. CSIRO Publishing, Melbourne, pp 3–14

Bradstock RA, Williams JE, Gill MA (2002) Flammable Australia: the fire regimes and biodiversity of a continent. Cambridge University Press, Cambridge

Brito JC (2011) Efeitos do fogo sobre a vegetação em duas áreas de campo rupestre na Chapada Diamantina, Bahia, Brasil. Dissertation, Universidade Estadual de Feira de Santana, Brasil

Brooks ML, D'Antonio CM, Richardson DM, Grace JB, Keeley JE, DiTomaso JM, Hobbs RJ, Pellant M, Pyke D (2004) Effects of invasive alien plants on fire regimes. Bioscience 54:677–688

Buhk C, Meyn A, Jentsch A (2006) The challenge of plant regeneration after fire in the Mediterranean Basin: scientific gaps in our knowledge on plant strategies and evolution of traits. Plant Ecol 192:1–19

Cawson J, Muir A (2008) Flora monitoring protocols for planned burning: a user's guide. Victorian Government Dept. of Sustainability and Environment, Fire and Adaptive Management, Melbourne, Report No. 74

Clarke PJ (2002) Habitat islands in fire-prone vegetation: do landscape features influence community composition? J Biogeogr 29:677–684

Coelho FF, Capelo C, Neves ACO, Martins RP, Figueira JEC (2006) Seasonal Timing of Pseudoviviparous Reproduction of *Leiothrix* (Eriocaulaceae) Rupestrian Species in South-eastern Brazil. Ann Bot 98:1189–1195

Coelho FF, Capelo C, Ribeiro LC, Figueira JEC (2008) Reproductive modes in *Leiothrix* (Eriocaulaceae) in South-eastern Brazil: the role of microenvironmental heterogeneity. Ann Bot 101:353–360

Conceição AA, Orr BJ (2012) Post-fire flowering and fruiting in the caulescent rosette *Vellozia sincorana*, an endemic plant to the Northeast of Brazil. Acta Bot Brasil 26:94–100

Conceição AA, Pirani JR (2005) Delimitação de habitats em campos rupestres na Chapada Diamantina: substratos, composição florística e aspectos estruturais. Bol Bot Univ São Paulo 23:85–111

Conceição AA, Alencar TG, Souza JM, Moura ADC, Silva GA (2013) Massive post-fire flowering events in a tropical mountain region of Brazil: high episodic supply of floral resources. Acta Bot Brasil 27:847–850

Coutinho LM (1976) Contribuição ao conhecimento do papel ecológico das queimadas na floração de espécies do cerrado. Dissertation, Universidade de São Paulo, Brasil

Coutinho LM (1990) Fire in the ecology of Brazilian cerrado. In: Goldammer JG (ed) Fire in the tropical biota—ecosystem process and global challenges. Springer, Berlin, pp 82–105

Couto-Júnior AF, Souza VV, Carvalho-Junior, Martins ES, Santana OA, Freitas, LF, Gomes RAT (2010) Integração de parâmetros morfométricos e imagem aster para delimitação das fitofisionomias da Serra da Canastra, Parque Nacional da Serra da Canastra, MG. Rev Bras Geom 11:57–68

Cundill G, Cumming GS, Biggs D (2012) Soft systems thinking and social learning for adaptive management. Conserv Biol 26:13–20

ELAT (2015) http://www.inpe.br/webelat/homepage/. Accessed 5 Aug 2015

Fariñas MR, Silva JF (2009) Fire in neotropical savannas. In: Tropical biology and conservation management—encyclopedia of life support systems (EOLSS), developed under the Auspices of the UNESCO, vol X. Eolss Publishers, Oxford

Federal Law 12.727 (2012) http://www.planalto.gov.br/ccivil_03/_ato2011-2014/2012/lei/L12727.htm. Accessed 29 Aug 2015

Ferreira GHC (2013) A regularização fundiária do Parque Nacional da Serra da Canastra e a expropriação camponesa: da baioneta à ponta da caneta. Dissertation, Universidade de São Paulo, Brasil

Fidelis A, Pivello VR (2011) Deve-se usar o fogo como instrumento de manejo no Cerrado e Campos Sulinos? Biodiv Bras 2:12–25

Figueira JEC (1998) Dinâmica de Populações de *Paepalanthus polyanthus* (Eriocaulaceae) na Serra do Cipó, MG. Dissertation, Universidade Estadual de Campinas, Brasil

Figueira JEC, Del Sarto ML (2007) Clonal growth and dispersal potential of *Leiothrix flagellaris* (Eriocaulaceae) in the rocky grasslands of Southeastern Brazil. Rev Brasil Bot 30:679–686

Figueiredo L (2014) Seasonal population dynamics and demography in *Leiothrix spiralis* (Eriocaulaceae). Dissertation, Université de Rennes, France

França H, Ribeiro KT (2008) Mapeamento de queimadas no Parque Nacional da Serra do Cipó e na Área de Proteção Ambiental Morro da Pedreira, MG: 1984–2007. Ministério do Meio Ambiente, Brasília, DF

França H, Ramos-Neto MB, Setzer A (2007) O Fogo no Parque Nacional das Emas. Ministério do Meio Ambiente, Série Biodiversidade, Brasília, DF

Frizzo TLM, Bonizário C, Borges MP, Vasconcelos HL (2011) Revisão dos efeitos do fogo sobre a fauna de formações savânicas do Brasil. Oecol Australis 15:365–379

Ghermandi L, Guthmann N, Bran D (2004) Early post-fire succession in northwestern Patagonia grasslands. J Veg Sci 15:67–76

Giulietti AM, Menezes NL, Pirani JR, Meguro M, Wanderley MGL (1987) Flora da Serra do Cipó, Minas Gerais: caracterização e lista de espécies. Bol Bot Univ São Paulo 9:1–151

Giulietti AM, Pirani JR, Harley RM (1997) Espinhaço range region, eastern Brazil. In: Davis SD, Heywood VH, Herrera-Macbryde O, Villa-Lobos J, Hamilton AC (eds) Centres of plant diversity. A guide and strategy for their conservation. Information Press, Oxford, pp 397–404

Giulietti AM, Harley RM, Queiroz LP, Wanderley MGL, van den Berg C (2005) Biodiversidade e conservação das plantas no Brasil. Megadiversidade 1:52–61

Gonçalves CN, Mesquita FW, Lima NRG, Coslope LA, Lintomen BS (2011) Recorrência dos incêndios e fitossociologia da vegetação em áreas com diferentes regimes de queima no Parque Nacional da Chapada Diamantina. Biodiv Bras 1:161–179

Grissino-Mayer HD (1999) Modeling fire interval data from the American southwest with the Weibull distribution. Int J Wildland Fire 9:37–50

Hoffmann WA, Adasme R, Haridasan M, de Carvalho MT, Geiger EL, Pereira MA, Gotsch SG, Franco AC (2009) Tree top-kill, not mortality, governs the dynamics of alternate stable states at savanna-forest boundaries under frequent fire in central Brazil. Ecology 90:1326–1337

Howe H (1994) Managing species diversity in tallgrass prairie: assumptions and implications. Conserv Biol 8:691–704

IBAMA (2005) Plano de manejo do Parque Nacional da Serra da Canastra. Ministério do Meio Ambiente, Brasília, DF

IBGE (2015) http://mapas.ibge.gov.br/pt/bases-e-referenciais/bases-cartograficas/. Accessed 6 Aug 2015

ICMBio (2009) Plano de manejo do Parque Nacional da Serra do Cipó. Área de Proteção Ambiental Morro da Pedreira. Instituto Chico Mendes de Biodiversidade/Ministério do Meio Ambiente, Brasília, DF

INPE (2015) http://www.inpe.br/queimadas/. Accessed 5 Aug 2015

Instituto Terra Brasilis (1999) Projeto sempre-vivas: subsídios para seu uso sustentado. Relatório Técnico, Instituto Terra Brasilis, Belo Horizonte

Jacobi CM, Carmo FF, Vincent RC, Stehmann JR (2007) Plant communities on ironstone outcrops: a diverse and endangered Brazilian ecosystem. Biodivers Conserv 16:2185–2200

Johansson M, Granström A, Malmer A (2013) Traditional fire management in the Ethiopian highlands—What would happen if it ends? Forest Facts 9 Sciences. http://www.slu.se/PageFiles/33707/2013/ForestFacts_09_2013.pdf. Accessed 4 Aug 2015

Keeley JE, Zedler PH (1978) Reproduction of chaparral shrubs after fire: a comparison of sprouting and seeding strategies. Amer Midl Nat 99:142–161

Keith DA, Martin TG, McDonald-Madden E, Walters C (2011) Uncertainty and adaptive management for biodiversity conservation. Biol Conserv 144:1175–1178

Klein C (2000) Geochemistry and petrology of some Proterozoic banded iron-formations of the Quadrilátero Ferrífero, Minas Gerais, Brazil. Econ Geol 95:405–428

Klink CA, Machado RB (2005) Conservation of the Brazilian Cerrado. Conserv Biol 19:707–713

Kolbek J, Alves RJV (2008) Impacts of cattle, fire and wind in rocky savannas, Southeastern Brazil. Acta Univ Carol Environ 22:111–130

Koproski L, Batista AC, Soares RV (2006) Impactos do fogo sobre serpentes (Squamata) no Parque Nacional de Ilha Grande (PR/MS), Brasil. Arq Ciên Vet Zool Unipar 9:129–133

Le Stradic S, Buison E, Fernandes GW (2013) Restoration of Neotropical grasslands degraded by quarrying using hay transfer. Appl Veg Sci 17:482–492

Lousada JM, Borba EL, Ribeiro KT (2011) Genetic structure and variability of the endemic and vulnerable Vellozia gigantea (Velloziaceae) associated with the landscape in the Espinhaço Range, in southeastern Brazil: implications for conservation. Genetica 139:431–440

Lucio SLB, Pereira LEC, Ludewigs T (2014) O gado que circulava: desafios da gestão participativa de unidades de conservação nos gerais do norte de Minas. Biodiv Bras 1:130–156

Lyon LJ, Huff MH, Telfer ES, Schreiner DS, Smith JK (2000) Fire effects on animal populations. In: Smith JK (ed) Wildland fire in ecosystems: Effects of fire on Fauna Gen. Tech. Rep. RMRS-GTR 42-vol 1. US Department of Agriculture Forest Service Rocky Mountain Research Station, Ogden, UT, pp 25–34

Mabunda D, Pienaar DJ, Verhoef J (2003) The Krueger National Park: a century of management and research. In: du Toit J, Biggs H, Rogers K (eds) The Krueger experience: ecology and management of savanna heterogeneity. Island Press, Washington, pp 3–21

Manning J (2007) Field guide to fynbos. Struik Publishers, Cape Town

Medeiros MB, Miranda HS (2008) Post-fire resprouting and mortality in Cerrado wood plant species. Edinb J Bot 65:1–16

Mesquita FW, Lima NRG, Gonçalves CN, Berlinck CN, Lintomen BS (2011) Histórico dos incêndios na vegetação do Parque Nacional da Chapada Diamantina, entre 1973 e abril de 2010, com base em imagens Landsat. Biodiv Bras 1:237–256

Miranda CAK (2002) *Paepalanthus polyanthus* (Bong.) Kunth (Eriocaulaceae) como bioindicadora de biomassa vegetal aérea nos campos rupestres da Serra do Cipó. Dissertation, Universidade Federal de Minas Gerais, Brasil

Mistry J, Berardi A (2005) Assessing fire potential in a Brazilian savanna nature reserve. Biotropica 37:439–451

Mistry J, Berardi A (2006) Introduction. In: Mistry J, Berardi A (eds) Savannas and dry forests: linking people with nature. Aldershot, England, pp 1–18

Mistry J, Bizerril M (2011) Why it is important to understand the relationship between people, fire and protected areas. Biodiv Bras 2:40–49

Mistry J, Berardi A, Andrade V, Krahô T, Krahô P, Leonardos O (2005) Indigenous fire management in the Cerrado of Brazil: the case of the Krahô of Tocantíns. Hum Ecol 33:365–386

MMA (2011) Plano de Ação para prevenção e controle do desmatamento e das queimadas: Cerrado. Ministério do Meio Ambiente, Brasília, DF

Morcatty TQ, El Bizri HR, Carneiro HCS, Biasizzo RL, Alméri CRO, Silva ES, Rodrigues FHG, Figueira JEC (2013) Habitat loss and mammalian extinction patterns: are the reserves in the Quadrilátero Ferrífero, southeastern Brazil, effective in conserving mammals? Ecol Res 28:935–947

Myers R (2006) Living with fire. Sustaining ecosystems & livelihoods through integrated fire management. The global fire initiative. The Nature Conservancy, Tallahassee

Negreiros D, Le Stradic S, Fernandes GW, Rennó HC (2014) CSR analysis of plant functional types in highly diverse tropical grasslands of harsh environments. Plant Ecol 215:379–388

Neves ACO, Bedê LC, Martins RP (2011) Revisão sobre os efeitos do fogo em Eriocaulaceae como subsídio para a sua conservação. Biodiv Bras 2:50–66

Neves ACO (2012) História de vida e efeitos do fogo em espécies de Leiothrix (Eriocaulaceae) com diferentes estratégias de propagação na Serra do Cipó, MG. Dissertation, Universidade Federal de Minas Gerais, Brasil

Neves ACO, Nogueira FB, Assis LR, Paglia AP, Bedê LC, Martins RP (2014) Reproductive allocation in rhizomatous, seminiferous, and pseudoviviparous *Leiothrix* (Eriocaulaceae) species. Plant Ecol 215:987–996

Neves SPS, Conceição AA (2010) Campos rupestres recém-queimados na Chapada Diamantina, Bahia: plantas de rebrota e sementes, com espécies endêmicas na rocha. Acta Bot Brasil 24:697–707

Noble IR, Slatyer RO (1980) The use of vital attributes to predict successional changes in plant communities subject to recurrent disturbances. Vegetatio 43:5–21

Ojeda F, Brun FG, Vergara JJ (2005) Fire, rain and the selection of seeder and resprouter life-histories in fire-recruiting, woody plants. New Phytol 168:155–165

Oliveira VB (2004) Inventário das espécies de mamíferos de médio e grande porte presentes no Parque Nacional da Serra do Cipó, Minas Gerais. Dissertation, Pontifícia Universidade Católica de Minas Gerais, Brasil

Oliveras I, Meirelles ST, Hirakuri VL, Freitas CR, Miranda HS, Pivello VR (2012) Effects of fire regimes on herbaceous biomass and nutrient dynamics in the Brazilian savannah. Int J Wildland Fire 22:368–380

Oriani A, Scatena VL, Sano PT (2008) Morphological architecture of *Actinocephalus* (Koern.) Sano (Eriocaulaceae-Poales). Flora 203:341–349

Parr CL, Woinarski JCZ, Pienaar DJ (2009) Cornerstones of biodiversity conservation? Comparing the management effectiveness of Kruger and Kakadu National Parks, two key savanna reserves. Biodivers Conserv 18:3643–3662

Parra LR, Giulietti AM, Andrade MJG, van den Berg C (2010) Reestablishment and new circumscription of *Comanthera* (Eriocaulaceae). Taxon 59:1135–1146

Pausas JG, Keeley JE (2009) A burning story: the role of fire in the history of life. Bioscience 59:593–601

Penman TD, Christie FJ, Andersen AN, Bradstock RA, Cary GJ, Henderson MK, Price O, Tran C, Wardle GM, Williams RJ, York A (2011) Prescribed burning: how can it work to conserve the things we value? Int J Wildl Fire 20:721–733

Phillips JFV (1930) Fire: its influence on biotic communities and physical factors in South and East Africa. S Afr J Sci 27:352–367

Pignatti E, Pignatti S, Lucchese F (1993) Plant communities of Stirling Range, Western Australia. J Veg Sci 4:477–488

Prous A (1992) Arqueologia Brasileira. Editora da Universidade de Brasília, Brasília

Ramos-Neto MB, Pivello VR (2000) Lightning fires in a Brazilian savanna national park: rethinking management strategies. Environ Manage 26:675–684

Ribeiro MC, Figueira JEC (2011) Uma abordagem histórica do fogo no Parque Nacional da Serra do Cipó, Minas Gerais, Brasil. Biodiv Bras 2:212–227

Rizzini CT (1992) Tratado de Fitogeografia do Brasil: Aspectos Ecológicos e Florísticos. Âmbito Cultural Edições, São Paulo

Rosière CA, Chemale F Jr (2000) Brazilian iron formations and their geological setting. Rev Bras Geoc 30:274–278

Rossi RD, Martins CR, Viana PL, Rodrigues EL, Figueira JEC (2014) Impact of invasion by molasses grass (*Melinis minutiflora* P. Beauv.) on native species and on fires in areas of campo-cerrado in Brazil. Acta Bot Brasil 28:631–637

Santos PR, Silvério ACM, Pereira G, Rocha LC, Cardozo FS (2014) Analysis of fires incidence in South America for the period between 2000 and 2012. Rev Elet Geog Territ Terram 2:101–112

Schmidt IB, Figueiredo IB, Scariot A (2007) Ethnobotany and effects of harvesting on the population ecology of *Syngonanthus nitens* (Bong.) Ruhland (Eriocaulaceae), a NTFP from Jalapão Region. Central Brazil. Econ Bot 61:73–85

Schmidt IB, Sampaio MB, Figueiredo IB, Ticktin T (2011) Fogo e artesanato de capim-dourado no Jalapão - usos tradicionais e consequências ecológicas. Biodiv Bras 2:67–85

Silva JA, Machado RB, Azevedo AR, Drumond GM, Fonseca RL, Goulart MF, Moraes EA Jr, Martins CS, Ramos Neto MB (2008) Identificação de áreas insubstituíveis para conservação da Cadeia do Espinhaço, estados de Minas Gerais e Bahia, Brasil. Megadiversidade 4:272–309

Silveira L, Jácomo ATA, Diniz-Filho JAF, Rodrigues FH (1999) Impact of wildfires on the megafauna of Emas National Park, central Brazil. Oryx 33:108–114

Silveira FAO, Negreiros D, Barbosa NPU, Buisson E, Carmo FF, Carstensen DW, Conceição AA, Cornelissen TG, Echternacht L, Fernandes GW, Garcia QS, Guerra TJ, Jacobi CM, Lemos-Filho JP, Le Stradic S, Morellato LPC, Neves FS, Oliveira RS, Schaefer CE, Viana PL, Lambers H (2016) Ecology and evolution of plant diversity in the endangered campo rupestre: a neglected conservation priority. Plant Soil 1–24

Simon MF, Grether R, Queiroz LP, Skema C, Pennington RT, Hughes CE (2009) Recent assembly of the Cerrado, a neotropical plant diversity hotspot, by in situ evolution of adaptations of fire. Proc Natl Acad Sci USA 48:20359–20364

Souza JM (2011) Efeitos do fogo na fenologia reprodutiva de angiospermas em vegetações campestres na Chapada Diamantina, Brasil. Dissertation, Universidade Estadual de Feira de Santana, Brasil

Trauernicht C, Brook BW, Murphy BP, Williamson GJ, Bowman DM (2015) Local and global pyrogeographic evidence that indigenous fire management creates pyrodiversity. Ecol Evol 5:1908–1918

Valeriano MM, Rossetti DF (2012) Topodata: Brazilian full coverage refinement of SRTM data. Appl Geogr 32:300–309

Van Wilgen BW (2013) Fire management in species rich Cape Fynbos shrublands. Front Ecol Environ 11:35–44

Van Wilgen BW, Biggs HC, Potgieter ALF (1998) Fire management and research in the Kruger National Park, with suggestions on the detection of thresholds of potential concern. Koedoe 41:1–33

Van Wilgen BW, Govender N, Forsyth GG, Kraaij T (2011) Towards adaptive fire management for biodiversity conservation: experience in South African national parks. Koedoe 53:1–9

Vasconcelos MF (2011) O que são campos rupestres e campos de altitude nos topos de montanha no leste do Brasil? Rev Bras Bot 34:241–246

Veldman JW, Buisson E, Durigan G, Fernandes GW, Le Stradic S, Mahy G, Negreiros D, Overbeck GE, Veldman RG, Zaloumis NP, Putz FE, Bond WJ (2015) Toward an old-growth concept for grasslands, savannas, and woodlands. Front Ecol Environ 13:154–162

Viana PL, Lombardi J (2007) Florística e caracterização dos campos rupestres sobre canga na Serra da Calçada, Minas Gerais, Brasil. Rodriguesia 58:159–177

Vilela RA, Melo RJ, Costa TAV, Lagoeiro LE, Varajão CAC (2004) Petrografia do minério hematita compacta da Mina do Tamanduá (Quadrilátero Ferrífero, MG). Rev Esc Minas 57:157–164

Vitta FA (2002) Diversidade e conservação da flora dos campos rupestres da Cadeia do Espinhaço em Minas Gerais. In: Araújo EL, Moura AN, Sampaio EVSB, Gestinári LMS, Carneiro JMT (eds) Biodiversidade, Conservação e Uso Sustentável da Flora do Brasil. Imprensa Universitária, Recife, UFRPE, pp 90–94

Whelan RJ (1995) The ecology of fire. Cambridge University Press, Cambridge

Whelan RJ (2004) Adaptive management: what does it mean and how ca it be used in fire management? In: Halse S (ed) Bushfire: managing the risk. New South Wales Nature Conservation Council, Sydney, pp 1–19

Chapter 19
Challenges in the Restoration of Quartzitic and Ironstone Rupestrian Grasslands

G. Wilson Fernandes, Tiago Shizen Pacheco Toma, Patrícia Angrisano and Gerhard Overbeck

Abstract Actively restoring ecosystems that have "been degraded, damaged or destroyed" became imperative in face of worldwide human impacts on nature. For rupestrian grasslands which are so peculiar and restricted in range, but also subjected to strong impact pressures, this seems to be even more important. Making use of ecological knowledge is fundamental to cope with the many uncertainties inherent to the process of ecological restoration. The overview on the ecology of rupestrian grasslands provided by this book thus is of utmost importance for the progress on this ecosystem's restoration and conservation. We benefit from this by invoking other chapters to base our assumptions and then present extant and possible ways of applying the ecological knowledge gathered. We summarize the academic background on restoration related to rupestrian grasslands, including examples of scientific restoration experiments, plant species with potential for restoration, among other aspects. We then point out potential restoration techniques and potential indicators of functional recovery during the restoration process. The problems imposed by invasive species on the process of restoration are highlighted due to its striking importance for restoration success and sustainability over time. Finally, we outline current gaps and challenges and indicate future directions to the ecological restoration of these ecosystems. This chapter represents the first attempt to review the efforts towards the ecological restoration of rupestrian grasslands at both scientific and technical perspectives.

G.W. Fernandes (✉) · P. Angrisano
Ecologia Evolutiva & Biodiversidade/DBG, ICB/Universidade Federal de Minas Gerais, CP 486, Belo Horizonte, MG 31270-901, Brazil
e-mail: gw.fernandes@gmail.com

G.W. Fernandes
Department of Biology, Stanford University, Stanford, CA 94305, USA

T.S.P. Toma · G. Overbeck
Laboratory of Grassland Vegetation, Department of Botany, Universidade Federal do Rio Grande do Sul, Porto Alegre 91501-970, RS, Brazil

© Springer International Publishing Switzerland 2016

G.W. Fernandes (ed.), *Ecology and Conservation of Mountaintop Grasslands in Brazil*, DOI 10.1007/978-3-319-29808-5_19

Contents

19.1 Introduction

Human impacts worldwide have led to the notion that ecosystem restoration is necessary and that it is only possible by intentionally assisting the recovery of ecosystems that have "been degraded, damaged or destroyed" (SER 2004). According to a survey made by Laestadius et al. (2011) more than two billion hectares of degraded land in the world are in need of restoration (but see Veldman et al. 2015a). In spite of the many idiosyncrasies of some ecosystems, along with general scientific and practical challenges, ecological restoration has progressed immensely in the last few decades. However, the restoration of harsh ecosystems, such as rupestrian grasslands, is not an easy task due to their many unique characteristics.

The restoration of any ecosystem inevitably relies on the knowledge of its ecology. This first substantial attempt on gathering knowledge on the ecology of rupestrian grasslands in the form of a book immensely benefits the development of restoration strategies for this ecosystem. Therefore, throughout the present chapter, we frequently reference other chapters in the book that provide the ecological basis for restoration. Here we focus on how this ecological knowledge can be applied to the restoration of rupestrian grasslands, and we urge the reader to consult the referenced chapters for further details on the ecological aspects of this ecosystem.

Some general considerations must be made in order to set the scene for developing strategies for restoration of rupestrian grasslands. Brazilian rupestrian grasslands are primarily restricted to mountaintops, which are often spatially separated (see details and maps in Chaps. 2 and 23). They are characterized by ironstone, quartzite or sandstone rock outcrops with rocky and sandy soils (see Chaps. 2 and 23) and with limited water availability, which can be considered harsh or even extreme conditions for the organisms that live in there (Chap. 4). This setting probably led to the highly diverse and endemic vegetation (see Hopper 2009; Chap. 6), which is often dependent on mutualistic interactions with microorganisms for their existence (Chaps. 8 and 13). The need to cope with the harsh environment has led to a variety of adaptive strategies (see Chaps. 11 and 12). In addition, rupestrian grasslands are subjected to the occurrence of occasional fires (Chap. 18).

A considerable portion of the Rupestrian Grassland ecosystem is under direct and indirect influence by mineral exploitation. Open pit mines, for example, completely alter the terrain to reach minerals. Mining has a profound impact on rupestrian grasslands, altering irreversibly the natural conditions established in ancient times (Veldman et al. 2015a, b; Chap. 23). A recently published article by Fernandes et al. (2014) highlights the compounding threats and the lack of specific conservation goals for the conservation of the Rupestrian Grassland (see also Chap. 23). One thing is clear, that the impacts of mining, roads, and real estate development have left thousands of hectares of the Rupestrian Grassland ecosystem in need of restoration. Once the rupestrian grassland vegetation-environment link collapses, due to the typical dependence of plants on specific site conditions, plant reproductive output is limited and new plant establishment decreases, thereby making natural regeneration difficult (Negreiros et al. 2011; Le Stradic et al. 2014a). To our knowledge, no one has yet provided empirical information on the natural regeneration of degraded rupestrian grasslands. Therefore, active restoration measures are necessary in the recovery of rupestrian grasslands.

This chapter represents the first attempt to review the efforts specifically directed towards the ecological restoration of rupestrian grasslands. We bring to light results from scientific studies on rupestrian grassland restoration, plant species with potential for restoration, and potential indicators of functional recovery during the restoration process. We also highlight the problems posed by invasive species to the process of restoration and illustrate these points with results from field experiments on the restoration of rupestrian grasslands. Finally, we summarize current gaps in knowledge and other emerging challenges and outline future directions for ecological restoration of this ecosystem.

19.2 Why Restore Rupestrian Grasslands?

The restoration of the unique and rare habitats associated with the Rupestrian Grassland ecosystems is a legal obligation, and is supported by Brazilian federal laws (Decree 97.632/1989 and decree 4.339/2002 which focus on the principles of the national policy for biodiversity; law 6.938/1981 that specifies the restoration of areas exploited by mining operations), as well as many state and municipal laws. Hence, restoration is a legal obligation. There are also accessory regulations, such as instruction ABNT 13030/1999, that call for the restoration of mined areas with native species. This norm, however, is already outdated and in need of urgent revision, especially with regard to more precise techniques and the exclusive use of local native species.

A second reason for restoration is that disturbed areas represent serious sources of impact on the Rupestrian Grassland ecosystem, with its low resilience, which results in the loss of biodiverstity and a decrease in the provision of ecosystem services (see Resende et al. 2013; Nunes et al. 2015; Chaps. 21 and 23). Furthermore, degraded areas may impact pristine environments far away from the original site of impact, including the silting of springs and other water bodies.

19.3 Sound Scientific Knowledge for the Development of Restoration Know-How

For long lasting and effective ecological restoration of the Rupestrian Grasslands, sound scientific knowledge is imperative. While restoration of many Brazilian ecosystems focuses on planting of trees—sometimes using non-native species, which critically needs to be discussed—the restoration of the harsh Rupestrian Grassland is much more challenging. Restoration is made difficult by the low survival rate of unassisted plants in the field due to drought and unpredicatable rainfall, which are common in the Rupestrian Grassland. Widely used restoration practices seem not applicable to the Rupestrian Grassland, and creativity in the development of appropriate restoration methods is urgently needed.

There is an urgent need to acquire basic knowledge on soil properties, habitat and microsite conditions, and on vegetation (preferentially from thorough studies on the flora and the definition of a reference ecosystem) to push forward the development of restoration techniques for the Rupestrian Grassland. Although there has been some increase in the study of the basic biology of potential plant species to be used in restoration, restoration technology must benefit from that knowledge and further develop it. Seed germination studies of rupestrian plant species increased over the last decade (see review in Chap. 10). These studies have helped in the development of sound ecological theory to subsidize the conservation of threatened species, but are also useful for the development of ecological restoration strategies (e.g., Gomes and Fernandes 2002; Chap.10). But not all species are suitable for the rigors of initiating restoration in the harsh and nutrient poor Rupestrian Grassland; hence, we ought to increase the number of studies on seed germination as well as stimulate more focused studies on the species best suited for initial restoration.

19.3.1 Reference Ecosystem

Perhaps the first step towards ecological restoration of the Rupestrian Grassland is to establish a reference ecosystem. In general, a reference ecosystem can be defined as "one or more existing, former, or hypothetical ecosystems that serve as a guiding image for ecosystem restoration or mitigation projects" (Miller et al. 2012). A reference ecosystem represents a target, reference, standard model, or mold to which the biological integrity, structure, function, condition or relative health of the ecosystem(s) under restoration can be compared (Jensen et al. 2000). Reference is basically a point of comparison, which can be conceptual, spatial (an environment), conditional (ecological states) or functional. Reference ecosystems are especially important to evaluate restoration success (SER 2004; Steyer et al. 2006; Miller et al. 2012).

Information that could be used for the description of a reference ecosystem in the Rupestrian Grassland includes: ecological descriptions, species lists and maps of the project site prior to degradation; historic and current photographs of the area at ground level; remnants of the site to be restored indicating previous physical conditions and biota; ecological descriptions and species lists from similar intact ecosystems; information from herbariums and museums; palaeoecological data such as fossil pollen, charcoal, history from tree growth rings, among others; and oral stories from people familiar with the project site prior to the damage (SER 2004).

While the theoretical definition of a reference ecosystem is more or less clear, its application in practice is a large problem for Rupestrian Grasslands, given their extremely rich flora and fauna and their high beta-diversity (Chaps. 1, 6 and 23). Local variation in habitats within small areas is very large and dependent on the lithotypes, as nicely demonstrated by the study of Dorr (1969) in the Iron Quadrangule. Carvalho et al. (2012), working on a single small area at a single elevation, recorded the existence of many distinct habitats in the quartzitic rupestrian grassland, each with different soil microorganisms and distinct flora, and consequently different resilience and functioning. The same richness of habitats is also found in the ironstone grasslands in northern Brazil. Although no one has yet attempted to delve into the subject and address this key question, this knowledge is central to both academics and practitioners. The ecological restoration of the Rupestrian Grassland must incorporate the unique local variations that provide the true identity of the ecosystem as a whole. Therefore, urgent work is called for to unravel the environmental signature of the Rupestrian Grassland so that ecological restoration can be acheived at its maximum.

19.3.2 Background Knowledge on Plant Species Propagation

In order propagate a native species for use in restoration basic information on its reproductive mode, germination, growth, performance, natural history and interactions with other organisms in the community is needed. A survey on availability of scientific studies on the propagation of ironstone rupestrian plant species was conducted for plant species reported to occur in the Iron Quadrangule (Table 19.1) and in Carajás (Table 19.2) (Scielo and Web of Science—1945 to 2012). The list of plant species surveyed originated from a compilation of species names reported from publicly available unpublished environmental assessment reports. Although we cannot ascertain with precision the correct taxonomic identification of the plant species of the reports by specialists, these are official documents and important material for consultation. On the other hand, some caution is necessary to interpret such reports. For instance, some native Brazilian species found in both lists are also repeatedly reported as noxious weeds in cultures, as noted in the references listed in the tables.

Table 19.1 presents a list of only 38 out of the 500 species that have been cited for the rupestrian grasslands of the Iron Quadrangule for which information on propagation was found. Plant species known to be non-native to Brazil were removed, as they are mostly invasive species (*Bidens pilosa, Solanum americanun*). Scientific information on propagation was found for only 7.6 % of 500 species surveyed. Most species for which some information was found are also recorded in other ecosystems while others are cosmopolitan (based on information at http://floradobrasil.jbrj.gov.br/jabot/listaBrasil). Out of the 38 species, 11 (29 %) are known to be endemic to the Cerrado, Atlantic Forest and Caatinga biomes, which are the biomes with which the Iron Quadrangule ironstone grassland should have the greatest affinity. However, they represent only 2.2 % of the total 500 plant species considered.

For the 541 species listed to occur in the ironstone rupestrian grasslands of Carajás, only 43 (7.9 %) had some information on their propagation. Twelve plant species were removed from this initial list: 6 species were not listed in the Brazilian flora (*Calathea ornata, Cyathea delgadii, Gleichenella pectinata, Heliconia birrai, Lycopodiella cernua, Nephrolepis bisserata*); five species are known as exotic invasive species (*Crotalaria juncea, Digitalia horizontalis, Digitalia insularis, Eleucine indica, Urocloa brizantha*). *Mimosa pudica* (Fabaceae) (http://floradobrasil.jbrj.gov.br/jabot/listaBrasil), a native species, was removed because it has been observed invading many areas in Carajás, suppressing the growth of native species (GWF, pers. obs.). Most species listed are also found in other ecosystems while others are cosmopolitan. Information is provided for only 31 species (5.7 %). Out of these, seven species (23 %) are known to be endemic to the Amazon and/or Cerrado biomes, which are the biomes with which the ironstone grassland of Carajás should have the greatest affinity. This represents only 1.3 % of the total 541 plant species considered.

Table 19.1 Native plant species found in unpublished lists of rupestrian grassland plants in the Iron Quadrangule, Brazil, for which some scientific data on propagation is available (Web of Science 1945–2012)

Family	Species	Occurrence	Propagation	Reference (e.g.)
Asteraceae	*Baccharis dracunculifolia*	Ce, Ma, Pm	Seed	Gomes and Fernandes (2002)
	Chresta sphaerocephala	Ce	Seed	Cury et al. (2010)
	Conyza bonariensis	All	Seed	Vidal et al. (2007)
	Eremanthus erythropappus	Ce, Ma	Seed	Rosal et al. (2007)
	Eremanthus incanus	Ca, Ce, Ma	Seed	Velten and Garcia (2005)
	Lychnophora pinaster	Ce	Seed	Melo et al. (2007)
Bignoniaceae	*Pyrostegia venusta*	All	Seed	Rossato and Kolb (2010)
Bromeliae	*Tillandsia gardneri*	Ca, Ce, Ma, Pm	Seed	Scatena et al. (2006)
	Tillandsia geminifolia	Ca, Ce, Ma, Pm	Seed	Stringheta et al. (2005)
	Tillandsia stricta	Ca, Ce, Ma, Pm	Seed	Scatena et al. (2006)
Cactaceae	*Epiphyllum phylanthus*	Az, Ca, Ce, Ma, Pl	Seed	Simão et al. (2010)
	Hylocereus setaceus	Az, Ca, Ce, Ma	Seed	Simão et al. (2010)
Cecropiaceae	*Cecropia glaziouvii*	Ce, Ma	Seed	Godoi and Takaki (2005)
	Cecropia pachystachya	Az, Ca, Ce, Ma Pl	Seed	Valio and Scarpa (2001)
Commelinaceae	*Commelina erecta*	Az, Ca, Ce, Ma Pl	Seed	Nisensohn et al. (2011)
Fabaceae	*Chamaecrista desvauxii*	Az, Ca, Ce, Ma, Pl	Seed	Caldeira et al. (2013)
Fabaceae	*Senna macranthera*	Ca, Ce, Ma	Seed	Cruz et al. (2010)
Fabaceae	*Copaifera langsdorffii*	Az, Ca, Ce, Ma	Seed	Augusto et al. (2003)
Flacourtiaceae	*Casearia sylvestris*	All	Seed	Rosa and Ferreira (2001)
Gesneriaceae	*Sinningia allagophilla*	Ce, Ma, Pm	Seed	Gomes (2006)
Lythraceae	*Cuphea carthagenensis*	All	Seed	Rosa and Ferreira (1998)
Malpighiaceae	*Byrsonima intermedia*	Az, Ce, Ma, Pl	In vitro	Nogueira et al. (2004)
	Byrsonima verbascifolia	Az, Ca, Ce, Ma	Seed	Alberto et al. (2011)
Melastomataceae	*Marcetia taxifolia*	Az, Ca, Ce, Ma	Seed	Silveira et al. (2004)
	Miconia ligustroides	Ca, Ce, Ma	Seed	Chaves et al. (2011)
Myrtaceae	*Blepharocalyx salicifolius*	Ca, Ce, Ma, Pm	Seed	Rego et al. (2009)
Onagraceae	*Ludwigia octovalvis*	Az, Ca, Ce, Ma, Pl	Seed	Wulff and Briceño (1975)
Orchidaceae	*Cattleya bicolor*	Ce, Ma	In vitro	Suzuki et al. (2010)
	Epidendron secundun	Az, Ca, Ce, Ma	Seed	Pereira et al. (2011)
Palmae	*Geonoma schottiana*	Ma	Seed	Aguiar (1990)
Passifloraceae	*Passiflora alata*	Az, Ce, Ma	Seed	Osipi et al. (2011)
Poaceae	*Andropogon bicornis*	All	Seed	Figueiredo et al. (2012)
	Andropogon leucostachyus	All	Seed	Figueiredo et al. (2012)
	Echinolaena inflexa	Az, Ca, Ce, Ma	Seed	Figueiredo et al. (2012)

(continued)

Table 19.1 (continued)

Family	Species	Occurrence	Propagation	Reference (e.g.)
Polygolaceae	*Polygala paniculata*	Az Ca, Ce, Ma, Pm	In vitro	Nogueira et al. (2005)
Smilacaceae	*Smilax campestris*	Ca, Ce, Ma, Pm	Seed	Soares et al. (2011)
Verbenaceae	*Lippia gracilis*	Ca, Ce	Seed	Marinho et al. (2011)
Vochysiaceae	*Vochysia tucanorum*	Ce, Ma	Seed	Pereira et al. (2011)

Occurrence in Brazilian biomes, mode of propagation and the source reference are indicated for each species
Az Amazônia, *Ca* Caatinga, *Ce* Cerrado, *Ma* Mata Atlântica, *Pm* Pampa, *Pl* Pantanal

Table 19.2 Native plant species found in unpublished lists of botanical surveys in the ironstone rupestrian grassland in Carajás, Brazil, for which some scientific data on their propagation was found

Family	Species	Occurrence	Propagation	Reference
Anacardiaceae	*Myracrodruon urundeuva*	Ca, Ce, Ma	Seed	Guedes et al. (2009)
	Tapirira guianensis	All	Seed	Santos-Moura et al. (2012)
Arecaceae	*Euterpe oleracea*	Az, Ce	Seed	Gama et al. (2010)
	Mauritia flexuosa	Az, Ca, Ce	Seed	Spera et al. (2001)
	Socratea exorrhiza	Az	Seed	Potvin et al. (2003)
Bignoniaceae	*Tabebuia impetiginosa*	Az, Ca, Ce, Ma, Pl	Seed	Oliveira et al. (2005)
Bromeliaceae	*Ananas ananassoides*	Az, Ca, Ce, Ma	Seed	Silveira et al. (2010)
	Ananas comosus	Ma	In vitro	Barbosa et al. (2009)
	Tillandsia streptocarpa	Az, Ca, Ce, Ma, Pm	Seed	Scatena et al. (2006)
Cactaceae	*Epiphyllum phylanthus*	Az, Ca, Ce, Ma, Pl	Seed	Simão et al. (2010)
Fabaceae	*Chamaecrista desvauxii*	Az, Ca, Ce, Ma, Pl	Seed	Caldeira et al. (2013)
	Enterolobium schomburgkii	Az, Ce	Seed	Braga et al. (2009)
Leguminosae	*Hymenaea courbaril*	Az, Ca, Ce, Ma, Pl	Seed	Pierezan et al. (2012)
	Sclerolobium paniculatum	Az, Ca, Ce	Seed	Felfili et al. (1999)
Melastomataceae	*Miconia albicans*	Az, Ca, Ce, Ma	Seed	Carreira and Zaidan (2007)
	Miconia Chamissois	Ca, Ce, Ma	Seed	Valio and Scarpa (2001)
Meliaceae	*Cedrela odorata*	Az, Ca, Ce, Ma	Seed	Passos et al. (2008)
Mimosoideae	*Parkia platycephala*	Az, Ca, Ce	Seed	Nascimento et al. (2009)
Myrristicaceae	*Virola surinamensis*	Az, Ca	Seed	Limas et al. (2007)
Orchidaceae	*Encyclia randii*	Az	Seed	Gonçalves et al. (2012)

(continued)

Table 19.2 (continued)

Family	Species	Occurrence	Propagation	Reference
Piperaceae	*Piper aduncum*	All	Seed	Bergo et al. (2010)
	Piper arboreum	Az, Ca, Ce, Ma	Vegetative	Souza et al. (2009)
Rubiaceae	*Alibertia edulis*	Az, Ce	Seed	Silva et al. (2008)
Rutaceae	*Pilocarpus microphyllus*	Az, Ca	Seed	Sabá et al. (2002)
Simaroubaceae	*Simarouba amara*	Az, Ca, Ce, Ma	Seed	Azevedo et al. (2010)
Smilacaceae	*Smilax brasiliensis*	Ce	Seed	Martins et al. (2012)
	Smilax campestris	Ca, Ce, Ma, Pm	Seed	Martins et al. (2012)
Verbenaceae	*Lantana camara*	All	Seed	Affonso et al. (2007)
	Stachytarpheta cayennensis	Az, Ca, Ce, Ma, Pl	Seed	Dias-Filho (1996)
	Swietenia macrophylla	Az, Ce, Ma	Seed	Souza et al. (2010)

Occurrence in Brazilian biomes, mode of propagation and the source reference are indicated for each species
Az Amazônia, *Ca* Caatinga, *Ce* Cerrado, *Ma* Mata Atlântica, *Pm* Pampa, *Pl* Pantanal

This survey shows that the scientific knowledge on the propagation of plant species of ironstone rupestrian grasslands that could help in the development of sustainable restoration strategies is negligible. Even worse is the fact that most studies were performed under laboratory conditions, hence being of limited practical value or still far from field application, where large-scale production of saplings is needed. Studies are biased towards the germination of seeds, with very few studies on other modes of propagation such as in vitro propagation. If it is acknowledged that plant introduction to restored areas is one of the strategies with high potential to restore degraded rupestrian ecosystems, then the lack of information on propagation techniques represents a large void that needs to be closed and strong planning and efforts should be made towards that end. Some development has been underway for the propagation of plant species that occur in the quartzitic rupestrian grasslands (see bellow). While this is of great importance, much more is yet to be done to reach a level of industrial production of plants for restoration of the Rupestrian Grassland (e.g., Wagner et al. 2011).

19.3.3 Nutritional Requirements and Site Preparation

Site preparation represents an important initial step in most restoration activities. Soil amendments are often needed in order to recover the physicochemical conditions of the soil that were lost in the process of degradation. On the other hand, rupestrian grasslands are naturally characterized by nutrient poor soils (e.g., Rodarte et al. 1998; Ribeiro and Fernandes 2000; Medina and Fernandes 2007; Negreiros et al. 2009, 2011; Chap. 3). Fertilization, a practice common in the restoration of forest sites, may rather represent a risk to rupestrian grasslands, as shown by Barbosa et al. (2010) and by Hilário et al. (2011). Enrichment of soil by nutrients through fertilization can instead end up promoting biological invasion by ruderal and exotic species (Negreiros et al. 2011; Chap. 23). Plants native to these ecosystems evolved the ability to survive under severe stresses caused by the lack of nutrients (Negreiros et al. 2011, 2014; Oliveira et al. 2015; Chap. 11). Caution is then needed when preparing the degraded sites for restoration and knowledge is again mandatory.

Although adapted to survive under the extreme environments of rupestrian grasslands, some species are able to exploit additional nutrients, such as the endemic species *Baccharis concinna* and the more widespread *Baccharis dracunculifolia* (Fernandes et al. 2007; Negreiros et al. 2014). Curiously, *B. dracunculifolia* exhibits a growth-survival tradeoff depending on nutrient availability (i.e., having higher growth rates in fertile soils and higher survival rates in nutrient poor soils; Negreiros et al. 2014). The restricted distribution of many plant species may be related to soil nutrient deficiency to which they have adapted, and are now perhaps prisoners of, such as the endemic shrubs *Calliandra fasciculata*, *Chamaecrista ramosa*, *Collaea cipoensis*, and *Coccoloba cereifera,* among others. The studies by Ribeiro and Fernandes (2000), Negreiros et al. (2008), and by Barbosa et al. (2015) provide support for this hypothesis. Studies on the nutritional requirements of plant species used for restoration purposes, or even the overall nutritional quality of the substrate, are of major relevance (see Negreiros et al. 2008, 2009, 2011; Le Stradic et al. 2014a, b; Machado et al. 2013; Messias et al. 2013; Oliveira et al. 2015).

The practical importance of these academic studies to restoration is twofold. First, they show that native species present superior competitive abilities compared to exotic species in the nutrient poor soils of rupestrian grasslands (e.g., Barbosa et al. 2010). Second, they demonstrate that soil fertilization can result in a negative outcome since exotic plants can be favored and outcompete natives, as shown by Hilário et al. (2011) and by Fernandes et al. (2015). Limiting conditions such as the presence of iron can favor native species, but negative outcomes can occur when invasive plants tolerate iron, as reported for *Calotropis procera* grown experimentally in rupestrian grasslands on ironstone (Oliveira et al. 2009). Future efforts to this end could be directed to studies on the acceleration of the growth of plant root systems to capture nutrients in a more effective way, developing more drought resistant species, increased plant associations with microorganisms to promote better field performance, among others.

19.3.4 Scientific Pilot Restoration Experiments

Only five scientific experiments are reported so far on the restoration of rupestrian grasslands: two in quartzitic grasslands degraded by quarrying and three in ironstone grasslands degraded by mining activities (Table 19.3). Although it is disappointing to find so few studies on restoration in the rupestrian grasslands, they are recent and as such indicate that we are moving forward. In this chapter we did not

Table 19.3 Scientific experiments on restoration of rupestrian grasslands published in peer-reviewed journals

Question addressed	Plant/seedling translocation	Substrate testing	Fertilization	Experimental design and treatments	Duration (months)	Main positive results	Main negative results	Type of degradation/outcrop	Source
Translocation viability of the orchid *Oncidium warmingii*	Yes	No	No	Direct and indirect translocation to a pristine area	20	Relatively high tolerance to translocation	Indirect translocation was less efficient	Mining/ironstone	Arruda et al. (2010)
Seedling growth and survival to translocation	Yes	No	NPK 4:14:8[a]	Planting seedlings in two degraded areas	54	High survival rates for 50 % of the species	Intra-specific competition lead to high mortality	Quarrying/quartzitic	Le Stradic et al. (2014a)
Planting of *Eremanthus erythropappus*	Yes	Yes	Likely[a]	Planting in three substrates (topsoil, laterite mowed and exposed)	12	Higher annual growth rates and biomass values for topsoil, followed by mowed laterite; High survival rates for all substrates	Not processing the exposed laterite resulted in lower annual growth and biomass	Mining/ironstone	Machado et al. (2013)
Reintroduction of plants	Yes	Yes	Termalphosphate, $MgSO_4$ and NPK 20:0:8[b]	Planting in two topsoil thicknesses and four fertilizer levels	42	Thicker topsoil layer lead to higher vegetation cover	High mortality rates after 42 months	Mining/ironstone	Rezende et al. (2013)
Resilience through seed bank	No	Yes	No	Hay transfer to degraded areas with 3 types of substrate	14	Indication that seed germination is a potential limitation	No effect observed for hay transfer	Quarrying/quartzitic	Le Stradic et al. (2014b)

[a]greenhouse, [b]four levels (0.00, 0.33, 0.66 and 1.00, see reference for further details on kg/ha)

attempt to locate and analyze private company reports as these are not readily available and because, unfortunately, most of them generally lack scientific rigor (Toy and Griffith 2001). On the other hand, we strongly stress that this knowledge should be rescued and brought to light to aid in the search for better restoration practices and scientific development.

Overall, the available studies covered only the initial stages of restoration, since their maximum duration was of 54 months. All of the studies made use of native plant species. Four out of the five studies tested plant species translocation, either directly from pristine areas or from propagation at greenhouse facilities; while two of them did not include testing with different substrates (Table 19.3).

The translocation of a single orchid species (*Oncidium warmingii*) showed to be quite successful. Arruda et al. (2010) found that all *O. warmingii* orchids translocated directly to the restoration area from the rescue area survived during the 20 month study period, while a smaller number of orchids translocated indirectly from the rescue area survived (70 %). Moreover, plant relative growth was higher for those individuals directly taken to the restoration area.

Another experiment consisted of planting seedlings of a set of woody species (18 species) in degraded areas of quartzitic rupestrian grassland in Serra do Cipó (Le Stradic et al. 2014b). Half of the species exhibited high survival rates (>78 %) 4.5 years after planting, and some were able to reproduce and/or allow the development of an herbaceous understory. The authors suggested the species *Calliandra fasciculata*, *Collaea cipoensis*, *Jacaranda caroba*, *Dasyphyllum reticulatum* and *Diplusodon hirsutus* as potential candidates for restoration of rupestrian grasslands since they presented positive responses to all aspects evaluated. Interestingly, intra-specific competition increased mortality rates in some cases and should be avoided by spreading conspecifics apart in the field (Le Stradic et al. 2014a; Fig. 19.1).

Two experiments consisted of plant species translocation in combination with substrate testing (Table 19.3). Machado et al. (2013) tested planting *Eremanthus erythropappus* seedlings in topsoil and in exposed lower soil layers (laterite) with or without processing to finer grains. The use of topsoil provided the best results for seedling establishment, followed by processed laterite, while unprocessed laterite presented the worst results. The authors argued that this common species of rupestrian grasslands has the potential for planting when there is no topsoil available. Another study by Rezende et al. (2013) tested the planting of seedlings of 15 native species in topsoil with two levels of thickness, in combination with four fertilization treatments. They found that treatments with a thicker topsoil layer presented higher vegetation cover independently of fertilization levels. However, mortality rates after 42 months were higher than 50 % for eight out of the 15 planted species; one species reached 100 % mortality. The authors suggested that practitioners should plant more seedlings per area in order to cope with the high long-term mortality levels.

Le Stradic et al. (2014b) experimentally tested the viability of hay transfer as a technique to restore degraded rupestrian grasslands. Three factors were analyzed: substrate (stony, sandy, and latosol), amendment to the substrate (with or without

Fig. 19.1 Pilot restoration study on quartzitic rupestrian grasslands in Serra do Cipó. **a** Preparation of the site began with the digging of plant pits (20 × 20 cm). **b** Saplings of native species produced in a nearby greenhouse were introduced in 2002. **c** Site three months after planting. **d** Site after 1 year after planting. **e** Site 2 years after planting. **f** Site 2 years after planting showing the differential response of the various plant species (for details see Le Stradic et al. 2014a)

geotextile), and addition of hay (with or without). Regardless of any factor or combination of factors, only a limited number of seedlings emerged from the seeds contained in the hay; most established species were ruderals probably from other sources (Le Stradic et al. 2014b). The authors suggest that seed germination is a limiting factor for the use of hay transfer as a restoration technique in this ecosystem.

At least one study has addressed the survival of seedlings planted in a restored rupestrian grassland ecosystem (Gomes et al. 2015). The authors experimentally investigated the survival, growth, and reproductive phenology of eight native rupestrian grassland species introduced to restore degraded quartzitic areas. Saplings were monitored for 12–18 months and all species presented high survival (90–100 %) and significant growth in height, indicating their successful establishment in the degraded areas (Fig. 19.2). The endemic and threatened species *Collaea cipoensis* (Fabaceae) had the highest growth, reaching approximately 150 cm in 12 months, six times its initial height; followed by *Mimosa foliolosa* (Fabaceae) and *Baccharis dracunculifolia* (Asteraceae), both with heights of approximately 75 cm in 12 months, around 3.75 times their initial heights. The woody species *Eremanthus incanus* (Asteraceae) and *Dalbergia miscolobium* (Fabaceae) reached ca. three times their initial heights in 12 months (ca. 60 cm).

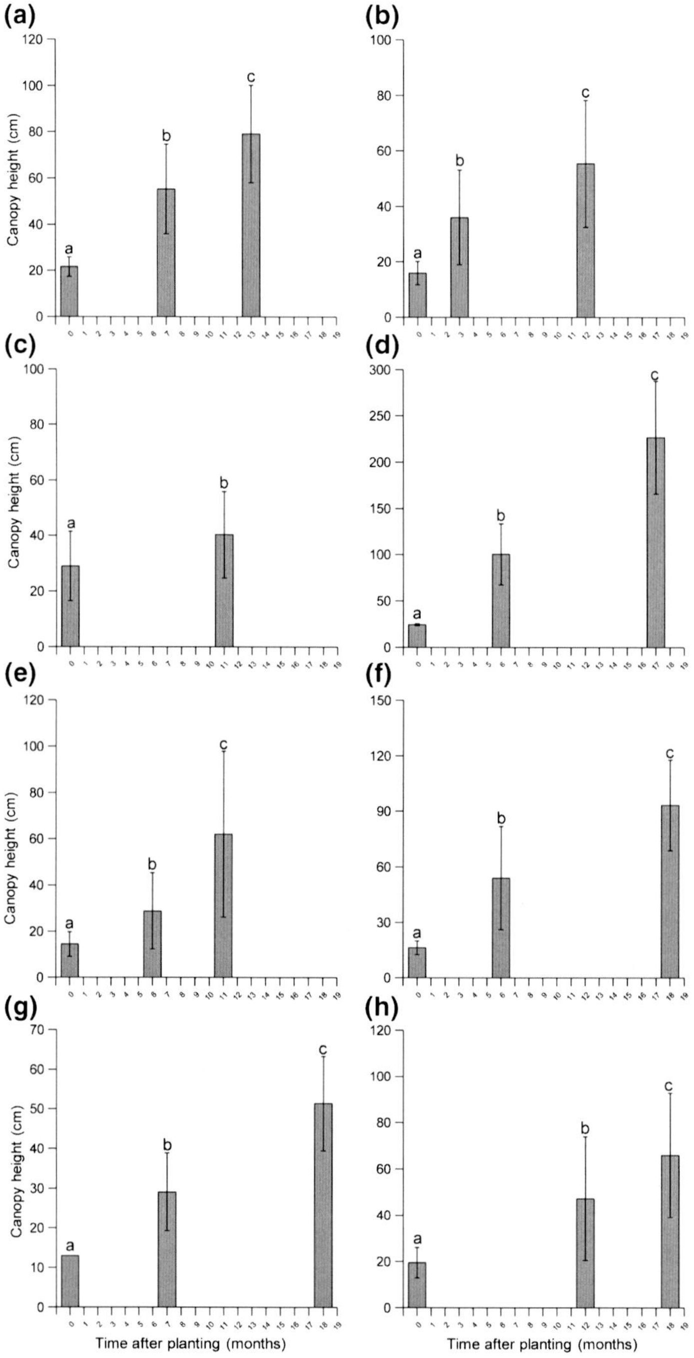
(a)
(b)
(c)
(d)
(e)
(f)
(g)
(h)
Canopy height (cm)
Time after planting (months)
a
b
c

◀ **Fig. 19.2** Sapling growth (height) of eight species native to rupestrian grasslands planted in degraded quartzitic areas in Serra do Cipó, MG, Brazil. For each graph, different letters indicate statistical difference between sampling dates ($p < 0.05$). *Vertical lines* show the standard deviation (modified from Gomes et al. 2015). **a** *B. dracunculifolia*. **b** *E. incanus*. **c** *H. ochraceus*. **d** *C. cipoensis*. **e** *D. miscolobium*. **f** *M. foliolosa*. **g** *L. campos-portoana*. **h** *T. heteromalla*

The species *Lavoisiera campos-portoana* (Melastomataceae) grew ca. 2.6 times its initial height in 12 months, reaching ca. 40 cm. In the same timespan, *Tibouchina heteromalla* (Melastomataceae) approximately doubled its initial size, reaching ca. 40 cm. The woody *Handroanthus ochraceus* (Bignoniaceae) grew 10 cm in about 11 months, approximately 1.3 times its initial size. These field data suggest the potential for successful establishment of native species and their ability to adaptat to the harsh conditions found in the degraded areas. Four species completed their reproductive cycle, reaching seed dispersion at an age of approximately two years. These results clearly indicate that planting saplings of native species represents an important tool for the restoration of degraded quartzitic areas in the Rupestrian Grassland.

19.4 Potential Techniques for the Ecological Restoration of Rupestrian Grasslands

A myriad of ecosystem restoration interventions has been suggested and tested for various ecosystems. Regardless of the intervention or set of interventions, implementing restoration is supposed to follow general guidelines as suggested, for example, by the Society for Ecological Restoration in their document "Guidelines for developing and managing ecological restoration projects" (SER 2005). The SER stresses the importance of planning for restoration, from concepts to implementation, going through preliminary and post-implementation tasks. Throughout the process of restoration, interventions should be carefully planned in order to maximize the chances of achieving a successful outcome. For the Rupestrian Grassland, as summarized in Table 19.3, only a few techniques have been scientifically tested. Here we present some potential restoration techniques for the Rupestrian Grassland.

19.4.1 Soil Preparation: Soil Fertility and the Use of Topsoil

Soil preparation is often the first step of a restoration project, consisting mostly of tractable physical and nutritional amendments. However, the peculiarities of Rupestrian Grasslands pose difficulties to this stage of restoration. The shallow and rocky soils characteristic of this ecosystem, with the marked presence of outcrops, constitute a complex mosaic of different soils with distinct successional stages of vegetation resulting from a myriad of ecological and geomorphological filters

(Chaps. 3 and 9). Hence, if superficial layers of soil (topsoil) are lost, it can take several years or even decades until any vegetation cover can be achieved.

Topsoil salvage is a restoration technique widely used worldwide and especially after degradation by mining (Bradshaw and Chadwick 1980) since it has the potential to allow the return of native plant species with lower technical and economic costs (Fowler et al. 2015, see Table 19.3 for two examples in rupestrian grasslands). Species present in the topsoil seed bank are of high importance, especially in such an endemic and diverse ecosystem as the Rupestrian Grassland (Toy and Griffith 2001; Medina and Fernandes 2007). The total knowledge on seed banks of this ecosystem includes only a single study that directly evaluated the seed bank (Medina and Fernandes 2007), and few others that make generealized references to the seed bank (e.g. Matias et al. 2009; Silveira et al. 2013, 2014). However, the often-needed storage of topsoil has not yet been evaluated regarding the viability of seeds (and microbiota) until use. Therefore, the recommendation is to directly transpose topsoil extracted from an active mining site to piles of sterile material or deactivated mining sites in the vicinity. In situations where no topsoil is available for restoring an area, a possibility could be, for instance, to process the available soil to finer grains, as tested for laterite (Machado et al. 2013; Table 19.3).

19.4.2 Seeding and Seedling Planting Techniques

Seeding and the planting of seedlings/saplings are widely used techniques in restoration around the world. On the other hand, the only experimental test of hay by Le Stradic et al. (2014b) was unsuccessful due to limited seed germination (see also Toy and Griffith 2001). More experiments must be performed in order to fully investigate ways of overcoming these limitations since seeding is a technique of great potential for application on a large scale.

Correctly choosing the plant community that will initiate the succession process in a degraded area is one of the most critical decisions in the recovering process (Corrêa 2007; Negreiros et al. 2009). After the appropriate establishment of species used in restoration plantings, success relies on the ability of the planted vegetation to self-regenerate, which requires studies on seedling development, natural regeneration, physiognomy, diversity, and seed rain, among others (Mandetta 2006).

19.4.3 Species Translocation

Translocation of plants from source areas or greenhouses to areas under restoration seems to be another viable option for rupestrian Grasslands. High survival rates of translocated individuals were found in rupestrian grasslands (Table 19.3). One experiment reported high mortality rates after 42 months regardless of substrate thickness or fertilization levels (Rezende et al. 2013). Another experiment on the rescue of almost 4000 individual plants of 43 species reported high survival rates for species belonging to Bromeliaceae, Velloziaceae and Orchidaceae; only six species presented survival bellow 5 % and two species showed 100 % mortality after four months (Mendonça et al. 2008). Le Stradic et al. (2014a) showed that the spatial distribution of plants in a restored area is an important consideration in order to avoid intraspecific competition; this finding is an indication on the importance of the reference ecosystem in the delineation of the restoration process.

Another example of the successes of translocation of species involves restoration attempts on former prospection sites. In planning mining activities, many areas are prospected, which involves the use of machinery that perforates the soils to precisely map the quality and quantity of minerals (Bradshaw and Chadwick 1980). However, not all probed areas are used due to the impurity of the minerals. While the extent and characteristics of these prospection areas has not been mapped, these areas usually are degraded and present risks to biodiversity, as they are generally suitable for the colonization by invasive species, while natural regeneration likely will not occur for decades or even centuries. In a pioneer study performed in an iron mine area of the company Vale in Carajás, dozens of prospection areas (Fig. 19.3a) were placed in the route of successional processes. In just six months after the planting of rescued plant individuals of a few selected species, the prospection areas showed high plant cover (30–70 %), while the control areas (no planting) had no colonization (Fig. 19.3b–d). This ongoing study clearly indicates the potential that

Fig. 19.3 Restoration of prospection areas in an ironstone rupestrian grassland in Carajás by translocation of native species. **a** Prospection areas after abandonement are likely to be uncolonized for decades. **b** and **c** Restored area 30 days after plant translocation. **d** Restored area six months after translocation

the application of scientific knowledge and the political will of mining companies can have on mitigating impacts resulting from prospection and promoting the return of biodiversity and environmental services. Nevertheless, a lot has yet to be done in this regard since prospection areas are often abandoned without any effort at reintegrating them into Rupestrian Grasslands. We still have no idea of the area and impacts of prospection, both direct and as potential sites for the establishment and spread of invasive species.

19.5 Potential Indicators of Restoration Progress

Monitoring ecological restoration is necessary to determine whether restoration interventions were effective. Monitoring constitutes an important part of any restoration initiative and should be considered early in the restoration planning phase (SER 2004, 2005). Restored areas are usually compared to references by using a set of indicators, but monitoring usually is performed for only a limited time (Ruiz-Jaen and Aide 2005; Wortley et al. 2013). However, evaluating restoration success is not an easy task and depends on a number of different aspects. For instance, the SER (2004) listed nine attributes of restored ecosystems in an attempt to provide a general guideline on what should be measured to determine that a restoration project was successful.

To date, very little is known regarding long term monitoring of rupestrian grassland restoration, making it difficult to present a synthesis. Yet some degraded rupestrian grassland areas seem to take a long time to recover even after restoration interventions (Le Stradic et al. 2014b, but see Gomes et al. 2015), and high mortality rates can strongly affect restoration in the midterm (Rezende et al. 2013; but see Gomes et al. 2015). A minimum of ten years after the restoration has begun seems to be a good criterion for monitoring until more research is developed on the subject. This is time period is based on the development of a pilot study done in the rupestrian grasslands (Fernandes and Negreiros, unpub. data). Additionally, we propose monitoring restoration progress rather than success (see Zedler 2007). For the rupestrian grasslands, measuring progress seems applicable and more reasonable since restoration success per se can take longer than the monitoring periods required by law.

We present here potential indicators for a set of aspects, from soil conditions to ecosystem functions, in an attempt to provide a starting point for the evaluation of the progress of rupestrian grasslands restoration. In each case, it is fundamental to perform a careful assessment of which indicators are to be use for determining restoration progress based on the objectives established in the original planning of the restoration. We focused on ecological indicators, but strongly recommend that other indicators, such as social and economic, also be considered (e.g. Wortley et al. 2013).

19.5.1 Soil Conditions

Adequate soil conditions at the restored area are crucial for plant establishment and development. As previously emphasized, rupestrian grassland soils are usually of low fertility and shallow (e.g., Benites et al. 2007). Therefore, even lower fertility soils are found in restored areas. The evolution of soil quality in restored areas must be compared with that of the reference area(s). Soil fertility analysis can then be considered a good indicator. Additionally, monitoring erosion is recommended as an indicator of soil stability in the restored sites. Erosive processes can lead to high rates of soil loss, which is highly detrimental to the restoration initiative of areas that inherently present shallow soils.

Although decomposition rates have been considered to be a good indicator of nutrient dynamics in restored sites (Meyer et al. 2015), studies on nutrient dynamics are rare in rupestrian grasslands. At this moment, it is probably difficult to determine the basic standards for evaluating nutrient dynamics, mostly due to the heterogeneity of rupestrian grasslands. The sole study found on this subject reported that litterfall dynamics varied greatly in time and among physiognomies within the same complex of ironstone rupestrian grassland (Valim et al. 2013). Yet another possible bioindicator of restoration in rupestrian grasslands is soil mycorrhizae. These organisms are highly diverse in rupestrian grasslands and are related to vegetation biodiversity and functioning (Carvalho et al. 2012; Chap. 8).

19.5.2 Plant Development and Performance

Some plant species native to rupestrian grasslands exhibit a growth-survival trade-off due to harsh conditions of water availability and soil fertility that can significantly limit plant growth (Negreiros et al. 2014). Thus, during the early stages of sucession such evaluations must be viewed with caution and can only be compared in relation to the reference ecosystem. Otherwise, plant growth in quartzitic rupestrian grasslands can in fact take place in a short time span (e.g. 12 months, Gomes et al. 2015). Additionally, some invasive species can present considerable growth even in degraded areas with important implications for the process of restoration (see following section). Therefore, monitoring vegetation through plant growth, survival and reproductive stage can provide good basic information on the progress of a restoration initiative.

19.5.3 Succession and Plant Cover

Ecological succession on rupestrian grasslands is limited by a number of factors. For instance, frequent fires in this ecosystem may represent a factor that strongly interferes with the succession process, causing it to regress once in a while and in very unpredictable ways. Furthermore, there is now some evidence that a long time is necessary for vegetation to exhibit clear changes between successional stages (Chap. 9), although the definition of stages for this ecosystem is also a matter of detailed studies. The slow rate of change in vegetation of rupestrian grasslands may be the result of the harsh environment. It is likely that some disturbances, such as fire, may be necessary from time to time to maintain the evolutionary pace of rupestrian grasslands and to achieve a condition of a mosaic of phytophysiognomies. At any rate, the classical successional model of a pioneer community to a climate stage, as commonly applied to forests, does not apply easily to rupestrian grassland, or to grasslands in general, for that matter. Nontheless, the development of vegetation cover—as long as it is composed of native species—is another important practical aspect since it protects soils, provides habitat for other organisms, is involved in many interaction networks between trophic levels, and is an indication of habitat productivity. On the other hand, we are not aware of detailed studies done regarding vegetation cover in rupestrian grasslands (see Le Stradic et al. 2014a, b).

19.5.4 Associated Organisms: Biodiversity

The return of biodiversity and ecosystem functioning to restored ecosystems is highly desirable since it is directly associated with sustainability through time. However, we are unaware of any study looking into this in the Rupestrian Grassland (see Jacobi et al. 2015 for a review of ecological interactions in ironstone rupestrian grasslands). Information on ants, in conjunction with information on the recovery of vegetation, was regarded as a reliable indicator in a restoration project in temperate grasslands by Fagan et al. (2010). In calcareous grasslands, Maccherini et al. (2009) revealed the potential use of different taxa of butterflies and vegetation in restoration evaluation. Galls induced by insects and their host plants have been shown to be good indicators of habitat quality in restored vegetation (Moreira et al. 2007; Fernandes et al. 2010; Toma et al. 2014), but all studies done so far are on forest ecosystems. These highly specific interactions might be even better indicators of habitat quality and health in rupestrian grasslands since galling herbivores are present in these ecosystems in great numbers and on a variety of host plants (Lara and Fernandes 1996; Lara et al. 2002). Pollinators, including bees, butterflies and hummingbirds also represent potential indicators of improvement in habitat quality, as these organisms tend to construct nests in the vicinity of such areas. At least in one example, pollinator behavior and frequency were considered in a restored rupestrian grassland area (Gelvez-Zúñiga et al. 2016).

19.5.5 Ecosystem Functions and Services

Parameters that provide information on ecosystem functions and processes must also be included in the evaluation of vegetation cover and functionality in restored areas of rupestrian grasslands. These include seed dispersal and pollination rates, plant recruitment, and the establishment of trophic structures similar to reference areas. A good starting point regarding seed dispersion and pollination could be the work of Jacobi and Carmo (2011) that lists plant species dispersion and pollination syndromes. Accounting for intra- and interspecific interactions and their associated ecosystem functions are of utmost importance, especially in such a diverse ecosystem. Evaluating ecosystem functions in the early stages of the restoration process of rupestrian grasslands can be a means of assessing the recovery of processes and services provided by the restored area. The recovery of some functions can help guide the management of the restoration process until vegetation is fully established. For the assessment of ecosystem functions, some methods that can be used to indicate restoration success are summarized in Meyer et al. (2015).

Research on ecosystem services has grown exponentially in the last ten years (see Guerry et al. 2015). Ecosystem service valuation was, for the first time, conducted for an area of quartzitic rupestrian grassland by focusing on the service provided by plant diversity storage by Resende et al. (2013). Such studies can help justify conservation and restoration initiatives based on services provided by a preserved or a restored area that are directly related to economic activities in rupestrian grassland areas. The restoration objectives must not just focus on soil protection, but also focus on the recovery of specific services related to biodiversity and ecosystem services (e.g., Bullock et al. 2011). With the increased focus on ecosystem services lately, this is a very promising venue for the advancement of ecological restoration in general.

19.6 The Threat of Invasive Plants to Rupestrian Grassland Restoration

The need to monitor the progress of restoration is not just related to the recovery of native vegetation, but also to the recognition of potential problems. One of the major problems in restored areas is that of invasive species. In addition to the fact that they represent a great threat to the identity of rupestrian grasslands (see Barbosa et al. 2010; Hilário et al. 2011; Fernandes and Barbosa 2013; Fernandes et al. 2014; Chap. 23), biological invasions are of major concern to the success of ecological restoration, and they may present a risk for adjacent well-conserved areas as well.

Fernandes et al. (2015) listed the non-native species invasions in restored quartzitic rupestrian grassland areas in Serra do Cipó and called attention to the aggressive behavior of some species. Among the highly competitive species that can come to dominate plant communities are the exotic African grasses *Urochloa brizantha* and *Melinis minutiflora*. These grasses have been spreading over huge areas and, consequently, have replaced native species in many areas of the Cerrado, as clearly shown by Pivello et al. (1999). Other non-native species of the rupestrian grasslands include *Cajanus cajan*, *Mimosa pigra*, *Crotalaria pallida*, *Crotalaria spectabilis*, *Achyrocline satureioides,* and *Ageratum fastigiatum,* among others. Furthermore, the native species *Stylosanthes guianensis* is also among those that might represent a threat to restoration of rupestrian grasslands. Therefore, wide and detailed evaluation of invasive species and their impact on restoration processes are urgently needed.

19.7 Summary of Current Gaps in Knowledge and Challenges for the Restoration of Rupestrian Grasslands

The acknowledged idiosyncrasies inherent to the Rupestrian Grassland call for a combination of techniques to restore degraded sites and, most importantly, long term monitoring. Success of ecological restoration of rupestriam grasslands must evolve under a sound scientific basis, as we cannot risk to follow wrong strategies in the field because of the serious impact they might cause. Negative results could promote biological invasions and the silting of springs and water basins. While we acknowledge an increase in scientific developments in the restoration efforts of some mining companies (some results have been presented above), the initiatives are very timid given the magnitude and importance of the area to be properly restored. The scenario is challenging because it involves governance, law enforcement, management, knowledge development, economic investments, know-how, well-trained human resources, and long term monitoring. Rupestrian grasslands must be restored as close as possible to reference ecosystems, as society will end up paying the costs of producing unsuitable habitats of low or zero ecosystem value. We must be able to figure out what does not work in the rupestrian grassland restoration and start working on alternatives. We must redirect restoration efforts towards being sustainable and ecologically oriented.

The accumulation of knowledge on the ecology of rupestrian grasslands is a good starting point from which practitioners can gain an appreciation of the complexity of this ecosystem. A true guide through the paths of restoring such complex ecosystems is still unavailable, mainly due to historical reasons. Neither were policies enacted to push conservation forward, nor were stakeholders made responsible for the ecological restoration of degraded areas. With very few exceptions, academia has for too long stayed away from this discussion, which has prevented advances. Rather, studies on conservation and restoration were developed in other ecosystems, such as the rainforests, traditionally in the focus of Brazilian conservation policy and research. The result was a profound gap in the knowledge on how to restore the harsh Rupestrian Grassland, or other open-type ecosystems in Brazil, for that matter (e.g. Overbeck et al. 2013). Therefore, a great effort must be made to equalize such knowledge and effectively develop some basal knowledge regarding the restoration of this old-growth grassland.

To achieve such know-how, we need to make use of the valuable knowledge of practitioners regarding their experience with restoration of rupestrian grasslands. We also need to overcome the unavailability of propagules in the market (i.e. seeds and seedlings) in order to apply large-scale restoration. However, this requires the will and action of stakeholders, and the constant inspection and pressure from governmental agencies. Only concerted efforts from all institutions involved can make true advances in restoration possible. Furthermore, more detailed and broadened research is needed on the ecology of seeds, and in the determination of key species to be used and directly managed in order to increase restoration success. Specific knowledge and technologies must also be developed for plant propagation, such as in vitro propagation, germination rates of the selected species, plant performance in the field, long-term survival of plants at restored sites, protection from invasions or severe disturbances, etc. Based on specific site conditions we ought to establish a variety of plant types including trees, shrubs, grasses and wildflowers that work together as a plant community, that hold the soil in place by slowing water runoff and facilitating its absorption, and that represent habitat for many other species and bring biodiversity back—and with that ecosystems services. Knowledge must be developed on species with deep, fibrous roots that solidly anchor each plant and help it withstand drought. Also, we must select tough, low-maintenance plants that need low staking, fertilizing and disease-control. In other words, resilient native plants need to be selected. Species that can naturally spread by underground suckering can quickly form thickets that protect a restored site and provide microsites for wildlife colonization. Similarly, herbs, shrubs and trees that regenerate easily by self-sowing, such as early and fruiting species, are also important for filling and stabilizing degraded space. Clearly, an important aspect is high survival.

More research on invasive species control is also necessary. Worldwide efforts are being made to advance this topic, but our knowledge for rupestrian grasslands is anectodal at best. For ecosystems that are complex and limited in range, such as the Rupestrian Grassland, the threat of invasive species is very substantial.

Restoration of the Rupestrian Grassland should also consider ecosystem functioning and the provision of services. It is essential to go beyond planting and monitoring vegetation structure in order to achieve some level of ecosystem sustainability and integration. Here we have presented some ways of achieving this, and we believe there is more yet to come. After filling these gaps in our knowledge, the expectation is that the exploitation of natural resources will be as sustainable as possible, allowing some of the Rupestrian Grassland to persist over time.

Acknowledgements We thank FAPEMIG and CNPq for the continuous support of this research and Reserva Vellozia for logistical support. We also thank Vale for invaluable assistance towards the development of ideas and support of some of the research. Finally, we deeply acknowledge the reading and criticisms provided by Daniel Negreiros.

References

Affonso VR, Bizzo HR, Lima SS, Esquibela MA, Sato A (2007) Solid microextraction (SPME) analysis of volatile compounds produced by in vitro shoots of *Lantana camara* L. under the influence of auxins and cytokinis. J Braz Chem Soc 18:1504–1508

Aguiar FFA (1990) Efeito de diferentes substratos e condições ambientais na germinação de sementes de *Euterpe edulis* Mart, e *Genoma schottiana* Mart. Acta Bot Bras 4:1–7

Alberto OS, Silva FG, Cabral JSR, Sales JF, Pereira F (2011) Métodos para superação da dormência de sementes de murici (*Byrsonima verbascifolia* Rich). Ciênc Agrar 32:1015–1020

Arruda LJ, Cheib AL, Ranieri BD, Negreiros D, Fernandes GW (2010) Resgate e translocação de *Oncidium warmingii* (Orchidaceae), espécie ameaçada de extinção de campo rupestre ferruginoso. Neotrop Biol Cons 5:10–15

Augusto DCC, Guerrini IA, Engel VL, Rousseau GX (2003) Utilização de esgotos domésticos tratados através de um sistema biológico na produção de mudas de *Croton floribundus* Spreng. (Capixingui) e *Copaifera langsdorffii* Desf. (Copaíba). Rev Arv 27:335–342

Azevedo IMG, Alencar RM, Barbosa AP, Almeida NO (2010) Estudo do crescimento e qualidade de mudas de marupá (*Simarouba amara* Aubl.) em viveiro. Acta Amazon 40:157–164

Barbosa SBSC, Teixeira JB, Portes TA, Copati LA, Lédo AS (2009) Cultivo inicial in vitro de gemas axilares de *Ananas comosus* (L.) Merr., em meio líquido/sólido, na presença/ausência de luz. Cienc Agrotec Lavras 33:1832–1836

Barbosa NPU, Fernandes GW, Carneiro MAA, Junior LAC (2010) Distribution of non-native invasive species and soil properties in proximity to paved roads and unpaved roads in a quartzitic mountainous grassland of southeastern Brazil (rupestrian fields). Biol Inv 12:3745–3755

Barbosa NPU, Fernandes GW, Sanchez-Azofeifa A (2015) A relict species restricted to a quartzitic mountain in tropical America: an example of microrefugium? Acta Bot Bras 29:299–309

Benites VM, Schaefer CEG, Simas FN, Santos HG (2007) Soils associated with rock outcrops in the Brazilian mountain ranges Mantiqueira and Espinhaço. Braz J Bot 30:569–577

Bergo CL, Ohlson OC, Biasi LA, Bianco MP (2010) Luz e temperatura na germinação de sementes de pimenta longa (*Piper hispidinervrum*) e pimenta-de-macaco (*Piper aduncum*) NC. Rev Bras Sem 32:170–176

Bradshaw AD, Chadwick MJ (1980) The restoration of land: the ecology and reclamation of derelict and degraded land. Univ of California Press, Los Angeles

Braga LF, Sousa MP, Almeida TA (2009) Germinação de sementes de *Enterolobium schomburgkii* (Benth.) Benth. submetidos a estresse salino e aplicação de poliamina. Rev Bras Pl Med 11:63–70

Bullock JM, Aronson J, Newton AC, Pywell RF, Rey-Benayas JM (2011) Restoration of ecosystem services and biodiversity: conflicts and opportunities. Tren Ecol Evol 26:541–549

Caldeira MVW, Delarmina WM, Juvanhol RS, Faria JCT (2013) Substratos alternativos na produção de mudas de *Chamaecrista desvauxii*. Rev Arv 37:31–39

Carreira C, Zaidan LBP (2007) Germinação de sementes de espécies de Melastomataceae de Cerrado sob condições controladas de luz e temperatura. Hoehnea 34:261–269

Carvalho F, Souza FA, Carrenho R, Moreira FMS, Jesus EC, Fernandes GW (2012) The hottest hospot for arbuscular micorrhyzal fungi: the habitat mosaic in the high altitude rupestrian fields. Appl Soil Ecol 52:9–19

Chaves IS, Alvarenga AA, Dousseau S, Soares GM, Arthur MAS (2011) Germination of *Miconia ligustroides* (Melastomataceae) diaspores submitted to different treatments for dormacy overcoming. Rev Bras Bot 34:335–341

Clewell A, Rieger J, Munro J (2005) Guidelines for developing and managing ecological restoration projects, 2nd edn. SER, Dec 2005. www.ser.org

Corrêa RS (2007) Recuperação de áreas degradadas pela mineração no cerrado. Manual para revegetação. Universa, Brasília

Cruz CAF, Paiva HN, Neves JCL, Cunha ACMCM (2010) Resposta de mudas de *Senna macranthera* (DC. Ex COLLAD) H.S.Irwin & Barnebi (fedegoso) cultivadas em latossolo vermelho-amarelo ditrófico e macronutrientes. Rev Arv 34:13–24

Cury G, Novembre ADDLC, Glória BAD (2010) Seed germination of *Chresta sphaerocephala* DC. and *Lessingianthus bardanoides* (Less.) H. Rob. (Asteraceae) from Cerrado. Braz Arch Biol Technol 53:1299–1308

Dias-Filho MB (1996) Germination and emergence of *Stachytarpheta cayennensis* and *Ipomoea asarifolia*. Pl Daninha 14:944–952

Dorr JVN (1969) Physiographic, stratigraphic, and structural development of the Quadrilátero Ferrífero Minas Gerais, Brazil. Geological survey professional paper 641A, United States Printing office, Washington, DC

Fagan KC, Pywell RF, Bullock JM, Marrs RH (2010) Are ants useful indicators of restoration success in temperate grasslands? Rest Ecol 18:373–379

Felfili JM, Hilgberg LF, Franco AC, Resende AV, Nogueira MVP (1999) Comportamento de plântulas de *Sclerobium paniculatum* Vog. var. *Rubiginosum* (Tul.) Benth. sob diferentes níveis de sombreamento em viveiro. Rev Bras Bot 22:297–301

Fernandes GW, Barbosa NPU (2013) Bombas relógio ameaçam a natureza em Minas Gerais. Sci Am Bras 135:60–61

Fernandes GW, Rodarte LHO, Negreiros D, Franco AC (2007) Aspectos nutricionais em *Baccharis concinna* (Asteraceae), espécie endêmica e ameaçada da Serra do Espinhaço, Brasil. Lundiana 8:83–88

Fernandes GW, Almada ED, Carneiro MAA (2010) Gall-inducing insect species richness as indicators of forest age and health. Environ Entomol 39:1134–1140

Fernandes GW, Barbosa NPU, Negreiros D, Paglia AP (2014) Challenges for the conservation of vanishing megadiverse rupestrian grasslands. Nat Cons 12:162–165

Fernandes GW, Santos RA, Barbosa NPU, Almeida HA, Carvalho VA, Angrisano P (2015) Ocorrência de plantas não-nativas e exóticas em áreas restauradas de campos rupestres. Pl Daninha 33:463–482

Figueiredo MA, Baeta HE, Kozovits AR (2012) Germination of native grasses with potential application in the recovery of degraded areas in Quadrilátero Brazil. Biota Neotrop 12:118–123

Fowler WM, Fontaine JB, Enright NJ, Veber WP (2015) Evaluating restoration potential of transferred topsoil. Appl Veg Sci 18:379–390

Gama JSN, Monte DMO, Alves EU, Bruno RLA, Braga Júnior JM (2010) Temperatura e substratos para germinação e vigor de sementes de *Euterpe oleraceae* Mart. Rer Cien Agron 41:664–670

Gelvez-Zúñiga I, Jaime AA, Martén-Rodríguez S, Gomes VM, Barbosa A, Bordignon L, Rocha R, Silva GG, Fernandes GW (2016) Necter robbing in *Collaea cipoensis* (Fabaceae): an endemic shrub of the Brazilian rupestrian grasslands. Rev Mex Biodiv (in press)

Godoi S, Takaki M (2005) Efeito da temperatura e a participação do fitocromo no controle da germinação de sementes de embaúba. Rev Bras Sem 27:87–90

Gomes MAN (2006) Cultura in vitro da espécie do cerrado *Sinningia allagophylla* (Martius) Wiehler (Gesneriaceae): enraizamento, tuberização e nutrição com ênfase em alumínio. Tese (Doutorado em Biologia Vegetal) Universidade Estadual de Campinas, Campinas SP

Gomes V, Fernandes GW (2002) Germinação de aquênios de *Baccharis dracunculifolia* D.C. (Asteraceae). Acta Bot Bras 16:421–427

Gomes V, Negreiros D, Fernandes GW (2015) Plasticidade no crescimento e alocação de biomassa em três espécies de campo rupestre. Neotrop Biol Conserv 33:159-168

Gonçalves LM, Prizão EC, Guitierre MAM, Magolin CA, Machado MFPS (2012) Use of complex supplements and ligth-differential effects for micropropagation of *Hadrolaelia purpurata* (=*Laelia purpurata*) and *Encyclia randii* orchids. Acta Sci 34:459–463

Guedes RS, Alves EU, Gonçalves EP, Colares PNQ, Medeiros MS, Silva KB (2009) Tratamento pré-germinativo em sementes de *Myracroduon urundeuva* Freire Allemão. Rev Arv 33:997–1003

Guerry AD, Polasky S, Lubchenco J, Chaplin-Kramer R, Daily GC, Griffin R, Ruckelshaus M, Bateman IJ, Duraiappah A, Elmqvist T, Feldman MW, Folke C, Hoekstra J, Kareiva PM,

Keeler BL, Li S, McKenzie E, Ouyang Z, Reyers B, Ricketts TH, Rockström J, Tallis H, Vira B (2015) Natural capital and ecosystem services informing decisions: from promise to practice. Proc Natl Acad Sci USA 112:7348–7355
Hilário RR, Castro SAB, Ker FTO, Fernandes GW (2011) Unexpected effects of pigeon-peas (*Cajanus cajan*) in the restoration of rupestrian fields. Planta Daninha 29:717–723
Hopper SD (2009) OCBIL theory: towards an integrated understanding of the evolution, ecology and conservation of biodiversity on old, climatically buffered, infertile landscapes. Plant Soil 322:49–86
Jacobi CM, Carmo FF (2011) Life-forms, pollination and seed dispersal syndromes in plant communities on ironstone outcrops, SE Brazil. Acta Bot Bras 25:395–412
Jacobi CM, Mourão FA, Antonini Y, Fagundes R, Kumagai AF (2015) Interações ecológicas em ambientes de canga do Quadrilátero Ferrífero, Minas Gerais. In FF Carmo, LHY Kamino (Orgs.). Geossistemas Ferruginosos do Brasil: Áreas Prioritárias para Conservação da Diversidade Geológica e Biológica, Patrimônio Cultural e Serviços Ambientais. Editora 3i, Belo Horizonte, pp 343–360
Jensen ME, Redmond RL, Dibenedetto JP, Bourgeron PS, Goodman IA (2000) Application of ecological classification and predictive vegetation modeling to broad-level assessments of ecosystem health. Environ Mod Asses 64:197–212
Laestadius L, Maginnis S, Minnemeyer S, Potapov P, Saint-Laurent C, Sizer N (2011) Mapping opportunities for forest landscape restoration. Unasylva 62:47–48
Lara ACF, Fernandes GW (1996) The highest diversity of galling insects: Serra do Cipó, Brazil. Biodiv Lett 3:111–114
Lara ACF, Fernandes GW, Gonçalves-Alvim SJ (2002) Tests of hypotheses on patterns of gall distribution along an altitudinal gradient. Trop Zool 15:219–232
Le Stradic S, Buisson E, Fernandes GW (2014a) Restoration of Neotropical grasslands degraded by quarrying using hay transfer. J Appl Veg Sci 17:482–492
Le Stradic S, Buisson E, Negreiros D, Campagne P, Fernandes GW (2014b) The role of native woody species in the restoration of Campos Rupestres in quarries. J Appl Veg Sci 17:109–120
Limas JD, Silva BMS, Moraes WS (2007) Germinação e armazenamento de sementes de *Virola surinamensis* (Rol.) Warb. (Myristicaceae). Rev Arv 31:37–42
Maccherini S, Bacaro G, Favilli L, Piazzini S, Santi E, Marignani M (2009) Congruence among vascular plants and butterflies in the evaluation of grassland restoration success. Acta Oecol 35:311–317
Machado NAM, Leite MGP, Figueiredo MA, Kozovtis AR (2013) Growing *Eremanthus erythropappus* in crushed laterite: a promising alternative to topsoil for bauxite-mine revegetation. J Environ Manage 129:149–156
Mandetta ECN (2006) Alternativas de RAD e importância da avaliação e monitoramento dos projetos de reflorestamento. In: Barbosa LM [Coord.]. Manual para recuperação de áreas degradadas do estado de São Paulo: matas ciliares no interior paulista. Instituto de Botânica, São Paulo, pp 106–118
Marinho MJM, Albuquerque CC, Morais MB, Souza MCG, Silva KMB (2011) Estabelecimento de protocolo para micropropagação de *Lippia gracilis* Schauer. Rev Bras Pl Med 13:246–252
Martins AR, Soares AN, Bombo AB, Fidelis A, Novembre ADLC, Glória BA (2012) Germination and seedling morphology of four American *Smilax* (Smilacaceae). Rev Biol Trop 60:495–504
Matias SR, Pegano MC, Muzzi FC, Oliveira CA, Carneiro AA, Horta SN, Scotti MR (2009) Effect of rhizobia, mycorrhizal fungi and phosphate-solubilizing microorganisms in the rhizosphere of native plants used to recover an iron ore area in Brazil. Eur J Soil Biol 45:259–266
Medina BMO, Fernandes GW (2007) The potential of natural regeneration of rocky outcrop vegetation on rupestrian field soils in "Serra do Cipó", Brazil. Rev Bras Bot 30:665–678
Melo PRBD, Oliveira JA, Pinto JEBP, Castro EMD, Vieira AR, Evangelista JRE (2007) Germinação de aquênios de arnica (*Lychnophora pinaster* Mart.) armazenados em diferentes condições. Ciênc Agrotec 31:75–82

Mendonça RC, Felfili JM, Walter BMT, Silva MCJR, Rezende TS, Filgueiras PE (2008) Nogueira FAGG (2008) Flora vascular do bioma cerrado: Checklist com 12.356 espécies. In: Sano SM, Almeida SP, Ribeiro JF (eds) Cerrado: ecologia e flora. Embrapa Cerrados, Brasília, pp 423–442

Messias MCTB, Leite MGP, Neto JAAM, Kozovits AR, Tavares R (2013) Soil-vegetation relationship in quartzitic and ferruginous Brazilian rocky outcrops. Folia Geobot 48:509–521

Meyer ST, Koch C, Weisser WW (2015) Towards a standardized Rapid Ecosystem Function Assessment (REFA). Trends Ecol Evol 30:390–397

Miller SJ, Pruitt BA, Theiling CH, Fischenich JC, Komlos SB (2012) Reference concepts in ecosystem restoration and environmental benefits analysis (EBA): principles and practices. ERDC TN-EMRRP-EBA-12

Moreira RG, Fernandes GW, Almada ED, Santos JC (2007) Galling insects as bioindicators of land restoration in an area of Brazilian Atlantic Forest. Lundiana 8:107–112

Nascimento IL, Alves EU, Bruno RLA, Gonçalves EP, Colares PNQ, Medeiros MS (2009) Superação da dormência em sementes de faveira (*Parkia platycephala*) Benth. Rev Arv 33:35–45

Negreiros D, Moraes MLB, Fernandes GW (2008) Caracterização da fertilidade dos solos de quatro leguminosas de campos rupestres, Serra do Cipó, MG, Brasil. Rev Cien Suelo Nutr Veg 8:30–39

Negreiros D, Fernandes GW, Silveira FAO, Chalub C (2009) Seedling growth and biomass allocation of endemic and threatened shrubs of rupestrian fields. Acta Oecol 35:301–310

Negreiros D, Fernandes GW, Berbara RLL, Rodarte LHO, Barbosa NPU (2011) Caracterização físico-química de solos quartzíticos degradados e áreas adjacentes de campo rupestre na Serra do Cipó, MG, Brasil. Neot Biol Cons 6:156–161

Negreiros D, Le Stradic S, Fernandes GW, Rennó HC (2014) CSR analysis of plant functional types in highly diverse tropical grasslands of harsh environments. Plant Ecol 215:379–388

Nisensohn LA, Tuesca DH, Vitta JI (2011) Características reproductivas de *Commelina erecta* L. asociadas com su propagación em sistemas agrícolas. Agriscien 28:51–60

Nogueira RC, Paiva R, Castro AH, Vieira CV, Abbade LC, Alvarenga AA (2004) Germinação In vitro de murici-pequeno (*Byrsonima intermedia* A. Juss.) Cienc Agrotez 28:1053–1059

Nogueira FLP, Fernandes SBO, Reis GM, Matheus ME, Fernandes PD, Lage CLS, Menezes FS (2005) Atividade analgésica e antiedematogênica de *Polygala paniculata* L. (Polygalaceae) selvagem e obtida por micropropagação. Rev Bras Farmacogn 15:310–315

Nunes FP, Fernandes GW, Negreiros D (2015) Campo rupestre. A restauração ecológica de um ecossistema ameaçado e megadiverso. Cien Hoje 327:24–27

Oliveira LM, Carvalho MLM, Borges DI (2005) Temperatura e regime de luz na germinação de sementes de *Tabebuia impetiginosa* (Martius ex A.P. De Candolle) Standley e *T. serratifolia* Vahl Nich. - Bignoniaceae. Ciênc Agrotec 29:642–648

Oliveira SH, Negreiros D, Fernandes GW, Barbosa NPU, Rocha R, Almeida-Cortez JS (2009) Seedling growth of the invader *Calotropis procera* in ironstone rupestrian field and seasonally dry forest soils. Neotrop Biol Cons 4:69–76

Oliveira RS, Galvão HC, Campos MCR, Eller CB, Pearse SJ, Lambers H (2015) Mineral nutrition of campos rupestres plant species on contrasting nutrient-impoverished soil types. New Phytol 205:1183–1194

Osipi EAF, Lima CB, Cossa CA (2011) Influência de métodos de remoção do arilo na qualidade fisiológica de sementes de *Passiflora alata* Curtis. Rev Bras Frutic E: 680–685

Overbeck GE, Hermann JM, Andrade BO, Boldrini II, Kiehl K, Kirmer A, Koch C, Kollmann J, Meyer ST, Müller SC, Nabinger C, Pilger GE, Trindade JPP, Vêlez-Martin E, Walker EA, Zimmermann D, Pillar VD (2013) Restoration ecology in Brazil time to step out of the forest. Nat Cons 11:92–95

Passos MAA, Silva FJBC, Silva ECA, Pessoa MML, Santos RC (2008) Luz, substrato e temperatura na germinação de sementes de cedro-vermelho. Pesq Agropec Bras 43:281–284

Pereira BF, Simão E, Filho JBT, Cardoso JC, Cardoso VJM (2011) Note on the germination of *Vochysia tucanorum* seeds treated with growth regulators. Rev Bra Bot 34:231–235

Pierezan L, Scalon SPQ, Pereira ZV (2012) Emergência de plântulas e crescimento de mudas de jatobá com uso de bioestimulantes e sombreamento. Cerne 18:127–133

Pivello VR, Shida CN, Meirelles ST (1999) Alien grasses in Brazilian savannas: a threat to the biodiversity. Biodivers Conserv 8:1281–1294

Potvin C, Cansari R, Hutton J (2003) Preparation for propagation: understanding germination of giwa (*Astrocaryum standleyanum*), wagara (*Sabal mauritiiformis*) and ba (*Socratea exorrhiza*) for future cultivation. Biodivers Conserv 12:2161–2171

Rego SS, Nogueira AC, Kuniyoshi YS, Santos AF (2009) Germinação de sementes de *Blepharocalyx sacilifolius* (H.B.K.) Berg. em diferentes substratos e condições de temperaturas, luz e umidade. Rev Bras Sem 31:212–220

Resende FM, Fernandes GW, Coelho MS (2013) Economic valuation of plant diversity storage service provided by Brazilian rupestrian grassland ecosystems. Braz J Biol 73:709–716

Rezende LAL, Dias LE, Assis IR, Braga R, Rezende ML (2013) Restoration of ironstones outcrops degraded by iron mining activity in Minas Gerais state-Brazil. J Am Soc Min Reclam 2:151–159

Ribeiro KT, Fernandes GW (2000) Patterns of abundance of a narrow endemic species in a tropical and infertile habitat. Plant Ecol 147:205–218

Rodarte LHO, Fernandes GW, Oliveira JD (1998) Características físico-químicas de solos naturais e em áreas degradadas por atividades de construção rodoviária em campo rupestre. Anais do 3° Encontro Íbero-Americano de Unidades Ambientais do Setor de Transportes, Florianópolis, Brazil

Rosa SGTD, Ferreira AG (1998) Germinação de sementes de espécies medicinais do Rio Grande do Sul: *Bromelia antiacantaha* Bert., *Cuphea carthagenensis* (Jacq.) Macbride e *Talinum patens* (Jacq.) Willdenow. Acta Bot Bras 12:515–522

Rosa SGTD, Ferreira AG (2001) Germinação de sementes de plantas medicinais lenhosas. Acta Bot Bras 15:147–154

Rosal LF, Pinto JEBP, Bertolucci SKV, Costa LCB, Corrêa RM (2007) Micropropagation of the medicinal plant *Eremanthus erythropappus* (DC.) MacLeish. HortScience 42:1420–1424

Rossato DV, Kolb RM (2010) Germinação de *Pyrostegia venusta* (Bignoniaceae), viabilidade de sementes e desenvolvimento pós-seminal. Rev Bras Bot 33:51–60

Ruiz-Jaen MC, Aide MT (2005) Restoration success: how is it being measured? Rest Ecol 13: 569–577

Sabá RT, Lameira OA, Luz JMQ, Gomes APR, Innecco R (2002) Micropropagação de jaborandi. Hortic Bras 20:106–109

Santos-Moura SS, Alves EU, Bruno RLA, Moura MF, Gondim PSS (2012) Influência de diferentes períodos de secagem na qualidade fisiológica de sementes de *Tapipira guianensis* Aublet. Revt Bras Frutic 34:382–389

Scatena VL, Segecin S, Coan AI (2006) Seed morphology and post-seminal development of *Tillandsia* L. (Bromeliaceae) from the "Campos Gerais", Paraná, southern Brazil. Braz Arch Bio Tech 49:945–951

SER (2004) The SER international primer on ecological restoration. Society for Ecological Restoration International Science and Policy Working Group, Version 2

Silva FAB, Pereira LAR, Silveira CES (2008) Micropropagation of *Alibertia edulis* Rich. Braz Arch Biol Tech 51:1103–1114

Silveira FAO, Negreiros D, Fernandes GW (2004) Influência da luz na temperatura na germinação de sementes de *Marcetia taxifolia* (A. St-Hil.) DC. (Melastomataceae). Acta Bot Bras 18:847–851

Silveira FAO, Santos JC, Fernandes GW (2010) Seed germination ecophysiology of the wild pineapple, *Ananas ananassoides* (Baker) L.B.Sm. (Bomeliaceae). Acta Bot Bras 24:1100–1103

Silveira FAO, Fernandes GW, Lemos JP (2013) Seed and seedling ecophysiology of Neotropical melastomataceae: implications for conservation and restoration of savannas and rainforests. Mo Bot Gard 99:82–99

Silveira FAO, Negreiros D, Ranieri BD, Silva CA, Araújo LM, Fernandes GW (2014) Effect of seed storage on germination, seedling growth and survival of *Mimosa foliolosa* (Fabaceae): implications for seed banks and restoration ecology Trop Ecol 55: 385–392

Simão E, Nakamura AT, Takaki M (2010) The germination of seeds of *Epiphyllum phylanthus* (L.) Haw. (Cactaceae) is controlled by phytochrome and by nonphytochrome related process. Biota Neotrop 10:115–119

Soares AN, Novembre ADDLC, Martins AR, Piedade SMDS, Appezzato-da-Glória B (2011) Propagation studies in *Smilax fluminensis* Steud, (Smilacaceae). Cien Rural 41:1762–1768

Souza LA, Moscheta IS, Mourão KSM, Albiero ALM, Iwasaki MC, Oliveira JHG, Rosa SM (2009) Vegetative propagation in Piperaceae species. Braz Arch Biol Tech 52:1357–1361

Souza CAS, Tucci CAF, Franco CA, Silva JF, Ortiz W (2010) Exigências nutricionais e crescimento de plantas de mogno (*Swietenia macrophylla* King). Acta Amaz 40:515–522

Spera MRN, Cunha R, Teixeira JB (2001) Quebra de dormência, viabilidade e conservação de sementes de buriti (*Mauritia flexuosa*) NC. Pesq Agropec Bras 36:1567–1572

Steyer GD, Twilley RR, Raynie RC (2006) An integrated monitoring approach using multiple reference sites to assess sustainable restoration in coastal Louisiana. In: Aguirre-Bravo C, Pellicane PJ, Burns DP, Draggan S (eds) Monitoring science and technology symposium: unifying knowledge for sustainability in the western hemisphere proceedings. U.S. Forest Service, Rocky Mountain Research Station—RMRS-P-42CD, Fort Collins, CO, pp 326–333

Stringheta ACO, Silva DJH, Cardoso AA, Luiz EFF, Barbosa JG (2005) Germinação de sementes e sobrevivência das plântulas de *Tillandsia geminifolia* Brongn, em diferentes substratos. Acta Sciet Agron 27:165–170

Suzuki RM, Almeida VD, Pescador R, Ferreira WDM (2010) Germinação e crecimento in vitro de *Cattleya bicolor* Lindley (Orchidaceae). Hoehnea 37:731–742

Toma TS, Fernandes GW, de Souza DG, Tabarelli M, Santos JC (2014) Galling insects as indicators of habitat quality. In: GW Fernandes, Santos JC (Orgs.). Neotropical insect galls. Springer, Dordretch, pp 143–150

Toy TJ, Griffith JJ (2001) Changing surface-mine reclamation practices in Minas Gerais, Brazil. Int J Surf Min Reclam Environ 15:33–51

Valim EAR, Nalini HA, Kozovits AR (2013) Litterfall dynamics in a iron-rich rock outcrop complex in the southeastern portion of the Iron Quadrangle of Brazil. Acta Bot Bras 27:286–293

Valio IFM, Scarpa FM (2001) Germination of seeds of tropical pioneer species under controlled and natural conditions. Rev Bras Bot 24:79–84

Veldman JW, Overbeck GE, Negreiros D, Mahy G, Le Stradic S, Fernandes GW, Durigan G, Buisson E, Putz FE, Bond WJ (2015a) Tyranny of trees in grassy biomes. Science 347:484–485

Veldman JW, Buisson E, Durigan G, Fernandes GW, Le Stradic S, Mahy G, Negreiros D, Overbeck GE, Veldman RG, Zaloumis NP, Putz FE, Bond WJ (2015b) Toward an old-growth concept for grasslands, savannas, and woodlands. Front Ecol Environ 13:154–162

Velten SB, Garcia QS (2005) Efeitos da luz e da temperatura na germinação de sementes de *Eremanthus* (Asteraceae), ocorrência na Serra do Cipó, MG, Brasil. Acta Bot Bras 19:753–761

Vidal RA, Kalsing A, Goulart ICGR, Lamego FP, Christoffoleti PJ (2007) Impacto da temperatura, irradiância e profundidade das sementes na emergência e germinação de *Conyza bonariensis* e *Conyza canadensis* resistentes ao glyphosate. Pl Daninha 25:309–315

Wagner M, Pywell RF, Knopp T, Bullock JM, Heard MS (2011) The germination niches of grassland species targeted for restoration: effects of seed pretreatments. Seed Sci Res 21:117–131

Wortley L, Hero JM, Howes M (2013) Evaluating ecological restoration success: a review of the literature. Rest Ecol 21:537–543

Wulff R, Briceño M (1975) Light and temperature interaction in the photocontrol of germination of seeds in *Ludwigia octovalvis*. Planta 128:195–199

Zedler JB (2007) Success: an unclear, subjective descriptor of restoration outcomes. Ecol Rest 25:162–168

Chapter 20
The People of the Mountains: The Biocultural Heritage of the Espinhaço Range in Minas Gerais State, Brazil

Emmanuel Duarte Almada, Felisa Cançado Anaya and Fernanda Testa Monteiro

Abstract The Espinhaço mountain range in Minas Gerais is characterized not only by its great biodiversity but also by its closely associated cultural diversity. Inhabited by groups of humans for over 12,000 years, the Espinhaço Range underwent significant change with European colonization, particularly due to the eighteenth century search for gold and diamonds, and Africans enslaved for this purpose. As a result of this process of occupation and interaction, several traditional rural communities were formed which inherited ecological knowledge and built highly complex ecosystem management systems. The interactions between these communities and rocky fields play a central role in the environmental history of the region. Being located at higher elevations these fields were generally used for grazing animals, plant extraction, and even agricultural crops and housing. The cultural and biological importance of the region due to existing socio-ecological processes have contributed to the formation of a diversity of people and forests which was made even more valuable with the creation of the Espinhaço Biosphere Reserve, in 2005. However, the current context of territorial disputes over this mountain range reflects the different meanings and social projects in the region. The advances of large capitalist enterprises based in the region contributes to a scenario of territorial and environmental conflicts where culturally differentiated communities try to defend their traditionally occupied lands against the advances of real estate speculation, overlapping full protection

---

E.D. Almada (✉)
Departamento de Ciências Biológicas, Laboratório de Estudos Bioculturais, Universidade do Estado de Minas Gerais (UEMG), Av. São Paulo, 3996, CEP 32400-000 Ibirité, Minas Gerais, Brazil
e-mail: almadaceae@gmail.com

F.C. Anaya
Departamento de Saúde Mental E Coletiva, Núcleo Interdisciplinar de Investigação Socioambiental, Universidade Estadual de Montes Claros, Montes Claros, Brazil

F.T. Monteiro
Faculdade de Filosofia, Letras E Ciências Humanas, Departamento de Geografia, Universidade de São Paulo, São Paulo, Brazil

© Springer International Publishing Switzerland 2016

G.W. Fernandes (ed.), *Ecology and Conservation of Mountaintop Grasslands in Brazil*, DOI 10.1007/978-3-319-29808-5_20

conservation units, eucalyptus monocultures and implementations of big industrial mining projects. In this context, the objective of this chapter is to describe the historical and political dimensions of the biocultural diversity of the Espinhaço mountain range, focusing especially on areas of rocky fields. The chapter critically reflects upon the naturalization of the political and economic processes that have come to threaten the cultural diversity and environment of this region due to ethnoecological and ecological policy.

Contents

20.1 Introduction

The Espinhaço mountain range starts in Minas Gerais, goes through Bahia and reaches the border of Piauí. It forms a great division between the watersheds of the Brazilian mid-east and the São Francisco River (COMIG/IGC/UFMG 1997). In Minas Gerais, the Espinhaço is a set of "highlands" forming a boomerang-shape, in a general north-south direction with the convex side to the west. Saadi (1995) considers the name "mountain range" to hide a physiographic reality that is best defined by the word "plateau". The two boomerang wings correspond to two plateau compartments—the southern and northern highlands—that differ in litho-structural and morphological aspects and are separated by a long depression in a SE-NW direction, passing Couto Magalhães Minas, north of Diamantina/MG. The southern portion extends from the area known as Cipó (north of the state capital) to the town of Olhos D'Água (north of Diamantina), and contains springs and tributaries that drain into different basins (River Doce, River Jequitinhonha and River São Francisco). These water sources are responsible for the water supply of important cities of the Southeast and Northeast.

It has been pointed out by Gontijo (2008) that "highlands" of the Espinhaço have a unique ecological function *sui generis*. It is also important to emphasize its

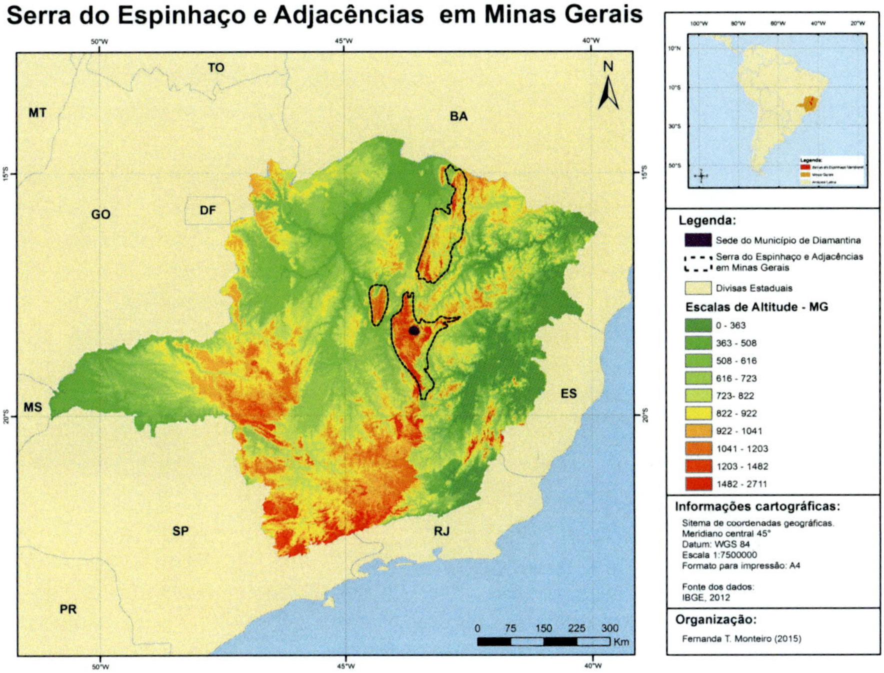

Fig. 20.1 Location of the Espinhaço range in the state of Minas Gerais

latitudinal position and distance from the Atlantic coast[1] (Fig. 20.1). These features attract studies by not only of geologists, but also by scientists focused on the ecology of the environments and biomes found in it.[2]

[1]"The vegetation that covers the region is represented by rocky fields and high-altitude grasslands, savannas and mainly semideciduous forests. The occurrence of a particular vegetation type is strongly influenced by climatic and morphological conditions. [...] There is a mosaic of vegetation and floristic landscapes along the mountain range and which is one of its greatest fascinations and is reflected in its western and eastern parts that define roughly altitudinal transitions, sometimes between rocky grasslands and savannas (San Francisco Basin), sometimes between rocky fields and Atlantic Forests (eastern basins) and sometimes between high-altitude grassland and savanna (lower latitudes). Interfaces with the Atlantic Forest, by the way, happen when the denser vegetation penetrates the eastern slope and persists along watercourses (riparian or gallery forests), and also in geological depressions and geomorphologically favorable mountain range tops (forest patches). The limestone outcrops, however, that occur in transition zones with the São Francisco Depression, on limestone rocks of the Bambuí Group, dominate the dry forest areas) (Gontijo 2008, pp. 10–11).

[2]This kind of mountain range caught the attention of the geologist Eschwege (1822) (Gontijo 2008), as well as several European naturalists, as mentioned in her studies of the nineteenth century like Auguste de Saint-Hilaire, Karl Friedrich von Martius and Ludwig Riedel (Costa 2005a, b).

The various scientists, travelers and naturalists who have visited the Espinhaço since the seventeenth century always extolled the beauty of the landscape and the uniqueness of the biodiversity found in the region.[3] It was in the Lagoa Santa region, near Serra do Cipó, that Eugene Warming conducted his studies that gave rise to the field of Plant Ecology.

However, the ecological processes of the Espinhaço, in addition to its biological dimension, needs to be described within the context of the social and cultural systems of which it is part. Since the early twentieth century, anthropological and more specifically ethno-ecological studies have shown a strong role of human actions in the configuration and determination of various structural and functional aspects of the ecosystems and terrestrial biomes (Baleé 1992; Descola 1996; Nazarea 2006; Nabhan 2009). Similarly, these mountain landscapes can be understood from different elevations, ranging from 600 m to approximately 2000 m above sea level, with their associated biomes and different social and cultural backgrounds (Acselrad 2004) that are conformed/produced by a process marked by territorial conflicts.

Without ignoring the unprecedented environmental crisis that the urban-industrial-capitalist system has led us into in this beginning of the millennium, human societies must be seen not only as consumers of natural resources, exerting an "anthropic pressure" in a naturalized and universal homogenizing way, but also as different groups of humans with a variety of social groups and, cultures with specific forms of meaning appropriate and related to the environment.

If on one hand the urban-industrial way of life that has become hegemonic, on the other hand, there is part of humanity represented by communities and people. Traditional groups[4] still perpetuate sociability models that are not based on the nature-society dichotomy (Viveiros de Castro 2002). These groups have contributed to the modeling and even increasing the biodiversity of terrestrial ecosystems from the genetic level to the landscape level (Maffi 2001; Toledo and Barrera-Bassols 2008). These connections between biological and cultural diversity have become in the last two decades a central theme for conservation policies and the search for ways to overcome the widespread environmental crisis (Gavin et al. 2015).

As clearly demonstrated by other chapters of this book, the Espinhaço is marked by an extraordinary diversity of environments, species and complex ecological processes. We understand, therefore, that all this biodiversity is the result of an

---

[3]The part of the Espinhaço range in Minas Gerais State is where most previous studies, especially geological, took place, emphasizing the importance of strengthening the understanding about its biota. In this sense, scientists and environmental NGOs, focused on biodiversity conservation research, with the support of government environmental agencies which gather information and perform studies on the Espinhaço Range and have already confirmed the existence of more than six thousand species in their biota.

[4]According to the Federal Decree 6.040/2007, Traditional Peoples and Communities are culturally differentiated groups who recognize themselves as such, which have their own forms of social organization, which occupy and use territories and natural resources as a condition for their cultural, social, religious, ancestral and economic way, using knowledge, innovations and practices generated and transmitted by tradition.

environmental history of over 10,000 years with a succession of cultures, which over time have shaped (and continue shaping) the "natural" landscapes of today, and of which environmental scientists dedicate themselves to understanding the ecological patterns and mechanisms.

From this perspective, and in the context of mega biocultural diversity, the rocky fields stand out as structural elements of local social ecological systems. Given their uniqueness, rocky fields, known as *Kapôt krã nhi mõk* by the Kayapó Indians (Posey 2002) are a central element of the cultural identity of local communities. Differing from what conservation scientists understand rocky fields to be (a good to be preserved) or from what capitalist agents believe (a good to be exploited), previous and current traditional societies of indigenous communities have always considered rocky fields or mountain tops as a dwelling place, a crossing, a sacred place, a place for harvest, cattle raising and agriculture, and a biocultural memory.

However, the disconnection between social, historical and political processes that represent the environment in this preservationist approach, contribute to a fragmented and decontextualized vision of the current problems related to the degradation and suppression of natural resources, the deterritorialization of the people, and contradictions among environmental policies that have been contributing to eucalyptus plantations and mining projects in the Espinhaço region in Minas Gerais. An emblematic case is the mining project in the region of Conceição do Mato Dentro/MG, and its advancement on other municipalities of the Espinhaço range, which has disregarded social groups and areas of rocky fields.

Thus, we briefly present in this chapter, the socio-cultural context operating on the biodiversity of the Espinhaço range in Minas Gerais, and how rocky fields are part of the cultural practices of communities and traditional peoples that reside there.

20.2 Environmental history of the Espinhaço and its People

The Espinhaço ranges one of the oldest geological structures of the planet, with formations of about 2.5 billion years (see Chap. 2). On the other hand, records of the first human occupation of the region date back to a period much earlier than the arrival of the first Europeans and enslaved Africans during the Brazilian colonial period. Throughout the entire length of the Espinhaço, archaeological remains are relatively abundant, with records of inscriptions, artifacts, organic debris and tools. Archaeological research in the southern portion of the Serra do Cipó indicates that human occupation may have started 20,000 to 12,000 years ago (Resende and Prous 1991; Prous 2000). These occupations occurred during the Pleistoscene, although there is no consensus about the origin of the migratory groups that arrived there.

During the first two centuries of the colonial period, the European occupation of Brazilian territory remained in regions near the coast. The penetration through *sertões*[5] seeking precious metals and diamonds begain in the seventeenth century and became more abundant in the eighteenth century with the promotion of *bandeiras* (flags). These great expeditions of exploration and occupation of the territories of the interior of Brazil, starting in the state of São Paulo, were based on conquest, enslavement and extermination of dozens of indigenous groups. With the beginning of colonization of the region in the late seventeenth century, the *Cerrado* biome,[6] and the Espinhaço Range were still largely occupied by indigenous groups of the linguistic stem Macro-Jê (Ribeiro 2005). With the arrival of the royal family in Brazil in 1808, war against these Indians began and led to almost complete decimation in a few decades.

With the discovery of gold and diamonds and the apogee of these activities between 1700 and 1850, the Espinhaço Range housed some of the main urban and economic centers of Latin America. The city of Ouro Preto and the municipality of Diamantina, in the southern portion of the Espinhaço, have long been the main economic centers of the state. The extraction of gold and diamonds, resulted in the development of rural areas for the provisioning of supplies to these and other cities. Between the seventeenth and nineteenth centuries, an estimated 500,000 Africans, ethnically represented by mainly Sudanese and Bantu, were brought to the regions of gold and diamond mining in Minas Gerais (Luna and Costa 2009).

Carney and Rosomoff (2009) point out that this African contingent brought with them a wide variety of agricultural species (e.g. rice, taro, millet, tamarind, balsam pear), and the complex knowledge and methods of traditional management of these people were adapted to the new environmental context. Soon these groups were finding cultural substitutes among the species cultivated for millennia by indigenous peoples, which were soon integrated into a new African-American diet, such as yams, cassava and maize. These authors conducted a rich and detailed rebuilding of the intense trade in knowledge and plants between the two continents and brought new light to understanding the role of the African diaspora in shaping the ecological knowledge of African descendants on American soil.

Given the harsh environments throughout the entire length of the Espinhaço, the presence of communities of African origin is remarkable. Throughout the period of gold and diamond exploitation, slaves that had escaped from the plantations and mines established diverse communities that would become the so-called *Quilombos*. The *Quilombos*, though often harassed and attacked by State forces,

[5]Ribeiro (2005) states that the origin of the term *sertã*, from etymological studies, is derived from *desertão* (large desert) and possibly express this notion in European expansionism throughout the planet, to establish itself as a center that radiates civilization for several "wildernesses" to be conquered. The term was already used in Portugal, certainly since the fourteenth century, to refer to areas within its territory and far from Lisbon.

[6]Space set of right order and territorial greatness under certain morphoclimatic and phytogeographical domain. These characteristics refer to the concept of *core area*—where the physiographic and biogeographic conditions form a relatively homogeneous and extensive complex (Ab'Sáber 2003).

often played an important role in the regional economy, with areas of food production supplying cities (Arruti 2006; Almeida 1996).

Among agricultural and management systems of fields, grasslands (savannas) and forests employed during this period, hybrids between indigenous, European and African systems certainly played a central role in the construction of the ecological knowledge of current communities. After the official end of the slavery system, in 1888, hundreds of other communities were formed by freed blacks, which led to the remaining current quilombo communities. These groups historically maintained the trajectory of their cultures and defended their ancestral territories, currently recognized as in the social category of *quilombolas* communities (Leite 2000). Since the Constitution of 1988, these traditional communities have been guaranteed the right to the territories, although they have often been threatened by political and economic groups.

The flow of production from the gold and diamonds mines was mainly through a path known as the *Estrada Real* (Royal Road) that connected the region of Diamantina/MG to the city of Paraty on the coast of Rio de Janeiro state (Machado Filho 1980). The decline of gold and diamond mining in the late nineteenth century, along with the official end of the slavery system in 1888, brought significant impacts to the social configuration of the entire region that houses the Espinhaço in Minas Gerais. The depletion of gold and diamond deposits has given way to an increase in agricultural and livestock activities, with the expansion of large farms. The Royal Road became the path of *tropeiros* (mule trains) carrying agricultural products—like coffee and *cachaça*- to the city of Rio de Janeiro, then the capital of the country. Until the 1970s, agriculture developed in the region and was based on the use of animals; virtually all plantation activities, including growing and harvesting were performed manually (Goulart 2009).

20.3 The Management of Rocky Fields

Agrarian societies that formed along the Espinhaço Range in Minas Gerais are characterized by their extensive cattle ranching and multiple uses of plant resources. Throughout its extension, the Espinhaço Range is comprised of large, heterogeneous environments that contact areas of Atlantic Forest on the east slope, Cerrado in the west slope and Caatinga on its northern portion (Chap. 7) (see Fig. 20.2).

Although there is significant diversity of cultural identities among groups that inhabit the Espinhaço, rocky fields[7] may be considered a common element throughout the range and possesses unique relationships with rural and traditional communities. This is a result of the ecological and biological characteristics of

[7]The rocky fields are found in mountain environments and correspond to 2.5 % of the initial length of the cerrado biome (Dias 1996 cited Mazzetto 2006). The region here in focus, is located in the Espinhaço Range and surroundings.

Fig. 20.2 The main vegetation types managed by rural and traditional communities throughout the Espinhaço range. *Below* we can see images of the Sempre-Vivas National Park. *Source* Andre Dib

rocky fields that contrast with other floral species of the surroundings (Atlantic Forest and Cerrado). Communities have developed specific ecological knowledge for the management of rocky fields, and especially for plant extraction and traditional livestock.

These groups interacted historically with the Espinhaço and are skillful in describing the location of resources—a detailed "mental map"—as well in the as *traditional*[8] uses associated with them. The native flora and fauna are widely recognized—habits, habitat and occurrence—as well as their many uses and meanings, including feeding strategies, housing, cooking utensils, medical and religious practices as well as income generation, of which the "*sempre-vivas*" flowers (dry flowers) are a key component.

Traditional livestock is one of the main economic activities in much of the Espinhaço of Minas. By the early twentieth century, cattle were raised with the use of natural pastures either in rocky fields on mountains or in the surrounding Cerrado environments locally known as "*chapadas*" (plateaus) (Almada 2012; Carvalho 2014).

In some regions, such as Serra do Cipó, there are records of the development of livestock breeds adapted to the peculiar conditions of the rocky fields. These more rustic races, in many places called "*pé-duro*" or "*curraleiro*", are adapted to the types of food available and the severe climatic conditions of these higher altitudes. Even today, this breed of cattle is used in the Espinhaço and is farmed on the "loose", staying free in the pastures for months, with no clear boundaries between properties as these areas are considered common use for the community. The use of fire in the management of these pastures was also a common practice, originally observed and inherited by indigenous peoples, who performed the cleaning of areas for planting with the controlled burning of certain spaces, a practice known by them as "*coivara*". Fire was used as a form of pasture renovation, and was performed, in general, in the period preceding the rainy season, with little wind and at intervals of up to three years (Almada 2012).

Another economic activity of great importance associated with the rocky fields in many traditionaland rural communities is the collection and commercialization of various ornamental species. In the second half of the twentieth century, especially with the increase of tourist activities in the region, species of Bromeliaceae and Orquidaceae, as well as Cyperaceae, Xyridaceae and Eriocaulaceae, became widely collected and marketed by Espinhaço communities. Before the advancement of conservation policies, this traditional activity was an important element of environmental conflict between the state and local communities, as will be discussed below (Figs. 20.3, 20.4 and 20.5). Many species of Velloziaceae, known locally as "*canelas-de-ema*" were collected for use as fuel. Indeed, the stalk of these species have high combustive power, and their morphology is adapted to fire, making them a key element of the dynamics of the plant community of rocky fields.

[8]The tradition here is not seen as the past surviving until the present, but as the past, which in the present, builds the possibilities of the future (Woortmann 1990).

Fig. 20.3 *Above*, the transport of objects and goods using pack animals dates back to the colonial period when *tropeiros* crossed great distances along the Espinhaço. The tropeiro's image is an extremely important cultural element throughout the region. *Below*, the use of plateaus for housing and livestock just bellow the rupestrian grassland vegetation. *Source* Elisa Cotta

Several species of Cyperaceae, Xyridaceae and Eriocaulaceae are locally called "sempre-vivas" flowers (dryflowers) (see Chap. 17). The inflorescences of these species have a long life, even after being harvested, which explains their name translated as "always alive". The traditional management of these species involves special knowledge developed by these communities concerning their taxonomy,

Fig. 20.4 *Above*, an area of the Sempre-Viva National Park with a *Vereda* habitat and the mountains of the Espinhaço range in the background. *Below*, two artisans collecting *sempre-vivas* and producing handicrafts, an important source of income for local communities. *Source* Andre Dib

phenology, response to fire on the population ecology and spatial distribution. These communities also have developed system for the management of these species involving rules of use and access to collecting areas in common use lands (Monteiro et al. 2012; Giulietti et al. 1988). The ecological effects of the collection

Fig. 20.5 Extractivism in Cerrado areas associated with the Espinhaço. **a** Araticum (*Annona crassiflora*) gathering. **b** Pequi oil production (*Caryocar brasiliense*). **c** Coquinho azedo gathering (*Butia capitata*). *Source* Igor Simoni Homem de Carvalho

of these species are not entirely clear, which has caused, among other factors, important debates between conservation perspectives and territorial identity and the rights of communities that still retains this extractive practice (Schmidt et al. 2007; Monteiro et al. 2012).

In the surroundings of Diamantina/MG, the collection of ornamental plants, including *sempre-vivas flowers* (dryflowers), is an important source of income for families in local communities. In surveys conducted by UFVJM researchers in the region, more than 200 species collected/managed plants were recorded (considering different elements of the plant: flowers, nuts, leaves, etc.), for residents of communities in 58 municipalities in the region. There is expectation, now that the activity is recognized and standardized by the State, beyond the collection permit in fields traditionally used by traditional communities involved. On the other hand, there is pressure from scientists and technicians from environmental non-governmental organizations to maintain the parks and define criteria for sustainable use of these resources external to these conservation these units. The impasse persists and generates tension and everyday conflicts, wherein the extractive economies that are part of the local culture find themselves challenged/threatened by an environmental agenda that focuses on the region (Fávero and Monteiro 2014).

Rocky fields are extensively used for agriculture for home consumption and, in general, little for industrial agriculture. Traditionally, agriculture is practiced near houses and, depending on the environments, employs rotation with areas left fallow for natural replenishment of soil fertility; the practice of "*roça-de-toco*" may also be used.

Commonly, planting focuses on the areas of Cerrado and Atlantic Forest in lower elevations or in areas of fertile soil close to rivers and lakes. Amid the rocky fields, there are enclaves of afforestation formations; the "*capões de mata*" (see Chap. 7). These areas of more fertile soil, locally known as "*terras de cultura*" (crop land) and are used for planting to supply families or even for regional trade.

After the decline of gold and diamonds exploitation, most workers remained in the villages near the mines. So, many families perform artisanal mining as one of their economic activities nowadays. During the dry season, they dedicate themselves to seeking diamonds, crystals and collecting flowers and during the rainy season, they exclusively perform agricultural and livestock activities. Although the environmental impacts of mining are undeniable, on the artisanal scale mining is not only an economic activity, but as an important part of cultural identity in the Espinhaço of Minas. Over decades of management of mines, these populations have built an important body of knowledge on local ecosystems, types of rocks and soil, and water regime (Ribeiro 2013).

In this context, the above examples have demonstrated that the traditional management of rocky fields is based on the multiple use of the landscape, according to the special and temporal variation of plant resources. However, it can be said that the main uses of rocky fields deal with plant harvest and traditional livestock. The importance of each one of these activities varies considerably over the Espinhaço of Minas depending on local histories and territorial dynamics resulting from preservationist conservation policies, space occupation by economic groups (mainly mining conglomerates and eucalyptus plantations) and the very action of the State.

20.4 A Biocultural Diversity Hotspot

It is not possible to define a unique cultural identity associated with the Espinhaço Range in Minas Gerais, due to the diversity of social groups therein, however, they do have in common cultural characteristics and specific ways of relating to the environment.

Communities and people of Espinhaço have various territorial practices, which operate as an identification factor and in environmental protection, which they have appropriated. The Espinhaço, including the rocky fields, are among various sources of resources—material and symbolic—used by the communities living there. This includes a complex system of traditional management of the fields, savannas and forests that occur on the eastern and west slopes. Throughout the environmental history of the Espinhaço of Minas, many communities have formed culturally distinct groups, especially regarding the need for policy coordination in defense of their territories and traditional ways of life.

The form of material and symbolic appropriation of the environment held by these communities is marked by the existence of a collective and familiar agrarian regime, which is based on a set of rules and consuetudinary values, corresponding to the diversity of ownership arrangements of natural resources included in legal category of "traditionally occupied lands", present in the Brazilian Constitution of 1988, as provided in Article 231, by granting rights to indigenous peoples.

The category of traditionally occupied lands recognizes different modalities and combined common use of forests, water resources, fields and pastures, involving different productive activities carried out by family labor units, such as plant extraction, agriculture, fishing, hunting, crafts and livestock. Therefore, this categorys also includes indigenous, quilombolas (descendants of runaway African slaves), ribeirinhos (riverside), seringueiros (rubber extractors), quebradeiras-de-côco babaçú (babaçú coconut-breakers), castanheiras (Brazilian nut collectors), among others (Almeida 2008), each with specific ways to appropriate and relate to their natural environments in different parts of Brazil.

Referring only to the portion of Espinhaço in the state of Minas Gerais and its surroundings, it is possible to identify, besides quilombolas and indigenous communities, the following self-determined cultural identities that have as an ethnic distinction some practice, form of management or environmental landscape with which they maintain a close relationship:

- *Dry-flower gatherers*: These communities inhabit the plains and have as one of their main activities the collection of dry flowers (sempre-vivas) in the rocky fields. Currently one of the main regions where this activity occurs represents the southern portion of the Espinhaço in the higher portions of the landscape.
- *Veredeiros*: communities living close to the *veredas*. The *vereda* areas are the Cerrado phytophysiognomies that occur in higher, flat areas with a shallow water table and a remarkable presence of buriti palm (*Mauritia flexuosa*).

- Geraizeiros: general name for the communities living in Cerrado areas, including the various biome phytophysiognomies therein, and use greatly the *chapadas* (plateaus).
- *Caatingueiros*: are the communities who live mainly in transition zones between Caatinga and Cerrado and make multiple use of these ecosystems of the Espinhaço and its surroundings.

Rocky fields, in a more central or peripheral way, comprise a set of landscape elements that are part of traditional management systems of all of these communities. Since they occupy the areas of higher altitude in these regions, rocky fields represent historical places of passage and communication between the two major river basins separated by the Espinhaço Range. In many communities, the highest points are also places of religious pilgrimages and sacred sites.

In addition, the Espinhaço also stands out in relation to food culture in a special way with various areas of artisanal cheese production. One of the main cultural elements of Minas Gerais state is *Serro* cheese that is artisanally produced in Serro district in the southern portion of the mountain chain of Espinhaço, between the cities of Diamantina and Conceição do Mato Dentro. The manufacturing recipe of this cheese would have come from the Serra da Estrela region of Portugal. In the nineteenth century, the region of Serro, as did a large part of the Espinhaço, stood out for livestock production, but it also included cheese production. In 2008, the Instituto do Patrimônio Histórico e Artístico Nacional (National Institute of Historical and Artistic Heritage) recognized the Serro cheese production technique as intangible cultural heritage of Brazil (Menezes 2009).

Each community along the Espinhaço is a result of centuries of hybridization of indigenous, African and European cultures. The popular cultures of these communities are extremely rich, marked by parties and religious manifestations of popular Catholicism, such as the Congado Guards and Folias de Reis. Ecosystems and native species are an important part of these cultural practices, whether as food, elements of magic, raw material for musical instruments or pilgrimage sites and celebrations (Almada 2012).

Two occurrences of species in rocky fields exemplify the role of the in popular religiosity of communities: *Baccharis dracunculifolia* and *Achyrocline satureioides*, two species of Asteraceae. *Baccharis* is used to manufacture brooms used to sweep the yards during the period of Lent, as a protection against evil spirits. This species is also frequently used by *benzedeiras* (healers) in their healing rituals for physical and spiritual ailments. The inflorescences of *Achyrocline* are usually harvested during the Holy week and used to adorn houses and as pillow filling (Almada 2012).

As for agro-biodiversity, there are few studies in the Espinhaço Range, although communities have inherited and developed countless agricultural varieties adapted to the climate and soil conditions of the region. Regarding the use of biodiversity of local natural ecosystems, however, several ethno-ecological studies have shown that communities use a large number of species of all the ecosystems associated with the Espinhaço. (Dayrrel 1998; Monteiro et al. 2012; Almada et al. 2013).

20.5 Territorial Environmental Conflicts in the Espinhaço of Minas

When the bandeirantes (pioneers) began the occupation of Espinhaço lands, it was done at a cost of genocide and expulsion of the indigenous people living there. The decline of gold and diamond mining in the nineteenth century was accompanied by a territorialization of Espinhaço's ecosystems by the remaining groups that led the current rural communities, traditional quilombolas and indigenous (e.g. Martins 2008).

On the one hand, historically, territories (Little 2002) of these groups combine different environments of the Espinhaço and its surroundings. Life strategies and complex knowledge were developed, permeated by meanings and understandings contextualized by the places where they are—*geographical knowledge* (Claval 2009), transmitted and reinvented over many generations. They also feature a family-based work organization for self-support and income through sale of what they produce; with each family having their own codes of ownership and land use, representations and social practices of interaction with nature, which is seen as the creator of life and as a whole to which they belong.

On the other hand, this region of significant biocultural diversity is now one of the regions of Brazil characterized by important environmental conflicts of a territorial nature. These conflicts are mainly between the modes of use and appropriation of land by local communities and the great undertakings of mining and monoculture led by companies and the often-authoritarian deployment of fully protected conservation units as a compensatory measure. This relationship is mediated by the State and tends to legitimize the actions of companies, thereby restating the asymmetric conditions of power and deepening social inequality and promoting social exclusion.

From a conceptual perspective, grounded in the notions of *field and habitus* of Pierre Bourdieu, the analytical tradition present in the sociology of environmental conflicts, Zhouri and Laschefski (2010) conducted a critical reflection of the *ecological modernization paradigm*, which guides speeches and actions in the environmental field, bringing a reading of environmental conflicts as being composed by a diversity and heterogeneity of actors with different ways of perceiving the world and projecting on it their actions. They view environmental conflicts as originating from different practices of technical appropriation, social and cultural, of the material world where the cognitive basis for the speeches and actions of the individuals involved are in conflict and are in accordance with their hegemonic views about the use of space. These conflicts show situations of *environmental injustice*, highlighting the unequal distribution of natural resources and economic development with consequences that fall asymmetrically on social groups of workers, low-income populations, racially discriminated segments, marginalized and most vulnerable portions of citizenship.

Perhaps the most dramatic cases currently refer to conflicts caused by the implementation of mega-projects of mineral extraction. The most emblematic is the undertaking of Anglo-American, in the town of Conceição do Mato Dentro, in the southern portion of the Espinhaço in Minas Gerais. The project started its implementation in 2007 and, since then, has led to the mobilization of dozens of groups and social movements (Zhouri 2014).

It is an iron ore-mining project whose deposit is located in areas of rocky fields of high biological and cultural value. With the implementation of mining, hundreds of employees were living in Conceição do Mato Dentro, causing serious social impacts such as increased cases of violence and disability in public education and health services. Furthermore, there have been several reports of violence in the purchase of land by the company, with threats to communities and families who resisted. Pollution of watercourses and soil due to emission of liquid effluents by the company has also been observed.

However, one of the most controversial elements of the project is the construction of a pipeline of about 525 km that will take the ore to the Açu port of Barra, in Rio de Janeiro, using water withdrawn from areas of rocky fields. Although it has gained great visibility, the case of mining in Conceição do Mato Dentro is not an isolated situation. There are numerous other ongoing or planned mining projects for the Espinhaço. For example, there is an attempt by Vale to deploy a mine in the range of Gandarela located in the metropolitan region of Belo Horizonte, the capital of Minas Gerais (Marent et al. 2011). This region is a major source of drinking water for the urban population of the capital and has caused several conflicts.

Another aspect of environmental conflicts involving local and traditional communities is the implementation of full-protection conservation units overlapping traditional community territories. The establishment of protected areas in the Espinhaço, in general, followed the same authoritarian model that has occurred throughout Brazil and, more generally, in Latin America. Given its high biological value, large areas of Espinhaço, especially the area dominated by rocky fields were turned into natural parks between years of 1970 and 1980. These areas were considered uninhabited rocky fields, representing large empty areas for the preservation of biodiversity; however, these units were created on lands of intense common use by traditional communities for both plant extraction and livestock.

A case of conflict between conservation and land rights in Espinhaço is the Parque Nacional das Sempre-Vivas (National Park of Sempre-vivas Flowers), in the region of Diamantina, Minas Gerais. The park was created in 2002 illegally by not holding a public hearing, but being justified in order to protect a high biological interest area. Additionally, the traditional use of these areas by dozens of "*apanhadores de flores*" (flowers gatherers) communities was dismissed, putting them in great difficulties to survive (Monteiro et al. 2012). Conflicts have been a factor in the mobilization of local communities to search for guarantee of their territorial rights. Currently, communities seek to change the category of the park for a sustainable use conservation unit, a category that permits the traditional use of natural resources essential to their material and symbolic existence linked to nature conservation. This struggle has preservationists' resistance who reaffirm the modern myth of untouched nature as presented by Diegues (1996).

Another interesting case is the implementation of the Serra do Cipó National Park in 1984, located 100 km to the north of Belo Horizonte. During the deployment process, many families were violently evicted from their land and to this day most families and their descendants have never received any compensation. Only recently did the community mobilization associated with anthropological and ethno-ecological researchers permit this story to be finally revealed and allow the families affected by the park to seek a guarantee to their right to territory and memory (Anaya and Souza 2014; Almada 2012).

Associated with the implementation of full-protection conservation areas, tourism activities also often represent a threat to the territorial rights of communities.

In the Mountain Range of Cipó, for example, the implementation of the Park has intensified commercial tourism activities and real estate speculation. In the last twenty years, local communities, pressured by economic and political power, have lost much of their land to owners coming from large urban centers to build hotels, condos and resorts. Besides the social impact, generally, these undertakings also bring significant ecological impacts and limit the access of communities to their areas of traditional use (Almada 2012).

20.6 The Espinhaço as Biocultural Heritage of Humanity

In 2005 UNESCO established the Biosphere Reserve of the Espinhaço Range (Reserva da Biosfera Serra do Espinhaço—RBSE), which covers the entire portion of the chain in the state of Minas Gerais. The RBSE currently covers 53 municipalities, with a total area of 3,070,000 ha. Although internationally known for its biodiversity and scenic beauty, the Espinhaço is primarily a cultural heritage of humanity. Biodiversity conservation policies will only be effective as long as they consider traditional knowledge and management modes of populations who have endured centuries of colonial exploitation, and recently neoliberalism and nationalist neo-developmentism of the Brazilian State.

The signaling by the Brazilian State the willingness to differentiate and recognize different social groups reflects the current concern about the conflicts arising from territorial disputes that have been developing over the overlapping of full-protection conservation units of nature with traditional territories, the construction of dams and hydroelectric power plants, large projects of eucalyptus plantations, mining, among others.

This situation has given visibility to these groups and the defense of place from the four fundamental right statements: its identity, its territory, political autonomy and its own development vision (Escobar 2003). The emerging self-determinations, through categories that claim a collective existence, has brought the complexity of identity elements to the environmental level, registering a profound break with the homogenizing colonialist attitude, which historically turned out ethnic differences and cultural diversity (Almeida 2008).

Thus, the struggles of these people of Espinhaço is strengthened by the process of reconstruction of their historical origins, checking the condition of the difference of their identities (sempre-viva flower pickers, veredeiros, quilombolas, geraizeiros, among others) in order to facilitate their stay in a territory before the restrictive processes of large economic enterprises take over their territory.

In the name of economic progress, the ecosystems of the Espinhaço Range, its peoples and the traditional communities living there, are constantly threatened by the expansion of mining projects, eucalyptus monocultures, tourism enterprises and the urban sprawl resulted from real estate speculation. At the same time, communities have organized politically to defend their territories and ways of life and, therefore, the entire local biodiversity.

To recognize the Espinhaço Range as a biocultural heritage is an essential step in overcoming false dichotomies, which consider nature in a dehumanized form. For traditional communities of Espinhaço, the mountain range, plants and animals have never been just natural resources, but non-human elements that comprise their sociability and constitution of sociological systems in which they live.

Acknowledgements The authors would like to thank all the traditional communities of Espinhaço collaboration with our research. His study received financial support from the Brazilian *Conselho Nacional de Desenvolvimento Científico e Tecnológico* (CNPq, National Council for Scientific and Technological Development; research grant no. 563304/2010-3.

References

Ab'Sáber NA (2003) Os domínios de natureza no Brasil: potencialidades paisagísticas. Ateliê Editorial, São Paulo

Acselrad H (2004) Conflitos ambientais no Brasil. Fundação Heinrich Böll, Rio de Janeiro, Relume Dumará

Almada ED (2012) Entre as serras: etnoecologia de duas comunidades no sudeste brasileiro, Thesis. Universidade Estadual de Campinas, Brazil

Almada ED, Coelho MS, Fernandes GW et al (2013) Traditional ecological knowledge of rural communities in areas of seasonally dry tropical forest in Serra do Cipó, Brazil. In: Sánchez-Azoifefa A, Powers JS, Fernandes GW, Quesada M (eds) Tropical dry forests in the americas: ecology, conservation, and management. CRC Press, London, pp 429–451

Almeida AWB (1996) Os quilombos e as novas etnias. In: O'Dwyer EC (ed) Quilombos: identidade étnica e territorialidade. ABA, FGV, Rio de Janeiro, pp 43–81

Almeida AWB (2008) Terra de quilombo, terras indígenas, "babaçuais livre", "castanhais do povo", faixinais e fundos de pasto: terras tradicionalmente ocupadas. 2nd edition. PGSCA—UFAM, Manaus

Anaya FC, Souza MCF (2014) Conflitos ambientais territoriais no Parque Nacional da Serra do Cipó. Rev Des Social. Special edition

Arruti JM (2006) Mocambo: antropologia e história do processo de formação quilombola. EDUSC, Florianópolis

Baleé WL (1992) Indigenous history and Amazonian biodiversity. In: Steen HK, Tucker RP (eds) Changing tropical forests: historical perspectives on today's challenges in Central and South America. Forest History Society, Durham, pp 185–197

Carney JA, Rosomoff RN (2009) In the shadow of slavery: Africa's botanical legacy in the Atlantic world. University of California Press, Los Angeles, USA

Carvalho ISH (2014) Back country cattle ranching and the conservation of biodiversity in the savanna region in the North of the state of Minas Gerais. Sust Deb 5:19–36

Claval P (2009) Globalização, migrações, inclusão e exclusão: algumas reflexões. In: Almeida MG, Cruz BN (eds) Território e cultura: inclusão e exclusão nas dinâmicas socioespaciais. UFG, Goiânia, pp 10–25

COMIG/IGC/UFMG (1997) Projeto Espinhaço. Belo Horizonte/MG (CD-ROM)

Costa FN (2005a) Campos rupestres. In: Silva AC, Pedreira LCVSF, Abreu PA (eds) Serra do Espinhaço meridional: paisagens e ambientes. O Lutador, Belo Horizonte, pp 139–145

Costa JBA (2005b) Cerrados norte mineiro: populaçõas tradicionais e suas identidades territoriais. In: Almeida MG (ed) Tantos Cerrados: múltiplas abordagens sobre a biogeodiversidade e singularidade sociocultural. Editora Vieira, Goiânia, pp 295–319

Dayrrell CA (1998) Geraizeiros e biodiversidade no Norte de Minas Gerais: a contribuição da agroecologia e da etnecologia nos estudos dos agroecossistema tradicionais. Dissertation. Universidad Internacional de Andalucía, Brazil

Descola P (1996) Constructing nature: symbolic ecology and social practice. In: Descola P, Pálsson G (eds) Nature and society: anthropological perspectives. Routledge, New York, pp 82–102

Dias BFS (1996) Cerrados: uma caracterização. In: Dias BFS (ed) Alternativas de desenvolvimento dos Cerrados: manejo e conservação dos recursos naturais renováveis. Fundação Pró-Natureza, Brasilia, pp 11–25

Diegues AC (1996) O mito do paraíso desabitado: as áreas naturais protegidas. In: Viola E (ed) Ferreira LC. Incertezas de sustentabilidade na globalização. Editora da UNICAMP, Campinas, pp 279–313

Escobar A (2003) Atores, redes e novos produtores de conhecimento: os movimentos sociais e a transição paradigmática na ciências. In: Santos BS (ed) Conhecimento prudente para uma vida decente: um discurso sobre as ciências. Editora Cortez, São Paulo, pp 639–666

Fávero C, Monteiro FT (2014) Disputas territoriais no Vale do Jequitinhonha: uma leitura pelas transformações nas paisagens. Rev Agric 11:7–15

Gavin MC, McCarter J, Mead A et al (2015) Defining biocultural approaches to conservation. Trend Ecol Evol 30:140–145

Giulietti N, Giulietti AM, Pirani JR, Menezes NL (1988) Estudos de sempre-vivas: importância econômica do extrativismo em Minas Gerais, Brasil. Acta Bot Bras 1:179–194

Gontijo BM (2008) Uma geografia para a Cadeia do Espinhaço. Megadiv 4:7–16

Goulart EMA (2009) O caminho dos currais do Rio das Velhas: a Estada Real do sertão. Coopmed, Belo Horizonte

Leite IB (2000) Os quilombos no Brasil: questões conceituais e normativas. Etnogr 4:333–354

Little PE (2002) Territórios sociais e povos tradicionais no Brasil: por uma antropologia da territorialidade. Ser Antropol 322:1–32

Luna FV, Costa IN (2009) Algumas características do contingente de cativos em Minas Gerais. In: Luna FV, Costa IN, Klein HS (eds) Escravismo em São Paulo e Minas Gerais, Imprensa Oficial/Edusp, São Paulo, pp 17–32

Machado Filho AM (1980) Arraial do Tijuco: cidade Diamantina. Itatiaia, Belo Horizonte

Maffi L (ed) (2001) On biocultural diversity: linking language, knowledge and the environment. Smithsonian Institution Press, Washington

Marent BR, Lamounier WL, Gontijo BM (2011) Conflitos ambientais na Serra do Gandarela, Quadrilátero Ferrífero—MG: mineração x preservação. Rev Geogr 7:99–113

Martins ML (2008) The voyagers' Jequitinhonha, nineteenth and twentieth centuries. Var Hist 24:707–728

Mazzetto CE (2006) Os cerrados e a sustentabilidade: territorialidades em tensão, Thesis. Universidade Federal Fluminense, Brazil

Menezes JNC (2009) Modos de fazer e a materialidade da cultura "immaterial": o caso do queijo artisanal de Minas Gerais. Patrim Mem 5:19–33

Monteiro FT, Pereira DB, Gaudio RS (2012) Os(as) apanhadores(as) de flores e o Parque Nacional das Sempre-vivas: entre ideologias e territorialidades. Soc Natur 24:134–419

Nabhan GP (2009) Ethnoecology: bridging disciplines, cultures and species. J Ethnobiol 29:3–7

Nazarea V (2006) Local knowledge and memory in biodiversity conservation. Annu Rev Anthrop 35:317–333

Posey DA (2002) Kayapo ethnoecology and Culture. Routledge, London

Prous A (2000) El poblamiento de América. Un debate sin fin. Cien Hoy 10:34–41

Resende ET, Prous A (1991) Os vestígios vegetais do Grande Abrigo de Santana do Riacho. Arq Mus Hist Nat UFMG, v. XII

Ribeiro RF (2005) Florestas anãs do sertão: o cerrado na história de Minas, Gerais edn. Autêntica, Belo Horizonte

Ribeiro L (2013) Cativos do diamante. Etnoarqueologia, garimpo e capitalismo. Rev Espinhaço 2:153–167

Saadi AA (1995) A geomorfologia da Serra do Espinhaço em Minas Gerais e de suas margens. Geonomos 3:41–63

Schmidt IB, Figueiredo IB, Scariot A (2007) Ethnobotany and effects of harvesting on the population ecology of *Syngonanthus nitens* (Bong.) Ruhland (Eriocaulaceae), a NTFP from Jalapão region, central Brazil. Econ Botany 61:73–85

Toledo V, Barrera-Bassols N (2008) La memoria biocultural: la importância ecológica de las sabidurias tradicionales. Icaria Editorial, Barcelona

Viveiros de Castro E (2002) A inconstância da alma selvagem. Cosac Naify, São Paulo

Woortman K (1990) Com parente não se "neguceia": o campesinato como ordem moral. An Antropol 11–73

Zhouri A (2014) Mapeando Desigualdades Ambientais: mineração e desregulação ambiental. In: Zhouri A, Valencio N (eds) Formas de Matar, de Morrer e de Resistir: limites da resolução negociada de conflitos ambientais. UFMG, Belo Horizonte, pp 111–141

Zhouri A, Laschefski K (eds) (2010) Desenvolvimento e conflitos ambientais. Editora UFMG, Belo Horizonte

Chapter 21
The Human Dimension in the Espinhaço Mountains: Land Conversion and Ecosystem Services

Ana Carolina de Oliveira Neves, Alisson Flávio Barbieri, André Aroeira Pacheco, Fernando de Moura Resende, Rodrigo Fagundes Braga, Alexsander Araujo Azevedo and G. Wilson Fernandes

Abstract The Espinhaço Mountains and their rupestrian grasslands hold significant historical, cultural, and economic value. The discovery of large gold deposits in Espinhaço in the 1700s started an enduring extractive tradition that persists until today. Since then, other important extractive-economic cycles took place in the region; for example, gold (18th and 19th centuries), diamond (19th and 20th), iron ore mining, gemstones, ornamental stones, sand, and plant extractivism (20th and 21st). Mining generated wealth for the Portuguese Crown and Brazil at a considerable environmental cost. However, in the 20th century, the Espinhaço Mountains developed additional

A.C.d.O. Neves (✉) · A.A. Pacheco · G.W. Fernandes
Evolutionary Ecology & Biodiversity, sala 172, Departamento de Biologia Geral, ICB, Universidade Federal de Minas Gerais, Caixa Postal 486, Belo Horizonte 30161970, MG, Brazil
e-mail: ananeves@gmail.com

A.A. Pacheco
e-mail: andrearoeirap@gmail.com

G.W. Fernandes
e-mail: gw.fernandes@gmail.com

A.F. Barbieri
Department of Demography and Center for Regional Development and Planning (Cedeplar), Universidade Federal de Minas Gerais, Av. Antônio Carlos, 6627, Belo Horizonte 31270-901, MG, Brazil
e-mail: barbieri@cedeplar.ufmg.br

F.d.M. Resende
Conservation Biogeography Lab, Departamento de Ecologia, Universidade Federal de Goiás, Caixa Postal 131, Goiânia 74001-970, GO, Brazil
e-mail: fermresende@gmail.com

R.F. Braga
Laboratório de Ecologia e Conservação de Invertebrados, Universidade Federal de Lavras, Caixa Postal 3037, Lavras 37200-000, MG, Brazil
e-mail: rodrigo.fagundes@yahoo.com.br

© Springer International Publishing Switzerland 2016

G.W. Fernandes (ed.), *Ecology and Conservation of Mountaintop Grasslands in Brazil*, DOI 10.1007/978-3-319-29808-5_21

values, focused on treasures of another kind. An astonishing and unique biodiversity occurs (with some of the world's highest richness values and several endemic species) over the colossal mineral deposits, especially in the rupestrian grasslands. This biodiversity contributes to cultural activities, provides people with medicines, raw materials and water, and maintains three major Brazilian river basins. Recent studies have translated into monetary metrics some of the services that these mountain ecosystems deliver to humans and encouraged more sustainable practices. Here, we offer conservation mechanisms to maintain biodiversity, as well as a proposal for land use management to promote sustainable using the wealth generated by mining.

Contents

21.1 Introduction

The Espinhaço Mountains form the longest mountain range in Brazil, stretching over 1,200 km from the states of Minas Gerais to Bahia. It has significant historical, economical, cultural, and biological importance (Fig. 21.1). 'Espinhaço' means large spine, referring to the topographical appearance of the mountain belt. This name was coined by the German geologist, geographer and metallurgist Wilhelm Ludwig von Eschwege, known as Eschwege Baron, in an article published in 1822 in Germany (see Gontijo 2008):

A.A. Azevedo
Instituto Biotrópicos, Praça JK, 25, Diamantina 39100-000, MG, Brazil
e-mail: alex@biotropicos.org.br

G.W. Fernandes
Department of Biology, Stanford University, P O Box 94305, Stanford, CA, USA

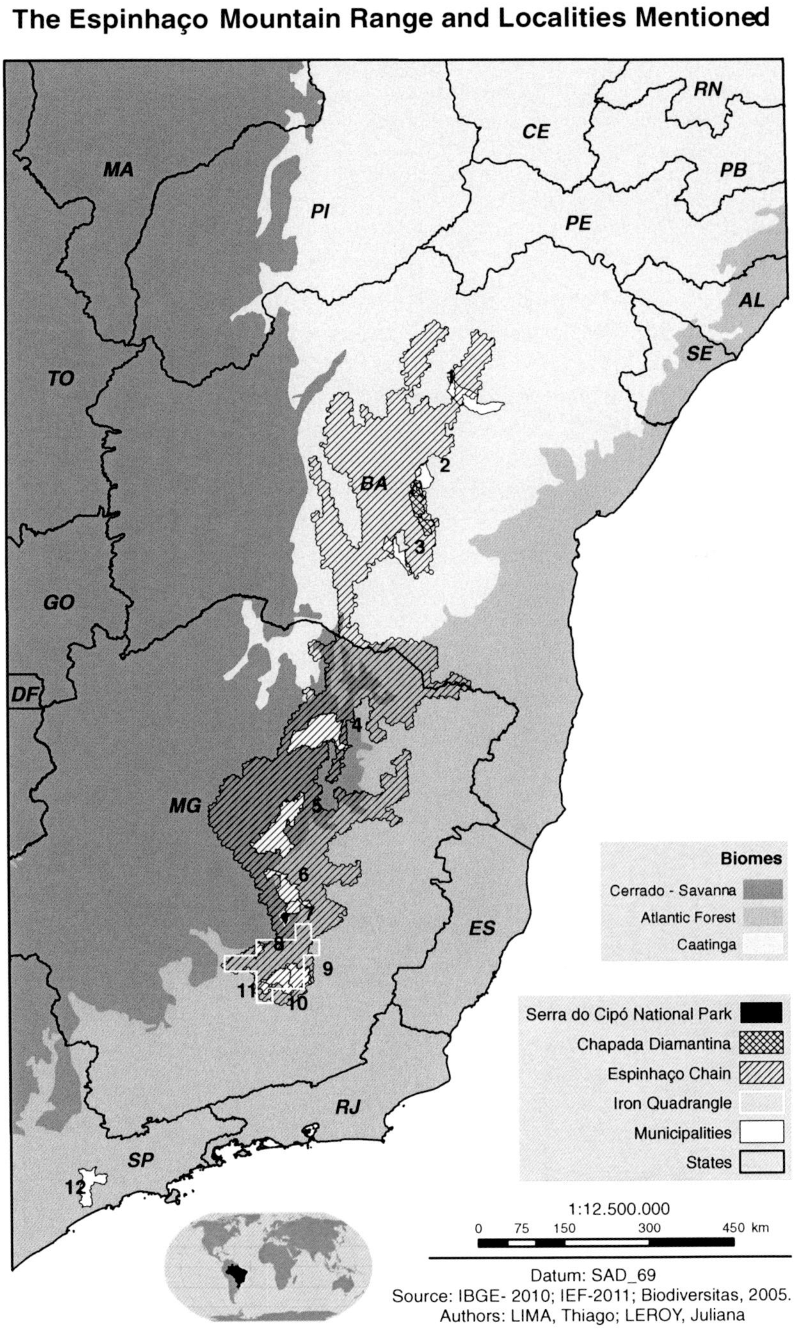

Fig. 21.1 Espinhaço Mountains: thresholds, biomes, and localities (see text for details). Municipalities: *1* Jacobina, *2* Lençóis, *3* Rio de Contas, *4* Grão Mogol, *5* Diamantina, *6* Conceição do Mato Dentro, *7* Morro do Pilar, *8* Belo Horizonte, *9* Mariana, *10* Ouro Preto, *11* Congonhas, *12* São Paulo

> I named it Espinhaço Mountains (*Rückenknochengebirge*) because it is not only the highest mountain belt, but also a remarkable one, especially for the naturalist, as it forms an important divider from the geognostical perspective, and also of huge importance due to its fauna and flora. (…) The regions from the east of this mountain belt to the sea are covered with the most exuberant forests. The western side forms a corrugated terrain and shows bare hills and open landscapes, covered by grasses and gnarled trees, or the grasslands whose valleys harbor dense vegetation only sporadically. The botanist finds in its virgin forests plants completely different from those in the grasslands, and the zoologist finds one or other animal, especially birds, as soon as he or she passes from the forests, through the Espinhaço Mountains, to the grasslands (von Eschwege 1822; Renger 2005).

In this passage, Eschwege mentions an important aspect of the Espinhaço Mountains: their role as a natural divider of two Brazilian biomes currently considered biodiversity hotspots: the Atlantic Forest and Cerrado. The Atlantic forest is located to the east, where forests predominate, and the Cerrado to the west, where savannas predominate. There is also the Caatinga to the north, a seasonally dry tropical vegetation. In addition, along the entire mountain belt, usually above 900 m a.s.l., there is a predominance of grassland ecosystems that grow on rocky outcrops, which are locally known as *campos rupestres* (rupestrian grasslands) (Fig. 21.2).

The history of this region is closely related to extractivism, especially mining. Gold was first discovered in Brazil in 1698, in the southern region of the Espinhaço Mountains in the state of Minas Gerais (literally translated as "General Mines"), in a region that later was named Vila Rica and then Ouro Preto. Many other mines were discovered in the Espinhaço Mountains, such as the gold alluvial deposits found in 1718 in the region of Jacobina and Rio de Contas (Bahia State)

Fig. 21.2 Typical landscape of the rupestrian grasslands in the Espinhaço Mountains. *Photo* Fernando Resende

(Machado and Figueirôa 2001). The discovery of gold in the Espinhaço Mountains attracted waves of miners and resulted in the first major Portuguese immigration flow to Brazil. In the first sixty years of the 18th century approximately 600 thousand people arrived from Portugal and the Atlantic islands (Fausto 2008). These discoveries had huge regional importance, as they encouraged the foundation of the first villages and roads in the mining areas. They were also important for the history of Brazil, as they represent a milestone in the development of the countryside and its urbanization in a previously agrarian and coastal land, which was based on the cultivation of sugarcane.

Diamonds were first found in Brazil in the rupestrian grasslands of the Espinhaço Mountains. The discovery was made public by the Royal House of Portugal in 1729. Five years later, the diamond-yielding areas around the colony of Tejuco (currently Diamantina) were declared to be a special district owned by the Crown (*Demarcação Diamantina*) (Machado and Figueirôa 2001; Furtado 2009). Later, diamond deposits started being explored in the Itacambiruçu valley (Grão Mogol, Minas Gerais) in 1771, in Chapada Diamantina (Bahia State) in 1830, and also outside the Espinhaço Mountains (Machado and Figueirôa 2001). The importance of Chapada Diamantina relies on its large *carbonado* deposits: thick and porous aggregates of very small gray or black diamonds, as hard as pure diamonds, currently used exclusively for industrial purposes. There was great demand for *carbonados* after the First World War (Machado and Figueirôa 2001). For 150 years, Brazil held the monopoly of diamond production. Then, in 1870 South Africa started to take the lead (Machado and Figueirôa 2001).

The discovery of gold and diamonds in the rupestrian grasslands was the most significant find of its type in the New Colonial World. At that time Brazil contributed approximately 50 % of the world's production (Machado and Figueirôa 2001). For example, from 1700 to 1801, 715 tons of gold were extracted from Minas Gerais (Machado and Figueirôa 2001), the equivalent to US$27.51 billion according to the current price of gold. During this period approximately one thousand tons of gold has escaped the royal taxation (Dean 1996). Brazil's cumulative production of gold and diamonds from colonial times to 2000 is estimated at 2,972 tons and 22,438 thousand carats, respectively (Machado and Figueirôa 2001). The value of gold alone, considering its current price, is estimated at US$114.33 billion (current dollar exchange rate of US$1 = R$3.09).

Some mining cities in the Espinhaço Mountains were important national political and cultural centers at their peak, where architecture and baroque art flourished and important events occurred. Among those events was the Emboabas War in 1708, a mining rights conflict between *bandeirantes* (explorers) from the state of São Paulo, who found gold in the region, and others, who arrived after the discovery. Another important event was the *Inconfidência Mineira*, in 1789, the most significant insurrection against Portuguese domination in Minas Gerais. At the peak of the gold cycle, Vila Rica (capital of the state of Minas Gerais at the time) supported 80 thousand people, and was considered the largest city in the Americas. At the time New York had fewer than 40 thousand people and the village of São Paulo approximately eight thousand people (Goulart 2009). Diamantina also grew rapidly:

from 884 free residents (i.e., non-slaves) in the third quarter of the 18th century to 6 thousand inhabitants in the early 19th century (Furtado 2009). With the decline of gold and diamond exploitation at the second half of the 18 and 19th centuries, respectively, due to the depletion of alluvial deposits, most of the mining community began devoting themselves to the extraction of gemstones, sand, rocks, ornamental plants, and subsistence farming.

After decades of stagnation, the rupestrian grasslands of the Espinhaço Mountains became the stage for a new period of economic growth and colonization. The exploitation of iron ore and the transference of the capital of Minas Gerais from the old gold region of Vila Rica to the region of iron ore of Belo Horizonte in 1897 sparked this change (Roeser and Roeser 2010). During the First and Second World Wars (1914–1918 and 1939–1945), Brazil played an important role in the supply of strategic minerals to the Allies, such as manganese and iron ore from the Espinhaço Mountains (Machado and Figueirôa 2001). The exploitation of iron ore grew, especially after the Second World War. Today over 160 million ton/year are produced (IBRAM 2012).

Currently, mining is the main economic activity of Minas Gerais, contributing 57.9 % of its exports (IBRAM 2013). The state is the most important mining province of Brazil (IBRAM 2013), in part due to minerals explored in the rupestrian grasslands of the Espinhaço Mountains. Two thirds of the inhabitants of the Espinhaço Mountains live in Belo Horizonte, the sixth most populous city of Brazil, with 2.48 million people (IBGE 2015). The region where Belo Horizonte is located, known as the Iron Quadrangle, is one of the most important and well-studied geological regions of the world.

Brazil presently has the strongest economy in Latin America and leads the regional trade block, known as Mercado Común del Cono Sur (MERCOSUR). In 2012, the country marketed approximately 85 mineral commodities (USGS 2012). Brazil is the world's second largest iron ore producer. However, the Brazilian economy became so diversified that in 2013 the value of mineral extraction accounted only for ca. 4.2 % of Brazil GDP (Gross Domestic Product), in spite of the growth of the Brazilian mining industry over the past decade (IBRAM 2012; IBGE 2014).

21.2 Natural Heritage of the Espinhaço Mountains

Study of the geological and biological heritage of the Espinhaço Mountains began in the 1800s by foreign naturalists, such as the Eschwege Baron, John Mawe, Auguste de Saint-Hilaire, Carl Friedrich Philipp von Martius, Johann Emanuel Pohl, George Langsdorff and George Gardner (Leite 1996). Their explorations were motivated by the discovery of precious metals in the previous century (Gontijo 2008). For example, the Eschwege Baron was hired by the Portuguese Crown to investigate the mining potential of Brazil, provide support for the declining gold mining production, and support the rising steel industry. In 1812, he founded the iron factory of the Prata, in the district of Congonhas do Campo, at that time located within the municipality of Ouro Preto (Gontijo 2008; IBGE 2014).

These naturalists made the first species lists for the region. They also described its scenic beauty, and ecological, and anthropological value (Fig. 21.3). The fundamental studies made by Peter Wilhelm Lund and Johannes Eugenius Bülow Warming lead these naturalists to be described as the "father of paleontology in the Americas" and "father of plant ecology", respectively. They worked in the ecotones between rupestrian grasslands, savannas, and dry forests on the limestone of Lagoa Santa (Minas Gerais). In the 20th century, especially from the 1970s on, the biodiversity of the Espinhaço Mountains started to be better known. Research revealed that the region supported one of the most spectacular biotas of Brazil and the world, both in the number of species and endemics, with extremely high local and regional diversity (Jacobi and Carmo 2008a, b; Rapini et al. 2008; see Chap. 6).

The location of the Espinhaço Mountains within three biomes, with extensive altitudinal and latitudinal variation, and isolated mountaintops (insularization) helped contribute to their high biodiversity (Fernandes et al. 2014). Although the Espinhaço area (18 million hectares; Silva et al. 2008) corresponds to only ca. 2 % of Brazil's territory, it hosts about ¼ of the country's bird and mammal species and 14 % of the amphibian species. Espinhaço ridge is also considered a center of plant diversity, as it hosts more than 4,000 vascular plant species, the equivalent to 7 % of the national flora (Table 21.1). But this value is certainly outdated and

Fig. 21.3 Illustration taken from *Flora Brasiliensis* (published between 1840 and 1906 by Carl Friedrich Philipp von Martius, August Wilhelm Eichler, and Ignatz Urban) showing a mountain landscape in the 'Minarum province' (Minas Gerais), with typical elements of rupestrian grasslands, such as rocky outcrops, sparse shrubs, *Vellozia* treelets (*right-hand side*), and individual *Lavosiera* sp., *Cariama cristata* and *Actinocephalus* sp. (*on the bottom*) (*Source* http://florabrasiliensis.cria.org.br/)

Table 21.1 Number of species in some groups of vertebrates and vascular flora for Brazil and the Espinhaço Mountains, and the percentage of Brazilian species that are found in this mountain range

Fauna and flora groups	Number of species in Brazil[a]	Number of species in Espinhaço	% of Brazilian species found at Espinhaço
Vascular flora	56,000	4,000[b]	7.1
Freshwater fishes	2,000	162[c]	8.1
Anurans	775	105[d]	13.5
Birds	1,696	411[e]	24.2
Mammals	541	141[f]	26.1

Source [a]Lewinsohn (2006), [b]Guilietti et al. (1997), [c]Alves et al. (2008), [d]Leite et al. (2008), [e]Vasconcelos and D'angelo Neto (2007), [f]Lessa et al. (2008)

underestimated [for example, Silveira et al. (2015) refers to about 5,000 species in the rupestrian fields—an narrower area than the Espinhaço chain).

The number of endemic species in the region is remarkable (especially in the rupestrian grasslands) and it is probably the largest among the Brazilian vegetation types (Giulietti et al. 1987; Giulietti and Pirani 1988; Harley 1995; Giulietti et al. 1997). For example, the Espinhaço hosts 70 % of all Brazilian species in the Eriocaulaceae family, one of the most representative families of rupestrian grasslands, 85 % of which are endemic to the region (Costa et al. 2008). About 50 % of all species of bromeliads found in Espinhaço are also considered endemic (Versieux et al. 2008) (see also Chap. 6).

The biological importance of the Espinhaço Mountains was identified in several studies that determined priority areas for conservation (Drummond et al. 2005; Costa et al. 2008). For example, the region is considered a center of plant diversity by the World Wide Fund for Nature (WWF) and the International Union for Conservation of Nature (IUCN), as well as an area of high bird endemism (Vasconcelos 2008). It was also considered one of the 238 global ecoregions that harbor exceptional biodiversity according to The Global 200 (Olson and Dinerstein 2002). In recognition of the region's global importance, the United Nations Educational, Scientific and Cultural Organization (UNESCO) designated it the Espinhaço Biosphere Reserve in 2005, an area of about 3 million ha spread over 53 municipalities in Minas Gerais. Since then, efforts have been made to preserve at least the priority areas throughout the Espinhaço Mountains (Silva et al. 2008; Azevedo et al. 2009).

21.3 Ecosystem Services Delivered by the Espinhaço Mountains

Recent studies on the ecosystem services provided by the Espinhaço Mountains, mainly by its rupestrian grasslands, indicate the importance of its natural ecosystems to human well-being at all scales. Some of those services are related to consumptive use and the harvesting of wildflowers and other raw materials. Non-consumptive uses include hiking, cultural and spiritual values, and tourism. Indirect uses include slowing water runoff, flood prevention, soil maintenance, carbon sequestration, nutrient cycling, and crop pollination. Potential future use or option values include biodiversity storage (e.g. plant, animal, mycorrhizae, and other taxa). Finally, there is the existence value of landscapes. Below, we present some important ecosystem services delivered by the Espinhaço Mountains and, whenever possible, their monetary value.

21.3.1 Plant Diversity Storage Service

The flora of the Espinhaço Mountains is used extensively by local people: species such as *mangaba* (*Hancornia speciosa*), *araticum* (*Annona crassiflora*), and *macaúba* palm (*Acrocomia aculeata*) are traditionally used as food. *Macaúba* palm oil was extracted for decades for domestic and commercial use. Resistant woody species, such as the silk floss tree (*Chorisia speciosa*), *vinhático* (*Plathymenia* spp.), and the Spanish cedar (*Cedrela odorata*) were and still are used at minor levels for construction and furniture production. *Candeia* (*Eremanthus erythropappus*, Fig. 21.4a) was used for house lighting and it is still used to build

Fig. 21.4 **a** Candeia (*Eremanthus erythropappus*), used for its wood and high concentration of alpha-bisabolol; **b** dehydration of *sempre-vivas* (*Comanthera* sp.: Eriocaulaceae) for commercialization; **c** flowers made with dried parts of *Chamaecrista orbiculata* (Fabaceae); **d** topiary balls made with dried fruits. *Photos* **a** G. Wilson Fernandes; **b–d** Ana Carolina Neves

fences (although it is protected by law). In addition, this shrub produces high concentration of alpha-bisabolol, a substance with anti-inflammatory, antibacterial, soothing and cicatrizing medicinal properties, traded by large international cosmetic companies. *Pacari* (*Lafoensia pacari*), *espinheira-santa* (*Maytenus* sp.), and *barbatimão* (*Stryphnodendron adstringens*) are also used for medicinal purposes, and more than one hundred species are used as ornamentals (as described below; Instituto Terra Brasilis 1999; Almada et al. 2013). Almada et al. (2013) lists more than 218 plant species of 65 botanical families that are used for several purposes by rural communities located in the southern region of the Espinhaço Mountains. In addition, the genetic prospecting of plants resources and medicines constitutes a promising future research direction, but have been poorly investigated.

Considering the benefits for local human populations and Brazilians as a whole, in situ conservation of the flora is an important ecosystem service provided by the Espinhaço Mountains. Resende et al. (2013) estimated the monetary value of the service of flora storage delivered by the ecosystems of Serra do Cipó in the south of the mountain range. This well studied area supports at least 2,154 plant species, 10 % of which are endemic to the Espinhaço Mountains, 15 % are endemic to Serra do Cipó and 10 % are under extinction threat (Madeira et al. 2008). To estimate the monetary value of floristic diversity storage delivered by the Serra do Cipó, Resende et al. (2013) used the replacement cost method (see Maia et al. 2004), were the monetary value of this ecosystem service is estimated using the cost of maintaining specimens in live plant collections at the botanical gardens of the Zoo-Botanica Foundation of Belo Horizonte. This cost is then applied to the flora in the wild and the annual value that would be spent to maintain this floristic diversity as if it were in a botanical garden. For the species with higher conservation value, i.e. endangered or endemic, the cost of maintaining 500 individuals were considered, and for common species the amount of money needed to support 50 individuals was considered. For endangered or endemic species storage costs equalled US$12.01 million/year and, for the other species, US$8.36 million/year. The taxonomic groups with highest storage value were Dicotyledoneae (1,465 species) and Monocotyledoneae (609 species), due to their high species richness, with US$12.19 million and US$7.53 million per year, respectively. The total value for the floristic diversity storage provided by the ecosystems of Serra do Cipó was US$20.38 million/year. Resende et al. (2013) report that when 1 ha of the region is degraded, society loses about US$644.62/year, considering only the ecosystem service assessed in their study (current dollar exchange rate, values adjusted by the period inflation).

21.3.2 Trade of Dried Ornamental Plants

The collection of plants for sale as ornamentals is an important social and cultural practice (Instituto Terra Brasilis 1999; Monteiro et al. 2012). Hundreds of plant species are collected by rural, low-income families, who sell them either live (orchids, bromeliads, and cacti) or as dried ornamentals (e.g., lichens, moss, and several parts of higher plants, such as twigs, leaves, branches, palm bracts, inflorescences, and fruits; Fig. 21.4a–d). The latter usually have little water in their tissues, and for this reason they keep the shape and color for long periods after being dried in the sun. The group of dry ornamentals with the highest market value and historical significance is the *sempre-vivas* (everlasting plants): inflorescences of the families Cyperaceae, Eriocaulaceae, Poaceae, Rapateaceae, and Xyridaceae, especially of the genus *Comathera* (Eriocaulaceae), which look like tiny daisies (Fig. 21.4a) (Saturnino et al. 1977; Giulietti et al. 1996; Instituto Terra Brasilis 1999; Scatena et al. 2004; Parra et al. 2010).

There have been some small scale initiatives for cultivating plants (e.g., *Comanthera elegans*), whose collection in the wild is prohibited by law (Regulation # 6/2008 of the Ministry for the Environment—Official List of Endangered Species of the Brazilian Flora). However, most sold species continue to come from natural areas, especially from the rupestrian grasslands in the Espinhaço Mountains and the cerrados of the Central Plateau of Brazil. In the Espinhaço Mountains, collecting occurs in the Chapada Diamantina, Bahia, and mainly in the Diamantina Plateau, Minas Gerais. In the latter region collectors of ornamental dry plants are found in 58 municipalities adjacent to the city of Diamantina (Oliveira et al. 2014). Most of the collectors are former diamond prospectors and miners who resorted to flower harvesting after mining declined (Giulietti et al. 1996).

The first record of the trade of *sempre-vivas* dates back to 1930, in Diamantina, Minas Gerais (Instituto Terra Brasilis 1999), though the habit of harvesting and drying plants as ornamentals is older. By the 1940s, growing demand lead to the harvesting of plants in other districts and villages around Diamantina for export. Historical records show that the *sempre-vivas* export peaked in the late 1970s. During this period 900 tons valued at US$3 million were exported per year (see Instituto Terra Brasilis 1999). Today, this trade represents a small fraction of past activity. From 1989 to 2013, the maximum values of all wild derived ornamental dry plants (including *sempre-vivas*, dry leaves, mosses, etc.) exported by Brazil per annum amounted to 940 tons worth US$2.74 million (Neves et al. 2014). From 1989 to 2014, the main importer countries, the United States of America, the Netherlands, and Italy were responsible for the purchase of over US$22 million in ornamental dry plants from Brazil.

Although exports have decreased, the collection and production of ornamental dry plants still involves a network of collectors, processors, middlemen, artisans, traders, and exporters, who work at fairs, markets, shops, workshops, or export warehouses in cities close to the harvesting regions (Datas, Diamantina, São João da Chapada, etc.), and distant cities (e.g., Belo Horizonte, São Paulo, and Brasília). These plants are exported to 54 different countries. Collecting and producing dry plants contributes to the maintenance of rural human populations and the employment of people in the cities. Fieldworkers tend to be local people (Giulietti et al. 1988; Instituto Terra Brasilis 1999), whereas traders and artisans are frequently people who were born in the harvest regions and later migrated to cities, while exporters are mostly older-former collectors (A.C. Neves et al. unpublished data).

On the one hand, the collecting of ornamental dry plants has large social and cultural importance but this trade raises conservation concern due to the restricted geographic distribution of many plant species and possible overharvesting as there are no scientific data to determine the sustainability of use. To induce mass flowering, the collectors burn the grasslands and collect the capitula before the seeds are dispersed, which compromises future recruitment (Giulietti et al. 1996; Instituto Terra Brasilis 1999). Some of the species are in a precarious state, for example, 50 Eriocaulaceae species are on the Official List of Endangered Species of the State of Minas Gerais (Fundação Biodiversitas 2007), and 16 are on the List of the Brazilian Threatened Flora. Excessive collecting is the main reason for the threatened status of several species.

21.3.3 Tourism in Serra do Cipó National Park

Some of the main tourism centers of the Espinhaço Mountains are historical cities such as Mariana, Ouro Preto, Congonhas do Campo, and Diamantina, in Minas Gerais, and Catolé and Lençóis, in Bahia. The first two cities and the Sanctuary of Bom Jesus de Matosinhos, in Congonhas, are considered world heritage sites by UNESCO. In the Espinhaço Mountains, and especially in the rupestrian grasslands, rivers and waterfalls are tourisms hotspots surrounded by exuberant landscapes and unique biodiversity (Fig. 21.5a–d). The geomorphological complexity and diversity, with different rocky formations and topographies, as well as a rich historical-cultural heritage, make the Espinhaço Mountains particularly important for ecotourism. Likewise, the rich fauna of the region, in particular birds (Melo-Júnior et al. 2001), attracts bird watching tourists (Rodrigues et al. 2005; Albano 2010). With all those characteristics, the rupestrian grasslands stand out for attracting a diversity of tourists.

Serra do Cipó is considered a touristic center in the state of Minas Gerais (ICMBio 2009). It receives approximately 100 thousand tourists per year. The Serra do Cipó National Park receives approximately 20 thousand visitors per year. Tourism has been increasing, and amounts to approximately 0.5 % of all visits to Brazilian protected areas in 2008 (MMA 2009). The visitors of Serra do Cipó are attracted not only by its natural beauties, but also by the infrastructure in and around the park. The main reasons to visit the park are visiting waterfalls, trekking, and sightseeing (A.A. Pacheco, unpublished data). Serra do Cipó has more than 1,000 sites suitable for bathing (ICMBio 2009). Aquatic environments (86 %), followed by rupestrian grasslands (62 %), are the main attractions reported by the park's visitors (A.A. Pacheco, unpublished data). Most visitors (60 %) come from the metropolitan region of Belo Horizonte and show a broad variety of professional occupations, consistent with the urban profile of ecotourists and the ideal of 'escaping' from large cities (A.A. Pacheco, unpublished data).

Fig. 21.5 The natural heritage of the Espinhaço Mountains attracts millions of tourists each year. Some attractions are **a** the beautiful landscapes of rupestrian grasslands; **b** Tabuleiro waterfall in Serra Geral do Intendente State Park, Minas Gerais, the third highest waterfall in Brazil (273 m); **c** viewpoints and peaks with up to 2,062 m in height (in this photo: Lapinha peak at Santana do Riacho); **d** the opportunity of practicing cultural/spiritual/adventure/ecotourism visitation. *Photos* **a** Heron Hilário; **b–d** Fernando Resende

21.3.4 Ecosystem Services Associated with Water

Another critical ecosystem service provided by the Espinhaço Mountains is water. This mountain chain harbors myriad springs and contributes to three large Brazilian hydrogeographic regions (Fig. 21.6 and Table 21.2), the São Francisco, Eastern Atlantic, and Southeastern Atlantic. Together, these river basins occupy an area of 1.21 million km^2 and provide water for mining, industry, agriculture, aquaculture, fishing, transportation, and 16.4 GW of hydropower generation (ANA 2013). The rivers support a population of approximately 43 million people with sanitation services, potable water supply, and leisure (IBGE 2010). This includes the traditional local populations of the Espinhaço Mountains, which are particularly dependent on water resources from the mountains (Domingues et al. 2012b). These river basins also have strategic importance, as they supply the population of northern Minas Gerais and other northeastern states, all located in a region that

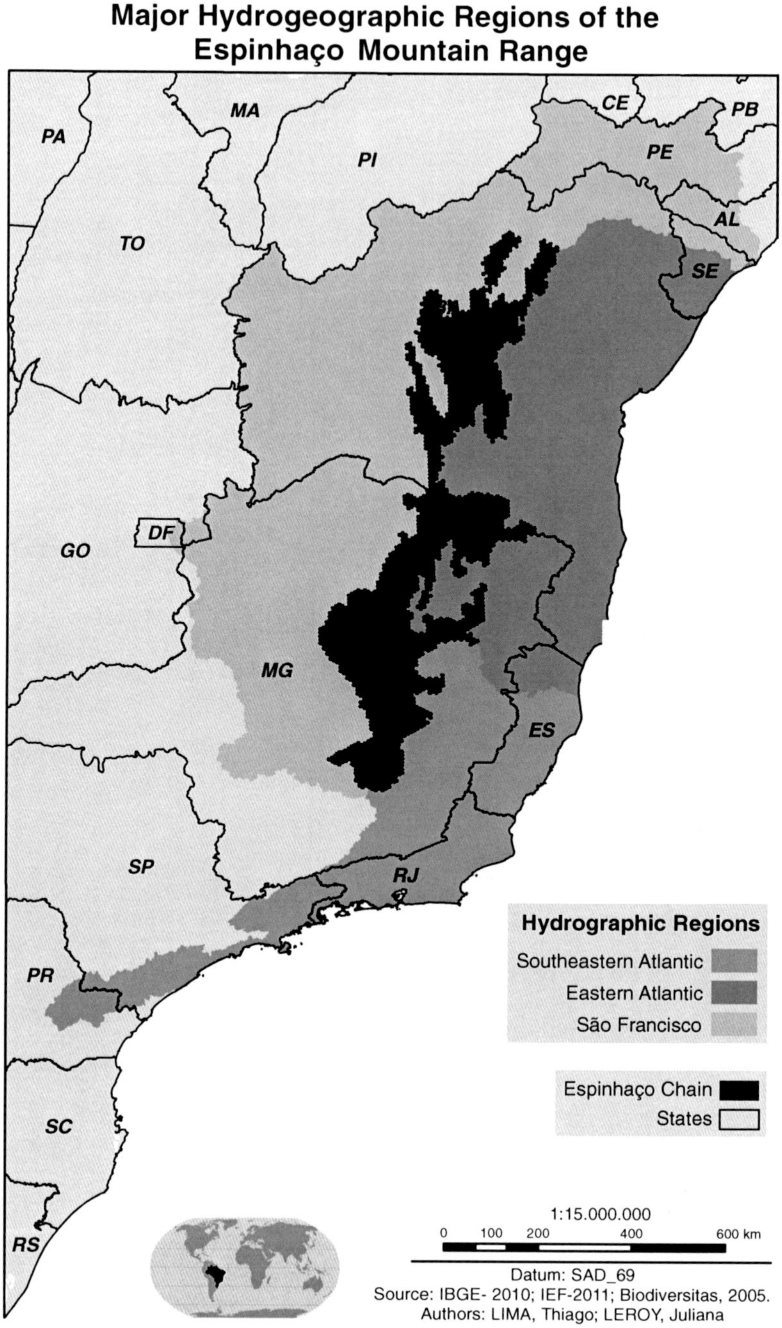

Fig. 21.6 Three major hydrographic regions that drain the Espinhaço mountains. Each one includes several river basins

Table 21.2 Main river basins that receive water from the springs and rivers of the Espinhaço Mountains and the ecosystems services provided

Hydrographic regions	Area ($\times 10^3$ km^2)	Population (millions)	States	Main rivers with springs in Espinhaço Mountains	Hydropower generation (GW)	Number of hydropower plants	Fluvial transport
São Francisco	640	14	MG, BA, AL, PE and SE	Velhas, Verde Grande, Paramirim, Vereda do Jacaré	10.7	63	National importance
Eastern Atlantic	390	15	ES, MG, BA, SE	Jequitinhonha, rio de Contas, Pardo, Itapicuru, Paraguaçu, Jacuípe	1.1	20	Port activities; local navigation
Southeastern Atlantic	187	13.4	MG, ES, RJ, SP, PR	Santo Antônio	5.3	163	Port activities; local navigation
Total	1,217	42.4	ES, MG, SP, PR, RJ, BA, AL, PE, SE	–	17.1	246	–

AL Alagoas, *BA* Bahia, *ES* Espírito Santo, *MG* Minas Gerais, *PR* Paraná, *PE* Pernambuco, *RJ* Rio de Janeiro, *SP* São Paulo, *SE* Sergipe experiences the highest water shortages in the country, where there is a high water demand/availability ratio (ANA 2013). Finally, they supply directly the metropolitan regions of Belo Horizonte, Salvador, and Aracaju, important state capitals that harbor industrial, mining, and agricultural regions and over 10 million inhabitants.

The São Francisco Hydrographic Region is formed by the São Francisco River, considered a river of national unity, as it connects the *sertão* region to the coast, and integrates ethnic and political entities of Brazil. The source of this river is in rupestrian grassland at Serra da Canastra mountains, Minas Gerais state, and it is fed by tributaries from the entire extension of the Espinhaço Mountains. The São Francisco river basin occupies an area of almost 640 thousand km^2, and supports more than 14 million people. It also feeds 63 hydropower plants (ANA 2013, 2015a). The waterway is also very important for transportation in Brazil as it is navigable for almost its entire length. The Middle São Francisco River (from northern Minas Gerais to the border between Pernambuco and Bahia) is particularly important (ANA 2013). The largest tributary of the São Francisco is the Velhas River, which is 761 km long and also fed by tributaries originating in the Espinhaço Mountains. The origin of the Velhas River is in the rupestrian grasslands near the town of Ouro Preto. It supplies water for over 4 million people (Camargos 2005), including inhabitants of the metropolitan region of Belo Horizonte, from where it receives a huge load of effluents that severely pollutes its waters (see also Chap. 5).

The Eastern Atlantic hydrographic region consists of the entire eastern part of the Espinhaço Mountains. It covers an area of 390 thousand km^2, with a population estimated as over 15 million inhabitants (IBGE 2010; ANA 2015b). Most of these people live in the states of Bahia and Minas Gerais. The basin includes two state capitals, Aracaju and Salvador. Human activities are denuding much of the native vegetation here, in particular in the Atlantic Forest biome. The eastern Atlantic hydrographic region is composed of several river basins, whose main rivers and tributaries originate in the rupestrian grasslands of the Espinhaço Mountains.

The southern Atlantic hydrographic region is partially supplied by waters coming from the eastern side of the Espinhaço Mountains. This region extends over 187 thousand km^2 and its rivers supplies water for a population of 13.4 million people. The main population benefitting from these waters is located in the Vale do Aço Metropolitan Region (Steel Valley), Minas Gerais, where approximately 500 thousand people live, in the Doce River Basin. If in the past the waters of the region were important for gold production. Today the waters of the Santo Antônio River Basin are used mainly for fishing, leisure, tourism, domestic purposes, and sand mining (ECOPLAN-LUME 2010).

21.4 Industry and Ecosystem Services: A Conflict of Interests

The main threat to ecosystem services of the region is the conversion of multifunctional into monofunctional landscapes. Such conversion occurs because the cost of ecosystem service loss is not internalized, private/short term gain is prioritized over public-long term needs, and the cost-benefit ratio is distorted by tax or governmental incentives (MEA 2005).

For example, tropical forests are multifunctional systems that provide benefits, such as non-timber forest products, sedimentation control, flood prevention, carbon storage, intrinsic values, and biodiversity maintenance. The conversion of tropical forests to agriculture for food production results in a mono-functional landscape and the loss of many ecosystem functions. In Mount Cameroon, Cameroon, the conversion of tropical forest to small-scale agriculture yielded private benefits (food production) of about $2,000 per hectare, but low-impact logging and maintenance of the forest yielded $3,400 per hectare (Yaron 2001).

In the Espinhaço Mountains, the most important trade-off between economic activities and ecosystem services is by mining. Mining benefits one product (minerals) at the expense of many other ecosystem services. Mining provides large private and short-term benefits. However, it changes or destroys landscapes and important landmarks, such as the Itabirito Peak—an obelisk of compact hematite with 1,568 m in height that attracted the attention of naturalists such as Carl Friedrich Philipp von Martius (Fig. 21.7a). In spite of being located in a protected area at the south of the metropolitan region of Belo Horizonte (APA SUL) and being a natural historical heritage site listed in the constitution of the state of Minas Gerais since 1989, its base was completely altered due to mining for iron ore (Fig. 21.7b). Mining also has other strong and irreversible direct and indirect impacts on biodiversity and ecosystem services. Mining proposals ought to be carefully evaluated through environmental perspectives before projects begin. The growth of urban areas, construction of roads, and silviculture projects are other examples of activities that will include trade-offs between ecosystem services and natural landscapes. In the following section we present some case studies in detail.

21.4.1 Conflicts Over Water Use

An important conflict of interests related to the use of natural resources involves water use by mining activities. The use of water from springs, for iron ore quarrying, processing, and transporting, is very controversial. In addition, complaints against mining companies are frequent. They include accusations of spring destruction, drainage and siltation of streams, and water contamination (GESTA 2015).

(a)
D. Stephan ad nat.
A. Brandmeyer in lap. del
E Septentrione visum.
CACUMEN MONTIS PICO D'ITABIRA DO CAMPO,
IN PROV. MINARUM.
(b)

◂ **Fig. 21.7** The Itabirito Peak, in the municipality of Itabirito (Minas Gerais State), is an iconic example of the destruction of a geographical/historical/identity landmark by mining. **a** The peak, as it was seen by the naturalist Carl Friedrich Philipp von Martius; **b** Its contemporary landscape completely transformed by iron ore mining, in spite of it being located in a protected area (APA SUL) and designation as a natural historical heritage site in the constitution of the state of Minas Gerais since 1989. *Source* **a** http://florabrasiliensis.cria.org.br/; **b** Google Maps

The use of large pipes that transport iron ore in aqueous medium is one of the most controversial issues. In the state of Minas Gerais there are several small-sized and four large-sized ore pipelines already in operation. They transport iron ore from quarries in the Espinhaço Mountains (Conceição do Mato Dentro, Congonhas, Mariana, and Morro do Pilar) to harbors in the states of Rio de Janeiro and Espírito Santo. According to Siqueira (2015), the water volume used by all these projects together reaches 19,350 m^3/hour. It is enough to supply approximately 2.9 million people, or to meet the demand of almost 50 % of the metropolitan region of Belo Horizonte.

The building of the largest ore pipeline in the world, close to the Serra do Cipó region, the so-called Minas-Rio ore pipeline is causing great controversy. With a transportation capacity of over 25 million tons of ore per year in an aqueous medium, this ore pipeline would extend over 529 km from Conceição do Mato Dentro (Minas Gerais state) to the harbor of Açu (RJ) (Fig. 21.8a, b). The operation of the ore pipeline and the mine involves removing 2.5 million liters of water per hour from the Peixe River, enough water to supply a city of over 350 thousand people. However, the company responsible for the enterprise was accused of irregularities and denounced several times in public hearings of the Extraordinary Commission of Waters of the Legislative Assembly of Minas Gerais, and in complaints by the Federal Public Ministry and by farmers made to the Supreme Federal Court (Becker and Pereira 2011; Tempo 2014; GESTA 2015). In spite of those societal, legal and procedural challenges, the Operation Permit for the ore pipeline was granted by the IBAMA the federal regulating authority at the end of 2014.

Other ongoing conflicts include water use by eucalyptus plantations in northern Minas Gerais (see Chap. 17). Here the goal is to supply charcoal to the steel industry of Minas Gerais. This activity is thought to dry out watercourses in a region with natural water deficit. There are also large ongoing irrigation projects, such as the one in Brumado (BA) (see Almeida and Pinto 2012) that will also drawdown water levels. Perhaps the most controversial enterprise is a project for diverting the water of the São Francisco River to the drainage basins of other rivers in several states of northeastern Brazil (see IPEA 2011). Lastly, additional dams are planned for the generation of hydropower.

In 2014 and 2015, the metropolitan regions of southeastern Brazil experienced a severe water crisis, due to the emptying of water reservoirs in the states of Minas Gerais, Rio de Janeiro, and São Paulo (Dobrovolski and Rattis 2015, Coutinho et al. 2015). In the middle of the summer, usually a rainy period, approximately 140 municipalities in Minas Gerais were forced to adopt restriction measures to preserve their water supply. The metropolitan region of Belo Horizonte was one of the most

Fig. 21.8 The installations of the largest ore pipeline in the world in the region of Conceição do Mato Dentro (municipality of Santo Antônio do Rio Abaixo, Minas Gerais) caused many social and environmental damages. **a** Silting of springs and creeks in a rain forest and riverine vegetation; **b** large landslides with active erosions draining into springs and rivers. *Photos* G. Wilson Fernandes

affected regions. The conflicts between land use and water provisioning for the human population inflame all discussion about the tradeoffs regarding ecosystem services.

21.4.2 *Social Impacts of Mining on the Region of Conceição do Mato Dentro*

The municipality of Conceição do Mato Dentro, near Serra do Cipó, is commited to the development of eco-, adventure-, and cultural-historical tourism. Due to its scenic beauty, waterfalls, baroque architectural heritage, and rich culture, this region is considered the ecotourism capital of Minas Gerais (Barbieri et al. 2014). The area from Conceição do Mato Dentro to Serro is also an active mineral reserve for itabirite, bauxite, and chromite (Barbieri et al. 2014). Conflict in the tourism and mining occurs because new mining investments have been made in the region, generating high economic expectations but also concern that the natural and cultural heritage will be destroyed.

Although nearly the entire region is already the object of mining decrees (Fig. 21.9), only one enterprise is currently in operation. However, mining activities are also planned for the municipalities of Alvorada de Minas (pellet plant), Dom

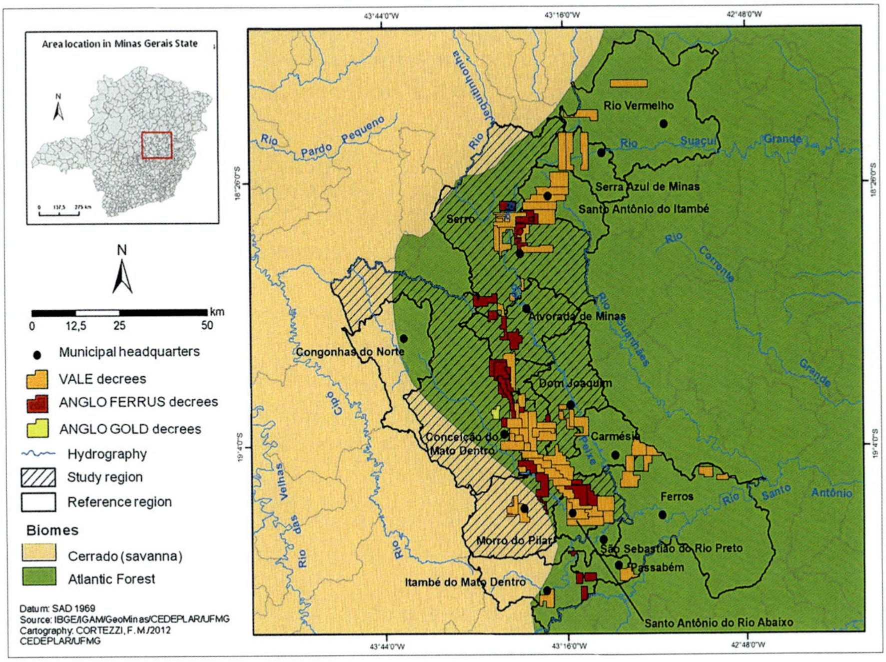

Fig. 21.9 Mining claims by the companies Vale, Anglo Ferrous and Anglo Gold in the region of Conceição do Mato Dentro, Minas Gerais, southern part of the Espinhaço Mountains (adapted from Barbieri et al. 2014)

Joaquim (water supply), and Serro (ore pipelines). The distribution infrastructure to transport mineral slur to harbors, though, will cross and additional 33 municipalities in the states of Minas Gerais and Rio de Janeiro, increasing mining impacts over an even larger territory. Part of this development corresponds to a mining decree presently obtaining a pre-operation permit. This project will establish a mine and operation center in the municipality of Morro do Pilar. Finally, there are further potential mining developments planned for Conceição do Mato Dentro and Morro do Pilar (Barbieri et al. 2014).

The impacts of the mining projects will affect an underdeveloped region, especially its economy and infrastructure. The fragile basis of a local economy that relies on agriculture, livestock farming, and tourism, with a precarious road network and a small population is in conflict with the infrastructure needed to support mining activities. The region's population, as of 2010, was 97,888 people, with 17,908 in Conceição de Mato Dentro and 3,399 in Morro do Pilar. During the operational phase the labor demand will probably not generate a significant population increase in the region (Barbieri et al. 2014); however, the increase in the floating population during the building phase may have dramatic social, environmental, and economical impacts. Conceição de Mato Dentro may have its population almost doubled during the peak period. This scenario tends to be repeat when other mining enterprises develop in the region.

On one hand, mining enterprises should increase the income and revenues but, there are important challenges for the infrastructure of the region during the first few years. According to Barbieri et al. (2014), mining enterprises in the middle Espinhaço generate a lot of capital but employ a small workforce. In addition, most of the requisitioning of goods and services is from outside the region, which means few possibilities for economic and productive diversification within the region.

The greatest challenge is how to prevent the region from becoming exclusively dependent on mining income and becoming a "company town". There is a need to improve and consolidate the existing agroindustry, manufacturing industry, tourism, and other extant productive services, as well as to remunerate Municipalities for ecosystem services.

Finally, it is worth mentioning that mining in the region has worsened conflicts related to land use and water resources. Historically these resources have been associated with systems of extreme concentration of land ownership and extensive livestock farming. According to Barbieri et al. (2014), the arrival of large-scale mining triggers a processes of land ownership concentration and displacement of traditional and rural populations, in particular in regions where these populations are already victims of intensive socio-economical and land exclusion processes.

21.4.3 Environmental Losses in the Iron Quadrangle

The Iron Quadrangle is an area of approximately 7,200 km^2, that includes part of the metropolitan region of Belo Horizonte (with 4.8 million people according to the

Demographic Census of 2010). With a history of 300 years of mining, this is one of the most important mineral region in the world, from which approximately 75 % of the iron ore extracted in Brazil is derived, as well as aluminum, manganese, gold, and other minerals. The main environmental problems associated with open-pit mining are the removal and the discharge of ironstone outcrops and their associated biota to reach deep deposits (up to 300 m). This digging frequently exposes ground water (Jacobi and Carmo 2008a). Excavation also destroys springs, drainage systems, and causes the siltation of streams. Mining wastes can also contaminate ground and surface water bodies with heavy metals and toxic elements. This contamination can occur when waste dams collapse, as has repeatedly occurred in the region. Associated urban growth has also led to habitat loss and the complete transformation of the landscape of the Iron Quadrangle.

Recent research (e.g. Viana and Lombardi 2006; Jacobi et al. 2007; Mourão and Stehmann 2007; see Jacobi and Carmo 2008b) has revealed remarkable species richness and endemism in the region, as well as extremely high local and regional diversity, in particularly in ecosystems associated with ironstone outcrops (locally known as *cangas*). For example, a floristic inventory of four disjunct *cangas* covering an area less than 300 ha in extent produced a list of 458 species of vascular plants from 86 families, with only 5 % of plants common to all *cangas* (Jacobi and Carmo 2008a). The *cangas* also supported several metallophytes (plants able to grow in the presence of toxic metals). These can deliver ecological services, such as phytoextraction, phytostabilization, and phytoprospection (Jacobi and Carmo 2008a).

Nevertheless, before being fully discovered, this heritage is being lost. Approximately 90 % of the ironstone outcrops are owned by mining companies and some of the few areas whose flora was inventoried have already been destroyed by mining (Jacobi and Carmo 2008b). There are only two areas of ironstone outcrops in fully protected areas in the Iron Quadrangle: Serra do Rola Moça State Park, with 3.94 thousand ha, created in 1994, and Serra do Gandarela State Park, with 34.40 thousand ha, created in 2014. There are also some located in private protected areas and areas of sustainable use, such as the Environmental Protection Area of Metropolitan Belo Horizonte (APA Sul RMBH), which protects very limitedly the natural heritage in terms of restricting land use forms.

Hence, without any knowledge of the biodiversity and ecosystem services prior to the arrival of mining and urban expansion in the region, we can only guess at what was lost due to the historical destruction of many ironstone outcrops.

21.5 Resource Generation by Mining and Investment in Natural Capital

The projections of the fiscal capacity of the mining municipalities in the Espinhaço Mountains, together with the future economic and demographic scenario for the region, suggest increasing income, consumption, and employment growth related to mining. This will occur in an area currently projected to grow at below the state average (Barbieri et al. 2014). However, job creation, increased income, and improved domestic trade will come at the expense of ecosystem function and services and biodiversity, creating a need for tax systems to compensate inhabitants and local governments for the depletion of environmental resources.

One possible tax mechanism is the Compensation for the Exploitation of Mineral Resources (Compensação Financeira pela Exploração Mineral, CFEM), by which mining companies financially compensate the government for the extraction of natural resources. These mining royalties are calculated based on the net amount obtained from product sales (i.e., after deduction of all marketing, transportation, and insurance costs). However, this corresponds to a mere 0.2 % for precious, colored, and carbonated stones, and precious metals; 1.0 % for gold; 2.0 % for iron; and 3.0 % for aluminum, manganese, rock salt, and potassium (Domingues et al. 2012a). To better compensate for the loss of ecosystem services and biodiversity a more realistic picture of the counterpart value for the exploitation of mineral goods in Brazil, might be similar to the highest tax charged in South Africa (0,5–7,0 %; PWC 2012) and Bolivarian Republic of Venezuela (3 %; Otto et al. 2006), especially when one consider that the exploitation of iron ore in Brazil is extremely cheap in comparison with other countries (Santos 2012). Only two-thirds (65 %) of the mining royalties collected by the federal government is allocated to the municipality from which minerals were extracted. An additional 23 % go to the state and 12 % remains with the federal government (Domingues et al. 2012a). For example, the annual collection of mining royalties in the Conceição do Mato Dentro region, during the operation phase, may reach an annual amount of approximately US$32.36 million (current dollar exchange rate) (Barbieri et al. 2014).

Barbieri et al. (2014) proposed that this amount could be used to start a regional development fund aimed at promoting sustainable development in the region and for funding projects for the conservation and management of natural areas. However, for this proposal to be considered requires drastic change in the mentality and common sense of power bases in terms of land use management. Such change would involve the creation of a planning structure that anticipates impacts and manages the territory, and also that integrates several planning sectors. Planning should anticipate any development interventions in the territory, be they private or public. A clear example of this approach is seen in the process of licensing large business ventures, which is detached from local and regional planning.

From this perspective, it is possible to create strategies for using mining income not only to minimize its impacts, but also to generate positive synergies between this activity and other factors (social, economical, cultural, environmental, etc.) to produce sustainable "green" regional development. In this context, planning depends on the creation of a capacity for institutional linking, and the integration of sectoral planning practices (Santos and Barbieri 2012).

A *sine qua non* condition for the accomplishment of this alternative approach is for public authorities and other institutions to coordinate between agents with multiple interests (companies, civil society, etc.). The objectives of the strategy could involve *ex-ante* planning processes, the correction of power asymmetries among agents, guarantee of rights of vulnerable social groups (minority or not), regulation and conservation of public or common properties and ecosystem services to prevent misappropriation by agents, and, finally, guarantee of internalization by entrepreneurs of the negative externalities generated during intervention in the territory (Barbieri et al. 2014).

Besides mining royalties, there is also the ecological ICMS (Tax over the Circulation of Goods and Services). This ecological tax was created to compensate municipalities for the economic restrictions they face for maintaining large protected areas, and to encourage them to invest in conservation. A municipality may benefit from this tax by delimiting or extending a protected area, or by investing in an existing area to improve its conservation. The share of revenue to which each municipality is entitled is defined by an index derived from the municipality's biodiversity conservation coefficient. In Minas Gerais, protected areas have grown considerably in size and number since this ecological tax was first implemented. Two-thirds (2.09 million ha) of the 3.08 million ha covered by the Serra do Espinhaço Biosphere Reserve are now protected. A successful example of the benefits of this system is exemplified by São Gonçalo do Rio Preto: the annual revenue for this municipality increased 3,691 % after the creation of the ecological tax. A total of US$10.29 million were transferred to the municipalities in 2009. This represents an average of ~US$4.53 per ha/year (current dollar exchange rate, values adjusted for inflation) (Domingues et al. 2012b). However, Resende et al. (2013) pointed out that the provision of a single ecosystem service (floristic diversity stock) corresponds to US$644.64 per ha/year in Serra do Cipó. Domingues et al. (2012a, b) discussed the potential of these two mechanisms for promoting sustainable development in the Espinhaço Biosphere Reserve. Thus, the revenue generated from the tax remains modest in comparison to the value of the services delivered by the natural ecosystems of the Espinhaço Biosphere Reserve. In reality neither the mining royalties nor the ecological tax is sufficient for compensating for the value of ecosystem services used and degraded (Domingues et al. 2012a, b).

Currently, there are no mandatory environmental criteria for the investment of royalties from mining. As a result, the protected areas receive inadequate financial benefits from the mining industry for their use of land and water (Domingues et al. 2012a, b). In addition, the distribution of mining royalties is extremely concentrated in few hands. In 2009, more than half of the royalties went to three municipalities: Nova Lima (US$36.88 million), Itabira (USR33.80 million), and Mariana (US $29.63; current dollar exchange rate, values adjusted for inflation) (Domingues et al. 2012a, b). It would be extremely beneficial for these municipalities to team up with the mining sector and invest the royalties for promoting 'green municipalities' and to cover the cost of maintaining and restoring damaged ecosystems (Domingues et al. 2012a, b). Mining companies should also develop protected areas of their own. This would help create a more extensive mosaic of private and public protected areas. They should also restore mining areas and degraded adjacent rivers. These activities would help improve the relationship between mining companies and local communities. Municipalities that do not earn mining royalties because they do not have protected areas could benefit by developing protected areas to increase revenue derived from the ecological tax (Domingues et al. 2012a, b).

There is a pressing need for mining companies to increase overall royalty rates, and for municipalities to receive higher rates from the ecological tax for the maintenance of protected areas. Studies of the valuation of ecosystem services can contribute to the establishment of fairer rates, by taking into account the multitude of ecosystem services provided by natural ecosystems. Hence, studies that estimate the value of ecosystem services per area/per year should be encouraged.

The cruel paradox behind the mechanism of mining royalties and the ecological tax is that their revenue is indexed on net invoicing obtained from product sales, and on consumer spending, respectively. That is, more mineral exploitation and increased consumption of resources, means that municipalities receive more compensation for environmental losses from mining royalties and the ecological ICMS. At the same time, the government will only be able to reduce the pressure on ecosystems by adopting more sustainable patterns of consumption for goods and services. Fixing this conundrum will take much more than the adoption of a new management system or new laws (Domingues et al. 2012b).

The objective of this discussion is not to establish a 'mining towards preservation' strategy. On the contrary, the intention is to establish proper compensation to communities, local populations, and nature for the conversion and degradation of multifunctional ecosystems to mining areas. Without mining impacts these areas benefit society by providing ecosystem services freely. A path towards long term sustainable socio-economic and environmental development requires a critical analysis on the hegemonic model of development through unlimited economic growth, flowed by a deep shift in paradigms (Daly and Farley 2004; Coelho et al. 2013). Such a change implies reducing society's consumption levels, the transformation of the Brazilian production matrix (which is largely based in commodities), and the internalization of the ecosystem services in the calculation of compensation values for the use and degradation of ecosystems and biodiversity. In the medium term, the measures suggested in the present study can contribute to the reduction of mining impacts.

Acknowledgments We thank CAPES, FAPEMIG and CNPq several types of grants to the authors. We also thank Dr. José M. V. Fragoso for careful reading the manuscript and for his helpful comments and suggestions.

References

Albano C (2010) Birding in north-east Brazil, part 2: the vast state of Bahia. Neotrop Bird 7:49–61

Almada ED, Coelho MS, Quintino AV, Fernandes GW, Sánches-Azofeifa A (2013) Traditional ecological knowledge of rural communities in areas of seasonally dry tropical forest in Serra do Cipó, Brazil. In: Sánchez-Azofeifa A, Powers JS, Fernandes GW, Quesada M (eds) Tropical dry forests in the Americas: ecology, conservation, and management, 1st edn. CRC, Boca Raton, pp 429–451

Almeida NO, Pinto JESS (2012) Seca e política de recursos hídricos na bacia do rio Brumado, Bahia. Rev GeoNorte 2:1133–1145

Alves CBM, Leal CG, Brito MFG, Santos AKA (2008) Biodiversidade e conservação de peixes do Complexo do Espinhaço. Megadiv 4:201–220

ANA (2013) Atlas geográfico digital de recursos hídricos do Brasil. http://www2.snirh.gov.br/atlasrh2013/. Accessed 2 Feb 2015

ANA (2015a) Região hidrográfica do São Francisco. http://www2.ana.gov.br/Paginas/portais/bacias/SaoFrancisco.aspx. Accessed 3 Feb 2015

ANA (2015b) Região hidrográfica atlântico leste. http://www2.ana.gov.br/Paginas/portais/bacias/AtlanticoLeste.aspx. Accessed 3 Feb 2015

Azevedo AA, Goulart M, Silva JA, Vilhena CF (2009) Mosaico de unidades de conservação do espinhaço: alto Jequitinhonha-Serra do Cabral: processo de criação e implantação. Instituto Biotrópicos, Diamantina 55p

Barbieri AF, Domingues EP, Monte-Mór RLM, Santos FBT, Rigotti JIR, Rodrigues BHM, Tonucci J, Neto OB, Cortezzi F (2014) Plano regional estratégico em torno de grandes projetos minerários no médio Espinhaço. FUNDEP, Belo Horizonte 712p

Becker LC, Pereira DC (2011) O projeto Minas Rio e o desafio do desenvolvimento territorial integrado e sustentado: a grande mina em Conceição do Mato Dentro (MG). In: Fernandes FRC, da Enríquez MARS, de Alamino RCJ (eds) Recursos minerais & sustentabilidade territorial. CETEM/MCTI, Rio de Janeiro, pp 229–258

Camargos LMM (2005) Plano diretor de recursos hídricos da bacia hidrográfica do rio das Velhas. IGAM, Belo Horizonte 228p

Coelho MS, Resende FM, Almada ED, Wilson Fernandes G (2013) Crescimento econômico e a moderna crise ambiental: uma análise crítica. Neotrop Biol Cons 8:53–62

Costa FN, Trovó M, Sano PT (2008) Eriocaulaceae na cadeia do espinhaço: riqueza, endemismo e ameaças. Megadiv 4:117–125

Coutinho RM, Kraenkel RA, Prado PI (2015) Catastrophic regime shift in water reservoirs and São Paulo water supply crisis. PLoS ONE 10(9):1–14

Daly HE, Farley J (2004) Ecological economics: principles and applications. Island Press, Washington

Dean W (1996) A ferro e fogo. Companhia das Letras, São Paulo

Dobrovolski R, Rattis L (2015) Water collapse in Brazil: the danger of relying on what you neglect. Natureza & Conservação 13(1):80–83

Domingues SA, Karez CS, Biondini IVF, Andrade MA, Fernandes GW (2012a) Economic environmental management tools in the Serra do Espinhaço biosphere reserve. J Sustain Develop 5:180–191

Domingues SA, Karez CS, Biondini IVF, Andrade MA (2012b) Putting a price on conservation. A World Sci Nat Sci Q Newsl 10:20–23

Drummond GM, Martins CS, Machado ABM, Sebaio FA, Antonini Y (2005) Biodiversidade em Minas Gerais: um Atlas para sua Conservação, 2nd ed. Fundação Biodiversitas, Belo Horizonte, 222p

ECOPLAN-LUME (2010) Plano integrado de recursos hídricos da bacia hidrográfica do rio Doce. IGAM, Belo Horizonte, 478p

Fausto B (2008) História concisa do Brasil. Editora da Universidade de São Paulo, São Paulo

Fernandes GW, Barbosa NPU, Negreiros D, Paglia AP (2014) Challenges for the conservation of vanishing megadiverse rupestrian grasslands. Nat Cons 12:162–165

FUNDAÇÃO BIODIVERSITAS (2007) Revisão das listas das espécies da flora e da fauna ameaçadas de extinção do estado de Minas Gerais. Biodiversitas, Belo Horizonte, 104p

Furtado JF (2009) Chica da Silva e o contratador dos diamantes - O outro lado do mito. Editora Companhia das Letras, São Paulo

GESTA (2015) Observatório dos conflitos ambientais de Minas Gerais, resistência contra a construção do mineroduto minas – Rio. http://conflitosambientaismg.lcc.ufmg.br/conflito/?id=549. Accessed 27 Feb 2015

Giulietti AM, Pirani JR (1988) Patterns of geographic distribution of some plant species from the Espinhaço range, Minas Gerais and Bahia, Brazil. In: Vanzolini P, Heyer W (eds) Proceedings of a workshop on neotropical distribution patterns. Academia Brasileira de Ciências, Rio de Janeiro, pp 39–69

Giulietti AM, Menezes NL, Pirani JR, Meguro M, Wanderley MGL (1987) Flora da Serra do Cipó, Minas Gerais: caracterização e lista das espécies. Bol Bot Univ São Paulo 9:1–151

Giulietti N, Giulietti AM, Pirani JR, Menezes NL (1988) Estudos em sempre-vivas: importância econômica do extrativismo em Minas Gerais, Brasil. Acta Bot Bras 1:179–193

Giulietti AM, Wanderley MDGL, Longhi-Wagner HM, Pirani JR, Parra LR (1996) Estudos em "sempre-vivas": taxonomia com ênfase nas espécies de Minas Gerais, Brasil. Acta Bot Bras 10:329–377

Giulietti AM, Pirani J, Harley RM (1997) Espinhaço range region. Eastern Brazil. In: Davis SD, Heywood VH, Herrera-MacBryde O, Villa-Lobos O, Hamilton AC (eds) Centres of plant diversity. A guide and strategies for the conservation, vol. 3. The Americas. IUCN Publication Unity, Cambridge, pp 397–404

Gontijo BM (2008) Uma geografia para a Cadeia do Espinhaço. Megadiv 4:7–15

Goulart EMA (2009) O Caminho dos currais do Rio das Velhas: a Estrada Real do Sertão. COOPMED, Belo Horizonte

Harley R (1995) Introduction. In: Stannard B (ed) Flora of the Pico das Almas, Chapada Diamantina, Bahia, Brazil. Royal Botanic Gardens, Kew, pp 1–40

IBGE (2010) Censo Populacional 2010. http://www.ibge.gov.br/home/. Accessed 18 Feb 2015

IBGE (2014) Indicadores IBGE - Contas nacionais trimestrais indicadores de volume e valores correntes. ftp://ibge.gov.br/Contas_Nacionais/Contas_Nacionais_Trimestrais/Fasciculo_Indicadores_IBGE/pib-vol-val_201403caderno.pdf. Accessed 18 Feb 2015

IBGE (2015) Instituto Brasileiro de Geografia e Estatística. http://www.cidades.ibge.gov.br/xtras/home.php. Accessed 4 May 2015

IBRAM (2012) Informações e análises da economia mineral Brasileira. http://www.ibram.org.br/sites/1300/1382/00002806.pdf. Accessed 5 Aug 2015

IBRAM (2013) Informações sobre a economia mineral do estado de Minas Gerais. http://www.ibram.org.br/sites/1300/1382/00003791.pdf. Accessed 5 Aug 2015

ICMBio (2009) Plano de manejo do Parque Nacional da Serra do Cipó. ICMBio, Brasília, 102 p

INSTITUTO TERRA BRASILIS (1999) Projeto sempre-vivas: subsídios para seu uso sustentado. Instituto Terra Brasilis, Belo Horizonte, 123 p

IPEA (2011) Transposição do Rio São Francisco: Análise de oportunidade de projeto. http://repositorio.ipea.gov.br/bitstream/11058/1418/1/TD_1577.pdf. Accessed 27 Feb 2015

Jacobi CM, Carmo FF (2008a) Diversidade dos campos rupestres ferruginosos no Quadrilátero Ferrífero, MG. Megadiv 4:25–33

Jacobi CM, Carmo FF (2008b) The contribution of ironstone outcrops to plant diversity in the Iron Quadrangle, a threatened Brazilian landscape. Ambio 37:324–326

Jacobi CM, Carmo FF, Vincent RC, Stehmann JR (2007) Plant communities on ironstone outcrops: A diverse and endangered Brazilian ecosystem. Biodiv Cons 16:2185–2200

Leite IB (1996) Antropologia da viagem: escravos e libertos em Minas Gerais no século XIX. Editora UFMG, Belo Horizonte

Leite FF, Juncá FA, Eterovick PC (2008) Status do conhecimento, endemismo e conservação de anfíbios anuros da Cadeia do Espinhaço, Brasil. Megadiv 4:158–176

Lessa LG, Costa BMA, Rossoni DM, Tavares VC, Dias LG, Moraes Jr EA, Silva JA (2008) Mamíferos da Cadeia do Espinhaço: riqueza, ameaças e estratégias para conservação. Megadiv 4:218–232

Lewinsohn TM (2006) Avaliação do estado do conhecimento da biodiversidade brasileira – vol. I and II. Ministério do Meio Ambiente, Brasília 520p

Machado IF, Figueirôa SFM (2001) 500 years of mining in Brazil: a brief review. Res Pol 27:9–24

Madeira JA, Ribeiro KT, Oliveira MJR et al (2008) Distribuição espacial do esforço de pesquisa biológica na Serra do Cipó, Minas Gerais: subsídios ao manejo das unidades de conservação da região. Megadiv 4:257–271

Maia AG, Romeiro AR, Reydon BP (2004) Valoração de recursos ambientais – metodologias e recomendações. IE/Unicamp, Campinas 116 p
MEA (2005) Ecosystems and human well-being: biodiversity synthesis. World Resources Institute, Washington
Melo-Júnior TA, De Vasconcelos MF, Fernandes GW, Marini MA (2001) Bird species distribution and conservation in Serra do Cipó, Minas Gerais, Brazil. Bird Cons Int 11:189–204
MMA (2009) Pilares para a sustentabilidade financeira do sistema nacional de unidades de conservação. Ministério do Meio Ambiente, Brasília, 72 p
Monteiro FT, Pereira DB, Gaudio RS (2012) Os(as) apanhadores(as) de flores e o Parque Nacional das Sempre-Vivas: entre ideologias e territorialidades. Soc Nat 3:419–433
Mourão A, Stehmann JR (2007) Levantamento da flora do campo rupestre sobre canga hematítica couraçada remanescente na Mina do Brucutu, Barão de Cocais, Minas Gerais, Brasil. Rodriguésia 58:775–786
Neves ACON, Negreiros D, Fazito D, Goulart MF, Echternacht L, Silva SR, Fernandes GW (2014) Brazilian dried ornamental plants: provisioning ecosystem services and livelihoods. In: Anais do I Simpósio Internacional de Ecologia da Conservação, Universidade Federal de Minas Gerais, Belo Horizonte, 25–27 Aug 2014
Oliveira MNS, Tanaka MK, Cabrini EC (2014) "Apanha de flores" no espinhaço meridional: importância, abrangência e diversidade. In anais do III Colóquio Internacional Sobre Povos e Comunidades Tradicionais, Unimontes, Montes Claros, 22–35 Apr 2014
Olson DM, Dinerstein E (2002) The global 200: priority ecoregions for global conservation. Ann Miss Bot Gard 89:199–224
Otto J, Andrews C, Cawood F, Doggett M, Guj P, Stermole F, Stermole J, Tilton J (2006) Mining royalties – a global study of their impact on investors, government, and civil society. The International Bank for Reconstruction and Development/The World Bank, Washington 296p
Parra LR, Giulietti AM, de Andrade MJG, van den Berg C (2010) Reestablishment and new circumscription of *Comanthera* (Eriocaulaceae). Taxon 59:1135–1146
PWC (2012) Corporate income taxes, mining royalties and other mining taxes – A summary of rates and rules in selected countries. https://www.pwc.com/gx/en/energy-utilities-mining/publications/pdf/pwc-gx-miining-taxes-and-royalties.pdf. Accessed 1 Dec 2015
Rapini A, Ribeiro P, Lambert S, Pirani J (2008) A flora dos campos rupestres da Cadeia do Espinhaço. Megadiv 4:16–24
Renger FE (2005) O "quadro geognóstico do Brasil" de Wilhelm Ludwig Von Eschwege : breves comentários à sua visão da geologia no Brasil. Phil J 13:91–95
Resende F, Fernandes G, Coelho M (2013) Economic valuation of plant diversity storage service provided by Brazilian rupestrian grassland ecosystems. Braz J Biol 73:709–716
Rodrigues M, Carrara LA, Faria LP, Gomes HB (2005) Aves do parque nacional da Serra do Cipó: o vale do rio Cipó, Minas Gerais, Brasil. Rev Bras Zool 22:326–338
Roeser HMP, Roeser PA (2010) O Quadrilátero Ferrífero - MG, Brasil: aspectos sobre sua história, seus recursos minerais e problemas. Geonomos 18:33–37
Santos RSP (2012) Fundamentos para a criação de um fundo social e comunitário da mineração no Brasil. http://www.observatoriodopresal.com.br/wp-content/uploads/2012/04/CadernoObservatorio21.pdf. Accessed 16 Feb 2016
Santos RO, Barbieri AF (2012) Reflexões sobre população, migrações e planejamento regional para o desenvolvimento. In: Guedes GR (ed) Território, mobilidade populacional e ambiente. Univale, Governador Valadares, pp 135–156
Saturnino HM, Saturnino MAC, Brandão M (1977) Algumas considerações sobre exportação e importação de plantas ornamentais em Minas Gerais. In: Anais da Sociedade Botânica do Brasil: XXIII Congresso Nacional de Botânica, Sociedade Botânica do Brasil, Garanhuns, 16–23 Jan 1972
Scatena VL, Vich DV, Parra LR (2004) Anatomia de escapos, folhas e brácteas de *Syngonanthus* sect. *Eulepis* (Bong. ex Koern.) Ruhland (Eriocaulaceae). Acta Bot Bras 18:825–837

Silva JDA, Machado RB, Azevedo AA, Drumond GM, Fonseca RL, Goulart MF, Moraes Júnior EA, Martins CS, Ramos Neto MB (2008) Identificação de áreas insubstituíveis para conservação da Cadeia do Espinhaço, estados de Minas Gerais e Bahia, Brasil. Megadiv 4:248–270

Silveira FAO, Negreiros D, Barbosa NPU, Buisson E, CarSmo FF, Carstensen DW, Conceição AA, Cornelissen TG, Echternacht L, Fernandes GW, Garcia QS, Guerra TJ, Jacobi CM, Lemos-Filho JP, Le Stradic S, Morellato LPC, Neves FS, Oliveira RS, chaefer CE, Viana PL, Lambers H (2015) Ecology and evolution of plant diversity in the endangered campo rupestre: a neglected conservation priority. Plant Soil. doi:10.1007/s11104-015-2637-8

Siqueira LPG (2015) Em Minas Gerais: e aí, governador Pimentel, é água ou mineroduto? http://www.viomundo.com.br/denuncias/em-minas-e-ai-governador-e-agua-ou-mineroduto.html. Accessed 20 Apr 2015

Tempo O (2014) Um rastro de insatisfação. http://www.otempo.com.br/cmlink/hotsites/especial-mineroduto. Accessed 20 Apr 2015

USGS (2012) Minerals yearbook U.S. Department of the interior U.S. geological survey Brazil. http://minerals.usgs.gov/minerals/pubs/country/2012/myb3-2012-br.pdf. Accessed 18 Feb 2015

Vasconcelos MF (2008) Mountaintop endemism in eastern Brazil: why some bird species from campos rupestres of the espinhaço range are not endemic to the cerrado region? Rev Bras Ornitol 16:348–362

Vasconcelos MF, D'Angelo Neto S (2007) Padrões de distribuição e conservação da avifauna na região central da Cadeia do Espinhaço e áreas adjacentes, Minas Gerais, Brasil. Cotinga 28:27–44

Versieux LM, Wendt T, Louzada RB, Wanderley MGL (2008) Bromeliaceae da Cadeia do Espinhaço. Megadiv 4:99–110

Viana PL, Lombardi JA (2006) Florística e caracterização dos campos rupestres sobre canga na Serra da Calçada, Minas Gerais. Rodriguésia 58:159–177

von Eschwege WL (1822) Geognostisches gemälde von brasilien und wahrscheinliches muttergestein der diamanten. Weimar, Berlin

Yaron G (2001) Forest, plantation crops or small-scale agriculture? An economic analysis of alternative land use options in the mount Cameroon area. J Environ Plan Manag 44:85–108

Chapter 22
Rupestrian Grassland: Past, Present and Future Distribution

Newton P.U. Barbosa and G. Wilson Fernandes

Abstract Despite the existing evidence about climate change in South America during the Pleistocene, ancient climate and vegetation data for the mountainous regions of eastern Brazil are still scarce. Thus, little is known about the possible changes that took place in the distribution of mountainous ecosystems, such as the rupestrian grasslands. Additionally, the extent of the Rupestrian Grassland has not yet been completely mapped. The main objective of this chapter is to present the results of distribution modeling to investigate the potential distribution of the Rupestrian Grassland for the present (1950–2001), for the last glacial maximum ($\sim$21,000 years B.P.), and for the middle Holocene ($\sim$6000 years B.P.). Since we expect that mountainous ecosystems should be more sensitive to climate change than other ecosystems, we also investigated the potential distribution of the Rupestrian Grassland into the future (for the decades of 2020, 2050, and 2080), under two different scenarios (pessimistic and optimistic). We also identify historically stable areas of the Rupestrian Grassland from a paleohistorical viewpoint and also from predictions based on general circulation models of the future. According to our results, there was no significant expansion of the Rupestrian Grassland during the middle Holocene and during the last glacial maximum, probably due to the strong edaphic specificity of this ecosystem. The historically stable areas for the rupestrian grasslands were basically the mountains of Espinhaço and Canastra, some isolated mountains in southern/southeastern Minas Gerais State, and central Brazil. The future models indicate a large loss of areas environmentally suitable for rupestrian grasslands into the 2080s, which is linked to various human impacts already existing in these regions and which may lead to unpredictable catastrophic results.

N.P.U. Barbosa (✉) · G.W. Fernandes
Ecologia Evolutiva & Biodiversidade/DBG, ICB/Universidade Federal de Minas Gerais, CP 486, Belo Horizonte, MG 31270-901, Brazil
e-mail: newtonulhoa@gmail.com

G.W. Fernandes
Department of Biology, Stanford University, Stanford, CA 94305, USA
e-mail: gw.fernandes@gmail.com

© Springer International Publishing Switzerland 2016

G.W. Fernandes (ed.), *Ecology and Conservation of Mountaintop Grasslands in Brazil*, DOI 10.1007/978-3-319-29808-5_22

Contents

22.1 Introduction

The Rupestrian Grassland we observe today is the result of extensive biological/evolutionary processes that have developed in association with specific geological formations, mainly in eastern Brazil, over the last 70–100 million years. The primary lithological basis for these ecological processes were formed 1.2 to 1.8 billion years ago in the Proterozoic, when a regimen of extensive forces caused the rupturing and fracturing of the upper parts of the fragile continental crust of the supercontinent Columbia (Dussin and Dussin 1995; Almeida-Abreu and Renger 2002; Chap. 2). These distensive processes, in association with rifting, sedimentation and active anorogenic magmatism, culminated in the fragmentation of this ancient supercontinent (Rogers 1996). Different geological basins were formed in this period, varying from sediments to volcanic rocks, which became metamorphosed overtime. Africa and South America collided to form west Gondwana around 900–1000 million years ago, and the impact induced great changes in these ancient rocks, building orogens and eroding others. The result was the liberation of magma in some parts of southeastern Brazil around the São Francisco Craton, and the formation of new geological basins that were integrated into the existing old matrix. At about 200 million years ago, Africa and South America began their separation, eroding old orogens and closing old oceans. From these ancient ocean floors many felsic rocks emerged, which had been greatly modified by the ocean water (for a review see Clapperton 1993). On this ancient rock matrix evolved one of the most diverse and intriguing ecosystems of our planet, the Rupestrian Grassland (see Chap. 1).

22.2 The Last Glacial Maximum in the Espinhaço Mountain Range

Paleoclimate data for the mountainous regions of eastern South America, which today include most of the Rupestrian Grassland, are often superficial. Dynesius and Jansson (2000) argued that the typical climate fluctuations that occurred in the southern hemisphere in the Pleistocene were responsible for the establishment, and the consequent differentiation, of many species in mountainous grasslands around the planet. There seems to be a consensus that the mountainous regions of south-eastern Brazil (with the exception of the northern part which is actually surrounded by the highly xeric Caatinga biome) were significantly cooler and drier around the last glacial maximum (ca. 21,000 years B.P.; LGM heretofore) and only became hotter and more humid with the progressive arrival of the Holocene (Vuilleumier 1971; Ledru 1993; Stute et al. 1995; Behling and Lichte 1997; Behling 1998, 2002; Ab'Saber 2000; Behling and Hooghiemstra 2001) (Fig. 22.1). However, these historical portraits were based mostly on non-empirical knowledge and palynological studies conducted in relatively lower adjacent areas (e.g. Behling and Hooghiemstra 2001). Fossil pollen data for the tops of the mountain regions of eastern South America are almost nonexistent.

There are many constraints to the development of palynological studies in the rupestrian grasslands, such as the shallow soils and the lack of high altitude lakes where such records are usually preserved. Furthermore, fire represents a strong threat to the more common deposits in peat bogs as they can be totally destroyed by below ground fires that consume the organic matter. Consequently, inferences about paleovegetation, and on the paleoclimate, of these mountainous regions are still anecdotal and much more research is needed in order to construct and test rigorous historical hypotheses. Not surprisingly, existing studies not only point out great limitations, but they often boil down to the cold/wet and warm/dry dualism. Behling and Lichte (1997), for instance, found evidence that areas currently dominated by Cerrado and Seasonal Deciduous Forest in the region of Catas Altas (surrounding the southern part of the Espinhaço mountain range in the state of Minas Gerais), were covered by grassland vegetation (mostly formed by Poaceae and Cyperaceae) and small areas of gallery forest during the LGM. Despite the current existence of rupestrian grasslands near the city of Catas Altas, the absence of typical Rupestrian Grassland taxa in these studies makes it impossible to propose at this time a definitive conclusion about the possibility that this mountain vegetation had migrated to lower areas during the LGM. According to some authors, many South American ecosystems had their geographic distributions shifted downhill to ca. 1 km during the last glacial maximum (e.g. Rind and Peteet 1985; Clapperton 1990; Stute et al. 1995). Barbosa et al. (2015) argued that the actual area of distribution of a relictual species (*Coccoloba cereifera*: Polygonaceae) in the southern Espinhaço functions as an interglacial microrefugium, whose ancestral species may have had a much wider and lower elevation distribution due to higher environmental suitability and less environmental restrictions (e.g., cold air pooling areas such as areas with a high incidence of frost). During the cold phases of the Pleistocene in South America (e.g. Vuilleumier 1971), microrefugia may have facilitated the persistence of some plant species in mountain ecosystems (see McLachlan et al. 2005; Holderegger and Thiel-Egenter 2009; Rull 2009). It is a more reasonable suggestion that only some groups of plants present today in the Rupestrian Grassland were able to migrate to lower regions, such as some shrubs for instance. Temperatures in the highlands may have reached below −5 °C according to Behling and Lichte (1997), and many thermophillous species

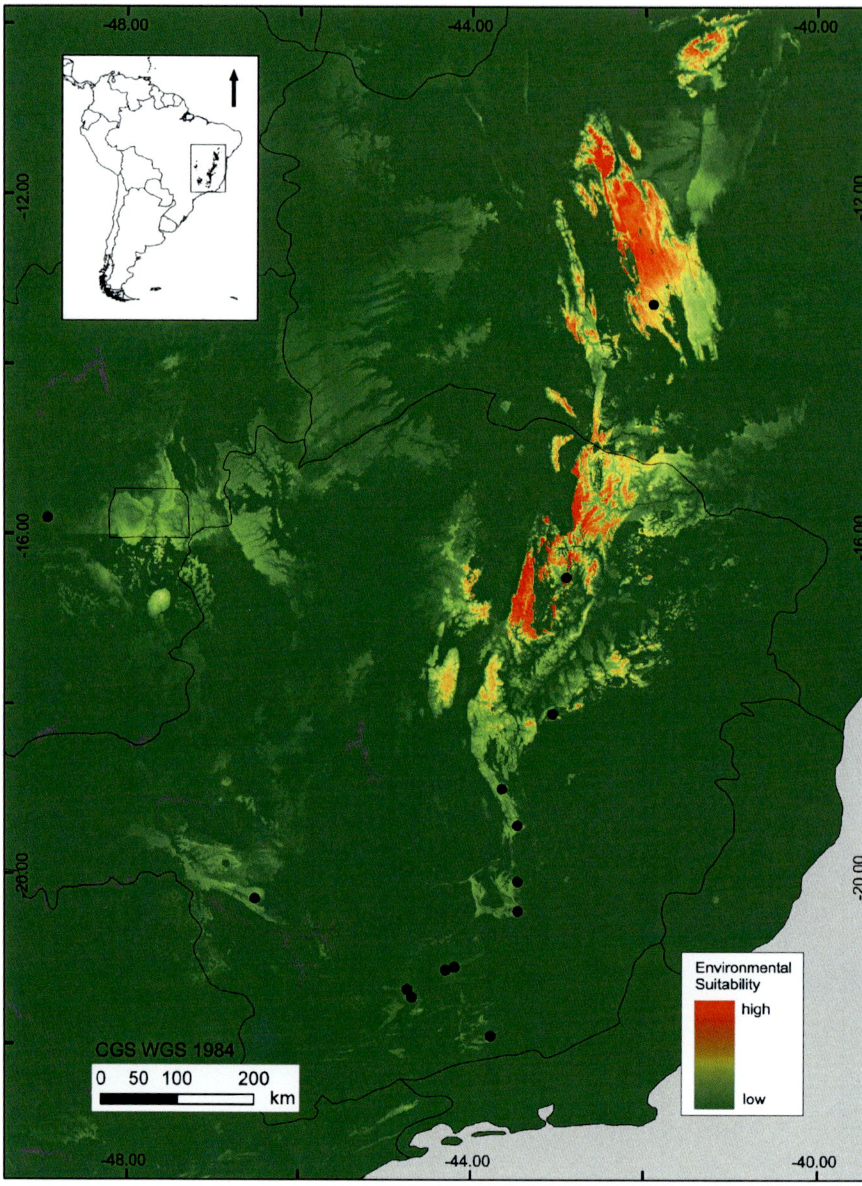

Fig. 22.1 Areas environmentally suitabile for rupestrian grasslands during the last glacial maximum (ca. 21,000 years BP) in southeast Brazil

(shrubs and trees) may have become locally extinct at higher elevations, which were covered solely by cold adapted grassland vegetation, similar to that of the steppes (see Behling 2002).

Many climate reconstructions for the LGM suggest that regions in latitudes above 40°N had temperatures 10–20 °C lower than present, while lower latitudes were significantly drier with temperatures 2.5 °C lower (Petit et al. 1999; Barron and Pollard 2002; Willis and van Andel 2004; Birks and Willis 2008). It is argued that a major change in weather patterns in southeastern South America occurred during the LGM. The current atmospheric dynamics in the eastern part of South America seems to be very similar to those found during most interglacial periods (Damuth and Fairbridge 1970). Currently, the center of the South Atlantic anticyclone is positioned below the Tropic of Capricorn and the warm currents coming from the Atlantic Ocean (currents of Brazil and Gulf) are quite influential on the continent, resulting in masses of warm and moist air that are extremely important for maintaining a mostly humid climate in southeastern Brazil. Thus, the influence of the cold Falkland Current is limited to southern Brazil and Argentina (Damuth and Fairbridge 1970). In addition, the annual variation of the intertropical convergence zone is a crucial factor in the formation of a rainy season in southeastern Brazil in the summer and a dry winter season (Damuth and Fairbridge 1970; Behling 2002). However, during glacial stages the center of the South Atlantic anticyclone was moved northward to the vicinity of the Equator, preventing the rise of moisture and creating a zone of semi-arid cold that could have largely encompassed southeastern Brazil (Damuth and Fairbridge 1970; Ledru 1993; Behling and Lichte 1997; Ab'Saber 2000; Behling 2002).

Thus, during the glacial periods of the Pleistocene, the southern ridge of the Espinhaço mountain range may have experienced something like a Frostwechselklima (Troll 1943), albeit in a milder form. Such conditions are characterized by small, though daily, periods of frost whose temperature is negative (Beck et al. 1982; Azócar et al. 1988) and require a moderate degree of frost tolerance by plants in parallel with an efficient mechanism for retarding the cooling (efficient insulation). According to Behling and Lichte (1997) and Behling and Hooghiemstra (2001), sediment samples dated to the period of 50,000 ^{14}C years B.P. to 40,000 ^{14}C years B.P. from the mountainous region of Salitre (west of the southern part of the Espinhaço mountain range), contained little arboreal pollen, indicating a very arid climate and that the region was already cold and dry even before the LGM. From the LGM, a myriad of forest formations were identified in the same region. Between 16,000 ^{14}C years B.P. and 11,000 ^{14}C years B.P. (late Pleistocene through early Holocene period), Behling and Hooghiemstra (2001) and Behling and Lichte (1997) found that a high proportion of forest pollen belonged to a mosaic of forest types, thus indicating a gradual increase in humidity. However, Ledru (1993) and Ledru et al. (1996) suggest a drier and colder climate at the beginning of this period, due to the low proportion of arboreal pollen between 14,000 and 13,000 ^{14}C years B.P.. Then, the increased presence of pollen in *Araucaria* sediments dated 12,000 ^{14}C years B.P., would be indicative of cold and wet weather (Ledru 1993; Ledru et al. 1996; Behling and Lichte 1997; Behling and Hooghiemstra 2001). Between 12,000 and 10,000 ^{14}C years B.P., there was an apparent decline in the presence of forest formations (Behling and Lichte 1997; Behling and Hooghiemstra 2001). This decline, also found by Ledru (1993), is probably related to "Recent Dryas", a brief period of cold weather that occurred during this time (e.g. Alley et al. 1993). After this short period, forest formations reappeared, and gradually from 8500 ^{14}C years B.P. began to be replaced by mesophytic vegetation. Around 5000 ^{14}C years B.P., sediment samples show large percentages of Poaceae, a clear indication of habitat aridity. After 4000 ^{14}C years B. P., the climate became more humid, similar to current levels (followed by aridity for only a short period around 1000 ^{14}C years B.P., as evidenced by the high percentage of Cyperaceae and Poaceae) (Behling and Lichte 1997; Behling and Hooghiemstra 2001). Similar data, showing a gradual increase in humidity from 5000 ^{14}C years B.P., were found in Lagoa Santa at the base of the region of Serra do Cipó. Between 5400 and 4600 ^{14}C years B.P. the presence of marshland vegetation reflects drier climatic conditions compared to present time. After that, a mosaic of Cerrado and semideciduous and gallery forests were present, reflecting conditions similar to the current climate (Parizzi et al. 1998).

The plant species of the rupestrian grasslands of the southern Espinhaço may have benefited from their various adaptations to fire and a dry climate in surviving the intense cold of this period. Most of these species have likely survived due to low precipitation rates during the winter. If these rates were high, the ice formed during the winter would be constant and pose a greater obstacle to the establishment of different taxa in these areas. Cell freezing poses a great threat to plant survival and largely influences their distribution (Silvertown and Charlesworth 2001). A relationship between the structural and molecular adaptations to cold tolerance and the evolution of desiccation tolerance in plants has been suggested by Larcher (1981). Mesleárd and Lepart (1989) suggest that the close relationship between fire and Mediterranean vegetation should be reconsidered, since they observed that the lignotubers of several species adapted to these regions could sprout in the absence of any other stimulus (and not only in response to fire), including intense cold. Carbohydrates and nutrients stored in these structures support the growth of meristems after a disturbance (e.g., fire, drought or extreme cold events) (Canadell et al. 1999). This is an example of a known efficient adaptation to fire present in many plant taxa of savanna and rupestrian grasslands (e.g. Ribeiro and Fernandes 1999; Barbosa et al. 2014, Chaps. 11 and 18). These ecosystems share the existence of many clonal systems and other types of geofitisms globally unique to, or typical of, savannas with a continental climate (rainy summer, dry winter).

22.3 The Actual Distribution of the Rupestrian Grassland

The Rupestrian Grassland is comprised of old growth herbaceous/shrub vegetation (Fernandes et al. 2014; Veldman et al. 2015), which is highly heterogeneous and intimately related to the rocky outcrops present in the upper parts of mountains, mainly on quartzite, sandstone and iron ore formations, among others (e.g. Joly 1970; Giulietti et al. 1997, Carvalho et al. 2012; Chaps. 2, 6). The Rupestrian Grassland ecosystem is distributed mainly along the Espinhaço mountain range, mostly above 900 m a.s.l., and along adjacent mountain ranges and isolated mountains such as Serra do Cabral (Alves and Kolbek 1994; Giulietti et al. 1997; Kolbek and Alves 2008). Moreover, rupestrian grasslands can also be found in other discrete regions such as the mountains of central Brazil (e.g., the Veadeiros Plateau and Serra dos Pirineus, both in Goiás State, and Serra da Canastra in southwest Minas Gerais State) or the mountains in the region of São João del Rei (Serra do Lenheiro), Tiradentes (Serra de São José) and Itutinga, beyond the region of Ibitipoca (Benites et al. 2007) in the state of Minas Gerais (see Vasconcelos 2011). Taxa typical of rupestrian grasslands, such as *Vellozia*, can also be found in the upper parts of the mountain range of Carajás, in the state of Pará (Silva et al. 1996; Piló et al. 2015; Chap. 2), and in some isolated mountains in the extreme west of Brazil, in the state of Mato Grosso do Sul (Fig. 22.2).

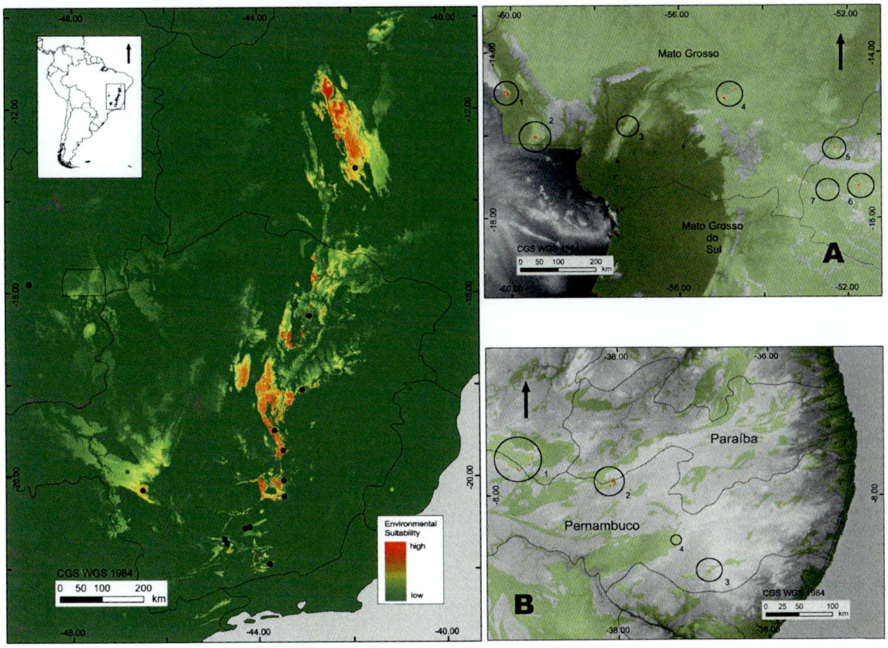

Fig. 22.2 Areas environmentally suitabile for rupestrian grasslands (years 1950–2000) in southeast Brazil. In detail, isolated environmentally suitabile areas for rupestrian grasslands in western Brazil (*1* municipality of Vila da Bela Santíssima Trindade—MT, *2* municipality of Pontes e Lacerda—MT, *3* municipality of Porto Estrela—MT, *4* Border between the municipalities of Nova Brasilândia—MT and Campo Verde—MT, *5* municipality of Baliza—GO, *6* municipality of Caiapônia—GO, *7* municipality of Mineiros—GO); and northeastern Brazilian (*1* municipalities of Exu—PE, Sítio dos Moreiras—PE, Serrita—PE and Jardim—CE, *2* municipalities of Triunfo—PE and Santa Cruz da Baixa Verde—PE, *3* boundary between the municipalities of Iati—PE and Saloá—PE, *4* Buíque region). The values were calculated using a threshold of 0.09 (MaxEnt, "equal test sensitivity and specificity")

22.3.1 Isolated Mountains with Rupestrian Grasslands

The region of Ibitipoca is located at the southern edge of the area environmentally suitabile for rupestrian grasslands, in the Mantiqueira Mountains (Benites et al. 2007). However, while some authors refer to this region as a region with a high prevalence of rupestrian grasslands (e.g. Rodela 1998; Monteiro and Forzza 2008; Silva et al. 2009), others refer to them as *campos de altitude* (Dias et al. 2002; Ladeira et al. 2007). This region presents a mosaic of quartzite rock formations and igneous or metamorphic rocks, such as granite or gneiss. The presence of "vestigial" quartzite formations (and other lithologies favorable to rupestrian grasslands, such as silts and metaconglomerates) in southern/southeastern parts of the state of Minas Gerais reflects the isolation of rupestrian grasslands refuges in these regions and also escarpments of the mountains of Lenheiro and São José (e.g., Vasconcelos 2011). Regions below this limit mostly have a combination of a very humid climate and low temperature seasonality, and a predominantly basaltic and/or granitic geology, a feature that strongly hinders the existence of rupestrian grasslands.

In central Brazil, there are some mountains with high environmental suitability for rupestrian grasslands. The Caiapó Mountain, located in the western state of Goiás (municipalities of Caiapônia and Mineiros), is on the geological formation of Irati and possesses sandstone and limestone outcrops (Araújo-Barberena et al. 2002; Chahud and Petri 2008). Typical species from the rupestrian grasslands of the Espinhaçomountain range, such as *Epidendrum campestre*, can also be found in this region (Barros 2002). The Ricardo Franco Mountain, in the municipality of Vila da Bela Santíssima Trindade, state of Mato Grosso, is also recognized as a potential area of distribution of rupestrian grasslands (Martinelli 2007). Areas of environmental suitability for rupestrian grasslands were also identified in the mountains located to the south of the city of Nova Brasilândia, south of the municipality of Porto Estrela and also south of the city of Pontes e Lacerda, both in the state of Mato Grosso. However, so far, there are no reports of collections or other studies that indicate the presence of rupestrian grasslands in these regions.

In northeast Brazil, there are other potential areas of suitability for rupestrian grasslands. The municipalities of Exu, Sítio dos Moreiras and Serrita Jardim, located in the south of the state of Ceará and in the west of the state of Pernambuco, feature promising areas for the development of rupestrian grassland species. Likewise, the municipalities of Triunfo, Santa Cruz da Baixa Verde, Manaíra and São José de Princesa, on the border of the state of Pernambuco with the western part of the state of Paraíba, also have small environmentally suitabile areas for rupestrian grasslands. Between the municipalities of Iati and Saloá in Pernambuco, there are also some environmentally suitabile areas for rupestrian grasslands, as well as the nearby Buíque, also in Pernambuco, where rupestrian grasslands are recognizably present (Rodal et al. 1998) (Fig. 22.2).

Outside Brazil, the region of Cerro Parabanó (located on the Macharetí y Mandiyutí formation with sandy outcrops), in Bolivia, also has high environmental suitability for the presence of rupestrian grasslands. According Saravia (2008), the region is covered by "*cerrado de llanura*", a typical white, sandy soil grassland ecosystem, covering mountains from 650 m a.s.l. upward, a description that closely resembles rupestrian grasslands.

22.4 Historically Stable Areas of the Rupestrian Grassland

In general, the Rupestrian Grassland has remained stable for at least the last 20,000 years, suffering contractions mainly on the borders of core areas: the mountains of central Brazil, the region of Canastra and the Espinhaçomountain range (which together comprise the core area of the Rupestrian Grassland) (Barbosa 2012). However, some regions, such as the Ricardo Franco Mountains in western Mato Grosso State, were also relatively stable during this period. There is evidence that the Espinhaço mountain range, in particular, has several areas of endemism, as do the vast majority of the mountainous regions of the planet (Janzen 1967; Kruckeberger and Rabinowitz 1985; Wiens and Graham 2005). Rapini et al. (2002) qualitatively defined four areas of endemism for the Espinhaço: the southern part, the Serra do Cipó, the Diamantina Plateau, and the northern region. Echternacht et al. (2011) defined six macro-areas of endemism through a parsimony analysis of endemism ("PAE", Rosen 1988) conducted in ten areas and with 178 species belonging to 17 families of vascular plants. Some of these areas correspond to the areas discussed by Rapini et al. (2002) and were considered priority areas for conservation (Silva et al. 2008).

22.5 The Future of the Rupestrian Grassland

Predictive models proposed by Barbosa (2012) for the 2020s, 2050s, and 2080s produced different results based on the scenarios and climate models used (Fig. 22.3). Under the pessimistic scenario (A2), it is expected that areas that are environmentally suitabile for rupestrian grasslands would loose more than 90 % of their extension by the end of the 2080s, with the largest loss of area being in northern Minas Gerais and southern Bahia. Indeed, public authorities expect scenarios of desertification and rising temperatures for these regions for the next 20 years (PAN-Brazil 2004). These scenarios are highly worrisome alerts for current conservation practitioners. According to the model, the regions that are likely to remain stable until the end of this century are the mountains of southern Minas Gerais, Ouro Preto, Caraça, Serra do Cipó, Canastra and part of the Diamantina Plateau, making them invaluable for conservation. The rest of the mountains of central Brazil, as well as the mountains located in northern Minas Gerais (e.g., Diamantina and Grão Mogol municipalities), are highly endangered according to

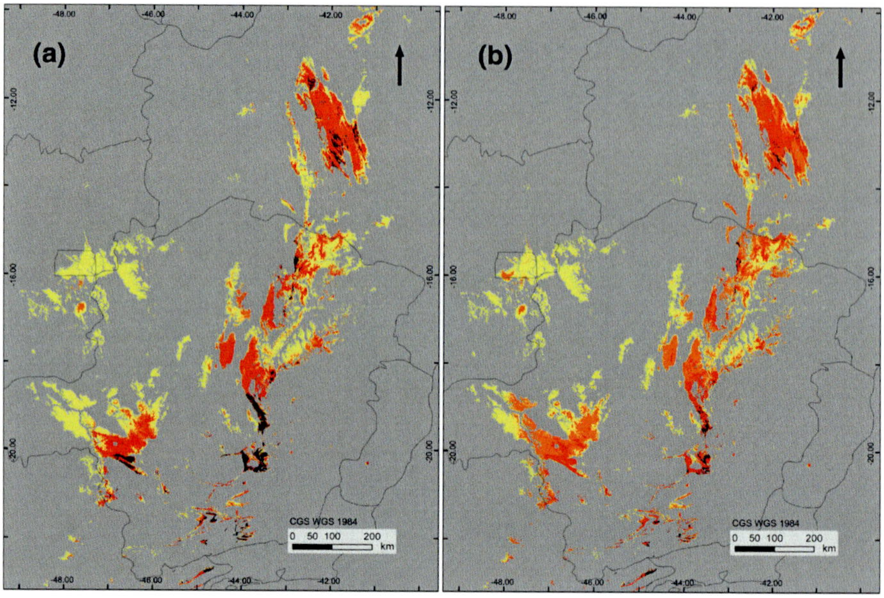

Fig. 22.3 Predicted loss of areas environmentally suitabile for rupestrian grasslands until late 2080 (general circulation model CCCma-CGCm2 in (**a**) and HADCM3 in (**b**), in A2A scenario. The areas in *yellow* show the losses current areas (1950–2000) by the decade of 2020; the areas in *orange* correspond to losses between the decades of 2020 and 2050, and the areas in *red* correspond to losses between the decades of 2050 and 2080. The areas highlighted in *black* are the predicted remaining areas environmentally suitabile for rupestrian grasslands by late 2080

the results obtained by both models (CCCma-CGCm2 and HadCM3). However, it is worth remembering that, under a more optimistic scenario (B2), the losses are less severe than those presented by the most pessimistic scenario. Mountains are regarded as the first places on earth to be impacted by global climate change (IPCC 2007; Kohler and Maselli 2009; Spehn et al. 2010). Other authors have discussed the fact that mountains are important refuges for species conservation in the future because of their heterogeneity, with possible areas for exchange (e.g., Epps et al. 2006). On the other hand, these studies did not examine in detail the resulting impacts of the immigration of new species to the mountaintop habitats. Furthermore, other elements such as the extensive cattle ranching, tourism, fire, and the entry of invasive species (Barbosa et al. 2010; Hilário et al. 2011; Fernandes et al. 2014; Bitencourt et al. 2016) are important threats that endanger the preservation of these systems, which in combination with climate change may causing unpredictable and catastrophic impacts on biodiversity and ecosystem services (Chap. 23). In 2005 the austral portion of the Espinhaço region was named a Biosphere Reserve by UNESCO, thus ensuring the recognition of the need for its conservation. On the other hand, there remain many challenges facing civil society and the government to implementing sound conservation strategies and policies to preserve this fragile ecosystem.

Acknowledgements We thank FAPEMIG, CAPES and CNPq for the continuous support of our research that has led to this synthesis. We also thank Neil Cobb and Daniel Negreiros for reading the chapter and for suggestions.

References

Ab'Saber AN (2000) Spaces occupied by the expansion of dry climates in South America during the Quaternary ice ages. Rev Inst Geol 21:71–78

Alley RB, Meese DA, Shuman CA, Gow AJ, Taylor KC, Grootes PM, White JWC, Ram M, Waddington ED, Mayewski PA, Zielinski GA (1993) Abrupt increase in snow accumulation at the end of the Younger Dryas event. Nature 362:527–529

Almeida-Abreu PA, Renger FE (2002) Serra do espinhaço meridional: um orógeno de colisão do mesoproterozóico. Rev Bras Geocien 32:1–14

Alves RJV, Kolbek J (1994) Plant species endemism in savanna vegetation on table mountains (campo rupestre) in Brazil. Vegetatio 113:125–139

Araújo-Barberena DC, Lacerda-Filho JV, Timm LL (2002) Mesossauro da Serra do Caiapó (Montividiu), GO: Um vertebrado fóssil típico do Paleozóico superior, importante na história da deriva continental. In: Schobbenhaus C, Campo DA, Queiroz ET, Winge M, Bebert-Born MLC (eds) Sítios geológicos e paleontológicos do Brasil. DNPM/CPRM - Comissão brasileira de sítios geológicos e paleobiológicos (SIGEP), Brasília, pp 81–85

Azócar A, Rada F, Goldstein G (1988) Freezing tolerance in *Draba chionophila*, a 'miniature' caulescent rosette species. Oecol 75:156–160

Barbosa NPU (2012) Modelagem de distribuição aplicado aos Campos Rupestres. Thesis. Universidade Federal de Minas Gerais, Belo Horizonte

Barbosa NPU, Fernandes GW, Carneiro MAA, Júnior LAC (2010) Distribution of non-native invasive species and soil properties in proximity to paved roads and unpaved roads in a quartzitic mountainous grassland of south-eastern Brazil (rupestrian fields). Biol Inv 12:3745–3755

Barbosa BC, Cappi VS, Ribeiro SP, Fernandes GW (2014) Avaliação da capacidade de rebrotamento pós-distúrbio das plantas lenhosas típicas dos campos rupestres. Ecol Aust 24:350–355

Barbosa NPU, Fernandes GW, Sanchez-Azofeita A (2015) A relict species restricted to a quartzitic mountain in tropical America: an example of microrefugium? Acta Bot Bras 29:299–309

Barron EJ, Pollard D (2002) High-resolution climate simulations of oxygen isotope stage 3 in Europe. Quatn Res 58:296–309

Barros F (2002) Notas nomenclaturais em Pleurothallidinae (Orchidaceae), principalmente brasileiras. Bradea 8:293–297

Beck E, Senser M, Scheibe R, Steiger H-M, Pongratz P (1982) Frost avoidance and freezing tolerance in afroalpine 'giant rosette' plants. Plant, Cell Environ 5:215–222

Behling H (1998) Late Quaternary vegetational and climatic changes in Brazil. Rev Paleobot Palyno 99:143–156

Behling H (2002) South and southeast Brazilian grasslands during late quaternary times: a synthesis. Palaeogeogr Palaeoclimatol Palaeoecol 177:19–27

Behling H, Lichte M (1997) Evidence of dry and cold climatic conditions at glacial times in tropical southeastern Brazil. Quat Res 48:348–358

Behling H, Hooghiemstra H (2001) Neotropical savanna environments in space and time: late Quaternary interhemispheric comparisons. In: Margraf V (ed) Interhemispheric climate linkages. Academic press, Cambridge, pp 307–323

Benites VM, Schaefer CEGR, Simas FNB, Santos HG (2007) Soils associated with rock outcrops in the Brazilian mountain ranges Mantiqueira and Espinhaço. Rev Bras Bot 30:569–577

Birks HJB, Willis KJ (2008) Alpines, trees, and refugia in Europe. Plant Ecol Div 1:147–160

Bitencourt C, Rapini A, Damascena LS, Junior PDM (2016) The worrying future of the endemic flora of a tropical mountain range under climate change. Flora 218:1–10

Canadell J, Djema A, López B, Lloret F, Sabaté S, Siscart D, Gracia CA (1999) Structure and dynamics of root system. In: Rodà F, Retana J, Gracia CA, Bellot J (eds) Ecology of Mediterranean Evergreen Oak Forests. Springer, Heidelberg, p 51

Carvalho F, Souza FA, Carrenho R, Moreira FMS, Jesus EC, Fernandes GW (2012) The mosaic of habitats in the high-altitude Brazilian rupestrian fields is a hotspot for arbuscular mycorrhizal fungi. Appl Soil Ecol 52:9–19

Chahud A, Petri S (2008) Chondrichthyes no membro Taquaral, base da formação irati, no centro leste do estado de São Paulo, Brasil. Rev Geol 21:169–179

Clapperton C (1990) Quaternary glaciation in the southern ocean and antarctic peninsula area. Quater Sci Rev 9:229–252

Clapperton C (1993) Quaternary geology and geomorphology of South America. Elsevier, Amsterdam, p 779

Damuth JE, Fairbridge RW (1970) Equatorial atlantic deep-sea arkosi sands and ice-age aridity in tropical South America. Geol Soc Am Bull 81:189–206

Dias HCT, Filho EIF, Schaefer CEGR, Fontes LEF, Ventorim LB (2002) Geoambientes do parque estadual do Ibitipoca, município de Lima Duarte-MG. Rev Arv 26:777–786

Dussin IA, Dussin TM (1995) Supergrupo espinhaço: modelo de evolução geodinâmica. Geonomos 3:19–26

Dynesius M, Jansson R (2000) Evolutionary consequences of changes in species geographical distributions driven by Milankovitch climate oscillations. Proc Natl Acad Sci USA 97: 9115–9120

Echternacht L, Trovó M, Oliveira CT, Pirani JR (2011) Areas of endemism in the espinhaço range in Minas Gerais, Brazil. Flora 206:782–791

Epps CW, PalsbØll PJ, Wehausen JD, Roderick GK, McCullough DR (2006) Elevation and connectivity define genetic refugia for mountain sheep as climate warms. Mol Ecol 15: 4295–4302

Fernandes GW, Barbosa NPU, Negreiros D, Paglia AP (2014) Challenges for the conservation of vanishing megadiverse rupestrian grasslands. Nat Cons 12:162–165

Giulietti AM, Pirani JR, Harley RM (1997) Espinhaço range region, eastern Brazil. In: Davis SD, Heywood VH, Herrera-MacBryde O, Villa-Lobos J, Hamilton AC (eds) Centres of plant diversity: a guide and strategy for they conservation. WWF/IUCN, Cambridge, pp 397–404

Hilário RR, Castro SAB, Ker FTO, Fernandes GW (2011) Unexpected effects of pigeon-peas (*Cajanus cajan*) in the restoration of rupestrian fields. Planta Daninha 29:717–723

Holderegger R, Thiel-Egenter C (2009) A discussion of different types of glacial refugia used in mountain biogeography and phylogeography. J Biogeogr 36:476–480

Intergovernmental Panel on Climate Change (IPCC) (2007) The physical science basis. In: Solomon S, Qin D, Manning M, Chen Z, Marquis M, Averyt KB, Tignor M, Miller HL (eds) Contribution of working group I to the fourth assessment report of the IPCC. Cambridge University Press, New York, 996 p

Janzen DH (1967) Why mountain passes are higher in the tropics? Am Nat 101:233–249

Joly AB (1970) Conheça a vegetação brasileira. Editora da Universidade de São Paulo e Polígono, São Paulo

Kohler T, Maselli D (eds) (2009) Mountains and climate change—from understanding to action. Geographica Bernensia/Swiss Agency for Development and Cooperation (SDC), Bern

Kolbek J, Alves RJV (2008) Impacts of cattle, fire and wind in rocky savannas, southeastern Brazil. Acta Univ Carol Environ 22:111–130

Kruckeberger AR, Rabinowitz D (1985) Biological aspects of endemism in higher plants. Annu Rev Ecol Syst 16:447–479

Ladeira AS, Ribeiro GA, Dias HCT, Schaefer CEGR, Filho EF, Filho ATO (2007) O perfil dos visitantes do parque estadual do Ibitipoca (PEIb), Lima Duarte, MG. Rev Arv 31:1091–1098

Larcher W (1981) Low temperature effects on Mediterranean sclerophylls: an unconventional viewpoint. In: Margaris NS, Mooney HA (eds) Components of productivity of

Mediterranean-climate regions. Basic and applied aspects. Junk Publishers, The Hague, pp 259–266
Ledru MP (1993) Late quaternary environmental and climatic changes in central Brazil. Quat Res 39:90–98
Ledru MP, Braga PIS, Soubiès F, Fournier M, Martin L, Suguio K, Turcq B (1996) The last 50,000 years in the Neotropics (Southern Brazil): evolution of vegetation and climate. Palaeogeogr Palaeoclimatol Palaeoecol 123:239–257
Martinelli G (2007) Mountain biodiversity in Brazil. Rev Bras Bot 30:587–597
McLachlan JS, Clark JS, Manos PS (2005) Molecular indicators of tree migration capacity under rapid climate change. Ecology 86:2088–2098
Mesléard J, Lepart J (1989) Continuous basal sprouting from a lignotuber: Arbutus unedo L. and Erica arborea L., as woody Mediterranean examples. Oecologia 80:127–131
Monteiro RF, Forzza RC (2008) A família Bromeliaceae no parque estadual do Ibitipoca, Minas Gerais, Brasil. Bol Bot Univ Sao Paulo 26:7–33
PAN-BRASIL (2004) Programa de ação nacional de combate à desertificação e mitigação dos efeitos da seca. MMA Secretaria de Recursos Hídricos, Brasília, p 242
Parizzi MG, Salgado-Labouriau ML, Kholer HC (1998) Genesis and environmental history of Lagoa Santa, Southeastern Brazil. Holocene 8:311–321
Petit JR, Jouzel J, Raynaud D, Barkov NI, Barnola J-M, Basile I, Bender M, Chappellaz J, Davis M, Delaygue G, Delmotte M, Kotlyakov VM, Legrand M, Lipenkov VY, Lorius C, Pepin L, Ritz C, Saltzman E, Stievenard M (1999) Climate and atmospheric history of the past 420,000 years from the Vostok ice core, Antarctica. Nature 399:429–436
Piló LB, Auler AS, Martins F (2015) Carajás National Forest: iron ore plateaus and caves in southeastern Amazon. In: Vieira et al. (eds) Landscapes and landforms of Brazil. Springer, Heidelberg, pp 273–283
Rapini A, Mello-Silva R, Kawasaki ML (2002) Richness and endemism in Asclepiadoideae (Apocynaceae) from the Espinhaço range of Minas Gerais, Brazil—a conservationist view. Biodiv Cons 11:1733–1746
Ribeiro KT, Fernandes GW (1999) Geographic distribution of *Coccoloba cereifera* Schw. (Polygonaceae), a narrow endemic plant from Serra do Cipó. Brazil BIOS 7:7–12
Rind D, Peteet D (1985) Terrestrial conditions at the last glacial maximum and CLIMAP sea-surface temperature estimates: are they consistent? Quat Res 24:1–22
Rodal MJN, Andrade KVS, Sales MF, Gomes APS (1998) Fitossociologia do componente lenhoso de um refúgio vegetacional no município de Buíque, Pernambuco. Rev Bras Biol 58:517–526
Rodela LG (1998) Cerrados de altitude e campos rupestres do parque estadual do Ibitipoca, sudeste de Minas Gerais: distribuição e florística por subfisionomias da vegetação. Rev Dep Geogr 12:163–189
Rogers JJW (1996) A history of continents in the past three billion years. J Geol 104:91–107
Rosen BR (1988) From fossils to earth history: applied historical biogeography. In: Myers A, Giller P (eds) Analytical biogeography: an integrated approach to the study of animal and plant distribution. Chapman and Hall, London, pp 437–481
Rull V (2009) Microrefugia J Biogeogr 36:481–484
Saravia ESR (2008) El costo de la conservación de los bosques tropicales. Thesis. Universidad Autónoma G. René Moreno, Santa Cruz
Silva MFF, Secco RS, Lobo MGA (1996) Aspectos ecológicos da vegetação rupestre da serra dos carajás, estado do Pará, Brasil. Acta Amaz 26:17–44
Silva JA, Machado RB, Azevedo AA, Drumond GM, Fonseca RL, Goulart MF, Júnior EAM, Martins CS, Ramos-Neto MB (2008) Identificação de áreas insubstituíveis para conservação da Cadeia do Espinhaço, estados de Minas Gerais e Bahia, Brasil. Megadiv 4:270–309
Silva DB, Forzza RC, Alves M (2009) Cyperaceae e Juncaceae no parque estadual de Ibitipoca, Minas Gerais, Brasil. Bol Bot Univ Sao Paulo 27:219–234
Silvertown J, Charlesworth D (2001) Introduction to plant population biology. Blackwell, London 360p

Spehn EM, Rudmann-Maurer K, Körner C, Maselli D (eds) (2010) Mountain biodiversity and global change. GMBA-DIVERSITAS, Basel

Stute M, Forster M, Frischkorn H, Serejo A, Clark JF, Schlosser P, Broecker WS, Bonani G (1995) Cooling of tropical Brazil (5 C) during the last glacial maximum. Science 269:379–383

Troll C (1943) Die Frostwechselhäufigkeit in den Luft- und Bodenklimaten der Erde. Metereol Zeits 60:161–171

Vasconcelos MF (2011) O que são campos rupestres e campos de altitude nos topos de montanha do leste do Brasil? Rev Bras Bot 34:241–246

Veldman JW, Buisson E, Durigan G, Fernandes GW, Le Stradic S, Mahy G, Negreiros D, Overbeck GE, Veldman RG, Zaloumis NP, Putz FE, Bond WJ (2015) Toward an oldgrowth concept for grasslands, savannas, and woodlands. Front Ecol Environ 13:154–162

Vuilleumier BS (1971) Pleistocene changes in the fauna and flora of South America. Science 173:771–780

Wiens JJ, Graham CH (2005) Niche conservatism: integrating evolution, ecol, and conservation biology. Annu Rev Ecol Evol Syst 36:519–539

Willis KJ, van Andel TH (2004) Trees or no trees? The environments of central and eastern Europe during the Last Glaciation. Quat Sci Rev 23:2369–2387

Chapter 23
The Shady Future of the Rupestrian Grassland: Major Threats to Conservation and Challenges in the Anthropocene

G. Wilson Fernandes

Abstract In spite of the increasing number of studies on the Rupestrian Grassland, the knowledge generated is insufficient to promote further conservation due to the rapid pace of impacts caused by human disturbances. These disturbances include biological invasions, afforestation, mining, road construction, tourism, fire, plant extraction, cattle ranching, and agriculture, among others. They vary from region to region but, in combination with climate change, these impacts could lead to an unseen wave of extinction of species and habitats and thereby impacting biodiversity and ecosystem health. Specific planning and conservation actions should be designed urgently in order to save the most speciose Brazilian ecosystem.

Contents

G.W. Fernandes (✉)
Ecologia Evolutiva & Biodiversidade/DBG, ICB/Universidade Federal de Minas Gerais, CP 486, Belo Horizonte, MG 31270-901, Brazil
e-mail: gw.fernandes@gmail.com

G.W. Fernandes
Department of Biology, Stanford University, Stanford, CA 94305, USA

© Springer International Publishing Switzerland 2016

G.W. Fernandes (ed.), *Ecology and Conservation of Mountaintop Grasslands in Brazil*, DOI 10.1007/978-3-319-29808-5_23

23.1 Ecological Illiteracy About the Rupestrian Grassland

The Rupestrian Grassland is a complex, ancient, and rare ecosystem that has been reported from only a few mountaintop environments in Brazil, although similar ecosystems occur in Africa (Congo) and Bolivia (see Chap. 21). The Rupestrian Grassland is mainly associated with the Espinhaço Mountain Range in southeastern and northeastern Brazil, although it is also found at lower elevations in isolated areas in the Amazonian state of Pará and in Morraria de Urucum in the state of Mato Grosso do Sul. In these latter two regions, rupestrian grasslands occur primarily on an iron rock matrix of significant economic importance (see Chaps. 2 and 18).

As repeatedly described in this book, and elsewhere, the Rupestrian Grassland ecosystem is composed of a mosaic of habitats shaped by relief, climate, and ancient geological history. The great heterogeneity found in this ecosystem is associated with the origin of the rock, which is comprised mostly of quartz, sand, or iron (see Chap. 2 and references there in). Many different types of herbaceous and shrubby vegetation developed on top of the nutrient poor soils that originated from the decomposition of the mother rock. In unique landscape habitats such as those formed by rock outcrops, stony and sandy grasslands, peat bogs, gallery forests, and many other transitional physiognomies, such as altitudinal cerrado and relict and hillside Atlantic Rain Forests, occur side-by-side (e.g., Giulietti et al. 1997; Medina and Fernandes 2007; Carvalho et al. 2012, 2014; Negreiros et al. 2014; Le Stradic et al. 2015; Chaps. 2, 6 and 7). The grassland physiognomies associated with the quartzitic soils are markedly water stressed due to shallow soils with low nutrient levels and extreme aluminum toxicity (Carvalho et al. 2012; Negreiros et al. 2014). These soils are also known for having a low water holding capacity. Harshness is amplified by a high incidence of radiation in the mountain environment and a long annual period of water deficit (Lüttge et al. 2007). Like other ancient ecosystems with extremely poor soils (Veldman et al. 2015a, b; see also Hopper 2009), rupestrian grasslands are amongst the most species-rich floras in the World. Approximately one third of its species are endemic, with varying degrees of rarity and life strategies (Giulietti et al. 1997; Rapini et al. 2008; Jacobi et al. 2011; Chap. 6). Silveira et al. (2016) provided a review of the literature in this regard, reinforcing the fine-tuned interactions of this vegetation with environmental stresses (see also Negreiros et al. 2014; Chap. 11). Plants of the Rupestrian Grassland exhibit several types of physiological, biochemical, morphological, demographic, structural, and phenological adjustments for surviving the various environmental stresses (e.g., Coelho et al. 2005, 2006, 2008; Lüttge et al. 2007; Negreiros et al. 2009, 2014; Belo et al. 2013; Neves et al. 2014; Oliveira et al. 2015; Chaps. 11 and 12).

In spite of recent efforts, more advances in the study of ecosystem functioning and resilience are needed for the Rupestrian Grassland. Present knowledge may not be sufficient to understand in detail large-scale projections of impacts on an ecosystem with such fine-tuned interactions among plants and animals as this one. This lack of knowledge may reflect a historically greater interest in forest type ecosystems (Sanchez-Azofeifa et al. 2013; Silveira et al. 2016). Government agencies, as well as

NGO's, have been primarily involved with studies of tropical rain forests and wetlands, and only recently have they been able to push forward with some scientific projects on other important ecosystems, including the Rupestrian Grassland. The Rupestrian Grassland presents a challenge to those interested in understanding conservation of target species because many species are extremely rare, and, in many cases, are found on inaccessible mountaintop environments. Many such species may go extinct without having even been recorded by science. An example of the magnitude of the problem is the discovery, from photos shared on Facebook, of a new species of carnivorous plant, 1.5 m tall (*Drosera magnifica*) and found above 1500 m in elevation in a rupestrian grassland environment (Gonella et al. 2015). Another difficulty that hampers more focused conservation planning is the lack of experimental and long-term studies to decipher the cause and effect of adaptations. On the other hand, this scenario can be viewed as a unique opportunity to generate more knowledge and to develop and innovate in science and conservation. The Rupestrian Grassland hides many new opportunities yet to be revealed.

23.2 Biological Invasions: More Than Meets the Eye

In spite of the many types of threats imposed on the rare and unique rupestrian grasslands, perhaps the most serious is that of biological invasion by alien plant species. This threat was not targeted by studies until recently, when Barbosa et al. (2010) clearly reported the impacts caused by invasive plant species associated with the asphalt paving of highway MG10, an important tourist road (part of the Royal Road—*Estrada Real*) that crosses rupestrian grasslands for ca. 500 km (see Chap. 21). Other studies have followed, and although far from being enough to provide a regional perspective, they are increasingly identifying potential invading species and their effects on biodiversity and ecosystem losses.

Barbosa et al. (2010) also unveiled an important mechanism by which invasion is facilitated in the nutrient poor soils of the Rupestrian Grassland. A high content of calcium was found along the asphalted road as a result of the paving process, in which limestone gravel is used in one of the several paving phases. This calcium made the roadside soils suitable for the invasive species. In these newly created habitats the toxicity of aluminum was drastically reduced and the nutrients enriched, hence producing sites favorable to the colonization and growth of non-native species. Populations of these alien species remain along the road for undetermined periods of time, shedding seeds that are carried away for several generations. Disturbance events, such as repeated fires, may provide the chance for some of the aliens to invade the adjacent pristine habitats.

Unfortunately, the imminent invasion reported by Barbosa et al. (2010) and Fernandes and Barbosa (2013) was not taken seriously by local authorities, and has now resulted in a wave of invasive species spreading through the rupestrian grasslands from this road margin (Fig. 23.1a). Instead of increasing the degree of protection and conservation of rupestrian grasslands, the situation has worsened in

(a)
(b)
(c)
(d)
(e)
(f)
(g)
(h)
ESTA·ENCHORRADA
ESTÁ ENTUPINDO
O MINADOR
AJUDE·NÓS

◂ **Fig. 23.1** Major threats to the Rupestrian Grassland in the Espinhaço range. **a** Biological invasion by *Ageratum fastigiatum* (Asteraceae) (white flower species) on a spring at 1200 m a.s.l. **b** Restoration of a road landscape by a mining company within a federal conservation unit, Environmetal protected area Morro da Pedreira, with *Melinis minutiflora*, a highly aggressive African grass. **c** Restoration of a rupestrian site with the exotic legume *Cajanus cajan* (Fabaceae). Note the high accumulation of organic material on the soil, atypical in the rupestrian grassland. **d** sapling of *Baccharis dracunculifolia* planted on a restored site being overtaken by a legume invasive to the rupestrian grassland. **e** *Eucalyptus* sp. that escaped a Cenibra plantation near the town of Itabira (MG) invades the remaining patches of Atlantic rain forest in valleys and canyons. **f** Afforestation with eucalyptus threatens high elevation rupestrian grasslands within a federal conservation unit in Serra do Cipó. Fire is a component of the Rupestrian Grassland, and a threat to forest type environments. **g** The construction of a road and slurry pipeline without proper knowledge of the threats to the land or the entire rupestrian grassland ecosystem [slurry pipeline construction by Anglo American in Morro do Pilar (MG)]. **h** A sign posted by a landowner during asphalt paving of highway MG-424 asks for help to save his water spring since during road construction debris and water were intentionally driven towards this sole source of water. No action was taken, even after authorities were warned

the face of erroneous management strategies. A clear example in the region is the deliberate introduction of exotic species to restore degraded areas in still otherwise pristine environments, be it in rupestrian grassland, Atlantic forest, or even Cerrado (see Fernandes and Barbosa 2013) (Fig. 23.1b). Many plant species have become invasive in the region after being mistakenly used for revegetation purposes, such as *Urochloa* spp., *Melinis minutiflora,* and *Cajanus cajan*, among others (Oliveira et al. 2009; Hilário et al. 2011). The use of a known and highly invasive exotic species (*Cajanus cajan*) to protect exposed soils and improve their nutritional quality—through interactions with nitrogen-fixing bacteria—to endemic species adapted to the nutrient-poor soils of rupestrian grasslands (see Hilário et al. 2011) clearly indicates the profound impacts misleading restoration programs can have (Fig. 23.1c).

Fernandes et al. (2015a) have recently provided the first attempt to list the exotic plants that can lead to the failure of restoration of quartzitic rupestrian grasslands along roadsides. Some of these alien species can now be found among native vegetation more than 2 km away from the roads where they were planted. The invaders threaten the native species by altering the natural food chain of the region and by bringing exotic species of fungi and insects, which can cause disease among the native plants. In disturbed sites, or those under restoration, the exotics outperform the native plants (Fernandes et al. 2015a) (Fig. 23.1d). These events might, unfortunately, lead to an unseen wave of serious impacts on the function, biodiversity, and ecosystem services of this ecosystem (see Fernandes and Barbosa 2013). In summary, biological invasion by alien plant species is increasing in the Rupestrian Grassland, yet our knowledge of it remains extremely poor. The sole list of threatening exotic plants developed so far is of limited regional value, and we are unaware of any empirical study on the influence of invasive species on the functioning of the Rupestrian Grassland ecosystem. With regard to biological invasions by animals, the situation is of complete ignorance, unfortunately, but again of major importance. Therefore, long lasting plans must be designed to study and eradicate the invasive species that are now spreading and placing the Rupestrian Grassland at a great risk.

Strong regulations on the processes involved in the construction of roads in rupestrian grasslands, on the mitigation of environmental impacts, and on restoration are urgently needed due to the special geology, landscape, fauna and flora of this ecosystem. Increasing mining activity and tourism in the Rupestrian Grassland will surely promote the construction of more roads which, unfortunately, will lead to further environmental impacts (Fig. 23.1e–h).

23.3 The Ill-Chosen Practice of Alien Forest Plantation

Afforestation (planting of tree species where they do not occur naturally, see Veldman et al. 2015a, b, c) in the Rupestrian Grassland started in the 1970s with exotic and invasive species of *Eucalyptus* spp. Afforestation has many impacts on biodiversity and ecosystem services and therefore cannot be considered as a conservation strategy, as advocated by some NGOs (see Veldman et al. 2015c). Although, afforestation by *Eucalyptus* in the Rupestrian Grassland was thoroughly reviewed in Chap. 16, a scientific evaluation on the effects of afforestation on the diverse rupestrian grasslands ought to be conducted before a strategic plan is pushed forward by government organizations (see also Fernandes et al. 2015b). Both government and industry once said that *Eucalyptus* does invade areas beyond the area of planting, but it has indeed been found invading what remains of gallery and Atlantic rain forests on the eastern slope of the Espinhaço mountains (Fig. 23.1e, f). The invasion of the Rupestrian Grassland by species of *Pinus* is also worrisome, and has increased dramatically in recent years. Many *Pinus* species are fire adapted, and as such could benefit from the increasing frequency of fire in this landscape (see Chap. 17). The extent of afforestation on the Rupestrian Grassland must be analyzed urgently, and policies created to protect this ecosystem from this source of disturbance. Sound science should also be developed to evaluate the effects that afforestation might have on the functioning of this ecosystem, already under stress from water shortage (see Chaps. 4, 5 and 21).

23.4 Mining Effects on Ecosystem Health

Mining represents a historical economic activity in the Rupestrian Grassland and remains a powerful and growing industry in the mountains of southeastern Brazil and in the irreplaceable islands of ironstone rupestrian grasslands in the Amazonian region (see Fernandes et al. 2014; Chaps. 18 and 21). The likely future scenario for conservation becomes disturbing when one considers the intensification of mining in this ecosystem (see SIGMINE/DNPM at http://sigmine.dnpm.gov.br/webmap, see Fernandes et al. 2014 for further details). Besides the visual impact, which causes an enormous economical depreciation of the value of the land (see Resende et al. 2013), many other problems occur due to direct and indirect effects of mining, now and long into the future. Mining causes local habitat degradation by the direct effect of removal of native vegetation and soil. Mining also causes indirect changes at the landscape level, such as the opening of roads and secondary access, and urbanization, deforestation for charcoal production, and intentional introduction of exotic species in rehabilitation projects, as already discussed above (e.g., Fernandes et al. 2014; Sonter et al. 2014). In addition, the effects of mining on the ground water table may be devastating to the fauna and flora at both local and regional scales (e.g. Daniel et al. 2015; Piló et al. 2015). Goulart et al. (2016) were the first to focus on the direct and indirect impacts of mining on amphibians of rupestrian mountaintops and called attention to the amplification of conservation threats if no detailed studies are initiated immediately. From an environmental perspective, fragmentation of the environment, extinction of populations and species, lowering of the ground water table, and pollution and siltation of water bodies are almost inevitable impacts, while the introduction of invasive alien species could be totally avoided.

We cannot argue against the fact that, at least for the rupestrian grasslands of the Espinhaço Range, historical use and regional socioeconomic development are related to the extraction of minerals; first gold, then diamonds and other precious gemstones, and later iron and manganese (see Chap. 21 for a review). In the Amazonian rupestrian grasslands, iron mining is comparatively recent but it is very intense in the small area where the ironstone rupestrian grasslands are located (see Piló et al. 2015). A myriad of impacts results from the mining process in rupestrian grasslands. One example, not normally evaluated from an environmental perspective, is that of the indisputable association between human immigration and demand for natural resources. Thousands of people with varying levels of knowledge and appreciation for nature are generally brought to a single region without any education about the new land where they will now live and work. For the rupestrian grasslands of the Espinhaço Range, 600,000 people arrived from Portugal and the Atlantic islands (as nicely illustrated in Chap. 21), completely transforming the environment. An evaluation of such pressure on biodiversity and ecosystem services has never been attempted in detail to my knowledge.

While geological studies have been undertaken in the Iron Quadrangle in Minas Gerais and in the Carajás region in Pará using the most modern technologies and the latest generation of machines, the same cannot be said about the ecology of such a fragile, low-resilient ecosystem. Negreiros et al. (2011) have argued that once broken, the link between this delicate vegetation and the environment seems to have little chance of being spontaneously regenerated (see also Le Stradic et al. 2014a, b; Chap. 19). Areas impacted from historical use remain devoid of vegetation, even after having been abandoned for many centuries.

There is intense pressure by the mining industry to exploit the grasslands in the Iron Quadrangle, a region where the existing network of strictly protected areas is not sufficient to guarantee the conservation of this ecosystem and its biodiversity. Given the high beta diversity inherent in the Iron Quadrangle—largely different plant communities among mountains and sites (e.g. Le Stratric et al. 2015; Goulart et al. 2016; Chap. 6)—unique elements of biodiversity and endemism are under-represented in the current system of protected areas primarily in the ironstone rupestrian grasslands (Jacobi et al. 2011). A similar situation can be found in the northern ironstone rupestrian grasslands in Pará since the region where this ecosystem is found is very small. Therefore, although Brazil has historically and economically relied on the extraction of minerals, mining activities should be carefully planned and presented with sound conservation strategies to conserve species and ecosystem services of the Rupestrian Grassland. Historically, the Brazilian industrial mining has left profound scars on the rupestrian grasslands with a drastically altered natural environment with many negative impacts, both social and environmental.

Although no attempt has been made in this chapter to investigate the environmental impacts of uncontrolled artisanal mining, one can easily picture in their mind the situation of a land with loose law enforcement. Although governmental agencies are aware of the critical points, this type of mining is generally aggressive to the environment, with no control on the impacts and no measures for restoration, among other factors. In many cases, the *garimpo* (artisanal mining) in the Rupestrian Grassland is surrounded by many sad shadows that span from social injustice, to health, to political and environmental issues. But sadly, this problem is not particular to the Rupestrian Grassland, but is a world wide problem in need of an urgent solution (for some details see Castilhos et al. 2015; Piló et al. 2015). No synthesis of the impact of artisanal mining and clandestine mining has yet been attempted for the Rupestrian Grassland. A possible explanation may involve the illegality of the activity or even the criminality associated with the underworld global gemstone market.

23.5 Tourism in the Rupestrian Grassland: Is There a Threshold Limit?

Since the 1980s, tourism has intensified in Brazil. Although Brazil was, and continues to be, mostly known to the general public for its coastal region (mostly due to the beach environment), and for the Amazon and Pantanal (forests and fauna), tourism has also increased in the mountains, primarily where rupestrian grasslands prevail. In recent years we have been experiencing a great increase in advertisement for outdoor activities and adventure tourism in the mountains. This growth was probably already likely to have happened in some areas because of improvements in road systems. Serra do Cipó, only 100 km away from the large city of Belo Horizonte, is a classic example. After paving the MG-010 highway, tourism multiplied by many times, but, unfortunately, in an uncontrolled fashion. Currently the region cannot support the growing number of tourists. The infrastructure and education needed for the conservation of natural resources is incipient, causing various types of impacts. This floating population of tourists with little understanding of the ecosystem is responsible for the deposition of a huge amount of waste and contaminating rivers and aquifers. This picture is further complicated by the critical lack of sanitation in these areas. While an evaluation of tourism in the region of Serra do Cipó is under way (Resende et al. in prep), a deeper understanding of the impact of tourism on this ecosystem has yet to be undertaken (see also Chap. 21).

A large number of orchids and bromeliads are extracted from the rupestrian grasslands of the Chapada de Diamantia region. Accidental fire can also be linked to tourism in many areas in Chapada de Diamantina and in Serra do Cipó. Unfortunately, human feces are often found along the trail leading to the touristic Fumaça waterfall in Chapada de Diamantina. Very often, tour guides allow tourists to collect plants. Therefore, due to the increasing threat posed by ecotourism in the Rupestrian Grassland, a detailed understanding of the pressures imposed by tourism is urgently needed along with the development of mitigation strategies so that the use of this ecosystem is maintained for future generations.

23.6 Fire and the Rupestrian Grassland

Fire is a common component of old growth grasslands (Veldman et al. 2015a, b, c). On the other hand, not all the fires in the Rupestrian Grassland are natural: non-natural rupestrian grassland fires are caused by tourists, outdated agricultural practices used for the management of harvested plant species and livestock, and, unfortunately, disputes involving those who lost their land to protected areas. The presence of the widespread invasive exotic grasses, brought to the region primarily for livestock feed (Kolbek and Alves 2008) and recently for the revegetation of degraded areas, facilitates the spread of fire. Although many of the Rupestrian Grassland plant species are adapted to fire, they are not adapted to the short fire regime imposed by human induced fire. Chap. 17 provides a sound review of the history, frequency and effects of fire on the rupestrian grasslands in the Serra do Cipó region (see also Fernandes et al. 2014).

Livestock activity in the rupestrian grasslands of the Espinhaço mountains, is rooted in the history of human occupation of the region, usually as small farmsteads, and is responsible for land degradation in many areas and also contributes to the spread of weed plants of great invasive potential, such as *Urochloa* spp., *Melinis minutiflora*, as well as *Melinis repens*. These grasses are highly combustible, far surpassing the native species, which have a much lower biomass. Yet an understanding of the affects of fire on the dynamics and evolution of the Rupestrian

Grassland is still anecdotal in spite of recent advances (see Chap. 17 and references therein). Much effort, however, has been otherwise directed to the other Cerrado ecosystems (e.g., Mistry 1998; Rossi et al. 2014; see also Parr et al. 2014) but these are not easily extrapolated to understanding the unique mountaintop vegetation of rupestrian grasslands, which have different and perhaps stronger environmental filters (see Negreiros et al. 2014). Therefore, detailed observational and experimental studies are urgently needed to understand the ecological effects of fire on the Rupestrian Grassland species and habitats.

23.7 The Extraction of Vegetal Resources

Uncontrolled harvesting of ornamental plants to produce handicrafts is very common in the Rupestrian Grassland, as well as in its surrounding environments, at least in the Espinhaço (e.g., Domingues et al. 2012; Almada et al. 2013, see Chaps. 20 and 21). New studies are now emerging from the ironstone rupestrian grasslands in northern Brazil (e.g., Skirycz et al. 2014). From a conservation perspective, the uncontrolled harvesting of plant products, be it for handicrafts or for medicinal purposes, may increase extinction risk of many threatened species. The impacts caused by extractivism on species of Eriocaulaceae and Xyridaceae represent classic cases in the quartzitic rupestrian grasslands. The extraction of orchids in the Espinhaço rupestrian grasslands is an old, but detrimental, activity with little scientific information available on the impact to populations, including extinction. People living in poor local communities depend on such activity to improve their income, even though these plant products are sold locally at very low prices (Domingues et al. 2012). The use of *Vellozia* aff. *sincorana* (Velloziaceae) in Chapada de Diamantina is another excellent example of the popular use of an endemic species of the Rupestrian Grasslands (Oliveira 2013). In relation to medicinal plants, I am unaware of any attempt to summarize their use and the effects of extraction on plant populations in this ecosystem (but see Almada et al. 2013). Although the individuals involved in extractivism in quartzitic rupestrian grasslands have become more knowledgeable over the last decade (see Chap. 21), much is yet to be understood.

The potential for providing the world with new plant species of economic value is tremendous and could bring to the poor people of the rupestrian grasslands important economical alternatives. On the other hand, such entrepreneurship has not yet been developed, notwithstanding the increasing knowledge on plant propagation for some species (see Skirycz et al. 2014, Chaps. 19 and 21). Detailed scientific investigation on flora with ornamental potential could provide the people of the rupestrian grasslands a new perspective and a prospect for welfare and, if intelligently coordinated, could lead to higher appreciation for such ecosystems, which would eventually lead to greater conservation efforts.

23.8 The Synergy of Climate Change and Human Induced Filters

The Rupestrian Grassland is perhaps among those ecosystems that shall first suffer the impacts of global climate change due to their location on mountaintops. The geographical location at higher elevations with extensive cattle ranching, uncontrolled tourism, frequent intense burning episodes, and the entry of invasive species can act synergistically with climate change and irreversibly change the Rupestrian Grassland. The loss of area of this already rare ecosystem is a reality (due to the several factors described above) and climate change will exacerbate it, leaving some species nowhere to go. Many species of the Rupestrian Grassland are endangered due to their small area of occurrence. Goulart et al. (2016) caution that the impacts of climate change and of mining on birds and amphibians are very important in this regard.

Predictive models are not favorable for the Rupestrian Grassland, and suggest a catastrophic future for this ecosystem in new climatic scenarios (e.g., Fernandes et al. 2014, Chap. 21). Models indicate that the regions which are likely to remain climatically stable until the end of this century will be those just south of the Espinhaço Mountains in Minas Gerais, including the region of the Iron Quadrangle and Serra do Cipó, and the Serra da Canastra and part of the Chapada Diamantina in the northern Espinhaço. These regions are therefore of enormous value for conservation. The remaining rupestrian grasslands on smaller mountains in central Brazil, as well as the mountains located in the north of Minas Gerais State, are also are highly endangered. According to models, by the end of the century, losses may exceed 90 % of the area presently recognized as being environmentally suitable for rupestrian grasslands (which corresponds to approximately 66,500 km^2).

23.9 Ecological Research on the Rupestrian Grassland in the Anthropocene

Despite recent advances, the knowledge generated in recent decades is insufficient to address the myriad of threats that face the enormous diversity of rupestrian grasslands. There is a need for a major effort to broaden and deepen observational and experimental knowledge on the ecology of native and invasive species and land use conversion. This will enable robust conservation and management strategies, like the control of biological invasions, the creation of new protected areas considering future climate model scenarios, and sustainable use. The conundrums facing scientists, policy makers, and society with regards to the conservation and use of the Rupestrian Grasslands is of paramount relevance to untangle.

The anthropogenic pressure on the Rupestrian Grassland ecosystem and its functioning is causing profound damage, which we have not yet been able to fully evaluate (Fig. 23.2). This is a rare ecosystem, where at least 50 % of the

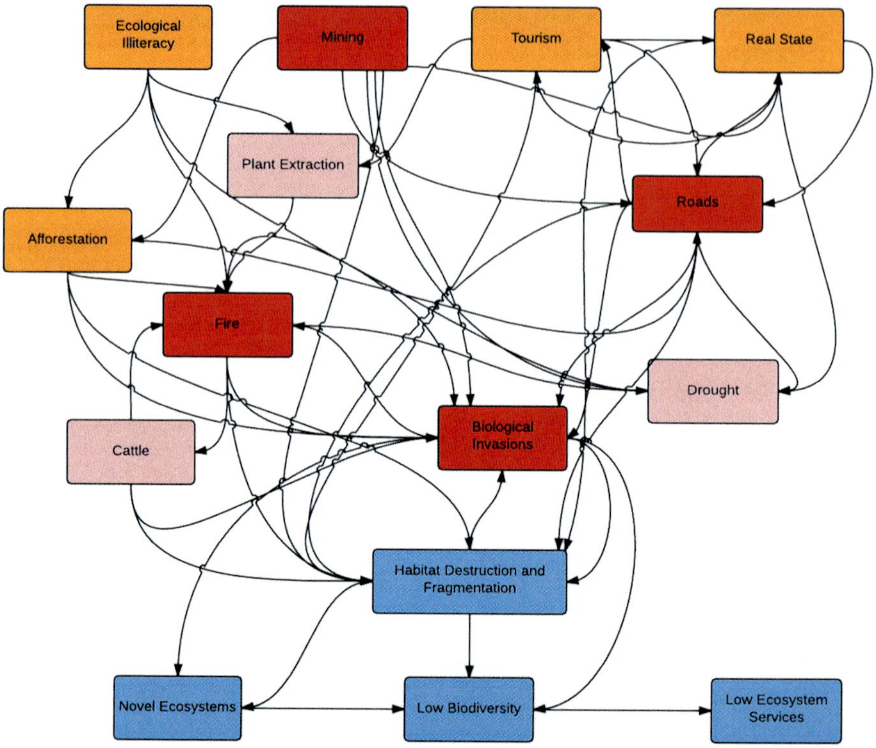

Fig. 23.2 The Rupestrian Grasslands evolved under strong environmental filters and has been stable until recently in evolutionary time. On the other hand, modern man has impacted the natural capital of the Rupestrian Grassland in several ways and in different intensities. These factors interact in different ways, often resulting in amplification of the impacts. Perhaps the factors most threatening to the Rupestrian Grassland are those related to mining, biological invasion, non-natural fires and road construction (*red squares*), followed by real state development, afforestation, tourism and ecological illiteracy (*orange squares*), while cattle, drought and plant extraction (*pink squares*) represent those threats with lesser impacts. The complex combination of these threats results in high habitat fragmentation and destruction and very often to novel ecosystems of very low public appreciation and ecosystem value (*blue squares*), with lower biodiversity and strong negative impacts on ecosystem services and identity

biodiversity of the most diverse savanna of the world (the Cerrado) is confined. Unfortunately, there is no replaceable area if these are lost. Local extinction in the Rupestrian Grassland means global extinction for many of these species. Hence, in addition to keeping up with basic scientific studies such as species surveys, and descriptions of plant structure and composition, natural history and ecology of these species, we must design and implement detailed studies to understand plant species longevity and population dynamics, decomposition, herbivore impacts on ecosystem productivity, resiliency and time required to reassemble in response to disturbances, and multitrophic interactions, among many others. Of practical importance is the development of ecological studies to improve our understanding of the roles of time in the development of these old-growth grassland plant communities, as well as other aspects of biodiversity and ecosystem services. Although it is clear that many decades of intense research are required to understand the functioning of this ecosystem, we should rather see this as an opportunity for more innovative studies. One such rare opportunity is the study of the filters leading to the astonishing rates of evolution and speciation, as well as levels of endemism, experienced by some genera and families. Such studies could start unlocking the mysteries involved in such central questions that have haunted evolutionary biologists since Darwin and Eugen Warming.

One area in which such innovation is urgently needed is restoration. Present knowledge is limited or insufficient for understanding natural regeneration processes, propagation of target plant species, and implementation of effective restoration or even conservation programs. Therefore, a collaborative strategy developed by both the private sector, that need to fulfill legislation requirements, and by society, must be constructed as urgently as possible since the danger associated with active erosion and areas wrongly restored with exotic species represent a major risk to the Rupestrian Grassland. All previous attempts to restore the Rupestrian Grassland were done with the wrong species, and are now environmental liabilities in need of correction.

The loss of habitat represents a major problem to be overcome in the Rupestrian Grassland and a sound strategy would be the full implementation of the conservation units already created. Many of the so called "conservation units" may be better described as "paper conservation units" as they are only legal parks created on decrees. This situation is no longer acceptable and investments for their full operation and payment to stake holders from whom the lands were taken must be enforced at all cost. The creation of new conservation areas is also desirable and strong efforts should be addressed to that end. Perhaps most relevant is the deep analysis of the functioning of those conservation units that have already been created. Rupestrian grasslands require specific planning in which people living in this ecosystem are called on to be partners in conservation. The management plans already established are failing to guarantee the effective conservation of the protected areas, and perhaps none have become drivers of sustainable development for local communities. Only a multidisciplinary understanding of the problem and wise planning could lead to a solution given the relevance of the rupestrian grasslands and conservation goals. While some recent studies have now emerged to this end (e.g., Silva et al. 2008; Sonter et al. 2014), these cannot be only subjects of academic interest.

The conservation of the Rupestrian Grassland represents a huge challenge to civil society and government owing to the current rates of conversion of their natural resources, the lack of understanding of their intrinsic value to human well-being, the approaching thresholds in their identity and the illiteracy regarding its biodiversity, functioning, and restoration. The solution of these challenges will need time, mighty planning, education and actions to implement sound and effective conservation plans for the preservation of this very fragile ecosystem.

23.10 Conclusions and Final Remarks

The very first challenge in the conservation of the Rupestrian Grassland is to overcome inertia and override the ecological illiteracy regarding this ecosystem. Strong planning to eradicate exotic plants must be elaborated and applied with the training of land managers and conservation agencies that neglect current knowledge. Non-native invasive species should be recognized as a serious threat to the biodiversity of the Rupestrian Grassland and a major hurdle to restoration. The creation of specific policies regarding the Rupestrian Grassland are needed so that we can control the risks and mitigate past obtuse management practices. More conservation units are urgently needed, and those already created need to be made fully operational with a long lasting solution to mitigate social impacts they may cause on society. Conservation advocacy groups should engage the public to seek legal protection for the Rupestrian Grassland and support educational outreach to raise awareness of this fragile ecosystem. Policies designed to promote afforestation in the Rupestrian Grassland for any of a variety of reasons, including climate change mitigation, should be abandoned due the mismatch of the forest with the structure and functioning of this old growth ecosystem. The destruction of the Rupestrian Grassland has deleterious environmental consequences similar to deforestation. Financial mechanisms (e.g., ecosystem service payments) should be developed that promote the protection and maintenance of the Rupestrian Grassland. Finally, joint strategies between the government and the private sector should be implemented to promote and disseminate knowledge regarding the Rupestrian Grassland and its conservation.

More specifically: (i) comprehensive monitoring approaches linking ground observation of biophysical properties with remote sensing can provide quick feedback on land use conversion and ecosystem health; (ii) better understanding of how changes in climate regimes of mountaintop rupestrian grasslands affect plant phenology and structure, composition of plant and animal communities, and biological invasions can provide strong tools for monitoring and elaboration of mitigation and adaptation strategies; (iii) we need to understand the ecological processes and functionality of the Rupestrian Grassland, and the primary processes such as herbivory, pollination, succession, dispersion, and interactions with micro-organism as they mediate many of the processes governing biodiversity and ecosystem services; (iv) we need also to understand the ability of the Rupestrian Grassland to sequester carbon and produce water. Rupestrian grasslands are true upside down forests and headwater areas. Carbon sequestration can provide insights on ecosystem function and water production; (v) develop a synthesis of the knowledge of the biology of the plant species capable of being used in restoration so we can better design sound programs for restoration; (vi) begin a strong program to evaluate the interaction of biodiversity and ecosystem services in the many habitats of the ironstone and quartzitic rupestrian grasslands; (vii) last, but not least, bioprospect for potential products that can be provided by the species and habitats in the rupestrian grassland, so we can gain a higher appreciation of this ecosystem.

From a more human perspective we also need equal development, among: (i) the promotion of conservation; (ii) the promotion of regular alternative economic activities; (iii) better understanding of land use history of rupestrian grasslands, and their use by rural communities; (iv) provide stimuli for students to work on rupestrian grasslands, as a chance for them to return to their home land and work on conservation and developmental issues of such ecosystems; (v) finally, we need to educate decision-making people about the Rupestrian Grassland via bio-literacy in a manner similar to many programs developed in other regions of the world. But for these to occur we need to create the opportunity and space for the development of strong integrated ties among all social actors involved such that they can contribute their knowledge and perspectives. Solutions coming from a pact will have a much higher likelihood of being effective and long lasting.

Acknowledgements I thank FAPEMIG, and CNPq for the continuous support of this research and Reserva Vellozia for logistical support. For reading and providing criticisms on an earlier version of this chapter I must acknowledge Benjamin Wilder, Abel Conceição, Hernani Almeida, Christina Lara and Daniel Negreiros.

References

Almada ED, Coelho MS, Quintino AV, Fernandes GW, Sanchez-Azofeif a A (2013) Traditional ecological knowledge of rural communities in areas of seasonally dry tropical forest in Serra do Cipó. In: Sanchez-Azofeifa A, Powers JS, Fernandes GW, Quesada M (eds) Tropical dry forests in the Americas: ecology, conservation, and management. CRC Press, Boca Raton, pp 429–451

Barbosa NPU, Fernandes GW, Carneiro MAA, Júnior LAC (2010) Distribution of non-native invasive species and soil properties in proximity to paved roads and unpaved roads in a quartzitic mountains grassland of southeastern Brazil (rupestrian fields). Biol Inv 12: 3745–3755

Belo MB, Negreiros D, Fernandes GW, Silveira FAO, Ranieri BD, Morellato PC (2013) Fenologia reprodutiva e vegetativa de arbustos endêmicos de campo rupestre na Serra do Cipó, Sudeste do Brasil. Rodriguesia 64:817–828

Carvalho F, Souza FA, Carrenho R, Moreira FMS, Jesus EC, Fernandes GW (2012) The mosaic of habitats in the high-altitude Brazilian rupestrian fields is a hotspot for arbuscular mycorrhizal fungi. Appl Soil Ecol 52:9–19

Carvalho F, Godoy EL, Lisboa FJG, Moreira FMS, Souza FA, Berbara RLL, Fernandes GW (2014) Relationship between physical and chemical soil attributes and plant species diversity in tropical mountain ecosystems from Brazil. J Mt Sci 11:875–883

Castilhos Z, Rodrigues-Filho S, Cesar R, Rodrigues AP, Villas-Bôas R, Jesus I, Lima M, Faial K, Miranda A, Brabo E, Beinhoff C, Santos E (2015) Human exposure and risk assessment associated with mercury contamination in artisanal gold mining areas in the Brazilian Amazon. Environ Sci Poll Res 22:11255–11264

Coelho FF, Neves ANO, Capelo C, Figueira JEC (2005) Pseudovivipary in two rupestrian endemic species (*Leiothrix spiralis* and *Leiothrix vivipara*). Cur Sci 88:1225–1226

Coelho FF, Capelo C, Neves ACO, Martins RP, Figueira JEC (2006) Seasonal timing of pseudoviviparous reproduction of *Leiothrix* (Eriocaulaceae) rupestrian species in Southeastern Brazil. Ann Bot 98:1189–1195

Coelho FF, Capelo C, Ribeiro LC, Figueira JEC (2008) Reproductive modes in *Leiothrix* (Eriocaulaceae) in South-eastern Brazil: the role of microenvironmental heterogeneity. Ann Bot 1010:353–360

Daniel W, Infante DM, Hughes RM, Tsanga Y, Esselman PC, Wieferich D. Herreman K, Cooper AR, Wang L, Taylor WW (2015) Characterizing coal and mineral mines as a regional source of stress to stream fish assemblages. Ecol Indic 50: 50–61

Domingues SA, Karez CS, Biondini IVF, Andrade MA, Fernandes GW (2012) Economic environmental management tools in the Serra do Espinhaço Biosphere Reserve. J Sust Develop 5:180–191

Fernandes GW, Barbosa NPU (2013) Bombas relógio que ameaçam a natureza em Minas Gerais. Sci Amer Bras 135:60–61

Fernandes GW, Barbosa NPU, Negreiros D, Paglia AP (2014) Challenges for the conservation of vanishing megadiverse rupestrian grasslands. Nat Cons 12:162–165

Fernandes GW, Santos RA, Barbosa NPU, Almeida HA, Carvalho VA, Angrisano P (2015a) Ocorrência de plantas não-nativas e exóticas em áreas restauradas de campos rupestres. Planta Daninha 33:463–482

Fernandes GW, Machado R, Ferreira M, Coelho MS, Aguiar L, Scariot A, Dirzo R (2015b) Afforestation of savannas: an impending ecological disaster. Cons Biol (in review)

Giulietti AM, Pirani JR, Harley RM (1997) Espinhaço range region, eastern Brazil. In: Davis SD, Heywood VH (eds) Centres of plant diversity: a guide and strategy for their conservation, vol 3. Cambridge, pp 397–404

Gonella PM, Rivadavia F, Fleischmann A (2015) *Drosera magnifica* (Droseraceae): the largest new world sundew, discovered on Facebook. Phytotaxa 220:257–267

Goulart F, Santos NB, Pena JC, Hoffmann D, Leite FSF, Fernandes GW, Soares-Filho B, Ziv G, Rodrigues M (2016) Habitat degradation by mining threatens endemic mountaintop anurans in Brazil. Biotropica (in review)

Hilário RR, Castro SAB, Ker FTO, Fernandes GW (2011) Unexpected effects of pigeon-peas (*Cajanus cajan*) in the restoration of rupestrian fields. Planta Daninha 29:717–723

Hopper SD (2009) OCBIL theory: towards an integrated understanding of the evolution, ecology and conservation of biodiversity on old, climatically buffered, infertile landscapes. Plant Soil 322:49–86

Jacobi CM, Carmo FF, Campos IC (2011) Soaring extinction threats to endemic plants in Brazilian metal-rich regions. Ambio 40:540–543

Kolbek J, Alves RJV (2008) Impacts of cattle, fire and wind in rocky savannas, southeastern Brazil. Acta Univ Carol Environ 22:111–130

Le Stradic S, Buisson E, Negreiros D, Campagne P, Fernandes GW (2014a) The role of native woody species in the restoration of campos rupestres in quarries. Appl Veg Sci 17:109–120

Le Stradic S, Buisson E, Fernandes GW (2014b) Restoration of Neotropical grasslands degraded by quarrying using hay transfer. Appl Veg Sci 17:482–492

Le Stradic S, Buisson E, Fernandes GW (2015) Vegetation composition and structure of some Neotropical mountain grasslands in Brazil. J Mt Sci 12:864–877

Lüttge U, Duarte HM, Scarano FR, Mattos EA, Cavalin OP, Franco AC, Fernandes GW (2007) Physiological ecology of photosynthesis of five sympatric species of Velloziaceae in the rupestrian fields of Serra do Cipó, Minas Gerais, Brazil. Flora 202:637–646

Medina BMO, Fernandes GW (2007) The potential of natural regeneration of rocky outcrop vegetation on rupestrian field soils in "Serra do Cipó" Brazil. Rev Bras Bot 30:665–678

Mistry J (1998) Fire in the cerrado (savannas) of Brazil: an ecological review. Prog Phys Geogr 22:425–448

Negreiros D, Fernandes GW, Silveira FAO, Chalub C (2009) Seedling growth and biomass allocation of endemic and threatened shrubs of rupestrian fields. Acta Oecol 35:301–310

Negreiros D, Fernandes GW, Berbara RLL, Rodarte LHO, Barbosa NPU (2011) Caracterização físico-química de solos quartzíticos degradados e áreas adjacentes de campo rupestre na Serra do Cipó, MG, Brasil. Neotrop Biol Cons 6:156–161

Negreiros D, Le Stradic S, Fernandes GW, Rennó HC (2014) CSR analysis of plant functional types in highly diverse tropical grasslands of harsh environments. Plant Ecol 215:379–388
Neves AC, Nogueira FB, Assis LR, Paglia AP, Bedê LC, Martins RP (2014) Reproductive allocation in rhizomatous, seminiferous, and pseudoviviparous *Leiothrix* (Eriocaulaceae) species. Plant Ecol 215:987–996
Oliveira RCS (ed) (2013) Uso e Conhecimento do Candombá. Universidade Federal de Feria de Santana, Feria de Santana
Oliveira SHF, Negreiros D, Fernandes GW, Barbosa NPU, Rocha R, Cortez JSA (2009) Seedling growth of the invader *Calotropis procera* in ironstone rupestrian field and seasonally dry forest soils. Neotrop Biol Cons 4:69–76
Oliveira RS, Galvão HC, Campos MCR, Eller CB, Pearse SJ, Lambers H (2015) Mineral nutrition of campos rupestres plant species on contrasting nutrient-impoverished soil types. New Phytol 205:1183–1194
Parr CL, Lehmann CER, Bond WJ, Hoffmann WA, Andersen AN (2014) Tropical grassy biomes: misunderstood, neglected, and under threat. Trend Ecol Evol 29:205–213
Piló LB, Auler AS, Martins F (2015) Carajás national forest: iron ore plateaus and caves in southeastern Amazon. In: Vieira BC, Salgado A, Santos L (eds) Landscapes and landforms of Brazil—world geomorphological landscapes. Springer, Dordrecht, pp 273–283
Rapini A, Ribeiro PL, Lambert S, Pirani RJ (2008) A flora dos campos rupestres da Cadeia do Espinhaço. Megadiv 4:15–23
Resende FM, Fernandes GW, Coelho MS (2013) Economic valuation of plant diversity storage service provided by Brazilian rupestrian grassland ecosystems. Braz J Biol 73:709–716
Rossi RD, Martins CR, Viana PL, Rodrigues EL, Figueira JEC (2014) Impact of invasion by molasses grass (*Melinis minutiflora* P. Beauv.) on native species and on fires in areas of campo-cerrado in Brazil. Acta Bot Bras 28:631–637
Sanchez-Azofeifa A, Powers JS, Fernandes GW, Quesada M (eds) (2013) Tropical dry forests in the Americas: ecology, conservation, and management. CRC Press, Boca Raton
SIGMINE/DNPM (2015) The ISSN register. http://sigmine.dnpm.gov.br/webmap. Accessed 20 Jun 2015
Silva JA, Machado RB, Azevedo AA, Drumond GM, Fonseca RL, Goulart MF, Júnior EAM, Martins CS, Neto MBR (2008) Identificação de áreas insubstituíveis para a conservação da Cadeia do Espinhaço, estados de Minas Gerais e Bahia, Brasil. Megadiv 4:248–270
Silveira FAO, Negreiros D, Barbosa NPU, Buisson E, Carmo F, Carstensen DW, Conceição AA, Cornelissen TG, Echternacht L, Fernandes GW, Garcia QS, Guerra TJ, Jacobi CM, Lemos-Filho JP, Le Stradic S, Morellato LPC, Neves FS, Oliveira RS, Schaefer CE, Viana PL, Lambers H (2016) Ecology and evolution of plant diversity in the endangered campo rupestre: a neglected conservation priority. Plant soil. doi: 10.1007/s11104-015-2637-8
Skirycz A, Castilho A, Chaparro C, Carvalho N, Tzotzos Siqueira JO (2014) Canga biodiversity, a matter of mining. Fron Plant Sci 5:653. doi:10.3389/fpls.2014.00653
Sonter LJ, Barrett DJ, Soares-Filho BS, Moran CJ (2014) Global demand for steel drives extensive land-use change in Brazil's iron quadrangle. Glob Environ Chan 26:63–72
Veldman JW, Buisson E, Durigan G, Fernandes GW, Le Stradic S, Mahy G, Negreiros D, Overbeck GE, Veldman RG, Zaloumis NP, Putz FE, Bond WJ (2015a) Toward an old-growth concept for grasslands, savannas, and woodlands. Front Ecol Environ 13:154–162
Veldman JW, Overbeck GE, Negreiros D, Mahy G, Le Stradic S, Fernandes GW, Durigan G, Buisson E, Putz FE, Bond WJ (2015b) Tyranny of trees in grassy biomes. Science 347:484–485
Veldman JW, Overbeck GE, Negreiros D, Mahy G, Le Stradic S, Fernandes GW, Durigan G, Buisson E, Putz FE, Bond WJ (2015c) Where tree planting and forest expansion are bad for biodiversity and ecosystem services. Bio Science doi: 10.1093/biosci/biv118

Index

© Springer International Publishing Switzerland 2016
G.W. Fernandes (ed.), *Ecology and Conservation of Mountaintop Grasslands in Brazil*, DOI 10.1007/978-3-319-29808-5